Handbook of Aerospace Electromagnetic Compatibility

IEEE Press
445 Hoes Lane
Piscataway, NJ 08854

Handbook of Aerospace Electromagnetic Compatibility

Edited by

Reinaldo J. Perez

EMC SOCIETY®

IEEE PRESS

WILEY

Published by John Wiley & Sons, Inc., Hoboken, New Jersey.
Published simultaneously in Canada.

Library of Congress Cataloging-in-Publication Data is available.

ISBN: 9781118910511

Cover design: Wiley
Cover images: (left) © Jag_cz/ Shutterstock; (right) © Andrey Armyagov/ Shutterstock

Printed in the United States of America.

10 9 8 7 6 5 4 3 2 1

Contents

Preface

Many of the early beginnings of electromagnetic compatibility (EMC) are coincidental with the technological advances in the aerospace industry, going as far back as the 1960s. Since the 1960s, EMC has been an essential part in the development of aerospace components. Furthermore, EMC has become an essential part in the development of aerospace subsystems and systems. Therefore, the role of EMC in the aerospace business is over 50 years old, probably older than any other technology-related business. The *Handbook of Aerospace Electromagnetic Compatibility* is an up-to-date snapshot of where EMC is today in the aerospace business. It is the first book of its kind. To that end, it was decided to bring into this book different aspects of EMC as these are applied today in the aerospace business. Each chapter in the handbook is independent and self-contained, but, in a convergent way, the ensemble of these chapters allows the reader to become aware of the big picture concerning EMC in aerospace electronics for both aircraft and space systems. The handbook is designed in such a way that it will be mostly of applied nature, as it is tailored to the practicing EMC engineer. The goal is to primarily provide practicing EMC engineers and even newly graduated engineers, who are involved or would like to be involved in aerospace EMC, with a good overall background in aerospace EMC, so they can "hit the ground running" in the business. The chapters cover both aircraft EMC and spacecraft EMC, and an effort has been made to develop a good cross-section of both. Because the field of aerospace EMC is evolving, mostly by new technological applications in aerospace (e.g., autonomous unmanned aerial vehicles), it is expected that future editions of the handbook will address newer aerospace EMC applications or existing applications that have yet to be covered.

The handbook has been divided into 14 chapters. The organization of the chapters is such that the first two chapters are more theoretical in nature while the remaining chapters are of applied nature. Although the first two chapters are theoretical, the theory is blended to aerospace applications. Chapter 1, titled "Introduction to E3 Models and Techniques in Aerospace Systems," addresses some theories and the accompanied mathematics for autonomous systems and coupled air and space systems. The theory of chaos is introduced as applied to EMC, and a mathematical approach to the effects of EMC in aerospace systems is also discussed. Chapter 2, titled "Deterministic and Statistical EMC Models for Field-to-Wire Coupling and Crosstalk in Wire Harness," addresses EMC in cables. This chapter is useful in developing a theoretical framework for one of the major problems in aerospace EMC—the large amount of cabling used in aerospace systems, which are responsible for generating a large number of EMC problems in aerospace. Chapter 3, titled "EMP Protection and Verification," looks at a subject, in detail, whose resurgence is now esteemed very important. Electromagnetic pulse (EMP) caused by either the space environment or nuclear explosions can cause incalculable damage to our aerospace systems and other technological areas such as electrical power, transportations, and informational infrastructures, just to name a few. Chapter 4, titled "HIRF and Lightning Effects and Testing," covers another topic of particular importance to the aircraft industry. The chapter is very extensive and discusses the theory, such as coupling analysis and the electromagnetic environment and it effects; it also includes considerable information concerning testing and protection against electromagnetic effects for lightning effects on aircraft. Chapter 5, titled "Techniques to Design Robust Lightning Protection Circuits for Avionics Equipment," discusses lightning effects at the box level and its electronics. The chapter combines both theory and testing methods for helping provide protection and hardening of avionics electronics. Chapter 6, titled "Pyrotechnic Systems in Aerospace Applications," addresses a particular type of aerospace component that is uniquely susceptible to EMC problems. Pyrotechnic devices (or pyros) are present in both aircraft and spacecraft vehicles. Pyros execute one-time functions when activated. Therefore, the need exists to protect such devices from EMC problems, which could cause some of those one-time functions to be activated prematurely or inadvertently with potential catastrophic consequences. Chapter 7, titled "Assembly-Level EMC Testing of Space Components/Subsystems," is dedicated to EMC testing of spacecraft components at the box or assembly level, while Chapter 8, titled "System-Level Testing of Spacecraft, thoroughly discusses the EMC testing of spacecraft at the subsystem level (i.e., the multiple spacecraft boxes that make up a subsystem). Chapter 9, titled "Subsystem EMC for Aircraft," discusses EMC testing of electronic components and subsystem that are intended for integration onto aircraft. The chapter covers central aspects of some of the commonly encountered technical tests performed under aviation EMC standards. Chapter 10, titled "EMI Effects in Flight Control Systems and Their Mitigations," is dedicated to EMC problems in aircraft control systems. The control systems of an aircraft are the most susceptible system to EMC problems because that is where most of the aircraft electronics resides, and these control systems manage all the control surfaces of the aircraft, which allow the aircraft to flight. The chapters cover both theoretical aspects of EMI control and EMC testing. Chapter 11, titled "EMC

Considerations for Unmanned Aerial Vehicles," presents a first-time introduction to EMC issues in unmanned aerial vehicles (also known as drones). We are in the early stages of this fascinating field, and this book provides interested parties with the early knowledge need to pursue this subject further in the future. Magnetic cleanliness is important not only for space systems but also for aircraft systems, which is the topic of Chapter 12, titled "DC Magnetic Cleanliness Description for Spaceflight Programs." Space systems must perform very sensitive scientific measurements as dictated by their mission profile. These measurements are made by highly sensitive payload instruments and sensors onboard the spacecraft. The science measurements can be easily corrupted by noise from DC magnetic fields, hence the need to provide such cleanliness. The chapter addresses both theoretical and testing methods. Chapter 13, titled "Spacecraft Charging," addresses a major issue in space systems that can cause significant EMC problems. Space vehicles are highly susceptible to charging due to the highly charged space environment. The accumulated charges can cause electrostatic discharge events in space, which can cripple satellites and many other types of space vehicles. Many examples of such space losses concerning spacecraft have been documented over the years. This chapter addresses the spacecraft charging phenomena and ways to protect a such charging. The final chapter, titled "Analysis and Simulations of Space Radiation-Induced Single-Event Effects and Transients," addresses the detrimental impact of electronics circuits in space vehicles due to transient currents caused by the impact of highly energetic particles produced in the highly charged space environment. The chapter describes the theory and impact of such transient events and provides ways to decrease the damage done to space electronics.

Each chapter of this handbook has been independently reviewed by a member of the IEEE Electromagnetic Compatibility Society (see Acknowledgments), most of whom have worked in the aerospace industry for many years. We thank the IEEE EMC Society, IEEE Press, and Wiley for their support of this handbook. I thank specially all the authors who dedicated their talent to the development of this handbook for a period of almost four years. The handbook is dedicated to each one of them.

Reinaldo J. Perez

Acknowledgments

The Editor and authors of the *Handbook of Aerospace Electromagnetic Compatibility* would like to express their sincere thanks to multiple individuals of the IEEE Electromagnetic Compatibility Society and others from the aerospace industry who contributed to the review of each of the chapters in the handbook. We would like first to express our thanks to Elya Joffe who reviewed a substantial number of the chapters in the handbook and provided very variable technical support and recommendations to multiple authors. We would like also to thank the following reviewers each of which reviewed one chapter in the handbook. Our gratitude to Albert Whittlesey from JPL/Caltech, John Norgard from NASA, James Lukash from Lockheed Martin, and Flavia Grassi from Politecnico di Milano.

List of Contributors

Karen Burnham
GM, USA

Martin Gabrisak
EMCC Germany

Paul Kay
Royal Australian Air Force, Australia

Ira Kohlberg
Kohlberg Associates Inc.

Irfan Majid
Institute of Space Technology

Dr. Clay McCreary
Rockwell Collins, USA

Pablo S. Narvaez
JPL, NASA

Reinaldo J. Perez
JPL, NASA

Sergio Pignari
Politecnico di Milano

William D. Prather
USAF AFMC AFRL, USA

Robert C. Scully
NASA, USA

Leslie R. Warboys
Consultant, USA

Johannes Wolf
European Space Agency

In Memoriam

As editor, I worked with Dr. Ira Kohlberg in the writing of his chapter. Soon after he delivered the fourth version of the chapter, he passed away, now over two years ago. I reviewed his chapter and edited it. Dr. Kohlberg was one of the first authors who enthusiastically accepted the call to contribute to the handbook. I was very impressed by his enthusiasm and kindness. We exchanged many ideas over the phone and he was the first author who delivered the first version of his chapter. Dr. Kohlberg had authored five books of his own, the first in 1976. Dr. Kohlberg was a physicist and mathematician by training and started his own consultant company in 1985, which provided consulting services to companies and the US government.

1

Introduction to E^3 Models and Techniques in Aerospace Systems

Ira Kohlberg

1.1 Introduction and Topics of Interest

1.1.1 Background

This chapter renders an overview and perspective of electromagnetic compatibility (EMC) and electromagnetic environmental effects' (E^3) theoretical considerations for current, near-term, and future aerospace systems. Our starting point extends in part from the current MIL-STD-464C, MIL-STD-461, and MIL-STD-3023 (all discussed in Section 1.5) baseline threats and technologies that include analytical models, testing techniques' measurements, shielding, numerical techniques, transients, antennas, power modulators, printed circuit board, cables, subsystems, individual limited-size systems, etc. By providing relevant academic and industrial educational resources (conferences, courses, journals, etc.), the EMC and E^3 community supports the creation of standards and the development of new components and systems.

The role of EMC and E^3 for aerospace systems is to not only maintain high standards, but also to improve the reliability and survivability of complex time-dependent networks that are susceptible to catastrophic failures. Survivability is enhanced by increasing the timeliness of delivery and reliability of long messages, improving reception for multiple multicast networks during link failure, and increasing robustness of heterogeneous networks through advanced communication techniques such as network coding.

It is absolutely essential that certified aerospace EMC and E^3 hardware and software are trustworthy. In addition to this mainstream bread-and-butter activity, it is also desirable and necessary to probe in the near term and future and prepare the theoretical foundations upon which to build robust analytical and testing techniques to meet new requirements. The need for new EMC and

Handbook of Aerospace Electromagnetic Compatibility, First Edition. Edited by Reinaldo J. Perez.

E^3 techniques arises because of increasingly complex emerging systems that, in large measure, will focus on:

(a) Testing techniques for autonomous systems that rely on dimensional analyses and stability theory
(b) Air-to-ground and space-to-ground and space-to-atmosphere communication and radar networks that contain combinations of random nondirect and direct communication electromagnetic propagation links
(c) New-generation aircraft that employ increasingly susceptible electronic components that may require advanced nonlinear control analyses and control systems, and advanced signal analyses that can identify and manage chaos signals and high-power microwaves (HPM) waveforms.

We assert that the EMC and E^3 technologies discussed in detail in Sections 1.2–1.5 meet the aforementioned "a-to-d" challenges.

1.1.2 Autonomous Systems

As pointed out in Section 1.2, an autonomous system is composed of three major functions: *perception*—sensing the environment, *decision-making*—selecting the best course of action, and *execution*—implementing the best course of action. The ultimate goal of an autonomous system is to provide the operating conditions and performance standards for each decision-making algorithm. Complex adaptive systems (CAS) that will include both human and software components need to be developed and tested to ensure that the autonomous system will perform its function as required within the system as a whole. Related control theory is routinely used to design electromechanical algorithms employed in the system while ensuring that the algorithms work effectively in a stable manner in response to their stimulation and interactions with other components in the overall system.

Figure 1.1 shows models of autonomous systems currently under development. The elements of these autonomous systems are discussed in Section 1.2. These systems must operate in the presence of strong electromagnetic interference, which plays a critical role in system performance since they are often the principal generator of unwanted electromagnetic signals. The creation of unwanted signals due to interference may require nonlinear control functions, which create undesirable complications.

1.1.3 Networks of Coupled Air and Space Systems

Today, the number of nodes in a network can be in the hundreds of thousands to over a million. Current networks have great connectivity for large numbers of nodes and include a mixture of deterministically positioned nodes, and

Figure 1.1 Models of the autonomous system.

randomly located nodes spread out over land, sea, atmosphere, and space. A key example is the class of mobil *ad hoc* networks (MANETs) involving directed and nondirected random graphs. Variables of modern networks include number of nodes and links, the probability density function of links connected to nodes, the probability density function of distance between node pairs, and selected system-unique parameters. Figure 1.2 illustrates a model of a MANET-type land system coupled to an airborne system having a relatively small (e.g., 1–100) number of platforms and possibly a few satellites.

Intentional electromagnetic interference effects (IEMIs) range from direct jamming and interference of nodes to the creation of fading dispersive channels

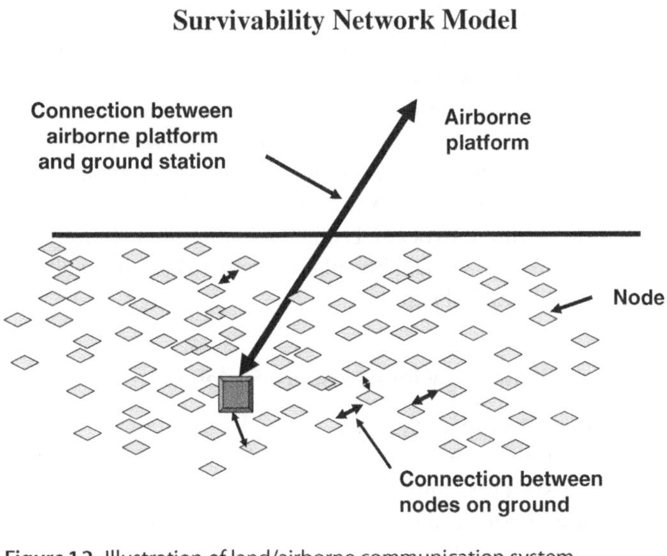

Figure 1.2 Illustration of land/airborne communication system.

that can limit both the coherence time and the coherence bandwidth. All these effects can reduce the information rate and, if bad enough, can cause the network to break into isolated clusters—possibly the most serious adverse effect. It is necessary to define the mathematical structure of these networks and the mathematical tools that are necessary to perform the communication survivability assessments. Section 1.3 discusses these issues in more detail.

1.1.4 EMC Considerations of Chaos and Related Waveforms

A technology that needs to be better understood in the aerospace EMC and E^3 community is a spectrum of anomalous waveforms that are embraced by the term "chaos," discussed in Section 1.4.1.2. There is currently much interest in quantifying, predicting, and measuring these effects on electronic and hybrid control systems, power electronics and power supplies generated by high-power electronics (HPE), and high power microwaves (HPM), which can cause temporary upset to permanent damage. These unwanted signals arise in multidimensional nonlinear dynamic systems and can manifest themselves as being extremely sensitive to initial conditions, splitting apart (unbounded), exhibiting unanticipated periodic/quasiperiodic motion and strange attractors, creating power spectrum with continuous parts, etc. Unfortunately, in many laboratory cases, the unwanted signals are not recognized as chaos but as "not understood" interference.

Researchers attempt to explain observations in terms of system variables such as microwave frequency, pulse duration, pulse repletion rate, peak power, average power, type of equipment, computer clock rate, and the number of components used in the experiment. Proposed connections between chaos and stochastic processes have been suggested with only modest success. As systems become more complicated, it will be necessary to have measurement algorithms that predict aerospace performance, especially for autonomous systems.

Much of what we have sketched out in this section can be visualized with the aid of Figure 1.3. This figure depicts a single system under attack by an HPE threat such as HPM, IEME, and chaos. The Xs are the assumed spots where the significant interactions are assumed to occur. Using the mathematical tools and theories developed in Section 1.5, we will be able to predict the output signal and the effect on critical aeronautical systems.

1.1.5 EMC Effects on and Technology for Aerospace Systems

1.1.5.1 Testing and Hardening Aerospace Systems

Section 1.5 embraces traditional advanced topics used in aerospace systems such as nonlinear control systems, multiple input multiple output (MIMO) control systems, and scaling theory and testing. While these are not necessarily new topics, their technologies need to be kept current in lieu of advancements

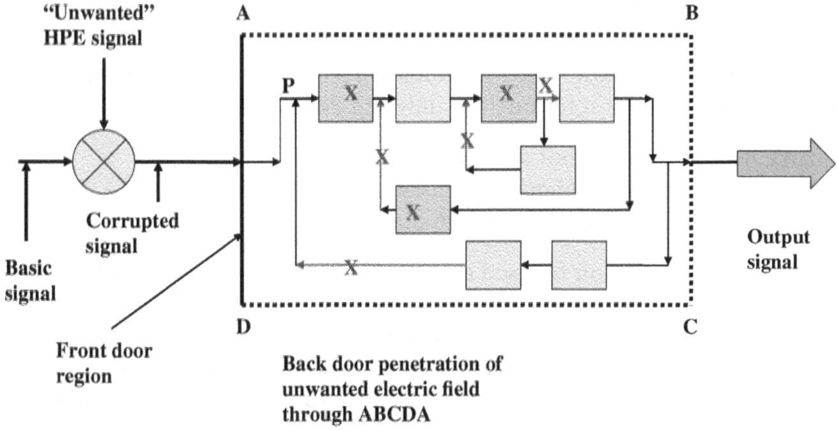

Figure 1.3 Example of electronic threat on the electronic system.

being made in information theory, nanomaterials, metasurfaces, etc. Our goal is that given a testing resource of several facilities that collectively cover the spectrum of threat waveforms applied to different targets over defined time interval, what is the most effective way by which we can assess the hardness, survivability, and reliability of the system. As a starting point, we assume that we know (1) the baseline cost and testing time for testing a single hypothetical object against a specified threat waveform in every relevant facility and (2) the total time for testing in a specified facility. The baseline threats include but are not limited to those identified in MIL-STD-461, MIL-STD-464C, and MIL-STD-3023, but do not include chaotic waveforms. In addition, special testing consideration may need to be given for simulating aircraft stability under threat conditions.

The primary goal of Section 1.5 is to provide techniques that will reduce testing and hardening costs while retaining required technology levels. Over the past 20 years, numerous studies were conducted to improve the quality of testing and reducing the cost that were consistent with the DoD budget. This needs to be examined more critically, and a rigorous analytical, practical, and realistic model that establishes the minimum testing cost for obtaining correct system assessment needs to be developed.

This chapter also begs the question "How do the technologies of this chapter relate to each other and to other chapters of this Handbook"? It does so in several seamless ways as we demonstrate in the following hypothetical scenarios. These scenarios reflect the kinds of issues that may need to be addressed by the aerospace EMC and E^3 community beginning now and extending to the near-term and the future. As we will show using three hypothetical emerging scenarios, new problems may cut across a spectrum of aerospace EMC and E^3 technologies.

For *illustrative purposes only*, we initially recommend the following three areas as topics of interest for further study and assessment:

- Large networks spread out over great distances,
- The inclusion of autonomous systems at all ranges,
- Emerging unwanted waveforms including chaos.

The foregoing selection is based on current international technical publications and conferences, but, by no means, is it meant to be comprehensive and, most likely, will change over time. It is anticipated that most of the core technologies outlined in this Handbook will remain over several decades.

For each of the three topics we have constructed a scenario that illustrates the connection between aerospace EMC and E^3 technologies. Scenario 1 emphasizes the extension of the current EMC and E^3 baseline to large systems, but does not include autonomous aspects. Scenario 2 includes an arbitrary combination of today's level of human control combined with autonomous systems. Scenario 3 recognizes that the electromagnetic environment will get more difficult to describe and quantify. The old standby of using Gaussian noise models will probably not suffice.

Scenario 1

Currently, the vast majority of testing and certification is conducted on individual newly developed electronic black boxes and/or modest-size systems. For example, a (large) system in today's vernacular might pertain to single aircraft, single space platform, single ground communication node, tank, HPM source or target, etc. All of the foregoing items usually fit into an anechoic chamber or a large free field simulator such as that used for simulating electromagnetic pulse (EMP) effects. Precious few electronic and communication systems that extend over large distances have been EMC and/or E^3-tested. Such tests may be difficult if not near impossible to perform in real-time and/or real-space (viz., operational) scenarios because of physical constraints and cost associated with acquiring enough reliable statistical data in real-world situations.

Consider, for example, a multiradar/multiaircraft scenario using current and/or near-term advanced communications and detection algorithms. Hypothetically, we might need to answer such questions as the following in order to translate test results into a real system:

- Are aircraft antenna transmitter and receiver gains adequate?
- Are detection, discrimination, and navigation algorithms adequate?
- Is the system number-of-aircraft limited?
- Are individual aircraft subject to unstable maneuvers?
- Is the system stable?

Although similar scenarios may have already have been tested for a limited number of aircraft and for a limited range and area, to our knowledge, the full potential of this and/or similar systems does not appear to have been explored. The analytical tools necessary to answer the aforementioned questions are rendered throughout the chapter.

The theoretical foundations of EMC and E^3 discussed in Section 1.4 provide the starting point since this is the place where baseline electromagnetic threats and state-of-the-art EMC technology sets the stage for expanding laboratory-scale results into real systems. This can be followed, for example, using the random graph models of Section 1.3, which show how degraded networks can even provide useful network performance. Ultimately, the ability to extend laboratory-scale results to realistic networks will likely require a variant of dimensional analysis theory, which is initially introduced to the reader in Section 1.2.4.1.

Scenario 2

This hypothetical scenario is a combination of scenario 1 combined with autonomous system considerations. The introduction of autonomous elements makes this problem a "complex" problem, as contrasted to a "complicated" problem. As pointed out in Section 1.2, by "complex" we mean that there may be significant uncertainties in performance due to the interaction of machine learning algorithms connected with autonomous behavior and natural unpredictable environmental effects. This environment is a relatively uncharted territory. By contrast, a "complicated" problem might include a hierarchical scheme containing known options—that is, even though there may be a large but finite number of outcomes, their characteristics and features are defined.

Currently, individual autonomous systems such as robots are making huge gains in capability, and in the near term we can easily imagine scenarios composed of hosts of robots and/or humans. Any time when test that involve large number of platforms (human and/or autonomous) in real operational situations are required the developmental cost increase dramatically. Here again, dimensional analysis provides a way to obtain meaningful test results. But as discussed in Section 1.2.4, dimensional analysis must be worked in harmony with the rigorous theoretical conditions of Sections 1.2.1–1.2.3 imposed on autonomous systems to get meaningful results.

Scenario 3

Scenario 3 is of much smaller scale than scenarios 1 and 2, but can be important in special cases. As the electromagnetic fields become more complicated, new waveforms emerge, and the methods for working with them in signal processing pose new challenges. The analysis of the chaos class of signals and their effects on aerospace EMC and E^3 is discussed throughout Section 1.4. For example, the theoretical signal processing community is debating whether the longtime behavior of chaos signals can be incorporated as a traditional stochastic process of classical noise theory.

1.2 Autonomous Systems

This section renders an overview and outline of EMC theoretical considerations for current and future aerospace systems. Our viewpoint extends from the current baseline threats and technologies to near-term and future EMC issues. In this chapter, we focus more heavily on (1) testing techniques on autonomous systems that rely on dimensional analyses and stability theory, (2) military air-to-ground and space-to-ground systems that employ networks with combinations of random nondirect and direct communication and sensor links, (3) new-generation aircraft that employ more susceptible electronic components requiring advanced nonlinear control analyses and control systems, and advanced signal analyses that can identify and manage chaos signals and HPM waveforms, and (4) efficient life-cycle cost-effective management systems. These issues are discussed respectively in Sections 1.2–1.5.

As recently pointed out by Weiss [1], there are great differences between manned systems and unmanned autonomous systems—"the main difference lies in the unmanned autonomous system's role in the decision process." Weiss also points out the need for a decision process that does not default into a human solution and also has a lower tolerance for errors as compared with robots. Additional definitions for autonomous systems also suggest that they operate and perform tasks without explicit human control over planned environmental conditions and make emergent and adaptive decisions [2–5]. Autonomous robots are also considered intelligent machines [6].

As developed principally by Roske [3, 4], and shown in Figure 1.4, there are three main functions of an autonomous systems that reside on the platform.

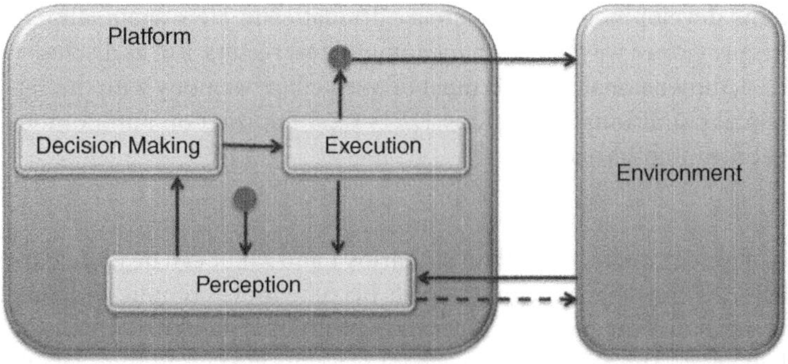

Figure 1.4 Perception, decision-making, and execution functions of autonomous systems [4].

These are *perception*—sensing the environment, *decision-making*—selecting the best course of action, and *execution*—implementing the selected best course of action. Each of these actions requires a combination and integration of self-consistent software and hardware. Moreover, much of the software will use machine learning [7]. These kinds of issues will place great demands on EMC. Not only will new ways to approach autonomous systems and subsystems need to be developed, but new testing paradigms, hardware, and software will also be essential [3, 4]. For example, Section 1.2.4 illustrates the potential role of dimensional analysis; implementation of this technique may also need new advanced EMC systems and analyses. There are numerous studies underway that address the development of autonomous systems (e.g., [8–11]).

The ultimate goal of an autonomous system is to provide the operating conditions and performance standards for each decision-making algorithm. Complex adaptive algorithms need to be developed and tested to ensure it will perform its function as required within the system as a whole. Related control theory is used to design electromechanical algorithms employed in the system while ensuring that the algorithms work effectively in a stable manner in response to their stimulation and interactions with other components in the overall system.

The underlying assumption in the illustration of Figure 1.5 is that autonomous decision-making algorithms will increasingly rely on CAS that include part both human and software components. The challenge to designers and testers will be to establish the operating conditions and performance objectives for these autonomous decision-making algorithms. A summary of the overall approach is:

1. Determine the extent of each human decision-making activity in a system as to maximize the overall effectiveness and minimize the overall risk based on the cost of the system.
2. Apply control theory to translate the mission environment and objectives into an expression of the required operating conditions and performance standards for each human decision-making activity.
3. Use the purpose, risk, and cost considerations to select human decision-making activities for conversion to CAS-based algorithms.
4. Define selected decision-making activities for system design-specific test conditions and performance objectives for test and execution (T&E) of the algorithm. Four key elements of testing will be dimensional analysis (particularly with regard to time-critical actions), nonlinear control actions, degree of human involvement, and projected cost.

Figure 1.5 is a model of a near-term autonomous system. The environment plays a critical role inasmuch as it is the principal generator of unwanted electromagnetic signals that may require nonlinear control functions.

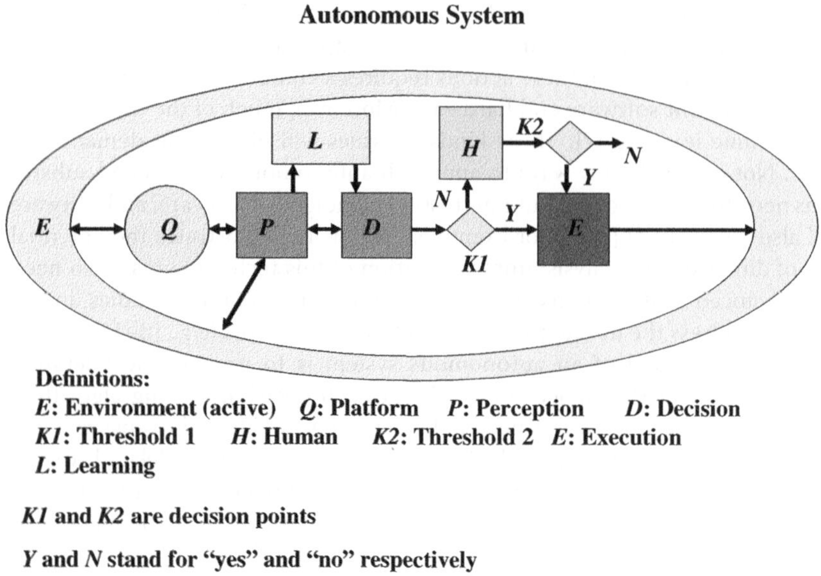

Figure 1.5 Illustrative model of a near-term autonomous system.

1.2.1 Classification of Autonomous Systems

1.2.1.1 Ideal Autonomous Systems

The benefits for developing robust autonomous systems for land, sea, air, space, and cyber applications are well established. Developing and testing these systems to required levels of capability is the challenge of the future. In addition to the basic issues of reliability, susceptibility, survivability, etc., future testing will need to address uncertainties in performance due to machine learning. For example, Nillson [7] has proposed "that a machine learns whenever it changes its structure, program, or data (based on its inputs or response to external information) in such a manner that it is that expected future performance improves." This statement covers a vast territory extending from those cases where machine learning is essentially just an efficient computational algorithm to cases where complex decisions regarding performance are rendered in domains never before encountered. Machine learning may or may not involve human participation in the decision process.

Interestingly enough, there is also some ambiguity regarding autonomy. In the extreme, we define an ideal autonomous system (IAS) as one that responds to its local immediate environment, and independently makes changes in accordance with some established overriding principles, rules, or criteria. These may in fact include programmed time-varying activity, but according to our

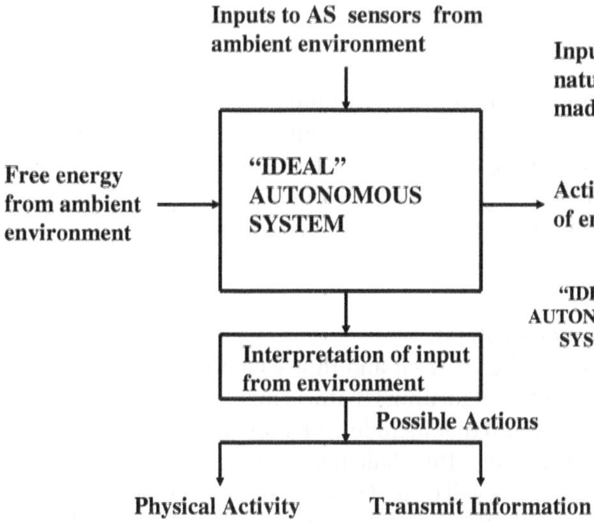

Figure 1.6 Concepts for ideal Autonomous System (AS).

definitions, autonomous systems receive no further instructions from a higher authority once set in motion. By ideal we mean it works perfectly, although we do not necessarily assume that the decisions are correct. There are some benefits in setting up the concept of an IAS. It establishes a baseline in which performance can be measured and quantified.

Figure 1.6 is a conceptual sketch for an IAS. The details of the internal wiring within the box labeled "Ideal Autonomous System" are not displayed at this top level. Indeed, the IAS of Figure 1.6 may be composed of several smaller autonomous systems, but there are generally two distinct classes of autonomous systems: (1) deterministic and normal stochastic control systems and (2) collections of artificial intelligence (AI) units (e.g., machine learning algorithms, and decision aids). What is important to note in Figure 1.6 is that all decisions and actions do not involve external guidance and updates, and end up with outputs that involve physical actions and/or the transmission (no further reception is allowed in the ideal case) of information to other systems.

The ideal system of Figure 1.6 may be interpreted as an eventual goal for an autonomous system. In the foreseeable future, we may need human intervention to at least prevent low-probability-of-event /huge catastrophic consequences. The IAS of Figure 1.6 inherently assumes that the system has enough intelligence to determine the best techniques to counter adversities. Today, only relatively simple systems approach true autonomous operation, and these typically have minor consequences for functional failure. Today's complex autonomous systems are essentially versions of Figure 1.6 with three major changes:

- The AI portions of Figure 1.6 are replaced by humans.
- The traditional feedback control systems of Figure 1.6 are not as robust as those that will be needed in the fully autonomous regime.
- The relationship between consequences of system failure and operating cost is easier to define because there is a vast experience base for assessing human operating cost.

1.2.1.2 Composite Autonomous Systems

The underlying assumption is that current and future composite autonomous systems will use decision-making algorithms that will increasingly rely on composite autonomous systems (CAS) that include both human and software components for their decision-making. The challenge to designers and testers of these decision-making algorithms will be to define those environments, operating conditions, and performance objectives where CAS can work and to design and test these autonomous decision-making algorithms.

The potential sequence of steps for the overall approach is as follows:

- Determine the extent of each human decision-making activity in a system design so as to minimize the overall "effectiveness and risk"-based cost of the system. This identifies the location, purpose, relative importance and enables the calculation of the cost of each human decision-making activity in the system design.
- Apply the mathematics of control theory to translate the mission environment and objectives through the system design and into an expression of the required operating conditions and performance standards for each human decision-making activity. This provides a disciplined and cost-effective approach to defining the purpose, functions, operating conditions, and performance objectives for each instance of human decision-making activity to be employed in the system.
- Use the purpose, risk, and cost considerations to select human decision-making activities for conversion to CAS-based algorithms as the CAS technology allows.
- Use the operating conditions and performance standards for a selected decision-making activity as the mission and system design specific test conditions and performance objectives for test and evaluation (T&E) of the algorithm designed to perform that activity to those standards under those conditions. Note that the opportunity for the human experience in performing a decision-making activity is to enhance the definition of effective operating conditions and performance standards for the CAS algorithm replacing the human.

To illustrate the first step in the overall approach, the extent of human involvement in a system will be determined to minimize the system's overall cost. This approach recognizes that cost is a consequence of designing the system not only to achieve its mission objectives but also to minimize cost of associated risks. Human involvement is currently the mechanism for best achieving the mission objectives and reducing the cost from unintended consequences.

It is the algorithms that define the characteristics and robustness of autonomous systems. It is also essential that we acknowledge that the autonomous systems we are considering are fundamentally all control systems. We are at liberty to call humans control systems if we recognize that humans have the capability of perception, decision-making, and execution embedded in a single integrated control system. To be sure, humans have limitations; at times they perform better than machines and at other times not as good. Human control systems are essentially good (but not perfect) at dealing with unpredicted events that are out of the usual domain of machine learning.

Operating conditions and performance standards for a decision-making activity provide a "template," a framework, in which an algorithm can be developed and tested to ensure it will perform its function as required within the system as a whole. Control theory is routinely used to define operating conditions and performance standards to guide the design of electromechanical algorithms employed in a system, ensuring the algorithms works effectively in a stable manner in response to their stimulation and interactions with other components in the overall system.

By characterizing the human decision-making activities in the form of a set of decision-making algorithms in a control theory framework, we develop an approach for expressing the operating conditions and performance standards for each human decision-making activity within in the context of the whole system performing its mission. Each activity's operating conditions and performance standards provide the framework to guide the design of a complex adaptive algorithm to perform that specific decision-making activity.

1.2.2 Combined Autonomy with Human Factors

1.2.2.1 Tradeoffs between Autonomy and Human Capability [3, 4]

By putting humans and traditional control systems on the same mathematical footing, we more clearly define their capabilities in T&E, and the resources necessary to sustain each component. In summary, we propose:

- All autonomous systems should be viewed as a combination of a physical part composed of N_P subsystems and a complex adaptive part comprised of N_{CA} human and human-related subsystems. We will shortly define N_P and N_{CA}

in more detail. The total number of subsystems embodied in an autonomous system is then

$$N_T = N_P + N_{CA} \tag{1.1}$$

and the fraction of physical subsystems is defined as

$$\gamma \equiv \frac{N_P}{N_P + N_{CA}} \tag{1.2}$$

We also have

$$N_{CA} = \frac{(1-\gamma)N_P}{\gamma} \tag{1.3}$$

$$\gamma = \frac{1}{(1 + N_{CA}/N_P)} = \sum_{n=0}^{n=\infty} (-1)^n \left(\frac{N_{CA}}{N_P}\right)^n \tag{1.4}$$

- As long as there is any subsystem within the entire autonomous system that has a human component: $N_{CA} > 1$ we have $\gamma < 1$. Future technology will push autonomous systems to larger values of γ.
- There is a conceptual limit of $\gamma \rightarrow 1$, but it may never be reached—"close, but no cigar."

Until recently, there has been a tendency to label control systems with some nonhuman elements as autonomous. Instead, we suggest the following taxonomy based on two major algorithms previously identified: Automated system algorithms and autonomous system algorithms. Each of these major groups is composed of two subgroups as indicated below. A large autonomous system may include all four groups.

Automated System Algorithms [3,4]

- Unmanned automatic system: This is the most basic kind of autonomous system and requires no human involvement. We define N_A to be the number of *physical* systems of this kind.

  Example: The positive crankcase vent (spring ball valve mechanical decision-making algorithm) in your car. It opens when the crankcase pressure gets too high and vents the vapor to the intake manifold to be reburned in the engine, and if it breaks, you get worse mileage. There is no human monitoring its performance; consequences from its failure are just not severe enough to worry about.
- Manned-automatic system: Similar to the aforementioned unmanned-automatic system, this system requires no human involvement except for performance approval. We define N_B to be the number of *physical* systems of this kind.

- Example: The autopilot and automatic landing system on an airliner; the stop and open door buttons on an elevator. The system works fine almost all of the time. The conditions are known and understood, and its performance is well and strictly defined and predictable. If something unforetold occurs that requires altering the performance, the human can step in.

Autonomous System Algorithms [3, 4]

- Manned autonomous systems: These are defined to be the most advanced autonomous systems and are based on a combination of machine learning algorithms, leading AI technology, experienced-based rule-base systems, operational experience, etc. They are constructed from integrated human experience with hardware but can make decisions within a broad range. We define N_C to be the number of *physical* systems of this kind.

  Example: An integrated air defense control system. It monitors and, on its own, generates recommendations (and can actually manage/execute) how best to employ the air defense sensors and weapons for a collection of ships, air, and various ground vehicles. The human can override it at any time, and may choose to just have it provide suggestions that the human can consider and then execute as it sees best.
- CAS: These are systems that are still in the formative stage. They represent the final few steps taken with human involvement, and the related decisions may have major impact. We expect a few subsystems of this type. We define N_{CA} to be the number of subsystems of this kind.

  Example: These come in two flavors: human and software. Complex adaptive algorithms are a class of autonomous algorithm software. The various expressions describing a system as manned, unmanned, autonomous, or automated are simply referring to the extent to which each form of decision-making algorithm is employed in the system, any use of a human autonomous algorithm yields a manned system (the man can be remote or onboard), and any use of software decision-making can be either automated or autonomous in the respective forms of rule-based or complex adaptive algorithms. Automated decision-making can also be in the form of electromechanical algorithms.

Combining the foregoing comments, we have

$$N_P = N_A + N_B + N_C \tag{1.5}$$

From equation (1.2) we can now also interpret γ (gamma) as a measure of autonomy. For example, if $N_{CA} = 0$, we get $\gamma = 1$, corresponding to a completely autonomous system.

The underlying assumption in this illustration is that autonomous decision-making algorithms will increasingly displace the human involvement. The challenge to designers and testers of these decision-making algorithms will be to establish the operating conditions and performance objectives for designing and testing these autonomous decision-making algorithms. For the simplest system, there is but one value of γ. However, for the class and level of autonomous systems we are concerned with—platform, perception, decision-making, and execution—there may be many physical control systems as well as significant human components.

It is not obvious that equations (1.1)–(1.5) are the best choices for evaluating the partition of physical and human components because it does not weight their importance. We are not ready to address this issue right now but need to point out that this is an important consideration. On the other hand, it is interesting to explore the use of equations (1.1)–(1.5) to better understand the balance between human factors and the degree of autonomy.

By applying control theory based system design, the characteristics of the operating conditions and performance standards associated with each subsystem function can be calculated from the detailed system design and the operational conditions and performance standards for the system as a whole. The operating conditions and performance standards associated with each subsystem provide the design goals which are also the test conditions and test performance objectives for an autonomous decision-making algorithm that would replace the human currently performing the associated decision-making function.

This illustration of a control theory approach to design of an autonomous system, so as to inform on the operating conditions and performance standards for the autonomous decision-making algorithms, suggests a basis for early and persistent collaboration among designers and testers of autonomous systems. The purpose of this collaboration would be to ensure that the system design provides the performance-, risk-, and cost-derived operating conditions and performance standards necessary to effectively test the system's autonomous decision-making functions.

Every autonomous systems we consider is an integrated set of electrical, electronic, electromechanical, mechanical, and fluid control systems working in harmony with human: control actions, inputs, and decisions. For reasons discussed in the previous sections, there is always some level of human involvement that we have incorporated in the term "complex adaptive system." The balance between the number and complexity of control systems that comprise a specific autonomous system is derived from decision-making technology and consideration of the operational objectives and risks. It is possible to determine from the autonomous system design the number of physical control subsystems N_P, and the number of subsystems attributed to human involvement, N_{CA}, the total number of subsystems, N_T, and γ.

There are two major categories of costs connected with developing an autonomous system: reducing the life-cycle cost, C_T, and the risk associated with mission failure, C_F. For illustrative purposes, we briefly discuss both aspects. We first identify some factors that are involved in finding the minimum value of C_T for autonomous systems. We assign a value C_B to represent the basic cost for creating, operating, and maintaining the system, including yearly testing, repairs, replacement of parts, and upgrades. This is the same as it is always been. We assume that C_B is a function f_B of its constituent parts N_P, γ, and a normal (usual) set of environmental and system parameters $\vec{\Omega} = \{\varpi_1, \varpi_2, \ldots, \varpi_m\}$ (e.g., temperature, altitude, etc.) that the system must operate in; ϖ_i stands for the ith environmental and system parameter. We have

$$C_B \equiv f_B(N_P, \gamma, \vec{\Omega}) \tag{1.6}$$

C_B is not in general simply the sum of its human and control parts. For brevity and without loss of generality, we treat N_P as a single variable instead of the three rendered in equation (1.6).

For a given system, let us now assume there is at least one critical variable or parameter of the system $Q(N_P, \gamma, \vec{\Omega})$, which needs to be maintained above or equal to a value Q_0. For example, Q_0 could be a minimum information rate in a communication system, or the desired best resolution in an optical image system, or the firing rate in a weapon system, etc. In theory, we can even make Q_0 multidimensional; for this illustrative example, we pick Q_0 to be a single parameter. We also select the condition

$$Q(N_P, \gamma, \vec{\Omega}) = Q_0 \tag{1.7}$$

Substituting equation (1.7) into equation (1.6) gives

$$C_B = f_B(Q_0, \gamma, \vec{\Omega}) \tag{1.8}$$

Equation (1.8) shows that in any real system in which all the variables and parameters are known, we can determine C_B as a function of γ. If we are interested only in minimizing the basic cost we can determine the specific value $\gamma = \gamma_c$ from the equation

$$\left(\frac{\partial C_B}{\partial \gamma}\right)_{\gamma_c} = 0 \tag{1.9}$$

However, there are two other risk-related costs that affect the life-cycle cost. These are denoted by $C_K(\gamma)$ and $C_U(\gamma)$, respectively. They are defined as follows:

- $C_K(\gamma)$ is the collective cost associated with *known* negative consequences and is a sum of all events in this category that can be identified when expressed as a function of γ. For example, it could include severe weather effects, normal accidents, hostile attacks, etc. We define $C_K(\gamma)$ as the risk. We let "i" be the symbol for the ith risk and C_i its cost; this is given by

$$C_i(\gamma) = (\text{probability of } i\text{th risk occurring}) \times (\text{loss for } i\text{th risk}) \tag{1.10}$$

There results

$$C_K(\gamma) \equiv \text{Risk} = \sum_i C_i \tag{1.11}$$

- $C_U(\gamma)$ is the collective cost (risk) for events and consequences that have yet to be defined.

Using the foregoing definitions, the life-cycle cost, $C_T(\gamma)$, can be expressed as a sum of three terms shown in the following equation:

$$C_T(\gamma) = C_B(\gamma) + C_K(\gamma) + C_U(\gamma) \tag{1.12}$$

Formulating a comprehensive quantitative framework for evaluating $C_U(\gamma)$ has yet to be developed. We conjecture it will be based in part on operational experience, learning algorithms, EMC technology and models, and other AI tools that model capability. However, at this preliminary stage, it is not necessary or possible to include $C_U(\gamma)$ in the factors relating the degree of autonomy to risk. The goal of the analysis is then to minimize

$$C_T(\gamma) = C_B(\gamma) + C_K(\gamma) \tag{1.13}$$

which is subject to the constraint: $0 \leq \gamma \leq 1.0$. There are two limiting cases: (1) $\gamma = 0$ is for nonautonomous systems, all control functions are known and deterministic, the ultimate control is human control; (2) $\gamma = 1$ is for a completely autonomous system—no human control.

Increasing the level of autonomy for an autonomous system is often based on the life-cycle cost and other factors. Using equation (1.13) as an example the minimum cost is determined from the equation

$$\frac{dC_T(\gamma)}{d\gamma} = \frac{dC_B(\gamma)}{d\gamma} + \frac{dC_K(\gamma)}{d\gamma} = 0 \tag{1.14}$$

The value of γ that is the solution of equation (1.14) is denoted as

$$\gamma = \gamma_c \tag{1.15}$$

and the minimum cost is

$$C_{T,\min} = C_T(\gamma_c) \tag{1.16}$$

As an illustrative example, let us consider how the analytical dependence of γ_c might depend on the human contribution. A plausible case is the one where the basic cost $C_B(\gamma)$ decreases with an increase in γ. This is consistent with historical evidence that supports the observation that as the level of automation goes up for elementary systems, the labor costs decrease. For example, let us assume

$$C_B(\gamma) = A - B\gamma \tag{1.17}$$

In the above example, A and B are constants determined from the autonomous system design. Equation (1.17) is a sample expression that demonstrates the fact that automation often reduces system cost for the usual kinds of activities. In this case, we assume that even at 100% automation ($\gamma = 1$), the basic cost is always finite: $A - B > 0$. A more general case is where the basic cost $C_B(\gamma)$ is of the form

$$C_B(\gamma) = \sum_n a_n \gamma^n \tag{1.18}$$

and extends up to $\gamma = 1$. Figure 1.7 is an example of a hypothetical basic cost C_B as a function of γ which shows the occurrence of a minimum.

As shown in Figure 1.7, if the basic components of life-cycle cost were the only consideration, there would be a definite value of γ that provides the minimum cost, and it would make no sense to build-in a greater degree of autonomy. But in military, space, and civilian operations, there are often other factors that lead us to determine the appropriate value of γ. This issue is further amplified when we need to account for the collected cost of known and unknown events and consequences. This means minimizing $C_T(\gamma)$ from equation (1.19)

$$C_T = \sum_n a_n \gamma^n \tag{1.19}$$

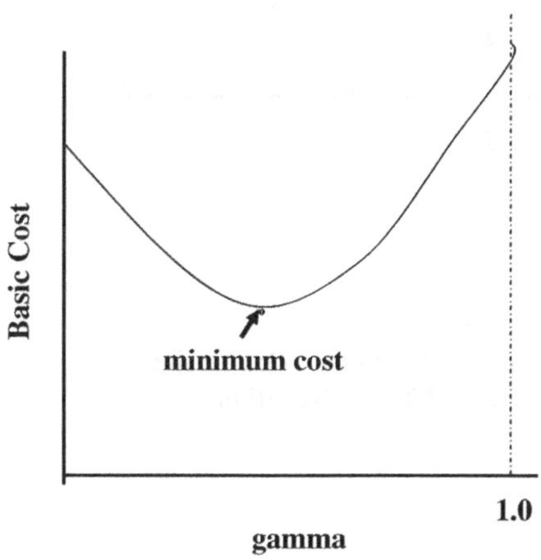

Figure 1.7 Hypothetical cost as a function of gamma.

The human involvement part of an autonomous system is often the factor that really determines its design and T&E. A major attribute of a human-in-the-loop is the inherent ability is to limit and/or large negative consequences. These are controlled by the CAS components and are connected with large autonomy. This was previously shown in equations (1.2) and (1.4) and is now rendered in equation (1.20):

$$\gamma = \frac{N_P}{N_P + N_{CA}} = \frac{1}{(1 + N_{CA}/N_P)} = \sum_{n=0}^{n=\infty} (-1)^n \left(\frac{N_{CA}}{N_P}\right)^n \tag{1.20}$$

For our interests, large autonomous systems in which the number of physical subsystems, N_P, greatly outnumbers the human and software based CAS, N_{CA}: $N_P \gg N_{CA}$ we have the approximation from equation (1.21):

$$\gamma \cong 1 - (N_{CA}/N_P) \tag{1.21}$$

Equation (1.21) shows that for the modern systems of interest, the autonomy is close to unity.

This is elucidated in Figures 1.8 and 1.9. The simplest case is that shown in Figure 1.8, as shown again, which applies to the automatic system case. The model could apply for the known events and consequence cases. Using equation (1.18), we could generate a diagram as in Figure 1.7, and corresponding

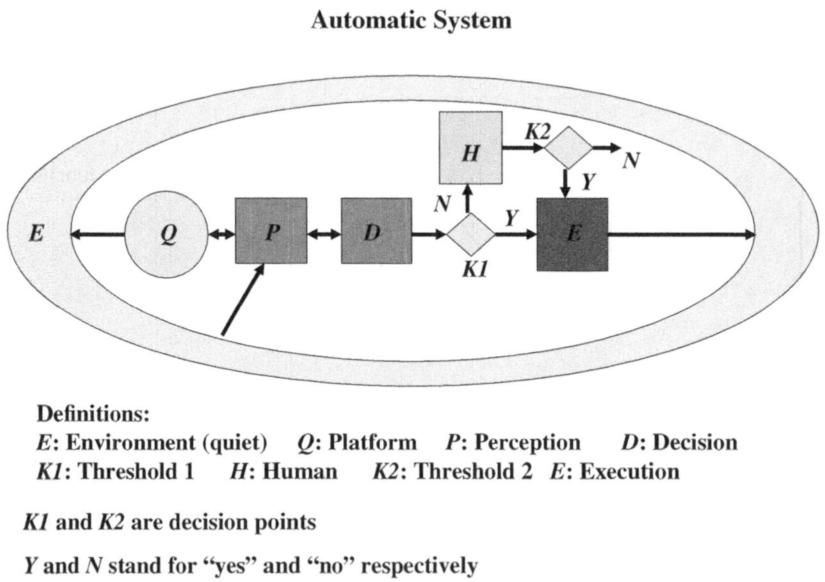

Figure 1.8 Model for automatic system.

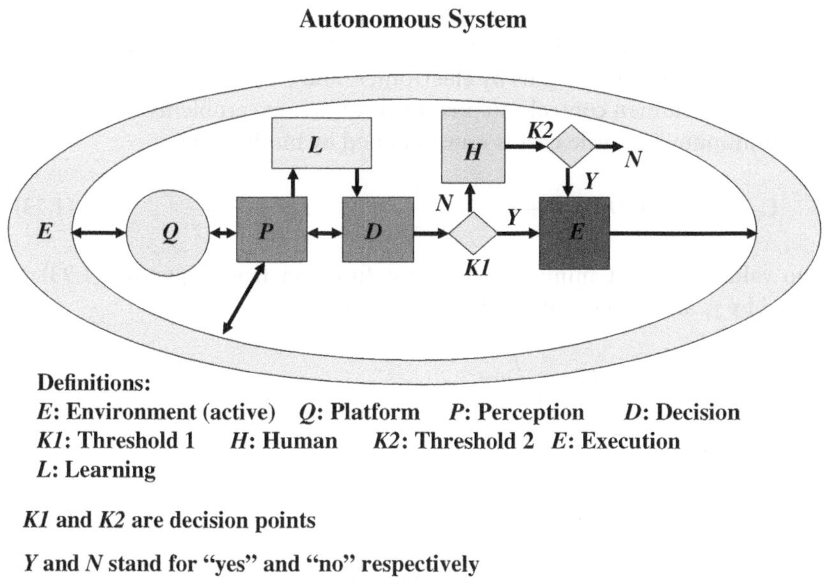

Figure 1.9 Model for autonomous system.

γ and minimum life-cycle cost. But this is not sufficient. Along with minimum life-cycle cost, we must ensure that the final system meets its functional requirements. This critical step is shown in the loop of Figure 1.8 originating with the diamond shaped object labeled *K1*. Conceptually, *K1* is the threshold corresponding to the requirement of equation (1.13). If the requirement of equation (1.13) is met, the execution of the decision occurs. On the other hand, if it is not met, the human steps in and can override the decision for various reasons using a modified threshold, *K2*.

Figure 1.9 is an expansion to the automatic case. It differs from Figure 1.8 in two significant ways: (1) the changes in the environment are random and (2) learning algorithms are introduced. Some suitable statistical methods for dealing with undefined events and consequents need to be developed. We see that the human plays a central role in autonomous systems. At the more basic level, humans compete with certain aspects of autonomous control, but it is at the final stages of decision-making and judgment that humans have the greatest impact. The total absence of human involvement as $\gamma \to 1.0$ could be severe. This point has been demonstrated in such movies as *Dr. Strangelove* and *West World*.

An *illustrative* hypothetical expression for known consequence, C_K, that emphasizes the sensitivity of a system to autonomy could be

$$C_K(\gamma) = D - E\gamma + \frac{\gamma F}{1-\gamma}, \tag{1.22}$$

where D, E, and F are event parameters. The consequence is composed of two parts: $D - E\gamma$ is that part which emphasizes the ability of the system to counter the threat by replacing humans by electronics, and $\gamma F/(1 - \gamma)$ is the part that the absence of human control can possibly lead to huge problems as $\gamma \rightarrow 1.0$.

The minimum life-cycle cost is now obtained by minimizing

$$C_T(\gamma) = (A + D) - (B + E)\gamma + \frac{\gamma F}{1 - \gamma} = G - H\gamma + \frac{\gamma F}{1 - \gamma} \tag{1.23}$$

The value of γ that minimizes the cost function from equation (1.23) is denoted by γ_c and is determined from the equation

$$\frac{dC_T(\gamma)}{d\gamma} = 0 = -H + \frac{F}{1 - \gamma_c^2} \tag{1.24}$$

The solution is

$$\gamma_c = 1 - \sqrt{\frac{F}{H}} \tag{1.25}$$

Equation (1.25) shows that when $(F/H) > 1$ there is no minimum, and the penalty for poor performance can be unbounded.

In actuality, human involvement can occur in many and various components of the system. The diversity of possible human involvement throughout the system motivates further expansion of this illustration. Control theory can then derive the required functional performance conditions and standards for each occurrence of human involvement in a specific system to enable that system to accomplish its performance objectives and thus provide a basis for defining the conditions and performance standards for designing and testing the autonomous decision-making algorithm that might replace the human decision maker in that system.

1.2.3 Stability of Autonomous Systems Including Nonlinear Theories

1.2.3.1 Linear Systems

This section explores the basic analytical methods and models that are necessary to create and evaluate usable autonomous systems. For illustrative purposes, this analysis is rendered in the context of deterministic feedback control model that also allows for human contribution. The question of how much human interaction is necessary is addressed. The feedback model of Figure 1.10 is shown for illustration.

In this model, $x(t)$ is the reference signal. This signal may be permanently fixed or time dependent in a purely autonomous system without human participation. Intelligence control comes into play when we allow $x(t)$ to be influenced by human intervention. The object of the control system is to generate

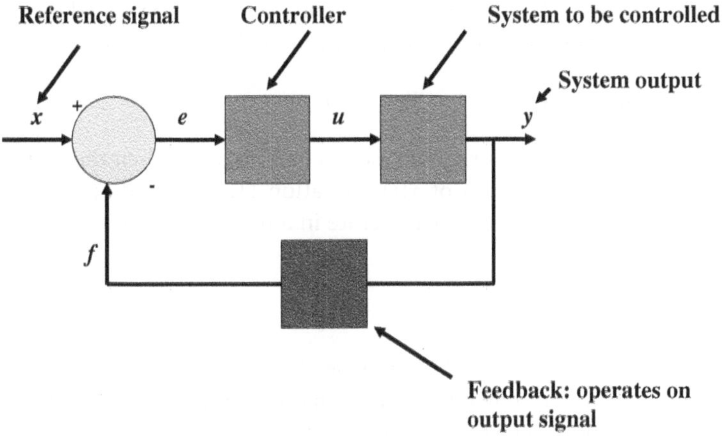

Figure 1.10 Deterministic feedback control model.

a desired prescribed system output $y(t)$. The controller operates on the error signal

$$e(t) = x(t) - f(t) \tag{1.26}$$

Its output $u(t)$ is the input to the system whose output is $y(t)$. The feedback function operates on $y(t)$, producing the function $f(t)$, which approximates the reference signal. Most feedback control systems operate in the linear range, which is characterized by small signals moving about some stable equilibrium points. The Laplace transform equations are

$$E(s) = X(s) - F(s)Y(s) \tag{1.27}$$

$$U(s) = C(s)E(s) \tag{1.28}$$

$$Y(s) = Q(s)U(s) = \left(\frac{Q(s)C(s)}{1 + Q(s)C(s)F(s)}\right) X(s) = H(s)X(s) \tag{1.29}$$

$$H(s) = \left(\frac{Q(s)C(s)}{1 + Q(s)C(s)F(s)}\right) \tag{1.30}$$

where $C(s)$, $F(s)$, $Q(s)$ are the transfer functions of the controller, feedback sensor, system to be controlled, closed-loop transfer function, $H(s)$, and output $Y(s)$, respectively.

For linear systems, the closed-loop transfer function, $H(s)$, will contain linear differential and integral operators, and in a strict sense this makes $y(t)$ a functional of $x(t)$ [12, 13]. However, the fundamental transfer functions $C(s)$, $F(s)$,

$Q(s)$ also depend on physical parameters $\alpha_1, \alpha_2 \ldots \alpha_N$, which are assumed fixed and thereby time invariant. The output is now given by the functional form

$$y(t) = \Psi(x(t), \alpha_1, \alpha_2, .\alpha_k \ldots \alpha_N) = \Psi(x(t), \vec{\alpha}) \tag{1.31}$$

where $\vec{\alpha} = \{\alpha_1, \alpha_2, .\alpha_k \ldots \alpha_N\}$ is the set of systems parameters. When these are specified, $y(t)$ is a unique function of $x(t)$. Equation (1.31) would apply for example when feedback control systems operate in a predictable regime often defined by the condition

$$\left|\Psi(x(t), \alpha_1, \alpha_2, .\alpha_k \ldots \alpha_N)\right| \leq M \tag{1.32}$$

For example, in equation (1.32), M could be the threshold for nonlinear effects. In the event of an unpredictable change in one or more of the parameters, e.g., $\alpha_k \to \hat{\alpha}_k$, we could get the unacceptable condition

$$\left|\Psi(x(t), \alpha_1, \alpha_2, .\hat{\alpha}_k \ldots \alpha_N)\right| > M \tag{1.33}$$

and some sort of troubleshooting corrective action would be necessary. In addition, there is also the possibility that the reference signal becomes unpredictable with or without cause, $x(t) \to \hat{x}(t)$. This also generates the unacceptable condition

$$\left|\Psi(\hat{x}(t), \alpha_1, \alpha_2, .\alpha_k \ldots \alpha_N)\right| > M \tag{1.34}$$

In serious cases leading to equation (1.33) or (1.34), the ultimate last resort of control is human control. For obvious reasons, this is not acceptable for autonomous systems. Deterministic feedback control systems must be more robust. Even though there may be good reason to select specific stable operating points, we must be assured that when the autonomous systems are driven way off the stable operating points, and possibly into the nonlinear range, we do not experience severe consequences. The object is to build autonomous systems that are *sufficiently* reliable and at *reasonable* cost.

The first step in building a robust autonomous system is to ensure that unanticipated disturbances that are capable of driving the deterministic feedback control system into the nonlinear region and *cause catastrophic damage* will occur infrequently. If $x(t)$ and $\vec{\alpha} = \{\alpha_1, \alpha_2, .\alpha_k \ldots \alpha_N\}$ are random variables characterized by the probability density function $\mu(x, \alpha_1, \alpha_2, .\alpha_k \ldots \alpha_N)$, and $\gamma(x, \alpha_1, \alpha_2, .\alpha_k \ldots \alpha_N)$ is the function space where catastrophic damage can occur, the probability that catastrophic damage will *not* occur is given by

$$P_N = 1 - \int\int\int \ldots \int \mu(x, \alpha_1, \alpha_2, .\alpha_k \ldots \alpha_N) dx d\alpha_1 d\alpha_2 .. d\alpha_k, .. d\alpha_N \tag{1.35}$$

The integration is carried out over the function space γ.

Today's deterministic feedback control models for are based on Lyapunov's stability theories. Establishing stability is the cornerstone of all control systems. There are two parts of Lyapunov's theories and deal with linear and nonlinear systems [14, 15]. The first method examines stability around a few selected operating points by rendering the nonlinear equations linear in a limited region around these points using small signal analysis. For this case, the functional analysis of equations (1.26)–(1.34) applies to the autonomous case. When the autonomous system gets forced into the nonlinear regime, the functional analysis of equations breaks down because the output of the system now depends on all other times [12–15].

In the linear time regime equation (1.29) becomes

$$y(t,\vec{\alpha}) = \int h(\tau,\vec{\alpha})x(t-\tau)d\tau \tag{1.36}$$

while for the nonlinear regime we can express $y(t,\vec{\alpha})$ in the Volterra series [14, 15]

$$y(t,\vec{\alpha}) = k_0(\vec{\alpha}) + \sum_{n=1}^{\infty} \frac{1}{n!} \int_{-\infty}^{+\infty} \dots \int_{-\infty}^{+\infty} k_n(t_1, t_2, ..t_n, \vec{\alpha})x(t-t_1)x(t-t_2)\dots.. \\ x(t-t_n)dt_1 dt_2 \dots dt_n \tag{1.37}$$

In the foregoing series, k_n are the Volterra kernels and $\vec{\alpha} = \{\alpha_1, \alpha_2, .\alpha_k, .\alpha_N\}$. An analogous formalism was developed by Norbert Wiener [12] that connects the output at any time to the system's history. Calculating the Volterra kernels, k_n, is difficult and in some cases the series of equation does not converge [14, 15]. Franz and Scholkoph have shown that significant improvement is possible by reformulating the Volterra and Wiener techniques as operators in a Reproducing Kernel Hilbert Space.

1.2.3.2 Volterra and Weiner Theories

Assuming that the computational difficulties for calculating k_n can be overcome, equation (1.37) may provide direct insight into the asymptotic time behavior of $y(t,\vec{\alpha})$ for the autonomous system. We would clearly like the result: $y(t \to \infty, \vec{\alpha}) \leq |\varepsilon|$, where $|\varepsilon|$ is a suitable bound determined from the system requirements. A result equivalent to equation (1.37) can also be derived by casting it in terms of state variables. This requires solving the nonlinear vector equation

$$\frac{d\vec{L}}{dt} = \tilde{G}(\vec{L}, \vec{u}, \vec{\alpha}) \tag{1.38}$$

where $\vec{L}$ is the state variable for the entire system and $\vec{u}$ is the control function vector. However, in contrast to the functional approach which involves a single input, the control vector may be more complex. The best method for designing a stable system with no catastrophic damage needs to be evaluated.

The linear case in which $P_N = 1$ defines the IAS. But, in reality perfection does not exist. We are always in the regime where P_N is less than but never exactly equal to 1.0. As the capability of autonomous systems increases so will its dependence on AI in the decision-making process. This is manifested in the input signal, which is now correctly recognized as a random variable, that is, $x(t) \to x(t, \beta)$. The random variable β represents the set of possible choices created by the AI decision-making systems. The possible choices may most likely be discrete but for brevity we treat it as continuous with a probability density function $\omega(\beta)$ properly normalized:

$$\int \omega(\beta)d\beta = 1.0 \tag{1.39}$$

If all the possible choices are known *a priori* from testing, the AI system is essentially a rule-based system. By using machine learning system new information gets added to the existing base to create an updated data base and new probability density function, $\bar{\omega}(\bar{\beta})$, which again satisfies the condition

$$\int \bar{\omega}(\bar{\beta})d\bar{\beta} = 1.0 \tag{1.40}$$

AI machine learning works best when new information has the opportunity to be digested in small bites.

This statement speaks to equation (1.37) in which case the output takes the form

$$y(t, \vec{\alpha}, \beta) = k_0(\vec{\alpha}) + \sum_{n=1}^{\infty} \frac{1}{n!} \int_{-\infty}^{+\infty} \cdots \int_{-\infty}^{+\infty} k_n(t_1, t_2, ..t_n, \vec{\alpha})x(t - t_1, \beta)x(t - t_2, \beta) \ldots.. \\ x(t - t_n, \beta)dt_1dt_2 \ldots dt_n \tag{1.41}$$

If $x(t, \beta)$ is an increasing function of β, equation (1.41) then shows an increasing possibility for system damage. Even though there is no guarantee that an abrupt change in the AI-derived input could be tolerated, the mathematical structure developed here should be integrated with machine learning to eliminate catastrophic damage and allow for system reconfiguration if necessary.

1.2.3.3 Functional Elements of Autonomous Systems

In Figure 1.1, we examined the functional elements that are embedded in autonomous systems. Roske and Kohlberg showed that the principal parts of an autonomous system can be modeled to include the following main sequential functions: *perception* →*decision-making* → and *execution* [3,4]. Figure 1.11 shows how these functions tie together. This figure shows that the perception

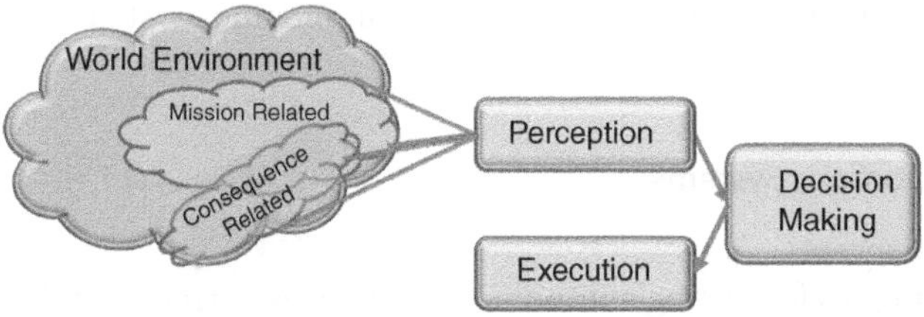

Figure 1.11 Perceptions for mission success and avoiding unwanted consequences [4].

function couples to the outside world and may also be the dominant source of both risk and unintended consequences via decision-making leading from the acquisition and sensing of the environment to the execution.

The functional blocks may contain distributed elements and there may in fact be many outputs from them. Equation (1.41) represents a hypothetical result for a distributed feedback control system that may exist in any of the major functional elements. In the simplest case, equation (1.41) could reside in the perception block where $x_P(t, \beta)$ originates in the *Mission-Related* cloud and its single output is $y_P(t, \vec{\alpha}_P, \beta)$. We rewrite equation (1.41) as (the subscript "*P*" refers to *perception*):

$$y_P(t, \vec{\alpha}_P, \beta) = k_0(\vec{\alpha}_P) + \sum_{n=1}^{\infty} \frac{1}{n!} \int_{-\infty}^{+\infty} \ldots \int_{-\infty}^{+\infty} k_n(t_1, t_2, ..t_n, \vec{\alpha}_P) x_P(t - t_1, \beta) x_P (t - t_2, \beta) \ldots x_P(t - t_n, \beta) dt_1 dt_2 \ldots dt_n \tag{1.42}$$

The input to the decision-making block is then $y_P(t, \vec{\alpha}_P, \beta)$ and the output is

$$y_{DM}(t, \vec{\alpha}_{DM}, \beta) = k_0(\vec{\alpha}_{DM}) + \sum_{n=1}^{\infty} \frac{1}{n!} \int_{-\infty}^{+\infty} \ldots \int_{-\infty}^{+\infty} k_n(t_1, t_2, ..t_n, \vec{\alpha}_{DM}) y_P (t - t_1, \vec{\alpha}_P, \beta) y_P(t - t_2, \vec{\alpha}_P, \beta) \ldots y_P(t - t_n, \vec{\alpha}_P, \beta) dt_1 dt_2 \ldots dt_n \tag{1.43}$$

In equation (1.43) the quantity $\vec{\alpha}_{DM}$ stands for the set of parameters contained in the decision-making functional block (same meaning as $\vec{\alpha}_P$). It is easy to see that the foregoing process demonstrated in equations (1.42) and (1.43) can be readily extended to include the execution functional block.

The point to be brought out here is that instabilities created by unanticipated consequences can propagate throughout an entire system. Pretesting and testing individual functional blocks is useful, but testing the entire system is essential to limit unanticipated catastrophic consequences. We should also reiterate a key point: interactions between perception, decision-making, and execution will, in general, be more complex than that shown in Figure 1.11. Structuring and evaluating testing methods for this new class of autonomous system is a key challenge.

1.2.4 Testing Autonomous Systems

1.2.4.1 Basic Testing Concepts

Developing the techniques for testing autonomous systems is currently in the formative stage. Initial structuring of an approach to this topic appears to have been addressed by Vince Roske of institute of defense analysis (IDA) between 2011 and 2012 (Roske and Kohlberg [4]). Numerous military and civilian personnel have urgent needs for autonomous systems: satellite systems, C4ISR, transportation systems, military, etc. The leaders in this field define an autonomous system as one that requires no human assistance in any function and can interpret and interact with their local environments.

Tests must precisely understand the difference between "automated systems," where alternative and known decisions are initially available, and "autonomous systems," where unpredictable conditions prevail—this is the domain of decision-making algorithms. A major difference between manned and autonomous systems lies in the decision-making algorithms, which are basically rooted in software [1, 3, 4]. At some point there is also the possibility that human-type adaptive reasoning may have been employed in the development of decision-making software algorithms, but at this point the human is out of the picture.

One may draw a sharp distinction between automatic systems and autonomous systems. While automatic systems might very well be embedded in the larger autonomous systems, the converse is not true; a larger automatic system that includes smaller embedded autonomous systems is still regarded as an automatic—*not* autonomous system. This distinction is a crucial aspect of testing. Automatic systems, no matter how complicated they are, deliver known well-defined results, whereas the outputs of autonomous systems inherently have a strong probabilistic flavor.

Even when probabilistic elements are present, it is often possible to quantify the behavior of systems through repeated tests conducted using standard statistical methods. However, this approach may not always be appropriate and/or feasible for autonomous systems because they may need to function for very long times in highly variable environments. Laboratory-scale testing for reasonable times may be required to extrapolate results to environments of interest. One possible tried-and-true method for rigorously extrapolating laboratory-scale results to true environments is dimensional analysis [16–20]. This technique has been used for about hundred years in numerous areas, such as fluid mechanics, heat transfer, shock physics, but not much in EMC. In the next section, we provide a brief summary of this technique in the hope that it will be examined by the EMC community in the near future.

1.2.4.2 Dimensional Analysis with EMC Considerations

The purpose of this technique is to show how laboratory-scale experiments could be scaled to real-life cases. For example, since it is expected that many autonomous systems may require very long operational lifetimes, it would be useful to show how laboratory tests conducted over hours or days could answer questions of autonomous system behavior in the months-to-years range, typical of NASA missions, drones that conduct C4ISR missions, environmental studies, etc. Unfortunately, we do not have specific examples at this time that show how laboratory-scale tests that use can complex hardware and software that can well simulate large-scale missions. However, the forthcoming discussion rendered here shows how to use dimensional analysis—more commonly known as "Buckingham's Π Theorem" [16] to accomplish our EMC goals.

As a starting point, consider, for example, the differential equation of motion for the velocity, v, of a particle of mass, m, initial velocity v_0, and frictional force $f_r = -\beta v^3$ with constant parameter friction coefficient β. The equation of motion is

$$m\frac{dv}{dt} = f_r = -\beta v^3 \tag{1.44}$$

The analytical solution to the foregoing equation is easily deduced, but if we could not solve it in closed form we could write the solution as

$$v = v(m, \beta, v_0, t) \tag{1.45}$$

The instantaneous velocity, v, is a function of three system parameters: m, β, v_0 and time, t. Introducing a dimensionless velocity, $y = v/v_0$ reduces equation (1.44) to

$$\frac{dy}{d\tau} + y^3 = 0 \tag{1.46}$$

where $\tau = t/\bar{t}$, $\bar{t} = m/\beta v_0$, and $y(\tau = 0) = 1.0$. The solution to equation (1.46) is

$$v = \frac{v_0}{\sqrt{1 + 2\tau}} \tag{1.47}$$

In the foregoing example, we have reduced greatly the dimensionality of the problem by knowing the differential equation. However, the foregoing achievement has not directly rendered insight into the technique that yields the connection between the laboratory and the real environment. This is accomplished via Buckingham's Π Theorem, which is summarized as follows. Assume a system can be described by "n" quantities composed of either physical constants and/or physical variables (e.g., electrical conductivity/permittivity/permeability, current/voltage, electrical or mechanical power, pressure, and temperature) and "m" dimensions. In systems composed of both fluid mechanics and electrical systems, the m dimensions are mass (M), length (L), time (T), temperature (θ), and electrical charge (Q).

Now, let $u_1, u_2, u_3 \ldots. u_n$ denote the aforementioned quantities which share the following relationship between themselves:

$$F(u_1, u_2, u_3 \ldots. u_n) = 0$$

Buckingham [16] proved that if each entity of the set $u_1, u_2, u_3 \ldots. u_n$ depended on one or all of the m dimensions, we could develop $n–m$ new dimensionless groupings of the set: $u_1, u_2, u_3 \ldots. u_n$. This new grouping, called Π–functions, is labeled: $\Pi_1, \Pi_2, \Pi_3 \ldots \Pi_{n-m}$ and satisfies the equation:

$$f(\Pi_1, \Pi_2, \Pi_3 \ldots \Pi_{n-m}) = 0 \tag{1.48}$$

Using equation (1.48) the dimensionality of the problem has been reduced from n to $n–m$. If all four basic dimensions (that is, $m = 4$) are used, Streeter (19) has shown how to reduce the computation to the form

$$\begin{aligned} \Pi_1 &= u_1^{w_1} u_2^{x_1} u_3^{y_1} u_4^{z_1} u_5 \\ \Pi_2 &= u_1^{w_2} u_2^{x_2} u_3^{y_2} u_4^{z_2} u_6 \\ \Pi_{n-4} &= u_1^{w_{n-4}} u_2^{x_{n-4}} u_3^{y_{n-4}} u_4^{z_{n-4}} u_n \end{aligned} \tag{1.49}$$

The final step in the method is to express the u's in their dimensions and then determine the w's, x's, y's, and z's so that all the Π's are dimensionless. The complete solution to a problem is to reintroduce the fewer variables into the fundamental equations of the system under study. This is not a trivial problem, but if the number of variables can be reduced significantly, the overall method is extremely simple since the same Π–functions establish a one-to-one connection between laboratory and real world. Kohlberg and Coffee [20] provide a detailed and comprehensive treatment of Buckingham's theorem. Application of Buckingham's theorem to EMC problems is the next step.

1.3 Coupled Air and Space Survivable Systems

1.3.1 Overview of Selected Systems

1.3.1.1 Types of Systems

We begin with a top-down aerospace point of view. There are about 1000 operational satellites in existence from all the nations, with about 3600 remaining in orbit at LEO, MEO, GEO, and HEO altitudes [21]. Their functions cover earth observation, communication, navigation, weather reporting, research, classified programs, and support of space station operations. Spacecraft that are not

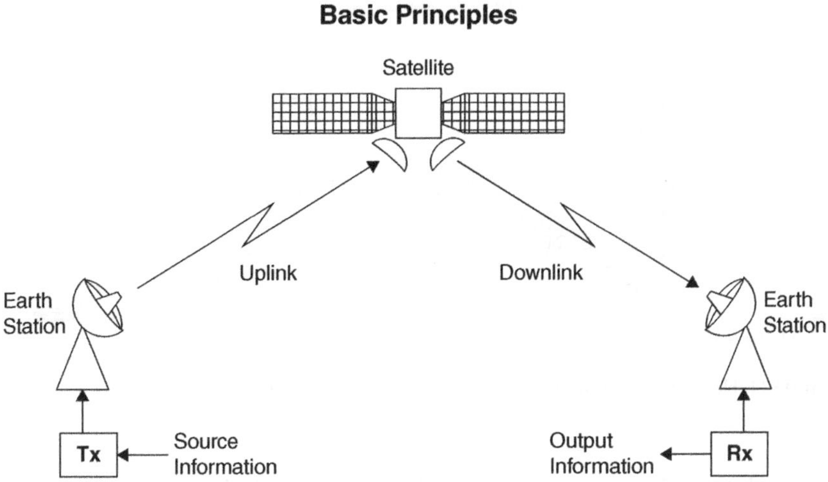

Figure 1.12 Basic links between satellite and ground station.

in orbit are not considered in this section. Beginning in the mid-1960s, the basic communications between a satellite and the earth was a single link that connected it to a single earth station. This is shown in Figure 1.12. Shortly thereafter, a single satellite could send messages to multiple ground targets, as shown in Figure 1.13. The concept of satellite-to-satellite communication, which was

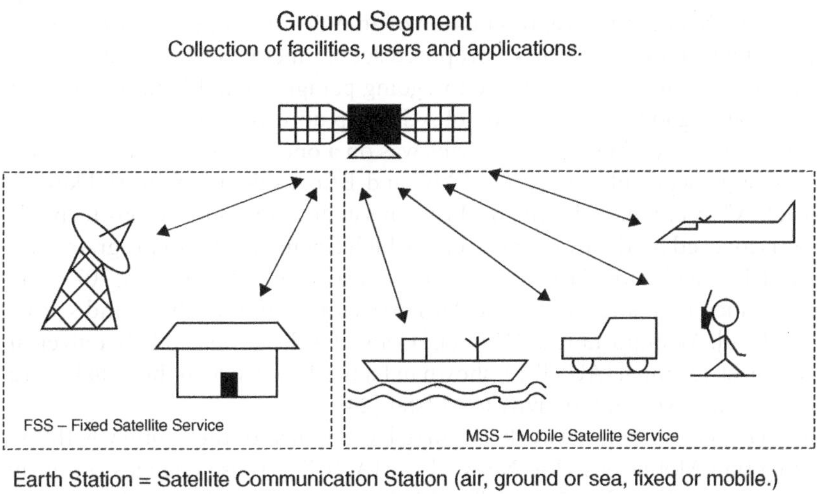

Figure 1.13 Sample of ground stations.

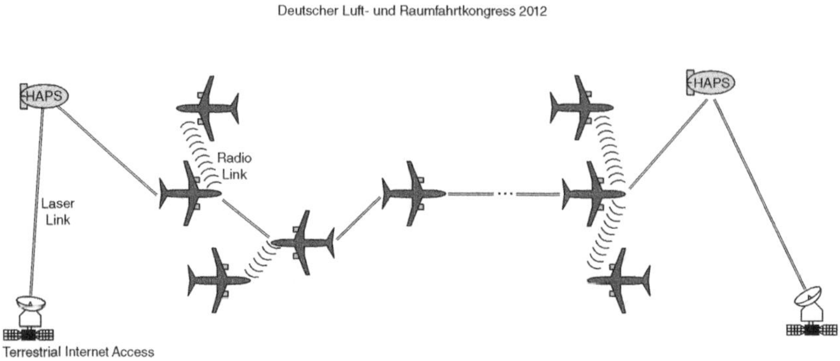

Figure 1.14 Networking between satellite, ground, and airborne systems.

helped along by laser communication, became strategically important during the Cold War. This capability enabled a global-wide picture of military events to become available to a few strategic command centers. It should be noted, however, that the number of satellites that communicated with each other was precious few.

Since the days of World War Two, aircraft have communicated with each other and ground stations via airborne radio, and in a sense this could be looked at as an airborne communication network (ACN). Earliest airborne communication in World War Two focused on dogfights. This capability translated into tactical air-to-ground support systems and then to large systems involving inter and intranetworks of satellites, airborne platforms, and land/sea platforms. Figure 1.14 is a hypothetical representation of an ACN involving the various platforms, while Figure 1.15 shows a proposed set of functions of an ACN [22].

The description rendered in the foregoing paragraph and Figures 1.14 and 1.15 is a fairly good representation of ACN's up until 2000. For the most part, the communication between platforms was on a one-to-one basis and did not involve large segments of ACMs. EMC and E^3 considerations were likewise "black box" oriented. A dominant thrust in satellite communications technology was focused on the Milstar extremely high frequency (EHF) program [23]. Figure 1.16 is a picture of the Paul Revere, "a heavily modified Boeing 707 operated by Lincoln Laboratory for the Air Force as a communications and sensor test bed [24]. An important EMC and E^3 element is the advanced EHF universal systems test-terminal (AUST-T), shown in Figure 1.17, used on the Paul Revere to provide airborne connectivity to Milstar [24].

Dramatic changes in networks occurred at the turn of the century with the introduction Mobile Ad-Hoc Networks (MANETs). These changes seriously affect the roles played by E^3. Initially, MANETs were focused on land networks, but now MANETs can also apply to airborne communications. In fact, complex

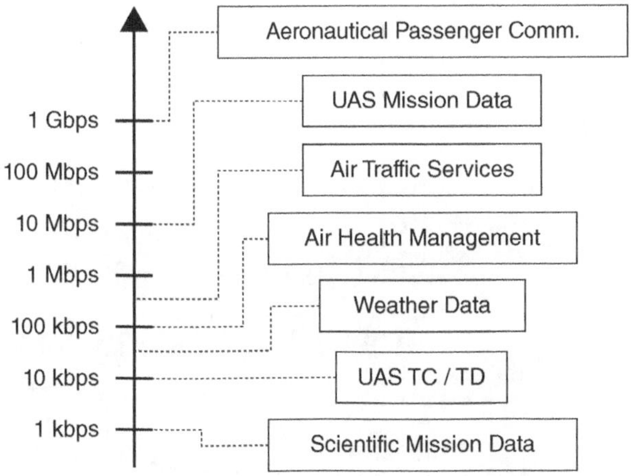

Figure 1.15 Functions of ACN.

systems concurrently involving land, air, and space networks are possible in the near future. As shown in Section 1.3.1.2, the demands for EMC and E^3 will grow substantially. In addition to novel hardware and software requirements, the philosophy of network connectivity and throughput will also change, moving in the direction of greater flexibility. This is attributed to the survivability *and* to the use of random graphs and percolation theory as rendered in Sections 1.3.2 and 1.3.3, respectively.

1.3.1.2 EMC and E^3 Perspective

EMC and E^3 have major roles in future networks, especially those involving aerospace systems. For example, Figure 1.18 shows possible communication requirements for ACNs that are in progress [25]. Following this figure is a list of anticipated technical areas for EMC and E^3.

- Development of satellite and airborne networks involving data management, protocols, information rate tradeoffs, jamming and countermeasure, etc.

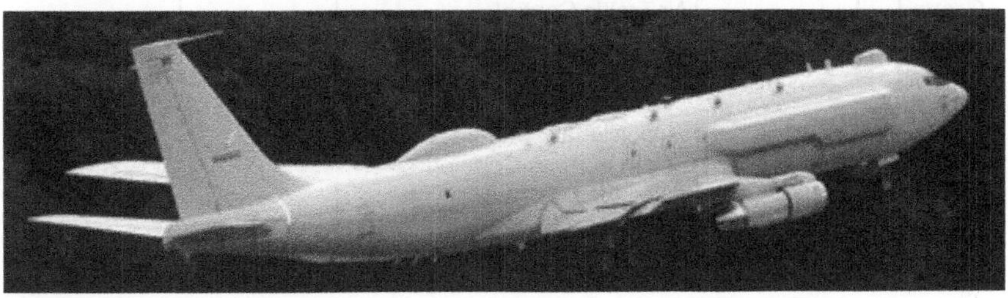

Figure 1.16 Paul Revere test bed.

Figure 1.17 AUST-T and accessories for airborne connectivity to Milstar.

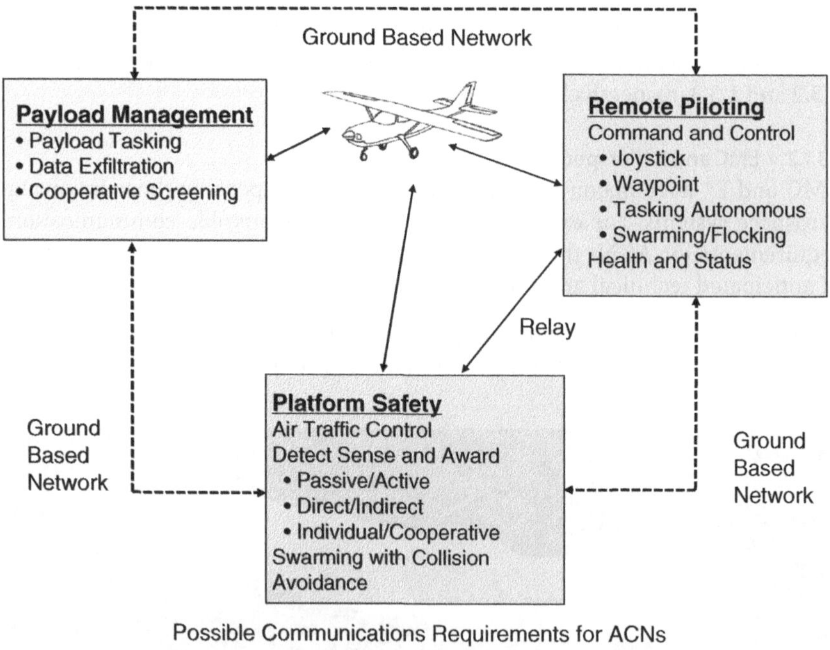

Figure 1.18 Possible communication requirements for ACNs.

- Modeling electromagnetic physical damage (upset, permanent damage) caused by voltage breakdown and/or heating of elements
- Testing large systems requiring a large number of elements and over large time intervals (most likely involving physical scaling theory)
- Making distinctions between issues for EMC and E^3 for spacecraft, airborne systems, and ground systems (where do the technical problems overlap)
- Developing innovative testing technologies for system-of-systems
- Advancing the theory for system-of-systems
- Modeling large autonomous airborne systems [26]
- Developing autonomous control meta-materials (ACMM) using 3D for microwave absorbers
- Developing system-level EMC for autonomous unmanned aircraft in all-weather conditions, intersystem interactions, platform integration, electromagnetic interference from HIRF and IEME
- Developing power electronics models with special emphasis on chaos
- Developing AI software for network protocol management in current and future IT environments

1.3.2 Evolution to Current Systems

1.3.2.1 Survivable Systems

Section 1.3.2 is concerned with the development of relatively new mathematical models of survivable communication, radar, and sensor networks for aerospace systems. Defining the mathematical structure of these networks is the starting point for introducing EMC and E^3 techniques and tools that will be used to build these networks. These new mathematical models, based on the theory of random graphs and percolation theory, took root around 2000 and continue to experience significant expansion. These models are based on the concept that communication, radar, and sensor networks are useful even if only a fraction (defined by an entity called the "percolation threshold") of them are operating, and if only a fraction of their maximum throughput capacity is available. This point of view is a departure from the initial idea of a survivable network introduced during the Cold War where a relatively few number of nodes controlled the strategic balance. For orientation purposes, a brief summary background of survivable systems is provided. The mathematical foundations are rendered in Section 1.3.2.2.

The Defense Threat Reduction Agency was one of the founders of "survivable communications." Its name arose during the Cold War when it became necessary to transmit the emergency action message (EAM) using an ensemble of frequencies from below VLF to EHF. The dominant threat was a multiburst nuclear laydown that included high-altitude EMP (HEMP). Communication platforms (nodes) ranged from ground stations to submarines to aircraft to satellites. The EAM was a very low data message that basically said "go to war." The NATO survivable communication model was developed in the late 1970s and was fundamentally an exercise in EMC—making the nodes hard. The number of nodes and links involved was minuscule by today's standards, but these were necessarily hardened against HEMP to very high probabilities of survival. Modest redundancy in communication paths helped raise the end-to-end probability of survival (getting the message through). Because the EAM data rate and message content were so low, it was assumed that connectedness between nodes alone was sufficient. Connectedness was equated with survivability. In retrospect this was the Cold War baseline.

The number of nodes in the Cold War was small and many were connected in series. Today, the number of nodes in a survivable network can be in the hundreds of thousands to over a million. The focus of interest in the Cold War era was command post-to-command post communication. This is radically different from modern tactical networks, where networks have great connectivity between large numbers of nodes.

The location of Cold War nodes (especially critical ones) was generally known due in part to their paucity combined with robust intelligence sources. The threat was the HEMP waveform generated by high-altitude nuclear detonations. Today's networks include a mixture of deterministically positioned and randomly located nodes. While the HEMP threat still remains (a low probability–high consequence event), IEMI waveforms are the most likely threats. These threat waveforms can be applied in large numbers and can be used to disrupt communication, radar, and sensor networks.

Today's networks that have an extremely large number of nodes make it possible and necessary to approach survivability on a statistical basis. These networks make extensive use of random graph theory, percolation theory, and wireless communications (Sections 1.3.2 and 1.3.3). The new network science tools of random graph theory and percolation theory are used to establish connectedness. These networks are a collection of nodes and links that connect the vertices with one another. In a fully connected network, all nodes are connected to each other. For most real networks a typical node is only connected to a few other nodes but it is sufficient to ensure good connectedness. IEMI can destroy nodes and/or links so as to reduce the connectedness to unacceptable values.

1.3.2.2 Current Survivable Random Systems

The random graph and percolation theories assume each node had a probability, p_s, of surviving and there are k links connected to it. Starting from this modest basis, one can compute the probability that the network will form a "network spanning cluster." These networks are essentially *ad hoc* networks that are composed of mobile and typically random nodes having just enough power to originate, receive, and relay packets by multihop transport. This cluster of nodes allows end-to-end communication at a minimum rate and thereby establishes baseline survivability. To illustrate the method, Sections 1.3.2 and 1.3.3

address the analytic foundations for determining the survivability for wireless communication, radar, and sensor system. The mathematical analysis applies to arbitrary combinations of satellite, airborne, land, and sea networks that can have nondirect and direct links. The EMC and E^3 hardware and software necessary to support the creation of these random networks will become evident when the networks are defined.

In the absence of IEMI, electromagnetic propagation occurs throughout the network. IEMI effects range from direct jamming and interference of nodes to creation of fading dispersive channels that can limit both the coherence time and the coherence bandwidth. All these effects can reduce the information rate and, if bad enough, can cause the network to break into isolated clusters—possibly the most serious adverse effect.

These networks are *random* networks because they are created in an *ad hoc* way and they satisfy the mathematical conditions for random graphs. It is necessary to define the mathematical structure of these networks and the mathematical tools that are necessary to perform the communication survivability assessments. A hypothetical model of a limited portion of a network under ideal and IEME attack conditions is shown in Figure 1.19. The black dots in this figure are the nodes and the solid black line are the propagation links between the nodes. The dotted lines are extensions (links) to additional nodes. For illustrative purposes, let our network consist of only the solid lines. Figure 1.19(a) shows the network in its normal operating state (no attack). There is communication between all nodes, the rates being consistent with network capacity.

Hypothetically, if we could imagine that the dotted lines of Figure 1.19 connect to lots of other nodes, then most likely we would not have clustering as shown in Figures 1.19(b) and 1.19(c) Nothing would happen to our

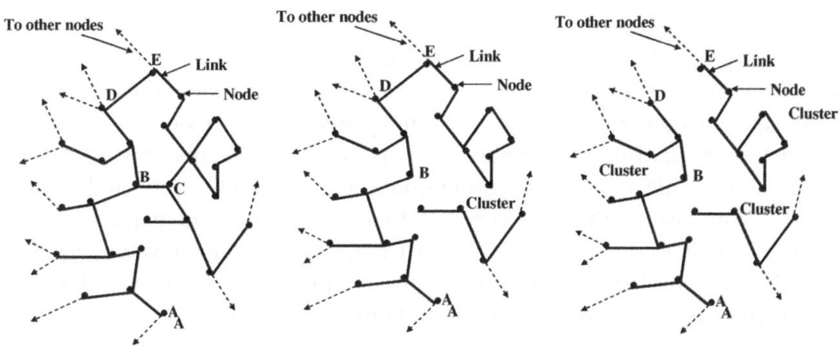

(a) No attack **(b) Attack node C** **(c) Attack node C and link between nodes D and E**

Figure 1.19 Network in absence and under attack.

hypothetical network! Large random networks are virtually immune to a few isolated attacks on nodes or links. We need to address the tolerance of these *ad hoc* wireless networks to network-wide IEMI attacks. What may be of great interest here is not the issue of information rate, but rather the ability of the network to communicate at any information rate over a significant extent—"a connected cluster of sites that spans the entire network even for an arbitrarily large fraction of sites, f, that are randomly removed."

1.3.3 Random Graphs and System Capacity

1.3.3.1 Random Graphs and Percolation Threshold

In this section, we (1) provide the theoretical principles based on random graph theory and percolation theory to evaluate the resilience of large random geometric *ad hoc* networks formed with nondirected and directed graphs, (2) provide a theoretical basis for expanding the theory to inhomogeneous networks that are more closely connected to those found in nature and in the military, and (3) propose that a metric for survivability is the network's ability to remain radio-frequency connected with a suitable surviving fraction of nodes [27–30].

For a network consisting of N nodes, which each node connected to every other node, the maximum number of links, N_L, is

$$N_L = \frac{1}{2}N(N-1) \tag{1.50}$$

For a given bandwidth, the maximum information rate that can be handled decreases as the number of nodes increases. We now have a tradeoff between survivability and communications capability that can be addressed from a random graph and percolation viewpoint. For a desired information rate, what is the minimum connectedness required that ensures end-to-end communication?

Random graph theory starts from a fully connected network; it demonstrates the conditions that must exist between the number of nodes and the number of links in order to generate a network spanning cluster, and provides estimates of the cluster size distribution. In particular, it shows that if a network-spanning cluster is not achieved, the remaining nodes are minimally connected to one another. Now assume that for every link there is the same independent probability, p, for that connection to actually exist. The smaller the value of p the fewer links are available to support network communication. In general, p can be a function of N. The average number of links, n, is then

$$n = pN_L \tag{1.51}$$

The starting point for percolation theory is the normalized degree distribution function, $P(k)$, the number of edges connected to a node—that is, the number of edges connected to a node can be a random variable. We have

$$1 = \sum_{k=2}^{k=k_{max}} P(k) \tag{1.52}$$

where k_{max} is the maximum number of edges considered. A key result of percolation theory is that a network-spanning cluster will be formed with a fraction, p_c, of the nodes given by

$$p_c = \frac{1}{K-1} \tag{1.53}$$

where

$$K = \frac{\langle k^2 \rangle}{\langle k \rangle} \tag{1.54}$$

$$\langle k \rangle = \sum_{k=0}^{k=k_{max}} kP(k) \tag{1.55}$$

$$\langle k^2 \rangle = \sum_{k=0}^{k=k_{max}} k^2 P(k) \tag{1.56}$$

What may be of great interest here is not only the issue of information rate, but rather the ability of the network to communicate at any information rate over a significant extent—"a connected cluster of sites that spans the entire network even for an arbitrarily large fraction of sites, f, that are randomly removed." The critical fraction, f, is defined as $f = 1 - p_c$, where p_c is the percolation threshold of the network—the critical probability that a giant cluster forms. "Above p_c the network is composed of isolated clusters, but below p_c a giant cluster appears that spans the entire network." The existence of a percolation property of networks is a consequence of random graph theory.

We seek to mitigate the loss of nodes by lowering the network's percolation threshold p_c. For a randomly connected network, such as the wireless meshed network, the critical entity is $P(k)$, with the normalizing condition is

$$\sum_{k=2}^{k=M} P(k) = 1 \tag{1.57}$$

The lower limit, $k = 2$, is a requirement that for a node to contribute to network connectivity, it must have as a minimum of 2 links. Nodes with only one link are end points. M is the maximum number of links connected to a node. The percolation threshold is given from equations (1.53) to (1.56) with

$$k_{max} = M \tag{1.58}$$

We have the condition $0 \le p_c \le 1$, as expected. For survivability, we would like to have p_c as close to zero as possible. The least survivable case occurs when $k = 2$ for all nodes. This is the case of string of nodes sequentially connected. From equations (1.53) to (1.56), we get $\langle k \rangle = 2$, $\langle k^2 \rangle = 4$, $K = 2$, and $p_c = 1$. What this means is that all the nodes need to be connected in order to maintain a network scanning cluster; conversely the network has zero tolerance for node failure. A more realistic illustrative example might be $\langle k \rangle = 4$, $\langle k^2 \rangle = 16$, $K = 4$, and $p_c = 0.33$. An attack would need to take out $f = 1 - p_c = 0.67$ fraction of the nodes to cause the network to fall apart.

1.3.3.2 IEMI and Throughput Capacity

Destroying a fraction of the nodes reduces the information rate and, if bad enough, can cause the network to break into isolated clusters. It is necessary to define the mathematical structure of MANETs under these conditions and the mathematical tools that are necessary to determine their survivability. Starting from this modest basis that each node had a probability of surviving, p, and there are k links connected to it, one can compute the probability that the network will form a "network spanning cluster." This cluster of nodes allows end-to-end communication at a minimum rate and thereby establishes a baseline level of survivability.

Under idealized conditions, Gupta and Kumar [31] show that for a *fully connected network*, the total network capacity, C_{NET}, measured in bit meters per second, is

$$C_{NET} = D(W\sqrt{n}) \tag{1.59}$$

where D is a constant in units meters and W is the channel capacity in bits/sec. If an IEMI attack removes $\hat{n}$ nodes or links, we can still compute the new network capacity. For example, if the n is the original number of nodes and a total of $\hat{n}$ nodes are lost to the creation of numerous small clusters, the capacity of the remaining fully connected network is

$$\hat{C}_{NET} = D(W\sqrt{n - \hat{n}}) \tag{1.60}$$

Another IEMI approach for disrupting the network is to reduce the coherence bandwidth B_c below W by introducing unwanted time delays. When this is combined with node attack, we get

$$\hat{C}_{NET} = D(B_c\sqrt{n - \hat{n}}) \tag{1.61}$$

If $\hat{C}_{NET}$ falls below an operational useful limit, the attack will have succeeded.

For orientation, we could also consider a wireless land system where the fraction of nodes having k nodes connected to it is the Poisson distribution

$$P(k) = \exp(-pN)\frac{(pN)^k}{k!} = \exp(-\langle k\rangle)\frac{\langle k\rangle^k}{k!} \tag{1.62}$$

It can be shown that the percolation threshold is

$$p_c = \frac{1}{\langle k\rangle} = \frac{D^2}{CR^2N} \tag{1.63}$$

where C is dimensionless constant of order unity, D^2 is the size of the domain, R is the range of a node, and N is the total number of nodes in the domain. We immediately see that the effect of spatially homogeneous random attack is to reduce the number of nodes that participate in the network—N decreases, thus increasing the percolation threshold. This makes the system less tolerant. The increase in p_c is easily calculated from the foregoing equation.

1.3.3.3 Directed and Nondirected Graphs

Some networks have directed links, and these must be combined with the undirected links to determine the percolation threshold for the network spanning cluster for combined systems. We have mathematically converted the undirected links to directed links and from this step have been able to determine the percolation threshold for a combined directed and undirected network.

Consider a system where there are N_D nodes that have directed edges and N_{ND} nodes that have nondirected edges. The total number of nodes is

$$N = N_D + N_{ND} \tag{1.64}$$

$$p_c = \frac{\langle k_0\rangle}{W_D\langle j_0\rangle\langle k_0\rangle + W_{ND}\left(\langle k_0^2\rangle - \langle k_0\rangle\right)} = \frac{1}{W_D\langle j_0\rangle + W_{ND}\left(K_0 - 1\right)} \tag{1.65}$$

$$W_D = \frac{N_D}{N},\ W_{ND} = \frac{N_{ND}}{N},\ W_D + W_{ND} = 1 \tag{1.66}$$

where $\langle j_0\rangle$ is the average in-degree for the directed nodes.

1.4 EMC Considerations of Chaos

1.4.1 Background

1.4.1.1 Brief History

Within the last 50 years, a number of excellent papers have been written to link the field of chaos to established disciplines of classical mechanics and non-conservative dynamical systems on one side, and set theory and stochastic processes on the other side. References [32] to [35] provide a limited tutorial sample of the available references. In this section, we show that many of these chaos-related topics (e.g., nonlinear dynamics, bifurcations, synchronization, and optical systems) can be folded into a roadmap that demonstrates chaos's strong dependence on statistical mechanics, electromagnetic theory, and power conversion when applied to EMC and E^3. Understanding and quantifying the nature of chaos waveforms may enable us in many cases to (1) predict the onset of the erratic phases of chaotic waveforms, (2) mitigate the effects, and (3) design systems that improve performance using chaos.

It is conjectured that chaos began when Newton's laws of mechanics started to accurately predict the motion of planetary bodies from mutual gravitational attraction. In retrospect, this was relatively easy to do for the two-body interaction because not only was the force between them known but one could also be confident that energy was conserved. The predicted orbits were well-defined and essentially periodic. Poincare's studies of the three-body problem around 1900 showed the existence of nonperiodic orbits and bounded in today's language as a limit cycle connected to a strange attractor. It was also noted that slight differences in initial positions of interacting bodies produced surprisingly huge differences in their orbits—this is one of the cornerstones of chaos theory.

Other significant observations include: (1) Maxwell's 1870 discovery of microscope randomness, (2) Henri Poincare's 1890 unexpected observation of the sensitive dependence on trajectories of the parameters for the three-body problem, (3) for the first half of the 20th century, unexpected randomness in classical mechanical systems was observed by a number of distinguished scientists, including Birkhoff, Kolmogorov, and Smale, and (4) in the last half of the 20th century, the term "chaos" was created to formally deal with this class of phenomena.

1.4.1.2 Evolution

A conclusion from the early studies demonstrated the need for physical models based on statistical mechanics where the focus shifted drastically from a few interacting bodies to gases and fluids. The classic application is Edward Lorenz's 1961 study of weather prediction [36]. His results showed an acute nonlinear sensitivity to initial conditions. There is also similarity between Lorenz's equations and those of the Rossler and Chua systems from the instability viewpoint and determining how long these systems can remain in quasiequilibrium before bifurcation begins. This issue appears to be of high interest in power electronics and power supplies where subharmonic bifurcations can connect a single point to a limit cycle. Chaos behavior can appear in a large number of power electronic circuits such as regulators, current limiter devices, and amplifiers [37].

Chaos can be a significant factor in power electronics and power supplies and can disrupt operation of systems for meaningful periods of time. For example, continuous operating electronic dynamic systems that operate at high gain can become unstable from nonlinear behavior, the most frequent source of nonlinearity being the switching element. Switching regulators often can generate subharmonics, which may start the bifurcation process. This problem may cause larger systems issues for the EMC designer and could cause initial design delays in understanding the observed chaotic behavior (random-like changes of state).

There are, however, indications of chaos that include ripple, subharmonic generation, and acoustic noise. These effects may be difficult to discern from usual random noise and can affect pulse-width-modulated regulators, ripple regulators, current-mode controllers, over-current protection circuits, various circuits, amplifiers, etc. A system whose asymptotic behavior is a limit cycle may exhibit a certain degree of immunity to chaos. Fortunately, chaos induced behavior can often be handled through clever EMC design because it is bounded and nondestructive. In addition, chaos signals can be useful in signal design.

1.4.1.3 Application for Power Conversion

Although chaos has many facets, medical analysis, communications, coding, those that deal with electronic circuits appear to be the most relevant to EMC. Within this context, power electronic converters are of great interest because of their anticipated increased use in the 21st century. For tutorial purposes, Figure 1.20 shows a diagram of a linear model of a basic open-loop buck dc/dc converter that converts a higher dc input to a lower dc voltage using a chopper circuit (see expanded discussion in text "Nonlinear phenomena in power electronics"); f_s is the switching frequency, d is the duty cycle, V_{in} is the driving voltage, D is the diode, and LC is the low pass filter that enables a smooth voltage, v, to appear across the load, R [38]. In practice, the output voltage may need to be regulated using a conventional model of the buck converter as shown in Figure 1.21.

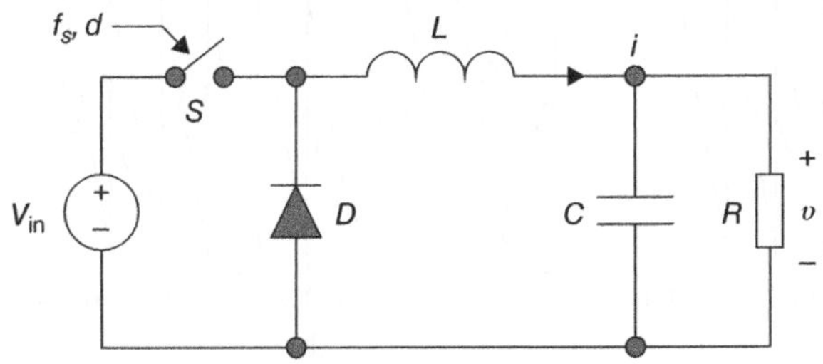

Figure 1.20 Open-loop buck dc/dc controller.

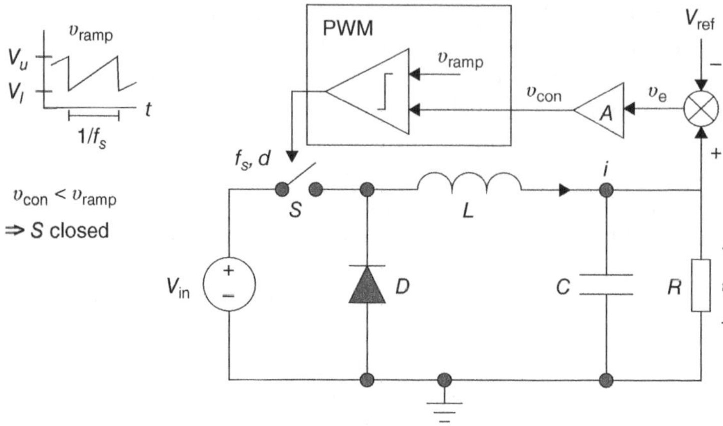

Figure 1.21 Buck converter with proportional closed-loop controller.

In both Figures 1.20 and 1.21, it is assumed that the converter operates in the linear range. This is analytically modeled using a basic linear model (obtained from perturbation theory); chaos is not present in this regime. As the driving ramp waveform increases slightly from its base value, the system moves into the nonlinear range, becoming unstable, and repeating every two cycles of the ramp instead of every cycle. This is one example of chaos. The initial stages of chaos, taken from a real case, are shown in Figure 1.22. The "(a)" figure is the normal nonchaotic state and the "(b)" figure is the initial stage of chaos [38]. As the driving ramp function becomes even larger, and pushes the system further into the nonlinear range, the chaotic behavior becomes even more dramatic.

It is not always obvious during testing systems in an EMC laboratory when chaotic behavior occurs, how we recognize it, can it be quantified in real time, how long it will it last, what the causes are (internal or external), and how to control it, etc. Despite continuing excellent insights on chaos, there appear to be a

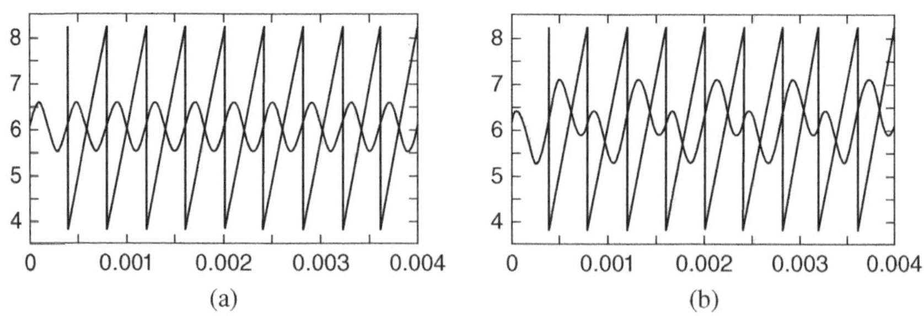

Figure 1.22 Onset of chaotic waveform.

number of issues which remain at least partially unresolved. A good example is quantifying the degree to which a chaos waveform can exhibit random behavior and concurrently have its autocorrelation function satisfy the traditional conditions of a stochastic process. Clarifying this point could significantly improve predict required signal strength and the efficiency of EMC testing.

1.4.2 Roadmap for General Solution

1.4.2.1 Fundamental Equations

The basic starting point is the general nonlinear autonomous equation for N dynamic variables:

$$\left(\frac{d\vec{x}}{dt}\right) = \vec{F}\left(x^{(1)}(t), x^{(2)}(t), \ldots . x^{(N)}(t)\right) \tag{1.67}$$

A critical feature of the foregoing system of equations is that they are not necessarily energy conserving or momentum conserving. The roadmap for the solution of the foregoing equation could, for example, begin with the following initial steps.

If possible, make an initial assessment as to whether the system of equations will need to accommodate random variables. Can we anticipate stochastic variables, chaotic effects, or a combination of both in the solution? What is the nature of the asymptotic solution? Can we make an initial assessment as to whether we are going to deal with positive Lyapunov exponents? Such exponents render long-term prediction questionable. If equation (1.67) only involves two variables, an exact solution can be handled exactly, as shown from the Poincare–Bendixson theorem.

Assuming there is no shortcut to solving equation (1.67), we could approach the solution by first determining whether equation (1.67) is a linear equation. If so, the solution is obtained from the matrix equation:

$$\left(\frac{d\vec{x}}{dt}\right) = M\vec{x}. \tag{1.68}$$

We have

$$\vec{x} = \exp(Mt)\vec{x}_0 \tag{1.69}$$

where $\vec{x}_0 = \vec{x}(t = 0)$ is the initial condition.

If the linear case is not operative, the next logical step is to search for time-independent solutions, characterized by the solution of the equation:

$$\left(\frac{d\vec{x}}{dt}\right) = \vec{F}\left(x^{(1)}(t), x^{(2)}(t), \ldots, x^{(N)}(t)\right) \!-\!\!\circ. \tag{1.70}$$

A time-independent solution is

$$F_k\left(x_\nu^{(1)}, x_\nu^{(2)}, \ldots, x_\nu^{(N)}\right) = 0, \tag{1.71}$$

where $k = 1, 2, \ldots, N$, $\nu = 1, 2, \ldots, N$, $x_\nu^k(x_{1,0}, x_{2,0}, \ldots, x_{N,0})$ is the νth equilibrium point for the kth dynamic variable, and $\{x_{1,0}, x_{2,0}, \ldots, x_{N,0}\}$ is the set of N initial conditions. It is not always clear that the equilibrium set can always be reached starting from an arbitrary set of initial conditions

The mathematical behavior of an N-dimensional nonlinear dynamic system can lead to the following cases:

- Coming to rest, as in a dissipative system
- System totally or partially splits apart (unbounded system)
- Periodic motion
- Quasiperiodic motion
- Chaotic motion; a chaotic system can have some of the following properties:
  - Extreme sensitivity of trajectories to initial conditions
  - Power spectrum with a continuous part
  - Solutions of differential equations being unstable
  - Topologically mixing of phase space
  - Dense periodic orbits

It is frequently of interest, especially for EMC laboratory testing, to explore whether a system that is assumed to be asymptotically stable for all initial condition is really so. This can be accomplished from perturbation theory starting with the approximation

$$\vec{x}_0 \to \vec{x}_0 + \delta\vec{x}_0 \tag{1.72}$$

where $\delta\vec{x}_0$ is a tiny perturbation of the initial conditions that satisfies the equation

$$\left|\delta\vec{x}_0\right| \ll \vec{x}_0 \tag{1.73}$$

Using the foregoing equations, it is easy to show that the initial separation will grow at a rate given by

$$\left|\delta\vec{x}(t)\right| \approx \exp(\lambda t)\left|\delta\vec{x}_0\right| \tag{1.74}$$

where λ is the Lyapanov exponent. In a finite dimensional system, there are a finite number of λ's equal to the dimensionality of the phase space. The largest value of λ from the set of Lyapunov exponents is called the maximum Lyapunov exponent (MLE). Should the MLE be positive, it could be likely that the system is unstable and/or chaotic. This observation needs to be addressed during EMC testing.

1.4.2.2 Interim Observations

Chaos deals with the behavior of dynamic systems that are extremely sensitive to time-independent initial conditions. The function $(\vec{F}(x^{(1)}(t), x^{(2)}(t), \ldots, x^{(N)}(t))$ does not include time-dependent random behavior. The time-dependent unpredictable behavior of the dynamic system is internally driven by its intrinsic nature once the initial conditions are known.

Chaotic behavior is often observed in large-scale physical, chemical, biological, meteorological, financial, computer systems, etc., as well as their laboratory counterparts and demonstrations. Unfortunately, in many of these laboratory cases, the unwanted signals are not recognized as chaos but as "not understood" interference. Chaos can accommodate a level of randomness (as in stochastic processes) provided that the stochastic time constants are much longer than those of chaos. On the other hand, it has been found that certain aspects of chaos behavior can be incorporated within the scope of stochastic processes. This can be very useful in predicting system behavior.

A good example of chaos chaotic behavior is the weather system modeled by Lorenz in 1963. He showed that his model of the complex weather system could be reduced from 12 equations to the following three coupled nonlinear equations:

$$\left(\frac{dx_1}{dt}\right) = \sigma(x_2 - x_1) \tag{1.75}$$

$$\left(\frac{dx_2}{dt}\right) = x_1(\rho - x_3) \tag{1.76}$$

$$\left(\frac{dx_3}{dt}\right) = x_1 x_2 - \beta x_3 \tag{1.77}$$

In the foregoing equations, x_1, x_2, x_3 are the dynamical variables and σ, ρ, β are parameters that define the physical features of the problem. In the Lorenz model, these parameters were connected with two precipitation states. The details of the Lorenz weather model are not critical for EMC, but what is critical is the nature of dynamics as it pertains for example to power electronics. At first glance, equations (1.75) to (1.77) look like any ordinary set of control theory sets, except for the fact that for some set of parameters, there is no closure; the system is chaotic, but bounded.

For example, Figure 1.23 depicts a solution of equations (1.75)–(1.77) for a particular set of parameters. This three-dimensional curve does not appear to ever terminate, although it is bounded. Figure 1.24 shows two sample chaotic time series (dashed and not-dashed) for the x_3 variable of equation (1.77). These curves differ because of slightly different initial conditions.

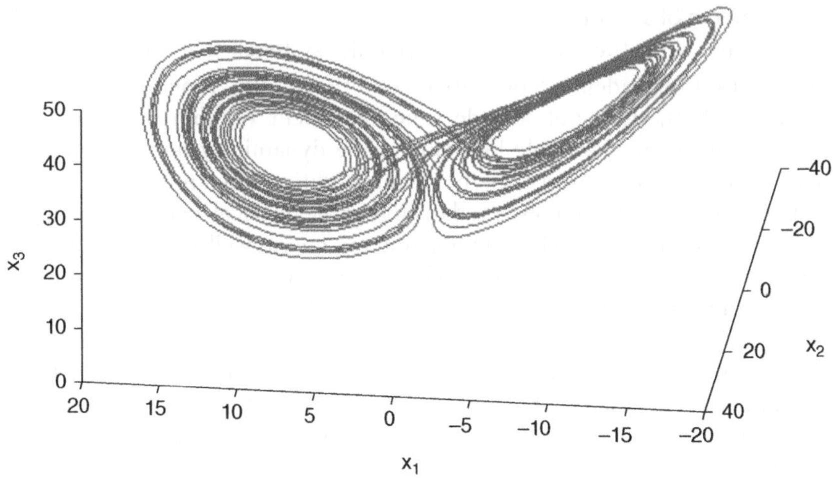

Figure 1.23 Sample phase space dynamics of Lorenz system.

1.4.3 Analytical Considerations of Chaotic Behavior

1.4.3.1 Connection with Stochastic Processes

Chaotic signals never end except at $t \to \infty$. On this basis, we make a connection with a wide-sense stationary random variable, $X(t)$, from the ergodic

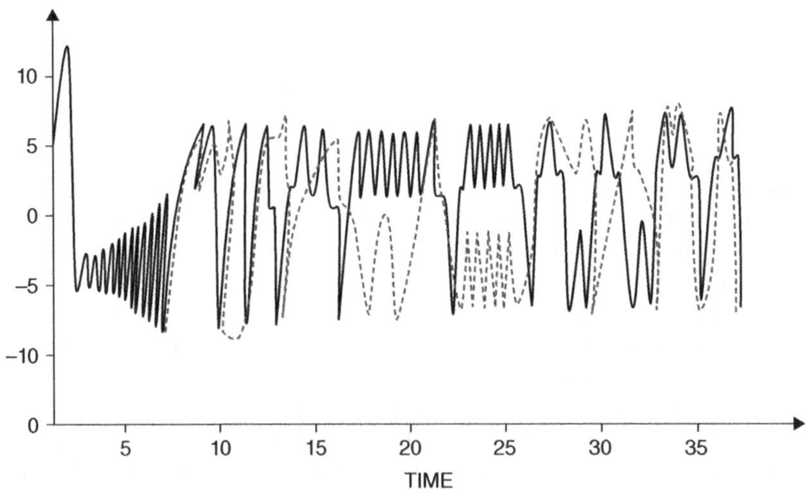

Figure 1.24 Sample of chaotic time series of Lorenz system.

theory viewpoint. The probability density function of $X(t)$ is determined from its moments ($n \geq 0$):

$$(X^n) = \lim_{T\to\infty} \frac{1}{2T} \int_{-T}^{T} X^n(t)dt. \tag{1.78}$$

The average power is the second moment of the distribution and is given by the foregoing equation for $n = 2$. Insight into the power spectrum of $X(t)$ is obtained from its autocorrelation function

$$R(\tau) = \lim_{T\to\infty} \frac{1}{2T} \int_{-T}^{T} X(t+\tau)X(t)dt. \tag{1.79}$$

From measurements, alone it would appear difficult to tell whether we are indeed seeing a random process or a chaotic waveform. The major difference between chaotic and random waveforms ones is that the former are totally deterministic once the initial conditions are precisely known while random process waveforms are probabilistic (for example a Markov process) at all times. It is the uncertainty in the physical specification (e.g., initial conditions) of chaotic waveforms that enables us to treat them as stochastic processes, and use the powerful techniques of random theory.

For illustrative purposes, now use the following form for the pulsed chaotic waveform as an example

$$\vec{\eta} = \vec{\chi}(t) \cos \omega_0 t. \tag{1.80}$$

In the foregoing equation, $\vec{\chi}(t)$ is the stochastic vector set of voltage waveforms,

$$\vec{\chi}(t) = \sum_k \chi_k(t) \tag{1.81}$$

$\cos \omega_0 t$ is a modulation term, and $T_0 = 2\pi/\omega_0 = 1/f_0$ is the pulse period. For our waveforms, we may imagine T_0 to be in the range of milliseconds or greater, while the HPM pulse width, T_P, is in the range of microseconds. The average energy of the kth component of $\vec{\eta}$ is

$$\left(\eta_k^2\right) = \lim_{T\to\infty} \frac{1}{2T} \int_{-T}^{T} \chi_k^2(t)\cos^2\omega_0 t dt. \tag{1.82}$$

and the autocorrelation function is

$$R(\tau) = \lim_{T\to\infty} \frac{1}{2T} \int_{-T}^{T} \chi_k(t+\tau)\chi_k(t) \cos \omega_0 t \cos \omega_0 (t+\tau) dt \tag{1.83}$$

We now use the assertion that $T_0 \gg T_P$ and therefore in the ranges where χ_k^2 is not negligible we have the approximation $\cos^2\omega_0 t \cong 1$. Thus,

$$\left\langle \eta_k^2 \right\rangle \cong \lim_{T\to\infty} \frac{1}{2T} \int_{-T}^{T} \chi_k^2(t)dt. \tag{1.84}$$

Correlation times will also be much less than T_0 and we get

$$R_k(\tau) \cong \lim_{T\to\infty} \frac{1}{2T} \int_{-T}^{T} \chi_k(t+\tau)\chi_k(t)dt, \tag{1.85}$$

$$R_k(\tau) = \left\langle \eta_k^2 \right\rangle \rho_k(\tau), \tag{1.86}$$

$$\rho_k(\tau) = \frac{\int_{-T}^{T} \chi_k(t+\tau)\chi_k(t)dt}{\int_{-T}^{T} \chi_k^2(t)dt}. \tag{1.87}$$

The quantities $\langle \eta_k^2 \rangle$ and $\rho_k(\tau)$ can both be measured for chaotic behavior and thus provide a means of predicting upset from these unwanted voltages. An approach to assess the susceptibility of electronic systems to chaos waveforms arising from HPM interaction with a system has been developed. Within certain limits related to ergodic theory, it is proposed that chaos could be treated as a stochastic process and that unwanted energy absorbed by the system could be determined from the Wiener–Khintchine theorem. Periodicity of HPM waveforms is also incorporated in the analysis.

There is currently much interest in the international EMC community in quantifying adverse anomalous effects on electronic and control systems produced by HPE and HPMs. These effects are incorporated in the realm of IEMI. They range from temporary upset to permanent damage. While a substantial amount of experimental data has been generated by HPE/HPM effects, a cohesive theory that links effects between different systems is missing.

The inherent complexity of the systems of interest has often led researchers to construct test programs that attempt to explain observations in terms of system variables such as microwave frequency, pulse duration, pulse repletion rate, peak power, average power, type of equipment, computer clock rate, and the number of components used in the experiment. These parameters may not be sufficient to allow extrapolation of results from one system to another or explain the same system's response for different waveforms. We propose that complementary sets of variables based on differential equations, stochastic control theory, and chaos could be used to explain heretofore puzzling and ambiguous observations. Should this approach be successful, it could lead to robust

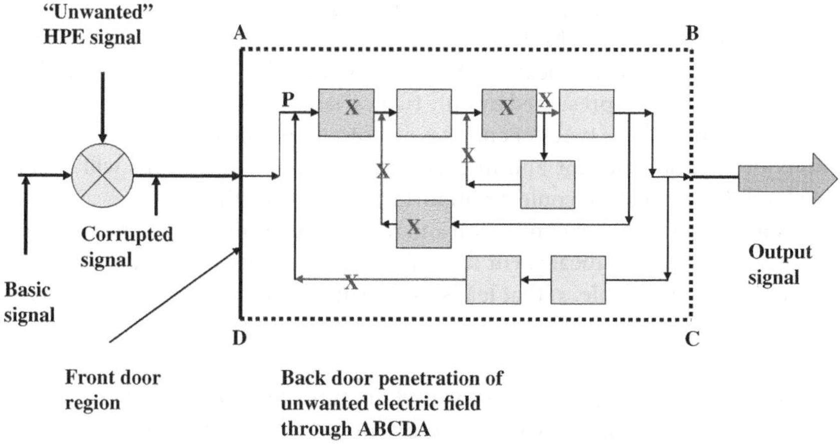

Figure 1.25 System under IEMI attack.

methods for designing electronic and control systems that are more resilient against HPE/HPM attacks.

Figure 1.25 shows the generic model of the class of systems we are addressing, emphasizing the front-door (FD) and back-door (BD) approach for the penetration of unwanted electromagnetic field. The system is enclosed by the rectangle *ABCDA* and the solid line, *DA*, denotes the FD—the place where information and/or desired control normally enters the system. The FD could, for example, be a receiving antenna or an intercomputer cable. The connected dashed lines *ABCD* define the BD. Unwanted signals in the form of electromagnetic waves enter through points of entry in the BD. It is these waves that create chaos signals in the volume.

Figure 1.25 is representative of a wide class of systems such as hybrid control systems, but it is not meant to be an all-inclusive representation. Unwanted signals can combine with the basic signal via the FD electronics and enter the first system component at point P. Unwanted system behavior results from corrupted information at point P. The Xs in Figure 1.25 depict places where adverse voltages are produced by the BD electromagnetic field. This field can produce misinformation on intrasystem cables and can also interact with hardware directly to produce nonlinear and random behavior [39].

In summary, experimental tests that ensure no BD coupling and simply examine FD effects using the aforementioned macroscopic parameters that avoid critical examination of its ordinary differential equations (ODE) may offer little insight into a system's physical behavior. Moreover, some theoretical studies suggest that common low level Gaussian noise can trigger instabilities in second-order systems. These kind of issues need to be examined as part of EMC pretest analyses.

1.4.3.2 Logistic Map and Statistical Considerations

Under certain conditions, by making a connection with turbulence theory, theoretical physics show that near a limit cycle a system involving a large number of equations can be represented by only two variables and sometimes even only a single variable. These types of analyses can often provide insight into the possibility of chaotic behavior and in particular whether bifurcations will occur. This kind of information could be important in EMC testing.

When applied to power conversion and related electronics, the dominant signals are sampled periodically. For illustrative purposes let us take the case of a dominant single variable, x, that leads to the solution of the discrete equation

$$x_{j+1} = f(x_j, \lambda) \tag{1.88}$$

In the foregoing equation, j is measured in units of the period and λ is a system dependent parameter determined by the bifurcation conditions and the multiplier μ—the numerical factor by which the frequency changes from the stable condition. At this transition point, x_*, it is required that $x_{j+1} = x_j \equiv x_*$ and $\mu = dx_{j+1}/dx_j$, which then provides the conditions for bifurcation. A common specific form for the solution of equation (1.88) is

$$x_{j+1} = f(x_j, \lambda) = rx_j(1 - x_j) \tag{1.89}$$

where r is a parameter in the range 0–4. The behavior of equation (1.89) is called a logistic map and is shown as a function of the parameter r in Figure 1.26. In the upper left, $r = 2.8$, there is only one fixed stable point; hence there is no chaos. In the upper right, $r = 3.4$, there is a stable two-cycle system (no chaos); in the and lower left, $r = 3.5$, there is a stable four-cycle system (no chaos); and in the lower right, $r = 3.8$ we have chaotic behavior [40].

Based on the foregoing illustration, it would appear that pretest EMC analyses of power conversion and related electronics are essential.

1.5 EMC Effects on and Technology for Aerospace Systems

1.5.1 Background and Overview

Beginning roughly in the year 2000, new and novel advances continue to be made in (1) developing new statistical theories for complex networks, (2) creating and modeling ACNs, (3) improving capabilities for robotics, (4) developing autonomous systems, (5) improving the ability of dynamic electronic and electromagnetic systems to cope with chaos, and (6) modeling the electromagnetic interactions between airborne and space systems. Alongside these aforementioned areas significant work continues to be made in the backbone topics of EMC; this information is rendered in a host of international journals and conferences. The aforementioned six new areas of R&D are not especially familiar

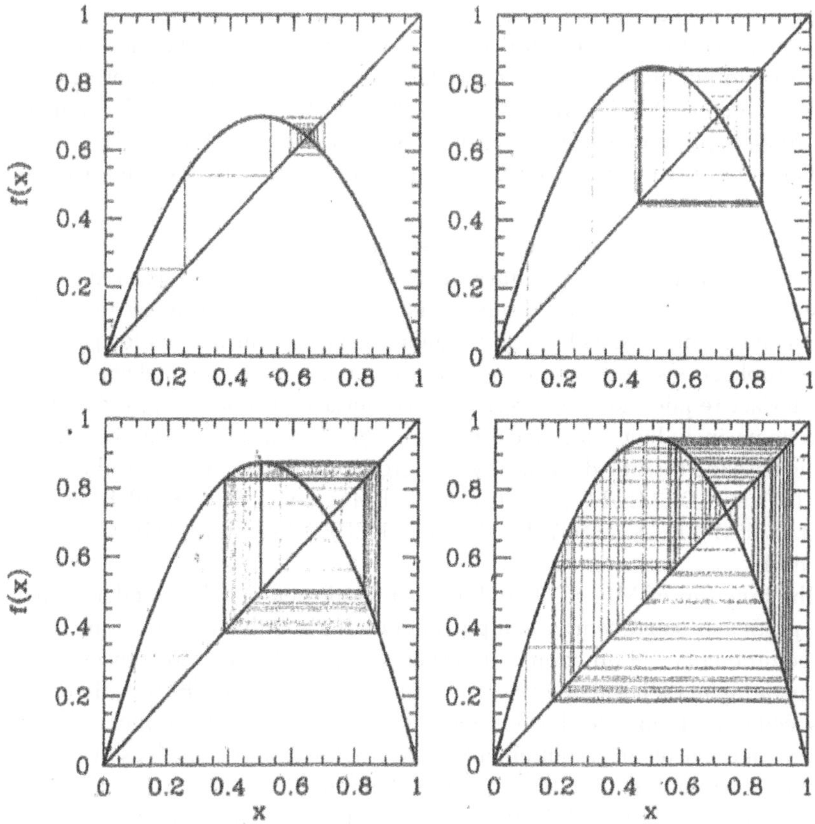

Figure 1.26 Cobweb diagram.

to most members of the EMC and E^3 communities. Sections 1.2 to 1.4 of this Handbook provide the general background for these new areas; and it is the intention of Sections 1.5.1 to 1.5.4 to provide a roadmap of advanced theoretical techniques that may be necessary for solving emerging problems in these new areas.

1.5.1.1 Brief Summary of Status of Current MIL-STD Issues and Threats

Until recently, the backbone for testing and ensuring that newly developed emerging equipment meets the proper international standards has rested in large part on the comprehensive set of standards: (1) MIL-STD-464C, electromagnetic environmental effects requirements for systems, December 1, 2010; (2) MIL-STD-461F, requirements for the control of electromagnetic interference characteristics of subsystems and equipment, December 10, 2007; and (3) MIL-STD-3023, HEMP protection for military aircraft, November 21, 2011.

Theoretical backup used in conjunction with the aforementioned standards continues to be provided in R.F. Gray's basic document, (4) Verification and validation of unified electromagnetic (UEM) design version 3.0, Contract No. DTRA01-03-D-0004; Delivery Order No: 0017, ATK, April 2009, and in numerous reports presented in international journals and conferences. The information rendered in these standards and Gray's document is summarized as follows:

1. MIL-STD-464C is a key source of information that is useful for establishing and understanding both military and civilian EMC between commonly used systems, subsystems, and components. This document "provides rationale, guidance, and lessons learned to tailor baseline requirements for particular applications; and establishes electromagnetic environmental effects (E^3) interface requirements and verification criteria for airborne, sea, and ground systems, including associated ordnance." This standard addresses the following baseline threats identified by the acronyms: The acronyms for the MIL-STD's are: DCMF, ECCM, ECM, EID, EMC, EME, EMI, EMP, EMRADHAZ, EMV, ESD, HEMP, HERF, HERO, HERP, HIRF, HPM, IMI, M&S, and NLEE.
2. The function of MIL-STD-461F is to "provide reasonable confidence that a particular subsystem or equipment complying with stated with interface requirements will function within their designated design tolerances when operating in their intended electromagnetic environment (EME)." There are specific limitations to the size and electromagnetic features of equipment that meet the testing requirements. Nevertheless, there is much useful information contained when testing to this standard.
3. The function of MIL-STD-3023 is to "provide HEMP protection for aircraft against functional upset or damage due to HEMP threat environments." At the unclassified level, it includes technical information, performance criteria, and test procedures for various EMP hardness levels.
4. Unified electromagnetic (UEM) design provides modeling support and analysis for a relatively large variety of electromagnetic environments, geometric configurations, waveforms, etc. The focus of this comprehensive document is the development of EM hardened systems with special emphasis on the EMP waveform. As pointed out, UEM does "provide critical tools and information for a program to build an EM hardened system, does not provide detailed calculations of system response in an EM environment helps decide on the level of protection needed for a system under design, and supports Combined Battlefield Environmental Effects (CBEE)."

1.5.1.2 Brief Summary of Status of Emerging System-Related Issues and Threats

Sections 1.2–1.4 proposed that vigorous IEEE-EMC programs begin to address three core specific electromagnetic system issues and threats: (1) aerospace autonomous systems, (2) airborne and aerospace networks, and (3) chaotic signals created by natural and manmade sources. The current set of theoretical and analytical techniques and experimental techniques in the EMC and E^3 communities are a necessary and excellent baseline from which to expand into large physical arenas.

For example, a publication by Frew and Brown [41] "explores the role of airborne communication networks in the operational performance of small unmanned aircraft systems." This paper suggests not only new system enhancing applications but also disruptive threats for airborne communication systems. A recent publication by Buchter et al. [42] proposes that future airborne communications include a host of applications including air traffic control, aircraft system integration, and weather and scientific data acquisition. A paper by Lee and Mark presents a cooperative control strategy for airborne networks that use aerial robotic vehicles [43]. Chaos enters the picture in an indirect way. If the chaos is caused by large-scale ambient environmental conditions, the correlation in information between airborne platforms may need to be addressed in new scenarios.

As the sophistication of future networks move in the direction of coupled platforms, autonomous behavior, and unexplored chaotic signals created in complex network configurations, more sophisticated analytical and experimental techniques need to be examined and evaluated for use in mainstream EMC and E^3 methodology.

1.5.1.3 EMC and E^3's Connection with Dynamical Systems

The mathematical foundations for EMC and E^3 are rooted in Maxwell's equation. Fundamentally, these are composed of time-dependent ODE and partial differential equations that involve spatial coordinates. If we then take the obvious step of breaking up continuum spatial variations into discrete elements, the most general class of system equations that EMC and E^3 workers will have to deal with is

$$\frac{d\vec{x}(t)}{dt} = \vec{f}(\vec{x}(t), \vec{u}(t), t) \tag{1.90}$$

In the foregoing equation, $\vec{x}(t)$ is the N-dimensional set of state variables (a column vector) of the system ($\vec{x}_i(t)$ is the ith component of $\vec{x}(t)$), $\vec{u}(t)$ is the M-dimensional set of external driving functions also treated as a column (some or all of which may be random), t is time, t_0 is the initial time, $\vec{x}_0$ are the initial conditions for the N state variables; there are exactly N initial conditions, and $\vec{f}(\vec{x}(t), \vec{u}(t), t)$ is an N-dimensional vector function. In component form, equation (1.90) can also be written as

$$\frac{dx_i(t)}{dt} = f_i(x_1(t), x_2(t) \ldots . x_N(t), u_1(t), u_2(t) \ldots . u_M(t), t) \tag{1.91}$$

Equation (1.91) is a nonlinear time-dependent equation for which there is no obvious universal solution. This does not however preclude the discovery of special-case solutions. The foregoing type of multifunction equation arises regularly in physics and engineering, and ingenuity is required to solve it, even approximately. On the other hand, there are substantially more cases that arise in practice where simpler versions of equation (1.91) occur, and these can be treated successfully using the current baseline techniques of references [44–47] combined with the recommended additional new techniques that are rendered in this document.

It is of interest to parse equation (1.91) into categories that can be solved either exactly or with good approximation. The first major simplification to equations (1.90/1.91) results when we delete the explicit dependence of "t" from consideration. Instead of equations (1.90) and (1.91) we then get

$$\frac{d\vec{x}(t)}{dt} = \vec{f}(\vec{x}(t), \vec{u}(t)) \tag{1.92}$$

$$\frac{dx_i(t)}{dt} = f_i(x_1(t), x_2(t) \ldots . x_N(t), u_1(t), u_2(t) \ldots . u_M(t)) \tag{1.93}$$

Special-case solutions to equations (1.92/1.93) can be found. Significant progress can be and has been made for those cases where equations (1.92/1.93) are further reduced to the familiar forms

$$\frac{d\vec{x}(t)}{dt} = \vec{f}(\vec{x}(t), \vec{u}(t)) = \vec{F}(\vec{x}(t)) + \vec{G}(\vec{u}(t)) \tag{1.94}$$

and its equivalent in component notation

$$\frac{dx_i(t)}{dt} = F_i(x_1(t), x_2(t), \ldots, x_N(t)) + G_i(u_1(t), u_2(t), \ldots, u_M(t)) \tag{1.95}$$

Equations (1.94/1.5) apply equally well to the areas of linear and nonlinear systems. For example, consider the case where equation (1.95) is written in matrix form. We have

$$\frac{dx_i(t)}{dt} = \widetilde{\mathrm{A}}_{ij}(x_1(t), x_2(t), \ldots x_N(t))x_j(t) + \widetilde{\mathrm{B}}_{ik}(u_1(t), u_2(t), \ldots u_M(t))u_k \tag{1.96}$$

$$F_i(x_1(t), x_2(t), \ldots x_N(t)) = (\widetilde{\mathrm{A}}_{ij}(x_1(t), x_2(t), \ldots x_N(t))x_j(t) \tag{1.97}$$

$$G_i(u_1(t), u_2(t), \ldots u_M(t))x_j(t) = \widetilde{\mathrm{B}}_{ik}(u_1(t), u_2(t), \ldots u_M(t))u_k \tag{1.98}$$

where the elements of a matrix $\widetilde{\mathrm{A}}_{ij}(x_1(t), x_2(t), \ldots x_N(t))$ are functions of the state variables, the elements of a matrix $\widetilde{\mathrm{B}}_{ik}(u_1(t), u_2(t), \ldots u_M(t))$ are functions of the driving functions; the matrix summation rule applies: $\widetilde{\mathrm{A}}_{ij}(x_1(t), x_2(t), \ldots x_N(t))x_j(t) = \widetilde{\mathrm{A}}_{i1}x_1 + \widetilde{\mathrm{A}}_{i2}x_2 \ldots . \widetilde{\mathrm{A}}_{iN}x_N$, $\widetilde{\mathrm{B}}_{ik}(u_1(t),$

$u_2(t), \ldots u_M(t))u_k(t) = \widetilde{B}_{i1}u_1 + \widetilde{B}_{i2}u_2 \ldots \widetilde{B}_{iM}u_M, \quad 1 \le i \le N, \quad 1 \le j \le N,$
$0 \le k \le M.$

In the case where there is no driving force, equation (1.96) reduces to the equation for autonomous systems,

$$\frac{dx_i(t)}{dt} = \widetilde{A}_{ij}(x_1(t), x_2(t), \ldots x_N(t))x_j(t) \tag{1.99}$$

Equation (1.99) is a coupled system of differential equations that does not depend on the independent variable. When the independent variable is time t, we call this set of equations a time-invariant system (LTI). These are the only ones that are covered in this document. A further parsing is made when the LTI systems are characterized as either linear or nonlinear. The linear system is the case where all the elements of $\widetilde{A}_{ij}$ are constants that are independent of the state variables, while in the nonlinear case the elements of $\widetilde{A}_{ij}$ depend on the state variables as in equation (1.99).

In summary, the EMC and E^3 systems considered in this document are LTI/linear—Section 1.5.2, and LTI/nonlinear—Section 1.5.3.

1.5.2 Linear Dynamical Systems

1.5.2.1 SISO and MIMO Models

The simplest kind of practical system that may be faced by EMC testing is the closed-loop single input single output (SISO) model which is shown in Figure 1.27. In this model, there is a single input, $u(t)$, and a single output $y(t)$. It is represented by a linear reduced form of equation (1.96). In those cases where

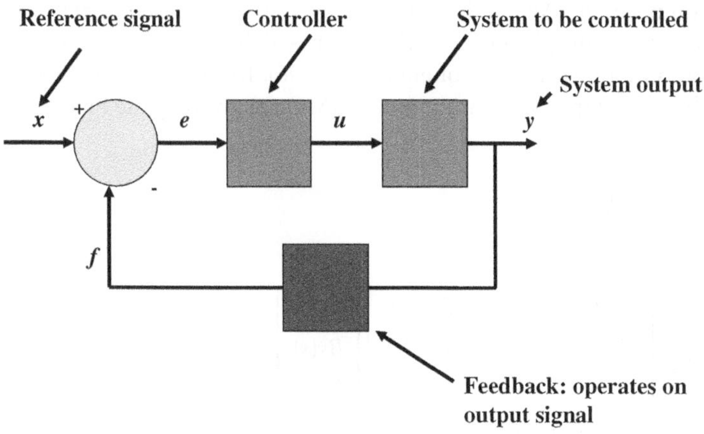

Figure 1.27 Closed-loop SISO model.

is no feedback we have an open-loop SISO model, which is used principally in noncritical situations. (Excellent discussions on this are rendered in [48].)

$$\frac{dy(t)}{dt} = \mathrm{A}y(t) + \mathrm{B}(u(t)) \tag{1.100}$$

In equation (1.100), A is a constant and B can be any well-behaved function of $u(t)$. If now s is the Laplace transform and a superscript "L" denotes the Laplace transform variables we have

$$Y^L(s) = \frac{y_0}{s - \mathrm{A}} + \frac{1}{s - \mathrm{A}}\mathrm{B}^L(s) \tag{1.101}$$

The modified new relationship between single input and single output is now

$$Y^L(s) = G_T(s)\mathrm{B}^L(s) \tag{1.102}$$

where $G_T(s)$ is the new scalar transfer function. A sample model of a closed-loop version has a feedback sensor $H(s)$. Feedback produces a "normal" error signal $E^L(s)$ delivered to the system transfer function. The well-known basic equation is

$$G_T(s) = \frac{G(s)}{(1 + H(s)G(s))} \tag{1.103}$$

where $G(s)$ is the open-loop transfer function. Should $G(s)$ have undesirable properties (e.g., small instability margin), equation (1.103) shows that the feedback sensor function $H(s)$ can in principle improve the overall transfer function.

As pointed out in [49], practical systems have more than a single variable and involve complicated EMC and E^3 than SISO systems. These systems are called multiple input multiple output (MIMO), in which the input and output are vectors and require the full use of equation (1.96). In the linear range, the relationships between input and output for a MIMO are given by the following form [49]:

$$Y_i^L(s) = G_{ij}(s)\mathrm{B}_j^L(s) \tag{1.104}$$

where

$$\begin{bmatrix} Y_1(s) \\ Y_2(s) \\ \vdots \\ Y_Q \end{bmatrix} = \begin{bmatrix} G_{11}(s) & G_{12}(s) \ldots G_{1P}(s) \\ G_{21}(s) & G_{22}(s) \ldots G_{2P}(s) \\ \vdots & \\ G_{Q1}(s) & G_{Q2}(s) \ldots G_{QP}(s) \end{bmatrix} \begin{bmatrix} B_1(s) \\ B_2(s) \\ \vdots \\ B_P(s) \end{bmatrix} \tag{1.105}$$

Equations (1.104) and (1.105) follow from equation (1.96). There are Q outputs and P inputs. The matrix elements G_{ij} of a transfer matrix function. For EMC test and evaluation it would be highly desirable to identify them. In the next section, we show how to determine the stability of the system under test as well as whether the system remains bounded.

1.5.2.2 Equilibrium Points, Initial Conditions, and Stability

At the fundamental level, the question of whether a MIMO system remains bounded and stable reduces to that for a linear SISO system, such as equation (1.102). For linear systems we can easily show [48]

$$G_T(s) = \sum_l \frac{\beta_l}{s + \alpha_l} \tag{1.106}$$

where the β_l and α_l are constants, "$-\alpha_l$" being the poles, and "l" the pole number designation. Using equation (1.106), the convolution Laplace transform to equation (1.102) the dependence of the output is

$$y(t) = \int_0^t g(t-\tau)b(\tau)d\tau = \int_{-\infty}^t g(t-\tau)b(\tau)d\tau \tag{1.107}$$

for causal systems and where $g(t)$ is the inverse Laplace transform of $G_T(s)$ and $b(t)$ the inverse Laplace transform of $B^L(s)$. From equation (1.106), we have

$$g(t) = \sum_l \beta_l \exp(-\alpha_l t) \tag{1.108}$$

Schetzen has pointed out that there are several ways to approach the question of stability and bounded behavior that applies to both linear and nonlinear systems [50]. The most popular discussions of stability, revert to Lyapunov's method—there are literally hundreds of thousands of articles on this subject. The method proposed by Schetzen is referred to as BIBO (bounded input–bounded output) and is well suited for EMC and E^3 testing of systems. For BIBO, a "stable system is one for which every bounded input gives rise to an output that also is bounded." We apply this concept to equations (1.107) and (1.108). By bounded we propose, in our notation, that the driving function $b(t)$ satisfy the condition (this is the basic approach)

$$|b(t)| < b_{\max} \tag{1.109}$$

for all t, and b_{max} is the absolute maximum value of $b(t)$. Using equations (1.107) and (1.109), we show that the system will be bounded only when

$$\int_{-\infty}^{\infty} \left|g(t-\tau)\right| d\tau < \infty \tag{1.110}$$

$$\left|y(t)\right| = \left|\int_{-\infty}^{\infty} g(t-\tau)b(\tau)d\tau\right|$$

$$\leq \int_{-\infty}^{\infty} \left|g(t-\tau)\right| \, |b(\tau| d\tau < b_{\max} \int_{-\infty}^{\infty} \left|g(t-\tau)\right| d\tau < \infty \tag{1.111}$$

The ability to ensure the validity of equation (1.111) requires equation (1.110) which in turn requires that all $\alpha_l \geq 0$.

Using equations (1.104)–(1.108), we can now show that the BIBO criteria for SISO can also apply to MIMO. We write equation (1.104) in the explicit form

$$Y_i^L(s) = G_{ij}(s)\mathrm{B}_j^L(s) = \sum_{j=1}^{j=P} G_{ij}(s)\mathrm{B}_j^L(s); \;\; 1 \leq i \leq Q \tag{1.112}$$

In order for equation (1.112) to satisfy the BIBO conditions, the only requirement is that each of its terms meets the BIBO requirements. This then requires that $G_{ij}(s)$ be of the form of equation (1.106). Although we have not explicitly demonstrated it, for brevity the reader can go through the individually steps and conclude that for a linear system the equilibrium points are stable if all the eigenvalues of the matrix "A" have negative real parts. This result is independent of the initial conditions.

1.5.2.3 Deterministic Force Response and Random Inputs

The analysis in the previous section implicitly dealt with the usual deterministic driving functions. Occasionally, selected inputs to parts of a system originate from external sources such as stochastic waveforms and chaotic signals. There is a long history of these problems, and fortunately these can be handled mathematically using the autocorrelation of the driving signal. Wilson Rugh [51] has provided a few compact expressions that make it easy to use in assessing the propagation of signals in nonlinear systems and also in the linear case considered here.

Following Rugh's approach, start with the linear expression of equation (1.107) now regarding $b(t)$ as a random function with expected value $\langle b(t) \rangle$ and autocorrelation function

$$R_{bb}(t_1, t_2) = \langle b(t_1) b(t_2) \rangle, \tag{1.113}$$

input/output cross-correlation function,

$$R_{yb}(t_1, t_2) = \langle y(t_1), b(t_2) \rangle \tag{1.114}$$

and output correlation function

$$R_{yy}(t_1, t_2) = \langle y(t_1) y(t_2) \rangle \tag{1.115}$$

Interchanging the order of integration of equation (1.107), we obtain the following results

$$\langle y(t) \rangle = \int_0^t g(\tau) \langle b(t-\tau) \rangle d\tau = \int_{-\infty}^{\infty} g(\tau) \langle b(t-\tau) \rangle d\tau \tag{1.116}$$

$$y(t_1) b(t_2) = \int_{-\infty}^{\infty} g(\tau) b(t_1 - \tau) b(t_2) d\tau \tag{1.117}$$

$$R_{yb}(t_1, t_2) = \int_{-\infty}^{\infty} g(\tau) R_{bb}(t_1 - \tau, t_2) d\tau \tag{1.118}$$

$$y(t_1) y(t_2) = \int_{-\infty}^{\infty} \int_{-\infty}^{\infty} g(\tau_1) g(\tau_2) b(t_1 - \tau_1) b(t_2 - \tau_2) d\tau_1 d\tau_2 \tag{1.119}$$

$$R_{yy}(t_1, t_2) = \int_{-\infty}^{\infty} \int_{-\infty}^{\infty} g(\tau_1) g(\tau_2) R_{bb}(t_1 - \tau_1, t_2 - \tau_2) d\tau_1 d\tau_2 \tag{1.120}$$

The EMC and E^3 analyst may find equations (1.113)–(1.120) common for the SISO case. On the other hand, a word of caution may be appropriate for MIMO systems. Potential difficulties arise when there is more than one random source. Let us consider for example when we have a single output and two random inputs denoted by α and β respectively. We have

$$y(t) = \int_{-\infty}^{\infty} g(\tau) b(t-\tau) d\tau \equiv \int_{-\infty}^{\infty} f(\tau) h(t-\tau) d\tau + \int_{-\infty}^{\infty} \bar{f}(\tau) \bar{h}(t-\tau) d\tau \tag{1.121}$$

$$y(t_1) = \int_{-\infty}^{\infty} g(\tau) b(t_1-\tau) d\tau \equiv \int_{-\infty}^{\infty} f(\tau) h(t_1-\tau) d\tau + \int_{-\infty}^{\infty} \bar{f}(\tau) \bar{h}(t_1-\tau) d\tau \tag{1.122}$$

$$y(t_2) = \int_{-\infty}^{\infty} g(\tau)b(t_2 - \tau)d\tau \equiv \int_{-\infty}^{\infty} f(\tau)h(t_2 - \tau)d\tau$$

$$+ \int_{-\infty}^{\infty} \bar{f}(\tau)\bar{h}(t_2 - \tau)d\tau \tag{1.123}$$

$$R_{yy}(t_1, t_2) = \int_{-\infty}^{\infty}\int_{-\infty}^{\infty} (f(\tau_1)h(t_1 - \tau_1) + \bar{f}(\tau_1)\bar{h}(t_1 - \tau_1))(f(\tau_2)h(t_2 - \tau_2)$$

$$+\bar{f}(\tau_2)\bar{h}(t_2 - \tau_2)d\tau_2 d\tau_1 \tag{1.124}$$

$$R_{yy}(t_1, t_2) = R_{yy}^{(1)}(t_1, t_2) + R_{yy}^{(2)}(t_1, t_2) + R_{yy}^{(3)}(t_1, t_2) + R_{yy}^{(4)}(t_1, t_2) \tag{1.125}$$

$$R_{yy}^{(1)}(t_1, t_2) = \int_{-\infty}^{\infty}\int_{-\infty}^{\infty} (f(\tau_1)h(t_1 - \tau_1)f(\tau_2)h(t_2 - \tau_2))d\tau_2 d\tau_1$$

$$= \int_{-\infty}^{\infty}\int_{-\infty}^{\infty} \left(f(\tau_1)f(\tau_2)R_{hh}(t_1 - \tau_1, t_2 - \tau_2)\right)d\tau_2 d\tau_1 \tag{1.126}$$

$$R_{yy}^{(2)}(t_1, t_2) = \int_{-\infty}^{\infty}\int_{-\infty}^{\infty} \left(f(\tau_1)h(t_1 - \tau_1)\bar{f}(\tau_2)\bar{h}(t_2 - \tau_2)\right)d\tau_2 d\tau_1$$

$$= \int_{-\infty}^{\infty}\int_{-\infty}^{\infty} \left(f(\tau_1)\bar{f}(\tau_2)R_{h\bar{h}}(t_1 - \tau_1, (t_2 - \tau_2)\right)d\tau_2 d\tau_1 \tag{1.127}$$

$$R_{yy}^{(3)}(t_1, t_2) = \int_{-\infty}^{\infty}\int_{-\infty}^{\infty} \left(\bar{f}(\tau_1)\bar{h}(t_1 - \tau_1)f(\tau_2)h(t_2 - \tau_2)\right)d\tau_2 d\tau_1$$

$$= \int_{-\infty}^{\infty}\int_{-\infty}^{\infty} \left(\bar{f}(\tau_1)f(\tau_2)R_{\bar{h}h}(t_1 - \tau_1, (t_2 - \tau_2)\right)d\tau_2 d\tau_1 \tag{1.128}$$

$$R_{yy}^{(4)}(t_1, t_2) = \int_{-\infty}^{\infty}\int_{-\infty}^{\infty} \left(\bar{f}(\tau_1)\bar{h}(t_1 - \tau_1)\bar{f}(\tau_2)\bar{h}(t_2 - \tau_2)\right)d\tau_2 d\tau_1$$

$$= \int_{-\infty}^{\infty}\int_{-\infty}^{\infty} \left(\bar{f}(\tau_1)\bar{f}(\tau_2)R_{\bar{h}\bar{h}}(t_1 - \tau_1, (t_2 - \tau_2)\right)d\tau_2 d\tau_1 \tag{1.129}$$

1.5.2.4 Example for SISO

The purpose of this section is to introduce selected mathematical aspects of EMC and E^3 for evaluating electromagnetic threats typical of those rendered in the MIL-STDs. Focusing on a few essential features can assist EMC and E^3 analysts in modeling and evaluating new equipment. The basic model is the linear model where we have a single input, $x(t)$, and a single output is $y(t)$. In the linear regime, the fundamental equation that relates $y(t)$ to $x(t)$ can be expressed by equation (1.130):

$$\sum_n A_n \int_0^t dt_1 \int_0^{t_1} dt_2 \ldots \int_0^{t_{n-1}} y(t_n)dt_n + \sum_m B_m \frac{d^m y(t)}{dt^m} = x(t) \tag{1.130}$$

The foregoing equation is a generic expression that connects an input to an output, and A_n and B_m are known physical constants of the system. At this point we have not yet stated precisely what $y(t)$ and $x(t)$ represent, other than being inputs and outputs. We are interested in relating the response of aerospace subsystems and components to unwanted electromagnetic threats. We define:

$$\widetilde{X}(s) = \text{Laplace transform of } x(t) \tag{1.131}$$

$$\widetilde{Y}(s) = \text{Laplace transformof } y(t) \tag{1.132}$$

$$G(s) = \left(\sum_{n=1} A_n \frac{1}{s^n} + \sum_{m=0} B_m s^m \right) = f(s) + h(s) \tag{1.133}$$

$$f(s) = \sum_{n=1}^{n=\gamma} A_n \frac{1}{s^n} = \frac{1}{s^\gamma}(A_\gamma + A_{\gamma-1}s \ldots . A_1 s^{\gamma-1}) \tag{1.134}$$

$$h(s) = \sum_{m=0}^{m=\sigma} B_m s^m = (B_0 + B_1 s \ldots . B_\sigma s^\sigma) \tag{1.135}$$

In the absence of unwanted EMPs, the characteristics of the driving function, $x(t)$, are assumed to be well known and appropriate for typical system actions. Even though the system can be rendered very complicated, we propose for tutorial purposes that its key physical features related to temporal response against the threat can be evaluated using a few parameters. As we will show the tradeoffs in performance are often governed by the ratio of threat pulse width to system response time constant. For relatively fast transients, the time constants buried in equation (1.130) are assumed to be much longer. For brevity, let us assume a simple form for equation (1.130):

$$\sum_n A_n \int_0^t dt_1 \int_0^{t_1} dt_2 \ldots \int_0^{t_{n-1}} y(t_n)dt_n + \sum_m B_m \frac{d^m y(t)}{dt^m} \to B_1 \frac{dy(t)}{dt} + B_0 y(t) \tag{1.136}$$

Inserting equation (1.136) into equation (1.130) gives

$$B_1 \frac{dy(t)}{dt} + B_0 y(t) = x(t) \tag{1.137}$$

The total solution to equation (1.137) is

$$y(t) = H \exp(-\lambda t) + \frac{\exp(-\lambda t)}{B_1} \int_0^t x(t') \exp(\lambda t') dt' \tag{1.138}$$

where H is an arbitrary constant and $\lambda = B_0/B_1$. In equation (1.138) there is no restriction on $x(t')$. For example, if $x(t') = C$, a constant, the solution to equation (1.138) is

$$y(t) = y_0 \exp(-\lambda t) + \frac{C}{B_0}(1 - \exp(-\lambda t)) \tag{1.139}$$

where

$$y(t = 0) = y_0 = H \tag{1.140}$$

In equation (1.138) now suppose that $x(t)$ is a "normal" operational waveform. We have not yet specified whether it is a voltage, current, electric and/or an electromagnetic field component; $y(t)$ is the output variable, and B_0 is a dimensional variable having the required form. For example, if $x(t)$ has the dimensions of voltage and $y(t)$ has the dimensions of meters then B_0 would have the dimensions of volts/meter.

Under normal conditions, we assume that $x(t)$ and y_0 will vary within defined suitable operational ranges:

$$-x_{\min} \leq x(t) \leq x_{\max} \tag{1.141}$$

$$-y_{0,\min} \leq y_0 \leq y_{0,\max} \tag{1.142}$$

By combining equations (1.138), (1.141), and (1.142), we can determine the maximum and minimum range of the variables. The maximum values occur when H and $x(t')$ of equation (1.138) are both positive. However, since $x(t') \leq x_{\max}$, we have

$$\frac{\exp(-\lambda t)}{B_1} \int_0^t x(t') \exp(\lambda t') dt' \leq \frac{\exp(-\lambda t) x_{\max}}{B_1} \int_0^t \exp(\lambda t') dt'$$
$$= \frac{x_{\max}}{B_0}(1 - \exp(-\lambda t) \tag{1.143}$$

Inserting equation (1.143) into equation (1.138) for positive $y(t)$ gives

$$\begin{aligned} y(t) &\le H\exp(-\lambda t) + \frac{x_{max}}{B_0}(1-\exp(-\lambda t) \\ &= \left(H - \frac{x_{max}}{B_0}\right)\exp(-\lambda t) + \frac{x_{max}}{B_0} \end{aligned} \tag{1.144}$$

The maximum of $y(t)$ occurs when $(dy(t)/dt) = 0$, which gives

$$\frac{dy(t)}{dt} = -\lambda\left(H - \frac{x_{max}}{B_0}\right)\exp(-\lambda t) = 0 \tag{1.145}$$

The maximum occurs for $H = x_{max}/B_0$ at which point $y_{0,max} = x_{max}/B_0$.

However, when $x(t)$ is derived from a threat waveform, the behavior of $y(t)$ is very different from that given by equation (1.144). Without getting into system specific details, the most general statement that can be made is that there will be a functional relationship between any of the types of voltages of the driving function $x(t)$.

We are concerned with threats where the domain of $x(t)$ and $y(t)$ is outside of the range given by equations (1.141)–(1.145), satisfying the condition

$$y_p(t) = \frac{\exp(-\lambda t)}{B_1}\int_0^t x_p(t')\exp(\lambda t')dt', \tag{1.146}$$

and $x_p(t)$ is the resulting action waveform produced by the voltage at the location of interest.

$$|x_p(t)| \ge \max\{x_{max} \text{ or } x_{min}\} \tag{1.147}$$

$$|y_p(t)| \ge \max\{y_{0,max} \text{ or } y_{0,min}\} \tag{1.148}$$

In general, we can write the following equation:

$$x_p(t) = \mathrm{O}_p(V_p(t)) \tag{1.149}$$

where O_p is a mathematical operation that transforms an unwanted pulse voltage waveform, $V_p(t)$ into the action waveform $x_p(t)$. The operator's action can range from basic algebraic functions, power series, and or differentiation and integration. For illustrative purposes, we consider the simple case

$$x_p(t) = \mathrm{O}_p(V_p(t)) \cong aV_p(t) \tag{1.150}$$

where a is a dimensional constant. We have

$$y_p(t) = \frac{a\exp(-\lambda t)}{B_1}\int_0^t V_p(t')\exp(\lambda t')dt' \tag{1.151}$$

Assume that the duration, t_p of $V_p(t)$ is substantially less than $\tau = 1/\lambda$ is by several orders of magnitude. Then the useful value of t' within the integral is t_p, $t_p\lambda = (t_p/\tau) \ll 1$, $\exp(t_p\lambda) \cong 1$, and we get

$$y_p(t) \cong \frac{a\exp(-\lambda t)}{B_1} \int_0^t V_p(t')dt' \tag{1.152}$$

For $t \geq t_p$, which is our range of interest, equation (1.152) becomes

$$y_p(t)_{t\geq t_p} \cong \frac{a\exp(-\lambda t)}{B_1} \int_0^{t_p} V_p(t')dt' = \frac{a\exp(-\lambda t)}{B_1} \bar{V}_p t_p \tag{1.153}$$

where $\bar{V}_p$ is a pulse-averaged voltage derived from

$$\bar{V}_p \equiv \frac{\int_0^{t_p} V_p(t')dt'}{t_p} \tag{1.154}$$

In the theoretical limit where

$$\begin{aligned} t_p &\to 0 \\ V_p &\to \infty \\ \bar{V}_p t_p &\to I_p = \text{finite voltage} - \text{time impulse} \end{aligned} \tag{1.155}$$

the threat pulse voltage waveform transforms approximately into a Delta function. This voltage waveform, $V_E(t)$, is represented by

$$V_E(t) = I_p\delta(t) \tag{1.156}$$

When equation (1.156) is combined with the normal operating voltage, $x(t)$, the starting point is now

$$B_1\frac{dy(t)}{dt} + B_0 y(t) = x(t) + I_p\delta(t) \tag{1.157}$$

The particular solution for equation (1.157) is

$$\hat{y}(t) = \frac{\exp(-\lambda t)}{B_1} \int^t \left(x(t') + I_p\delta(t')\right) \exp(\lambda t')dt', \tag{1.158}$$

the total solution is

$$y(t) = \hat{y}(t) + y_c(t) = H\exp(-\lambda t) + \frac{\exp(-\lambda t)}{B_1} \times \int_0^t (x(t') + aI_p\delta(t'))\exp(\lambda t')dt' \tag{1.159}$$

and H is determined from the initial condition: $y(t = 0) \equiv y_0$.

Replacing $x(t')$ by a local time average value: $x(t') = C$, we get

$$\hat{y}(t) = \frac{C\tau}{B_1}(1 - \exp(-\lambda t)) + \frac{\exp(-\lambda t)}{B_1} a\bar{V}_p t_p \tag{1.160}$$

$$\hat{y}(t) = \frac{C\tau}{B_1}\left(1 + \exp(-t/\tau)\left[\frac{\bar{V}_p t_p}{C\tau} - 1\right]\right) \tag{1.161}$$

$$y(t) = y_c(t) + \hat{y}(t) = y_0 \exp(-t/\tau) + \frac{C\tau}{B_1}\left(1 + \exp(-t/\tau)\left[\frac{a\bar{V}_p t_p}{C\tau} - 1\right]\right) \tag{1.162}$$

As observed from equation (1.162), the critical parameter is the ratio

$$R \equiv \frac{a\bar{V}_p t_p}{C\tau} = \frac{aI_p}{I_c} \tag{1.163}$$

where we define $I_c \equiv C\tau$ as the "normal impulse." When R is large the threat will produce a significant change in the system response, and when R is small the converse is true. But whether large or small, the duration of the effect is controlled by the normal impulse.

The threat impulse introduces a step function jump at $t = 0$. From equation (1.162) we get

$$y(t = 0+) = y_0 + \frac{C\tau}{B_1}\left(\left[\frac{a\bar{V}_p t_p}{C\tau}\right]\right) = y_0 + \frac{a\bar{V}_p t_p}{B_0 \tau} \tag{1.164}$$

If the system has been in steady-state operation for a long time (compared with τ) with a constant $x(t) = \hat{C}$ before the threat, then the initial condition just before the threat insult would be

$$y_0 = \frac{\hat{C}}{B_0} \tag{1.165}$$

We then have

$$B_1 y_0 \cong \frac{B_1 \hat{C}}{B_0} = \tau \hat{C} \tag{1.166}$$

$$y(t = 0) = y_0 + \frac{a\bar{V}_p t_p}{B_0 \tau} = y_0\left(1 + \frac{a\bar{V}_p t_p}{\hat{C}\tau}\right) \tag{1.167}$$

and the relative size of the jump at $t = 0+$ is $(a\bar{V}_p t_p/\hat{C}\tau)$.

Now let us consider a more complicated system model based on equation (1.130), and equation (1.156). Instead of equation (1.130), we now have

$$\sum_n A_n \int_0^t dt_1 \int_0^{t_1} dt_2 \ldots \int_0^{t_{n-1}} y(t_n)dt_n + \sum_m B_m \frac{d^m y(t)}{dt^m} = x(t) + aI_p\delta(t) \tag{1.168}$$

For brevity and without loss of generality, let's use $x(t') = C$, and assume that $y(t=0) = y_0$ and $(d^n y/dt^n)_{t=0} = 0$ for $n \geq 1$. Taking the Laplace transform of equation (1.168) using the aforementioned conditions gives

$$\left(\sum_{n=1} A_n \frac{1}{s^n} + \sum_{m=0} B_m s^m\right)\widetilde{Y}(s) - B_1 y_0 = \frac{C}{s} + aI_p \tag{1.169}$$

$$\left(\sum_{n=1} A_n \frac{1}{s^n} + \sum_{m=0} B_m s^m\right)\widetilde{Y}(s) = \frac{C}{s} + aI_p + B_1 y_0 \tag{1.170}$$

Let us assume that γ is the largest value for n and σ is the largest value of m in the summations of equation (1.170). We then define the functions

$$G(s) = \left(\sum_{n=1} A_n \frac{1}{s^n} + \sum_{m=0} B_m s^m\right) = f(s) + h(s) \tag{1.171}$$

$$f(s) = \sum_{n=1}^{n=\gamma} A_n \frac{1}{s^n} = \frac{1}{s^\gamma}(A_\gamma + A_{\gamma-1}s \ldots . A_1 s^{\gamma-1}) \tag{1.172}$$

$$h(s) = \sum_{m=0}^{m=\sigma} B_m s^m = (B_0 + B_1 s \ldots . B_\sigma s^\sigma) \tag{1.173}$$

$$G(s) = \frac{1}{s^\gamma}(A_\gamma + A_{\gamma-1}s \ldots . A_1 s^{\gamma-1} + B_0 s^\gamma + B_1 s^{\gamma+1} \ldots . B_\sigma s^{\gamma+\sigma}) \equiv \frac{P(s)}{s^\gamma} \tag{1.174}$$

$$P(s) = (A_\gamma + A_{\gamma-1}s \ldots . A_1 s^{\gamma-1} + B_0 s^\gamma + B_1 s^{\gamma+1} \ldots . B_\sigma s^{\gamma+\sigma}) \tag{1.175}$$

The polynomial $P(s)$ has $\eta = \gamma + \sigma$ roots and can be factored to produce the form

$$P(s) = \mathrm{K}(s+p_1)(s+p_2) \ldots .. (s+p_i) \ldots .. (s+p_\eta) \tag{1.176}$$

In the foregoing expression, K is a constant derived in the factoring process with the p_i's being the roots of $P(s)$. For brevity, we have assumed that all the roots are distinct. If some of the roots should turn out to be repeated the forthcoming analysis becomes more complicated but does not change the physical consequences.

Inserting equation (1.165) into equation (1.170) gives

$$\widetilde{Y}(s) = \frac{s^{\gamma}}{P(s)}\left(\frac{C}{s} + aI_p + B_1 y_0\right) = \frac{s^{\gamma}}{P(s)}\left(\frac{C}{s}\right) + \frac{s^{\gamma}}{P(s)}(aI_p + B_1 y_0) \quad (1.177)$$

Using the partial fraction expansion for $1/P(s)$ we get

$$\frac{1}{P(s)} = \frac{1}{\mathrm{K}}\sum_{i=1}^{i=\eta}\frac{E_i}{(s+p_i)} \quad (1.178)$$

The E_i's are constants derived in the partial fraction expansion from the p_i's. The time behavior of $y(t)$ is readily derived by first computing the inverse Laplace transforms of the functions $(s^{\gamma}/P(s))$ and $(s^{\gamma-1}/P(s))$.

We have

$$\widetilde{L}^{-1}\left(\frac{s^{\gamma}}{P(s)}\right) = \frac{1}{\mathrm{K}}\frac{d^{\gamma}}{dt^{\gamma}}\sum_{i=1}^{i=\eta}E_i\exp(-p_i t) = \frac{1}{\mathrm{K}}\sum_{i=1}^{i=\eta}(-p_i)^{\gamma}E_i\exp(-p_i t) \quad (1.179)$$

$$\widetilde{L}^{-1}\left(\frac{s^{\gamma}}{P(s)}\right) = \frac{1}{\mathrm{K}}\frac{d^{\gamma-1}}{dt^{\gamma-1}}\sum_{i=1}^{i=\eta}E_i\exp(-p_i t)$$

$$= \frac{1}{\mathrm{K}}\sum_{i=1}^{i=\eta}(-p_i)^{\gamma-1}E_i\exp(-p_i t) \quad (1.180)$$

$$\widetilde{Y}(s) = \frac{s^{\gamma}}{P(s)}\left(\frac{C}{s}\right) + \frac{s^{\gamma}}{P(s)}(aI_p + \tau C) = \frac{s^{\gamma}}{P(s)}\left(\frac{C}{s}\right) + \frac{s^{\gamma}\tau C}{P(s)}\left(\frac{aI_p}{\tau C} - 1\right) \quad (1.181)$$

$$y(t) = \frac{C}{\mathrm{K}}\sum_{i=1}^{i=\eta}(-p_i)^{\gamma-1}E_i\exp(-p_i t) + \frac{\tau C}{\mathrm{K}}\sum_{i=1}^{i=\eta}(-p_i)^{\gamma}\left(\frac{aI_p}{\tau C} - 1\right)E_i\exp(-p_i t) \quad (1.182)$$

$$y(t) = \frac{C}{\mathrm{K}}\sum_{i=1}^{i=\eta}(-p_i)^{\gamma-1}E_i\exp(-p_i t)\left\{1 - p_i\tau\left(\frac{aI_p}{\tau C} - 1\right)\right\} \quad (1.183)$$

1.5.3 Nonlinear Dynamical Systems

1.5.3.1 Autonomous and Nonautonomous Systems

In Section 1.5.1.3, we provided general overview of nonlinear autonomous (time-invariant) systems and nonlinear nonautonomous systems and proceeded at considerable length to examine linear dynamical systems that included equilibrium conditions, stability, forced response. In particular, we constructed a hypothetical dynamical model for a SISO system. Although the material presented was not intended to be comprehensive, the discussion combined with the cited references should allow the practicing EMC and E^3 scientists to expand their technology base and handle the analyses for a large number of linear systems.

If it is desired in the analyses to go beyond the basic criterion of BIBO, it is relatively simple to examine in more detail the conditions for: *stable, unstable, single, and multiple equilibrium points of autonomous nonlinear systems* [52]. Even for the basic model

$$\frac{d\vec{x}(t)}{dt} = \vec{f}(\vec{x}(t)) \tag{1.184}$$

the possibilities are enormous. For tutorial purposes, reference [52] has provided an excellent set of one-dimensional functions, $f(x(t))$, that may be useful in testing of EMC and E^3 systems, because they cover a wide range of functions including polynomials, rational functions of x, irrational functions of x, transcendental functions of x, and multiple finite-valued zeros, and become infinite at a number of points and have discontinuities. Useful examples given in [52] are

a. $\dot{x} = -x^3$ (1.185)

b. $\dot{x} = x^3$ (1.186)

c. $\dot{x} = -k\tanh(x)$ (1.187)

d. $\dot{x} = -\text{sgn}(x)x^2$ (1.188)

e. $\dot{x} = -kx(1-x)$ (1.189)

f. $\dot{x} = -x(16x^4 - 20x^2 + 5)$ (1.190)

g. $\dot{x} = -\sin(x)$ (1.191)

The material rendered in the foregoing discussion and in equation (1.185) should be regarded as "warm-up" for real systems. In fact, in some cases, it is possible to combine a few of the equations within equations (1.185)–(1.191) and develop unique solutions for actual situation. We need to also realize that the world of time-invariant nonlinear equations is enormous. The next logical steps in rendering time-invariant nonlinear useful for aerospace EMC and E^3 are the Volterra and Weiner theories discussed in Section 1.5.2.2.

1.5.3.2 Volterra and Weiner Theories

In this section, we present a brief mathematical background for the Volterra and Weiner theories that are useful for the EMC and E^3 scientist to proceed further into the complicated mathematical structure of the theories. An example of mathematical depth required to use these theories can be found in Schetzen's book [50].

The Volterra series is an extension of the Taylor series for nonlinear behavior. Whereas the Taylor series only applies when the output of a system at a particular at time t depends specifically on the input to the input at the same time. We say that the system is memory-less. The Volterra series applies for all times, which gives it memory capability [53]. The series originated in 1887 by the Italian mathematician Vito Volterra, and became of great interest in the United States in the 1920s due to Norbert Weiner. It became of broad interest in many branches of nonlinear systems in 1957. Franz and Scholkopf pointed out that the Volterra approach essentially extended the standard convolution description given by

$$y(t) = H_1(x(t)) = \int h^{(1)}(\tau)x(t-\tau)d\tau \tag{1.192}$$

where $y(t)$ is the output, $x(t)$ is the input, and $h^{(1)}(t)$ is the linear kernel defining the impulse response [54].

The Volterra series rendered in the forthcoming equations (1.104)–(1.105) is from [54] and is an extension of equation (1.103) to nonlinear systems given by

$$y(t) = H_0x(t) + H_1x(t) + H_2x(t) + \cdots + H_nx(t) + \tag{1.193}$$

where $H_0x(t) = h_0 = const.$ and

$$H_nx(t) = \int\int ..... \int h^{(n)}(\tau_1, \tau_2, \dots \tau_n)x(t-\tau_1) \dots . x(t-\tau_n)d\tau_1 \dots d\tau_n \tag{1.194}$$

is the nth-order Volterra operator and $h^{(n)}(\tau_1, ..\tau_n)$ are the Volterra kernels.

Lamnabhi–Lagarrigue also shows that the nonlinear is causal if and only if

$$h^{(n)}(\tau_1, \tau_2, ..\tau_n) = 0, \text{ for } \tau_j < 0, \tag{1.195}$$

and any kernel $h^{(n)}(\tau_1, \tau_2, ..\tau_n)$ can be replaced by a symmetric one using the formula [55]

$$h^{(n)}_{sym}(\tau_1, \tau_2, \dots, \tau_n) = \frac{1}{n!} \sum_{\substack{\text{Set of all} \\ \text{permutations} \\ \text{of}\,\tau_1,\tau_2,\dots,\tau_n}} h^{(n)}(\tau_1, \tau_2, \dots, \tau_n) \tag{1.196}$$

The point that we are bringing out with the few examples given here is that there is a large pool of resources that can be used to calculate the Volterra kernels once the physical models have been constructed. The Weiner series is an alternative way of computing the response of nonlinear systems [50, 54, 55].

The foundation for the Weiner series is to measure the system response for all possible input functions. Weiner has suggested that suitable random functions based on the random walk process are useful in this approach. When this is done, one can write

$$y(t) = G_0x(t) + G_1x(t) + G_2x(t) + \cdots + G_nx(t) + \cdots \tag{1.197}$$

where a Wiener series of operators G_n are mutually uncorrelated (orthogonal with respect to the Wiener process. The G_ns are linear combinations of Volterra operators up to order n [54].

In summary, we anticipate that systems having extensive nonlinear properties will become more prevalent in the future. A very small population of the EMC and E^3 community is familiar with the technical nonlinear techniques that have been available in related fields. By increasing the awareness of such techniques, we can dramatically increase our ability to test and evaluate aerospace systems.

1.5.4 Representation of an Autonomous Systems Dealing with EMI

A true autonomous system must identify and address its own EMI problems independently of any external input and driving function. It was previously stated in this section that the mathematical foundation of EMC problems is rooted in Maxwell's equations which are basically time dependent ordinary linear partial differential equations involving spatial coordinates. The most general class of system equations that EMC will have to deal, was shown previously in equation (1.90), and now shown again as equation (1.198)

$$\frac{d\vec{x}(t)}{dt} = \vec{f}(\vec{x}(t), \vec{u}(t), t) \tag{1.198}$$

In equation (1.198) when there is no driving force (i.e. u(t) = 0) equation (1.198) reduces to the previously shown equation (1.99) herein written again as equation (1.199) . Equation (1.199) represents the governing equation for an autonomous system.

$$\frac{dx_i(t)}{dt} = \tilde{A}_{ij}(x_1(t), x_2(t), \ldots x_N(t))x_j(t) \tag{1.199}$$

Equation (1.199) can represents a linear (A_{ij} matrix elements are constants) or a nonlinear (A_{ij} matrix elements depend on the state variables, x_i(t)) time invariant system.

Figure 1.28 shows an autonomous system composed of four independent units (Unit #1 through Unit # 4). Each unit communicates with another, as shown in the figure, where data is passed from Unit #1 through Unit # 4, and finally converted into a data stream output as shown in the figure. It is also shown in Figure 1.28 that Unit #1 has an EMI source, and the figure also shows

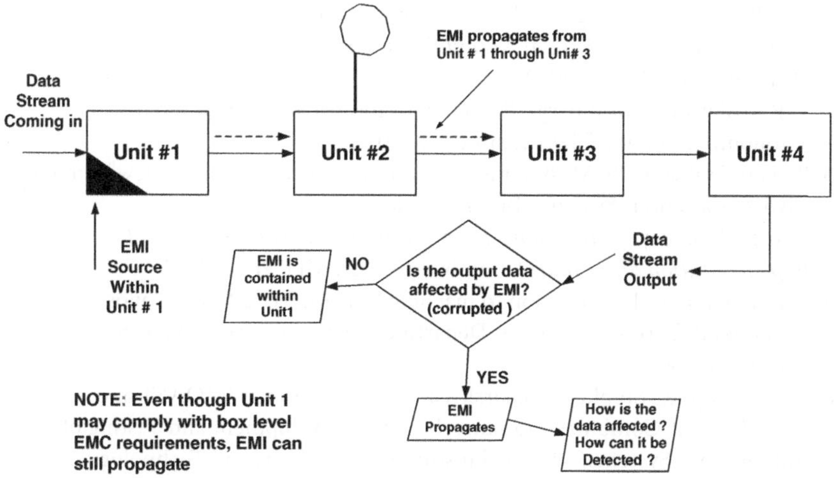

Figure 1.28 EMI Diagnosis within an autonomous system.

that the EMI propagates to Unit #2 and Unit #3 from its source in Unit #1. Notice that even though Unit #1 may have passed compliance with its EMC requirements, it is shown in the Figure that EMI can still propagate to other assemblies in the autonomous system. A true autonomous system must assess if its output data stream is corrupted by EMI, which will be an indication of EMI propagation as shown in the flowchart of Figure 1.28.

In equation (1.199) when a parameter $x_n(t)$ where $(n = 1 \ldots j)$ is affected by EMI, the parameter is physically distorted ("d"). A distorted parameter in equation (1.199) can be re-named as $x_{nd}(t)$. A true autonomous system must first recognize a distorted parameter $x_{nd}(t)$ and where does it come from. The answer will determine the source of the EMI in the autonomous system. Once the source of the EMI has been identified and the nature of the distorted parameter also identified, the autonomous system can assess how other parameters in equation (1.199) might be affected, and from such information the EMI propagation within the autonomous system can be predicted.

Our present technology in autonomous system does not allow for such unprecedented diagnostic capabilities by autonomous systems, but future progress in this area is envisioned as autonomous systems become more independent due to the future infusion of machine learning.

References

1 Lora G. Weiss, "Autonomous robots in the fog of war," *IEEE Spectrum*, Aug. 2011.
2 Thomas Tenorio, "Meeting the challenges of unmanned autonomous system test and evaluation," Mar. 10, 2010, USC.

3 V. P. Roske, I. Kohlberg, and R. Wagoner, "Autonomous systems challenges to T&E," in *NDIA 28th Annual National Test and Evaluation Conference*, Hilton Head, SC, Mar. 12–15, 2012.

4 V. P. Roske and I. Kohlberg, "Foundations for test and evaluation of autonomous systems," IDA Report, Jun. 2012.

5 Richard Williams, "BAE systems – Autonomous capability overview," Director Civil Autonomous Systems, Briefing Slides.

6 George A. Bekey, *Autonomous Robots, from Biological Inspiration to Implementation and Control*. Cambridge, MA: MIT Press, 2005.

7 Nils. J. Nilsson, "Introduction to machine learning," Draft of a proposed textbook, Robotics Laboratory, Department of Computer Science, Stanford University, 2005.

8 Chris Scrapper, Stephen Balakirsky, and Elena Messina, "MOAST and USARSim—A Combined Framework for the Development and Testing of Autonomous Systems," National Institute of Standards and Technology, Gaithersburg, MD 20899, USA, Proceedings of the SPIE Defense and Security Symposium, Orlando, FL. Apr. 17–21, 2006.

9 Adam Jacoff, Elena Messina, and John Evans, "Experiences in deploying test arenas for autonomous mobile robots," Intelligent Systems Division National Institute of Standards and Technology, Gaithersburg, MD 20899; Proceedings of the 2001 Performance Metrics for Intelligent Systems (PerMIS) Workshop, in association with the IEEE CCA and ISIC, Mexico City, Mexico, Sep. 4, 2001.

10 The Royal Academy of Engineering, *Autonomous Systems: Social, Legal and Ethical Issues*. London: The Royal Academy of Engineering, 2009.

11 Kris Cowart, Maj, USAF, Ricardo Valerdi, PhD, C. Robert Kenley, PhD, "Development, validation and implementation considerations of a decision support system for unmanned & autonomous system of systems test and evaluation," Test Week 2010, Jun. 16, 2010.

12 Norbert Wiener, "Nonlinear problems in random theory," New York: The Technology Press of MIT and John Wiley & Sons, Inc, 1958.

13 Vito Volterra, *Theory of Functionals and of Integral and Integro-Differential Equations*. New York: Dover Publications, 1959.

14 Martin Schetzen, *The Volterra and Wiener Theories of Nonlinear Systems*. New York: John Wiley & Sons, Inc., 1980.

15 Matthias O. Franz and Bernhard Scholkopf, "A unifying view of Wiener and Volterra theory and polynomial kernel regression," *Neural Comput.*, vol. 18, pp. 3097–3118, 2006.

16 E. Buckingham, "On physically similar systems: illustrations of the use of dimensional equations," *Phys. Rev.*, vol. 4, no. 4, p. 345, 1914.

17 L. I. Sedov, *Similarity and Dimensional Methods in Mechanics*. Moscow: MIR Publishers, 1982.

18 G. Birkhoff, *Hydrodynamics: A Study in Logic, Fact, and Similitude*. Princeton, NJ: Princeton University Press, 1960.

19 V. L. Streeter, *Fluid Mechanics*, 4th ed. New York: McGraw-Hill Book Company, 1996.

20 I. Kohlberg and T.P. Coffee, "Dimensional analysis and self-similarity theory for regenerative liquid propellant guns," ARL Report ARL-TR-2157, Jan. 2000.

21 http://en.wikpedia.org/wiki/Satellite

22 K.-D. Buchter, A. Reinhold, G. Stenz, and A. Sizmann, "Drivers and elements of future airborne communication networks," Deutscher Luft-und Raumfahrtkongress 2012, Document: 281323.

23 Milstar Satellite Communications System. March 22, 2017. https://www.afspc.af.mil/About-Us/Fact-Sheets/Display/Article/1012622/milstar-satellite-communications-system/

24 "New Photos Show the Rare Heavily Modified B-707 Operated by MIT as a Communications and Sensor Testbed for the U.S. Air Force." April 17, 2017, *The Aviationist*. https://theaviationist.com/2017/04/17/new-photos-show-the-rare-heavily-modified-b-707-operated-by-mit-as-a-communications-and-sensor-testbed-for-the-u-s-air-force/

25 E. W. Frew and T. X. Brown, "Airborne communication net for small unmanned aircraft," in *Proceedings of the IEEE*, vol. 96, December 2008.

26 Deok-Jin Lee and R. Mark, "Decentralized control of unmanned aerial robots for wireless airborne communication networks," *Int. J. Adv. Robot. Syst.*, vol. 7, no. 3, pp. 191–200, 2010.

27 I. Kohlberg, "Spatially inhomogeneous survivable communication networks with directed and non-directed graphs," in *Proceedings of the International Conference on Electromagnetics in Advanced Applications* (ICEAA 14), Palm Beach, Aruba, Aug. 3–9, 2014.

28 SDAN 40, "Random graph model for determining the survivability of spatially inhomogeneous communication and sensor networks," System Design and Assessment Note 40, University of New Mexico, Dec. 2012.

29 SDAN 39, "Survivable communication networks with non-directed and directed graphs," System Design and Assessment Note No. 39, University of New Mexico, Dec. 2011

30 L. Zeger, "Survivability and recovery of degraded communication networks", in *Proceedings of MILCOM 2011*, Baltimore, MD, Nov. 7–10, 2011.

31 P. Gupta and P. R. Kumar, *The Capacity of Wireless Networks*, IEEE Transactions on Information Theory, vol. 46, no. 2, March 2000.

32 S. H. Kellert, *In the Wake of Chaos: Unpredictable Order in Dynamic System*, Chicago: University of Chicago Press, 1993.

33 A. V. Oppenheim et al., "Signal processing in the context of chaotic Signals," *IEEE International Conference on Acoustics, Speech, and Signal Processing*, March 23–26, 1992.

34 L. O. Chua and R. Brown, "Clarifying chaos III. Chaotic and stochastic processes, chaotic resonance, and number theory," *International Journal of Bifurcation and Chaos*, vol. 9, no. 5, pp. 785–803, 1999.

35 J. Kasac and A. Stranjak, “The criteria of quantitative determination of chaotic behavior of non-linear dynamic systems,” in *Proc. 20th International Conference on Technology Interfaces*, Zagreb, 1998.

36 E. N. Lorenz, “Deterministic non-periodic flow,” *Journal of Atmospheric Sciences*, vol. 20, no. 2, pp. 130–141, 1963.

37 S. Banerjee and G. C. Verghese, eds., *Nonlinear Phenomena in Power Electronics*. Hoboken, NJ: John Wiley & Sons, 2001.

38 D. C. Hamill, S. Banerjee, and G. C. Verghese, “Introduction to power electronics,” in *Nonlinear Phenomena in Power Electronics*. Hoboken, NJ: John Wiley & Sons, 2001.

39 I. Kohlberg, “A stochastic process and chaos interpretation of HPM effects on electronic systems,” in *Proc. Asia-Pacific Symposium on Electromagnetic Compatibility*, Beijing, China, Apr. 12–16, 2010.

40 L. M. Berliner, “Statistics, probability, and chaos,” *Statistical Science Journal*, vol. 7, no. 1, pp. 69–90, 1992.

41 E. W. Frew and T. X. Brown, “Airborne communication networks for small unmanned aircraft systems,” in *Proc. IEEE*, vol., 96, no. 12, Dec. 2008.

42 K.-D. Buchter, A. Reinhold, G. Stenz, and A. Sizmann, “Drivers and elements of future airborne communication networks,” Deutcher Luft und Raumfahrtkongress 2012, Document: 281323.

43 Deok-Jin Lee and R. Mark, “Decentralized control of unmanned aerial robots for wireless airborne communication networks,” *Int. J. Adv. Robot. Syst.*, vol. 7, no. 3, 2010. https://doi.org/10.5772/9702

44 MIL-STD-464C, Electromagnetic environmental effects requirements for systems, Dec. 1, 2010.

45 MIL-STD-461F, Requirements for the control of electromagnetic interference characteristics of subsystems and equipment, Dec. 10, 2007.

46 MIL-STD-3023, High-altitude electromagnetic pulse (HEMP) protection for military aircraft, 21 Nov. 2011.

47 R. F. Gray, Verification and validation of unified electromagnetic (UEM) design, Version 3.0, Contract No. DTRA01-03-D-0004; Delivery Order No:0017, ATK, Apr. 2009.

48 R. Saucedo and E. E. Schiring, *Introduction to continuous and digital control systems*. New York: Macmillan Publishing Co., Inc., 1968.

49 H. Bessai, *MIMO Signals and Systems*, Berlin: Springer, 2006.

50 M. Schetzen, *The Volterra and Weiner Theories of Nonlinear Systems*. New York: John Wiley & Sons, Inc., 1980.

51 W. J. Rugh, *Nonlinear System Theory: The Volterra/Wiener Approach*. Baltimore: The Johns Hopkins University Press, 1981. Web version prepared in 2002.

52 “Nonlinear, time-invariant (autonomous systems),” Chapter 4, http://eecs.ceas.uc.edu/˜pramamoo/CourseWork/BookNonLinearAndAdaptiveSysPDFfiles/Chapter4_2Columns.pdf

53 "Volterra series," http://en.wikipedia.org/wiki/Volterra_series
54 M. O. Franz and Scholkopf, "A Unifying View of Wiener and Volterra Theory and Polynomial Kernel Regression," *Neural Comput.*, vol. 18, no. 12, pp. 3097–3118, Dec. 2006.
55 Francoise Lamabhi-Lagarrigue, "Volterra and Fliess series expansion," in *Control Systems, Robotics and Automation*, vol. XII, Encyclopedia of the Life Support Systems, H. Unbehauen, Ed. UNESCO, 2009.
56 E. Buckingham, "On physically similar systems: Illustrations of the use of dimensional equations," *Phys. Rev.*, vol. 4, no. 4, p. 345, 1914.
57 L. I. Sedov, *Similarity and Dimensional Methods in Mechanics*. Moscow, Russia: MIR Publishers, 1982.
58 G. Birkhoff, *Hydrodynamics: A Study in Logic, Fact, and Similitude*. Princeton, NJ: Princeton University Press, 1960.
59 V. L. Streeter, *Fluid Mechanics*, 4th ed. New York: McGraw-Hill Book Company, 1996.
60 I. Kohlberg and T. P. Coffee, "Dimensional analysis and self-similarity theory for regenerative liquid propellant guns," ARL Report ARL-TR-2157, Jan. 2000.

2

Deterministic and Statistical EMC Models for Field-to-Wire Coupling and Crosstalk in Wire Harness

Sergio Pignari

2.1 Introduction

This chapter focuses on the modeling of field-to-wire coupling and crosstalk in wiring harness. The fundamental phenomena and mechanisms of coupling are presented by considering elemental line configurations (e.g., the single-ended line above ground and the ideally twisted-wire pair). The modeling approach is then extended to complex wiring harness, and the influence of complexity on the wiring harness EMC behavior is shown. To this end, bundles of wires and bundles of twisted-wire pairs (TWPs) are here considered.

Both deterministic and statistical models based on multiconductor transmission line (MTL) theory and modal analysis are described. Deterministic models are used to explain the involved phenomena, provide simplified circuit representations, assess MTL model validity in controlled configurations, and show the effect of terminal load and wire harness imbalance on mode conversion and resulting immunity characteristics.

On the other side, statistical modeling is introduced as an extension of deterministic modeling, able to deal with unknown and/or uncontrolled parameters of the wiring harness and/or the terminal networks, as well as to recognize the random nature of the impinging electromagnetic disturbance.

2.2 Deterministic Modeling

2.2.1 MTL Models

In this section, the basic concepts of MTL theory are briefly reviewed as they are fundamental principles in the development of distributed-parameter

Handbook of Aerospace Electromagnetic Compatibility, First Edition. Edited by Reinaldo J. Perez.

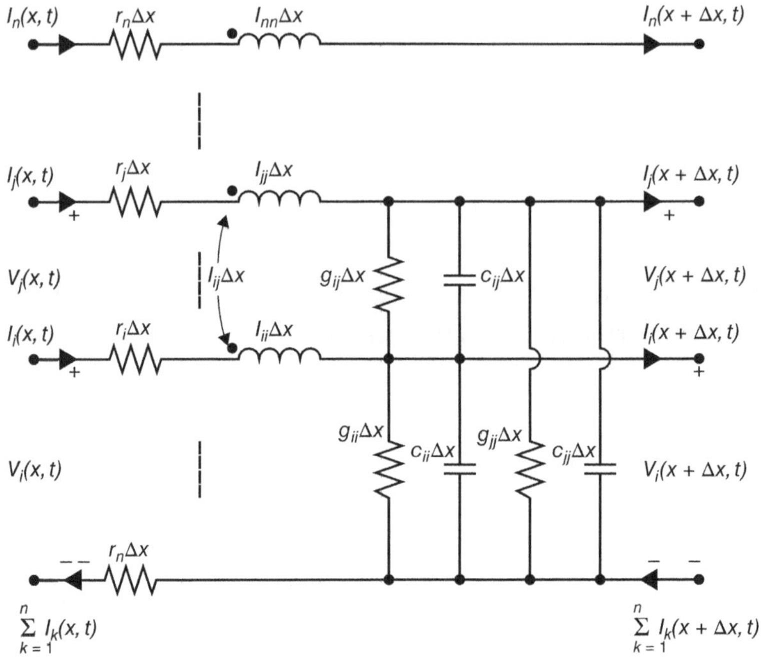

Figure 2.1 Infinitesimal line section of an MTL composed by $(n + 1)$ conductors.

models for the prediction of crosstalk and field-to-wire coupling in complex cable bundles.

2.2.1.1 MTL Equations

Given an MTL composed by $(n + 1)$ wires, and chosen the $(n + 1)$th conductor as reference, propagation of the n currents and n voltages along the line can be expressed by writing the Kirchhoff laws for the infinitesimal line section in Figure 2.1, and taking the limit for $\Delta x \to 0$. This leads to the following system of $2n$ *coupled first-order partial differential equations* (also known in the literature as *Telegrapher's equations*):

$$\frac{\partial}{\partial x}\boldsymbol{V}(x,t) = -\boldsymbol{R}\boldsymbol{I}(x,t) - \boldsymbol{L}\frac{\partial}{\partial t}\boldsymbol{I}(x,t) \tag{2.1}$$

$$\frac{\partial}{\partial x}\boldsymbol{I}(x,t) = -\boldsymbol{G}\,\boldsymbol{V}(x,t) - \boldsymbol{C}\frac{\partial}{\partial t}\boldsymbol{V}(x,t), \tag{2.2}$$

where $\boldsymbol{V}(x,t)$, $\boldsymbol{I}(x,t)$ denote the $n \times 1$ vectors of voltages and currents at line position x, and $\boldsymbol{R}$, $\boldsymbol{L}$, $\boldsymbol{G}$, $\boldsymbol{C}$ are the $n \times n$ matrices of the per-unit-length (p.u.l.) resistance, inductance, admittance, and capacitance parameters associated with the MTL cross-section. For different cable categories, approximate analytical expressions for such p.u.l. parameters, based on the wide-separation approximation can be found in [1, Ch 5], but their validity is strictly limited to the case of homogeneous media surrounding the wires. Hence, in the majority of practical applications, the evaluation of the line p.u.l. parameters is carried out resorting to numerical methods involving electrostatic and magnetostatic 2D simulation of the MTL cross-section.

2.2.1.2 Frequency-Domain Solution

As long as the frequency-domain solution is the target, sinusoidal excitation of the line is considered, the line voltages and currents are replaced by their phasor form, and the system in (2.1) and (2.2) is rewritten as

$$\frac{d\hat{\boldsymbol{V}}(x)}{dx} = -\hat{\boldsymbol{Z}}\hat{\boldsymbol{I}}(x), \quad \hat{\boldsymbol{Z}} = \boldsymbol{R} + j\omega\boldsymbol{L}, \tag{2.3}$$

$$\frac{d\hat{\boldsymbol{I}}(x)}{dx} = -\hat{\boldsymbol{Y}}\hat{\boldsymbol{V}}(x), \quad \hat{\boldsymbol{Y}} = \boldsymbol{G} + j\omega\boldsymbol{C}, \tag{2.4}$$

which represent a system of $2n$ (coupled) *first-order ordinary differential equations with complex coefficients* (in the following denoted as *first-order MTL equations*), where $\omega = 2\pi f$ is the radian frequency, and the caret is used to denote complex quantities. By differentiating (2.3) with respect to x and substituting in (2.4), and vice versa, the following systems of *second-order differential equations* (in the following denoted as *second-order MTL equations*) are obtained:

$$\frac{d^2\hat{\boldsymbol{V}}(x)}{dx^2} = (\hat{\boldsymbol{Z}}\hat{\boldsymbol{Y}})\hat{\boldsymbol{V}}(x), \tag{2.5}$$

$$\frac{d^2\hat{\boldsymbol{I}}(x)}{dx^2} = (\hat{\boldsymbol{Y}}\hat{\boldsymbol{Z}})\hat{\boldsymbol{I}}(x), \tag{2.6}$$

With respect to (2.3) and (2.4), these sets of equations are independent, since (2.5) only involves the line voltages, whereas (2.6) only involves the line currents. Therefore, it is possible to solve the equations in (2.5) for the line voltages and then retrieve the line currents from (2.3), or equivalently to solve (2.6) with respect to line currents and retrieve line voltages from (2.4). As a matter of fact, since propagation constants are obtained as the eigenvalues of the matrix products $\hat{\boldsymbol{Z}}\hat{\boldsymbol{Y}}$ and $\hat{\boldsymbol{Y}}\hat{\boldsymbol{Z}}$, respectively, consistency of the obtained results is ensured by the fact that the eigenvalues of $\hat{\boldsymbol{Z}}\hat{\boldsymbol{Y}}$ and $\hat{\boldsymbol{Y}}\hat{\boldsymbol{Z}}$ are the same, since physical lines are always characterized by symmetric matrices of p.u.l. parameters, thus satisfying the property $(\hat{\boldsymbol{Z}}\hat{\boldsymbol{Y}})^T = \hat{\boldsymbol{Y}}^T\hat{\boldsymbol{Z}}^T = \hat{\boldsymbol{Y}}\hat{\boldsymbol{Z}}$, [2].

Although the two systems in (2.5) and (2.6) are uncoupled, the n equations of each system are still coupled, as matrix products $\hat{\boldsymbol{Z}}\hat{\boldsymbol{Y}}$ and $\hat{\boldsymbol{Y}}\hat{\boldsymbol{Z}}$ are, at least in the general case, full matrices. Therefore, for the solution of (2.5) or, equivalently, of (2.6), decoupling techniques are usually exploited [2], which are based on the introduction of similarity transformation matrices to decompose the physical voltages or/and currents into a fictitious set of modal quantities (in the following denoted by subscript "*m*") as

$$\hat{\boldsymbol{V}}(x) = \hat{\boldsymbol{T}}_V \hat{\boldsymbol{V}}_m(x), \quad \hat{\boldsymbol{I}}(x) = \hat{\boldsymbol{T}}_I \hat{\boldsymbol{I}}_m(x). \tag{2.7}$$

In terms of modal quantities, (2.5) and (2.6) can be therefore rewritten as

$$\frac{d^2 \hat{\boldsymbol{V}}_m(x)}{dx^2} = \hat{\boldsymbol{T}}_V^{-1} (\hat{\boldsymbol{Z}}\hat{\boldsymbol{Y}}) \hat{\boldsymbol{T}}_V \hat{\boldsymbol{V}}_m(x) = \boldsymbol{\Gamma}^2 \hat{\boldsymbol{V}}_m(x), \tag{2.8}$$

$$\frac{d^2 \hat{\boldsymbol{I}}_m(x)}{dx^2} = \hat{\boldsymbol{T}}_I^{-1} (\hat{\boldsymbol{Y}}\hat{\boldsymbol{Z}}) \hat{\boldsymbol{T}}_I \hat{\boldsymbol{I}}_m(x) = \boldsymbol{\Gamma}^2 \hat{\boldsymbol{I}}_m(x), \tag{2.9}$$

where $\boldsymbol{\Gamma}^2 = \hat{\boldsymbol{T}}_V^{-1} (\hat{\boldsymbol{Z}}\hat{\boldsymbol{Y}}) \hat{\boldsymbol{T}}_V = \hat{\boldsymbol{T}}_I^{-1} (\hat{\boldsymbol{Y}}\hat{\boldsymbol{Z}}) \hat{\boldsymbol{T}}_I$ is a $n \times n$ diagonal matrix, the root square of its main diagonal entries being the propagation constants $\hat{\gamma}_1, \hat{\gamma}_2, \ldots, \hat{\gamma}_n$. Therefore, the columns of the similarity transformation matrices $\hat{\boldsymbol{T}}_V, \hat{\boldsymbol{T}}_I$ are the eigenvectors associated with the eigenvalues of the matrix products $\hat{\boldsymbol{Z}}\hat{\boldsymbol{Y}}, \hat{\boldsymbol{Y}}\hat{\boldsymbol{Z}}$. Based on this result, the line characteristic impedance matrix (full matrix) can be equivalently expressed as

$$\hat{\boldsymbol{Z}}_C = \hat{\boldsymbol{T}}_V \hat{\boldsymbol{\Gamma}}^{-1} \hat{\boldsymbol{T}}_V^{-1} \hat{\boldsymbol{Z}}, \quad \hat{\boldsymbol{Z}}_C = \hat{\boldsymbol{T}}_V \hat{\boldsymbol{\Gamma}} \hat{\boldsymbol{T}}_V^{-1} \hat{\boldsymbol{Y}}^{-1}, \tag{2.10}$$

$$\hat{\boldsymbol{Z}}_C = \hat{\boldsymbol{Z}} \hat{\boldsymbol{T}}_I \hat{\boldsymbol{\Gamma}}^{-1} \hat{\boldsymbol{T}}_I^{-1}, \quad \hat{\boldsymbol{Z}}_C = \hat{\boldsymbol{Y}}^{-1} \hat{\boldsymbol{T}}_I \hat{\boldsymbol{\Gamma}} \hat{\boldsymbol{T}}_I^{-1}, \tag{2.11}$$

depending on the choice to solve the MTL equations for the line voltages (2.10) or the line currents (2.11). It is important to stress here that since $\hat{\boldsymbol{Z}}\hat{\boldsymbol{Y}}, \hat{\boldsymbol{Y}}\hat{\boldsymbol{Z}}$ are square matrices, solution of the *second-order MTL equations* can be always accomplished, since it is always possible to find a similarity transformation matrix (i.e., $\hat{\boldsymbol{T}}_V$ or $\hat{\boldsymbol{T}}_I$) able to make the matrix products $\hat{\boldsymbol{Z}}\hat{\boldsymbol{Y}}, \hat{\boldsymbol{Y}}\hat{\boldsymbol{Z}}$ diagonal. However, restrictions apply if one attempts to uncouple the *first-order equations* in (2.5) and (2.6). Indeed, in that case, the two matrices $\hat{\boldsymbol{T}}_V, \hat{\boldsymbol{T}}_I$ were simultaneously involved, and the possibility to make diagonal also the p.u.l. parameter matrices $\hat{\boldsymbol{Z}}, \hat{\boldsymbol{Y}}$ is strictly related to special symmetry properties of the line cross-section [2].

For lossless MTLs in homogeneous medium, the previous analysis significantly simplifies. Indeed, as long as losses are negligible, the p.u.l. impedance and admittance matrices simplify to $\hat{\boldsymbol{Z}} = j\omega \boldsymbol{L}$, $\hat{\boldsymbol{Y}} = j\omega \boldsymbol{C}$. Additionally, if the medium surrounding the line conductors is homogeneous, the fundamental relationship $\boldsymbol{LC} = \boldsymbol{CL} = \mu\varepsilon \mathbf{1}_n$ holds between the p.u.l. inductance and capacitance matrices, where $\mu = \mu_r \mu_0$ and $\varepsilon = \varepsilon_r \varepsilon_0$ are the permeability and permittivity[1] of the medium, respectively, and $\mathbf{1}_n$ is the $n \times n$ identity matrix. This special case is known as *degenerate mode* of propagation, since the n modes of the line travel with the same propagation velocity $v = c_0/\sqrt{\varepsilon_r \mu_r}$, where c_0 denotes the speed of light in free space. In this case, the similarity transformation matrices are identity matrices, i.e., $\hat{\boldsymbol{T}}_V = \hat{\boldsymbol{T}}_I = \mathbf{1}_n$, and the characteristic impedance matrix, which is real-valued and frequency independent, can be simply evaluated as $\hat{\boldsymbol{Z}}_C = v\boldsymbol{L} = \boldsymbol{C}^{-1}/v$.

2.2.1.3 Chain-Parameter Representation

Once matrix $\hat{\boldsymbol{\Gamma}} = diag\{\hat{\gamma}_1, \hat{\gamma}_2, \dots, \hat{\gamma}_n\}$ and the characteristic impedance matrix are known, the standard approach is to resort to the chain-parameter representation. Accordingly, the MTL is interpreted as a $2n$-port network, whose voltages and currents at the terminal ports are related by the $2n \times 2n$ chain-parameter matrix $\hat{\boldsymbol{\Phi}}_{2n\times 2n}(L)$:

$$\begin{bmatrix} \hat{\boldsymbol{V}}(L) \\ \hat{\boldsymbol{I}}(L) \end{bmatrix} = \hat{\boldsymbol{\Phi}}_{2n\times 2n}(L) \cdot \begin{bmatrix} \hat{\boldsymbol{V}}(0) \\ \hat{\boldsymbol{I}}(0) \end{bmatrix} = \begin{bmatrix} \hat{\boldsymbol{\Phi}}_{11}(L) & \hat{\boldsymbol{\Phi}}_{12}(L) \\ \hat{\boldsymbol{\Phi}}_{21}(L) & \hat{\boldsymbol{\Phi}}_{22}(L) \end{bmatrix} \cdot \begin{bmatrix} \hat{\boldsymbol{V}}(0) \\ \hat{\boldsymbol{I}}(0) \end{bmatrix}, \tag{2.12}$$

where L denotes the line length, and the $n \times n$ submatrices $\hat{\boldsymbol{\Phi}}_{11}(L)$, $\hat{\boldsymbol{\Phi}}_{12}(L)$, $\hat{\boldsymbol{\Phi}}_{21}(L)$, and $\hat{\boldsymbol{\Phi}}_{22}(L)$ take the form:

$$\hat{\boldsymbol{\Phi}}_{11}(L) = \frac{1}{2}\hat{\boldsymbol{T}}_V(\mathrm{e}^{\hat{\boldsymbol{\Gamma}}L} + \mathrm{e}^{-\hat{\boldsymbol{\Gamma}}L})\hat{\boldsymbol{T}}_V^{-1} = \frac{1}{2}\hat{\boldsymbol{Y}}^{-1}\hat{\boldsymbol{T}}_I(\mathrm{e}^{\hat{\boldsymbol{\Gamma}}L} + \mathrm{e}^{-\hat{\boldsymbol{\Gamma}}L})\hat{\boldsymbol{T}}_I^{-1}\hat{\boldsymbol{Y}}, \tag{2.13}$$

$$\hat{\boldsymbol{\Phi}}_{12}(L) = -\frac{1}{2}\hat{\boldsymbol{T}}_V(\mathrm{e}^{\hat{\boldsymbol{\Gamma}}L} - \mathrm{e}^{-\hat{\boldsymbol{\Gamma}}L})\hat{\boldsymbol{\Gamma}}^{-1}\hat{\boldsymbol{T}}_V^{-1}\hat{\boldsymbol{Z}} = -\frac{1}{2}\hat{\boldsymbol{Y}}^{-1}\hat{\boldsymbol{T}}_I\hat{\boldsymbol{\Gamma}}(\mathrm{e}^{\hat{\boldsymbol{\Gamma}}L} - \mathrm{e}^{-\hat{\boldsymbol{\Gamma}}L})\hat{\boldsymbol{T}}_I^{-1}, \tag{2.14}$$

$$\hat{\boldsymbol{\Phi}}_{21}(L) = -\frac{1}{2}\hat{\boldsymbol{Z}}^{-1}\hat{\boldsymbol{T}}_V\hat{\boldsymbol{\Gamma}}(\mathrm{e}^{\hat{\boldsymbol{\Gamma}}L} - \mathrm{e}^{-\hat{\boldsymbol{\Gamma}}L})\hat{\boldsymbol{T}}_V^{-1} = -\frac{1}{2}\hat{\boldsymbol{T}}_I(\mathrm{e}^{\hat{\boldsymbol{\Gamma}}L} - \mathrm{e}^{-\hat{\boldsymbol{\Gamma}}L})\hat{\boldsymbol{\Gamma}}^{-1}\hat{\boldsymbol{T}}_I^{-1}\hat{\boldsymbol{Y}}, \tag{2.15}$$

$$\hat{\boldsymbol{\Phi}}_{22}(L) = \frac{1}{2}\hat{\boldsymbol{Z}}^{-1}\hat{\boldsymbol{T}}_V(\mathrm{e}^{\hat{\boldsymbol{\Gamma}}L} + \mathrm{e}^{-\hat{\boldsymbol{\Gamma}}L})\hat{\boldsymbol{T}}_V^{-1}\hat{\boldsymbol{Z}} = \frac{1}{2}\hat{\boldsymbol{T}}_I(\mathrm{e}^{\hat{\boldsymbol{\Gamma}}L} + \mathrm{e}^{-\hat{\boldsymbol{\Gamma}}L})\hat{\boldsymbol{T}}_I^{-1}. \tag{2.16}$$

Based on such a $2n$-port representation, the voltages and currents at the line terminals (and, subsequently, at an arbitrary position x along the line) can be predicted by combining (2.12) with the port constraints enforced by the terminal sections, that can be either expressed in terms of Thevenin or Norton equivalent circuits as shown in [1].

1 In the expressions of μ and ε, subscript "r" is used to denote the relative permeability/permittivity of the material with respect to free space, whose properties are denoted by subscript "0."

2.2.1.4 Nonuniform MTLs

For nonuniform MTLs (i.e., wiring harness exhibiting variable cross-section along the longitudinal direction), the simplest, though inherently approximate, way to extend the previously described approach is to represent the MTL as the cascade connection of uniform line sections [3,4], each of them characterized by nearly constant cross-section. The resulting chain-parameter matrix of the entire MTL is therefore obtained by multiplying the chain-parameter matrices associated with each line section as

$$\begin{bmatrix} \hat{V}(L) \\ \hat{I}(L) \end{bmatrix} = \hat{\mathbf{\Phi}}_{2n\times 2n}(L) \cdot \begin{bmatrix} \hat{V}(0) \\ \hat{I}(0) \end{bmatrix} = \\ = \hat{\mathbf{\Phi}}_n(L_n) \cdot \hat{\mathbf{\Phi}}_{n-1}(L_{n-1}) \cdot \ldots \cdot \hat{\mathbf{\Phi}}_1(L_1) \cdot \begin{bmatrix} \hat{V}(0) \\ \hat{I}(0) \end{bmatrix}, \tag{2.17}$$

where $\hat{\mathbf{\Phi}}_n(L_n)$ and $\hat{\mathbf{\Phi}}_1(L_1)$ are the chain-parameter matrices associated with the rightmost ($x = \mathrm{L}$) and leftmost ($x = 0$) line sections, respectively.

Optimal segmentation fineness is strictly dependent on the spatial scale of variation of the line cross-section. In the modeling of complex wiring harness, this choice is usually a trade-off between prediction accuracy and computational burden (or, equivalently, simulation time).

2.2.2 Field-to-Wire Coupling

2.2.2.1 Distributed Line Excitation

The first attempt to model the coupling between a transmission line and an external electromagnetic field dates back to 1965 [5], when Taylor firstly resorted to transmission line (TL) theory to predict the terminal response of a two-wire line structure in free space, driven by a plane-wave electromagnetic field. In the Taylor model work, the effect of the interfering field was included into the distributed-parameter model of the victim circuit by means of distributed voltage and current sources, as shown in (2.18). According to this formulation (known in the literature as *total voltage formulation*), the telegrapher's equations for an infinitesimal line section can be written as

$$\frac{d}{d\xi}\begin{pmatrix} V(\xi) \\ I(\xi) \end{pmatrix} + j\omega \begin{bmatrix} 0 & \ell \\ c & 0 \end{bmatrix} \cdot \begin{pmatrix} V(\xi) \\ I(\xi) \end{pmatrix} = \begin{pmatrix} V_F(\xi) \\ I_F(\xi) \end{pmatrix}, \tag{2.18}$$

where ξ denotes the line coordinate along the line, ℓ, c are the line p.u.l. inductance and capacitance, and the infinitesimal voltage and current sources $V_F(\xi)$, $I_F(\xi)$ are, respectively, related to the normal component of the incident magnetic field and to the longitudinal component of the incident electric field.

Subsequently, several other scientists contributed to refine and extend the Taylor model. In particular, Paul [6] extended the model to the case of uniform MTLs. Agrawal [7] reformulated the field-to-wire coupling problem in terms of electric-field components only, whereas Rachidi [8] proposed a dual formulation, where the induced sources involve magnetic-field components only.

In particular, as regards the model proposed by Agrawal (known in the literature as *scattered voltage formulation*), the main difference with respect to the *total voltage formulation* proposed by Taylor can be appreciated by comparing the telegrapher's equations for an infinitesimal line section. Namely, instead of the total line voltage $V(\xi)$, the telegrapher's equations formulated according to the Agrawal approach

$$\frac{d}{d\xi}\begin{pmatrix} V^{sca}(\xi) \\ I(\xi) \end{pmatrix} + j\omega \begin{bmatrix} 0 & \ell \\ c & 0 \end{bmatrix} \cdot \begin{pmatrix} V^{sca}(\xi) \\ I(\xi) \end{pmatrix} = \begin{pmatrix} V_\ell(\xi) \\ 0 \end{pmatrix} \tag{2.19}$$

involve the "scattered" component $V^{sca}(\xi)$ of the total line voltage, that is, the component of $V(\xi)$ associated with the scattered electric field. Circuit interpretation of the Agrawal model is provided in Figure 2.2(b), where the effects due to the external EM field are included into the distributed-parameter model of the victim TL by

(a) a distributed voltage source $V_\ell(\xi)$, related to the tangential component of the electric field evaluated along the line length;

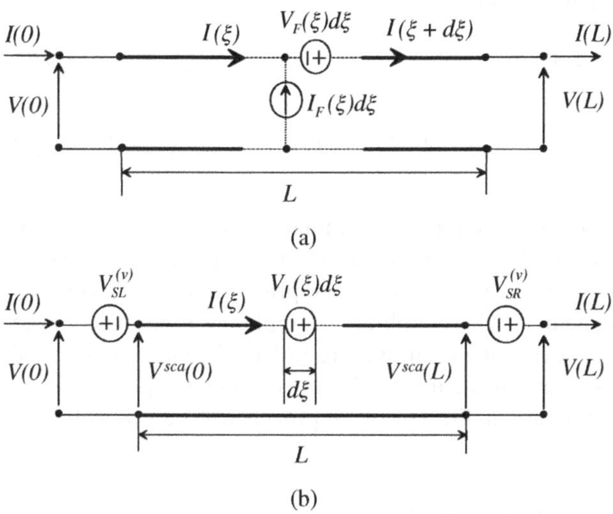

Figure 2.2 Circuit interpretation of the coupling between an external EM field and a two-conductor TL according to the (a) Taylor and (b) Agrawal formulation.

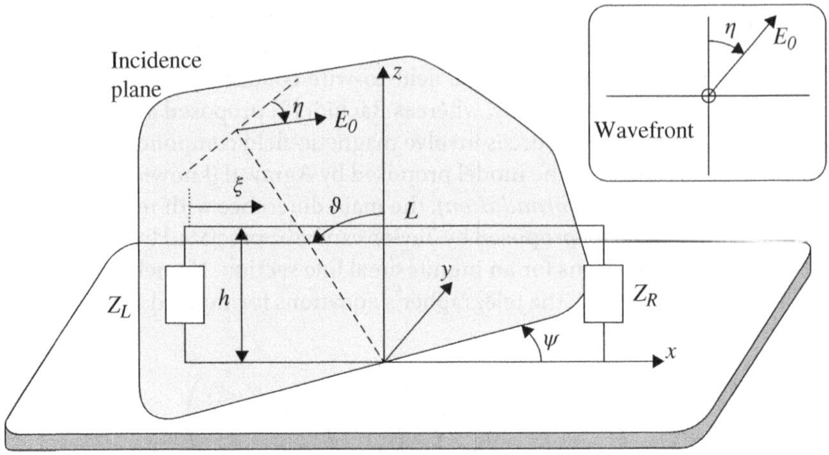

Figure 2.3 Single-ended interconnection above ground illuminated by a plane-wave field, with electric-field strength E_0 and incidence and polarization angles ϑ, ψ, and η, respectively.

(b) two lumped voltage sources, $V_{SL}^{(v)}$, $V_{SR}^{(v)}$, related to the vertical component of the electric field evaluated at the line ends.

Indeed, with reference to the single-ended interconnection above ground shown in Figure 2.3, the involved distributed, $V_\ell(\xi)$, and lumped, $V_{SL}^{(v)}$, $V_{SR}^{(v)}$, sources take the general expressions

$$V_\ell(\xi) = E_x^{inc}(x_0 + \xi, y_0, h), \tag{2.20}$$

$$V_{SL}^{(v)} = -\int_0^h E_z^{inc}(x_0, y_0, z)dz, \quad V_{SR}^{(v)} = -\int_0^h E_z^{inc}(x_0 + L, y_0, z)dz. \tag{2.21}$$

Since the Agrawal model only requires (a) evaluation of the E-field component tangential to the line path and (b) evaluation of the vertical E-field components at line terminals, this formulation is often preferred to the Taylor formulation for predicting field-to-wire coupling in complex systems, where high nonuniformity and complexity of the impinging EM field require numerical evaluation (by 3D EM solvers) of the involved field components [9–12]. More details on this hybrid "numerical-TL theory" approach will be given in Section 2.2.4.

2.2.2.2 Equivalent Circuits at Line Terminals

In spite of the distributed nature of the field-to-wire coupling phenomenon, equivalent circuits involving lumped voltage/current sources at line ends can

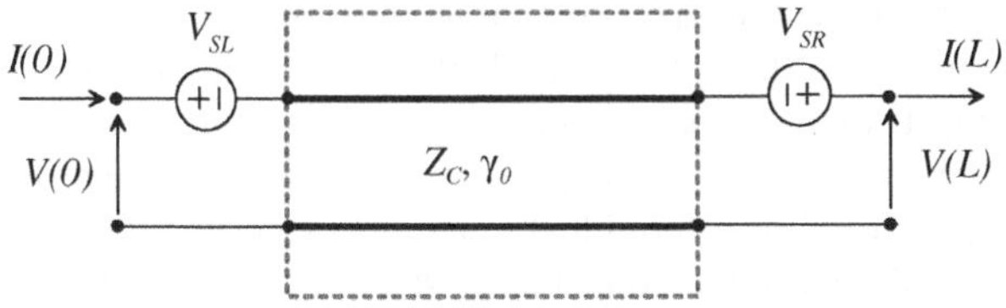

Figure 2.4 Equivalent circuit at line ends of a two-conductor TL driven by an external EM field.

be derived and combined with the port constraints enforced by the terminal networks in order to predict the voltages and currents induced by the impinging field at line ends.

An example is shown in Figure 2.4, where the effects due to the external electromagnetic wave are modeled by voltage sources connected at the line terminals. For the canonical case of a single-ended interconnection above ground, illuminated by the uniform plane-wave field, shown in Figure 2.3, analytical expressions of the induced sources V_{SL}, V_{SR} due to field-to-wire coupling can be obtained and cast in closed form as[2] [13]:

$$V_{SL} = 2E_0 h \left\{ \frac{F(\vartheta, \psi, \eta)}{\gamma_0 \sinh(\gamma_0 L)} [\hat{\gamma}_0 \sinh(\gamma_0 L) - \gamma_0 \cosh(\gamma_0 L) + \gamma_0 \mathrm{e}^{-\hat{\gamma}_0 L}] - G(\vartheta, \eta) \right\} \tag{2.22}$$

$$V_{SR} = V_{SL}^* \mathrm{e}^{-\hat{\gamma}_0 L} \tag{2.23}$$

where

$$\hat{\gamma}_0 = \gamma_0 \sin\vartheta \cos\psi, \quad G(\vartheta, \eta) = \cos\eta \sin\vartheta, \tag{2.24}$$

$$F(\vartheta, \psi, \eta) = \frac{\cos\vartheta(\cos\eta\cos\vartheta\cos\psi + \sin\eta\sin\psi)}{1 - \sin^2\vartheta\cos^2\psi} \tag{2.25}$$

and $\gamma_0 = j\omega/c_0$ denotes the line propagation constant, E_0 is the electric-field amplitude, and ϑ, ψ, and η are the incidence angles and the polarization angles of the impinging plane-wave field, respectively. These expressions, valid for a lossless line in free space, can be readily extended to bundles of wires as in [14].

2 Without loss of generality, (22)–(25) were evaluated by assuming the left termination of the TL in Figure 2.3 coincident with the origin of the coordinate system, i.e., $x_0 = y_0 = 0$.

2.2.2.3 Limitation of Field-to-Wire Coupling Models based on TL Theory

While modeling field-to-wire coupling through TL theory, it should always be kept in mind that only transverse-electromagnetic (TEM) currents can be predicted. Indeed, TL theory inherently neglects the so-called antenna or common-mode currents [15, 16]. These "nonideal" currents can only be modeled through 3D full-wave simulation of the wiring structure under analysis, since they can be generally ascribed to imperfections and/or asymmetries affecting the wiring structure and/or the ground plane [1], e.g., finiteness and special shape of the ground plane, presence of holes, and neighboring metallic objects.

For a two-conductor line in free space, the contribution to the overall current distribution due to antenna-mode current can be sensibly larger than the one predicted by TL theory [17]. However, such a contribution vanishes at the line ends, and TL-based models can therefore provide accurate prediction of currents/voltages induced at the input ports of terminal networks [1]. The situation is less critical if the line is placed in proximity to a ground plane. Namely, in this case, TL theory can be successfully applied for field-to-wire coupling prediction since the contribution predicted by TL theory is proven to be dominant with respect to the antenna-mode one also in terms of current distribution [8, 17].

2.2.3 Twisted-Wire Pairs

First introduced at the end of 19th century by Bell with the objective to prevent interference between telegraph and telephone lines [18, 19], the TWP technology is nowadays widespread in a variety of other sectors, spanning from electrical and electronic systems. Indeed, since disturbance theoretically cancel out within the twist, wire twisting allows increasing immunity to electromagnetic interference from external sources as well as reducing crosstalk coupling with surrounding wires within cable bundles. Additionally, it is a valuable remedy to reduce radiated emissions [20].

The standard approach to model EMC problems involving TWPs is based on the "click model," originally proposed by Paul and McKnight in [21, 22]. According to this model, the twisted line—inherently nonuniform with respect to ground due to wire twisting—is modeled as the cascade connection of uniform transmission-line sections with abrupt wires' interchanges.

Circuit representations of TWPs based on the click model allowed putting in evidence the fundamental role that nonideal implementation of the twisting as well as of the terminal loads may play on TWP performance. As a matter of fact, although differential-mode (DM) noise currents entering the terminal units were theoretically null, TWP's ability in mitigating electromagnetic interference (EMI) and crosstalk coupling may be seriously jeopardized by imperfections affecting the bifilar helix (e.g., twist nonuniformity, presence of residual twists, and untwisted line sections at the terminal sections of the TWP) as well as by possible unbalance affecting the terminal networks, which are responsible for common-mode (CM) into DM conversion.

2.2.3.1 Field-to-Wire Coupling Models

The first attempt to model TWP interaction with external electromagnetic fields was made by Taylor et al. [23], who derived closed-form expressions of the frequency-domain response of a TWP driven by a plane-wave field with electric field parallel to the TWP axis. Extension of such a model to account for arbitrary conditions of incidence and polarization of the impinging EM field was subsequently proposed by Tesche [24] and Stolle [25]. Then, Armenta and Sarris [26] restated the model by solving the Taylor inhomogeneous TL equations [27] in closed form.

All the above-mentioned models assume TWPs in free space. However, since in several practical applications, such as in systems for the aerospace sector, TWPs are installed in close proximity to planar metallic structures, the field-to-TWP coupling problem was recently reformulated by Pignari and Spadacini [28] to manage TWP lines running above metallic ground planes. As a matter of fact, the presence of a ground plane makes the structure behaving as a nonuniform three-conductor TL, where immunity characteristics are significantly influenced by nonuniformity of the p.u.l. parameters, coexistence of CM and DM disturbance, as well as field reflection due to the ground plane.

In [28], an accurate geometrical representation of the TWP helix above ground was proposed [see Figure 2.5(a)], and used to solve in closed-form the resulting MTL equations. The obtained distributed-parameter model is shown in Figure 2.5(b), and comprises (a) a passive part (matrix $\mathbf{\Phi}$), whose entries are evaluated starting from inductance and capacitance matrices of p.u.l. parameters averaged over the twist pitch, that is,

$$\bar{\mathbf{L}} = \begin{bmatrix} \bar{l} & \bar{l}_m \\ \bar{l}_m & \bar{l} \end{bmatrix}, \quad \bar{\mathbf{C}} = \frac{1}{c_0^2} \tag{2.26}$$

with

$$\bar{l} \cong \frac{\mu_0}{2\pi} \ln\left(\frac{2h}{r_w}\right), \quad \bar{l}_m \cong \frac{\mu_0}{2\pi}\left[\ln\left(\frac{2h}{s}\right) + \frac{s^2}{16h^2}\right], \tag{2.27}$$

c_0 denoting the light speed in free space, and (b) an active part, consisting of two voltage and two current sources lumped at the right TWP end. Such a model allows accurate prediction of CM and DM disturbance induced at line terminals and proved to be in satisfactory agreement with numerical simulation based on the method of moments (MoM) in a wide frequency range and in spite of the absence of dielectric coating (Figure 2.6a). Additionally, it allows pointing out the large sensitivity (even tens of decibels at low frequency) to incomplete

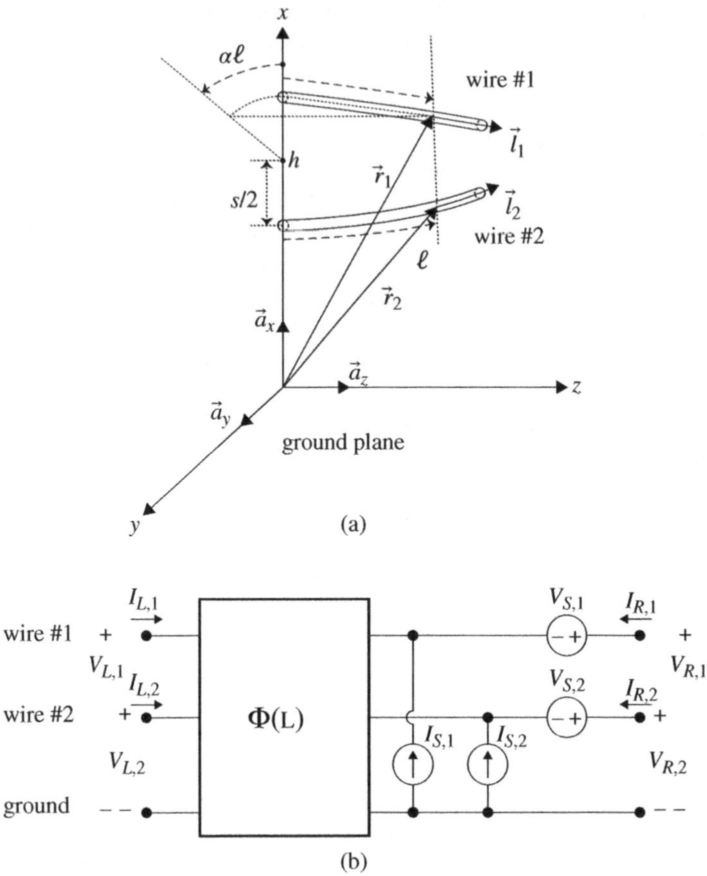

Figure 2.5 (a) Geometrical representation of the TWP bifilar helix above ground; (b) distributed-parameter model of the TWP driven by an external wave [28].

terminal twists of DM disturbances induced at the TWP terminals. As shown in Figure 2.6(b), worst-case DM noise levels are observed in TWPs made of $N + 1/4$ twists.

For better understanding the involved phenomena, an approximate lumped-parameter circuit model was also developed in [28] for electrical short TWPs. For such a low-frequency (LF) model, neglecting propagation along the TWP, approximate expressions for the CM and DM components of the induced sources $V_{S1} = V_{S,CM} + V_{S,DM}$, $V_{S2} = V_{S,CM} - V_{S,DM}$, $I_{S1} = I_{S,CM} + I_{S,DM}$, $I_{S2} = I_{S,CM} - I_{S,DM}$ can be cast as

$$V_{S,CM} \cong j\omega\mu_0 2H(\cos\eta\cos\psi + \cos\vartheta\sin\eta\sin\psi)hL_z \tag{2.28}$$

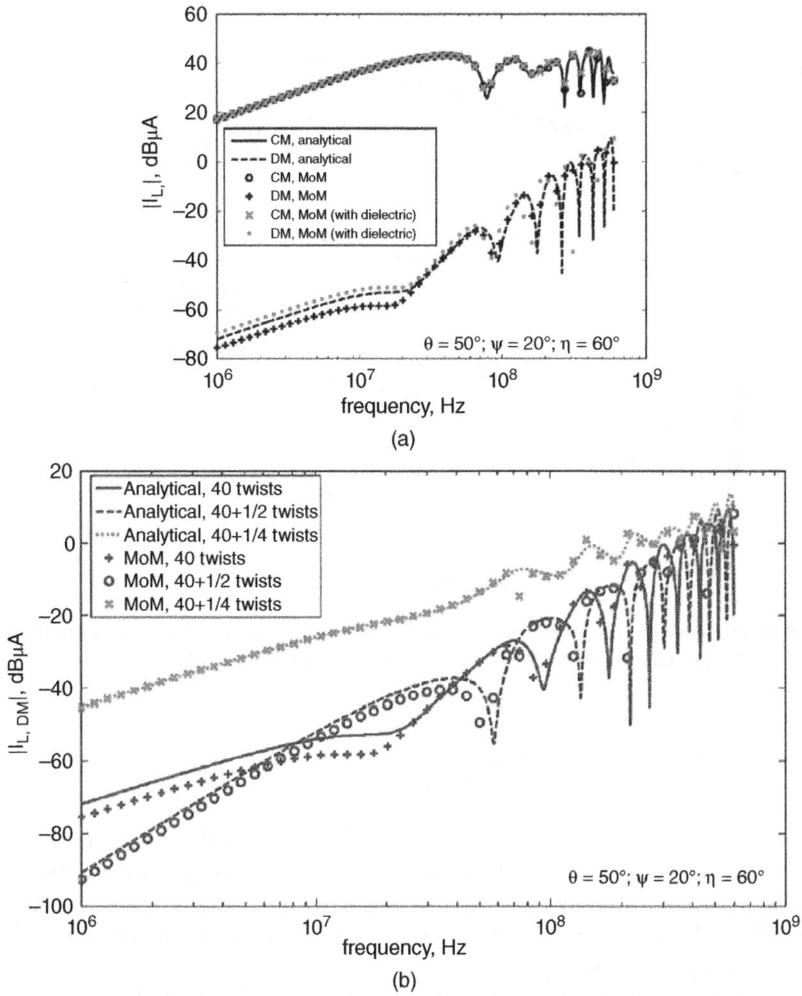

Figure 2.6 (a) CM and DM currents induced at the left TWP termination: Prediction obtained by the proposed model (analytical) and numerical simulation carried out by the MoM; (b) influence of the residual twist on DM currents induced at the left TWP termination.

$$I_{S,CM} = -j\omega 2E\cos\eta\sin\vartheta\frac{hL}{c_0^2(\bar{l}+\bar{l}_m)} \tag{2.29}$$

$$V_{S,DM} = j\omega\mu_0 H\begin{bmatrix}\dfrac{p}{2\pi}(\cos\eta\cos\psi+\cos\vartheta\sin\eta\sin\psi)+\\ -h(\cos\eta\sin\psi-\cos\vartheta\sin\eta\cos\psi)\end{bmatrix}s\sin(\alpha L) \tag{2.30}$$

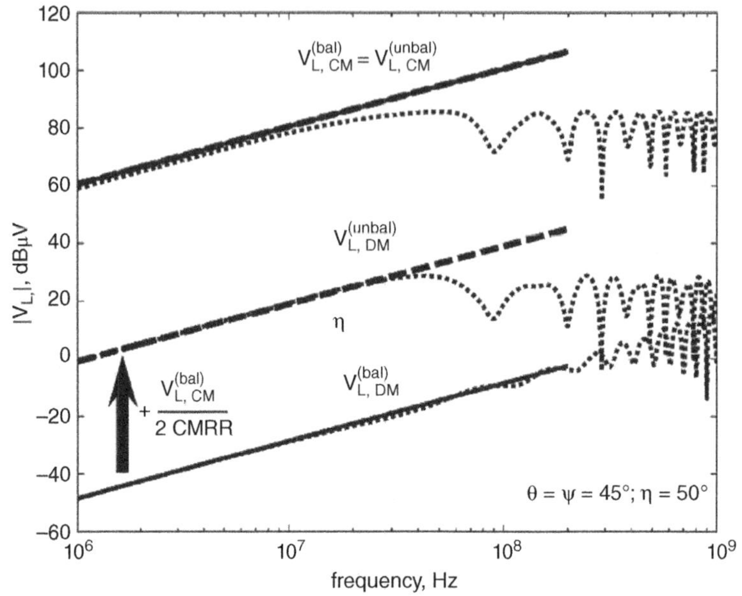

Figure 2.7 Influence of imbalance affecting the terminal networks on CM and DM voltages induced at the terminations of a TWP circuit driven by a plane-wave field.

$$I_{S,DM} = -j\omega E \cos\eta \sin\vartheta \frac{s\sin(\alpha L)}{\alpha c_0^2(\bar{l} - \bar{l}_m)}. \tag{2.31}$$

In (2.28)–(2.31), $E = E_0 e^{j\varphi}$, $H = (E_0/\sqrt{\mu_0/\varepsilon_0})e^{j\varphi}$, and angles ϑ, ψ, η (defined according to Figure 2.3) characterize the impinging plane-wave field, whereas L denotes the actual length of the wires in the TWP (untwisted), $L_z = \alpha pL/(2\pi) < L$ is the total distance the TWP extends in the longitudinal direction, and p is the twist pitch.

Based on the approximate expressions in (2.28)–(2.31), the following conclusions are drawn. As regards CM, the induced sources cannot be reduced by twisting the line. Conversely, DM-induced sources are theoretically reduced to zero in electrically short TWPs comprising either an even or an odd number of half twists. In this case, CM-to-DM conversion due to possible imbalance affecting the terminal networks plays a fundamental role on TWP immunity performance. As a matter of fact, generation of DM disturbance is strictly dependent on CM-rejection properties of the TWP terminal networks, which, according to common practice, are qualified in terms of *common mode rejection ratio* (CMRR) [29]. This is made evident in Figure 2.7(a), where CM and DM voltages induced by field coupling at the unbalanced termination of a

TWP circuit are plotted for two different values of CMMR. Circuit interpretation of the large sensitivity to termination imbalance exhibited by induced DM voltages is provided in [30–32], where the mechanism of CM-to-DM conversion is included into the DM equivalent circuit of the TWP by means of controlled voltage sources lumped at line terminals and proportional to (a) CM voltages induced by the external wave at the line ends (nearly unaffected by termination imbalance) and (b) the inverse of the CMMR of each terminal network.

2.2.3.2 Crosstalk Models

The standard procedure to investigate crosstalk in wiring structures involving TWPs is based on seminal works by Paul, who first modeled the TWP as the cascade connection of uniform TL sections [21] and provided an LF interpretation of crosstalk coupling to TWPs in terms of inductive and capacitive contributions [22]. The model was used to highlight the impact of wire twisting on the voltage induced at TWP terminals [22] and to explain the large sensitivity to slight rotation of one end of the twisted pair experimentally observed in circuits involving low termination impedances [33].

The results and modeling approach proposed by Paul were subsequently extended and refined by several other researchers, who focused on specific TWP features possibly influencing TWP performance.

For instance, in [34] and [35] the influence of twist nonuniformity [34] and CM-to-DM conversion [35] on undesired voltages induced by crosstalk at the terminations of a TWP circuit in close proximity to a nearby wire is studied by suitable prediction models.

In [36], an advanced model based on the cascade connection of uniform line sections (see Figure 2.8) and aimed at predicting crosstalk between a TWP, acting as generator circuit, and a nearby wire, acting as victim circuit, has been developed. That model allowed proving that the current distribution along a nearby wire is due to different variations of the mutual capacitance and inductance terms, and, in the absence of actual asymmetries, is influenced by angle α (see Figure 2.8) between the TWP and the ground plane. Moreover, it is shown that in practical installations, crosstalk is dominated by the coupling due to untwisted-wire sections at the TWP ends, that is, where the bifilar helix is usually untwisted in order to allow the TWP terminals to be plugged into a connector. Based on this finding, an approximate prediction model where untwisted line sections are included and the twisted line is replaced with a uniform TL with averaged p.u.l. parameters can be readily cast. Such an approximate model yields predictions of the current distribution along the victim wire in good agreement with those obtained by the original model, but with significant advantages in terms of computational and modeling efforts.

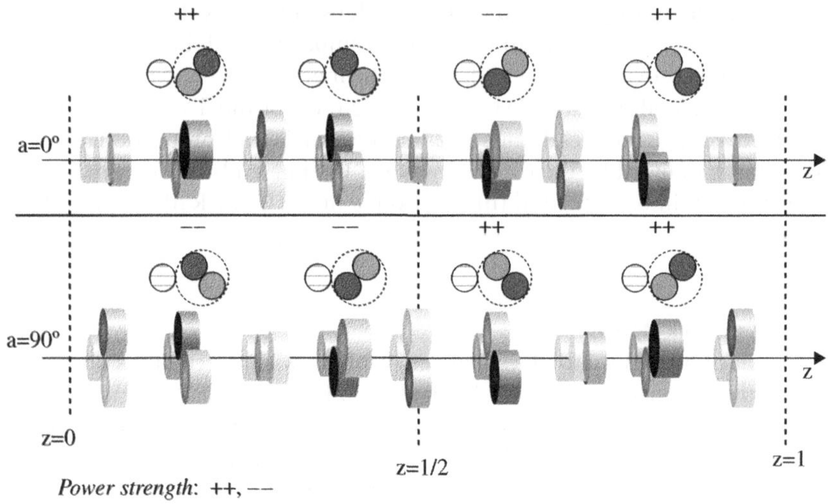

Figure 2.8 Modeling TWP-to-wire crosstalk coupling by cascade-connecting uniform TL sections: Example of discretization of the cable harness in section with nearly uniform cross-section for two different angles α between the TWP and the ground plane [36].

2.2.4 Multipair Bundles

When prediction of crosstalk and/or field-to-wire coupling in multipair bundles is addressed, closed-form analytical solutions cannot be obtained due to wiring harness complexity, and prediction models based on numerical solution of the MTL equations represent the typical approach. Particularly, as regards field-to-wire coupling, a twofold problem may arise in practical applications, which is related not only to modeling of the cable harness itself (usually requiring a stochastic representation, as will be discussed in the following section), but also to the complexity of the interfering field. For this reason, the prediction models available in the literature usually exploit the cascade connection of several line sections to account for possible nonuniformity of the cable cross-section and/or possible nonuniformity of the interfering EM field. Suitable approximations are usually introduced with the objective to minimize the required segmentation (significantly influencing the simulation time), while retaining satisfactory levels of prediction accuracy.

An example of cascade-connecting several line sections to account for nonuniformity of the cable cross-section can be found in [37], where this strategy is used to predict the terminal response of a bundle of TWPs illuminated by a plane-wave field. According to the Taylor formulation of field-to-wire coupling, the effect due to the interfering field is included in each line section by voltage and current sources, whose expressions are derived starting from an analytical description of the plane-wave external field. The wires in the bundle are surrounded by an inhomogeneous medium, and dielectric losses are considered. The adopted segmentation is very fine, with the shortest full twist contains several tens of line segments.

An example in which bundle segmentation is exploited to account for nonuniformity of the interfering field is the hybrid "numerical-TL theory" approach described in [11]. In this work, 3D numerical simulations and MTL theory are combined to predict the response of complex cable bundles installed inside a car. The EM field impinging onto the cable harness is evaluated by FDTD-based numerical simulation carried out inside a 3D CAD model of the car chassis, sampled along the cable axis, and then included into the MTL model of the cable harness by distributed voltage sources. The main advantage of this hybrid approach is that, according to the Agrawal formulation, numerical evaluation of the field is carried out in the absence of the cable harness, simplification that significantly increases simulation efficiency. Moreover, for a given cable path, a database with impinging field data can be computed once and for all for different emission sources, and then used to predict field-to-wire coupling in cable harness with different geometrical and electrical characteristics.

This approach was successfully applied in [38] for field-coupling prediction in a bundle of TWPs running above ground, and, subsequently, in [12], where immunity of a TWP bundle excited by the nonuniform EM field inside a small satellite was considered.

With respect to [37], in [38], inhomogeneity of the surrounding medium is neglected, and cable nonuniformity due to wire twisting is overcome by averaging the TWPs' p.u.l. parameters over the twist pitch [28]. These simplifying assumptions are of paramount importance as they allow for modeling a multipair bundle composed of N TWPs running above ground (see Figure 2.9a) as a

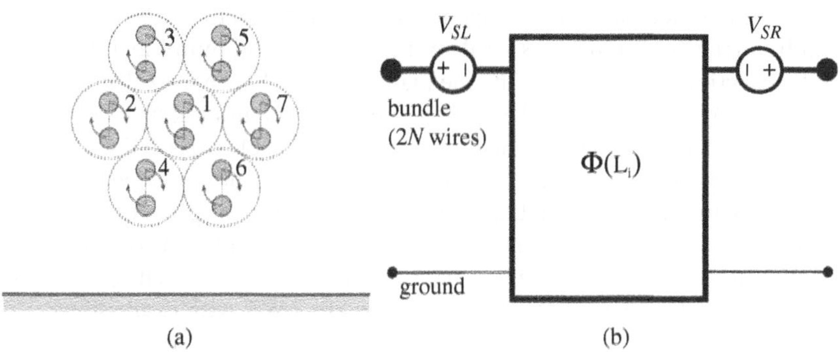

Figure 2.9 (a) Example of cross-section of a multipair bundles above ground (in this example, seven TWPs are arranged in a pseudocircular shape); (b) equivalent circuit model of the *i*th bundle section.

uniform MTL with p.u.l. inductance and capacitance matrices ($2N \times 2N$ sized) in the form:

$$
\mathbf{L} = \begin{array}{c}
\begin{matrix} \overbrace{\qquad\qquad}^{1} & \overbrace{\qquad\qquad}^{2} & \overbrace{\qquad\qquad}^{\cdots} & \overbrace{\qquad\qquad}^{N} \end{matrix} \\
\begin{bmatrix}
\bar{l}_1 & \bar{l}_{M,1} & \bar{l}_{12} & \bar{l}_{12} & \cdots & \cdots & \bar{l}_{1N} & \bar{l}_{1N} \\
\bar{l}_{M,1} & \bar{l}_1 & \bar{l}_{12} & \bar{l}_{12} & \cdots & \cdots & \bar{l}_{1N} & \bar{l}_{1N} \\
\bar{l}_{12} & \bar{l}_{12} & \bar{l}_2 & \bar{l}_{M,2} & \cdots & \cdots & \bar{l}_{2N} & \bar{l}_{2N} \\
\bar{l}_{12} & \bar{l}_{12} & \bar{l}_{M,2} & \bar{l}_2 & \cdots & \cdots & \bar{l}_{2N} & \bar{l}_{2N} \\
\cdots & \cdots & \cdots & \cdots & \cdots & \cdots & \cdots & \cdots \\
\cdots & \cdots & \cdots & \cdots & \cdots & \cdots & \cdots & \cdots \\
\bar{l}_{1N} & \bar{l}_{1N} & \bar{l}_{2N} & \bar{l}_{2N} & \cdots & \cdots & \bar{l}_N & \bar{l}_{M,N} \\
\bar{l}_{1N} & \bar{l}_{1N} & \bar{l}_{2N} & \bar{l}_{2N} & \cdots & \cdots & \bar{l}_{M,N} & \bar{l}_N
\end{bmatrix}
\end{array}
\begin{matrix} \left.\vphantom{\begin{matrix}a\\a\end{matrix}}\right\} 1 \\ \left.\vphantom{\begin{matrix}a\\a\end{matrix}}\right\} 2 \\ \left.\vphantom{\begin{matrix}a\\a\end{matrix}}\right\} \cdots \\ \left.\vphantom{\begin{matrix}a\\a\end{matrix}}\right\} N \end{matrix}
\quad \mathbf{C} = c_0^{-2}\mathbf{L}^{-1} \tag{2.32}
$$

$$
\bar{l}_n \cong \frac{\mu_0}{2\pi}\left[\ln\left(\frac{2h_n}{r_w}\right) - \frac{s^2}{16h_n^2}\right], \ \bar{l}_{M,n} \cong \frac{\mu_0}{2\pi}\left[\ln\left(\frac{2h_n}{s}\right) + \frac{s^2}{16h_n^2}\right],
$$

$$
\bar{l}_{nk} \cong \frac{\mu_0}{4\pi}\ln\left(1 + \frac{h_n h_k}{d_{hk}^2}\right) \tag{2.33}
$$

where r_w denotes the wire radius, s is the separation of wires in the same TWP, h_n is the distance to ground of the axis of the nth TWP, and d_{hk} is the distance between the axis of the hth and kth TWPs.

Likewise in [11], line segmentation is here exploited to account for the nonuniformity of the external field. To this end, a suitable number, N_S, of line sections is identified, based on the spatial nonuniformity of the field at the maximum frequency of interest. As a rule of thumb, the (minimum) length, L_i, of each line section should satisfy the inequality $L_i < \lambda/20$, where λ denotes the wavelength. According to the Agrawal formulation, the impinging EM field is then evaluated (numerically or analytically) along the cable axis (in the absence of the cable harness), sampled for each line segment, and included into the model by means of induced voltage sources connected at the ports of each single wire segment as shown in Figure 2.9(b). In particular, since the two wires in the TWP lie on average at the same height above ground, equal voltage sources are assumed for the two wires belonging to the nth TWP, with expressions

$$
V_{SL,n} = \int_0^{L_i} \frac{\sin[\beta_0(z - L_i)]}{\sin(\beta_0 L_i)} E_z(h_n, 0, z)dz - \int_0^{h_n} E_x(x, 0, 0)dx \tag{2.34}
$$

$$V_{SR,n} = \int_0^{L_i} \frac{\sin(\beta_0 z)}{\sin(\beta_0 L_i)} E_z(h_n, 0, z)dz - \int_0^{h_n} E_x(x, 0\ , L_i)dx, \tag{2.35}$$

where E_x and E_z denote the vertical and longitudinal component, respectively, of the electric field. The above integrals can be numerically evaluated by approximating E_x with a constant value, and E_z with a piece-wise linear function of z.

Cascades connecting the N_S bundle sections and enforcing the port constraints imposed by the terminal networks allow for evaluating the CM currents at the terminations of each TWP in the bundle. As a matter of fact, by enforcing equal sources on the two wires in the TWP implies assuming the CM excitation to be dominant in the field-coupling mechanism. Accordingly, the nonnull DM noise induced at the terminations of each TWP in the bundle can be mainly ascribed, and therefore *a posteriori* predicted, to CM-to-DM conversion due to imbalance affecting the terminal networks [30–32].

Prediction accuracy of the above-described model was assessed versus full-wave 3D simulation based on the MoM. Figure 2.10 shows model validation examples involving specific TWPs in the bundle cross-section: (a) CM currents induced at TWP terminals by an Hertzian dipole and (b) DM currents induced at TWP terminals by a plane-wave field.

2.2.5 The Equivalent Cable Bundle Method

The hybrid "numerical-TL theory" method described in the previous section represents a valuable attempt to reduce the computational burden involved in full-wave 3D numerical simulation of a complex setup comprising multiwire bundles as long as the fundamental assumptions of TL theory are satisfied. However, at high frequencies, i.e., when the high of the cable axis does not satisfy the inequality $h \ll \lambda$ (as a rule of thumb, $h < \lambda/5$ [39]), the assumption of TEM propagation is no more satisfied due to the occurrence of higher-order propagation modes (transverse electric and magnetic modes), which are not accounted for by TL theory. Therefore, resorting to full-wave 3D numerical simulation of the cable harness installed in its EM environment (e.g., the metallic body of a vehicle) is the only solution to get reliable prediction of the line response, in spite of the involved discretization issues and consequent computational burden.

The "equivalent cable bundle method" was proposed in this context, with the aim to reduce the computational time and memory requirements of 3D full-wave simulation of an entire car with the involved multiwire bundles [39]. The method allows approximate prediction of the CM current induced by field-to-wire coupling into cable harness by replacing it with an equivalent bundle, whose cross-section is composed of four equivalent conductors obtained by

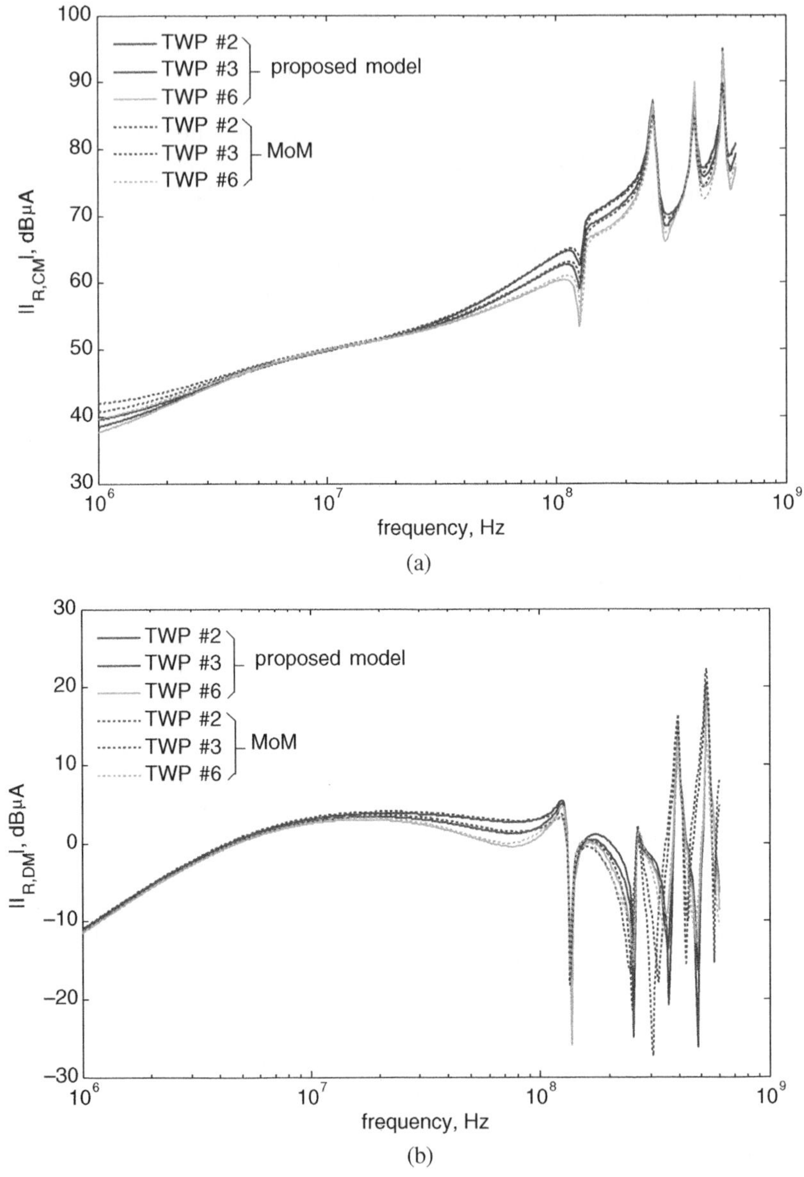

Figure 2.10 (a) CM and (b) DM currents induced at the (right) terminations of three TWPs in the bundle in Figure 2.9(a). In (a), the bundle is illuminated by an Hertzian dipole, and in (b), by a plane-wave field.

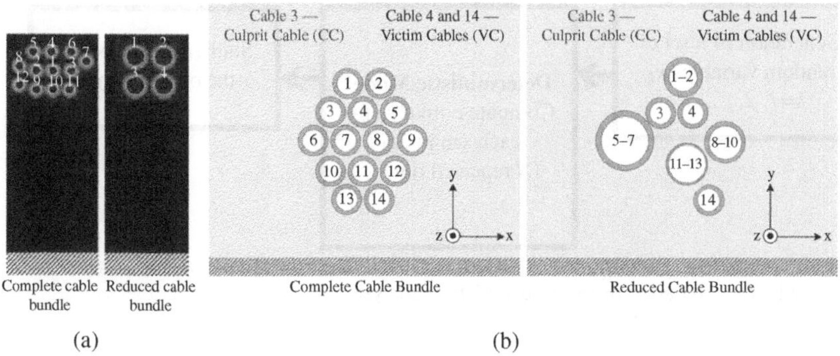

Figure 2.11 Examples of complex cable harness reduction into equivalent cross-sections for radiated susceptibility [39] and crosstalk [42] prediction.

sorting the N wires in the original bundle in four groups. The sorting criterion is based on the fact that the left and right termination impedances of each wire in the original bundle are smaller or larger than the CM impedance of the bundle. The p.u.l. inductance and capacitance matrices of the equivalent bundle are determined starting from those of the original bundle, by assuming the wires belonging to the same group connected in parallel. Starting from the p.u.l. parameters of the reduced bundles, geometrical and electric characteristics of the equivalent conductors are retrieved resorting to analytical expressions of the p.u.l. inductances in [1] and electrostatic simulation, respectively. Finally, parallel connection of the wires belonging to the same group is exploited to evaluate the equivalent loads at the terminations of the conductors in the equivalent cross-section. Reduction of a cable bundle comprising 12 conductors is exemplified in Figure 2.11(a).

The method was experimentally validated versus measurements carried out into an anechoic environment [39] and subsequently extended to the case of bundles running on complex ground structures [40]. The method was successfully exploited also for the prediction of radiated emissions from complex cable harnesses [41].

More recently, the equivalent cable bundle method was reformulated in [42] with the objective to reduce complexity and time required for crosstalk-coupling prediction between two arbitrary conductors packed into a multiwire cable bundle (or into two adjacent bundles), as exemplified in Figure 2.11(b).

2.3 Statistical Modeling

The use of deterministic models presupposes system parameters (geometrical, electrical, etc.) are known to a large extent. Unfortunately, this is not always the case for EMC, often dealing with uncontrolled parasitic effects (e.g.,

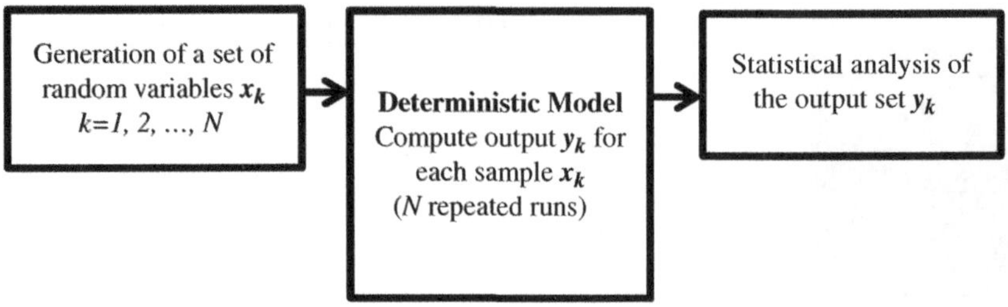

Figure 2.12 Monte Carlo approach to statistical analysis.

high-frequency behavior of electronic components), uncontrolled geometries (e.g., cable routing and layout), uncertainty due to tolerances, large spread of the parameters of interest, etc. This kind of input parameters needs description in terms of random variables, characterized by probability distributions and associated statistical moments (expected value, variance, skewness, kurtosis, etc.). As a consequence, the target quantities of the analysis (e.g., an induced voltage or current) have also to be regarded as random variables.

A statistical EMC model can be defined as a transformation of random variables, that is, a modeling procedure which allows computation of the statistics (distribution, moments) of predicted output quantities, knowing the statistics of input model parameters. Unless the transformation can be treated with closed-form mathematical derivations (this is possible only for problems of limited complexity), the common analysis technique is based on the Monte Carlo method, that is, the use of numerical repeated-run simulations.

According to Figure 2.12, the Monte Carlo approach consists of three steps. The first step is the use of pseudorandom number generators to construct a set of samples $\boldsymbol{x}_k$, $k=1,2,\ldots, N$ of the random-variable vector $\boldsymbol{x}$representing input quantities, according to known statistical distributions. Then, a deterministic model (implemented by a computer code) is run N times, to evaluate each output sample $\boldsymbol{y}_k$ corresponding to the input sample $\boldsymbol{x}_k$. Eventually, a statistical analysis is performed on the output set $\boldsymbol{y}_k$, $k=1,2,\ldots, N$, to investigate the empirical cumulative distribution function, the empirical probability density function (pdf), and the estimates of statistical moments (sample mean, sample variance, etc.).

Modeling of field-to-wire coupling and crosstalk in wire harness benefits from the use of statistical techniques, basically for two main reasons: (a) the possibility to account for random geometry and (b) the possibility to account for random sources of interference. On the one hand, a random geometry can be introduced to account for uncertainty in the description of wire paths (e.g., distance, height above ground) which is typical of cable interconnections used at system level to interconnect functional units in complex systems (e.g., in the automotive or aerospace industry). Similarly, a random geometry is inherently involved for the description of hand-made bundles of wires. On the other hand, random sources of interference can be introduced to model random electromagnetic fields and their coupling with cables. Both aspects (random geometry and random sources, as well as their possible combination) are presented in the following paragraphs. Eventually, an alternative approach to time-consuming Monte Carlo simulations based on the polynomial-chaos expansion technique is presented.

2.3.1 Crosstalk in Random Wire Harness

A canonical structure (composed of a three-conductor uniform and lossless TL) can be first considered to introduce the crosstalk problem in wiring structures having random geometries, and to illustrate basic statistical concepts and modeling approaches that can then be extended to more complex wire interconnections [43, 44]. Namely, let us consider the simplest TL configuration for crosstalk analysis [45] shown in Figure 2.13(a), and composed of two signal conductors and a reference conductor. Letters G and R denote generator and receptor (i.e., victim) circuit, respectively. The line length is L and the generator wire is fed by voltage source V_S at the near end. In order to ease the mathematical derivations, the assumption of a lossless TL immersed in homogeneous medium is adopted. For sake of simplicity, line terminations are connected to resistive loads R_L, R_S, R_{NE}, R_{FE}; however, extension to frequency-dependent loads is straightforward. For the characterization of crosstalk, voltages V_{R0} and V_{RL} induced by the generator circuit at the near- and far-end terminations of the receptor circuit are taken as the reference quantities.

The TL cross-section is shown in Figure 2.13(b) and defines a structure where the victim wire R is place in a fixed position, whereas the generator wire is placed in a random position within a circular sector, defined by boundary values

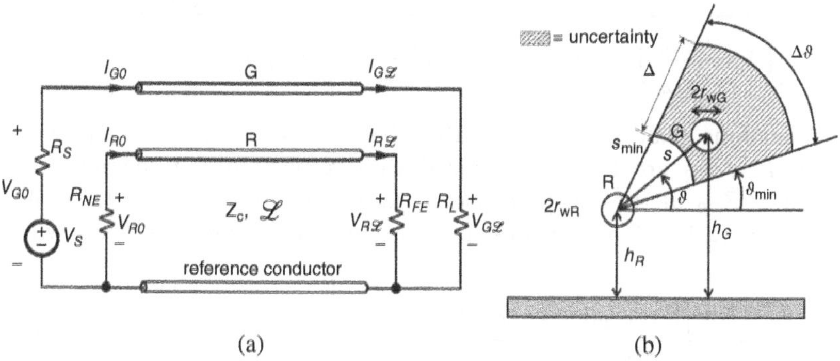

Figure 2.13 (a) Three-conductor TL for crosstalk analysis and (b) reference cross-section allowing for tracking random fluctuations of generator wire G around receptor wire R in an uncertainty region [44].

of parameters s and ϑ, defining a separation distance and an angle, respectively. Conversely, the height above ground of the receptor wire h_R and the wire radii r_{wR}, r_{wG} are known deterministic parameters.

The analysis of such a structure is presented in [44] and foresees the preliminary derivation of a deterministic circuit model in which the victim circuit embeds all the interference effects due to the presence of the generator circuit. Such effects are represented by a voltage and a current lumped noise source and a passive two-port distributed parameter. Under the assumptions of weak coupling [45] and matched generator circuit, it is then shown that the passive part of the model can be neglected. Results are valid without frequency limitations (within the frequency range of validity of the TL model), thus extending previous results in [43] limited to electrically short TLs.

The deterministic model is then exploited to introduce parameters s and ϑ treated as independent random variables (RVs) distributed within intervals $[s_{\min}, s_{\max}]$ and $[\vartheta_{\min}, \vartheta_{\max}]$, and to compute closed-form expressions for the mean value and variance of near-end (NE) and far-end (FE) crosstalk transfer ratios NEXT $= |V_{R0}/V_S|$ and FEXT $= |V_{RL}/V_S|$, respectively. These expression can be cast as

$$\mu_{XEXT} = F_{XE}(\beta L; \alpha_{RNE}, \alpha_{RFE})\ \mu_\gamma \tag{2.36}$$

$$\sigma^2_{XEXT} = F_{XE}(\beta L; \alpha_{RNE}, \alpha_{RFE})^2\ \sigma^2_\gamma \tag{2.37}$$

where $X = N, F$ and

$$F_{NE} = \frac{|\sin(\beta L)|}{\left|\left(1+\dfrac{\alpha_{RFE}}{\alpha_{RNE}}\right)\cos(\beta L) + j\sin(\beta L)\left(\dfrac{1}{\alpha_{RNE}} + \alpha_{RFE}\right)\right|}\ \frac{1+\alpha_{RFE}}{2} \tag{2.38}$$

$$F_{FE} = \frac{|\sin(\beta L)|}{\left|\left(1+\dfrac{\alpha_{RNE}}{\alpha_{RFE}}\right)\cos(\beta L) + j\sin(\beta L)\left(\dfrac{1}{\alpha_{RFE}} + \alpha_{RNE}\right)\right|}\ \frac{|\alpha_{RNE}-1|}{2} \tag{2.39}$$

In equations (2.38)–(2.39), β is the phase constant, $\alpha_{RNE} = R_{NE}/Z_{CR}$, $\alpha_{RFE} = R_{FE}/Z_{CR}$ are loads normalized with respect to the characteristic impedance Z_{CR} of the receptor circuit (in absence of wire G). Quantities μ_γ and σ^2_γ in (2.36)–(2.37) are the expected value and variance of the following RV:

$$\gamma = \frac{1}{2}\frac{\log\left[1 + 4h_R(h_R + s\sin\vartheta)/s^2\right]}{\log\left[2(h_R + s\sin\vartheta)/r_{wG}\right]} \tag{2.40}$$

As a specific example, Figure 2.14 illustrates NEXT statistical estimates (mean value and standard deviation) for $L = 3$ m, $r_{wR} = r_{wG} = 0.1$ mm,

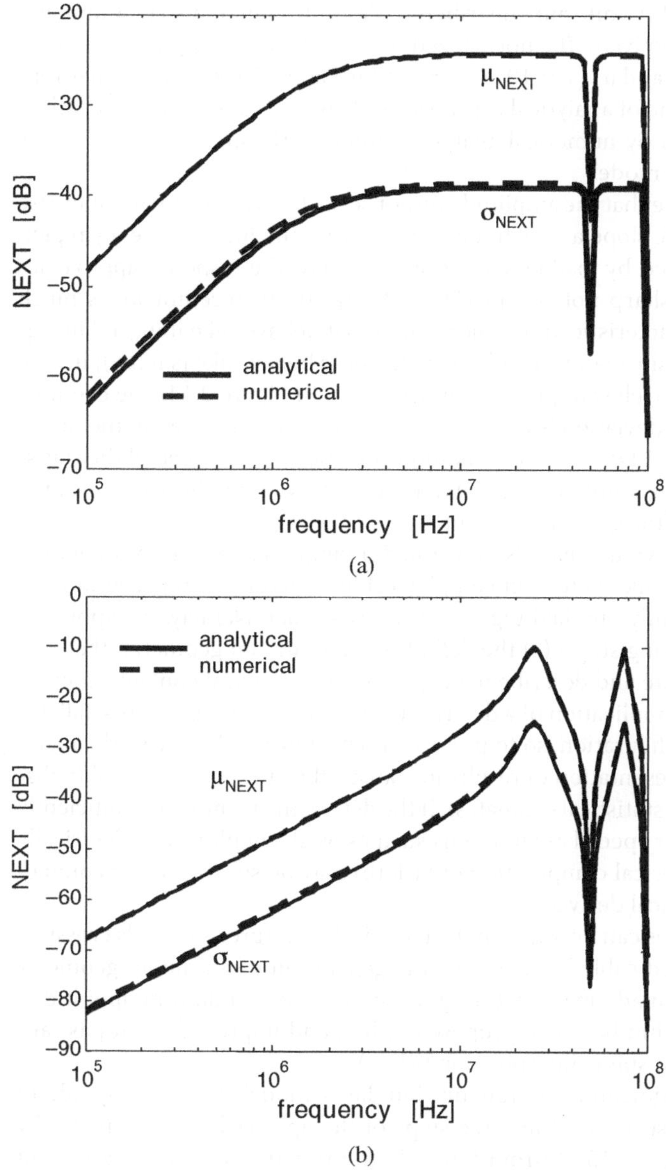

Figure 2.14 Mean value and standard deviation of NEXT: (a) for $\alpha_{RNE} = \alpha_{RFE} = 20$, i.e., both receptor loads much greater than characteristic impedance; (b) for $\alpha_{RNE} = 10$, $\alpha_{RFE} = 0.1$, i.e., opposite loading conditions [45].

$h_R = 1$ mm, s and ϑ uniformly distributed RVs in the intervals [1, 3]mm and [45°, 135°], respectively. The normalized receptor loads are $\alpha_{RNE} = \alpha_{RFE} = 20$ for Figure 2.14(a) and $\alpha_{RNE} = 10$, $\alpha_{RFE} = 0.1$ for Figure 2.14(b). Solid lines represent the outcome of analytical expressions (2.36) and (2.37), whereas dashed lines are obtained by numerical analysis (Monte Carlo simulations) to assess the validity of the model.

One can observe that the amplitude response of statistical estimates increase with 20 dB/decade slope at low frequency, whereas the high-frequency region is highly influenced by loading conditions. Namely, the response approaches a flat level with sharp notches in Figure 2.14(a) (both receptor loads much greater than characteristic impedance) whereas it behaves like $|\tan(\beta \mathrm{L})|$ in Figure 2.14(b) (opposite loading conditions). It would behave like $|\sin(\beta \mathrm{L})|$ in case of matched loads (relevant plot not shown here). In Figure 2.14, one can note that the standard deviation is not much lower than the mean value (the difference is only about 14 dB). This observation corroborates the need of the statistical approach to account for large spread of crosstalk values due to uncertainty in the position of the generator circuit.

Though the previous analysis concerned a simple canonical example, the rationale of crosstalk analysis in complex cable structures composed of multiple wires randomly bundled together is fairly similar. Namely, the approach follows the following steps: (a) the definition of a random geometry, that is, a method to generate and describe many possible geometrical configurations of the cable, (b) the application of a deterministic model to compute crosstalk for each random configuration, so to perform a repeated-run Monte Carlo analysis, and (c) the presentation of results in statistical terms (e.g., probability distributions and/or statistical estimates). If the deterministic model is sufficiently simple (e.g., under special assumptions such as weak coupling and electrically short lines) numerical computation of task (b) may be substituted or complemented by analytical derivations [46].

Several examples can be found in the technical literature as regards crosstalk in hand-assembled cable bundles. For the generation of a random geometry mimicking hand-made manufacturing of bundles, the random midpoint displacement algorithm has been proposed in [47] and improved in [48] as random displacement spline interpolation (RDSI).

In the RDSI algorithm, the random bundle is modeled as the cascade of n uniform MTL sections. The three steps of the approach are schematically described in Figure 2.15(a) (from top to bottom) with reference to a wire of the bundle. Namely, the x coordinate in the cross-section is considered (the y following a similar fashion), and a first set of few points are evaluated by linear interpolation between the known starting and ending coordinates corresponding to cable terminals. In a second step, points are randomly displaced by adding random numbers following a Gaussian distribution with expected value equal to zero and a given standard deviation (which is the key parameter to

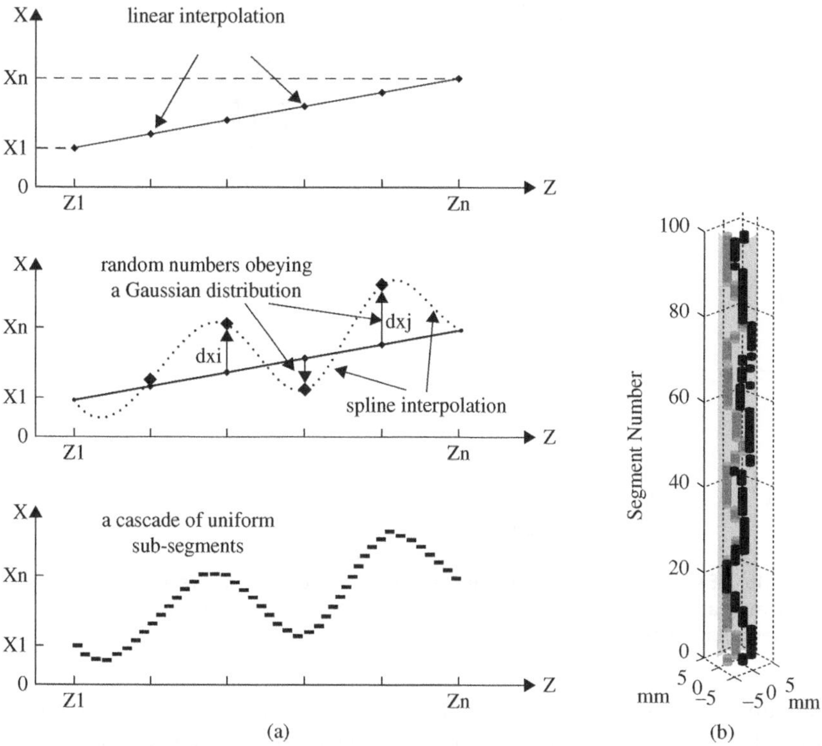

Figure 2.15 The RDSI algorithm: (a) modeling approach and (b) three-dimensional visualization of the random paths of two wires, determined via RDSI. Figures are reprinted from [48].

control the randomness of the cable). Finally, the displaced points are interpolated by a spline function to find a larger number of n points required by the model. This three-step procedure is applied for all wires of the bundle. Subsequently, the generated wire coordinates are mapped into fixed and predetermined position of the reference cross-section required by the MTL model. The proposed algorithm and, in particular, the use of spline functions, proved to be efficient for the construction of realistic geometries of hand-made cable bundles, characterized by smooth, continuous transitions of wire paths. An example is shown in Figure 1.15(b), [48].

With some modifications, the aforementioned RDSI algorithm has been used in [49] to model crosstalk in hand-made cables composed of wires randomly bundled together and tightened using lacing cords distributed along the cable length. The presence of lacing cords along the bundle determines expansions and contractions (at lacing positions) of the cross-section, thus increasing the nonuniformity and randomness of the MTL structure. An example of NEXT

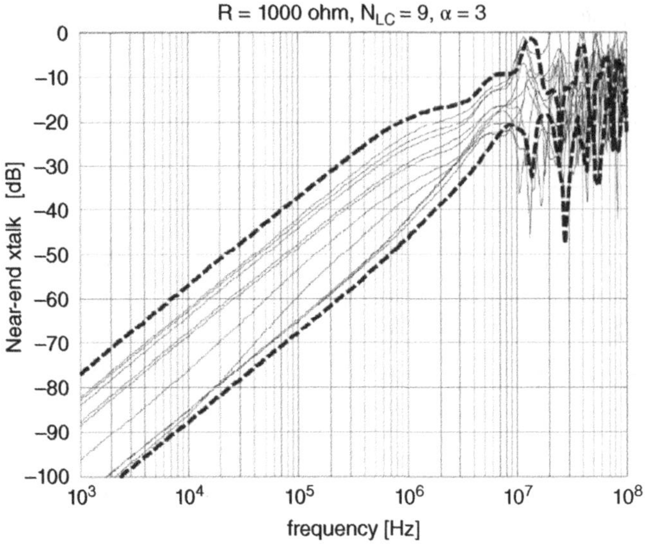

Figure 2.16 NEXT in a random bundle with lacing cords. Each solid line refers to a different random realization of the bundle (10 in total), whereas dashed lines are approximate upper and lower bounds defined in [49].

prediction is shown in Figure 2.16, referring to a bundle of length 4 m, composed of 25 AWG #22 stranded wires, running at an average height of 2.4 cm above a metallic ground (reference conductor), with nine lacing cords and high-impedance terminal loads (one of the 25 wires is considered as generator and another is considered as receptor).

Since Monte Carlo simulations are time-consuming, simpler analysis techniques can be developed in case a comprehensive statistical characterization is out of the scope, but only simple figures of merit are of interest. For instance, for the determination of an approximate estimator of worst-case crosstalk, the work in [50] proposed to analyze the probability distributions of p.u.l. parameters (inductances and capacitances) from a single harness cross-section and to use an approximate crosstalk model (for electrically short lines) to infer crosstalk properties. Results proved good estimation of worst-case crosstalk (error within 5 dB) compared with the use of the RDSI algorithm via repeated-run simulations.

2.3.2 Coupling of Deterministic Wire Harness with Random Electromagnetic Fields

Coupling of random electromagnetic fields to deterministic cables can be introduced by considering at first the elemental configuration consisting of a wire running above ground and illuminated by a uniform plane wave. This circuit structure was introduced in Section 2.2.2 (see Figure 2.3), where h is the line height, $E_0$is the electric-field strength, ϑ, ψ are the incidence angles, and η is the polarization angle. Unlike what was previously done, here the electromagnetic wave is assumed to be described by RVs. For instance, one can assume to know the field amplitude E_0 but to ignore the exact direction of incidence, as well as the polarization of the field. Hence, the analysis requires to define wave angles as RVs and to consequently evaluate the pdf and/or statistical estimates of the target quantity of interest, e.g., the current induced in a terminal load.

Generally, this problem can be readily solved by using an appropriate deterministic model of field-to-wire coupling and applying the Monte Carlo technique, based on processing repeated-run simulations [51]. Several contributions on this subject are now available in the scientific literature. Those papers report analytical results of field-to-wire coupling problem, cast in probabilistic terms and derived under certain simplifying assumptions. For instance, some pdfs of the induced current were derived in [52] for low frequency (i.e., for lines electrically short with respect to the free-space wavelength).

As a specific result, for a line length $\text{L} = 1$ m, $h = 10$ cm, wire radius 0.5 mm, matched terminal loads, E_0 beta-distributed in [0, 1] V/m (with parameters 2, 3), frequency 50 MHz, the pdf of the current induced in the right terminal load is shown in Figure 2.17 as evaluated by the analytical expression reported in [52] (solid line) and by 10^4 repeated runs (histogram bars). Specifically, Figure 2.17(a) refers to known angles $\eta = \psi = 0°$ and random elevation ϑ, uniformly distributed in [0,90°]; Figure 2.17(b) refers to known angles $\vartheta = 30°$, $\psi = 135°$, and random polarization η uniformly distributed in [0,360°]. By comparing the plots, one can clearly appreciate that the worst-case current (about 50 dBμA) is the same, but the distribution is significantly different, with larger dispersion in Figure 2.17(a).

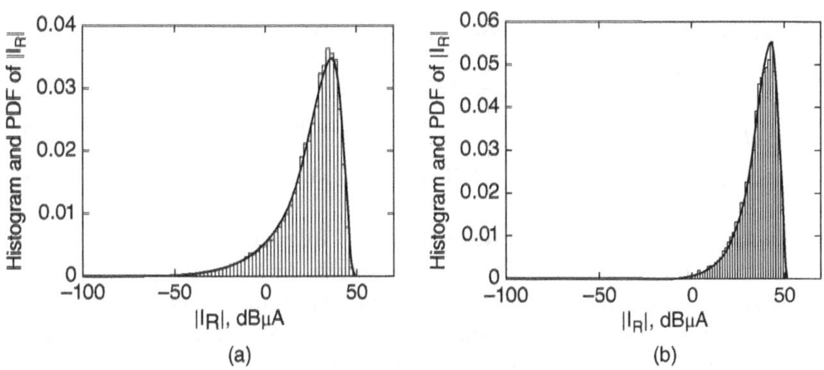

Figure 2.17 Pdf of the current induced by a random plane wave in the right terminal resistor for (a) random electric-field amplitude E_0 and elevation angle ϑ, (b) random electric-field amplitude E_0 and polarization angleη. Reprinted from [52].

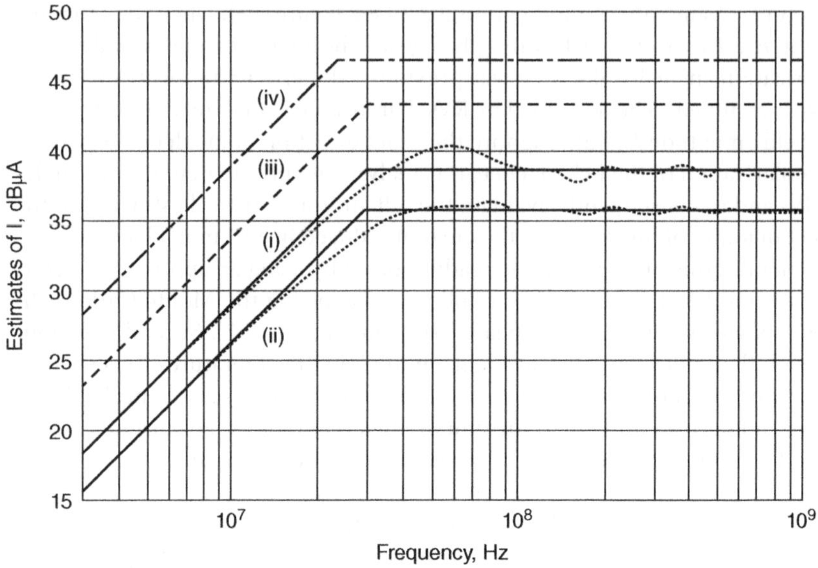

Figure 2.18 Statistical estimates of the current induced in the terminal load of a TL composed of a wire running above ground, illuminated by an electromagnetic field with random incidence and polarization angles.

For a line of arbitrary length and with general loading conditions (i.e., not necessarily matched) approximate closed-form expressions of statistical estimates are provided in [53]. As an example, for a matched line, one obtains the following approximate expression of the expected value and standard deviation of the current magnitude (expressed in dBμA):

$$\mu_{LF} \cong 6 + 20\log_{10}E_0 + 20\log_{10}(2h/Z_C) + 20\log 10(L/\lambda) \tag{2.41}$$

$$\mu_{HF} \cong -7.8 + 20\log_{10}E_0 + 20\log_{10}(2h/Z_C) \tag{2.42}$$

$$\sigma_{LF} \cong 3.4 + 20\log_{10}E_0 + 20\log_{10}(2h/Z_C) + 20\log 10(L/\lambda) \tag{2.43}$$

$$\sigma_{HF} \cong -10.7 + 20\log_{10}E_0 + 20\log_{10}(2h/Z_C) \tag{2.44}$$

where subscript LF (HF) refers to high- (low-) frequency interval, Z_C is the characteristic impedance, and λ is the wavelength. As a specific example, Figure 2.18 reports plots of estimates (2.41)–(2.44) of the current magnitude in the line loads of a matched TL with length 3 m, $h = 3$ cm, wire radius 0.5 mm, illuminated by an electric field with known amplitude $E_0 = 1$ V, and random angles uniformly distributed in $\vartheta \in [0, 90°]$, $\psi \in [0, 360°]$, and $\eta \in [0, 360°]$. Plotted curves represent (i) the expected value, (ii) the standard deviation, (iii) the expected value plus a standard deviation, and (iv) the worst-case envelope. Dotted curves are obtained by 10^4 repeated runs for validation purposes. The difference between the worst case and the expected value amounts to about 10 dB.

The plane-wave field with random amplitude, incidence, and polarization is not a good representation of a random electromagnetic field for closed and resonating environments, where multiple reflections from boundaries may occur. An extreme case of great practical interest is the reverberation chamber used for radiated susceptibility testing, where the electromagnetic field can be properly represented by a plane-wave integral [54], that is, the superposition of infinite plane waves, coming from all possible directions (the full solid angle $\Omega = 4\pi$), with random amplitude, polarization, and phase. In mathematical terms, the electric field at position $\boldsymbol{r}$ is written as

$$\mathbf{E}(\mathbf{r}) = \int_{\Omega} \mathbf{F}(\Omega)\, e^{-j\mathbf{k}\cdot\mathbf{r}} d\Omega \tag{2.45}$$

where $\boldsymbol{F}(\Omega)$ is a complex random variable modeling the wave-spectrum angular density, $\mathbf{k}$ is the vector wave number. The plane-wave integral representation of fields was implemented in [55] for an electrically short two-conductor TL (i.e., at low frequencies), leading to closed-form expressions of the distribution of the induced current. For a reverberation-chamber environment, the distribution proved to be a Rayleigh, whose parameter σ (mode) is the square root of

$$\sigma^2 = \frac{4\pi}{3}\left(\omega \frac{hL}{c_0 Z_C (a_L + a_R)}\right)^2 D_\Gamma \left(1 + a_L^2\right) \tag{2.46}$$

where D_Γ is the average intensity of the plane-wave amplitude density [in an ideal reverberation chamber, $D_\Gamma = E_0^2/(4\pi)$, where E_0^2 is the electric-field mean-square value], ω is the angular frequency, h is the line height above ground, L is the line length, c_0 is the speed of light in free space, Z_C is the characteristic impedance, $a_L = R_L/Z_C$, $a_R = R_R/Z_C$, where R_L (R_R) is the left (right) terminal resistor. An example of two Rayleigh distributions of the induced current is shown in Figure 2.19. It is interesting noticing that, unlike distributions for the single plane-wave excitation (see Figure 2.17), a worst-case current does not exist for the random plane-wave spectrum, since the tail of the Rayleigh distribution extends to infinite values.

These results were generalized in [56] for electrically long lines by implementing a numerical procedure, and closed-form results were also obtained in the special case of matched terminal loads. Additionally, simulation and comparison with measurements allowed for investigating the statistical distribution and correlation of the coupled current along the TL. Finally, TWPs illuminated by a random plane-wave spectrum were investigated in [57, 58].

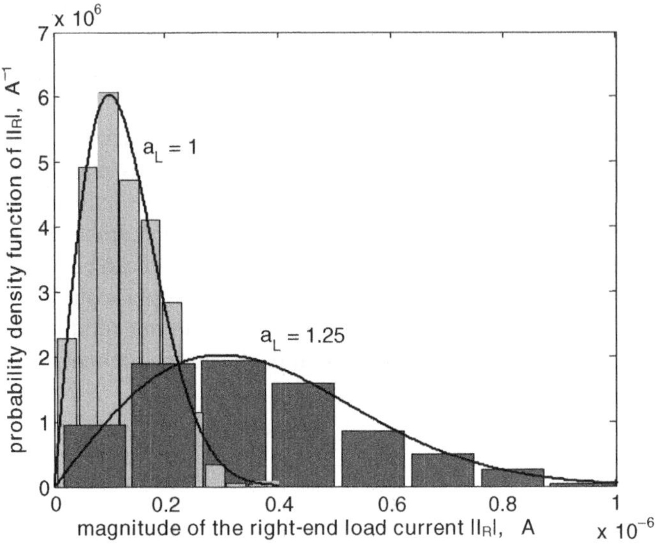

Figure 2.19 Normalized histogram (1000 repeated runs) and analytical pdf (solid lines) of the right-end load current magnitude, with angular extension of the excitation equal to 30°, for two different values of the normalized left-end load a_L. Reprinted from [55].

2.3.3 Coupling of Random Wire Harness with Deterministic Electromagnetic Fields

Random wire bundles analyzed in Section 2.3.1 for crosstalk are of interest also as concerns coupling to deterministic electromagnetic fields. Specifically, an LF model for a random bundle illuminated by a plane-wave field was presented in [59], whereas an approximate CM model for random bundles of TWPs illuminated by an arbitrary (uniform or nonuniform) electromagnetic field was presented in [38, 60].

In those works, a new algorithm for the generation of a statistical population of random bundles was based on the definition of a reference cross-section and random cycles of wires. Namely, the bundle can be approximated by cascading *Ns* uniform sections (as for the previous RDSI method). All cross-sections are similar and obtained by allowing minimum-distance interchanges of wires in a reference cross-section. Any set of such position interchanges is called *cycle*. For a given reference cross-section, the determination of all possible cycles is a so-called NP complete problem in the framework of graph theory. Namely, one can define a graph whose nodes are wire positions and whose branches connect adjacent nodes (i.e., adjacent wire positions in the reference cross-section). The solution consists in finding all possible routes in the graph, including the possibility to exclude from a route the positions occupied by those wires which do

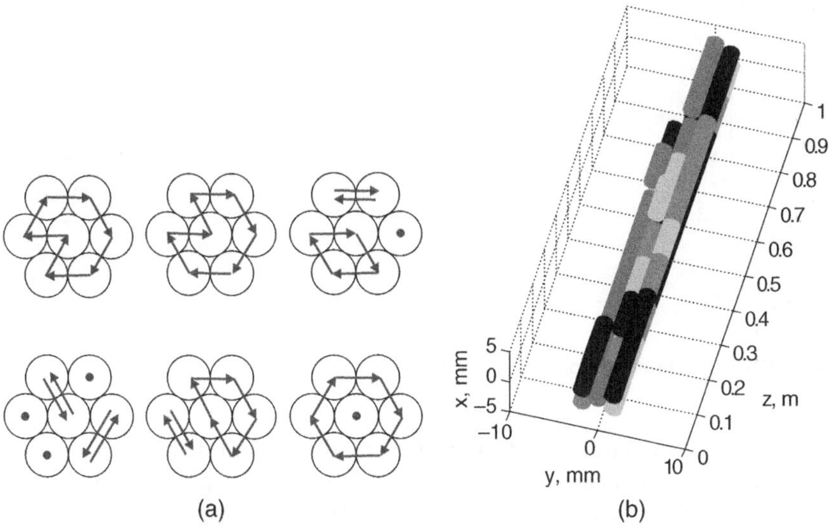

Figure 2.20 (a) Six possible cycles in a reference cross-section composed of six wires. (b) Example of a random bundle (cascade of 10 uniform sections) constructed by selecting 10 random cycles. Figure reprinted from [60].

not move, and including disjoint routes. As an illustrative example, six cycles for $N = 7$ wires are shown in Figure 2.20 (arrows represent shifts between adjacent positions, whereas dots denote wires which do not move).

Once the population of all possible cycles is available, a bundle sample can be constructed by randomly selecting a set of Ns cycles according to a uniform-discrete probability distribution. Each cycle is applied to a bundle section to obtain the arrangement of wires for the subsequent bundle section (the first cycle being applied to the reference cross-section). By this approach, one can construct several samples of random bundles. An example is shown in Figure 2.20(b), for a bundle composed of 5 wires, $Ns = 10$ sections with equal length.

The algorithm described above for bundle of wires can be equally applied for random bundles of TWPs (in this case, each circle in Figure 2.20a represents the space occupied by a pair) as done in [60] and experimentally assessed in [12]. The experiments consisted in using a full-scale mock-up reproducing the metallic chassis of the satellite AGILE satellite (sent into orbit in 2007 in the framework of an Italian scientific mission). Mock-up base is hexagonal with side length ~40 cm and height ~120 cm. The whole structure does not present apertures, with the exception of the upper base, which is equipped with a large square window bordered by additional small holes, which—in the real satellite—are used to fasten the payload. A picture of the test setup is shown in Figure 2.21.

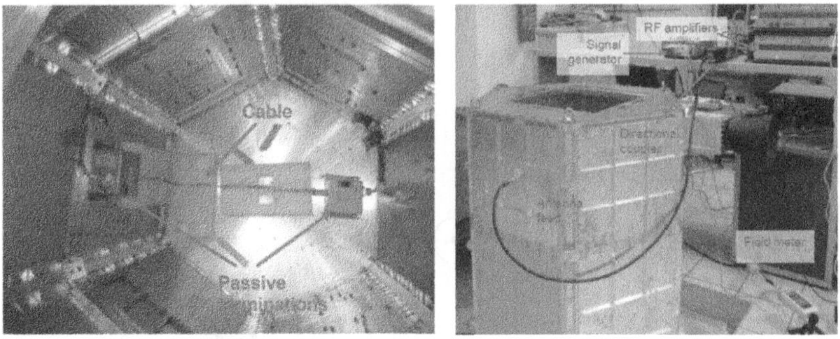

Figure 2.21 AGILE satellite mock-up and test setup.

To assess the susceptibility of hand-made random bundles of TWPs to radio frequency (RF) EM fields inside the satellite, the mock-up was equipped as follows. A random bundle composed of six TWPs ran inside the structure at a constant distance of 50 mm from the inner metallic surface (by the use of nonconductive holders). The bundle was connected to terminal units for measuring the common mode voltage induced in terminal loads of each twisted-wire pair by the electromagnetic field generated by a short monopole antenna (with wire radius 0.5 mm and length 120 mm) installed inside the mock-up. Several hand-made samples of random bundle were constructed and tested in the frequency range 250 MHz–1 GHz. Measurements are reported with colored lines in Figure 2.22, where the gray shaded area represents a prediction interval (between minimum and maximum predicted coupling) obtained by repeated-run (Monte Carlo) simulations.

2.3.4 The Polynomial-Chaos Expansion Technique

In previous paragraphs, the Monte Carlo method based on repeated-run simulations was proposed whenever the complexity of the electromagnetic problem prevents finding of analytical solutions. Actually, the Monte Carlo method is nowadays implemented by virtually all available commercial-design software tools. Though it is an easy and robust algorithm, it should be recognized that the repeated evaluation of output quantities while randomly varying the input quantities is a time-consuming task (often prohibitive even for modern computers). Hence, the search for computationally efficient alternatives to repeated runs is currently of great interest. As far as transmission-line analysis is concerned, a potential candidate known as "polynomial-chaos" (PC) is recently emerging.

The theory of PC for the stochastic simulation of multiwire cables is presented in [61]. That work provides an effective solution for the simulation

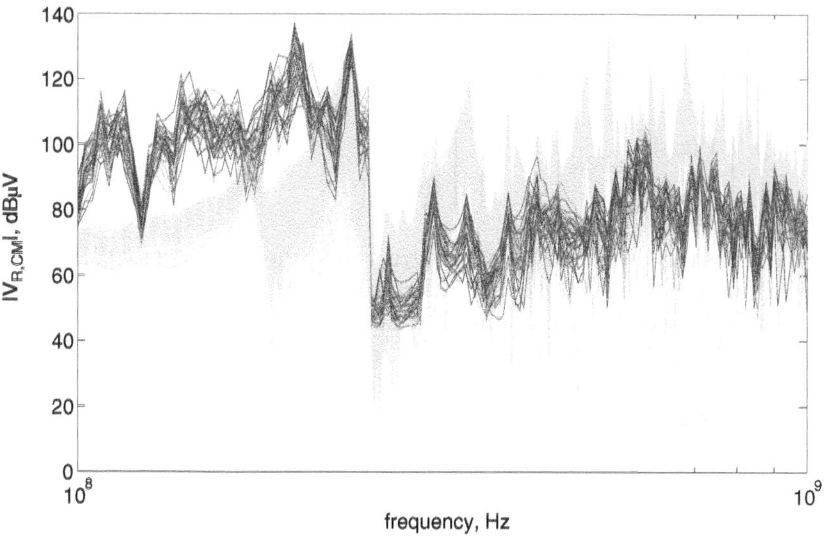

Figure 2.22 Colored lines: measured CM terminal voltages for each TWPs in several random bundle samples installed in the AGILE satellite mock-up. Shaded area: prediction interval (between minimum and maximum levels at each frequency) obtained by processing several Monte Carlo simulations.

of cables and interconnects with the inclusion of the effects of parameter uncertainties. The problem formulation is based on the telegrapher's equations with stochastic per-unit-length parameters. These parameters are expanded in terms of orthogonal polynomials of RVs. For instance, the per-unit-length capacitance matrix $\boldsymbol{C}$and the per-unit-length inductance matrix $\boldsymbol{L}$ can be expanded as

$$\boldsymbol{C} = \sum_{k=0}^{P} \boldsymbol{C}_k \phi_k(\xi), \quad \boldsymbol{L} = \sum_{k=0}^{P} \boldsymbol{L}_k \phi_k(\xi) \tag{2.47}$$

where $\{\boldsymbol{C}_k\}$ and $\{\boldsymbol{L}_k\}$ are expansion-coefficient matrices with respect to orthogonal components $\{\phi_k\}$, which are Legendre polynomials of ξ representing a vector of RVs involved in the specific problem under analysis (e.g., geometrical parameters defining the height of a wire above ground, and the separation between wires).

The randomness of the p.u.l. parameters reflects into stochastic values of the voltage and current distribution in the TL. Hence, similar expansions in orthogonal polynomial are also used to express voltages and currents (in this case, involving unknown expansion coefficients). By inserting the expanded p.u.l. parameters, voltage, and currents in the TL equations, and by the projection of each equation on the basis functions through the inner product in the Hilbert space of ξ, so to eliminate the functional dependence on ξ, the following system of differential equations is found:

$$\frac{d}{dz}\begin{bmatrix}\tilde{V}(z,s)\\ \tilde{I}(z,s)\end{bmatrix} = -s\begin{bmatrix}\mathbf{0} & \tilde{\boldsymbol{L}}\\ \tilde{\boldsymbol{C}} & \mathbf{0}\end{bmatrix}\begin{bmatrix}\tilde{V}(z,s)\\ \tilde{I}(z,s)\end{bmatrix} \tag{2.48}$$

which retains the formal mathematical structure of the TL model (with an augmented number of equations), and where vectors $\tilde{\boldsymbol{V}}$ and $\tilde{\boldsymbol{I}}$ collect the coefficients of the PC expansion of voltage and current, respectively, s is the complex frequency, $\tilde{\boldsymbol{L}}$ and $\tilde{\boldsymbol{C}}$ are matrices of coefficients. Once $\tilde{\boldsymbol{V}}$ and $\tilde{\boldsymbol{I}}$ are computed by solving (2.48), the PC expansions of voltage and currents (now determined) will express in analytical form the dependence on RVs ξ, leading to undoubted

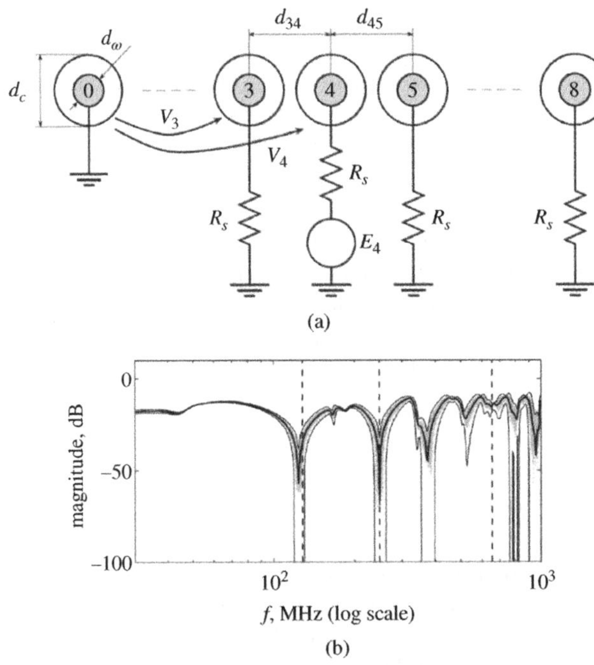

Figure 2.23 (a) 80-cm-long commercial flex cable ("0.050" High Flex Life Cable, 28 AWG Standard, PVC, 9-wire configuration). $R_S = 50\ \Omega$, $d_w = 15$ mils, $d_c = 35$ mils. The nominal value of the distance between adjacent wires (e.g., d_{34} and d_{45}) is 50 mils. (b) NEXT transfer function $H(j\omega)$ of the test case. Solid black thick line: deterministic response; solid black thin lines: 3σ tolerance interval of the third-order PC expansion; gray lines: a sample of responses obtained by means of the MC method (limited to 100 curves, for graph readability). Figures reprinted from [61].

computational advantages and allowing insight on the involved transformation of RVs.

Hence, the advantage of the PC method over Monte Carlo techniques resides in the fact that finding once and for all the coefficients of the aforementioned expansions in Legendre polynomials (which describe a closed-form approximate dependence on RVs) is more computationally efficient than solving several times the electromagnetic problem. Actually, this method offers accuracy and improved efficiency in computing parameter-variability effects on system responses. As a specific example, for the crosstalk problem reported in Figure 2.23, the Monte Carlo method required 3 m 10 s versus 3 s of the PC method [61].

References

1 C. Paul, *Analysis of Multiconductor Transmission Lines*. New York: Wiley Interscience, 1994.

2 C. R. Paul, "Decoupling the multiconductor transmission line equations," *IEEE Trans. Microwave Theory Tech.*, vol. 44, no. 8, pp. 1429–1440, Aug. 1996.

3 C. R. Paul and J. W. McKnight, "Prediction of crosstalk involving twisted pairs of wires—Part I: A transmission-line model for twisted-wire pairs," *IEEE Trans. Electromagn. Compat.*, vol. EMC-21, no. 2, pp. 92–105, May 1979.

4 A. Shoory, M. Rubinstein, A. Rubinstein, C. Romero, N. Mora, and F. Rachidi, "Application of the cascaded transmission line theory of Paul and McKnight to the evaluation of NEXT and FEXT in twisted wire pair bundles," *IEEE Trans. Electromagn. Compat.*, vol. 55, no. 4, pp. 648–656, Aug. 2013.

5 C. D. Taylor, R. S. Satterwhite, and C. H. Harrison, "The response of a terminated two-wire transmission line excited by a nonuniform electromagnetic field," *IEEE Trans. Antennas Propag.*, vol. 13, no. 6, pp. 987–989, Nov. 1965.

6 C. R. Paul, "Frequency response of multiconductor transmission lines illuminated by an electromagnetic field," *IEEE Trans. Electromagn. Compat.*, vol. 18, no. 4, pp. 183–190, Nov. 1976.

7 A. K. Agrawal, H. J. Price, and S. H. Gurbaxani, "Transient response of multiconductor transmission-lines excited by a nonuniform electromagnetic field," *IEEE Trans. Electromagn. Compat.*, vol. 22, no. 2, pp. 119–129, May 1980.

8 C. A. Nucci, F. Rachidi, M. Ianoz, and C. Mazzetti, "Lightning induced voltages on overhead lines," *IEEE Trans. Electromagn. Compat.*, vol. 35, no. 1, pp. 75–86, Feb. 1993.

9 S. Frei and R. Jobava, "Coupling of inhomogeneous fields into an automotive cable harness with arbitrary terminations," in *Proc. EMC Zurich Int. Symp. Electromagnetic Compatibility*, Zurich, Switzerland, Feb. 2001, pp. 87–92.

10 R. Neumayer, A. Stelzer, F. Haslinger, G. Steinmair, M. Troscher, J. Held, B. Unger, and R. Weigel, "Numerical EMC-simulation for automotive applications," in *Proc. EMC Zurich Int. Symp. Electromagnetic Compatibility*, Zurich, Switzerland, Feb. 2003, pp. 459–464.

11 L. Paletta, J. Parmantier, F. Issac, P. Dumas, and J. Alliot, "Susceptibility analysis of wiring in a complex system combining a 3-D solver and a transmission-line network simulation," *IEEE Trans. Electromagn. Compat.*, vol. 44, no. 2, pp. 309–317, May 2002.

12 F. Grassi, G. Spadacini, S. A. Pignari, and F. Marliani, "Combined MTL-fullwave statistical approach for fast estimation of radiated immunity of spacecraft cable assemblies involving multipair bundles," *IEICE Trans. Commun.*, vol. E98-B, no. 07, pp. 1204–1211, Jul. 2015.

13 F. Grassi, H. Abdollahi, G. Spadacini, S. A. Pignari, and P. Pelissou, "Radiated immunity test involving crosstalk and enforcing equivalence with field-to-wire coupling," *IEEE Trans. Electromagn. Compat.*, vol. EMC-58, no. 5, pp. 66–74, Feb. 2016.

14 S. Pignari and F. G. Canavero, "Theoretical assessment of bulk current injection versus radiation," *IEEE Trans. Electromagn. Compat.*, vol. 38, no. 3, Aug. 1996, pp. 469–477.

15 A. A. Smith, Jr., *Coupling of External Electro-Magnetic Fields to Transmission Lines.* New York: John Wiley and Sons, 1977.

16 K. S. H. Lee, "Two parallel terminated conductors in external fields," *IEEE Trans. Electromagn. Compat.*, vol. 20, no. 2, May 1978, pp. 288–296.

17 F. M. Tesche, "Comparison of the transmission line and scattering models for computing the HEMP response of overhead cables," *IEEE Trans. Electromagn. Compat.*, vol. 34, no. 2, pp. 93–99, May 1992.

18 A. G. Bell, "Research in electric telephony," *J. Soc. Telegraph Eng.*, vol. 6, no. 20, pp. 385–421, 1877.

19 A. G. Bell, "Telephone-circuit," U.S. patent 244 426, Jul. 19, 1881.

20 F. Grassi, S. A. Pignari, and J. Wolf, "Channel characterization and EMC assessment of a PLC system for spacecraft DC differential power buses," *IEEE Trans. Electromagn. Compat.*, vol. 53, no. 3, pp. 664–675, Aug. 2011.

21 C. R. Paul and J. W. McKnight, "Prediction of crosstalk involving twisted pairs of wires—Part I: A transmission-line model for twisted-wire pairs," *IEEE Trans. Electromagn. Compat.*, vol. 21, no. 2, pp. 92–105, May 1979.

22 C. R. Paul and J.W. McKnight, "Prediction of crosstalk involving twisted pairs of wires—Part II: A simplified low-frequency prediction model," *IEEE Trans. Electromagn. Compat.*, vol. 21, no. 2, pp. 105–114, May 1979.

23 C. D. Taylor and J. P. Castillo, "On the response of a terminated twisted wire cable excited by a plane-wave electromagnetic field," *IEEE Trans. Electromagn. Compat.*, vol. 22, pp. 16–19, Feb. 1980.

24 F. M. Tesche, "Plane wave coupling to cables," in *Handbook of Electromagnetic Compatibility*. San Diego, CA: Academic Press, 1995, pp. 67–115.

25 R. Stolle, "Electromagnetic coupling of twisted pair cables," *IEEE Jrnl. Sel. Areas Comm.*, vol. 20, no. 5, pp. 883–892, Jun. 2002.
26 R. B. Armenta and C. D. Sarris, "Efficient evaluation of the terminal response of a twisted-wire pair excited by a plane-wave electromagnetic field," *IEEE Trans. Electromagn. Compat.*, vol. 49, no. 3, pp. 698–707, Aug. 2007.
27 D. Taylor, R. S. Satterwhite, and C. W. Harrison, "The response of a terminated two-wire transmission line excited by a nonuniform electromagnetic field," *IEEE Trans. Antennas Propagat.*, vol. 13, pp. 987–989, 1965.
28 S. A. Pignari, G. Spadacini, "Plane-wave coupling to a twisted-wire pair above ground," *IEEE Trans. Electromagn. Compat.*, vol. 53, no. 2, pp. 508–523, May 2011.
29 H. W. Ott, *Noise Reduction Techniques in Electronic Systems*, 2nd ed. New York: J. Wiley and Sons, 1988.
30 F. Grassi and S. A. Pignari, "Bulk current injection in twisted-wire pairs with not perfectly balanced terminations," *IEEE Trans. Electromagn. Compat.*, vol. 55, no. 6, pp. 1293–1301, Dec. 2013.
31 F. Grassi, G. Spadacini, and S. A. Pignari, "The concept of weak imbalance and its role in the emissions and immunity of differential lines," *IEEE Trans. Electromagn. Compat.*, vol. 55, no. 6, pp. 1346–1349, Dec. 2013.
32 F. Grassi, X. Wu, Y. Yang, G. Spadacini, S. A. Pignari, "Modeling of imbalance in differential lines targeted to SPICE simulation," *Progress in Electromagnetics Research B*, vol. 62, pp. 225–239, 2015.
33 C. R. Paul and M. B. Jolly, "Sensitivity of crosstalk in twisted-pair circuits to line twists," *IEEE Trans. Electromagn. Compat.*, vol. 24, no. 3, pp. 359–364, Aug. 1982.
34 C. Jullien, P. Besnier, M. Dunand, and I. Junqua, "Advanced modeling of crosstalk between an unshielded twisted pair cable and an unshielded wire above a ground plane," *IEEE Trans. Electromagn. Compat.*, vol. 55, no. 1, pp. 183–194, Feb. 2013.
35 G. Spadacini, D. Bellan, and S. A. Pignari, "Impact of twist non-uniformity on crosstalk in twisted-wire pairs," in *Proc. IEEE Int. Symp. Electromagn. Compat.*, Aug. 2003, vol. 2, pp. 483–488.
36 G. Spadacini and S. A. Pignari, "Impact of common-to-differential mode conversion on crosstalk in balanced twisted pairs," in *Proc. 2006 IEEE Int. Symp. on Electromagn. Compat.*, Portland, OR, USA, Aug. 14–18, 2006, pp. 1–6.
37 R. B. Armenta, C. D. Sarris, "Modeling the terminal response of a bundle of twisted-wire pairs excited by a plane wave," *IEEE Trans. Electromagn. Compat.*, vol. 49, no. 4, pp. 901–913, Nov. 2007.
38 G. Spadacini, F. Grassi, F. Marliani and S. A. Pignari, "Transmission-line model for field-to-wire coupling in bundles of twisted-wire pairs above ground," *IEEE Trans. Electromagn. Compat.*, vol. 56, no. 6, pp. 1682–1690, Dec. 2014.

39 G. Andrieu, L. Koné, F. Bocquet, B. Démoulin and J. P. Parmantier, "Multiconductor reduction technique for modeling common-mode currents on cable bundles at high frequency for automotive applications," *IEEE Trans. Electromagn. Compat.*, vol. 50, no. 1, pp. 175–184, Feb. 2008.

40 G. Andrieu and A. Reineix, "On the application of the equivalent cable bundle method to cables bundles in presence of complex ground structure," *IEEE Trans. Electromagn. Compat.*, vol. 55, no. 4, pp. 798–801, Aug. 2013.

41 G. Andrieu, A. Reineix, X. Bunlon, J. P. Parmantier, L. Koné, and B. Démoulin, "Extension of the equivalent cable bundle method for modeling electromagnetic emissions of complex cable bundles," *IEEE Trans. Electromagn. Compat.*, vol. 51, no. 1, pp. 108–118, Feb. 2009.

42 Z. Li, Z. J. Shao, J. Ding, Z. Y. Niu and C. Q. Gu, "Extension of the 'equivalent cable bundle method' for modeling crosstalk of complex cable bundles," *IEEE Trans. Electromagn. Compat.*, vol. 53, no. 4, pp. 1040–1049, Nov. 2011.

43 S. Shiran, B. Reiser, and H. Cory, "A probabilistic model for the evaluation of coupling between transmission lines," *IEEE Trans. Electromagn. Compat.*, vol. 35, no. 3, pp. 387–393, 1993.

44 D. Bellan, S. A. Pignari, and G. Spadacini, "Characterization of crosstalk in terms of mean value and standard deviation," *IEEE Proc. Sci., Meas. Tech.*, vol. 150, no. 6, pp. 289–295, 2003.

45 C. R. Paul, *Introduction to Electromagnetic Compatibility*. New York: Wiley-Interscience, 1992.

46 S. Pignari, D. Bellan, and L. Di Rienzo, "Statistical estimates of crosstalk in three-conductor transmission lines," in *Proc. 2002 IEEE Int. Symp. on Electromagnetic Compatibility*, Minneapolis, MN, USA, Aug. 19–23, 2002, pp. 877–882.

47 S. Salio, F. Canavero, J. Lefebvre, and W. Tabbara, "Statistical description of signal propagation on random bundles of wires," in *Proc. 13th EMC Zurich Int. Symp. on Electromagnetic Compatibility*, Zurich, Switzerland, 1999.

48 S. Sun, G. Liu, J. L. Drewniak, and D. J. Pommerenke, "Hand-assembled cable bundle modeling for crosstalk and common-mode radiation prediction," *IEEE Trans. Electromagn. Compat.*, vol. 49, no. 3, pp. 708–718, Aug. 2007.

49 D. Bellan and S. Pignari, "Efficient estimation of crosstalk statistics in random wire bundles with lacing cords," *IEEE Trans. Electromagn. Compat.*, vol. 53, no. 1, pp. 209–218, Feb. 2011.

50 M. Wu, D. G. Beetner, T. H. Hubing, H. Ke, and S. Sun, "Statistical prediction of 'reasonable worst-case' crosstalk in cable bundles," *IEEE Trans. Electromagn. Compat.*, vol. 53, no. 1, pp. 842–851, Feb. 2011.

51 S. Pignari and D. Bellan, "Statistical characterization of multiconductor transmission lines illuminated by a random plane-wave field," in *Proc. 2000 IEEE Int. Symp. on Electromagn. Compat.*, Washington, D.C., USA, Aug. 21–25, 2000, pp. 605–609.

52 D. Bellan and S. Pignari, “A probabilistic model for the response of an electrically short two-conductor transmission line driven by a random plane-wave field,” *IEEE Trans. Electromagn. Compat.*, vol. 43, no. 2, pp. 130–139, May 2001.

53 S. A. Pignari, “Statistics and EMC,” *URSI Radio Science Bulletin*, no. 316, Mar. 2006, pp. 13–26.

54 D. A. Hill, “Plane wave integral representation for fields in reverberation chambers,” *IEEE Trans. Electromagn. Compat.*, vol. 40, no. 3, pp. 209–217, Aug. 1998.

55 D. Bellan and S. A. Pignari, “Complex random excitation of electrically-short transmission lines,” in *Proc. 2006 IEEE Int. Symp. on Electromagn. Compat.*, Portland, OR, USA, Aug. 14–18, 2006, pp. 663–668.

56 M. Magdowski, S. V. Tkachenko, and R. Vick, “Coupling of stochastic electromagnetic fields to a transmission line in a reverberation chamber,” *IEEE Trans. Electromagn. Compat.*, vol. 53, no. 2, pp. 308–317, May 2011.

57 G. Spadacini and S. A. Pignari, “Radiated susceptibility of a twisted-wire pair illuminated by a random plane-wave spectrum,” *IEICE Trans. on Communications*, vol. E93-B, no. 7, pp. 1781–1787, July 2010.

58 G. Spadacini, F. Grassi, and S. A. Pignari, “Influence of load imbalance on noise induced in a twisted-wire pair illuminated by a random plane-wave spectrum,” in *Proc. Asia-Pacific Microwave Conf.*, Sendai, Japan, Nov. 4–7, 2014, pp. 73–75.

59 G. Spadacini, F. Grassi, and S. A. Pignari, “Statistical properties of low frequency voltages induced by a plane-wave field across the terminal loads of a random wire-bundle,” in *Proc. 2015 IEEE Int. Symp. on Electromagn. Compat.*, Dresden, Germany, Aug. 16–22, 2015, pp. 824–829.

60 G. Spadacini, F. Grassi, and S. A. Pignari, “Field-to-wire coupling model for the common-mode in random bundles of twisted-wire pairs,” *IEEE Trans. Electromagn. Compat.*, vol. 57, no. 5, pp. 1246–1254, Oct. 2015.

61 I. S. Stievano, P. Manfredi, and F. G. Canavero, “Stochastic analysis of multiconductor cables and interconnects,” *IEEE Trans. Electromagn. Compat.*, vol. 53, no. 2, pp. 501–507, May 2011.

3

HEMP Protection and Verification[1]

William D. Prather

3.1 Introduction

High-altitude nuclear electromagnetic pulse (HEMP) has been recognized as a threat to electrical and electronic equipment since the early 1960s. It is becoming increasingly important today because of the growing global nuclear threat and the increased dependence of all of our forces on computers and electronic systems to carry out their missions. Though the HEMP threat has not changed dramatically in recent years, the magnitude of external radio frequency interference (RFI) threats, also known as high-intensity radiated fields (HIRF), in MIL-STD-464 have increased significantly, as have the required FAA certification levels. In addition, the FAA lightning protection requirements have become much more stringent, due in particular to the 1994 issuance of Federal Aviation Regulation (FAR) 25.1316, system lightning protection. As a result, aircraft are designed and built much differently today from the way they were in the 1970s, and during this same time period, our knowledge of HEMP effects, hardening, and testing has grown noticeably.

The purpose of this chapter is to describe the phenomenon of HEMP, the effect it has on aircraft, and the methods used for hardening and testing hardened aircraft.

[1] This work was supported by the Air Force Research Laboratory, the Defense Threat Reduction Agency, and the Oklahoma City Air Logistics Center.

Handbook of Aerospace Electromagnetic Compatibility, First Edition. Edited by Reinaldo J. Perez.

3.2 High-Altitude Electromagnetic Pulse

3.2.1 First Evidence of HEMP

On the evening of July 9, 1962, the United States detonated a 1.4 MT device known as Starfish Prime 400 km above Johnston Island in the Pacific Ocean. The scientists knew there would be some electromagnetic (EM) interference from the detonation, but they were very surprised by what they actually saw. Because there is almost no air at an altitude of 400 km, there was no fireball, no shockwave, and no sound, only a bright flash of light, as shown in Figure 3.1. There were, however, many other notable effects. In Hawaii, about 1500 km away, 300 street lights failed, television sets and radios malfunctioned, burglar alarms went off and power lines fused. On Kauai, telephone calls to the other islands were interrupted when the microwave relay station burned out [1–5]. Interestingly enough, the Soviet Union's experience was very similar to ours. In October 1962, they detonated a 300 kT warhead 300 km over Kazakhstan [6]. The HEMP it produced exceeded that of the Starfish because of the stronger

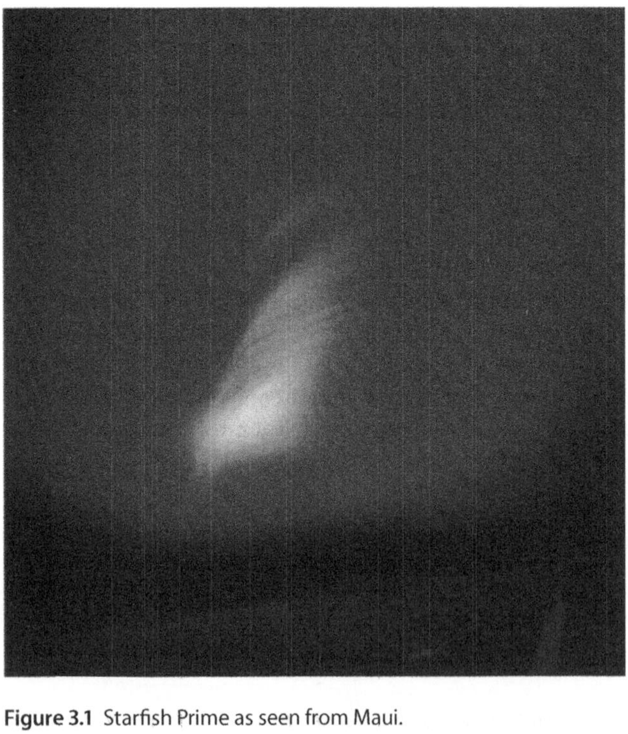

Figure 3.1 Starfish Prime as seen from Maui.

magnetic fields present in the more northerly latitude.[2] As a result, more severe impacts were noted in electrical systems including physical damage to power line insulators, outages of long communications lines (both buried and above-ground), damage to diesel power systems, and radar systems. Clearly, there was much to be learned about this new phenomenon.

As a side note, possibly the most famous victim of Starfish Prime was Telstar, the first telecommunications relay satellite, the shining star, a showcase of modern technological achievement, which had the misfortune of being launched the day after Starfish Prime. The large flux of high-energy electrons and charged particles produced by Starfish generated an Aurora Borealis that was magnificent to see. Unfortunately the electrons, being the lightest, did not fall down into the earth's atmosphere, but got caught up in the earth's magnetic field and remained aloft for several months. They rapidly degraded Telstar's electronics and rendered it silent within four months [1, 7]. At least six other satellites were also rendered useless within a few months, including one from the Soviet Union [6].

3.2.2 Source of the HEMP Field

HEMP is generated when the gamma rays from a nuclear detonation interact with the atmosphere. A perspective on the layers of the atmosphere and the relative height of a high-altitude burst (HAB) can be seen in Figure 3.2. The gamma rays illuminate a large area of the ionosphere as shown. The resulting HEMP is a short, but very intense pulse that reaches its peak in 1 to 10 nanoseconds and spreads outward at the speed of light. The intensity of the HEMP as it propagates outward is a function of the height of the burst as illustrated in Figure 3.3. As it spreads in all directions toward the horizon, it couples to any metal object in its path including antennas, cables, conduits, power lines, ships, aircraft, and missile bodies. The direction of the electric field is always normal to the radial vector originating at the weapon, and the polarization seen at the ground will vary with position. The HEMP threat consists of three parts:

E1 is the prompt or early-time pulse in the range of 0 to 1 μs. It has a high peak power, fast rise time, and extremely wide bandwidth across 4 decades of frequency as described in [2–4]. The waveform and spectrum from MIL-STD-464C [8] are illustrated in Figure 3.4. It is this part of the HEMP that is of primary concern to aircraft.

[2] If Starfish Prime had been detonated over the United States, the HEMP generated would have been about 2.4 times larger than that seen on Johnston Island, which is close to the equator, due to the difference in direction of the earth's magnetic field in the higher latitudes.

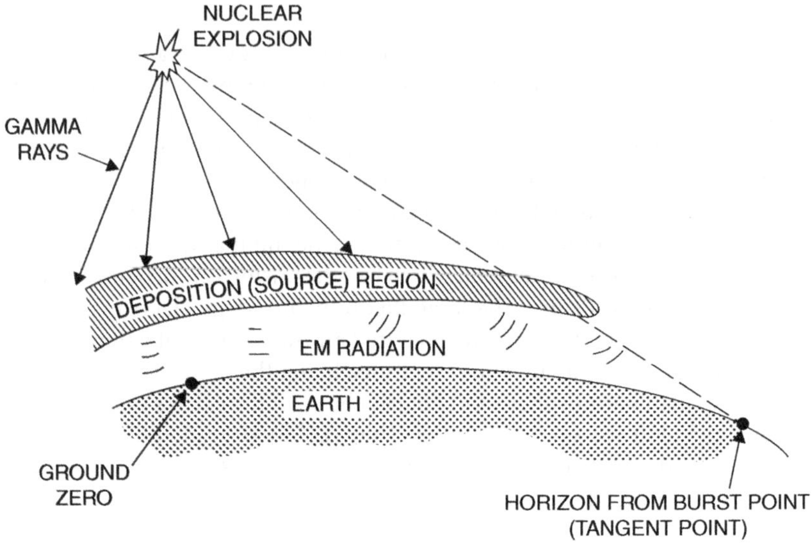

Figure 3.2 HEMP generation.

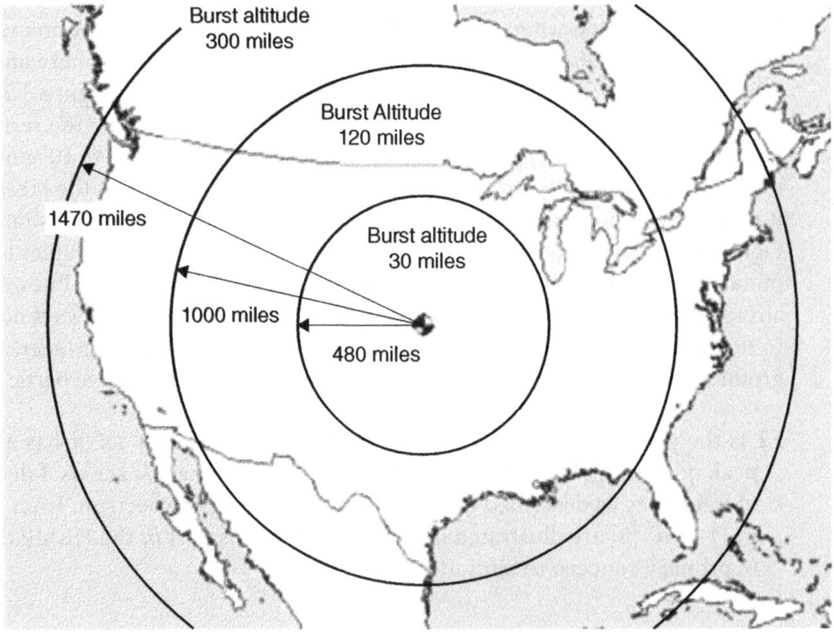

Figure 3.3 Effective range of a high-altitude EMP as a function of burst height.

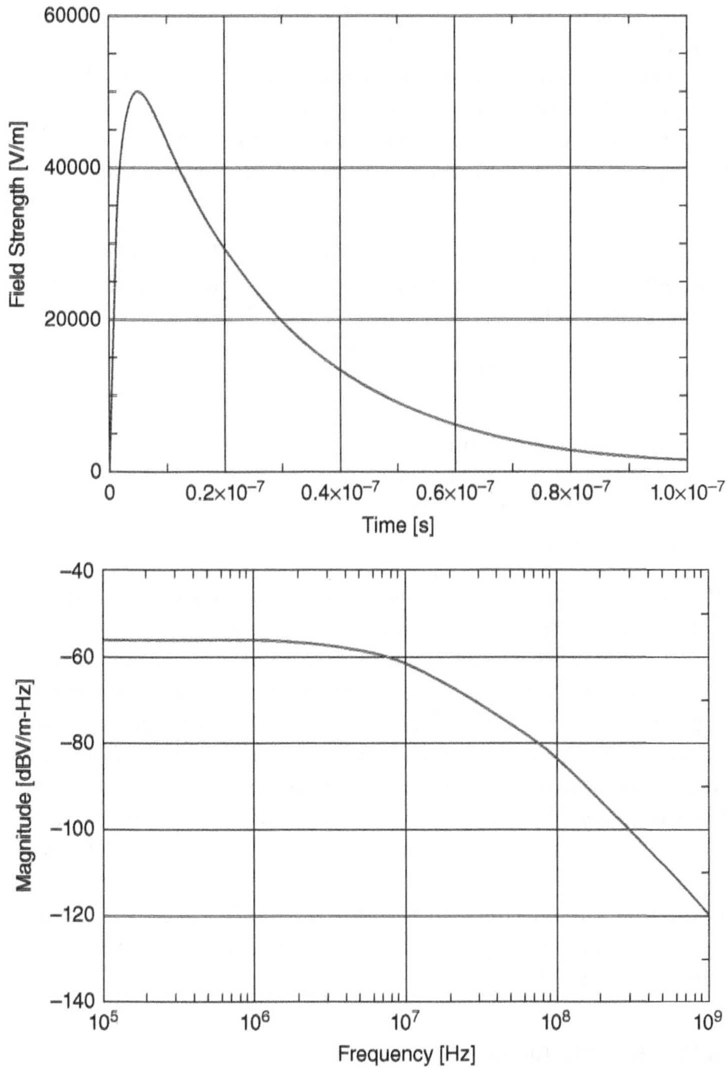

Figure 3.4 HEMP waveform and frequency spectrum from MIL-STD-464C.

E2 is the intermediate time pulse (1 μs to 1 s). This part of the HEMP field is comparable in amplitude and spectrum to the pulse produced by "nearby lightning," and is not nearly as much of a threat to aircraft as E1 or E3.

E3 is the magnetohydrodynamics (MHD) or late-time pulse (1 to 200 sec). During this time period, the magnitude of the E3 field is very small, only tens of mV/m, but the wavelengths are very long and thus couple to miles-long power and communication lines. As a result, the voltages induced in the long lines can be in the megavolt range, enough to cause serious damage, as exemplified by the experiences in Kauai and Kazakhstan. The effects caused by the very low frequency E3 fields is analogous in many ways to the ground currents induced by large sunspots (coronal mass ejections) like the one that caused the power blackout in Quebec in 1989 and the famous Carrington Event of 1889 [1].

For this discussion, we will be addressing the early-time (E1) pulse only, and for this we will refer to the unclassified HEMP threat waveform in MIL-STD-464C [8], which is illustrated in Figure 3.4. This is described by

$$E(t) = E_o(e^{-\alpha t} - e^{-\beta t})u(t) \text{ in V/m} \tag{3.1}$$

where

E_o = field intensity constant in V/m
α = decay constant in radians/s
β = rise time constant in radian/s
$u(t)$ = unit step function

The HEMP from a high-altitude detonation can cover thousands of square miles ranging much farther than the blast and radiation from the weapon itself. When the HEMP encounters metallic conductors, it causes currents to flow, and the conductors carry the energy into electrical and electronic equipment where it may cause upset or permanent damage. Equipment that operate at low currents, such as computers and solid-state systems without protection, cannot withstand a HEMP power surge and are likely to burn out. Because of the longer wavelengths involved, HEMP can present a serious threat to the systems of the size of an aircraft or larger that depend upon electronics to complete their mission.

HEMP is a potential threat to not only the military, but to the civilian infrastructure as well, so it has been reported several times to the U.S. Congress as well as the military leaders [9].

3.2.3 The HEMP Threat to Aircraft

The EM threats to aircraft can range from a small nuisance to something potentially very harmful, and it may be either natural or man-made. Some examples of threats are as follows:

- Natural: Lightning, p-static[3]

---

[3] P-static or precipitation static (also referred to as electrostatic discharge or ESD) is the continuous build-up and discharge of electrostatic charge on the aircraft in flight, in particular when flying through moisture. This can cause high levels of noise on radios, navigation equipment, etc.

- Induced: EMI, EMC, TEMPEST, HIRF
- Hostile: HEMP, HPM, jammers

The main EM threats during regular flight operations are from lightning strikes and HIRF. As a result, all aircraft must have effective EM protections designed in, and these are required by the FAA to obtain certification. Military aircraft must also survive in a hostile environment and be able to fight back, so additional protection is needed.

Developments in the aerospace industry over time have led to the replacement of classic flight and communications systems with solid-state and microchip-based electronics. In addition, flight controls are becoming more and more "fly-by-wire." The advantages of these new systems are numerous, but the disadvantage is that they are more sensitive to EM interference and more vulnerable to HEMP.

Observed damage from HEMP tests that have been observed include:

- Destruction of interface circuits and inadequate transient suppression devices
- Voltage breakdown of semiconductor junctions
- Destruction of the bonding of components
- Destruction of printed circuit board (PCB) traces or contacts

Upset effects observed have included:

- Intermittent operation of latches, thyristors, and TRIACs[4]
- Erasure or corruption of the memory in programmable logic controllers, computers
- Program errors or computer system crashes
- Malfunction of digitally controlled equipment such as fuel valves
- Data and transmission errors

3.2.4 Comparison with Other EMI Threats

The spectrum and waveform of HEMP differ from those of any other natural or common man-made sources. The spectrum is extremely broad, extending from extremely low frequencies into the UHF and even the L-band regions. The time-domain waveform has a higher amplitude and much faster rise time than, for example, the fields generated by a lightning stroke, referred to as "nearby lightning." Therefore, the protective components and practices for non-HEMP

---

[4] A thyristor is a solid-state device with four layers of alternating N- and P-type material. They act as bistable switches, conducting when their gate receives a current trigger, and continue to conduct as long as the voltage across the device is not reversed. A TRIAC is a subset of thyristor closely related to silicon-controlled rectifiers (SCRs). TRIACs are bidirectional and thus allow current flow in either direction.

environments—RFI, lightning, etc.—do not necessarily provide adequate protection against HEMP. The HEMP is a unique transient phenomenon—at once fast, powerful, and very broadband—and must be treated as such.

3.2.5 Comparison with Lightning

Natural lightning and HEMP are often compared, perhaps because they have waveforms of similar shape. However, they are fundamentally different phenomena and should not be confused in terms of their interactions with aerospace systems or the protection methods used.

A lightning strike is a direct current injection into the airframe and can cause physical damage (direct effect) or electronic effects by cross-coupling to the aircraft wiring (indirect effects). In the case of a nearby strike, the EM fields generated by the lightning current channel will illuminate the aircraft as an EM wave, which has a large amount of low-frequency energy and very little above 1 MHz. Since HEMP, HIRF, and "nearby strike" lightning are all radiated fields, they couple through the hull to the aircraft cables as the *time derivative* of the incident field. As a result, HEMP will produce *higher peak currents and voltages*, but of shorter duration than lightning. This was confirmed by a thorough study of the coupling to aircraft of lightning and HEMP conducted in 1973–1974 by the Air Force Weapons Laboratory (AFWL) and NASA using NASA's F-106 lightning research aircraft [10, 11]. Another major difference is that HEMP, originating a long distance away, is to the observer, essentially a plane wave, whereas the radiated field from a lightning channel is relatively localized. It is cylindrical and decreases as $1/r$ with distance.

The protection methods used for HEMP and lightning are not the same because of not only their different frequency content, but also because of the differences in the coupling and interaction mechanisms. Lightning has a slower rise time, a much longer duration, higher energy, and a much larger charge transfer than does an EM wave. HEMP coupling will generate higher peak currents of much shorter duration.

Lightning arrestors are built to be very robust, but may have slower reaction times than those required for EMP. Special hybrid surge arrestors, usually solid state, are available that will respond to the fast rise time of the HEMP and carry enough current to protect against lightning. On the other hand, low-frequency *shielding* (such as mu-metal), which is installed to protect sensitive subsystems against lightning, does provide some protection for HEMP, but may not be entirely sufficient.[5]

---

[5] Mu-metal is a soft alloy of nickel and iron which has a very high permeability, μ. It is often used in cable shields or cable shield foil wraps, it is effective in shielding components against low-frequency magnetic fields.

Lightning protection in aircraft includes: [12, 13]

- Current diversion through large ground straps and buses
- Filters and spark-gap arrestors
- Cable shields emphasizing metal thickness/current-carrying capability
- In composite aircraft, wire mesh, metallic coating, and diverter strips

whereas EMI/HIRF/HEMP protection relies mostly on

- Hull and rack shielding and shielding topology
- Shielded cables with peripherally bonded connectors
- Filters with fast-acting voltage clamps
- Aperture protection with gaskets, window screens, and honeycomb
- MIL-STD-461G immunity for equipment protection

3.3 HEMP Coupling to Aircraft

When HEMP is incident on an aircraft, it induces currents and charges on the surface. These currents and charges can penetrate to the interior through antennas, windows, doors, or unprotected conducting penetrations and find their way into the aircraft's electronic systems, where they can cause functional upset or physical damage. On an unprotected aircraft, HEMP can result in peak currents of 100s of amperes and peak voltages in the kilovolt range, enough to cause damage to the electronics.

As seen in the previous chapter, the HEMP from a high-altitude detonation contains an extremely wide band of frequencies. At the high end of the spectrum (1 GHz), the wavelength is only about 30 cm, whereas at the low end (100 kHz), it is kilometers in length. It is important to understand the wavelengths compared with the size of the aircraft and its components, because the coupling mechanisms will be different at different wavelengths.

3.3.1 External Coupling

An EM wave will induce currents and charges onto the external surface of the aircraft as a function of (1) the frequency and polarization of the incoming wave and (2) the coupling transfer function of the airframe, which is dependent on its size and shape, as well as the direction of incidence of the wave [14–16].

When the wavelength of the incident field is near the size of the airframe, it will excite the resonant modes on the surface. At the resonant frequencies, the surface current density J_s and surface charge density ρ_s will be at their highest values. As with a dipole antenna, the current will be the strongest in the center of the dipole and the charge strongest at the extremities. Therefore, in

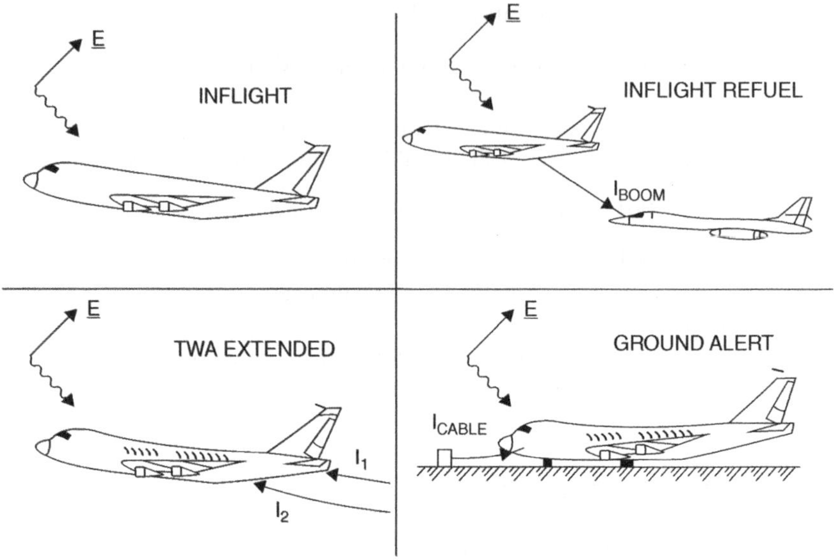

Figure 3.5 Aircraft EM coupling configurations.

an aircraft, there are *two* coupling mechanisms against which we must harden, magnetic field coupling through the apertures and electric field coupling from the extremities.

The largest currents on the aircraft occur in the HF and VHF bands, because at those frequencies the wavelength is roughly the same size as the aircraft and its components [14–16]. For example, the lowest resonant mode of a typical passenger aircraft like a Boeing 737 occurs around 3.2 MHz, and that of a Boeing 747 occurs around 2.1 MHz.

The resonant frequencies and the Q will be a function of the size and shape of the aircraft. That is to say, a long, narrow aircraft will have a Q around 15, so the charge enhancement on the extremities (nose, tail, and wingtips) will be high. From the aircraft scale model work done by the University of Michigan, we can see that at resonance the surface charge density (surface electric field) on the nose will be $E_{NOSE} \approx 15 \times E_{inc}$, an enhancement factor of about 24 dB [16]. The surface current density (magnetic field) $\boldsymbol{J}$ will be about 10 times the incident H field, an enhancement of 20 dB.[6]

An aircraft can be in several different EM coupling configurations during its mission: in-flight, ground alert, in-flight refueling, or with a trailing wire antenna (TWA) extended, as illustrated in Figure 3.5.

[6] The nose of the aircraft enhances the field in two dimensions (a hemisphere), whereas the side of the fuselage only enhances it in one dimension (a cylinder).

- **Ground-alert Mode.** In ground-alert mode, the engines are turned off, and the aircraft is operating on ground power. The presence of the ground is a realistic part of the configuration. The aircraft may have additional electrical connections for phone, Internet, etc.
- **Taxi.** In the taxi mode, the aircraft is powered up with engines running and crew on board, but still sitting on the ground. There are no external wires connected.
- **In-flight.** When the aircraft is in free flight, there is no ground reflection or interaction. Wheel well and weapons bay doors are closed.
- **In-flight Refueling.** During in-flight refueling, the aircraft is airborne and connected to a tanker. In this configuration, the aircraft and the tanker combine to form a much larger resonant structure, so the resonant modes and frequencies seen by the aircraft are quite different. The lowest resonant frequency is about a factor of 2 lower than that of the aircraft alone and the maximum surface current and charge densities will be higher by a factor of 2. In addition, the distribution of surface current and charge on the surfaces of the two aircraft is also very different. From Figure 3.5, we can see that the nose of the aircraft being refueled (in the rear) is now near the center of the resonant structure, so the net result is that there is a great increase in the surface current at the nose of the receiving aircraft, but no charge concentration, because it is being drawn off by the boom. In similar fashion, the charge and current on the tail of the tanker is also changed.
- **TWA Extended.** If the aircraft carries a VLF/LF TWA, then the aircraft is connected to two long wires, which will couple large amounts of energy from the HEMP.

3.3.2 Internal Coupling

The coupling of energy into the interior of an aircraft is often described as being either "front-door" or "back-door." Front-door coupling refers to EM energy coupling directly into the aircraft's antennas, which are of course designed to receive energy within a certain band. However, as we know, antennas can often have out-of-band responses that can inadvertently admit energy from HEMP, lightning, or other broadband sources. To avoid surprises, the antennas and their tuners should be tested across the entire threat band to make sure there are no out-of-band responses that might cause problems. Back-door coupling refers to coupling through apertures, seams, cracks, energy coming into the shielded volume on wiring or tubes, or through out-of-band antenna resonances. These are often referred to as "inadvertent penetrations."

3.3.2.1 Magnetic Field Coupling

The surface current J_S on the fuselage and wings will couple to any loops of wire and apertures and is best described in units of transfer impedance. The wiring in the leading and trailing edges of the wings is strongly excited by the current density flowing on the surface of the wing as well as the charge density on the wingtips. The coupling through an aperture is represented by

$$V_{OC}(\omega) = Z(\omega)\, I_{SC}(\omega) = (R + j\omega L)\, I_{SC}(\omega) \tag{3.2}$$

where

I_{SC} = short-circuit current flowing across the aperture.
V_{OC} = open-circuit voltage across the aperture
Z = aperture transfer impedance
R = aperture resistance
L = aperture inductance

The short-circuit current is the surface current induced by the incident field that would flow across the aperture if the aperture were not there. In our case, when making a measurement, it is the driven current across the aperture as described in Section 3.3.

Also,

$$\boldsymbol{I}_{SC} = \boldsymbol{J}_S W_{eff} \tag{3.3}$$

where

J_S = surface current density across the aperture = H_S, the surface magnetic field
W_{eff} = effective width of the aperture.

Note that the surface current on the fuselage will follow the phase of the incident electric field. However, the voltage appearing on the inside of the apertures, which is what drives the nearby cables is, according to (3.2), proportional to the *time derivative* of the surface current J_S. As a result, the current on the wires will be proportional to the *time derivative of the incident field.* In similar fashion, the current on wires such as those in the leading or trailing edges of the wings respond according to Faraday's Law to the time derivative of the magnetic field generated by the current on the wings. If the magnetic field passes through the loop formed by the cables and the edge of the wing, there will be a current induced on the wire that is proportional to the time derivative of the B field on the wing.

3.3.2.2 Electric Field Coupling

The surface electric field (or charge density) on the nose, tail, and wingtips will couple strongly to any wires, tubes, control cables, blade antennas, etc., exposed to the field. An electric field penetration can be represented as either a Thévenin equivalent circuit (with a transfer impedance) or a Norton equivalent (with a transfer admittance). It is also worth noting that even though transfer admittance Y_T and transfer capacitance C_T are technically correct quantities to use in this case, they are very difficult to measure on board the aircraft, so for practical reasons, electric field penetrations are usually specified in terms of Z_T.

A simple example is the coupling to a wingtip or fin-cap HF antenna, which responds to the charge density on the wingtip in the same fashion as a D-dot electric field sensor as is described in Section 3.4. In the frequency domain,

$$V_{OC} = Z_T I_{SC} = h_{eff} E_n \tag{3.4}$$

$$I_{SC} = j\omega\, \varepsilon_o E_n A_{eq} = j\omega\, D_n A_{eq} \tag{3.5}$$

$$D_n = \varepsilon_o E_n = \rho_s \tag{3.6}$$

where

I_{SC} = short-circuit entering the aircraft
V_{OC} = open-circuit voltage induced on the exposed element
A_{eq} = the effective area of the exposed metallic part in m^2
h_{eff} = the effective height of the exposed part in m
E_n = electric field normal to the surface in V/m
D_n = electric flux normal to the surface in coulombs/m^2
ρ_s = the surface charge density in coulombs/m^2

Notice from (3.5) that the cable currents induced by the electric field coupling to the nose, tail, etc., is proportional to the time derivative of the electric field. This may explain why the currents on the wires inside an aircraft often have larger amplitudes at the higher frequencies than one might expect. Note that on an aircraft, the electric field (charge) coupling is stronger than the aperture coupling. Note, too, that the wires on the leading and trailing edges of the wings are excited by both magnetic and electric fields and are driven very hard. In addition, since these wires are often exposed, they can receive direct coupling from the incident field and that is where we will often see high frequency energy penetrating.

3.4 Shielding and Shielding Topology

To be effective, the design of the shielding on an aircraft should follow the rules of shielding topology, which is the division of space into volumes and boundary surfaces so as to clearly define and maintain inside versus outside. In developing topology, there is only one question, "Does EM coupling occur or not." The *amount* of coupling is not a consideration in the design, only the configuration. The design of the shield layers and Points of Entry (POEs) is the next step.

The concept of shielding topology and its application to EM protection has been described by a number of authors [17–19]. The first step is to define the shield surfaces through which the HEMP energy must penetrate. The most obvious of such surfaces is the exterior skin (hull) of an all-metal aircraft with its windows, skin panel joints, and antennas being POEs. Smaller metal-enclosed areas, such as conduits, shielded cables, and equipment housings, are examples of second-layer surfaces.

HEMP protection topology is built into an aircraft in several ways as illustrated in Figure 3.6. It is often desirable to allocate part of the protection to the aircraft structure in order to provide a more benign interior environment, part to the wiring shields, and part to the strength of the boxes. Such an integrated design has been found to not only be effective, but to reduce weight, production cost, and downstream maintenance cost. In addition, a multilayer

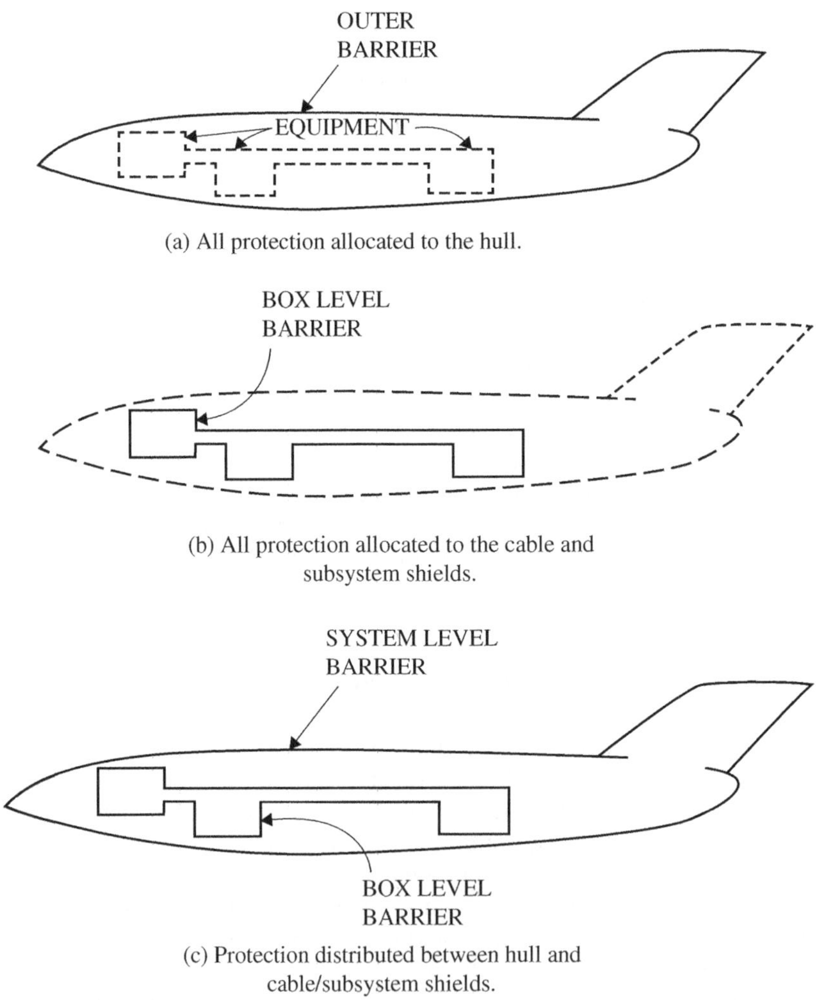

(a) All protection allocated to the hull.

(b) All protection allocated to the cable and subsystem shields.

(c) Protection distributed between hull and cable/subsystem shields.

Figure 3.6 Three ways of distributing HEMP shielding protection.

design provides a measure of insurance that as the components age (which they will), the aircraft will continue to meet its HIRF, HEMP, and lightning requirements.

3.5 EM Protection Technology

The differences between 1980 and 2018, in both aircraft construction and test technology, are significant. Around 1970, when this learning curve began, there were no HEMP standards at all and only a few basic EMI/EMC standards. Today, many robust standards do exist, and as a result, new aircraft are now delivered with a substantial amount of EM protection already built in. This makes upgrading an aircraft to meet HEMP standards a lot more straightforward that it was ever before. In addition, methods have been developed to measure the EM shielding as it is designed and as it is installed on board the airframe.

3.5.1 Standards and Technology Available Today

What do we have in place now that we did not have before? We have quite a lot, in fact.

- Threat-level HEMP simulators [20]
- Swept CW systems for pretest evaluation and hardness maintenance (HM) [21, 22]
- Tools for measuring hardening components on board the aircraft [21, 22]
  - Loop Resistance Tester (LRT) for testing cable shields on board
  - Cable Shield Tester (CST) for testing cable shields in the laboratory
  - Aperture Tester for measuring windows and doors
  - Single Point Excitation for Hardness Surveillance (SPEHS)—A system-level test method for HM/hardness surveillance (HM/HS).
  - Pulse Current Injection (PCI) equipment that will operate up to 1 GHz
- MIL-STDs for EMI/EMC/HIRF and HEMP [7] and [23–25]
- Knowledge of how to write specifications for aircraft protection
- Aircraft with *substantial* EMI and HIRF hardening *already built in*
- Hardening kits available from some of the manufacturers
- The ability to conduct HEMP verification tests in a few weeks at a reasonable cost
- Fast transient digitizers and CW data recording systems
- Fast computers and sophisticated data processing software

3.5.2 Integrated EM Environment Effects Programs

All of the EM hardening in a modern aircraft is built around an integrated electromagnetic environment effects (E3) design that includes electromagnetic

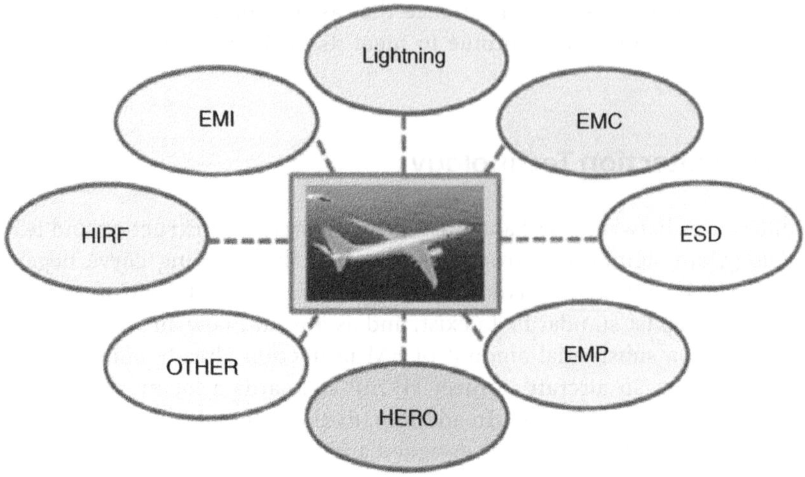

Figure 3.7 Integrated electromagnetic environment effects program illustrated.

interference (EMI), HIRF, lightning, and electrostatic discharge (ESD). On top of this, the Defense Ministry will place additional requirements that support their various military missions, such as Hazards of EM Radiation to Ordnance (HERO) and perhaps HEMP, as illustrated in Figure 3.7. This substantial base of hardening has a significant influence on the HEMP hardening design, making it more straightforward and less expensive to implement than it ever was in the past. The aircraft designer is now always working within an E3 program, so a new commercial aircraft will have already incorporated some EM protections into the design. To aid the designer, tools exist today to measure all of the EM hardening components *in situ*, something that did not exist when the first HEMP-hardened aircraft were built.

As a result of the increasing number of EM standards, aircraft manufacturers have developed improved protection strategies and hardening components, and since the new hardening approaches were developed to meet the FAA standards, the commercial aircraft manufacturers have invested a great deal of money and effort into making them low cost, reliable, and maintenance free. As a result, the military also benefits from these improvements.

3.5.3 Military and Industry: A Shared History

If we look at the history of aircraft hardening design, in some ways we have come full circle. That is to say, the basic topological shielding concepts and many of the hardening techniques used on aircraft today derive directly from the early HEMP hardening work done by the military services during the Cold War. The development of topological shielding concepts, knowledge of how EM energy couples into an aircraft, and the testing and evaluation of many of the early generation hardening fixes came directly from the work done at the AFWL, Oklahoma City Air Logistics Center, Aeronautical Systems Center, and Patuxent River NAS. In like manner, much of what is found in the current FAA regulations for qualifying new aircraft, in particular the CW illumination methods, came directly from AFWL's development of CW test methods and participation in the standards committees during the early 1990s [26, 27].

3.6 System-Level Specifications and Measurements

Aircraft HEMP hardening began about 40 years ago, and since that time, hardening designs and measurement techniques have evolved into a mature engineering discipline. This is due in large part to the significant increases in EM protection requirements for the commercial industry in the form of FAA and European standards. The airline industry especially requires effective and economical protection, and they have invested a great deal of money to engineer EM hardening that is effective, reliable, and easy to maintain. From this, the military departments have derived great benefit [28].

3.6.1 System-Level Specifications

In a HEMP-hardened aircraft, the system specification is written in terms of a design margin (DM), which is the ratio of the maximum HEMP-induced cable currents (stress) to the minimum equipment immunity (strength) at each cable connector interface as illustrated in Figure 3.8. At these interfaces, strength and stress are usually expressed in terms of three waveform norms:

- 1 Norm = Peak magnitude norm = the maximum of the absolute value of the waveform:

$$I_P = |I(t)|_{max} \text{in Amps} \tag{3.7}$$

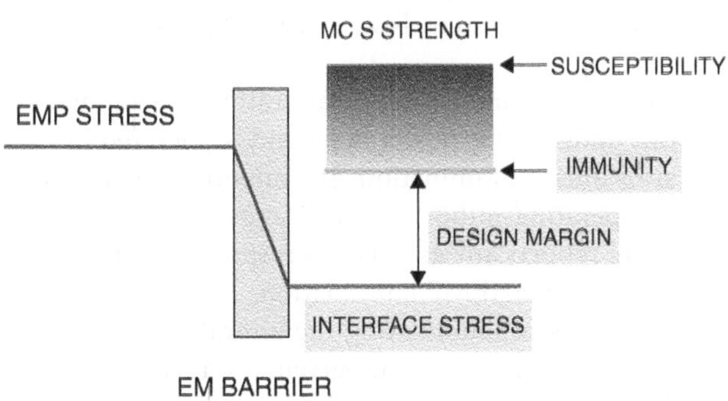

Figure 3.8 Design margin illustrated.

- 2 Norm = maximum derivative norm = the maximum rate of rise of the waveform:

$$PD = \left| \frac{dI(t)}{dt} \right|_{max} \quad \text{in Amps/sec} \tag{3.8}$$

- 3 Norm = Root action integral (RAI) norm is proportional to the energy in the waveform:

$$RAI = \sqrt{\int_0^{\infty} I^2(t)dt} \quad \text{in Amps} \cdot \text{sec}^{1/2} \tag{3.9}$$

These can be defined for each test frequency *k* (from MIL-STD-461, CS116) to which the boxes were qualified. Then, the DM is specified for each norm at each measurement interface *i* by

$$DM_{ki} = 20 \log \frac{\|I_k\|_i}{\|R\|_i} \tag{3.10}$$

where

i = norm index
k = test center frequency index from CS116
$\|I_k\|_i$ = immunity norms
$\|R\|_i$ = residual stress norms

The final measured DM is usually specified as the maximum of the minimum norm margins measured for each center frequency in CS116. The pass/fail criteria are determined by the sponsoring agency.

3.6.2 HEMP Simulator Tests

HEMP simulator hardness verification testing is conducted in 3 phases as illustrated in Figure 3.9, including an Active System Test, a Passive System Test, and a Direct Drive Test.

The Active System Test consists of illuminating the aircraft with a simulated HEMP waveform. The aircraft under test is powered on and is operated in flying simulated mission scenarios during illumination. The mission scenarios are determined by the procuring/operating authority. The air crew on board monitors the aircraft operations and reports any anomalous responses (functional upsets or damage) during and after the HEMP illumination.

The Passive Systems Test consists of illuminating the aircraft with pulsed, high-level EM fields while measuring the residual currents induced on some sample of cables inside the EM barrier(s). The aircraft under test is unpowered

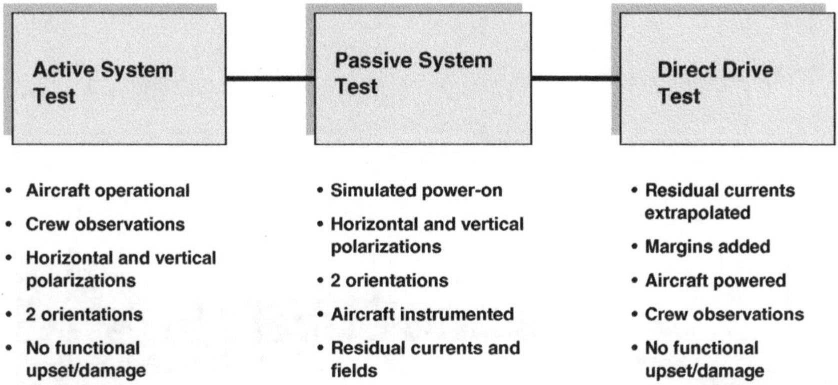

Figure 3.9 HEMP verification test approach for aircraft.

and is instrumented to measure the induced currents. Aircraft wiring is temporarily configured (jumpered) to simulate the power-on configuration.

The Direct Drive Test confirms that mission critical failure (due to both functional upset and damage) does not occur within the required DM. The direct drive testing is performed with Mission Critical Systems (MCS) operational. Crew observations during the direct drive testing are used to monitor MCS response. Direct drive levels determined from the HEMP-induced residuals, increased by the DM required for the appropriate aircraft hardness level, are used.

A HEMP verification test consists of measuring the HEMP-induced stress, usually the common-mode core current on the cables, and comparing that to the immunity of the electronic boxes, usually determined by test CS116 of MIL-STD-461, which is a test consisting of a series of damped sine waves at different frequencies.

The hardness evaluation is done by first measuring the HEMP-induced stress on the cables at a box cable connector. Then the same interface is driven with a waveform that best matches the measured waveform to a level of the HEMP-induced current plus some specified DM. This is done while the aircraft is powered up and operating.

The data systems available today make it possible to do an HEMP verification test in 4–6 weeks as opposed to the 4–6 months required in the early days. And when the test is completed, the results are graded according to the now-existing HEMP standards, so the determination as to whether a given aircraft meets its design requirements or not is not subject to debate.

3.6.2.1 Horizontally Polarized HEMP Simulator

An overhead incident, horizontally polarized simulated HEMP is produced by a Horizontally Polarized Dipole (HPD) simulator like that shown in Figure 3.10.

Figure 3.10 The horizontally polarized HEMP simulator at Patuxent River, MD.

3.6.2.1.1 Symmetric Excitation In the HPD or Ellipticus simulators, the excitation is said to be symmetric excitation when the *E*-field vector is arriving from overhead and parallel to the fuselage. This orientation excites the symmetric resonant modes on the fuselage and wings such that there is a high charge density on the nose and tail, and the charge induced on the wingtips is in phase. This is often referred to as $E_{//}$ (i.e., E-parallel).

3.6.2.1.2 Antisymmetric Excitation The aircraft excitation is antisymmetric when the electric field vector E is perpendicular to the fuselage. This excitation drives current from wingtip to wingtip. There will be very little current on the centerline of the aircraft and the charge induced on the wingtips will be out of phase. This is often referred to as $\mathrm{E}_{\perp}$ (i.e., E-perpendicular).

3.6.2.2 Vertically Polarized Simulator

For completeness, the aircraft must also be excited with a vertically polarized wave because the HEMP can also have a strong vertical component. The aircraft may be placed with the nose toward the simulator (nose-on) as shown in Figure 3.11 or tail-on or wing-on. If there is time and sufficient interest, tests are conducted in several orientations.

3.6.2.3 Transient Data Acquisition Systems

A typical data van currently used for transient data collection is the Data Acquisition and Processing System (DAPS). It consists of a PC-based data acquisition system called E3DAS, which controls up to 16 LeCroy 7700, dual-channel digitizing oscilloscopes and 32 wideband fiber-optic links (FOLs).

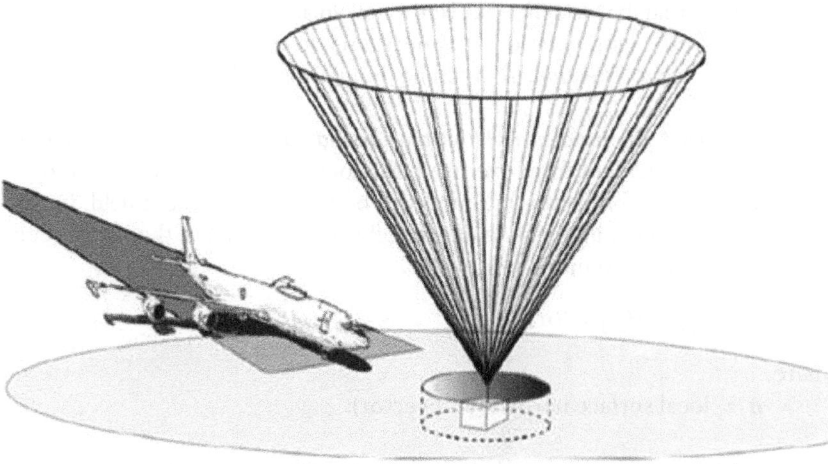

Figure 3.11 Vertically polarized HEMP simulator.

3.6.2.4 Transient Data Processing and Extrapolation to Threat

After the data are recorded, they are corrected for instrumentation effects using frequency-domain calibration data that have been loaded into the computer for each sensor and data link. The data are then extrapolated to the threat environment using the methods set forth by Dr. Carl Baum in Sensor and Simulation Note 222 [29].

The measured cable current is extrapolated to threat level in accordance with (3.11)

$$I_{\mathrm{HEMP}}(\omega) = \frac{I_{\mathrm{CW}}(\omega)}{E_{\mathrm{INC}}(\omega)} \cdot E_{\mathrm{HEMP}}(\omega) \tag{3.11}$$

where

$I_{\mathrm{CW}}(\omega)$ = cable current measured with the CW illumination
$E_{\mathrm{INC}}(\omega)$ = incident field from the illuminating antenna
$E_{\mathrm{HEMP}}(\omega)$ = threat-level HEMP specification from MIL-STD-464.

The extrapolated time-domain waveform is then simply the inverse Fourier Transform of $I_{\mathbf{HEMP}}(\omega)$.

There are several different ways to perform the extrapolation. One can use the field at the center of the parking pad or take an average of the fields over the working volume. The final answer, of course, will depend on what value is used for the incident field, $E_{\mathrm{inc}}(\omega)$. The HEMP waveform is an incident field without any ground reflection, so the simulator field map data must be the same.

3.6.2.5 Measuring the Incident Field: Horizontal Polarization

In order to do the extrapolation, it is necessary to measure the *incident field without the ground reflection*. This can be done in several ways.

3.6.2.5.1 Measure the Incident Field on the Ground Surface For a horizontally polarized wave, incident from overhead, the total magnetic field measured on the surface of the (conducting) ground will be twice the incident field. In this case, extraction of the incident field is straightforward. That is, the surface current density on the ground is

$$J_S = H_{inc} + H_{refl} = 2\boldsymbol{n} \times H_{inc} \tag{3.12}$$

where

$\boldsymbol{n}$ = local surface normal (unit vector).

This is true because $H_{inc} \approx H_{refl}$. Such measurements may be made on the surface of a concrete test pad even though the surface is not perfectly conducting. The reflection is not 100%, but the difference is only a few decibel.

3.6.2.5.2 Measuring E_{TOTAL} and H_{TOTAL} Simultaneously If we measure the E and H field components at the same point, we can then use this information to subtract out the ground-reflected field.

$$\begin{aligned} E_{total} &= E_{inc} + E_{refl} \\ H_{total} &= H_{inc} + H_{refl} \end{aligned} \tag{3.13}$$

Note that E_{refl} is negative. So we can then combine these to get

$$E_{inc} = 1/2\left(E_{total} + Z_0 H_{total}\right) = Z_0 H_{inc} \tag{3.14}$$

This works for the test point (TP) in the center of the test pad right under the balun and for the other TPs away from the balun, also. This works because, even though the arrival time changes and the ground bounce notch appears at a different frequency at each TP, both the *E* and the *H* fields arrive at the same time and have opposite polarity. So for the horizontal components, we can apply the formulas above to get rid of (or severely reduce) the notch at each TP. This is illustrated in Figure 3.12.

3.6.2.6 Measuring the Incident Field: Vertically Polarized Field

As in the case of the overhead incident, horizontally polarized field, it is necessary to remove the reflected fields from the measurement of the incident field in order to be able to accurately extrapolate the measured data.

If a reflection is coming from a long distance away such as the pole at the far end of the antenna, one way to remove the reflection is to transform the data into the time domain and truncate (time gate) the data file before the reflection

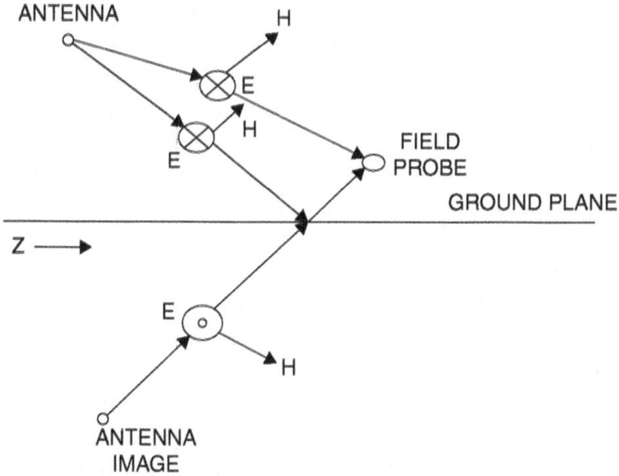

Figure 3.12 Fields away from the centerline of the antenna at oblique incidence.

appears. This can be seen by looking at Figure 3.13. This is described in Section 3.5.3.10 on data averaging.

3.6.3 Swept Frequency CW Measurements

Since there is currently only one HEMP test facility in the United States for fixed-wing aircraft, it is not possible to test every aircraft in every fleet. Therefore, in order to gain more knowledge of the hardness of an aircraft fleet, we

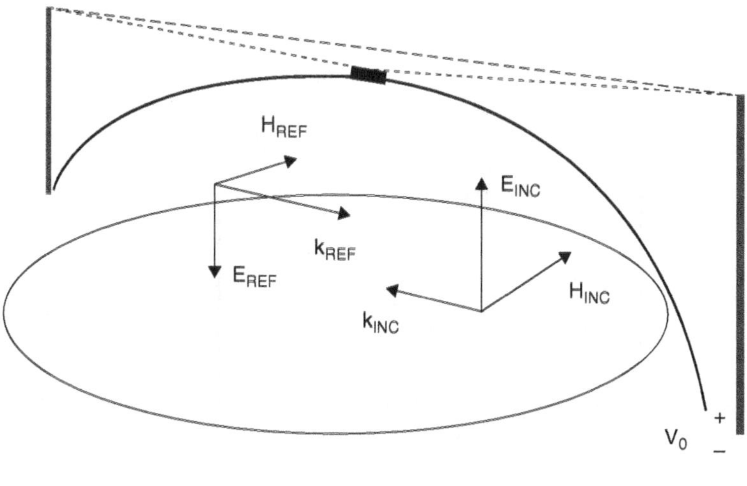

Figure 3.13 Incident and reflected fields in the Ellipticus CW illuminator, vertically polarized.

must reach out for other available tools and methods including analysis, low-level continuous wave (LLCW) testing, and PCI, which are discussed in the sections below.

Swept LLCW testing can be used as a first-level evaluation of the vulnerability of a system by measuring the currents and voltages induced on a test object as a function of frequency and then extrapolating and Fourier transforming the result to get a prediction of the time-domain waveform, the same that would be measured in a threat-level HEMP simulator. CW illumination testing, of course, does not replace threat-level transient testing, but it can be used to advantage in system design, prequalification, and HM/HS.

The measured CW transfer functions are scaled up to the threat-level HEMP spectrum, then transformed into the time-domain. These time-domain waveforms are a good prediction of the levels that will be seen in the high-level HEMP simulators. Recent side-by-side comparisons from tests run at Patuxent River have shown that for cable bundles and linear shields, the CW prediction is quite good.

LLCW tests of systems are used for several purposes:

1. System design and factory quality control
2. System-level shielding measurements once aircraft is fully assembled
3. Pretest evaluation before going into a threat-level HEMP test
4. HM/HS or reverification of the shielding after a major design modification.

The Ellipticus LLCW Illuminator shown in Figures 3.14 and 3.15 was designed to be a CW counterpart to the HPD HEMP simulator [21]. It was designed to be used for evaluating the HEMP hardness of shielded systems and performing HM/HS testing.

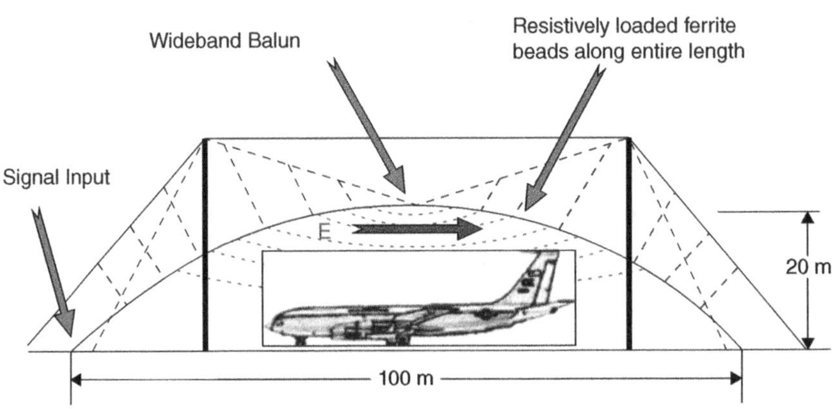

Figure 3.14 Ellipticus CW illuminator.

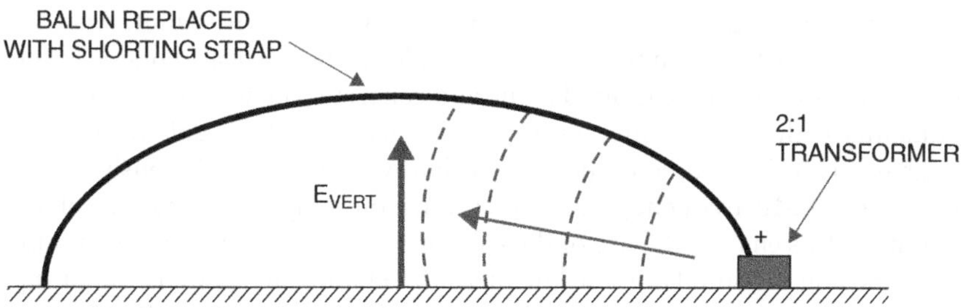

Figure 3.15 Ellipticus simulator in the vertical mode.

The Ellipticus CW radiator is relatively inexpensive, and the system response data have been shown to compare very favorably to the HPD and Vertically Polarized Bounded Wave (VPBW) simulators. The system has been modified to operate in both horizontal and vertical polarizations. The existing models are 20 m high in the center and 100 m from end to end. For testing larger aircraft, such as 747s and 767, a 30 m version is required.

The Ellipticus Swept CW simulator can be used for a first-level evaluation of the coupling to a system by measuring the currents and voltages induced on a test object as a function of frequency and then extrapolating and Fourier transforming the result to get a prediction of the time-domain waveform, the same that would be measured in a threat-level HEMP simulator. CW illumination testing, of course, does not replace threat-level transient testing, but it can be used to advantage in system design, prequalification, and HM/HS. The measured CW transfer functions are scaled up to the threat-level HEMP spectrum, and then transformed into the time-domain. These time-domain waveforms are a good prediction of the levels that will be seen in the high-voltage HEMP simulators. Side-by-side tests of aircraft have shown that the CW prediction of the penetration of the linear shielding is quite good.

The Ellipticus LLCW illuminator (LLCWI) was designed to produce the same field pattern and low-frequency wave impedance as the ATHAMAS II (HPD) HEMP simulator [21], which is why the comparison is so good. If an antenna with a different field distribution or impedance is used, the comparisons will suffer.

In recent years, aircraft CW testing capability has matured a great deal and has been demonstrated to provide an accurate measure of the shielding of a system. This experience has given valuable insight into ways to improve the approach to making HEMP measurements between 100 MHz and 1 GHz, where wavelength of the incident field ranges from 30 cm to 3 m. In this range, small features of the simulator and instrumentation system that could previously be ignored suddenly become important, such as the length and configuration of the instrumentation cables, the spacing between the ferrites on the antenna, and scattering objects near the vertical drive point. Another phenomenon that was immediately detectable in the vertical mode was the high frequency loss in the ground plane and concrete pad, which is impossible to quantify with transient measurements. We have also been able to see how valuable it is to have an antenna system that will sweep continuously across the entire HEMP frequency band, recording the amplitude and the phase accurately. This enables the use of the unwrapped phase and impulse response for diagnosing the simulator and the measurement quality as well as assuring a more accurate extrapolation to transient waveform such as HEMP [30].

3.6.3.1 The Ellipticus Antenna

The antenna is made from low-loss coaxial cable with a combination of ferrite and resistive loading on the outside. For horizontal polarization, the signal is fed through the coax from the bottom of one leg to a balun and radiating gap at the apex [41]. For vertical polarization, the outer shield of the cable is driven against the ground plane at one end with a matching transformer [42]. The antenna was designed to have the same resistive profile as the HPD HEMP Simulator (Designated by Carl Baum as ATHAMAS I) so the low-frequency fields in the working volume have a wave free space wave impedance of $Z_0 = 377\ \Omega$. With a 100 W amplifier, the system produces an electric field of at least 1 V/m in the working volume, which is enough to measure the wire and cable currents in a HEMP-hardened aircraft. (Most of the power is absorbed by the large number of ferrites and resistors on the antenna, which are necessary to generate the flat spectrum.)

3.6.3.2 Impedance-matching Transformers

For radiating in the horizontal polarization, the wideband, the low-inductance Quad Coaxial Balun (QCB) is installed at the apex of the antenna. The QCB is a remarkably ingenious device designed by Carl Baum and Gary Sower that employs 100 Ω solid-jacket cable, Moebius gaps, and wideband ferrite chokes. With the QCB, the Ellipticus operates seamlessly and spectrally flat ± 3 dB from 100 MHz to 1 GHz.

For vertical mode transmission, a 50–100 Ω matching transformer is installed on one end. The one we currently have uses the same design concepts as the QCB, and like the QCB, it is also spectrally flat ± 3 dB from 100 kHz to 1 GHz.

3.6.3.3 Ellipticus in Horizontal Polarization

When the system is operated in the horizontally polarized mode, the antenna gap is located at the apex and illuminates the aircraft from above in the same fashion as the HPD as shown in Fig. 3.14. The antenna wire, which is made out of coaxial cable, is loaded on its exterior with resistor/ferrite combinations that give it the same resistive loading profile as the HPD in order to achieve an E/H ratio of 377 Ω at low frequencies.

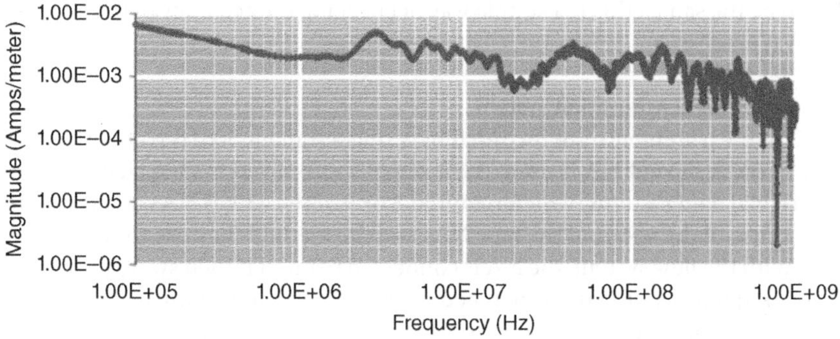

Figure 3.16 Magnetic field in center of working volume at $h = 3$ m, horizontal polarization with ground reflection removed.

3.6.3.4 Ellipticus in Vertical Polarization

The antenna can also be configured to operate in a vertically polarized mode by driving the outer shield of the coax against a ground plane using a 50–100 Ω wideband transformer mounted on the ground, as shown in Figure 3.15. The balun at the apex can be left in place or replaced by a shorting strap. We have found that the difference it makes is negligible.

3.6.3.5 Measured Simulator Fields

The measurements in Figures 3.16 and 3.17 show the measured field produced by the Ellipticus CW simulator. Figure 3.16 shows the H field measured at a height of 3 m with the ground reflection removed using the formula in (3.14). Figure 3.17 shows the H field measured on the ground (concrete pad), where $H_{meas} = 2\ H_{inc}$. Except for a slight difference in amplitude due to $1/r$, they

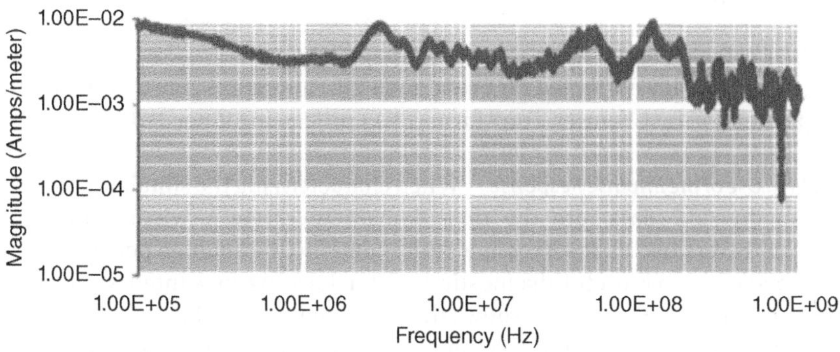

Figure 3.17 Magnetic field on the ground in center of the working volume at $h = 0$ m, horizontal polarization. $H_{meas} = 2\ H_{inc}$

should be the same. This is a simple way of checking the accuracy of the ground removal process.

3.6.3.6 CW Data Acquisition System

The CW measurements are recorded by a single Agilent 5061B Network Analyzer (NA). A second NA is used as the signal source. This is a big change from the original data systems, which required 8 or 9 of the 4396B NAs and is possible because of the speed of the new NAs and the method they use to record the data. With the new system, the NA is connected to one TP, then sweeps through all the frequencies. It is then switched to the next TP and so forth. Thus, with the new system, instead of having to have a bank of 9 NAs, only 2 are required along with as many coaxial switches as one would like, perhaps up to 32 or 64. The dwell time on each frequency is reduced from 300 ms to around 30 ms, and the total data acquisition time for 16 TP is reduced now to 10–15 minutes as opposed to 40 minutes required by the previous system.

The frequency steps are preprogrammed into the computer to radiate only on allowed frequencies in accordance with the local frequency permit. The software steps the system through up to 3000 frequency points in less than 15 minutes, depending on the receiver bandwidth used (typically 100 Hz). Instrumentation response variations and gain changes contributed by signal path components are removed automatically in real-time using stored frequency-domain calibration data for each component. Accurate phase data are acquired and corrected continuously to provide data quality suitable to be inverse-transformed into the time domain.

The Ellipticus antenna operates across the entire band with no changes in the antenna configuration, so the resulting phase is continuous and Fourier Transformable. It is not necessary to use a Hilbert Transform to reconstruct the phase as is done on other systems that have to use more than one antenna to cover the frequency range. As a result, the comparison with HPD aircraft cable current data is quite good.

3.6.3.7 CW Data Processing, Analysis, and Interpretation

The data are initially corrected for probe and FOL calibration. The data are then interpolated for Fast Fourier Transform (FFT) processing if needed.

3.6.3.7.1 Use of the Impulse Response When characterizing the simulator, transforming the CW data into the time domain to create a "virtual impulse response" is very useful. It serves as a sanity check on the antenna performance and provides a very useful diagnostic tool for locating unwanted scattering objects in the area. Figure 3.18 shows three different views of the electric field environment as measured at the volume center, at a height of $h = 3$ m, using an ACD-4, D-dot Free-field sensor with a DMB-3 balun. Figures 3.18a and b show the magnitude and phase of the measured field as a function of frequency. The

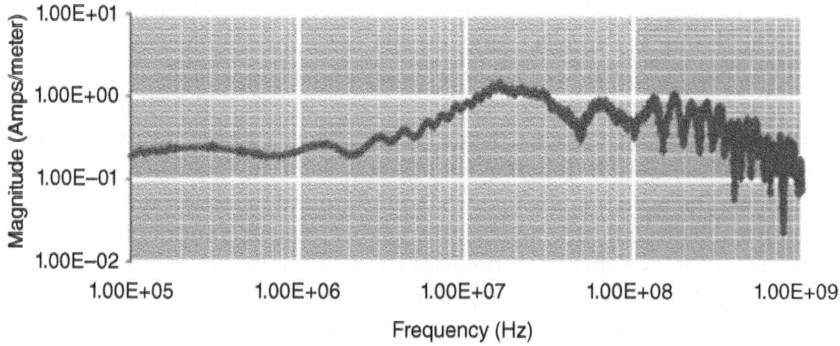

(a) Measured horizontal electric field.

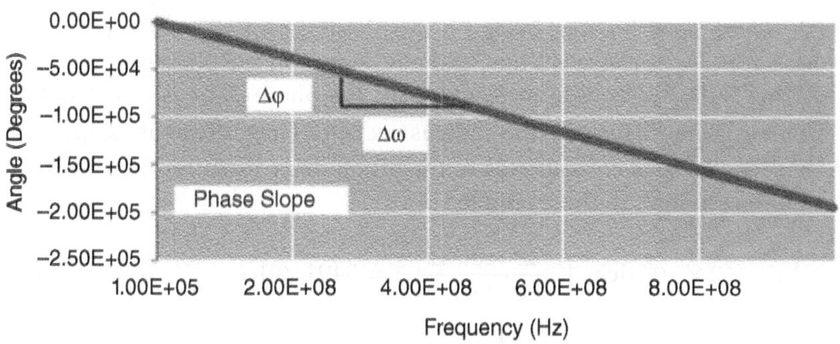

(b) Unwrapped phase on a linear scale.

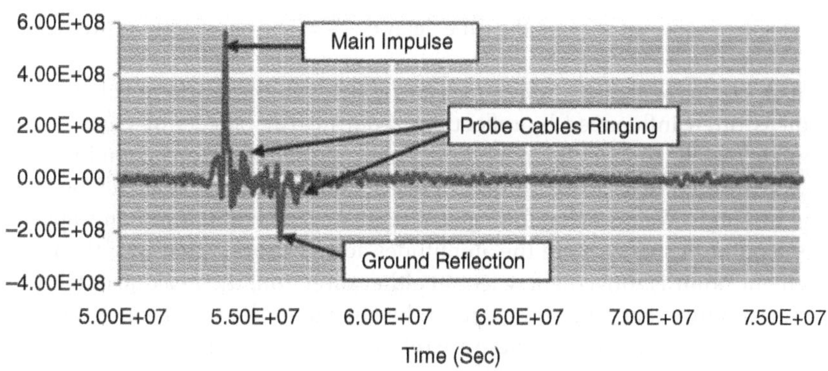

(c) Impulse response of electric field measurement.

Figure 3.18 Horizontal electric field in center of the working volume at $h = 3$ m.

frequency-domain data shown are the field at each frequency normalized to the incident field, so it is a transfer function and not a spectral density. Thus, the time-domain data shown here are in arbitrary units, but are sufficient to illustrate the use of the impulse response, and we can clearly see the initial impulse and the reflections in the data. Figure 3.18c shows a direct inverse Fourier Transform of the transfer function, which represents the impulse response. It clearly shows the main impulse followed by the ground reflection at 17 ns (6 m). Notice that the ground-reflected electric field is returning out of phase as we would expect. The smaller, high frequency reflections are a result of the field interacting with the fiber-optic transmitter case and the 24" long probe cable, something we do not see when testing only to 100 MHz. Nor will these details show up in the high-level HPD field measurements because of the higher noise level and much reduced dynamic range of the measurements.

3.6.3.7.2 Phase Slope and Time Delay If the phase is plotted on a linear scale, the phase appears as a straight line, and any change in the phase slope is easy to see. This is illustrated also in Figure 3.18. The phase slope is calculated as $\frac{\Delta\varphi}{\Delta\omega}$, where φ is in radians and ω in radians/sec. Therefore, the phase slope is in units of seconds. This corresponds to the group delay (turn-on time) seen in the impulse response. In this example,

$$\Delta\varphi = \left(2\times10^5\ \text{deg}\right)\left(\frac{2\pi\ \text{radians}}{360\ \text{degrees}}\right) = 3491\ \text{radians} \tag{3.15}$$

and

$$\Delta\omega = 2\pi\,\Delta f = 2\pi\times10^9\ \text{radians/sec} \tag{3.16}$$

In Figure 3.18c, we see that the time delay $= \frac{\Delta\varphi}{\Delta\omega} = 560$ ns, exactly as predicted.

3.6.3.7.3 Data Averaging When measuring fields in the working volume, the result is often influenced by reflections from nearby objects or parts of the simulator itself, such as the telephone poles. These reflections combine with the transmitted field at each point in space and cause a "spatially dispersed" peak or notch. We note that for a given frequency, each notch will only appear at one point in space where the amplitudes add or cancel. That is, if the measurement point is moved a few feet in either direction, the peak or notch will appear at a different frequency due to the change in the travel time from the reflecting object. As a result, these are referred to as "spatially dispersed" peaks or notches and can be removed by averaging the data across the space of the working volume.

Averaging the measured free-field data across the working volume has been found to be beneficial in two ways. First, it compensates for the nonuniformity of the fields in the working volume, which tend to become less at the outer edges. Since HEMP is a plane wave, we must compensate for this reduction in the field by simply performing an arithmetic average of the fields at all field map points. This was recommended by Dr. Carl Baum in Sensor and Simulation Note 222 on Extrapolation [29] for this reason. Secondly, we have found that when transmitting in vertical polarization, there is a noticeable amount of scatter from objects around the site, some of which cannot be moved, such as telephone poles and data trailers.

Figure 3.19 shows the vertical electric field measured at the near edge of the working volume at 3 m from the ground. Figure 3.19a clearly shows some wide notches beginning around 17 MHz due to the presence of the instrumentation van and several storage trailers (which could not be moved), located 20–35 m behind the launch point. There was also some scattering from the antenna support pole which was about 5 m behind the launch point. These objects

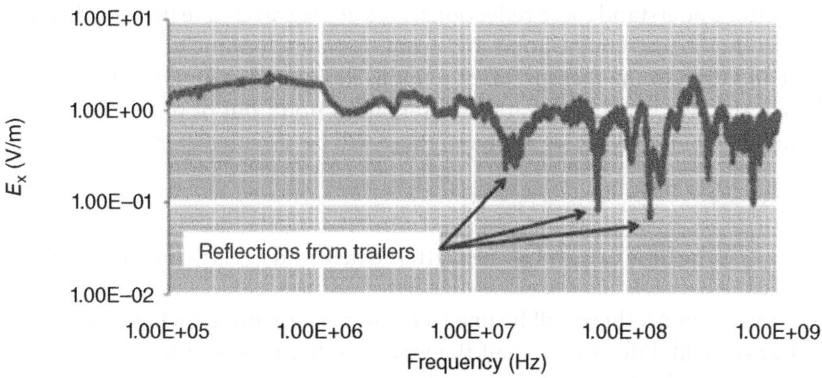

(a) Frequency sweep measurement showing reflections from nearby trailers.

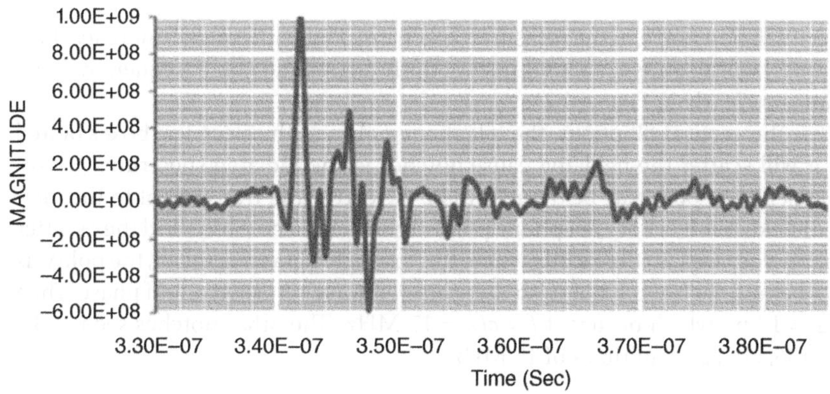

(b) Virtual impulse response.

Figure 3.19 Vertical electric field in center of working volume at height $h = 3$ m. Vertical polarization.

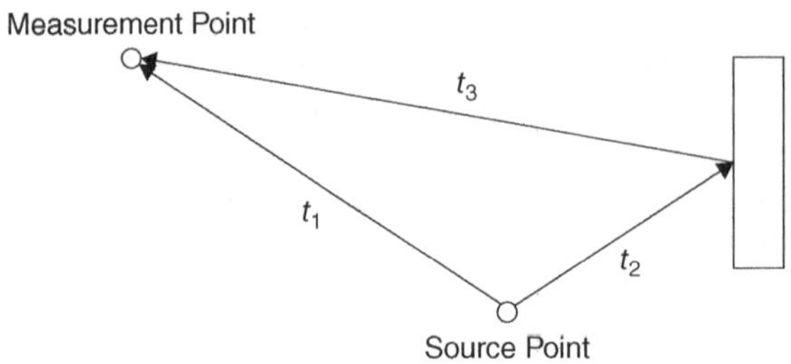

Figure 3.20 Clear time for a scattered signal.

had no noticeable effect in the horizontal mode, but were quite evident in the vertical mode. In Figure 3.19b, we can see this displayed in the time domain.

To better understand these reflections, look at the travel time from the launch point to the measurement point and from the launch point to the scattering object and then to the measurement point as illustrated in Figure 3.20.

When the two of these reach the observation point out of phase, there will be a notch; when they arrive in phase, there will be an enhancement.[7]

$$\Delta t = t_{\text{refl}} - t_{\text{inc}} = t_2 + t_3 - t_1 \tag{3.17}$$

where t_{refl} is the time of travel of the scattered field and t_{inc} is that of the incident field.

For any given Δt, there will be one frequency where there incident and scattered waves will differ by one half the wavelength and will cancel to form the notch. *This will be different for every TP.* That is, when

$$\Delta d = c\Delta t = \lambda/2 \tag{3.18}$$

there will be a notch in the measured spectrum. Higher frequency notches (harmonics) will also appear at multiples of the fundamental notch frequency. These show up clearly in Figure 3.21a.

For the vertical Ellipticus, the telephone pole and the electrical boxes bolted to it will scatter the field and cause notches in the radiated spectrum. A typical telephone pole may be thought to be dielectric and transparent to the incident field, but it is not. Telephone poles are soaked in creosote, and they scatter energy very well. In the facility where this example data were taken, the pole was 5 m behind the launch point, so the scattered wave returned out of phase when $\lambda/2 = 10$ m, which occurs at $f = c/\lambda = 17$ MHz. The other notches seen at 34, 67, 136, etc., are multiples of 17 MHz.

[7] Given that the speed of light ≈1 ft/ns, we can determine the time of flight using a tape measure.

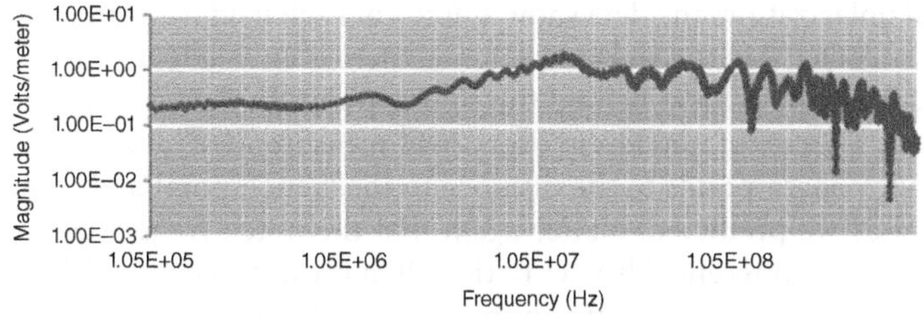

(a) Horizontally polarized E-field measured in the center of the working volume at h = 3 m.

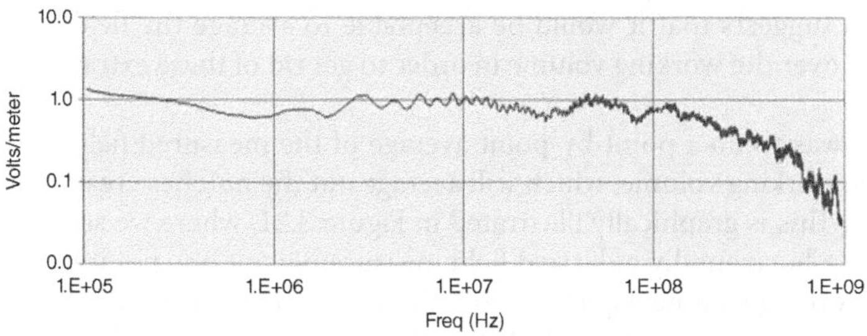

(b) Average of electric fields across the working volume along the antenna axis, horizontal polarization.

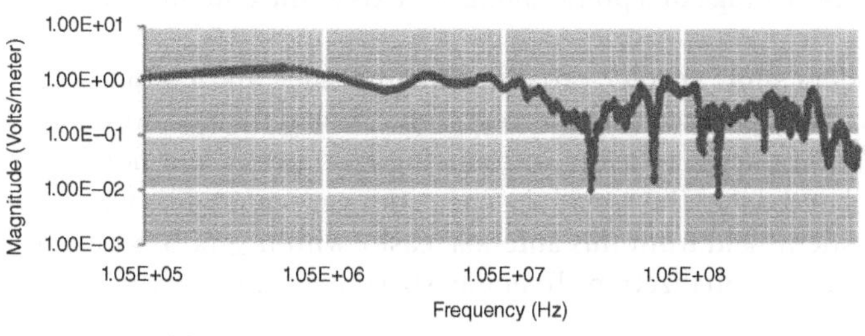

(c) Vertically polarized E-field in the center of the working volume at h = 3 m.

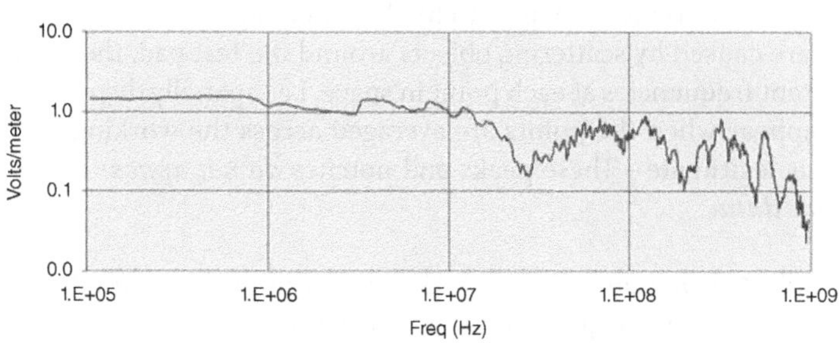

(d) Average of the vertical electric fields along a line parallel to the antenna through center of the WV, h = 3 m.

Figure 3.21 Illustration of the effect of spatially averaging on the spatially dispersed peaks and notches caused by reflections from outside the working volume.

In the vertical polarization, the data show a notch at each field measurement point. That is, for each TP in the working volume, there is an incident field and a reflected field from the trailers behind the source point. In the vertical field map data, there is a series of notches in every field measurement, and in each case, the frequency corresponds to the distance to the launch point and some scattering object as we predicted. Notice again that the notch frequencies at each point in space are different. Also, we noticed that these notches, which we see in the field map data, *do not* show up in the aircraft data. Why? Because the large metal fuselage of the aircraft has a spatial integrating effect across the working volume, the sharp notches do not appear in the aircraft cable current data. This suggests that it would be acceptable to average the field map measurements over the working volume in order to get rid of these extraneous notches.

The process was to do a point-by-point average of the measured field map data across the working volume, which will average out the notches created by the reflections. This is graphically illustrated in Figure 3.21, where we see first in Figure 3.21a a horizontally polarized field measurement at one point in the center of the working volume. Figure 3.21b is the average of 5 measured points along the axis of the antenna. Similarly, Figure 3.21c shows the vertically polarized field measured at one point in the center of the working volume, and Figure 3.21d shows the average of 5 points along the axis of the antenna (direction of current flow).

The averaged data show smaller variations in amplitude and no high peaks or deep notches. The measured cable currents, etc., should then be normalized to the *averaged field* curves to avoid introducing false peaks and notches in the final extrapolated data. The result is a spectrum very much like we would expect of the incident field from this antenna, easily within ± 6dB, which will disappear in the data normalization. To emphasize the meaning of this:

- If the notches are caused by resonances in the antenna, they will appear at the same frequencies at all TPs. These cannot be averaged out; they are there to stay. The only way to remove them is to fix the antenna.
- If the notches are caused by scattering objects around the test pad, they will appear at different frequencies at each point in space, i.e., spatially dispersed. These will disappear when the points are averaged across the working volume, and this is legitimate—These peaks and notches *do not appear in the internal aircraft data.*

3.6.3.7.4 CW-to-Pulse Comparison Figures 3.22 and 3.23 give some examples of the comparison between data acquired on the same aircraft, on the same TPs, with pulse and CW excitations [30]. For these figures,

- Blue = HPD measurement extrapolated to MIL-STD-464 HEMP.
- Red = Red = CW extrapolated to 464.
- Green = raw HPD measurement. Ignore this.

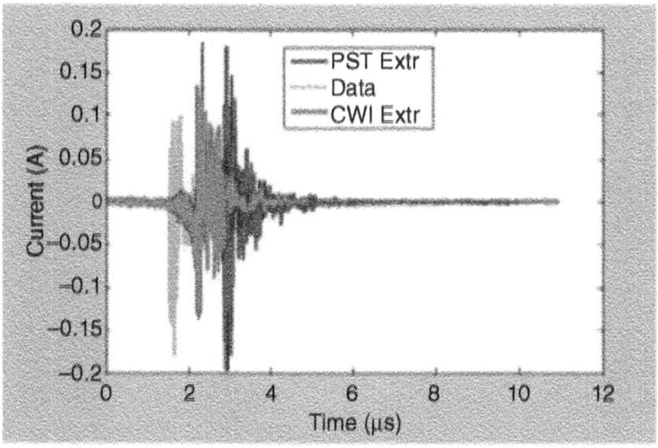

(a) Time-domain comparison.

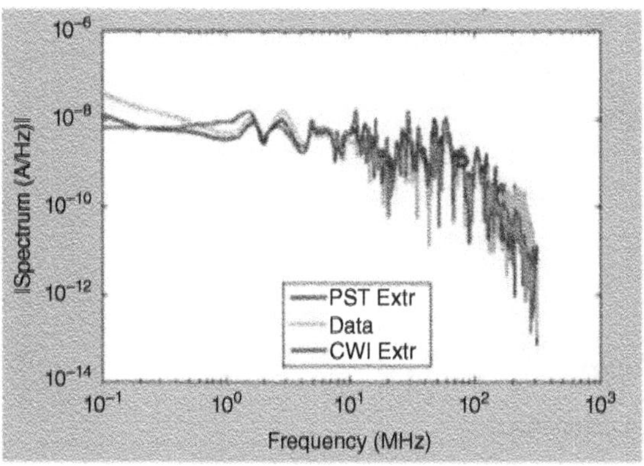

(b) Frequency-domain comparison.

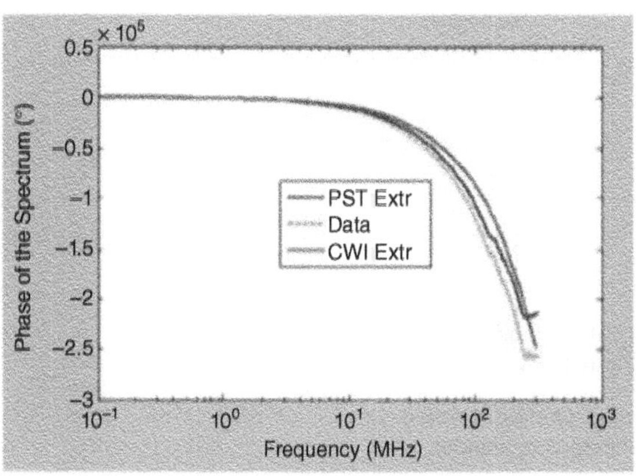

(c) Unwrapped phase comparison.

Figure 3.22 Example of a CW-to-pulse comparison where all of the signal was above the noise floor.

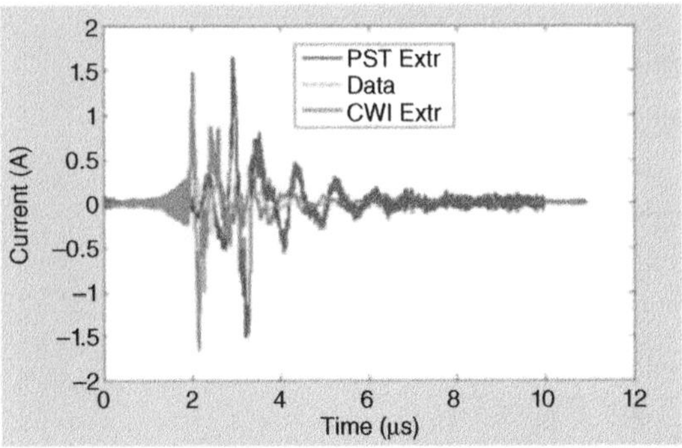

(a) Extrapolated time-domain comparison.

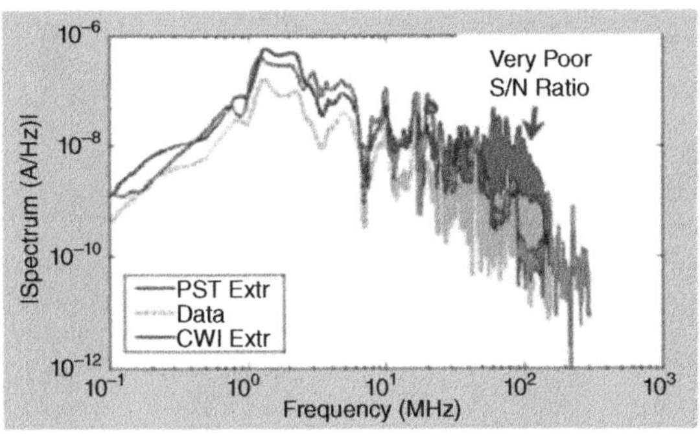

(b) Frequency-domain comparison.

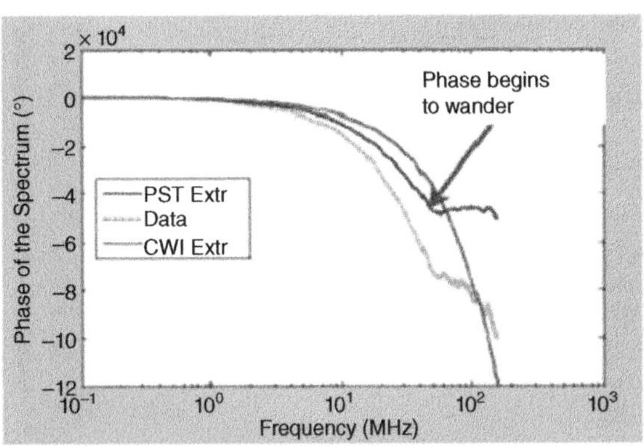

(c) Phase comparison showing where the phase of the time-domain signal begins to wander.

Figure 3.23 Comparison of CW-to-pulse measurements of a cable showing where the phase begins to wander, meaning the instrument has dropped below the noise floor.

Note that the blue trace has been offset by 1 μs for visual clarity. This delay is also evident in the phase data of Figure 3.22c.

The pulse data were measured with the aircraft in the HPD Simulator; the CW data were measured in the Ellipticus CW Illuminator. In both the cases, the aircraft was in the same configuration with the same instrumentation. The HPD was fitted with the new Navy pulser that has a rise time of about 2 ns, which corresponds to a maximum frequency content of around 500 MHz. The Ellipticus was sweeping to a full 1 GHz with a spectrum that is flat ± 3 dB over the entire range. For comparison, the measured cable currents from both the CW and HEMP simulators were extrapolated to MIL-STD-464.

3.6.3.7.5 Phase Response and the Noise Floor. Figure 3.23 is an example of a cable current that is dominated by the aircraft fuselage resonance, which creates a dynamic range problem for the higher frequency part of the measurement. As we can see, the frequency-domain data are very noisy above about 10 MHz, but it is almost impossible to determine from the frequency magnitude where the signal falls below the noise floor. However, if we look at the unwrapped phase, we get a clearer picture of what is happening. In Figure 3.23c, it can be seen that at around 50 MHz, the phase of the transient measurement begins to wander, which is an indication that the signal strength is getting close to the noise floor of the instrumentation. The measurement has some higher frequency content, but it is so small compared with the dominant resonant peak; the NA is having trouble measuring it. The dynamic range of the transient instrumentation is only about 32 dB. The CW system, on the other hand, which has a dynamic range of 120 dB, continues to record even the very small signal.

3.6.4 PCI Testing

PCI tests are conducted onboard the aircraft to establish the margin of protection over the coupled HEMP waveform. This is accomplished by driving the cable in the same location where the HEMP-induced current was measured with a waveform that represents the maximum norms of all of the possible waveforms that could be induced on that cable from an HEMP signal. That is, the amplitude of the various spectral components of the waveform may vary depending on the angle of incidence and polarization of the HEMP field rather than driving the cable x times with x different waveforms to cover all the possibilities. Figure 3.24 shows the direct drive probe and monitor probe hook-ups to the cable.

3.6.4.1 Current Transformers (Drive Probes)

For direct drive testing, the current is inductively coupled onto the cables with a toroidal current transformer or "drive probe." The drive probe is electrically the same as a receiver probe, but made more robust so they can be used to drive 10s

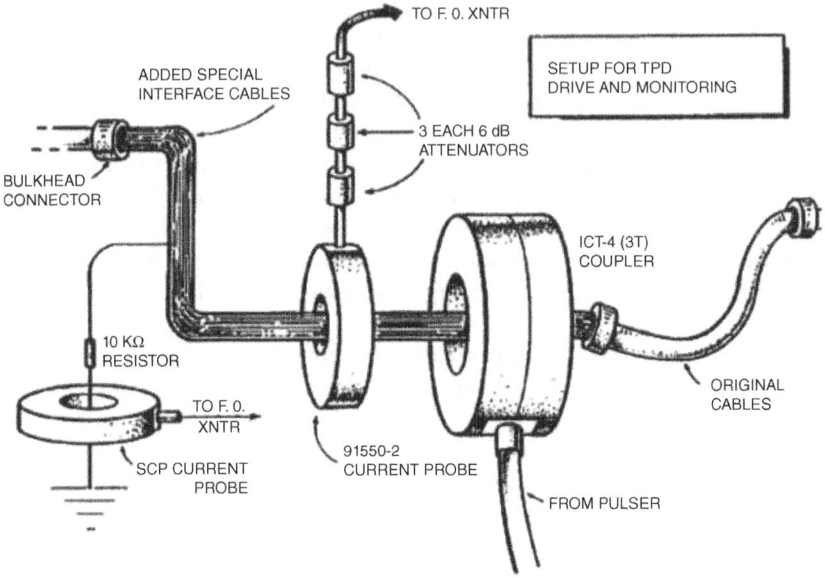

Figure 3.24 Typical cable direct drive configuration.

of amps in a repetitive mode. As a result, they will generally not operate over as broad a band as the receiver probes, and they take up more space. These drivers come in a variety of sizes, current ratings, and frequency ranges.

3.6.4.2 Waveform Generator

The PCI system uses an arbitrary waveform generator (AWG) and a power amplifier to achieve the desired waveform and a current coupler to drive the current onto the target cable. The AWGs generate the desired waveforms that are computed from norms of the data collected during the HEMP tests.

3.6.4.3 Power Amplifier

Power amplifiers used for the direct drive tests are typically 1 kW in CW mode and 1000 volts, 10 amps in pulse mode. There are usually two or more amplifiers in a direct drive system, so that they can be operated in series or parallel depending on the impedance and level required for the drive. When operated in cascade, a pair of amplifiers can usually deliver in excess of 2 kW in CW mode, and peaks of 1500 V, 15 A in the pulse mode into a 100 Ω load. The bandwidth of the amplifiers in the Patuxent River PCI system is from 1 kHz to 150 MHz. For tests to accompany the new generation of simulators, a broader bandwidth is needed.

Typically, these large power amplifiers require a pre-amp to drive them. The pre-amps develop 15 W, which is what is required to drive the power amplifiers to their maximum output. Maximum input to the preamps from the AWG is then 0 dBm at 50 Ω.

Alternatively, for low amplitude signals, amplifiers can be procured with a 50 W output and a bandwidth of 1 MHz–1 GHz. They can be used to drive low-level, high-frequency waveforms, and do not require a preamp.

3.6.4.4 Determining the Direct Drive Waveform

The equivalent waveform to be used for the PCI (i.e., direct drive) test may be derived in several ways. The main criterion is that for any given TP, the norms of the equivalent waveform must be greater than or equal to the worst case norms measured in various aircraft orientations. Norms considered for specification verification included the time-domain peak, the maximum rise time, and the RAI. The waveform measured at each connector in each orientation may be driven onto the equipment connector, or an equivalent damped cosine waveform can be computed that envelopes the maximum norms of all of the measurements for a given connector.

A damped sine waveform is described by

$$I(t) = I_0\,\mathrm{k}\,\mathrm{e}^{-\beta t}\sin(\omega_0 t) \tag{3.19}$$

where

I_0 = maximum amplitude
k = a constant.

The parameters that uniquely define the waveform are I_0 (the initial value of the exponential envelope at $t = 0$), the damping constant β (in radians/s), and the fundamental frequency $f_0 = 2\pi\,\omega_0$. The time-domain peak is given by

$$I_P = I_o\,\mathrm{e}^{-\frac{\beta}{4f_0}} \tag{3.20}$$

and the Fourier transform (spectral formula) is

$$I(f) = \frac{I_o\,\omega_0}{\left(\beta^2 + \omega_o^2 - \omega^2\right) + j2\beta\omega} \tag{3.21}$$

The magnitude of the spectrum can then be found from

$$|I(f)| = \frac{I_o\,\omega_0}{\sqrt{\left(\beta^2 + \omega_0^2 - \omega^2\right)^2 + (2\beta\omega)^2}} \tag{3.22}$$

Plots of the time-domain and frequency-domain waveforms are shown in Figure 3.25 for a damped sine wave with a moderate Q. The Q of the sine waves, usually in the range of 10 to 15 is similar to most waveforms measured inside an aircraft.

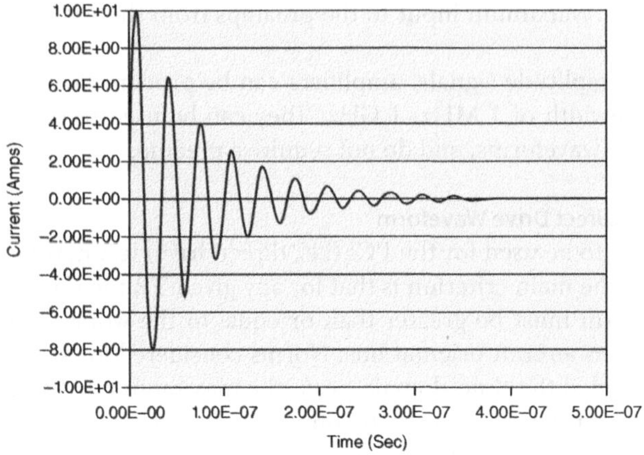

(a) Time domain waveform.

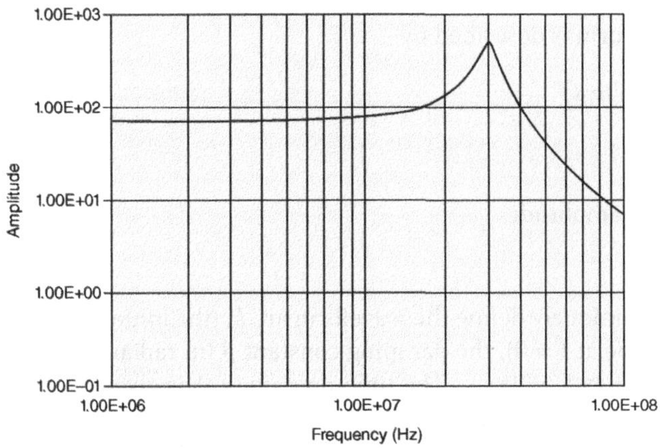

(b) Frequency domain waveform.

Figure 3.25 Example of a damped sine waveform.

3.6.5 Transient and CW Instrumentation

3.6.5.1 Current Probes

Current probes are used to measure the currents induced on a wire, cable bundle or other conductor inside the aircraft. The voltage output of the probe is proportional to the current input. The ratio is the transfer impedance of the probe, usually 1, 2, or 5 Ω. Current probes are made in a variety of sizes, impedances, frequency responses, and sensitivities.

$$V_{out} = I_0 Z_T \tag{3.23}$$

Figure 3.26 Snap-on current probes.

where

V_{out} = voltage out of the probe into 50 Ω
I_0 = current on the conductor under test
Z_T = probe transfer impedance.

Some typical sensors used in aircraft testing are shown in Figures 3.26 and 3.27.

Figure 3.27 Clip-on current probe.

3.6.5.1.1 Snap-on Current Probe (SCP) The SCP shown in Figure 3.26 is a wideband probe capable of measuring currents on conductors in an aircraft or other test object. This is a broadband probe with a response characteristic which is flat from below 100 kHz to above 100 MHz. It is a rugged, low-weight device which is easily installed over existing conductors due to its snap-on design. The SCP is designed to redirect output to a standard 50 Ω cable terminated in its characteristic impedance. A type N connector is standard; other connectors may be supplied on request. The SCP may have a transfer impedance or either 1 Ω or 5 Ω. It also comes with a 1, 2, or 3 cm aperture. It may also be ordered with a selection of frequency bands. The "STD" version is the standard probe. The "LF" version is designed for low-frequency measurements. The "HF" version is a special high frequency model which has a sensitivity that is within ± 6 dB of the specified value to above 1 GHz. The high frequency probe exhibits a pulse response with less than 0.5 ns rise time and less than 10% deviation from ideal response.

3.6.5.1.2 Clip-on Current Probe The Clip-on-Probe (CoP) is a small, broadband probe suitable for measuring current on small conductors inside a test object, where limited physical space is available. It is rugged, lightweight, and easy to install due to its clip-on design. It is designed to redirect output to a 50 Ω cable terminated. The output connector is a 3 mm SMA, as illustrated in Figure 3.27. The COP may be ordered with a transfer impedance of either a 1 Ω (COP-1) or 5 Ω (COP-5).

3.6.5.2 Voltage Probes

Voltage measurements on board require the use of a voltage probe with the proper physical and electrical characteristics. There are a number of acceptable versions available. Two of those commonly used in HEMP testing are shown here.

3.6.5.2.1 Voltage Pickoff Probe The Voltage pickoff probe (VPP), shown in Figure 3.28, is used to measure the voltage on a 50 Ω coaxial cable without perturbing the monitored signal. The output of the VPP is designed to drive a 50 Ω cable terminated in its characteristic impedance. The VPP is very useful for making Time-Domain Reflectometry (TDR) measurements. Both the incident and reflected voltage TDR signals can be accurately measured by the sensor.

3.6.5.2.2 van Lint Voltage Probe Measurement of EM coupling and electronics responses inside operating electronic assemblies during HEMP or HPM exposure requires probes that

- Do not load the circuit and change the response,
- Do not significantly alter the EM coupling by their presence, and
- Are sufficiently sensitive to measure the response at the frequencies of interest.

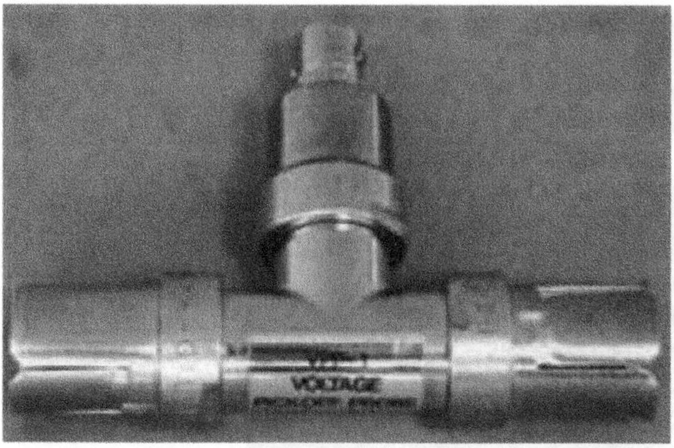

Figure 3.28 Voltage pickoff probe.

The voltage probe shown in Figure 3.29, designed by Victor van Lint, consists of a resistive voltage divider, and can be designed by suitably trading off between loading the circuit and obtaining enough sensitivity. In a typical probe, a 1 k Ω resistor is connected to the circuit node of interest. It is connected to a 50 Ω cable which is terminated at the fiber-optic transmitter. Solid-shield coax is used to shield the measurement from other fields in the vicinity. Such probes can be used to measure the voltages at nodes inside electronic assemblies up to frequencies of several GHz. The advantage of these probes is that they can be easily made to custom fit each application. They only need to be calibrated.

3.6.5.2.3 Breakout Boxes For bulk cable current measurements, the sensors are simply clipped around the cables. However, to measure the current on a wire or bundle inside a braided shield, one must use a breakout box or breakout cable in order to preserve the shielding. An example of one of these is shown in Figure 3.30.

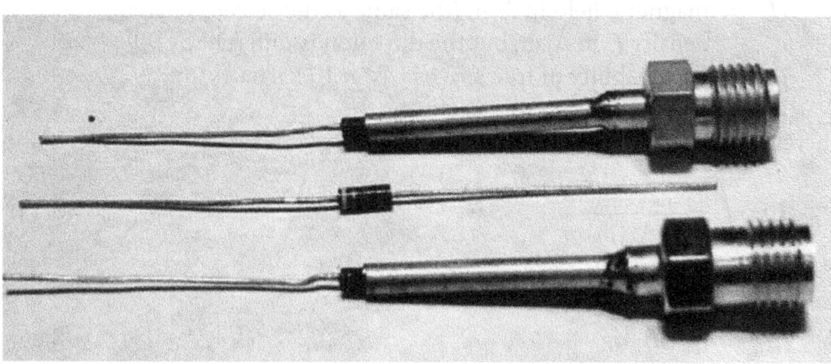

Figure 3.29 van Lint voltage probe.

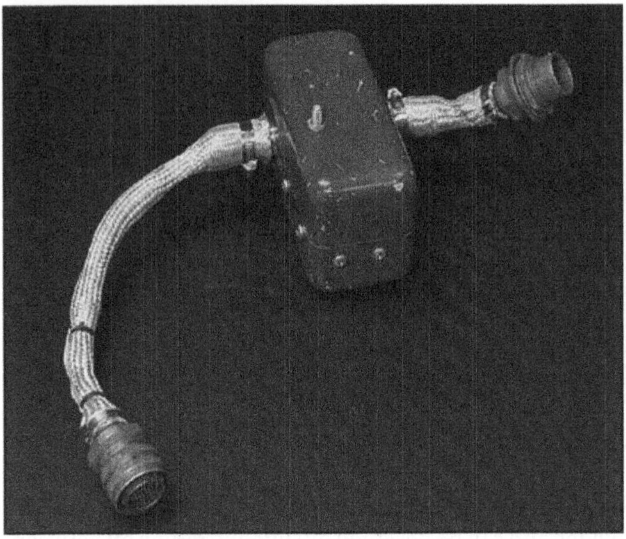

Figure 3.30 Typical break-out box.

3.6.5.2.4 Magnetic Field Sensors The output voltage from a magnetic field sensor is proportional to the time rate of change (derivative with respect to time) of the magnetic flux passing though the loops. Thus, they are often referred to as B-dot ($\dot{\mathbf{B}}$) sensors. The voltage produced by a B-dot sensor is given in MKS units by

$$V_{\text{out}} = A_{\text{eq}} \frac{dB}{dt} = A_{\text{eq}} \dot{\boldsymbol{B}} = A_{\text{eq}} \mu_0 \dot{\boldsymbol{H}} \tag{3.24}$$

where

$\boldsymbol{A}_{\text{eff}}$ = equivalent area of the sensor in m^2.
$\boldsymbol{B}$ = magnetic flux density in Teslas = Webers/m^2
$\boldsymbol{H}$ = magnetic field in A/m (the same as the surface current density J_s in A/m, but the direction is different by 90°)
μ_0 = permeability of free space = $4\pi \times 10^{-7}$ Henry/m.

Therefore,

$$\boldsymbol{H} = \int_0^T \frac{V_{\text{measured}}\, dt}{A_{\text{eq}} \mu_0} \tag{3.25}$$

and

$$V_{\text{out}} = A_{\text{eq}}\, d\boldsymbol{B}/dt \tag{3.26}$$

where

A_{eq} = effective area of the sensor, usually specified by the manufacturer in m^2

B = magnetic flux in Webers/m^2.

and

$$B = \mu_o H \tag{3.27}$$

where

μ_o = permeability of free space = $4\pi \times 10^{-7}$ Henry/m,

H = magnetic field in A/m.

On the surface, the surface current density $J = H$, but the two are at right angles to each other. When making a measurement, the MGL sensor should be taped down to the surface of the aircraft with conducting copper tape, but often this is not possible, so the sensor is simply placed on top of the paint, and the difference is usually small. A length of coaxial cable, usually no more than 2' long, is used to connect the sensor to a fiber-optic transmitter. The cable is usually loaded with ferrite beads to damp any disturbance introduced by the cable. The basic limiting factor of these types of sensors is their size, since the sensor must be electrically small in order for it to function properly. Figure 3.31 illustrates several different types of magnetic field sensors that can be employed in CW tests. Figure 3.31a shows the free-field B-dot sensors commonly used for measuring the incident field of a transient or CW antenna. The large MGL-1 sensor has a maximum frequency of 120 MHz and a rise time capability of about 3 ns. The smaller MGL-6 sensor has a maximum frequency of about 1.7 GHz and a rise time capability of about 0.5 ns. Figure 3.31b shows some examples of surface-mount B-dot sensor (half-loop over a ground plane) used to measure the B field (surface current density) on a ground plane or on the surface of an aircraft.

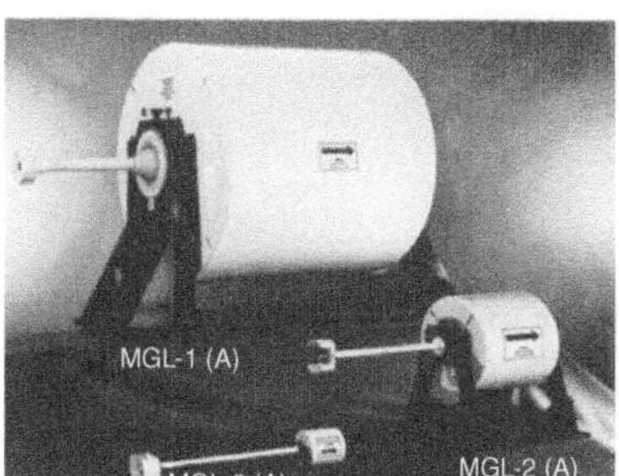

(a) Free field sensors.

(b) Surface field sensors.

Figure 3.31 Magnetic field sensors.

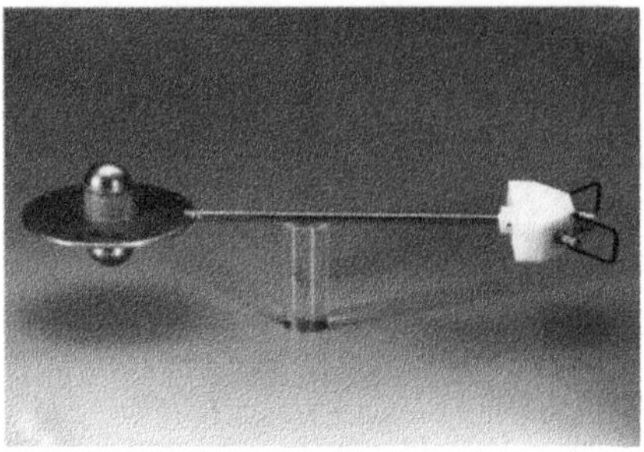

(a) Free field sensor

(b) Surface field sensor

Figure 3.32 ACD electric field sensors.

3.6.5.2.5 Electric Field Sensors Figure 3.32 shows several sensors for measuring the E-field. The sensors shown with the spherical shape are hollow spherical dipole (HSD) sensors. The voltage output from these sensors is proportional to the time rate of change of the electric flux D. The voltage produced by a D-dot sensor is given by

$$\begin{aligned} V &= RA_{eq}dD/dt = R\varepsilon_o A_{eq}dE/dt \\ &= h_{eff}dD/dt = h_{eff}\varepsilon_o \text{E-dot} \end{aligned} \tag{3.28}$$

or

$$I = A_{eq}dD/dt = A_{eq}\text{D-dot} = A_{eq}\varepsilon_o\text{E-dot} \tag{3.29}$$

where

h_{eff} = effective height of the sensor in m.
A_{eq} = equivalent area of the sensor in m^2
R = characteristic load impedance = 100 Ω for free-field sensors or 50 Ω for ground planes

Note that

$$D = \varepsilon_o E \tag{3.30}$$

where

D = electric displacement or flux density in coulombs/m^2,
E = electric field strength in V/m,
ε_o = permittivity of free space = $1/36\,\pi \times 10^{-9}$ Farads/m.

$$E = \int_0^T \frac{V_{measured}dt}{h_{eff}\varepsilon_0} = \int_0^T \frac{V_{measured}dt}{R\,A_{eq}\varepsilon_0} \tag{3.31}$$

3.6.5.2.6 ***Baluns*** When making free-field measurements, a differential-mode balun is required to transform the balanced 100 Ω signal from a differential (free-space) field sensor to an unbalanced 50 Ω signal for input to a 50 Ω coaxial cable, oscilloscope, or FOL. It converts the input signal of V_0 across 100 Ω to an output signal of $V_0/2$ across 50 Ω. Thus, the insertion loss is 6 dB. Differential baluns may come in a variety of sizes and frequency response characteristics to match the field sensors being used.

3.6.5.2.7 ***Integrators*** Since the data output from the field sensors is the time derivative of the incident field, it must be integrated in order to get the time-domain waveform. This may be done numerically in data processing, or it may be done using either a passive or an active hardware integrator.

Passive Integrators

Depending on the signal strength and frequency band being measured, a passive integrator may be used. These are commercially available with standard RC time constants of 1, 5, 10, or 100 μs. The transfer function of a passive integrator is

$$\frac{V_{\mathrm{out}}(s)}{V_{\mathrm{in}}(s)} = \frac{1}{1 + sRC} \tag{3.32}$$

where $s = j\omega$ = the Laplace operator.

This device will integrate the signal from a sinusoidal voltage, if the frequency is large compared with $1/(2\pi RC)$ or for transient voltages, where the pulse width is small compared with RC. If a field sensor is used with a passive integrator, the conversion factor should be multiplied by the integrator time constant (10^{-6} for a 1 μs integrator or 10^{-5} for a 10 μs integrator.

Active Integrators

Active integrators are commercially available with 1 and 10 μs time constants. It has a high impedance unity gain amplifier built in immediately after the RC integrator circuit to provide an output impedance of 50 Ω. Nowadays, these are usually built into the fiber-optic data systems, so all one has to do is switch them on and set the time constant τ and the gain.

$$V_{out}(t) = \frac{1}{\tau}\int_0^t e^{\frac{(t'-t)}{\tau}} \cdot V_{in}(t')\,dt' \tag{3.33}$$

where $\tau = 1$ μs. For $t \ll \tau$, then one has to a good approximation

$$V_{out}(t) = \int_0^t V_{in}(t')\,dt' \tag{3.34}$$

i.e., V_{out} is the time integral of V_{in}. For times larger than or comparable to τ, one has to use the more complicated relationship in the previous equation.

3.6.5.2.8 Fiber-Optic Links The signal transmission links for the reference and measurement sensors should not violate the shield topology surrounding the measurement equipment. A common way of insuring that the shielding is maintained is to use FOLs. This requires a conversion of the electrical signals at the sensors to optical signals by means of a fiber-optic transmitter, the transmission of the optical signals via an optical cable, and the reconstitution of the electrical signal within the equipment enclosure by a fiber-optic receiver. The units currently used for CW and HEMP tests have a bandwidth of 10 kHz to 1400 MHz with built-in integrators and automated control from the central data computer.

In order to expand the data acquisition capability, coaxial switches are often used. These will connect as many as 7 instrumentation cables to one FOL as illustrated in Figure 3.33. This saves a great deal of time when measuring a large test object, allowing the tester to remotely switch from one probe to the

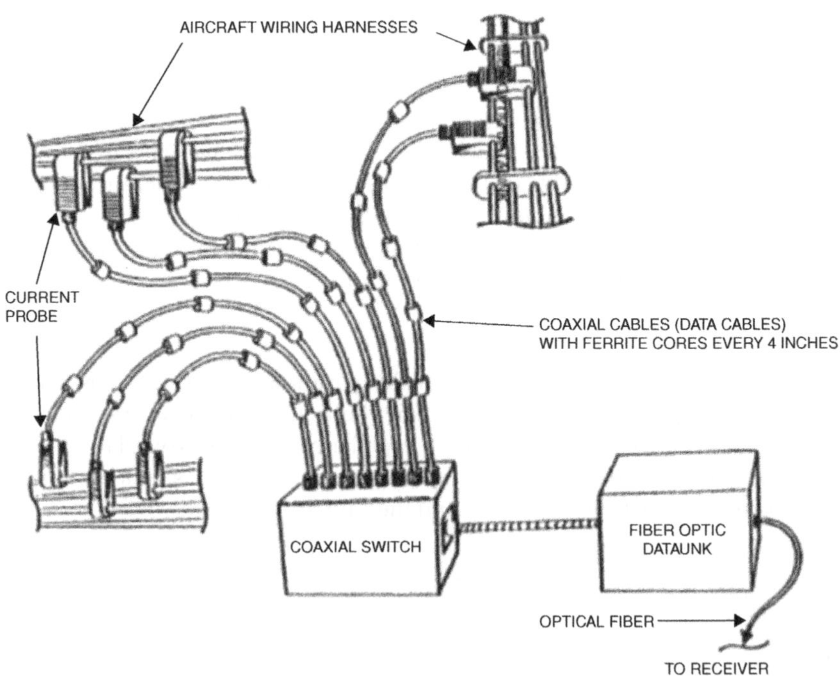

Figure 3.33 7-to-1 coaxial switch driving a single fiber-optic link.

other without a technician having to open up panels on the test object to move probes.

3.7 Hardening Component Specifications and Measurements

For a shielding specification to be effective, it must be written in real physical units that are measurable and that relate in an unambiguous way to the attenuation of EM fields on the object. (That is, merely specifying that an aircraft fuselage is to have X dB of attenuation is ambiguous and subject to interpretation by the builder.) Thus, shielding component specifications are best represented by their transfer impedance or transfer admittance as a function of frequency, depending on whether they are driven by magnetic field (surface current) or electric field (surface charge). Transfer impedance $Z_T(\omega)$ will in general be a complex quantity with both resistive and inductive parts.

$$Z_T(\omega) = \frac{V(\omega)}{I(\omega)} = R_T + j\omega L_T \tag{3.35}$$

where

$V(\omega)$ is some voltage on the interior of the shield
$I(\omega)$ is a current on the exterior
R_T is the resistive component
L_T is the inductive component

If a panel joint or gasket is making good electrical contact along its entire length, the inductance will be zero and only a resistance will be present. If, however, part of the gasket is not making good contact, there will be an inductive component as well, which indicates that there is essentially a hole in the shield, and high frequency fields can leak through. This is why there is so much emphasis on peripheral bonding of shields and elimination of pigtails on cables.

3.7.1 Seams, Joints, and Gaskets

Seams, joints, and gaskets, including the skin panel joints and the RF gaskets that are used on windows, doors, and hatches, can be characterized in terms of their transfer impedance. The voltage across a joint at a certain location divided by the current flowing across the joint at that point on the outside will yield a transfer function in units of impedance (the transfer impedance). The transfer impedance is the ratio of the internal voltage across the gasket to the external surface current density.

$$Z'_{\mathrm{T}}(\omega) = V(\omega)/J(\omega) = l \cdot V(\omega)/I(\omega) = R' + j\omega L' \text{ in } \Omega \cdot \mathrm{m} \qquad (3.36)$$

where the dimension l is measured transverse to the direction of current flow in m

R' is the resistance in ohm m
L' is the inductance in Henry m.

The resistance appears as the low-frequency portion of a measured curve. The inductance is the slope of the curve between 1 and 10 MHz.

$$L' = \Delta Z'/2\pi\Delta f \text{ in } \Omega\cdot\mathrm{m}\cdot\mathrm{sec (Henry}\cdot\mathrm{m)} \qquad (3.37)$$

where $\Delta Z'$ is the change in the transfer impedance over some interval of frequency Δf chosen to be read in the linear portion of the inductive slope.

If a seam is of finite-length, like a slot, we will know the total current flowing across it. Therefore, Z_{T} can be expressed in units of ohms. If, however, the measurement is being made at only one point along a seam, then while the voltage measured is definite, the current density across the slot must be expressed in amps-per-unit-length, where the length l is along the seam transverse to the flow of current. Then, the current is expressed in units of A·m. R is similarly expressed in units of Ω·m, L in units of Henry·m, and Z_{SEAM} (ω) in units of Ω·m [31–34].

Seams and joints on an aircraft include the skin panel joints, but more importantly, the RF gaskets that are used on windows, doors, and hatches. If the seams or joints are not making good electrical contract across the gap, then EM energy can leak through the joint and induce currents on the interior wires.

The transfer impedance can be easily measured in the laboratory using, for example, a small parallel plate transmission line to drive current across the seam and a voltage probe to measure the induced voltage across the aperture on the other side [35, 36]. Measurements can also be made on board the aircraft, especially during assembly before paint and surface coatings have been applied.

An example measurement of a door gasket is shown in Figure 3.34. The measured data have been divided by the length of the side of the door to convert the units into ohms. Note that the circumferential resonance of the door appears clearly as the dip in the curve at 75 MHz where the wavelength equals one circumference of the 4' × 6' passenger door. The inductance measurement is only valid below about 30 MHz, where the impedance begins to show the beginnings of the resonance. The resistance is about 30 mΩ, and the inductance is around 26 nH, which indicates a gasket that is making good peripheral contact.

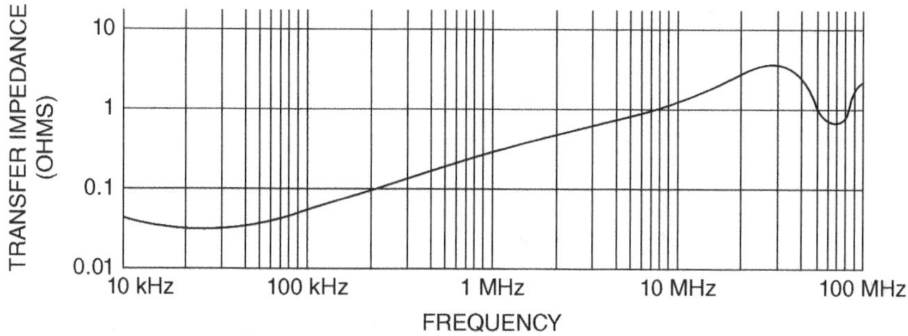

Figure 3.34 Example transfer impedance measurement of an RF gasket on an aircraft door.

3.7.2 Apertures

The transfer impedance of an aperture is represented in the same way as that of a slot or gasket, but with slightly different units. When an aperture is illuminated by an EM wave, the induced surface current on the outside induces a voltage across the aperture on the inside as illustrated in Figure 3.35. Therefore, the shielding specification for an aperture can be written in terms of Z_A (ω), the ratio of these two quantities as a function of frequency. Thus, by driving the aperture with a transmission line, as illustrated in Figure 3.36, and simultaneously measuring the current in the transmission line and the induced voltage across the aperture, one can compute the Z_A (ω) in ohms as described in (3.38), [35–36].

$$Z_A(\omega) = \frac{V_A(\omega)}{I_A(\omega)} = R_A + j\omega L_A \tag{3.38}$$

where V_A is the voltage measured across the aperture on the inside and I_A is the total short-circuit current flowing across the aperture.

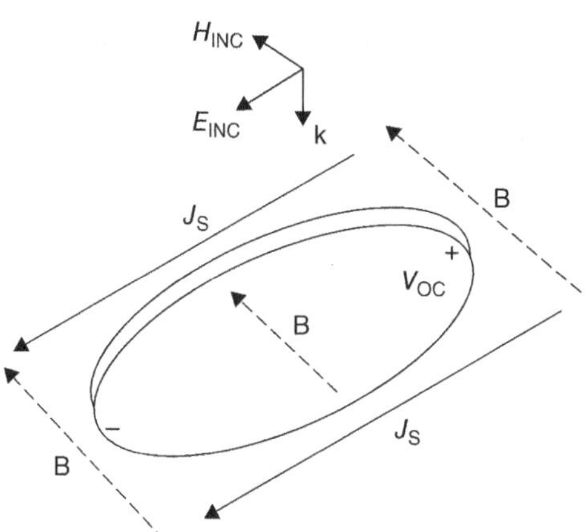

Figure 3.35 Aperture penetration illustrated.

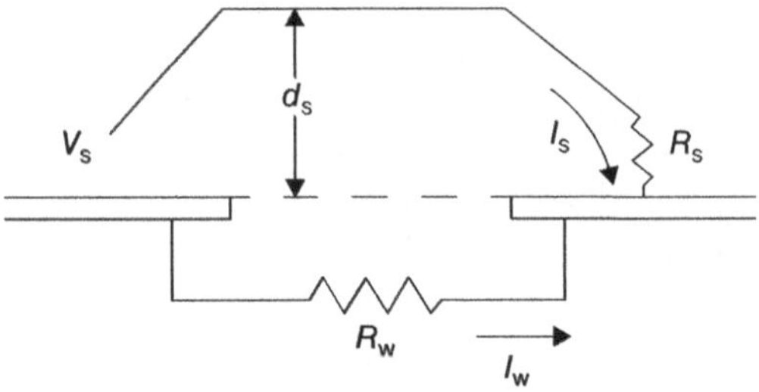

Figure 3.36 Aperture tester showing the transmission line driver and the voltage probe.

Sometimes it is easier to understand the behavior of a screened aperture, if we construct an equivalent circuit. For example;

(a) The open aperture is purely an inductance L_A whose value is determined by the size and shape of the opening. That is,

$$Z_A(\omega) = j\omega L_A \text{ in } \Omega \tag{3.39}$$

(b) The screen mesh is specified in terms of its surface transfer impedance in Ω/square [33–36].

$$Z_{SCR}(\omega) = R_{SCR} + j\omega L_{SCR} \text{ in } \Omega/\text{square} \tag{3.40}$$

This can be measured in the laboratory, but it is not usually measurable on the aircraft. However, that is not a hindrance, because the screen itself does not change with time. It is the RF gasket around the screen that is of interest, because that is what will degrade over time.

To get the surface transfer impedance of the entire window, we put these all together as shown in the equivalent circuit in Figure 3.37. Note that these quantities are specific to the direction of surface current flow. The joints and the screen in the direction of current flow combine in series. The voltage is

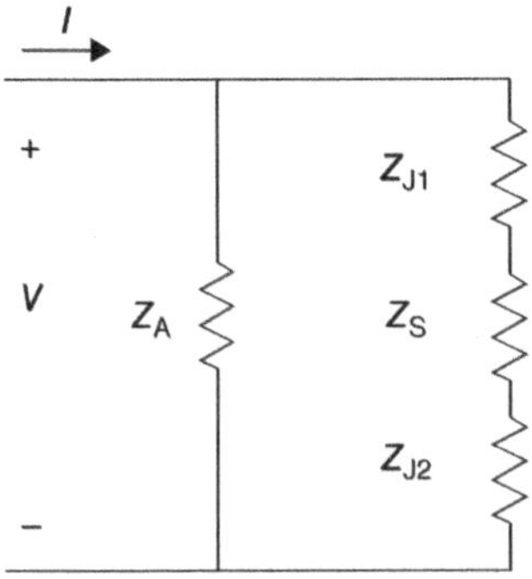

Figure 3.37 Screened aperture equivalent circuit.

common to these elements and the aperture inductance, so they combine in parallel. Thus,

$$Z_W(\omega) = Z_A(\omega)//\left[R_{J1} + GFZ_S(\omega) + R_{J2}\right] \text{ in } \Omega \tag{3.41}$$

where GF is a geometrical factor related to the shape of the aperture. That is, the slope of the Z_T curve is

$$L = \Delta Z_T/2\pi\Delta f \text{ in } \Omega\text{-sec (Henries)} \tag{3.42}$$

The aperture tester consists of a low-impedance transmission line placed on the outside of the aircraft window or door and a voltage probe connected across the aperture on the inside as illustrated in Figure 3.38. It can be used effectively to measure the resistance and inductance of window screens, window blanks, and doors with their associated gaskets. It can also be used to measure the $Z_T(\omega)$ of seams and joints in units of Ω/m, where the units of m are measured along the seam lateral to the flow of current [35, 36].

Apertures in an aircraft hull include window, personnel doors, and equipment bay doors. Hardening for windows is usually accomplished by covering them with screen mesh that is circumferentially bonded to the aircraft skin with an RF gasket. Doors and hatches are similarly connected to the hull with an RF gasket.

Figure 3.38 Measuring the transfer impedance of an over-wing hatch with the Aperture Tester.

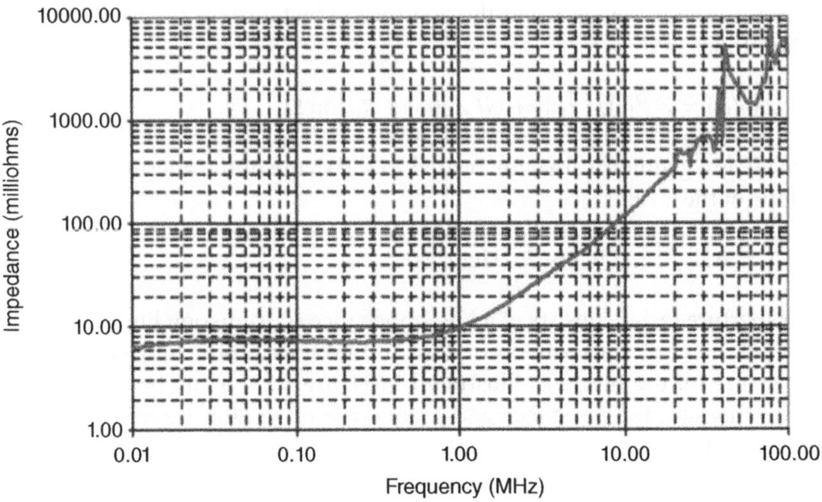

Figure 3.39 Measured transfer impedance of a screened aperture.

A typical measurement of the transfer impedance of a screened window on an aircraft is shown in Figure 3.39. The flat part of the curve at low frequencies is the series combination of the resistance of the gasket on each side of the window and that of the screen. The inductive part is a measure of the inductance of the screen itself and any unwanted gaps in the RF gasket. It is noted here that this apparatus yields a good, repeatable measurement. For large apertures, it is simply not possible to make an extremely accurate measurement of the inductance. The test method was designed to be used on board and aircraft to make a quality control measurement and to monitor the degradation of the aperture shielding over time, and for this purpose, it is quite adequate.

This can be measured in the laboratory or on board using the aircraft using the aperture tester shown in Figure 3.38. This consists of a lightweight transmission line to drive current on the outside surface and a voltage probe that measures the induced voltage across the aperture on the inside [35, 36]. The ratio of these two quantities is the transfer impedance.

Figure 3.40 shows the transfer impedance of a screened aircraft window in three configurations in order to show the range of variation from open to closed configuration. In the open configuration, we measured an inductance of 32 nH, which is reasonably close to the calculated value. Notice there is no resistance showing, only the inductive rise. The screened window measures 1.5 nH, not as small as it should be, but the window did not have a peripheral RF gasket, only spring clips holding it in. This too shows only an inductive rise and no resistance. The third curve shows the window covered with a metal cover (slug) with an RF gasket. As expected, all we see is the resistance of the gasket. Finally, in

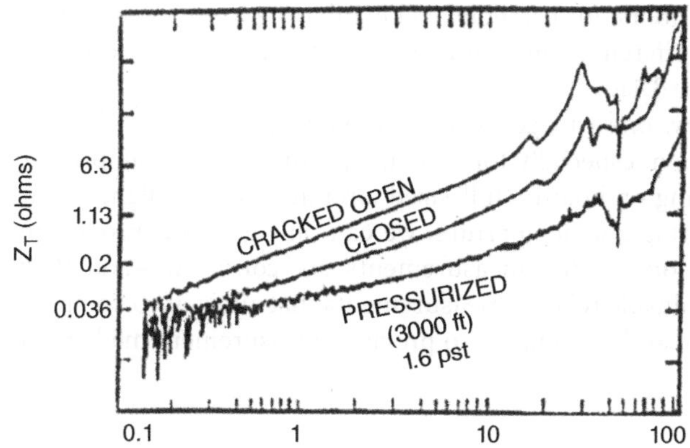

Figure 3.40 Z_T measurement of a 4′ × 6′ aircraft passenger door with RF gaskets.

Figure 3.41, we see a measurement of the Z_T of a screened window with a proper RF gasket. At low frequencies, the resistance of the RF gasket at about 7 mΩ and then beginning about 1 MHz, the rising curve indicates the inductance of the screen mesh. At around 40 MHz, the resonance of the fixture begins to distort the measurement and we cut it off. However, we need only go high enough in frequency to be able to clearly see the slope of the inductive curve, which will give us the measure of the aperture inductance that we are looking for.

The measured data from a door or hatch will have the same general shape as that of a window except it is not possible to go quite as high in frequency

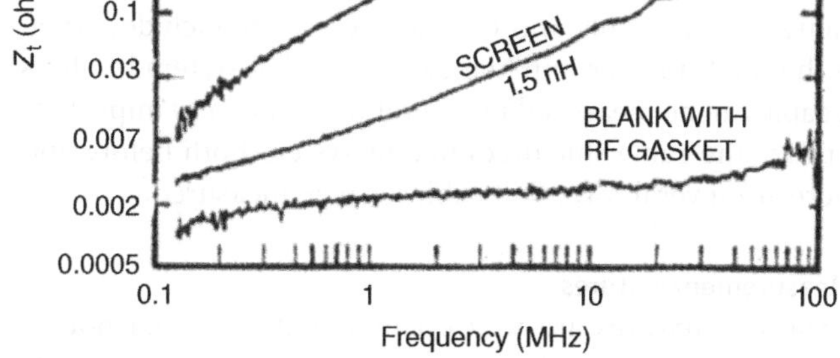

Figure 3.41 Z_T measurement of a 9″ × 12″ screened aircraft window.

because the passenger doors, for example, will exhibit their fundamental resonance (λ = one circumference) beginning around 20 MHz as shown in the measured data of Figure 3.41.

Note: In a laboratory, one can get a measurement that is quite close to the theoretical prediction, especially for a small aperture. That is simply not possible when measuring on board a full-sized aircraft. With the lightweight transmission line covering a large aperture, we can only get an approximate measurement, but as long as the measurements are consistent—and they are—we can get very valuable relative measurements that allow us to compare approximately to a spec and to compare to previous measurement made with the same setup.

3.7.3 Shielded Cables

On many aircraft, shielded cables serve as the primary defense against EMI or, if there is hull hardening, they form the second layer of protection, depending on the design of the system. They also provide grounding for the equipment and an alternate path to ground for lightning-induced currents. In like manner, overbraid shielding or conduits in the unpressurized areas or the aircraft contribute to meeting all of the hazardous EM requirements by *becoming a part of the shielding topology of the hull.* Shielded outboard wire bundles with their shields peripherally terminated at the hull significantly increase the effectiveness of hull hardening and often reduce cost and weight by reducing the need for terminal protection modules (TPMs) (i.e., filters and terminal protection devices (TPDs)).

The braided shield on a cable, including the connector joints and backshells, is also described by a resistance and inductance per unit length. In this case, the measurement of length is along the cable in the same direction as the current flows, so Z_T is expressed in ohms/m and Henries/m.

$$Z_T(\omega) = \frac{V_c(\omega)}{I_c(\omega)} = R_c + j\omega L_c \tag{3.43}$$

Thus, if the total length of the cable and the connectors are included, then the units will reflect the total transfer impedance of the cable system in ohms [37–40]. The Z_T of a cable can be measured in several ways. It is very important to be able to characterize the cable shield/connector system both before and after installation. There are several ways that cables can be measured.

3.7.3.1 Laboratory Measurement Fixtures

Triaxial and quadraxial test fixtures are well documented and need not be described here. Their limitation is that they can only measure a short sample of a cable design, perhaps 2 m, so they are used for verification of cable and connector designs. In order to measure full-length shielded cables as installed on board an aircraft, we must use a different instrument, either a LRT or a CST as described below.

3.7.3.1.1 ***Loop Resistance Tester*** The LRT is becoming a standard tool for measuring the resistance of the joints in shielded cables and connectors on board the aircraft. Originally built by Boeing and British Aerospace for maintenance on the Boeing 777, the LRT measures the resistance of a cable assembly by driving a low-frequency (quasi-DC) current around the loop formed by the electronic boxes, the interconnecting shielded cables, and the aircraft frame with a very low impedance drive coil as shown in Figure 3.42. With the current established, the voltage can be measured in one of the two ways. The

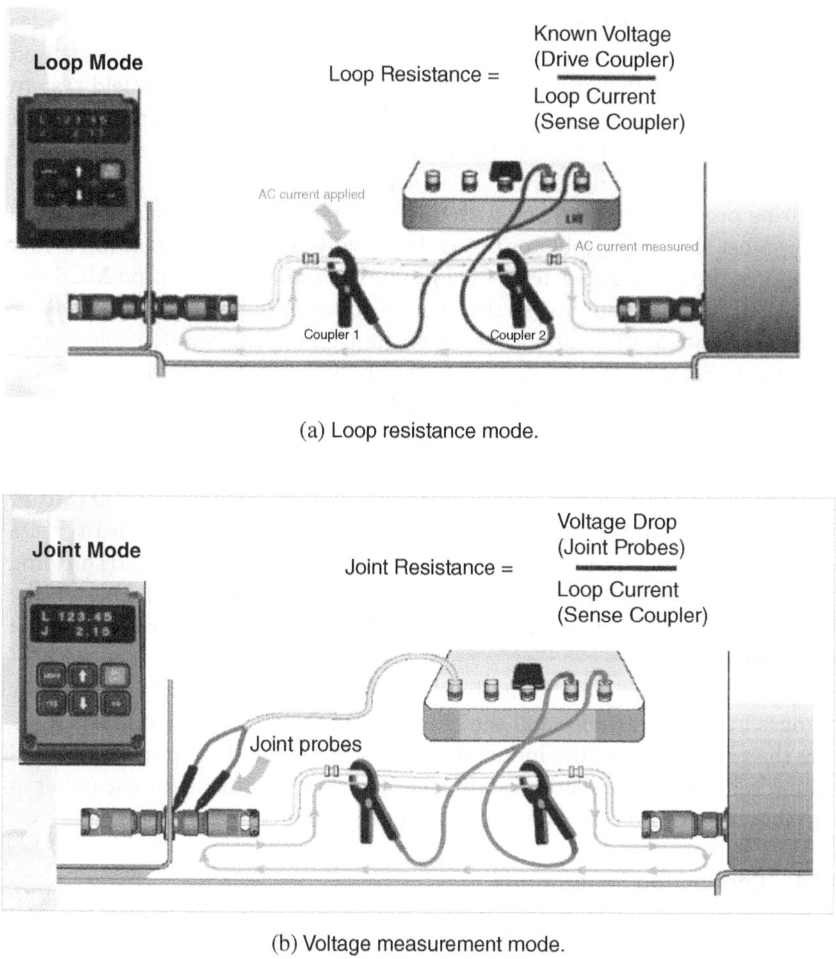

Figure 3.42 Loop resistance tester.

back-impedance on the drive coil, properly calibrated, will give a measure of the resistance around the entire loop, which can be used as a screening tool. If it is below some specified value, no additional measurements are needed. If the loop resistance is above the specification, then the included voltage probe can be used to measure across each joint in the cable loop to locate the bad connection [38]. The attractive feature of the LRT is that it does not require any of the connectors to be demated (disconnected) in order to make the measurement. This is very important on aircraft, because if a connector is "demated" for any reason, the entire system has to then be recertified before it can be flown. As a result, the LRT is finding increased use in flight-line and depot-level maintenance of military systems. It is nationally stock listed and is therefore available for use by the military.

3.7.3.1.2 Cable Shield Tester The CST is designed to measure the transfer impedance of shielded cable systems as a function of frequency. This method has the advantage that it measures the inductance of the cable shield system as well as the resistance and is thus a valuable tool for use during development and initial qualification tests of the cable system. It can also detect problems with the braided shield (such as a bad splice, a tear in the braid, or a manufacturing process problem) and with the backshell-to-braid interface, which is where most shielding problems occur. The disadvantage of this method is that it requires the cable under test (CUT) to be disconnected from the MCS and connected to a breakout box so the internal voltage can be measured.

The test configuration for measuring the transfer impedance of a cable system is shown in Figure 3.43. The current is driven by the RF OUT port on the NA. The cable shield drive current is recorded by the Reference (REF) channel. The core voltage is recorded on the A channel.

At low frequencies, the impedance measured will be the DC resistance of the cable shield and connection joints. At intermediate frequencies (100 kHz to 10 MHz), the impedance decreases because the skin depth is comparable to or less than the shield thickness (the curve dips). At higher frequencies, braid shielding becomes inductive, and the fields couple through the holes in the braid. The inductance is indicated by the 20 dB/decade rise in the impedance curve (on a log scale) [38].

Figure 3.44 shows how faults in shielded cables are readily identified by changes in their resistance and the inductance [39, 40]. It will accurately measure the resistance and the inductance of a cable shield, thus providing information for design, installation, and maintenance. It is suitable for use either in the laboratory or on board an aircraft or other shielded system.

3.7.4 Conducting Penetrations

Conducting penetrations can be categorized as being either electrical (wiring) or nonelectrical (metal tubing or control cable). Electrical penetrations into the

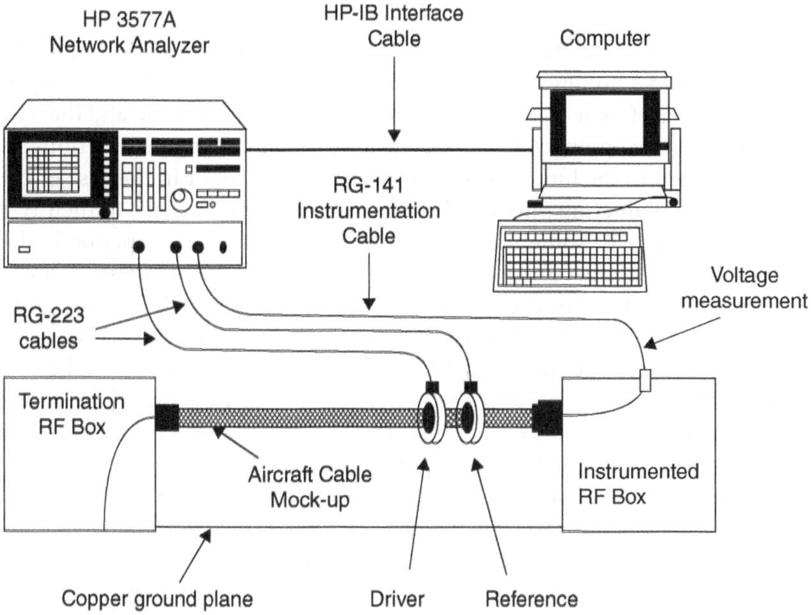

Figure 3.43 Cable shield test measurement set up.

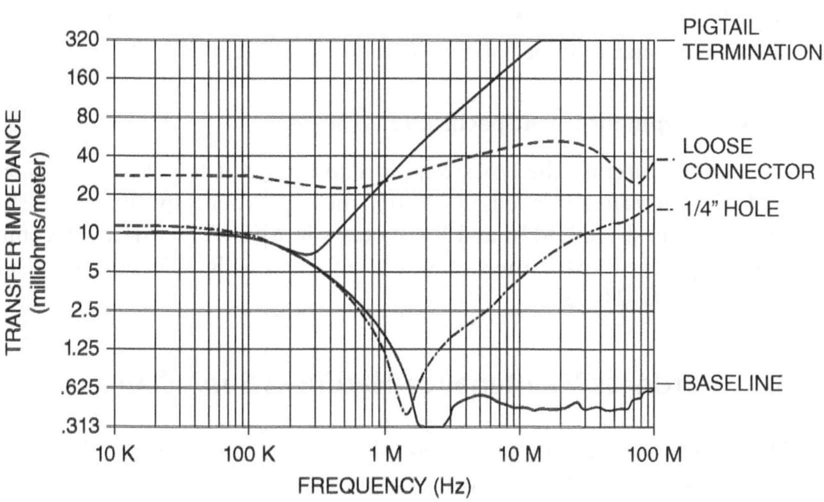

Figure 3.44 Z_T measurements of cable shields showing characteristic signatures of common cable faults.

hull, including antenna penetrations should be protected by a TPM consisting of a filter and a nonlinear transient voltage suppressor. The filters are specified in terms of $Z_T(\omega)$ or $Y_T(\omega)$ over the required frequency range and the TPDs in terms of their clipping voltage.[8] Nonelectrical conducting penetrations should be grounded to the hull by a collet (grounding flange) or in the case of control cables, by conducting pulleys. These can be specified in terms of their transfer impedance, Z_T, or sometimes by a simple current transfer function in dB. For feed- through connectors, tubes, and ducts, the quantity of interest is the transfer impedance from one side of the wall to the other, which is governed by the impedance between the penetrating connector and the shield wall (ground). As with gasketed apertures, the feed-through should have a peripheral bond to the shield wall to prevent the penetration of EM energy. If the connector is making contact at only one point, for instance, there is a hole in the shield, and the joint will be inductive, meaning it will leak high frequency energy. If a peripheral bond is achieved, the transfer function will be purely resistive.

Such penetrations can be expressed as either a transfer impedance $Z_T(\omega)$ in Ohms or a transfer admittance $Y_T(\omega)$ in Siemens (mhos or ℧) depending on the field driving the penetration. If it is being driven by a surface magnetic field $\boldsymbol{H}$ (surface current density $\boldsymbol{J}$) on the outside, then the specification should be written in terms of $Z_T(\omega) = \frac{V_{\text{Internal}}(\omega)}{H_{\text{Surface}}(\omega)}$ in ohms, but if the specified penetration is excited by a surface electric field $\boldsymbol{E}$ (surface charge density $\boldsymbol{Q}$), the penetration specification should be written as $Y_T(\omega) = \frac{I_{\text{Internal}}(\omega)}{E_{\text{Surface}}(\omega)}$ in Siemens, so that it is always normalized to the correct excitation field [31].

3.7.5 Equipment-Level Specifications (Boxes)

For military aircraft, the box strength requirement is found in MIL-STD-461 [23]. For aircraft hardened to HEMP, the equipment specifications should include at least the CS116 and RS105 tests in MIL-STD-461G. For commercial equipment (boxes), the strength is governed by RTCA DO-160 or one of the IEC STDs [24, 25].

3.8 Hardness Maintenance/Hardness Surveillance

Once the shielding components and the assembled system are characterized using CW, the information can form the basis for future HM/HS testing to monitor hardness degradation and to evaluate the system integrity after any major modification.

---

8 These can include nonlinear voltage clamps such as transient voltage suppressor diodes (transorbs). For higher voltages, metal-oxide varistors (MOVs), or neon discharge tubes.

3.8.1 Ellipticus Swept CW Tests

As discussed previously, the Ellipticus antenna can be used for detecting degradations in the overall shielding of the aircraft and easily reveal any areas where HEMP signal levels are becoming too high.

3.8.2 SPEHS

The SPEHS concept originated from the need to create a realistic surface charge distribution on the extremities of an aircraft for the purpose of making localized transfer admittance measurements of the hull penetrations. Whereas localized surface current can be easily generated using a transmission line, surface charge densities are more difficult. Some researchers had tried using a flat or curved plate transmission line driven in common mode against the skin with only moderate success. It is awkward and tends to distort the very thing you are trying to measure. In a laboratory setting, this is manageable, but on the fuselage of an aircraft, it proved to be extremely difficult. The SPEHS concept, however, offers a practical solution to this problem [41, 42]. To do this, a signal generator with a small amplifier is placed beneath the aircraft and grounded to one of the ground rods or a screen mesh. The output lead is connected to the bottom of the airframe somewhere near the center. The aircraft is then driven in common mode against the ground across some selected frequency band beginning at perhaps 1 MHz. As the frequency is increased and the airframe enters its first resonance, current and charge densities will be established on the surface similar to those which would result from an exposure to HEMP as illustrated in Figure 3.45. The fundamental mode resonance of a Boeing 707 is around 3.2 MHz and about 2.1 MHz for a 747. The SPEHS thus operates around the first few resonant modes of the airframe, up to about 30 MHz. These current and charge densities can then be used for localized transfer function measurements, as shown in Figure 3.45. Measurements of the surface

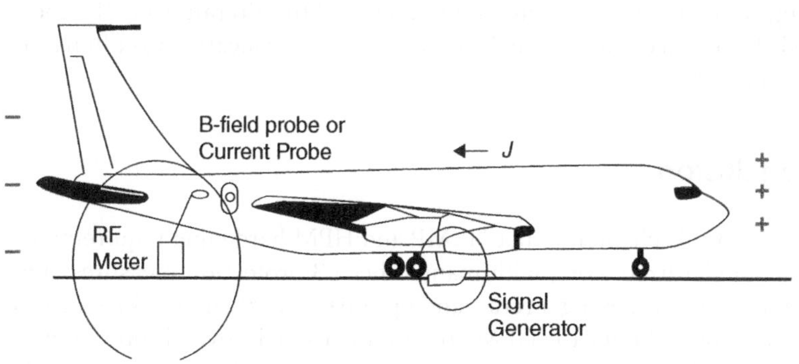

Figure 3.45 The SPEHS test concept.

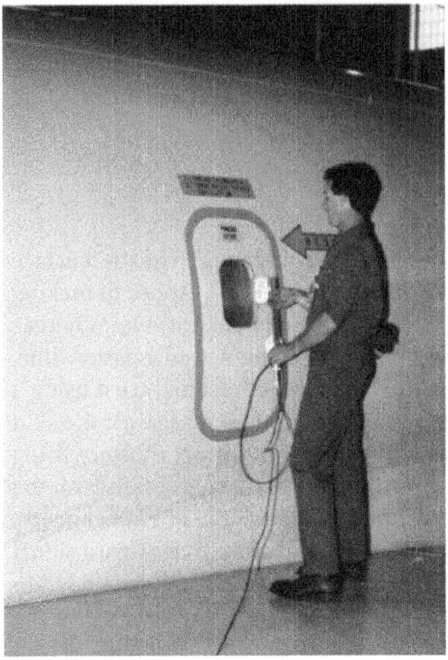

Figure 3.46 Using SPEHS to measure the SE of a hardened over-wing escape hatch.

magnetic fields can then easily be made with a hand-held B-dot sensor as illustrated in Figure 3.46. The system can be used at one frequency, as it often is for HM/HS, or the frequency can be swept through the first few resonant modes in order to generate a curve. Because the measurements are localized to the surface of the aircraft, this technique does not require an outdoor test range. The measurements can be done in a hangar with no effect on the accuracy.

3.8.3 Localized Hardening Component Tests

Hardening components can be measured on board the aircraft using the tools described above and compared against their design specifications as often as is deemed necessary.

3.9 Conclusion

The message of this chapter is that HEMP and HPM hardening requirements have been standardized along with many other E3 protection measures, and the technology for hardening aircraft and optimizing cost and performance is well advanced. By including the HEMP requirements in the overall aircraft integrated E3 design from the beginning, the services can meet their requirements with a minimum of extra costs and can blend the maintenance costs with the overall E3 Life Cycle Maintenance Program.

For HEMP hardness verification testing, there are facilities at Patuxent River NAS and White Sands Missile Range that will do the job. For those faced with designing and evaluating the hardening of a system, tools are now available that can measure the effectiveness of the hardening components in place and the overall system shielding performance, all using low-level CW.

References

The Theoretical Notes (TNs) and Sensor & Simulation Notes (SSNs) referenced herein are available through the University of New Mexico, Albu-querque NM, at http://www.ece.unm.edu/summa/HEMPseries.htm.

1 T. H. Lee, "The Carrington Event, H-bombs, Telstar, and the Great Geomagnetic Storm of 1989," keynote presentation at the *IEEE/EMC & SI Symposium*, Santa Clara CA, March 2015.

2 C. L. Longmire, "On the Electromagnetic Pulse Produced by Nuclear Explosions," *IEEE Trans. on EMC, Special Issue on the Nuclear Electromagnetic Pulse*, Vol. EMC-20, No. 1, February 1978.

3 K-D. Leuthäuser, "A Complete HEMP Environment Generated by High-Altitude Nuclear Bursts: Data and Standardization," Theoretical Note 364, Air Force Phillips Lab, Kirtland AFB, NM, February 1994.

4 Samuel Glasstone and Philip J. Dolan, *The Effects of Nuclear Weapons*, U.S. Government Printing Office, Washington DC, 2006.

5 C. Meng, "Numerical Simulation of the HEMP Environment," *IEEE Trans on EMC Special Issue on HEMP*, W.A. Radasky (ed.), Vol. 55, No. 1, May 2013.

6 Vladimir M. Loborev, "Up to Date State of the NHEMP Problems and Topical Research Directions," Electromagnetic Environments and Consequences: Proc. AMEREM 96 Int'l Symposium, Albuquerque NM, May 1996.

7 "Collateral Damage to Satellites from an HEMP Attack," DTRA-TR-10-22, Defense Threat Reduction Agency, Ft. Belvoir, VA, August 2010.

8 MIL-STD-464C, "Electromagnetic Environmental Effects Requirements for Systems," December 1, 2010.

9 W. R. Graham (Chairman, HEMP Commission) et al., "Report of the Commission to Assess the Threat to the United States from Electromagnetic Pulse (HEMP) Attack, Vol. 1: Executive Report," April 2004. Available online at http://www.HEMPcommission.org/reports.php.

10 D. J. Andersh, et al., "Comparison of Simulated HEMP and Natural Lightning Environments on a NASA F-106B Aircraft," *Air Force/Navy Scientific Engineering Symposium*, Norfolk NAS VA, November 1974.

11 T. F. Trost and Felix L. Pitts, "Analysis of EM Fields on an F-106B Aircraft during Lightning Strikes," *Proc. Int'l Aerospace Conf. on Lightning and Static Electricity*, St. Catherine's College, Oxford, England, March 1972.

12 Franklin A. Fischer and J.A. (Andy) Plumer, *Lightning Protection of Aircraft*, 2nd ed., Lightning Technologies Inc., 2004.

13 "Aircraft Lightning Environment and Related Test Waveforms," SAE Aerospace Recommended Practices SAE ARP5412A, Revised 2005.
14 Capt. C. E. Baum, "Interaction of Electromagnetic Fields with an Object which has an Electromagnetic Symmetry Plane," Interaction Note 63, Air Force Weapons Laboratory, March 1971.
15 C. E. Baum, "The Singularity Expansion Method," Chapter 3 in *Transient Electromagnetic Fields*, L.B. Felsen (ed.), Heidelberg, Springer-Verlag, 1976.
16 V. V. Liepa, "Surface Field Measurements on Scale Model EC-135 Aircraft," Interaction Application Memo 15, Air Force Weapons Laboratory, August 1977.
17 L. O. Marin, J. P. Castillo, and K. S. H. Lee, "Broad-band Analysis of the VLF/LF Aircraft Wire Antenna," *IEEE Trans on Antennas & Propagation*, Vol. AP-26, No. 1, January 1977.
18 F. M. Tesche, "Topological Concepts for Internal HEMP Interaction," *IEEE Trans. on EMC*, Vol. 20, No. 1, February 1977.
19 E. F. Vance, "Shielding and Grounding Topology for Interference Control," Interaction Note 306, Air Force Weapons Laboratory, Kirtland AFB NM, April 1977.
20 J. C. Giles and W. D. Prather, "Worldwide High Altitude Nuclear HEMP Simulators," *IEEE Trans on EMC Special Issue on HEMP*, W.A. Radasky (ed.), Vol. 55, No. 1, May 2013.
21 W. D. Prather, "The Ellipticus Illuminator," AFRL-PS-TR-2013-0009, Air Force Research Laboratory, Kirtland AFB NM, December 2013.
22 W. D. Prather, J. C. Cafferky, L. Ortiz, and G. Anderson, "CW Measurements of HEMP Shielding," *IEEE Trans on EMC Special Issue on HEMP*, W.A. Radasky (ed.), Vol. 55, No. 1, May 2013.
23 MIL-STD-461G, "Requirements for the Control of Electromagnetic Interference Characteristics of Subsystems and Equipment," December 11, 2015.
24 RTCA/DO-160G, Environmental Conditions and Test Procedures for Airborne Equipment, FAA, Dec 2010.
25 International Electrotechnical Commission (IEC), Subcommittee 77C. EMC: High power transient phenomena. Standards available online at http://www.iec.ch.
26 FAA Advisory Circular 20-157, "Certification of Aircraft Electrical/ Electronic Systems for Operation in a High Intensity Radiated Fields (HIRF) Environment," July 30, 2007.
27 SAE Aerospace Recommended Practice (ARP) 5573, "Guide to the Certification of Aircraft in the High Intensity Radiated Field (HIRF) Environment," SAE AE-4R Working Group, January 2003.
28 W. D. Prather, "Shielding Specification Techniques and Measurement Methods for Aircraft", *Proc. IEEE Symp. on EMC*, Detroit MI, August 2008.

29 C. E. Baum, "Extrapolation Techniques for Interpreting the Results of Tests in HEMP Simulators in Terms of HEMP Criteria," Sensor & Simulation Note 222, AFWL, Kirtland AFB NM, March 1977.

30 W. D. Prather, et al., "CW Measurements of Electromagnetic Shields," *Proc. EUROEM Symposium*, Toulouse, France, July 2012.

31 L. O. Hoeft and J. S. Hofstra, "Penetration Transfer Impedance and Admittance – the Intrinsic EM Parameters for Specifying Filters, Bonds, Isolators, and Other Devices for Treating Conductive Penetrations," *Proc. IEEE Int'l Symposium on EMC*, Atlanta, August 1977.

32 "Coaxial Test Procedure to Measure the RF Shielding Characteristics of EMI Gasket Materials," Aerospace Recommended Practice ARP1705, Rev A, Society of Aerospace Engineers (SAE), 2006.

33 K. F. Casey, "Electromagnetic shielding behavior of wire-mesh screens," *IEEE Trans. on EMC*, Vol. 30, No. 3, August 1977.

34 K. F. Casey, "Advanced composite materials and electromagnetic shielding," *IEEE Int'l Symposium on EMC*, Atlanta GA, June 1977.

35 W. D. Prather and C. D. Taylor, "Verification of the HEMP Hardening of Aircraft Windows and Doors," Hardness Surveillance Memo 2, Air Force Weapons Laboratory, Kirtland AFB NM, June 1977.

36 L. O. Hoeft, C. Herrmann, and W. D. Prather, "Measured EM Performance of Hardening Elements for A/C Windows and Doors," *IEEE Conf. on EMC*, Atlanta, August1977.

37 E. F. Vance, *Coupling to Shielded Cables*, John Wiley and Sons, Inc., New York, 1977.

38 E. L. Godo and B. Van Deventer, "Loop Resistance Tester: A non-intrusive method to measure connector and shield resistance," *Proceedings of the IEEE*, 1997. Additional information on the Loop Resistance Tester is available at www.boeing.com/commercial/aeromagazine/aero_10/loop, 2010.

39 L. O. Hoeft, P. J. Miller, and W. D. Prather, "Development of a Cable Shield Tester for In-situ Hardness Surveillance of Aircraft Cables," *IEEE Symposium on EMC*, 1977.

40 L. O. Hoeft and J. S. Hofstra, "Measured EM Shielding Performance of Commonly Used Cables and Connectors," *IEEE Trans on EMC*, Vol. 30, No. 3, August 1977.

41 L. O. Hoeft and W. D. Prather, "Single Point Excitation – A New Technique for Exciting Aircraft for Hardness Surveillance Measurements," *Int'l Conf. on EMC*, Institute of Electronic and Radio Engineers, University of York, England, September 1977.

42 L. O. Hoeft, D. McLemore, and W. D. Prather, "Swept Frequency Single Point Excitation Technique for Measuring the Shielding of Aircraft," *Proc. EUROEM 94 Conf.*, Bordeaux, France, May 1994.

4

HIRF and Lightning Effects and Testing

Martin Gabrisak

4.1 Introduction

All aircraft systems must be designed to ensure electromagnetic compatibility (EMC) with the electromagnetic environment (EME) in which they are intended to operate. This environment includes all electromagnetic phenomena present at the given installation location, i.e., superimposition of energies emanating from sources arising through natural phenomena as well as those generated by manmade activities.

Both lightning and high-intensity radiated fields (HIRF) are external EME elements, thus posing threats to the complete aircraft at the same time. Importance for the protection of avionic electrical and electronic systems against external electromagnetic energy has increased substantially in the last decades because of

1. replacement of mechanical/electromechanical/hydraulic and pneumatic devices and systems by electrical and electronic systems;
2. the reduced electromagnetic shielding provided by some composite materials used in aircraft designs;
3. the increase in susceptibility of electrical and electronic systems to electromagnetic energy because of higher data transmission speeds, higher density integrated circuits, and lover levels of signals.

Well-known effects of electromagnetic energy on electrical and electronic system are ranging from slight functional upsets to component failures and permanent damages.

Handbook of Aerospace Electromagnetic Compatibility, First Edition. Edited by Reinaldo J. Perez.

4.1.1 Certification Documents and Advisory Materials

There are four basic certification requirement documents issued by US Federal Aviation Agency (FAA)—Codes of Federal Regulations:

- 14CFR Part 23 applies to "normal, utility, acrobatic, and commuter category aeroplanes."
- 14CFR Part 25 is relevant to "transport category aeroplanes."
- 14CFR Part 27 applies to "normal category rotorcraft."
- 14CFR Part 29 is relevant to "transport category rotorcraft."

Their European Aviation Safety Agency (EASA) counterparts are Certification Specifications CS-23, CS-25, CS-27, and CS-29, where the same numbers denote the same aircraft categories as it is a case of FAA CFRs. All mentioned certification requirement documents contain paragraphs covering exposure of the aircraft to radio frequency energy and lightning.

However, these documents do not provide any guidance how the show compliance with the requirements. Therefore, both FAA and EASA issued advisory material outlining acceptable paths to compliance:

- FAA AC20-158A [4] and EASA AMC 20-158 [5] apply to HIRF protection.
- FAA AC20-136B [6] and EASA AMC 20-136 [7] cover protection of aircraft electrical and electronic system against lightning.
- Advisory Circular AC20-53B [8] deals with lightning protection of aircraft fuel systems.

Due to complicated nature and extent of topics involved in HIRF and lightning protection of aircraft, a number of additional guidance documents have been issued over the years by the non-profit industry organizations such as SAE International or EUROCAE. A reference to these documents will be made throughout this chapter.

4.1.2 General Approach to Analyses of HIRF/Lightning Effects

When analysing the HIRF and Lightning effects on electrical and electronic equipment and systems, the following general procedure for analysis of system/equipment EMC can be applied:

1. Functional analysis of the system and functional hazard analysis (FHA)
2. Definition of external sources of electromagnetic energy
3. Coupling analysis and definition of threats at interfaces
4. Verification of compliance—interference analysis and assessment of functional degradation
5. Introduction of corrective measures, if needed
6. Repeated verification after implementation of the corrective measures

The steps above will be outlined further on in the following sections.

Table 4.1 HIRF/lightning failure conditions and system certification levels

<table>
<tr><th colspan="2">Requirements excerpts</th><th rowspan="2">Failure condition</th><th rowspan="2">System HIRF/lightning certification level</th></tr>
<tr><th>HIRF</th><th>lightning</th></tr>
<tr><td colspan="2">Each electrical and electronic system that performs a function whose failure would prevent the continued safe flight and landing of the rotorcraft/aeroplane</td><td>Catastrophic</td><td>A</td></tr>
<tr><td>Each electrical and electronic system that performs a function whose failure would significantly reduce the capability of the rotorcraft/aeroplane or the ability of the flightcrew to respond to an adverse operating condition</td><td rowspan="2">Each electrical and electronic system that performs a function, for which failure would reduce the capability of the aeroplane or the ability of the flightcrew to respond to an adverse operating condition</td><td>Hazardous</td><td>B</td></tr>
<tr><td>Each electrical and electronic system that performs a function whose failure would reduce the capability of the rotorcraft/aeroplane or the ability of the flightcrew to respond to an adverse operating condition</td><td>Major</td><td>C</td></tr>
</table>

4.1.3 Functional Hazard Analysis

The first step in showing compliance is an identification of electrical or electronic systems those failures may cause or contribute to an adverse effect on the safety of flight of an aircraft. This involves classification of the equipment or system failure conditions and division of the systems into certification levels (groups) according to their criticality as defined in Table 4.1

System HIRF/lightning certification level A function systems (denoted further on as Level A systems) are further subdivided into two groups, depending on the pilot's presence in control loop [4–7]:

- Level A Non-Display Systems (formerly denoted as Level A Control Systems) involve functions that have some automated influence on a system (i.e., engine system, flight control system) and whose failure would prevent the continued safe flight and landing of the aircraft (Catastrophic failure). The pilot is not in the loop through pilot/system information exchange.
- Level A Display Systems are electronic instruments and associated systems sensors, which display critical information to the flight crew and whose failure would prevent the continued safe flight and landing of the aircraft (Catastrophic failure). The pilot is in the loop through pilot/system information exchange.

The actual procedures for system safety analyses are defined in several reference documents listed in all four relevant advisory documents [4–7]. Therefore, they will not be repeated here.

Summing up, all electrical or electronic systems shall be analyzed under all normal aircraft operating modes, stages of flight, and operating conditions to identify any failure due to HIRF or lightning that may cause or contribute to an adverse effect on the aircraft. It is important to know that when dealing with HIRF and lightning, redundancy alone cannot be considered as a mean of protection or a reduction in hazard, since the HIRF and lightning-generated electromagnetic effects can induce transients in all electrical wiring of the aircraft simultaneously.

4.2 Coupling Analysis

4.2.1 Introduction

The external lightning and HIRF electromagnetic environments are defined in Sections 4.3 and 4.4, respectively. The process, through which external electromagnetic energy couples to the internal electrical and electronic installation and possibly generates interferences in internal electronic circuitry, is an important problem for system design. In general, the victim system may be subjected to simultaneous interference from more than one source. However, for practical reasons, each source is evaluated individually, one at a time.

The electromagnetic energy is associated with an electromagnetic field characterizing the electric and magnetic conditions at the given location. It may include static components, i.e., an electrostatic field and a magnetostatic field, and time-varying components representing electromagnetic waves.

A perfectly conducting closed shield would prevent any penetration of the electromagnetic energy from external sources as indicated in Figure 4.1. However, an aircraft cannot be considered as an ideal shield since it has windows, doors, and other openings as well as external electrical/electronic installation (lights, antennas, etc.) connected via wiring to the internal circuitry. Therefore, some portion of the external electromagnetic energy can be transferred (electromagnetic coupling) to electrical/electronic circuits installed in aircraft and cause interference. In order to determine possible interaction of electromagnetic energy with the installation of interest, it is necessary to perform the following steps:

1. Define electromagnetic environment for the system/equipment.
2. Define possible coupling paths.
3. Determine the types of interference and their time dependence/frequency response at the interfaces of interest.

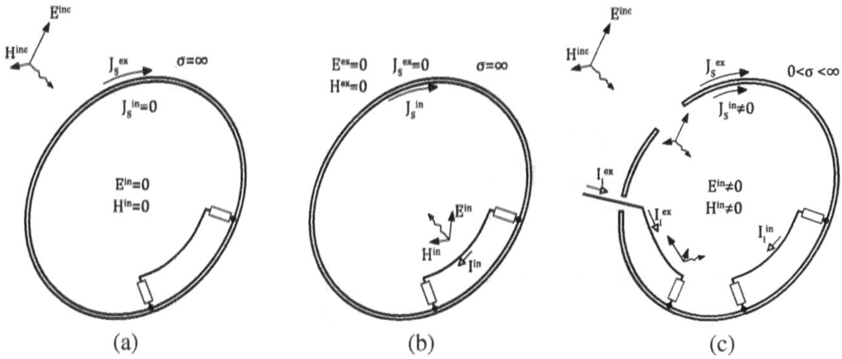

Figure 4.1 The concept and limitations of perfect shield. (a) perfectly conducting shield in external field, (b) perfectly conducting shield with internal excitation, and, (c) practical shield with penetrating wires and apertures.

The analysis can be carried out by one or more of the following methods:

- Approximate calculations;
- Estimation based on the similarities with the previous projects;
- Experimental methods (measurements);
- Numerical modeling.

The suitability of each approach depends on the design stage of the given project. At the early stages of aircraft development minimum design information available, the designer will often reach for the first two methods, or perform measurements or/and numerical simulation on rudimentary models. On the contrary, at the later stages with the most of design data or even real aircraft are available, the exact measurements and simulations are possible.

The possible methods to make approximate calculations are given in the following subsections. The experimental methods and numerical methods will be outlined later separately for HIRF and lightning, respectively.

4.2.2 Outline of Coupling Process

In order to perform an approximate analysis of a system response, it is useful to divide a complicated chain of processes into a number of simpler parts, starting from the knowledge of the incident field and proceeding step by step toward the internal component response. During each step, a simple problem is solved. The estimation of the system/equipment response is obtained by combining the results of individual steps.

An incident external RF electromagnetic field or the charge conducted by lightning channel induces surface current and charge densities on the conducting exterior (aircraft skin, antennas, lights, etc.) Such a process is referred to as

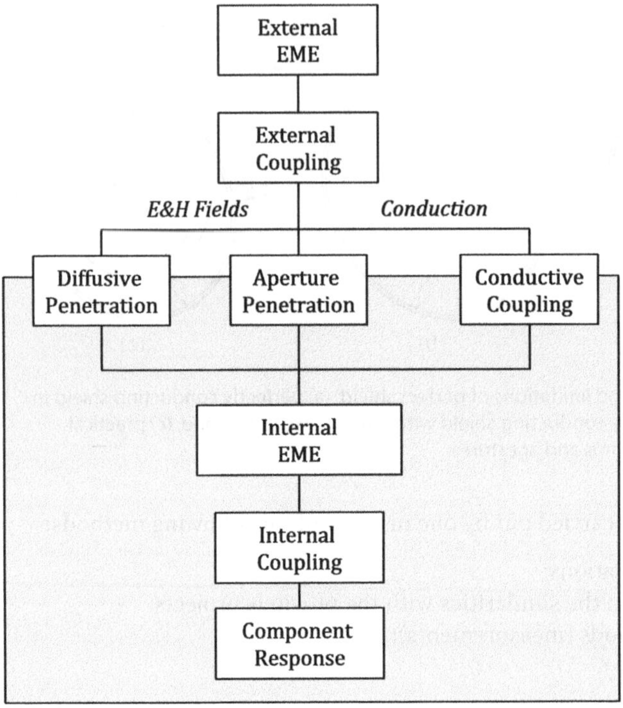

Figure 4.2 Coupling process.

external coupling and can be described by a transfer function from the external sources to these surface response quantities.

Once the surface responses are known, further calculations of the EM energy penetration into the interior are carried out. Several processes can be involved in the energy transfer, giving the rise to internal electromagnetic fields.

These internal electromagnetic fields once again give rise to interior surface currents and charge densities on the cables, shields, or equipment cases, as shown in Figure 4.2. This process of coupling, penetration, and propagation will be repeated until the component level is reached. Each step can be expressed by a transfer function. The complete chain of processes taking place during penetration of the external EM energy to an avionic system is depicted in Figure 4.3.

4.2.3 External and Internal Fields and Surface Currents

4.2.3.1 Basic Considerations

4.2.3.1.1 Time Domain and Frequency Domain Analyses Analysis of the electrical signals based on variations over time is known as Time Domain (TD)

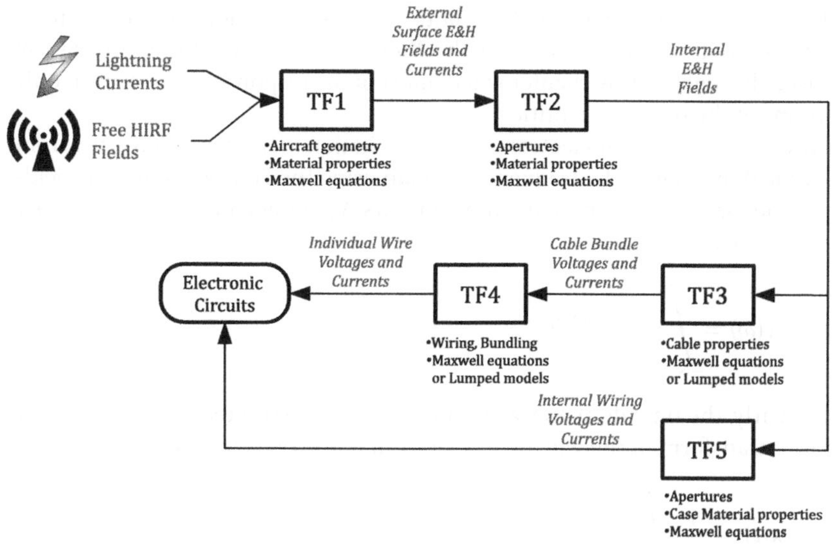

Figure 4.3 Aircraft coupling chain.

analysis. Another important form of analysis is Frequency Domain (FD) analysis wherein the time domain signal is converted into a frequency domain signal and its component frequencies are analyzed.

Depending on the nature of the source of the electromagnetic energy, the induced response in the victim circuitry can be either a time harmonic quantity (i.e., best expressed and depicted depending on frequency in the frequency domain) or a transient quantity (i.e., best expressed and depicted as a function of time in the time domain)

In the development of the coupling analysis in this chapter, the formulation and solution of problems in the frequency domain will be used where the excitation is assumed to be a sinusoidal waveform. In this notation, a time harmonic quantity $x(t)$ is described by an amplitude X_0, an angular frequency ω, and a phase angle φ as:

$$x(t) = X_0 \cos(\omega t + \varphi) \tag{4.1}$$

This can be expressed using the phasor notation as

$$x(t) = \mathrm{Real}[X(\omega)e^{j\omega t}] \tag{4.2}$$

where $X(\omega) = X_0 e^{j\varphi}$ is the complex phasor quantity describing the time response $x(t)$. The function Real[] extracts the real part of the quantity in the brackets, and j denotes the complex unity $j = \sqrt{-1}$.

In the frequency domain analysis, both the Real[] function and the $e^{j\omega t}$ term are for simplicity omitted and only the $X(\omega)$ is considered. This approach will be used throughout this chapter where most of the responses will be given only in terms of the phasor quantities.

Any general transient electromagnetic excitation $x(t)$ can be converted from the time domain into the frequency domain using the Fourier transform yielding a wideband spectrum defined by phasors $X(\omega)$ depending on the angular frequency ω:

$$X(\omega) = \int_{-\infty}^{\infty} x(t)e^{-j\omega t}dt \tag{4.3}$$

Similarly, the time domain waveform can be reconstructed using the inverse Fourier transform of the wideband spectrum of the phasors $X(\omega)$:

$$x(t) = \frac{1}{2\pi}\int_{-\infty}^{\infty} X(\omega)e^{-j\omega t}d\omega \tag{4.4}$$

One of the advantages of the phasor notation and thus frequency domain analysis can be shown in problems involving time derivatives of the response functions, where:

$$\frac{dx(t)}{dt} \leftrightarrow j\omega X(\omega) \tag{4.5}$$

i.e., the time derivatives are obtained simply by multiplying the phasor response by the expression $j\omega$. Higher-order derivatives are obtained similarly. In this way, the problems expressed in the time domain using partial differential equations differentiated both in space time can be in the frequency domain be solved using ordinary differential equations differentiated in space only.

Depending on the time and frequency variations, these Fourier transform pairs can be evaluated either analytically by direct integration or using the numerical discrete or fast Fourier transforms (DFT/FFT) algorithms. As it will be shown later in this chapter for standardized lightning environment, where the excitation waveforms are described analytically the transient responses can obtained by first analytically converting the excitation waveform from the time into the frequency domain, then computing the wideband spectrum of the response phasor $X(\omega)$ and finally obtaining the transient responses using the inverse DFT (IDFT) or inverse FFT (IFFT).

4.2.3.1.2 Low Frequency (LF) Approximation For electrically small objects, a simplified approach can be used, where the electrical quantity is considered to be constant along a dimension of interest. In the frequency domain, electrical size is determined by physical size in relation to wavelength. An object is electrically small when its physical size is small compared to a wavelength:

$$l \ll \frac{c}{f} = \lambda \tag{4.6}$$

where l is the dimension of the object (e.g., the length of the fuselage, the wing span, the diameter of the fuselage, the length of a slot, or the diameter of a circular hole) along which the given quantity should be considered constant, c is the speed of light and f and λ denote the frequency and the wavelength of the external field, respectively.

For transients, an object is considered being electrically small if the rate of change of the exciting field (E or H) is much smaller than the time $t = l/c$ it takes for a wave to cross the aperture, or

$$\frac{l}{c} \ll \frac{H}{max\left(\frac{dH(t)}{dt}\right)} \approx t_r \tag{4.7}$$

where t_r is the rise time of the transient.

For a HIRF and lightning electromagnetic pulse, an object may be electrically large for the high-frequency part of the spectrum and electrically small at the lower frequencies. For example, when setting the conditions for smallness to be

$$l < \frac{\lambda}{10} \quad \text{or} \quad \frac{l}{c} < \frac{t_r}{10} \tag{4.8}$$

a circular aperture with the diameter $d = 0.5$ m would be electrically small for frequencies below

$$f < \frac{c}{10d} = \frac{3 \times 10^8\ \text{m}\cdot\text{s}^{-1}}{10 \times 0.5\ \text{m}} = 6 \times 10^7\ \text{Hz} = 60\ \text{MHz}$$

or in time domain for pulses with the rise time more than:

$$t_r > 10 \times \frac{d}{c} = 10 \frac{0.5\ \text{m}}{3 \times 10^8\ \text{m}\cdot\text{s}^{-1}} = 1.67 \times 10^{-8}\ \text{s} = 16.7\ \text{ns}$$

On the other side, if the length of the fuselage is $l = 20$ m, then the aircraft is electrically small below:

$$f < \frac{c}{10l} = \frac{3 \times 10^8\ \text{m}\cdot s^{-1}}{10 \times 20\ \text{m}} = 1.5 \times 10^6\ \text{Hz} = 1.5\ \text{MHz}$$

At overlapping frequencies, it is possible to combine both approaches, first treat the fuselage as electrically large, find the distribution of the surface fields and currents at the location of the aperture and then locally treat the aperture as an electrically small one.

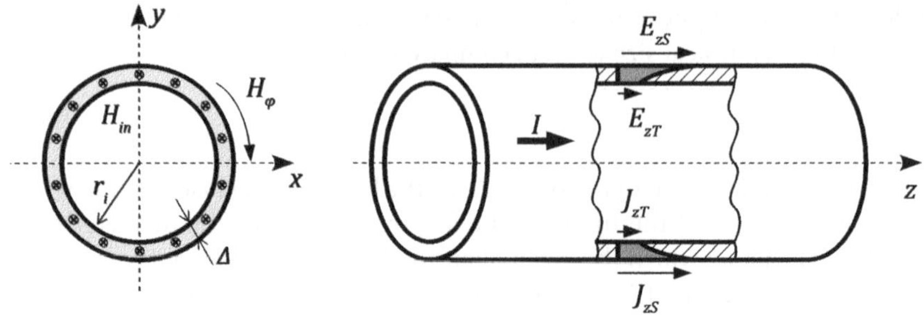

Figure 4.4 Problem geometry.

4.2.3.2 Direct Injection and Conductive Coupling

At the first approach, let's represent aircraft fuselage by solid tubular conductor of uniform cross-section and uniform wall thickness with electrical conductivity σ and permeability μ_0 (i.e., non-ferromagnetic material) as depicted in Figure 4.4. At points distant from the injection points, an even distribution of the current along the circumference can be assumed. The magnetic field density at the external surface of the skin can be then expressed as:

$$H_\varphi = \frac{I}{2\pi(r_i + \Delta)} \tag{4.9}$$

where r_i denotes the inner radius of the tube and Δ is the wall thickness.

When $r_i \gg \Delta$, the surface current density J_{zS} can be expressed as [14, 15]

$$E_{zS} = \frac{J_{zS}}{\sigma}$$

$$E_{zS} = I\frac{(1+j)\frac{\Delta}{\delta}}{2\pi r\sigma\Delta}\coth\left[(1+j)\frac{\Delta}{\delta}\right] = R'_{DC}I(1+j)\frac{\Delta}{\delta}\coth\left[(1+j)\frac{\Delta}{\delta}\right]$$

$$E_{zS} = Z'_S \cdot I \tag{4.10}$$

where r is the average radius of the tube $r = \sqrt{r_i \cdot (r_i + \Delta)}$, δ denotes the skin depth (see Figure 4.5)

$$\delta = \frac{1}{\sqrt{\pi f \mu_0 \sigma}} \tag{4.11}$$

and Z_S denotes the surface (also denoted as internal) impedance of the tube defined as

$$Z'_S = R'_{DC}(1+j)\frac{\Delta}{\delta}\coth\left[(1+j)\frac{\Delta}{\delta}\right] \tag{4.12}$$

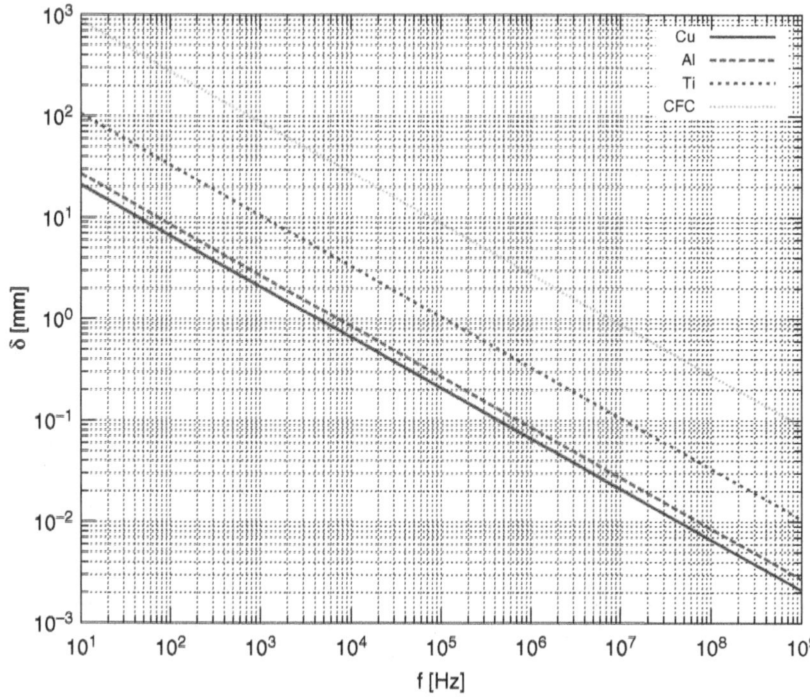

Figure 4.5 Skin depth for different materials.

The radial distribution of the current in the wall is governed by diffusion of the electromagnetic field [13]. The relation between the voltage gradient along the inner surface of the tube V_{in} and the impressed current I_{ex} can be defined as transfer impedance of the tube per unit of length:

$$Z'_T = \frac{1}{I_{ex}} \frac{dV_{in}}{dx} \tag{4.13}$$

Once again using the simplifications for a thin-walled tube ($r \gg \Delta$), the transfer impedance of a solid thin-walled cylinder can be expressed as [14, 15]

$$Z'_T = \frac{1}{2\pi r \sigma \Delta} \frac{(1+j)\frac{\Delta}{\delta}}{\sinh\left[(1+j)\frac{\Delta}{\delta}\right]} \tag{4.14}$$

At low frequencies, where $\Delta/\delta \ll 1$, the magnitude (modulus) of the transfer impedance reduces to

$$|Z'_T| \approx \frac{1}{2\pi r \sigma \Delta} = R_{DC} \tag{4.15}$$

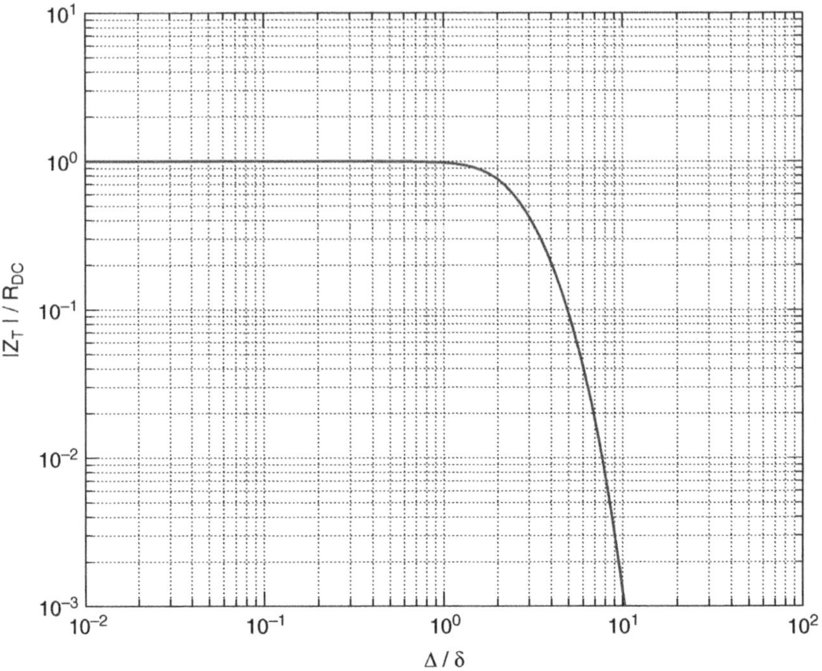

Figure 4.6 Normalized dependence of transfer impedance on relative skin depth.

where R'_{DC} is the DC resistance of the tube per unit length. The material-independent variation of the transfer impedance on the frequency is shown in Figure 4.6, where the modulus of the transfer impedance $|Z'_T|$ is normalized to R'_{DC} and the frequency-dependent skin depth (Figure 4.5) is normalized to the thickness, yielding the relative skin depth δ/Δ.

An alternative expression for the transfer impedance is obtained by letting [14]

$$(1+j)\frac{\Delta}{\delta} = \sqrt{j\omega\tau_s}$$

$$Z'_T = R_{DC}\frac{\sqrt{j\omega\tau_s}}{\sinh\left(\sqrt{j\omega\tau_s}\right)} \tag{4.16}$$

where τ_s is the wall diffusion time constant expressed as

$$\tau_s = \mu\sigma\Delta^2 \tag{4.17}$$

By rearranging the expression for skin depth (4.11) and inserting in (4.17), the wall diffusion time constant is then

$$\mu\sigma = \frac{1}{\pi f \delta^2}$$

$$\tau_s = \frac{1}{\pi f}\left(\frac{\Delta}{\delta}\right)^2 = \frac{1}{\pi f_\delta} \tag{4.18}$$

where f_δ is the frequency at which the skin depth is equal to the wall thickness.

Summing up, when a current flows along the aircraft structure, it gives rise to a voltage drop due to finite impedance of the structure. Depending on electrical properties, the main contributors to the resistive coupling are impedance of the aircraft structure (e.g., skin) and the contact resistances.

4.2.3.3 Multilayered Thin-Walled Cylindrical Structures

Tubular structures made of several layers of different conducting materials (see Figure 4.7) can be analyzed recursively by applying the two-layer formulae repeatedly until all the layers have been considered.

If the thin-walled condition applies to all layers ($r_{i(1\ldots n)} \gg \Delta_{(1\ldots n)}$), the transfer impedance of an n-layer shield with no insulation (no spaces) between individual layers can be expressed as [14]

$$Z_T'^{(n)} = \frac{Z_{Tn}' Z_T'^{(n-1)}}{Z_{Sn}' + Z_S'^{(n-1)}} \tag{4.19}$$

where $Z_T'^{(n-1)}$ and $Z_S'^{(n-1)}$ denote the equivalent transfer and surface impedances of the first $n-1$ layers while Z_{Tn}' and Z_{Sn}' indicate the transfer and

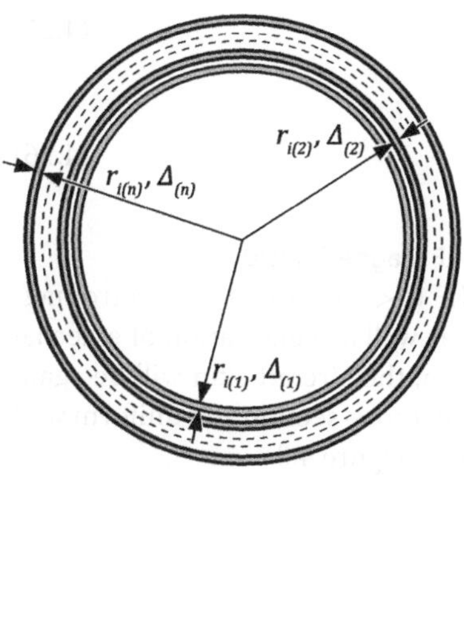

Figure 4.7 Multilayered cylinder.

the surface impedances of the n th layer alone, respectively. The "cumulative" equivalent surface $Z'^{(n)}_S$ impedance for n layers can be obtained as

$$Z'^{(n)}_S = Z'_{Sn} - \frac{(Z'_{Tn})^2}{Z'_{Sn} + Z'^{(n-1)}_S} \tag{4.20}$$

Then, for the first two layers (or for two-layer cylindrical structure), the transfer and surface impedances can be expressed as

$$Z'^{(2)}_T = \frac{Z'_{T1} Z'_{T2}}{Z'_{S1} + Z'_{S2}} \tag{4.21}$$

$$Z'^{(2)}_S = Z'_{S2} - \frac{(Z'_{T2})^2}{Z'_{S2} + Z'_{S1}} \tag{4.22}$$

These formulas can also be applied to electrically short insulated multilayer shields connected in parallel or long shields that are repeatedly connected together at electrically short spacings (i.e., there are no travelling wave effects between the shields). The influence of insulation between the n-th and $(n-1)$-th can be included by replacing the Z'_{Sn} by

$$Z'_{Sn-corr} = Z'_{Sn} + \frac{1}{G'_{n(n-1)} + j\omega C'_{n(n-1)} + \frac{1}{j\omega L'_{n(n-1)}}} \tag{4.23}$$

where $G'_{n(n-1)}$ is the per-unit insulation conductance, $C'_{n(n-1)}$ is the per-unit capacitance, and $L'_{n(n-1)}$ is the per-unit mutual inductance between layers n and $(n-1)$ given as

$$G'_{n(n-1)} = \frac{2\pi\sigma_{n(n-1)}}{\ln \frac{r_{i(n)}}{r_{i(n-1)}+\Delta_{(n-1)}}} \tag{4.24}$$

$$C'_{n(n-1)} = \frac{2\pi\varepsilon_{n(n-1)}}{\ln \frac{r_{i(n)}}{r_{i(n-1)}+\Delta_{(n-1)}}} \tag{4.25}$$

$$L'_{n(n-1)} = \frac{\mu_{n(n-1)}}{2\pi} \ln \frac{r_{i(n)}}{r_{i(n-1)} + \Delta_{(n-1)}} \tag{4.26}$$

4.2.3.4 Cylindrical Object Exposed to External LF Magnetic Field

In order to estimate the distribution of the LF ($l < \lambda/10$) surface electromagnetic fields and the surface current densities as well as penetration of external electromagnetic fields into aircraft by diffusion, the aircraft skin will be again represented by solid tubular cylinder of uniform cross-section and uniform wall thickness with a half-sphere covers, as shown in Figure 4.8.

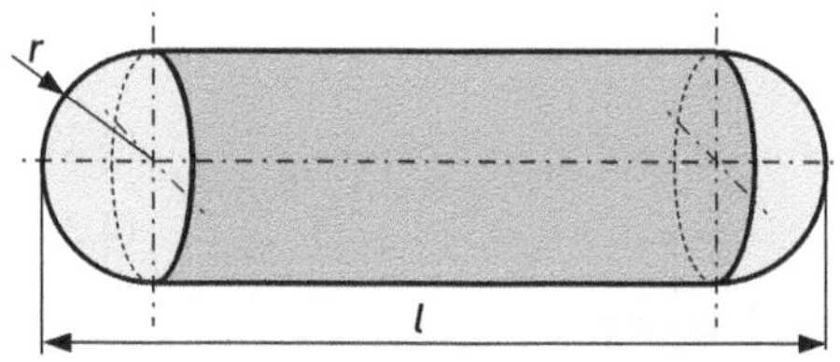

Figure 4.8 Model geometry.

A case of an aircraft exposed to an external transverse electromagnetic field (i.e., a plane wave) of arbitrary incident angle can be decomposed into two separate cases:

- Immersion in longitudinal magnetic field, where the vector of magnetic field intensity of the incident field is parallel to cylinder axis
- Immersion in a transversal magnetic field, i.e., the vector of magnetic field intensity of the incident field is perpendicular to cylinder axis

The total coupling can be calculated as a superimposition of these two separate solutions.

4.2.3.4.1 Thin-Walled Sphere Immersed in a Uniform External Magnetic Field The end caps of cylinder can be approximated as half-spheres. The surface magnetic field and current density distribution can be estimated by fields and currents on a sphere of a radius r_i, conductivity σ, and permeability μ_0 immersed in external LF magnetic field, as shown in Figure 4.9.

The radial and angular components outside the sphere ($r \geqq r_i + \Delta$) are described by [15]

$$H_r = H_e \left(1 - \frac{2(r_i + \Delta)^3}{r^3} W\right) \sin\xi \tag{4.27}$$

$$H_\xi = H_e \left(1 + \frac{(r_i + \Delta)^3}{r^3} W\right) \cos\xi \tag{4.28}$$

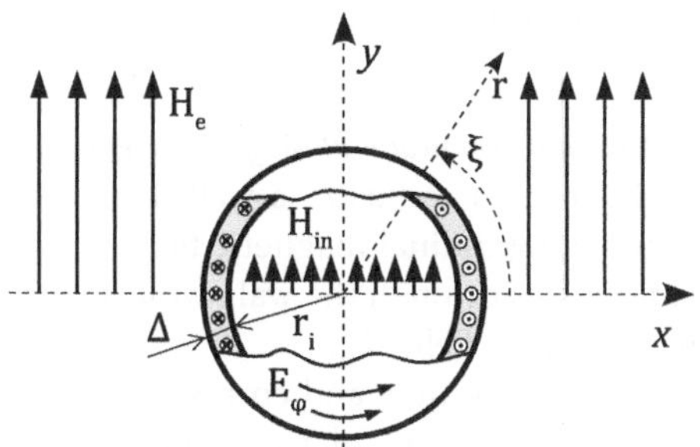

Figure 4.9 Sphere in external magnetic field.

and inside the sphere ($0 \leq r \leq r_i$)

$$H_r = H_{in} \cos \xi = QH_e \sin \xi \tag{4.29}$$

$$H_\xi = H_{in} \sin \xi = QH_e \sin \xi \tag{4.30}$$

with the parameters Q and W defined as follows:

$$Q = \frac{1}{\cosh \gamma\Delta + \frac{1}{3}\left(K + \frac{2}{K}\right) \sinh \gamma\Delta} \tag{4.31}$$

$$W = \frac{\frac{1}{3}\left(\frac{K}{2} - \frac{2}{K}\right) \sinh \gamma\Delta}{\cosh \gamma\Delta + \frac{1}{3}\left(K + \frac{2}{K}\right) \sinh \gamma\Delta} \tag{4.32}$$

with

$$K = \frac{\mu_0}{\mu} \gamma r_i \tag{4.33}$$

The propagation constant of the material γ is given as

$$\gamma = \sqrt{j\omega\mu_0\sigma} = \frac{1}{\delta} + j\frac{1}{\delta} \tag{4.34}$$

where δ denotes the skin depth:

$$\delta = \frac{1}{\sqrt{\pi f \mu_0 \sigma}} \tag{4.35}$$

The variation of field-enhancement coefficient W and attenuation coefficient Q with frequency is shown in Figure 4.10. It can be deduced that the external field strength is enhanced to maximum $1.5 \times H_e$ at high frequencies.

The angular eddy current density J_φ at the outer surface of the sphere is given as

$$J_\varphi(r_i + \Delta) = \sigma E_\varphi(r_i + \Delta) = -j\omega \cdot \mu_0 \cdot r_i \cdot \sigma \cdot H_e \left(\frac{1 - 2W}{2}\right) \cos \xi \tag{4.36}$$

4.2.3.4.2 Tubular Conductor in a Uniform External Longitudinal Magnetic Field A non-magnetic tubular conductor ($\mu_r = 1$) of inner radius r_i, a wall thickness Δ, and a finite conductivity σ is placed in a longitudinal magnetic field H_e, as shown in Figure 4.11. The magnetic field diffuses through the wall, while being attenuated while, producing the internal magnetic field H_{in}.

The resulting electric field on the surfaces of the tube and correspondingly the density of induced electric current will be perpendicular to the magnetic

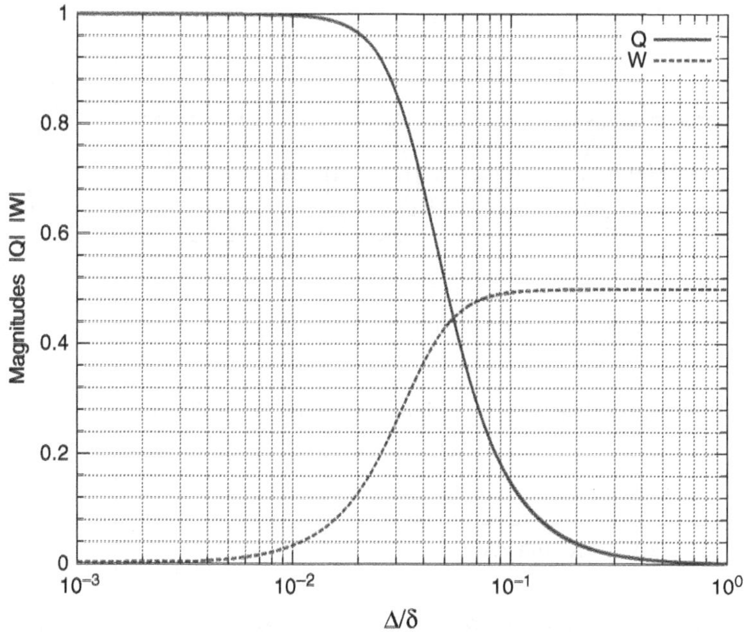

Figure 4.10 Frequency variation of the sphere field-enhancement/ attenuation coefficients.

field, i.e., flowing along the internal and external circumference of the tube, respectively [15, 16]:

$$J_{\varphi}(r_i + \Delta) = \sigma E_{\varphi}(r_i + \Delta) = -j\omega\sigma\mu_0 H_e \frac{r_i + \Delta}{2} \tag{4.37}$$

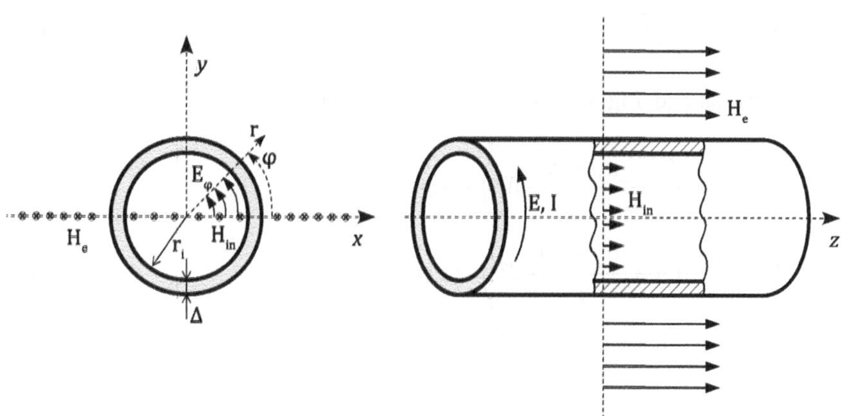

Figure 4.11 Cylindrical tube in a uniform external longitudinal magnetic field.

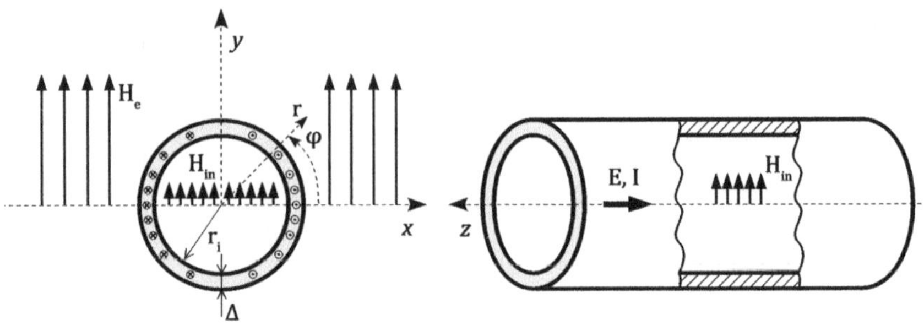

Figure 4.12 Cylindrical tube in a uniform external transversal magnetic field.

The internal magnetic field can be calculated using the following equation [15, 16]

$$H_{in} = \frac{H_e}{\cosh \gamma\Delta + \frac{1}{2}K \sinh \gamma\Delta} = QH_e \tag{4.38}$$

4.2.3.4.3 Tubular Conductor in a Uniform External Transverse Magnetic Field The same tube as defined in the previous subsection is placed in the transversal field as shown in Figure 4.12. The induced eddy currents in the tube produce magnetic fields that oppose the inducing field. The resulting external magnetic field H_e^{ex} outside the tube ($r \geqq r_i + \Delta$) is composed of the external electromagnetic H_e and the induced magnetic fields due to the eddy currents:

$$\vec{H}_e^{ex} = \vec{H}_e + \vec{H}_e^{eddy} \tag{4.39}$$

$$|H_e^{ex}| = \sqrt{H_r^2 + H_\varphi^2} \tag{4.40}$$

with the radial and angular components defined as [15]

$$H_r = H_e\left(1 - \frac{(r_i + \Delta)^2}{r^2}W\right)\sin\varphi \tag{4.41}$$

$$H_\varphi = H_e\left(1 + \frac{(r_i + \Delta)^2}{r^2}W\right)\cos\varphi \tag{4.42}$$

The diffused magnetic field inside the tube ($0 \leqq r \leqq r_i$) is defined as

$$H_r = H_{in}\sin\varphi = QH_e\sin\varphi \tag{4.43}$$

$$H_\varphi = H_{in}\cos\varphi = QH_e\cos\varphi \tag{4.44}$$

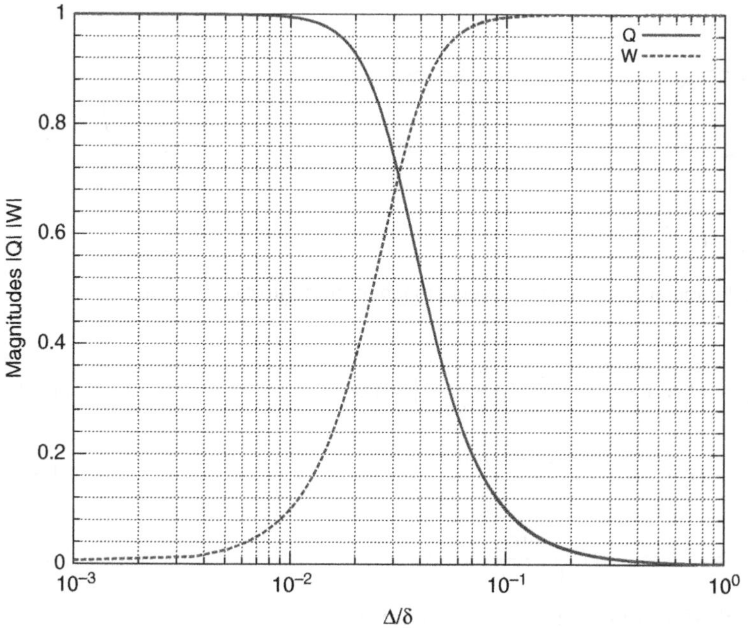

Figure 4.13 The cylinder field-enhancement / attenuation coefficients for transversal H-field.

with the parameters Q and W defined as follows:

$$Q = \frac{1}{\cosh \gamma\Delta + \frac{1}{2}\left(K + \frac{1}{K}\right)\sinh \gamma\Delta} \tag{4.45}$$

$$W = \frac{\frac{1}{2}\left(K - \frac{1}{K}\right)\sinh \gamma\Delta}{\cosh \gamma\Delta + \frac{1}{2}\left(K + \frac{1}{K}\right)\sinh \gamma\Delta} \tag{4.46}$$

The variation of field-enhancement coefficient W and attenuation coefficient Q with frequency is depicted in Figure 4.13. The electric field at the outer surface of the tube can be expressed as

$$J_z(r_i + \Delta) = \sigma E_z(r_i + \Delta) = -j\omega\cdot\mu_0\cdot r_i\cdot\sigma\cdot H_e\,(1 - W)\cos\varphi \tag{4.47}$$

Further simplification can be reached neglecting the skin effect, i.e., replacing the wall by thin sheet with constant current density ι_z [17–19].

$$H_{in} = \frac{H_e}{1 + j\omega\tau} = H_e\frac{1 - j\omega\tau}{1 + \omega^2\tau^2} \tag{4.48}$$

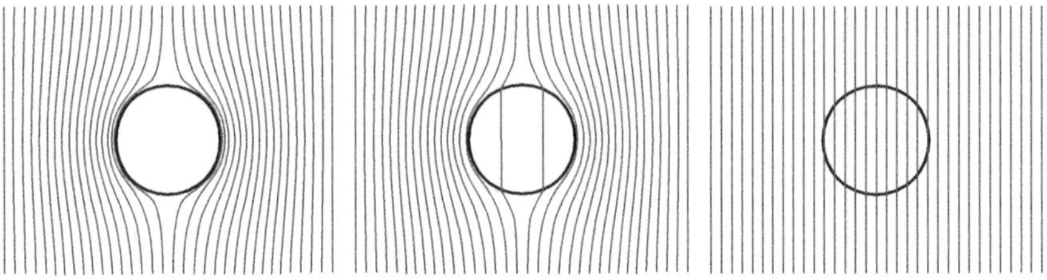

Figure 4.14 Temporal diffusion of transversal magnetic field into cylindrical tube.

$$H_r = H_e \left(1 - \frac{(r_i + \Delta)^2}{r^2} \frac{j\omega\tau}{1 + j\omega\tau}\right) \sin\varphi \tag{4.49}$$

$$H_\varphi = H_e \left(1 + \frac{(r_i + \Delta)^2}{r^2} \frac{j\omega\tau}{1 + j\omega\tau}\right) \cos\varphi \tag{4.50}$$

$$\iota_z = J_z \cdot \Delta = -2H_e \frac{j\omega\tau}{1 + j\omega\tau} \cos\varphi \tag{4.51}$$

with the diffusion time constant τ of the form:

$$\tau = \frac{\mu_0 \cdot r_i \cdot \Delta \cdot \sigma}{2} \tag{4.52}$$

The magnetic field distribution outside and inside the cylinder at high and low frequencies is shown in Figure 4.14. The diffusion time constant can also be expressed as [17]

$$\tau = \frac{L_S}{R_S} = \mu_0 \Delta\sigma \frac{A}{P} \tag{4.53}$$

where L_S corresponds to the internal inductance of the structure

$$L_s = \frac{\mu_0}{4\pi} \tag{4.54}$$

and R_S is the resistance of the structure

$$R_s = \frac{1}{2\pi r_i} \cdot \frac{\sigma}{\Delta} \tag{4.55}$$

with A denoting the area enclosed by the tube $A = \pi r_i^2$ and P the circumference of the structure, $P = 2\pi r_i$.

Looking at the equations, it can be said that the penetration of the field inside the cylinder may be understood and studied as an analogue low-pass filter shown in Figure 4.15 with the analogies $H_{in} \rightleftarrows V_{in}$ and $H_e \rightleftarrows V_e$. The low-pass

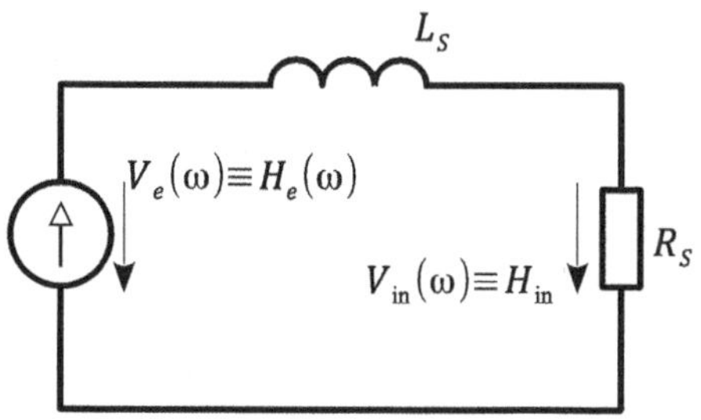

Figure 4.15 Analogous circuit for magnetic field penetration.

filter circuit is then defined by the cutoff frequency:

$$f_C = \frac{1}{2\pi\tau} \tag{4.56}$$

Comparing the maximum external and internal field intensities for original and simplified solutions, it can be deduced that at frequencies above the "cut-off" frequency, f_C is the shielding effect noticeable and above a frequency of $5f_C$, the cylinder can be treated as a perfect conductor, for which the radial and angular external field components are given by [17]:

$$H_r = H_e\left(1 - \frac{(r_i + \Delta)^2}{r^2}\right)\sin\varphi \tag{4.57}$$

$$H_\varphi = H_e\left(1 + \frac{(r_i + \Delta)^2}{r^2}\right)\cos\varphi \tag{4.58}$$

$$H_{in} = 0 \tag{4.59}$$

It can be easily shown that the external field strength is enhanced to maximum $2 \times H_e$ at high frequencies, as illustrated in Figure 4.16.

4.2.3.4.4 ***Layered Conducting Shells*** The more general expression for the spherical and cylindrical multilayer shells with all layers then compared to their respective radius ($r_i \gg \Delta_i$) is given by the expression [19, 20]:

$$\begin{aligned}\frac{H_e}{H_{in}} = S^{(n)} = 1 + \sum_{i=1}^{n} S_i + \sum_{i=1}^{n-1}\sum_{j>i}^{n} S_iS_j\left[1 - \left(\frac{r_i}{r_j}\right)^x\right] + \\ + \sum_{i=1}^{n-2}\sum_{j>i}^{n-1}\sum_{k>j}^{n} S_iS_jS_k\left[1 - \left(\frac{r_j}{r_k}\right)^x\right]\left[1 - \left(\frac{r_i}{r_j}\right)^x\right] + \cdots + \\ + S_1S_2\ldots S_n\left[1 - \left(\frac{r_{n-1}}{r_n}\right)^x\right]\ldots\left[1 - \left(\frac{r_1}{r_2}\right)^x\right]\end{aligned} \tag{4.60}$$

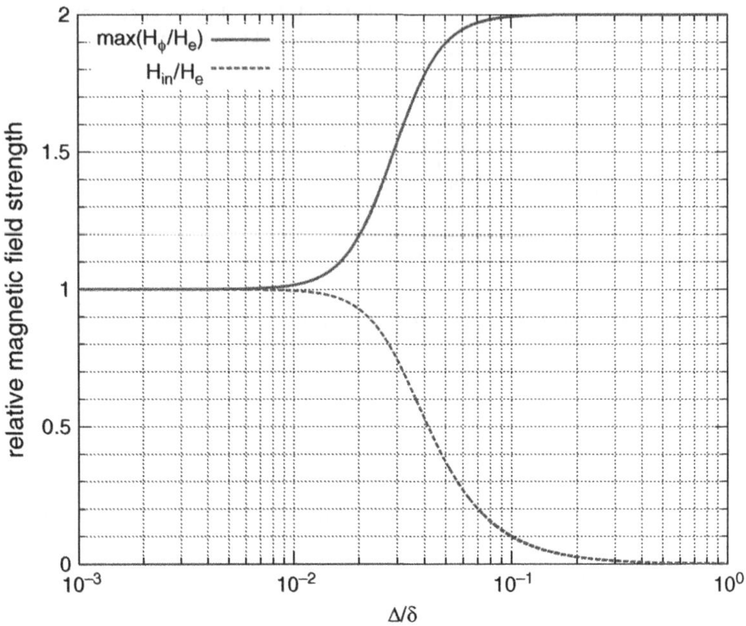

Figure 4.16 Variation of external and internal H-field strengths with frequency.

where x is the geometry factor, $x = 2$ for cylinders and $x = 3$ for spherical shields. The shielding factor S_i can be expressed as

$$S_i = \frac{1}{Q_i} - 1 \tag{4.61}$$

where Q_i is the corresponding shielding factor defined by equations (4.31), (4.38), or (4.45). For densely spaced thin shields (that means $r_1 \approx r_2 \approx \cdots \approx r_n$ and consequently $[1 - (r_1/r_2)^x] \approx 0$), the total shielding can be estimated as

$$\frac{H_e}{H_{in}} = S^{(n)} = 1 + \sum_{i=1}^{n} S_i \tag{4.62}$$

4.2.3.5 Extension to Non-circular Cylinders

Analytical solutions of field distribution for cylinders of non-circular cross-sections are limited to very special cases. For example, an oval fuselage or wing cross-section can be approximated by an ellipse with the major axis a and minor axis b (see Figure 4.17).

The area $A_{ellipse}$ enclosed by an ellipse can be calculated as

$$A_{ellipse} = \pi ab \tag{4.63}$$

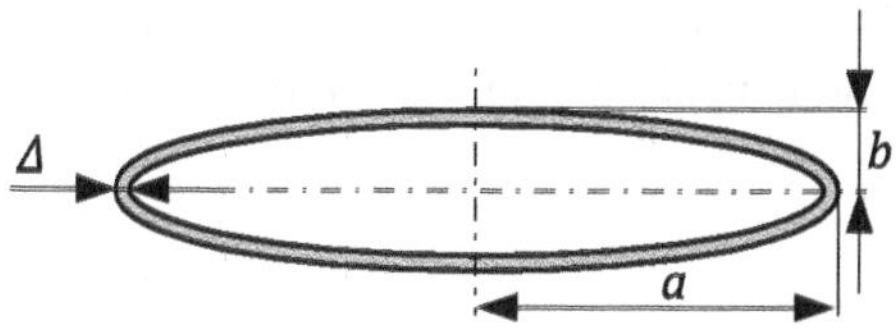

Figure 4.17 Elliptical shell.

The circumference (perimeter) of the ellipse cannot be exactly represented by elementary functions. An approximate formula is

$$C_{ellipse} \approx \pi \left[3(a+b) - \sqrt{(3a+b)(a+3b)}\right] \tag{4.64}$$

4.2.3.5.1 *Direct Injection of the Current* If a current I flows through the elliptical shell with a constant thickness Δ at very low frequencies and under DC conditions, the current distribution along the circumference will be governed by the resistance, i.e., for homogeneous shell, the current will distribute constantly over the shell. At high frequencies, the current distribution will be different since the penetration of magnetic field into the conductor will be opposed by the induced eddy currents and the total current will not be evenly distributed along the shell.

The DC magnetic field (i.e., no eddy current effect) of elliptic tube can be expressed using the magnetic field distribution for a full conductor of elliptic cross-section [21] where the magnetic field is expressed in dependence on complex variable $\bar{z}$:

$$\bar{z} = x + jy \tag{4.65}$$

The superimposition of the fields of two concentric elliptical conductors with axes (a, b) and $(a - \Delta, b - \Delta)$, respectively, yields the DC magnetic distribution for an elliptical shell:

Inside the shell

$$H_{in}^{DC}(\bar{z}) = \frac{I}{\pi\Delta(a+b-\Delta)}\left(\frac{bx - jay}{a+b} - \frac{(b-\Delta)x - j(a-\Delta)y}{a+b-2\Delta}\right) \tag{4.66}$$

and outside

$$H_{e}^{DC}(\bar{z}) = \frac{I}{\pi\Delta(a+b-\Delta)}\left(\frac{ab}{\bar{z}+\sqrt{\bar{z}^2-a^2+b^2}} - \frac{(a-\Delta)(b-\Delta)}{\bar{z}+\sqrt{\bar{z}^2-(a-\Delta)^2+(b-\Delta)^2}}\right) \tag{4.67}$$

At high frequencies, where due to eddy current effect, no magnetic field penetrates into the elliptical shell, the magnetic field outside is given by the equation [15]:

$$H_e^{HF}(\bar{z}) = \frac{I}{2\pi\sqrt{\bar{z}^2 - a^2 + b^2}} \tag{4.68}$$

Since for any point on the ellipse, the following is valid:

$$y = \pm b\sqrt{1 - \frac{x^2}{a^2}} \tag{4.69}$$

the magnetic field at the outer surface of the elliptic shell is given as

$$\left|H_e^{HF}\right| = \frac{I}{2\pi\sqrt{a^2 - \frac{a^2-b^2}{a^2}x^2}} \tag{4.70}$$

at center (x=0, y=b)

$$\left|H_e^{HF}\right|_{x=0,y=b} = \frac{I}{2\pi a} \tag{4.71}$$

and at the tip of the ellipse (x=a, y=0):

$$\left|H_e^{HF}\right|_{x=a,y=0} = \frac{I}{2\pi b} \tag{4.72}$$

The magnetic field distribution at DC and high frequencies is depicted in Figure 4.18.

The solution at low frequencies, where the magnetic field penetrates through the shell into the internal space requires application of numerical techniques. However, approximate response can be estimated using the low-pass filter

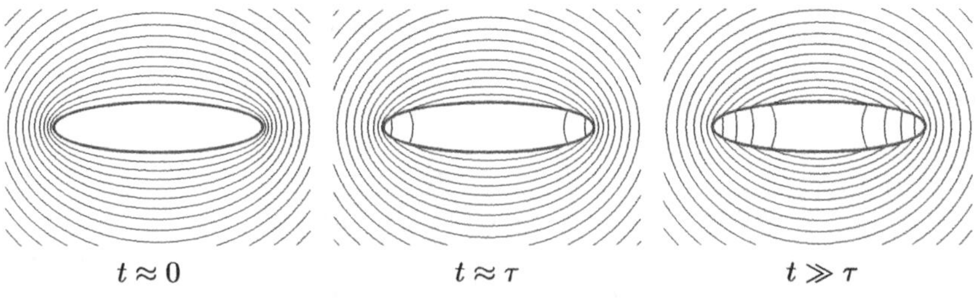

Figure 4.18 Temporal diffusion of magnetic field into elliptic tube.

approach described in subsection 4.2.3.4.3. The field inside then can be estimated to be

$$H_{in}^{LF}(\bar{z}) = \frac{H_{in}^{DC}(\bar{z})}{1 + j\omega\tau} \tag{4.73}$$

and the field outside:

$$H_{e}^{LF}(\bar{z}) = H_{e}^{DC}(\bar{z}) + \frac{j\omega\tau}{1 + j\omega\tau}\left[H_{e}^{HF}(\bar{z}) - H_{e}^{DC}(\bar{z})\right] \tag{4.74}$$

with the diffusion time constant expressed as

$$\tau = \mu_0 \Delta\sigma \frac{A_{ellipse}}{C_{ellipse}} \tag{4.75}$$

4.2.3.5.2 Elliptic Shell Immersed in External Field An analytical solution for an elliptical shell immersed in longitudinal field will be the same as in the case of cylindrical shell. The situation is different for the transversal field immersion. Analytical solution is not possible but one again the low-pass filter approach derived for cylindrical shell can be applied. The internal field can be estimated to be

$$H_{in} = \frac{H_e}{1 + j\omega\tau} = H_e \frac{1 - j\omega\tau}{1 + \omega^2\tau^2} \tag{4.76}$$

with the time constant:

$$\tau = \mu_0 \Delta\sigma \frac{A_{ellipse}}{C_{ellipse}} \tag{4.77}$$

4.2.3.6 Field Penetration Through Apertures

4.2.3.6.1 Open Apertures at Low Frequencies Aperture coupling denotes penetration of electromagnetic energy inside the aircraft by means of fields penetrating apertures, such as windows or canopies. The same process can take place at locations where doors, avionic compartments, wheel wells, etc., those covers are made of poorly conducting materials or/and are equipped with gaskets of poor electrical conductivity (poor electromagnetic seals), thus permitting significant penetration as shown in Figures 4.19(a) and 4.20(a).

The problems of apertures are generally solved in two steps applying the equivalence principle [22, 24, 25]. With the source of EM fields outside, we replace the internal region by a perfect conductor, so the fields inside are zero. This allows finding the field and surface current distributions that fulfill the

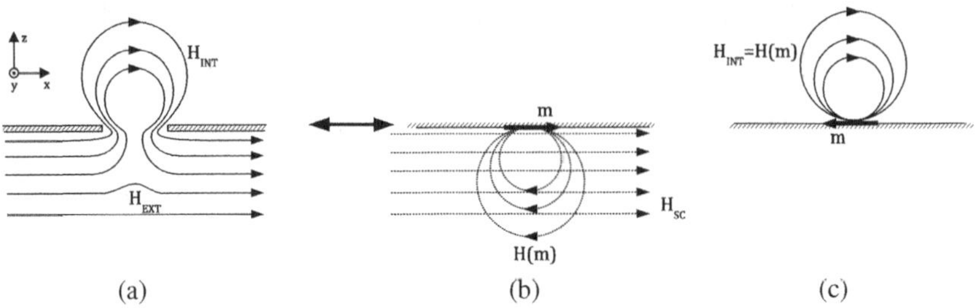

Figure 4.19 Magnetic field penetration through aperture and equivalent problems (a) problem outline, (b) external field, and (c) internal field.

boundary conditions. The apertures are in this step covered by perfect conductors and the normal electric field E_{SC} and the tangential magnetic field H_{SC} existing in the absence of the opening are calculated.

In the second step, a magnetic dipole m, whose moment is parallel to H_{SC}, is placed in the center of aperture. The resulting magnetic field outside (below the plane) is then superposition of the unperturbed H_{SC}, plus an opposing dipole field generated by the dipole m, $H(m)$ (see Figure 4.19b). Similarly, the electric field in outside region (below the plane) the appearance of the unperturbed normal field E_{SC}, plus the field $E(p)$ of a dipole p oriented oppositely to E_{SC} as depicted in Figure 4.20(b).

In the internal region, i.e., above the plane in Figures 4.19(c) and 4.20(c), the magnetic field appears to be generated by a magnetic dipole m whose moment is directed anti-parallel to H_{SC} and the electric field seems to originate from a vertical dipole moment p directed along E_{SC}. The electric and magnetic dipoles are defined by their respective moments given as

Magnetic dipole:

$$\vec{m} = -2\vec{\alpha}_m \vec{H}_{SC} \tag{4.78}$$

Electric dipole:

$$\vec{p} = 2\varepsilon_0 \vec{\alpha}_e \vec{E}_{SC} \tag{4.79}$$

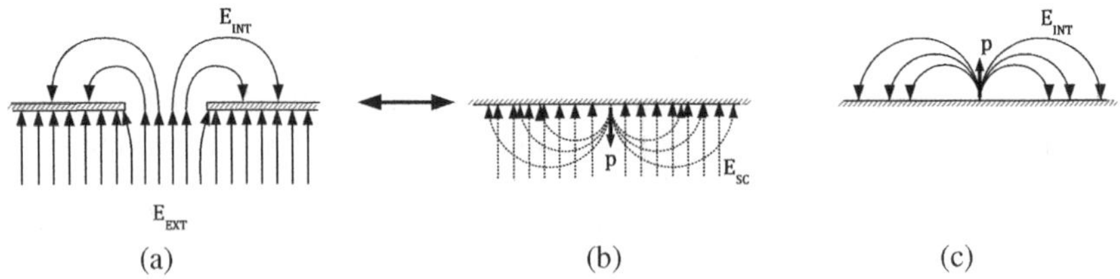

Figure 4.20 Electric field penetration through aperture and equivalent problems (a) problem outline, (b) external field, and (c) internal field.

Table 4.2 Polarizabilities of open apertures

Aperture shape	α_e	α_{mx}	α_{my}
Circle diameter d	$\frac{d^3}{12}$	$\frac{d^3}{6}$	$\frac{d^3}{6}$
Ellipse width w, length l	$\frac{\pi}{24}\frac{w^2 l}{E(e)}$	$\frac{\pi}{24}\frac{e^2 l^3}{K(e)-E(e)}$	$\frac{\pi}{24}\frac{e^2 l^3}{\frac{l^2}{w^2}E(e)-K(e)}$
Narrow ellipse ($w \ll l$)	$\frac{\pi}{24}w^2 l$	$\frac{\pi}{24}\frac{l^3}{\ln\frac{4l}{w}-1}$	$\frac{\pi}{24}w^2 l$
Narrow slit ($w \ll l$)	$\frac{\pi}{16}w^2 l$	$\frac{\pi}{24}\frac{l^3}{\ln\frac{4l}{w}-1}$	$\frac{\pi}{16}w^2 l$

$K(e)$ and $E(e)$ are the complete elliptical integrals of the first and the second kind

$$E(e) = \int_0^{\frac{\pi}{2}} \left(1 - e^2 \sin^2 \xi\right)^{\frac{1}{2}} d\xi \qquad K(e) = \int_0^{\frac{\pi}{2}} \left(1 - e^2 \sin^2 \xi\right)^{-\frac{1}{2}} d\xi$$

Ellipse eccentricity

$$e = \sqrt{1 - \left(\frac{w}{l}\right)^2}$$

where $\bar{\alpha}_m$ is the magnetic and $\bar{\alpha}_e$ is the electric polarizability of the aperture defined as

$$\vec{\alpha}_m = \alpha_{mx}\hat{x} + \alpha_{my}\hat{y} \tag{4.80}$$

$$\vec{\alpha}_e = \alpha_e\hat{z} \tag{4.81}$$

The polarizabilities for a few simply shaped apertures are given in Table 4.2 [22, 25]. Polarizabilities of other-shaped apertures can be either approximated using similarity with the simple shapes or found in [22]. The electric and magnetic fields of the dipoles given above are then given by the following set of equations (see Figure 4.21) [26]:

Electric dipole:

$$p = \frac{I}{j\omega} \cdot l \tag{4.82}$$

$$E_r = \frac{Il\cos\theta}{2\pi} e^{-j\frac{\omega}{c}r} \left(\frac{Z_0}{r^2} + \frac{1}{j\omega\varepsilon_0 r^3}\right) \tag{4.83}$$

$$E_\theta = \frac{Il\sin\theta}{4\pi} e^{-j\frac{\omega}{c}r} \left(\frac{j\omega\mu_0}{r} + \frac{Z_0}{r^2} + \frac{1}{j\omega\varepsilon_0 r^3}\right) \tag{4.84}$$

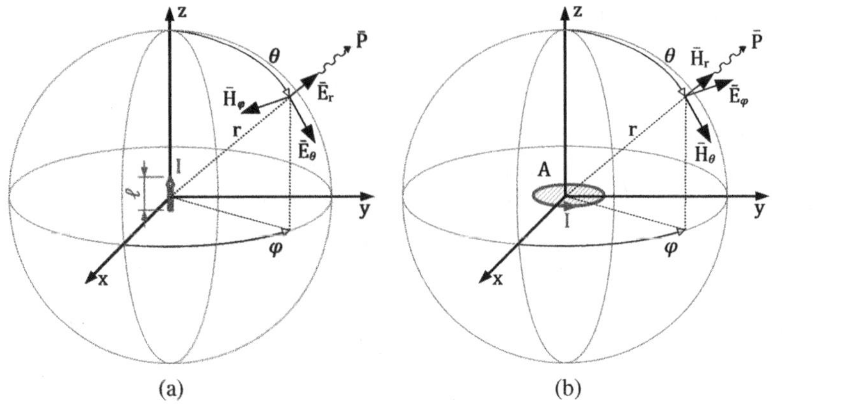

Figure 4.21 Electric and magnetic field components around dipoles (a) electric dipole, and (b) magnetic dipole.

$$H_\varphi = \frac{Il \sin\theta}{4\pi} e^{-j\frac{\omega}{c}r} \left(\frac{j\omega}{cr} + \frac{1}{r^2} \right) \tag{4.85}$$

Magnetic dipole:

$$m = I \cdot A \tag{4.86}$$

$$H_r = \frac{IA \cos\theta}{2\pi} e^{-j\frac{\omega}{c}r} \left(\frac{j\omega}{cr^2} + \frac{1}{r^3} \right) \tag{4.87}$$

$$H_\theta = \frac{IA \sin\theta}{4\pi} e^{-j\frac{\omega}{c}r} \left(-\frac{\omega^2}{c^2 r} + \frac{j\omega}{cr^2} + \frac{1}{r^3} \right) \tag{4.88}$$

$$E_\varphi = Z_0 \frac{IA \sin\theta}{4\pi} e^{-j\frac{\omega}{c}r} \left(\frac{\omega^2}{c^2 r} - \frac{j\omega}{cr^2} \right) \tag{4.89}$$

where Z_0 denotes the intrinsic impedance of free space (120π or approximately 377 Ω) and c is the speed of light.

A cover or door on an aperture, which is made of the same material as the aircraft skin and can be sealed by means of gaskets to reduce the penetration of electromagnetic energy, can be accounted for in the determination of the polarizabilities. The resulting polarizabilities of both unsealed and gasket-sealed covered apertures as well as hinged circular apertures can be found in [22].

Treatment of loaded apertures, i.e., apertures, where a non-conducting cover or door is provided with a conductive sheet, or the cover/door is made of poorly conducting material, can be found in [25].

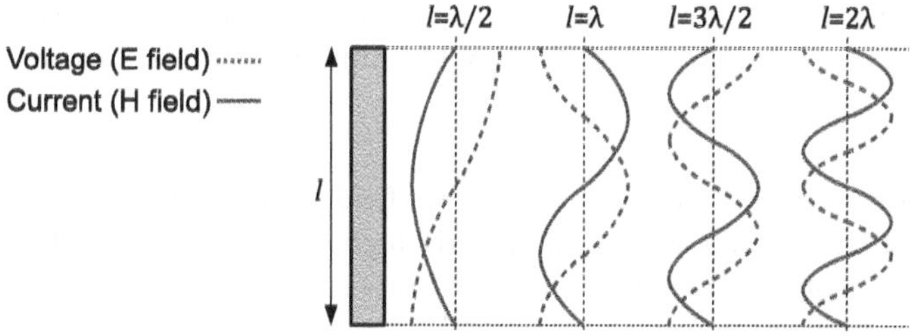

Figure 4.22 Modes of resonance for cylinders.

4.2.3.7 High-Frequency (HF) Behavior

4.2.3.7.1 External Resonances When the length of the aircraft is comparable with the wavelength of the incident electromagnetic field, the structure becomes resonant. At airframe fundamental resonances, the surface electromagnetic field is several times higher than the incident field at points not too close to the extremities of the airframe.

Since the aircraft is not a simple structure, an analytical solution is not possible and the solution requires implementation of numerical techniques. However, the basic behavior can be explained on the simplest example possible, i.e., representation of the aircraft fuselage by a conducting cylinder.

Let's consider the case of a cylindrical body as shown in Figure 4.22, immersed in a plane wave such that the electric field vector is parallel to the major axis of the cylinder. Longitudinal current will flow in the same way as current flow on a transmission line which is open circuited at each end, i.e., the longitudinal current distribution will be such that boundary conditions of zero current at the ends are satisfied.

The first (lowest) resonance frequency f_{res1} can be found from the condition using equation (4.6):

$$l = \frac{\lambda}{2} = \frac{c}{2f_{res1}} \Rightarrow f_{res1} = \frac{c}{2l} \tag{4.90}$$

If a 15 m-long aeroplane, the first resonance frequency can be estimated to be

$$f_{res1} = \frac{c}{2l} = \frac{3 \times 10^8\,\mathrm{m \cdot s^{-1}}}{2 \times 15\,\mathrm{m}} = 10\,\mathrm{MHz}$$

In practice, because of end capacitance, the current even at the end of cylinders is never zero. Additionally, on real aircraft, there will be superimposition of several resonance processes since every substructure (front fuselage, wings, aft fuselage, tail assembly) have different lengths and correspondingly different resonant frequencies. The amplitudes will depend on aircraft materials and the re-radiation of energy due to induced surface currents. Therefore, at high frequencies, measurements on real aircraft or complex numerical simulations are preferred.

4.2.3.7.2 Internal Resonances The electromagnetic field penetrating into the internal cavities of the aircraft will at sufficiently high frequencies give rise to resonances. The resonant frequencies of a cavity are simply the frequencies whose wavelengths correspond to its dimensions.

If the energy coupled is greater than the losses, the oscillating waves will increase in amplitude until the losses just equal the energy supplied. Losses take place at conducting surfaces, in any dielectric material present, and, open structures, through re-radiation. An important characteristic of a resonant mode is its quality factor Q, defined as [23, 24]:

$$Q = \omega_0 \frac{W_{AV}}{P_L} = 2\pi f_0 \frac{W_{AV}}{P_L} \tag{4.91}$$

where W_{AV} is the average energy stored, f_0 is the resonant frequency and P_L denotes power loss. The quality factor is a very important property of a cavity at a resonance mode because it influences both the maximum field amplitude and the bandwidth of the given mode, as depicted in Figure 4.23(a). The quantity $\Delta\omega$ is the separation of the frequencies where the power stored in the field drops to one half of the maximum (also denoted as the 3-dB bandwidth). It can be easily shown that very high quality (lossless) cavities are characterized by very high field strengths in very narrow frequency bands.

Influence of the lossy walls and objects in the cavity is outlined in depicted in Figure 4.23(b), where an increase in the loading of the cavity is accompanied by a decrease in the chamber Q, resulting in reduced maximum field strengths and more stable constant frequency behavior.

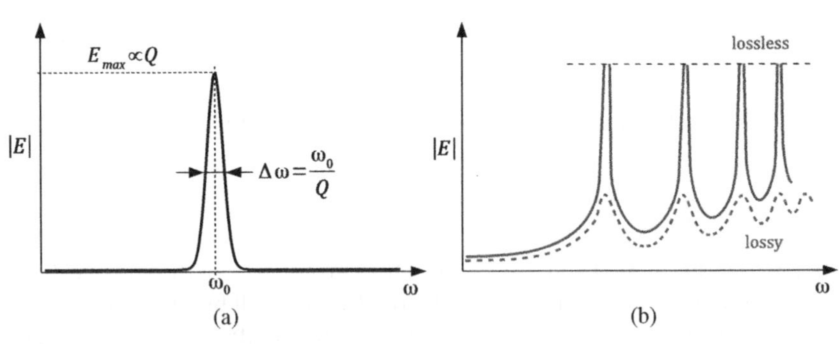

Figure 4.23 The frequency response curve of a resonant cavity (a) influence of the quality factor, and (b) damped and undamped cavity resonances.

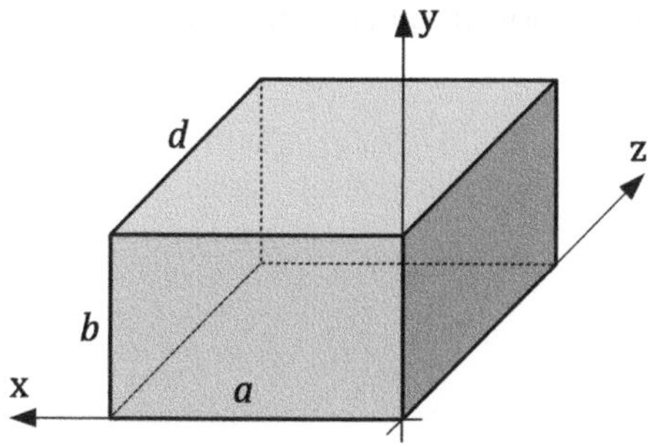

Figure 4.24 Geometry of rectangular cavity.

Rectangular cavities

A rectangular cavity with dimensions *a*, *b*, and *d* (see Figure 4.24) has the resonant frequencies defined by the well-known formula [24]:

$$f_{mnp} = \frac{1}{2\sqrt{\varepsilon_0 \mu_0}} \cdot \sqrt{\left(\frac{m}{a}\right)^2 + \left(\frac{n}{b}\right)^2 + \left(\frac{p}{d}\right)^2} \quad (4.92)$$

There are two fundamental cases:

1. Case 1: Transverse Magnetic TM_{mnp} mode, i.e., $H_z = 0$ and $E_z \neq 0$. For this mode $m = 1 \ldots \infty, n = 1 \ldots \infty, p = 0 \ldots \infty$. All other magnetic and electric field components (i.e., in directions of y and y axes, respectively) are non-zero.
2. Case 2: Transverse Electric TE_{mnp} mode, i.e., $E_z = 0$ and $H_z \neq 0$. For this mode, $p = 1 \ldots \infty, m = 0 \ldots \infty, n = 0 \ldots \infty$, except $m = n = 0$ at the same time. All other magnetic and electric field components are non-zero.

The mode that has the lowest resonant frequency for a given cavity size (a, b, d) is the dominant mode, governed by the two longest dimensions. For example: TE_{101} mode is the dominant mode of the rectangular resonator in case of $a > b < d$.

Circular cavities

An air-filled circular cylindrical cavity resonator of radius *a* and length *d* is depicted in Figure 4.25. The resonant frequencies of the TE_{mnp} modes are

$$f_{TE_{mnp}} = \frac{1}{2\pi\sqrt{\varepsilon_0 \mu_0}} \sqrt{\left(\frac{X'_{mn}}{a}\right)^2 + \left(\frac{p\pi}{d}\right)^2} \quad (4.93)$$

where the quantity X'_{mn} is the nth root of the derivative of the mth-order Bessel function:

$$J'_m(X) = 0$$

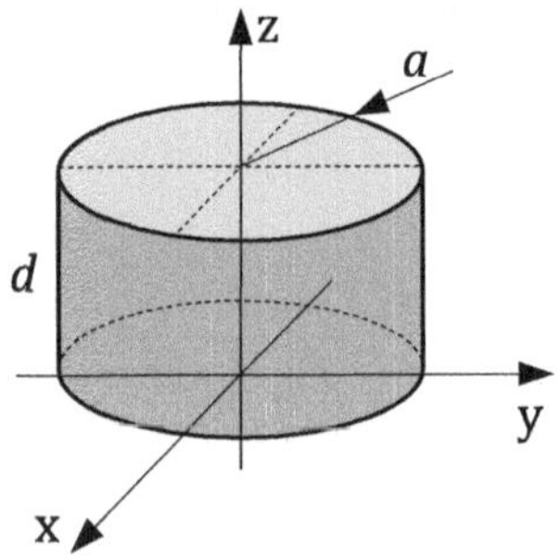

Figure 4.25 Circular cylindrical hollow cavity.

The resonant frequencies of the TM_{mnp} modes are different

$$f_{TM_{mnp}} = \frac{1}{2\pi\sqrt{\varepsilon_0\mu_0}}\sqrt{\left(\frac{X_{mn}}{a}\right)^2 + \left(\frac{p\pi}{d}\right)^2} \tag{4.94}$$

the quantity X_{mn} is the nth root of the mth-order Bessel function of the first kind:

$$J_m(X) = 0$$

Roots X_{mn} and X'_{mn} are calculated numerically and can be found tabulated, for example, in [24].

The lowest TE mode has $m = n = p = 1$, and is, therefore, denoted as TE_{111}. Root $X'_{11} = 1.841$ and the corresponding resonance frequency is

$$f_{TE_{111}} = \frac{1.841}{2\pi a\sqrt{\varepsilon_0\mu_0}}\sqrt{1 + 2.912\left(\frac{a}{d}\right)^2}$$

The lowest TM mode has $m = 0, n = 1, p = 0$, and so is designated by TM_{010}. Root $X_{01} = 2.405$ Its resonance frequency is

$$f_{TM_{010}} = \frac{2.405}{2\pi a\sqrt{\varepsilon_0\mu_0}}$$

When $d/a < 2$, the dominant mode is the TM_{010}, whereas for $d/a > 2$, the dominant mode is the TE_{111} mode. Both modes are outlined in Figure 4.26.

4.2.4 Estimation of Threat Levels at Equipment Interfaces

This subsection outlines methods to estimate the threat levels at the equipment interfaces. They are based on electromagnetic theory; however, a lot of

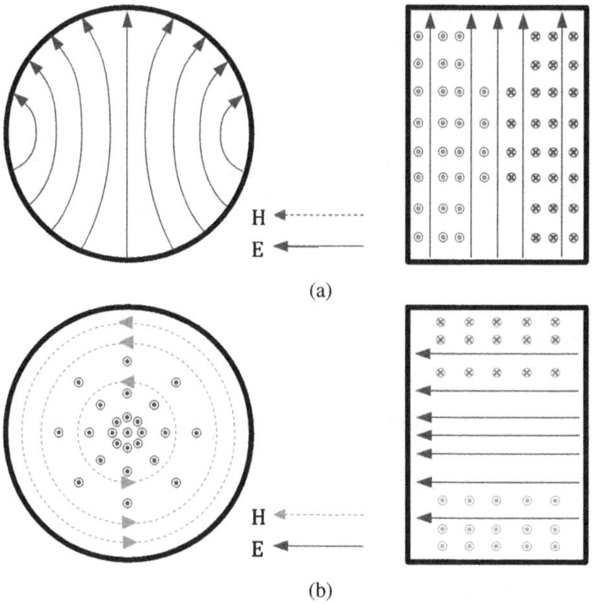

Figure 4.26 Field patterns in cylindrical cavity resonators (a) node pattern for the TE_{111} mode; (b) node pattern for the TM_{010} mode.

approximations and simplifications will be made to allow finding analytical solutions.

4.2.4.1 Resistive and Diffusion Coupling to Wires

Arbitrary two-dimensional (2D) structure can be approximated by an array of thin parallel conductors that are at each end short-circuited, so the voltage drop ΔV along of each conductor is the same. Using the circuit analysis, the current distribution in the structure can be calculated. Then, it is possible to calculate the field distributions in both the inner and outer regions.

Each of parallel conductors represents a segment of the structure, as illustrated in Figure 4.27(a). Therefore, the conductor's resistance shall be set to a value of the segment it represents. This is especially important for the structures with higher resistances. However, as it was mentioned in the previous sections, the time-varying penetrating magnetic field induces eddy current opposing the penetrating field. This action of circulating induced currents is taken into account by including both the wire self-inductance and mutual inductances between the wire filaments. The wire of interest routed parallel to the longitudinal axis can be then added to the structure as an additional filament. Then,

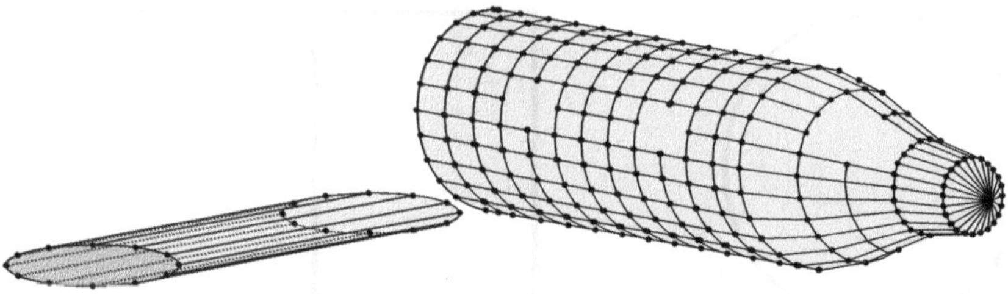

Figure 4.27 Thin wire model of aircraft structures (a) a parallel-filaments model of a wing; (b) 3D thin-wire model of a fuselage.

either the open-circuit voltage or the short-circuit current induced in the wire can be calculated.

The below-given formulas can be used to calculate the external partial self-inductance $L_{\{p,e\}}$ of a filament of length l and radius r_w [27]:

$$L_{\{p,e\}} = \frac{\mu_0}{2\pi} l \left(\ln \frac{2l}{r_w} - 1 \right) \quad l \gg r_w \tag{4.95}$$

For practical purposes or at high frequencies, the internal inductance of the wires can generally be neglected. However, for the sake of completeness, the value of the internal inductance of a wire is independent of the wire radius and can be expressed as

$$L_{\{p,i\}} = \frac{\mu_0}{8\pi} l \tag{4.96}$$

The total partial self-inductance is then given as

$$L_{\{p,total\}} = L_{\{p,e\}} + L_{\{p,i\}} \tag{4.97}$$

The mutual partial inductance between parallel wires of the same length l separated by distance d_{12} under condition $d_{12} \gg r_{w2}$ is then given by

$$M_{p12} = \frac{\mu_0}{2\pi} l \left[\ln \left(\frac{l}{d_{12}} + \sqrt{\left(\frac{l}{d_{12}} \right)^2 + 1} \right) - \sqrt{1 + \left(\frac{d_{12}}{l} \right)^2} + \frac{d_{12}}{l} \right] \tag{4.98}$$

Under condition $l \gg d$, this can be further simplified to

$$M_{p12} = \frac{\mu_0}{2\pi} l \left(\ln \frac{2l}{d_{12}} - 1 \right) \tag{4.99}$$

Then, for each of n wire filaments in the network, the following circuit equation can be written:

$$(R_{p,i} + j\omega L_{p,i})I_i + j\omega \sum_{\substack{k=1 \\ k \neq i}}^{n} M_{ik} I_k - \underbrace{(V_{node1} - V_{node2})}_{\Delta V} = 0 \tag{4.100}$$

When current I_{total} enters node 1 and exits node 2, then

$$\sum_{k=1}^{n} I_k = I_{total} \tag{4.101}$$

In the matrix form, the problem can be expressed as

$$\begin{bmatrix} [Z] & [A]^T \\ [A] & [0] \end{bmatrix} \begin{bmatrix} [I] \\ [\Delta V] \end{bmatrix} = \begin{bmatrix} [0] \\ I_{total} \end{bmatrix} \tag{4.102}$$

where [A] is the branch-node incidence matrix. Solving this set of equations, the current distribution between individual filaments can be obtained.

This approach can be expanded to 3D structures where the structure can be represented by a mesh of wires. Figure 4.27(b) illustrates this configuration. Mutual inductance for arbitrary oriented wire elements that are inclined with respect to each other at an angle θ is given by [27]

$$M_{p12} = \frac{\mu_0}{4\pi} \int_{l_2} \int_{l_1} \frac{dl_1 \cdot dl_2}{r_{12}} = \frac{\mu_0}{4\pi} \cos\theta \int_{l_2} \int_{l_1} \frac{1}{r_{12}} dl_1 dl_2 \tag{4.103}$$

where r_{12} is the distance between the differential segments dl_1 and dl_2. Analytical expression for several geometries can be found in [31]. When $l_1 \ll r_{12} \gg l_2$, then

$$M_{p12} = \frac{\mu_0}{4\pi} \int_{l_2} \int_{l_1} \frac{dl_1 \cdot dl_2}{r_{12}} \approx \frac{\mu_0}{4\pi} \frac{l_1 l_2}{r_{12}} \cos\theta \tag{4.104}$$

The model can be further expanded by adding the filament self and mutual capacitances.

Example The parallel-filaments model depicted in Figure 4.28 represents a 5 m-long aluminum wing with a composite leading edge, both of a thickness of 2 mm. A shielded cable of radius 2 mm and resistance of 0.1 Ω/m is routed at a height of 2.5 cm above the aluminum wall under the composite leading cover. The resistivity of the composite material is a hundred times higher than the one of aluminum.

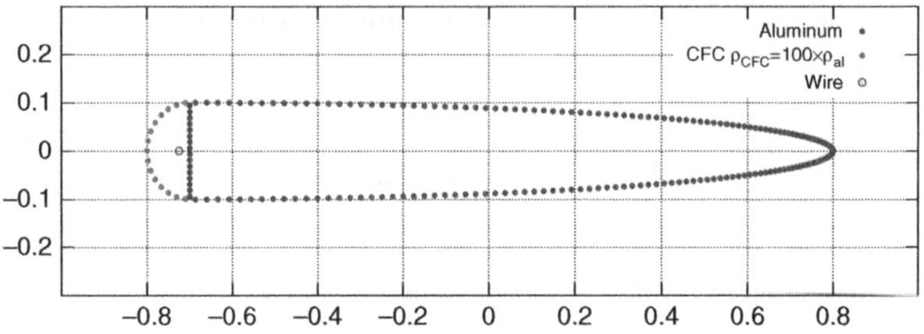

Figure 4.28 Thin wire model of an aluminum wing with a composite leading edge.

The injected impulse current and the resulting current in the shield of the cable are shown in Figure 4.29.

4.2.4.2 Aperture Coupling to Wires

A situation, where a wire bundle is located near a hole in conducting aircraft skin, can be represented as an aperture-wire configuration depicted in Figure 4.30(a) where a wire of radius *a* is routed at height *h* horizontally above the conducting wall at distance *S* from the center of the aperture with dimensions *w* and *l*, respectively.

This problem can be solved in two steps. First, the tangential magnetic field H_{SC} and the normal electric field E_{SC} are calculated at the location of the aperture but with the aperture absent, as described in subsection 4.2.3.6. Then, the

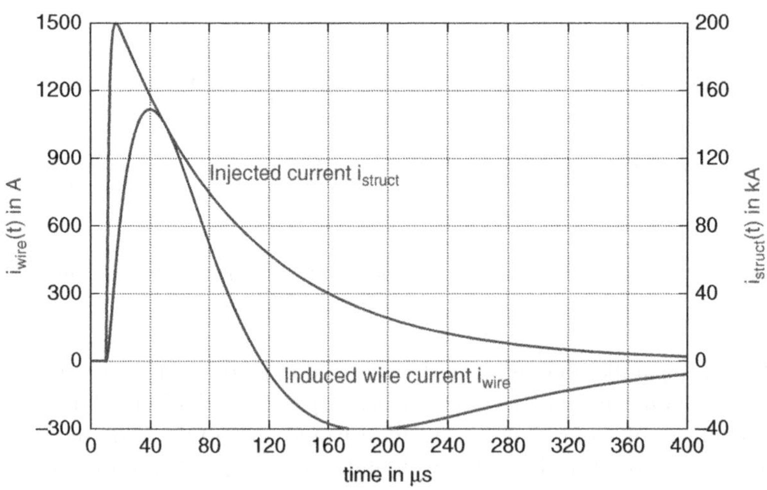

Figure 4.29 Thin wire model of an aluminum wing with a composite leading edge.

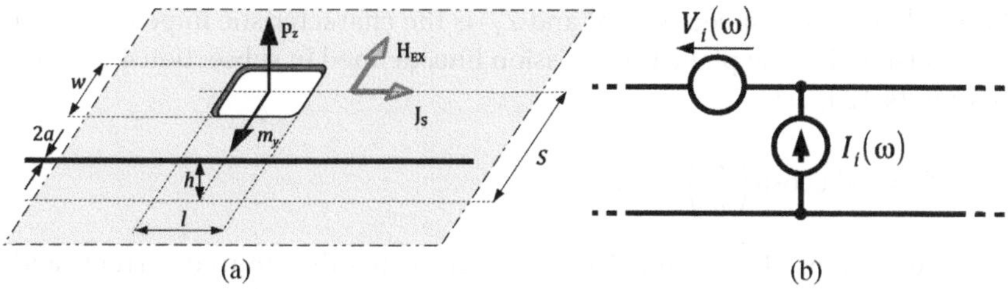

Figure 4.30 Aperture penetration and interaction with an interior wire (a) physical arrangement for simplified problem; (c) equivalent circuit.

coupling of fields penetrating through the aperture to a wire routed close to the aperture can be represented by ideal voltage and current sources according to the equivalent circuit diagram depicted in Figure 4.30(b) [22].

For a wire directly routed over aperture, the maximum short-circuit current induced by the electric field E_{SC} can be estimated as [28]:

$$I_i = j\omega Q = j\omega\varepsilon_0 A E_{SC} \tag{4.105}$$

where A is the aperture area. The maximum open-circuit voltage induced by the magnetic external field H_{SC} is approximately

$$V_i = j\omega\mu_0 \frac{A}{Pw} H_{SC} \tag{4.106}$$

where P denotes the perimeter of the aperture.

For a wire not too close to the edge of window ($w/R_0 < 1$) and close to the conducting wall ($h/R_0 \leq 0.1$), the voltage induced on the wire by a surface tangential magnetic field H_{SC} perpendicular to the wire (the worst-case coupling) is represented by the ideal voltage source with open-circuit voltage:

$$V_i = \mu_0 \alpha_m \frac{h}{\pi R_0^2} j\omega H_{SC} \tag{4.107}$$

where

$$R_0 = \sqrt{S^2 + h^2} \tag{4.108}$$

and α_m is the corresponding magnetic polarizability of the aperture defined in subsection 4.2.3.6. The short-circuit current of the current source representing the influence of the electric field is given as

$$I_i = \frac{\alpha_e}{cZ_C} \frac{h}{\pi R_0^2} j\omega E_{SC} \tag{4.109}$$

where c denotes the speed of light and Z_C is the characteristic impedance of a wire-over-conducting plane transmission line, defined in subsection 4.2.4.3 by equation (4.124) as:

$$Z_C = 60 \text{ acosh}\left(\frac{h}{a}\right)$$

From equations (4.107) and (4.109), it is clear that the induced currents and voltages are proportional to frequency, or in other words, that high-frequency interference fields couple more strongly to internal circuits than low frequency ones. In the time domain, the expression $j\omega$ can be replaced by the derivative operator d/dt. Then, the induced currents and voltages are proportional to the derivatives (the rate of change) of the magnetic and electric fields, respectively. This is the reason why aperture coupling is important for both fast-rising lightning transients and RF fields.

When R_0 is not large compared to the aperture size, a small-hole correction factor f_S must be applied:

$$V_{OC} = f_S \mu_0 \alpha_m \frac{h}{\pi R_0^2} j\omega H_{SC} \tag{4.110}$$

$$I_{SC} = f_S \frac{\alpha_e}{cZ_C} \frac{h}{\pi R_0^2} j\omega E_{SC} \tag{4.111}$$

The correction factor depends on the shape of the aperture as well as the position of the wire with respect to the aperture. The correction factors for different apertures can be found in [22, 29, 30].

Using the concept of representing the aperture coupled voltages and currents by the equivalent lumped ideal sources, the responses of arbitrary wires or cables in the form of the terminal voltages and currents can be calculated [25].

Example If fuselage of diameter D with a circular aperture of diameter d carries current $I(\omega)$ distributed evenly around the fuselage circumference, the magnetic field coupling to an internal wire routed at height h at a distance S from the center of the aperture can be estimated as follows:

1. An equivalent voltage source is given by equation (4.107)

$$V_i(\omega) = j\omega\mu_0 \frac{h}{\pi(S^2 + h^2)} \alpha_{my} H_{SC}$$

2. For circular aperture with diameter d is the corresponding magnetic polarizability α_{my} (see Table 4.2):

$$\alpha_{my} = \frac{d^3}{6} \tag{4.112}$$

3. Without the aperture present, the surface magnetic field due to current $I(\omega)$ distributed evenly along the circumference of the fuselage can be expressed by Ampère's Law as

$$H_{SC} = \frac{I(\omega)}{\pi D} \tag{4.113}$$

4. Inserting the expressions (4.112) and (4.113) into equation (1)

$$V_i(\omega) = j\omega\mu_0 \frac{h}{\pi(S^2+h^2)}\alpha_{my}H_{SC} = j\omega\,\underbrace{\mu_0\frac{h}{\pi(S^2+h^2)}\frac{d^3}{6}\frac{1}{\pi D}}_{M}I(\omega)$$

$$V_i(\omega) = j\omega M I(\omega) \tag{4.114}$$

where M denotes the mutual inductance between aperture and internal wiring. In the time domain (refer to subsection 4.2.3.1.1), the corresponding expression for the induced voltage is

$$v_i(t) = M\frac{di(t)}{dt} \tag{4.115}$$

In the case of a circular fuselage of diameter $D = 3$ m and the internal wire routed at height $h = 5$ cm at a distance $S = 0.5$ m from the centre of a circular aperture of diameter $d = 50$ cm, the mutual inductance M is

$$\begin{aligned} M &= \mu_0\frac{h}{\pi^2(S^2+h^2)}\frac{d^3}{6} = \frac{\mu_0}{\pi^2}\frac{0.05\,\text{m}}{(0.5\,\text{m})^2+(0.05\,\text{m})^2}\frac{(0.5\,\text{m})^3}{6}\frac{1}{3\,\text{m}} \\ &= 1.75\times10^{-10}\,\text{H} \end{aligned} \tag{4.116}$$

4.2.4.3 Load Responses for a Finite Line with Excitation at the Beginning

As an initial case, let's consider a finite line terminated at both ends with load impedances Z_1 and Z_2 and having a voltage excitation V_0 at the beginning of the line as shown in Figure 4.31. The voltage and current at the input of the transmission is described by the Telegrapher's equations [31]:

$$V(0) = V(l)\cosh(\gamma l) + Z_C I(l)\sinh(\gamma l) \tag{4.117}$$

$$I(0) = \frac{V(l)}{Z_C}\sinh(\gamma l) + I(l)\cosh(\gamma l) \tag{4.118}$$

where Z_C represents characteristic impedance of the line,

$$Z_C = \sqrt{\frac{R' + j\omega L'}{G' + j\omega C'}} \tag{4.119}$$

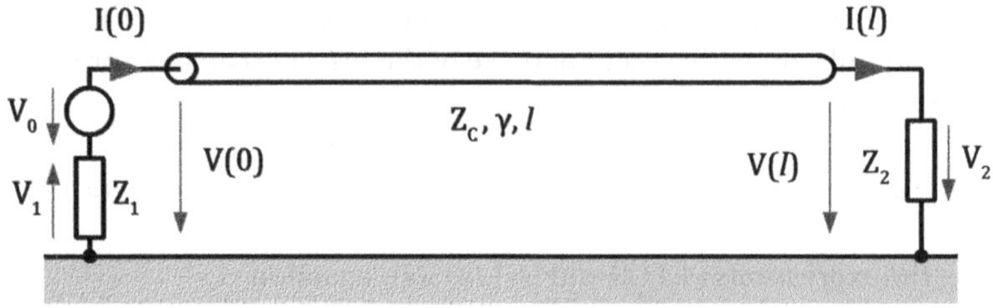

Figure 4.31 Finite line with excitation at the beginning.

l is the length and γ is the propagation constant:

$$\gamma = \sqrt{(R' + j\omega L')(G' + j\omega C')} \tag{4.120}$$

with R', L', G', and C' representing per-unit-length (p.u.l.) resistance, inductance, conductance and, capacitance of the line. For a horizontal wire or radius a routed in air at height h above an infinite ideal ground plane, the following expressions for L' and C' can be used:

$$L' = \frac{\mu_0}{2\pi}\text{acosh}\left(\frac{h}{a}\right) \approx \frac{\mu_0}{2\pi}\ln\left(\frac{2h}{a}\right) \tag{4.121}$$

$$C' = \frac{2\pi\varepsilon_0}{\text{acosh}\left(\frac{h}{a}\right)} \approx \frac{2\pi\varepsilon_0}{\ln\left(\frac{2h}{a}\right)} \tag{4.122}$$

with the approximate formulas valid for $h \gg a$. The p.u.l conductance can be generally neglected:

$$G' = \frac{2\pi\sigma_{\text{air}}}{\text{acosh}\left(\frac{h}{a}\right)} \approx 0 \tag{4.123}$$

The last variable to be defined is the p.u.l. resistance of the wire. It is often neglected, especially for the "worst-case" estimations. For LF considerations, where the penetration depth is much higher than the wire radius, the p.u.l. DC resistance can be used. At higher frequencies, the formulas for the surface impedance of a tube (4.12) or a woven shield (4.181) should be used.

Neglecting the p.u.l. resistance and conductance and inserting equations (4.121) and (4.122) into equation (4.119), one obtains formula for a characteristic impedance of a horizontal wire above an infinite ideal ground plane

$$Z_C = 60\ \text{acosh}\frac{h}{a} \approx 60\ln\frac{2h}{a} \tag{4.124}$$

The equations (4.117) and (4.118) can be rewritten as

$$V(0) = V(l)\left[\cosh(\gamma l) + \frac{Z_C}{Z_2}\sinh(\gamma l)\right] \quad (4.125)$$

$$\frac{V_0 - V(0)}{Z_1} = V(l)\left[\frac{1}{Z_C}\sinh(\gamma l) + \frac{1}{Z_2}\cosh(\gamma l)\right] \quad (4.126)$$

Rearrangement of equation (4.125) and its insertion into equation (4.126) yields

$$\frac{V_0 - V(0)}{Z_1} = V(0)\left[\frac{\frac{1}{Z_C}\sinh(\gamma l) + \frac{1}{Z_2}\cosh(\gamma l)}{\cosh(\gamma l) + \frac{Z_C}{Z_2}\sinh(\gamma l)}\right] = \frac{V(0)}{Z_{in1}} \quad (4.127)$$

The inverted expression in square brackets as can be interpreted as the input impedance of the transmission line:

$$Z_{in1} = Z_C\frac{Z_2 + Z_C\tanh(\gamma l)}{Z_C + Z_2\tanh(\gamma l)} = Z_C\frac{1 + \rho_2 e^{-2\gamma l}}{1 - \rho_2 e^{-2\gamma l}} \quad (4.128)$$

where ρ_2 denotes reflection coefficient for the end of the line

$$\rho_2 = \frac{Z_2 - Z_C}{Z_2 + Z_C} \quad (4.129)$$

Expressing the hyperbolic sine and cosine in equation (4.127) using the exponential form and subsequent rearrangement yields the voltage at the beginning:

$$V(0) = V_0\left(\frac{1-\rho_1}{2}\cdot\frac{1 + \rho_2 e^{-2\gamma l}}{1 - \rho_1\rho_2 e^{-2\gamma l}}\right) \quad (4.130)$$

with the reflection coefficient for the beginning of the line ρ_1 defined as

$$\rho_2 = \frac{Z_1 - Z_C}{Z_1 + Z_C} \quad (4.131)$$

The voltage across Z_1 can be obtained straightforwardly as follows:

$$V_1 = V_0 - V(0)$$

$$V_1 = V_0\frac{(1+\rho_1)(1-\rho_2 e^{-2\gamma l})}{2(1-\rho_1\rho_2 e^{-2\gamma l})} \quad (4.132)$$

The expression for the voltage at the end of the line is obtained by insertion of equation (4.130) into equation (4.125):

$$V_2 = V(l) = V_0 \frac{(1+\rho_2)\, e^{-\gamma l}\,(1-\rho_1)}{2\left(1-\rho_1\rho_2 e^{-2\gamma l}\right)} \tag{4.133}$$

The line voltage and current at arbitrary point x on the line can be also of interest. One way of finding them is by replacing the remaining section of the line between the point of interest and the remote terminal $l - x$ by the section's input impedance $Z_{in}(l-x)$ defined equivalently to equation (4.128) and correspondingly modifying expression for the reflection coefficient (4.129) before solving equation (4.133) for line length x. After some rearrangements, one obtains:

$$V(x) = \frac{Z_C}{Z_1 + Z_C} V_0 e^{-\gamma x} \frac{1+\rho_2 e^{-2\gamma(l-x)}}{1-\rho_1\rho_2 e^{-2\gamma l}} \tag{4.134}$$

$$I(x) = \frac{1}{Z_1 + Z_C} V_0 e^{-\gamma x} \frac{1-\rho_2 e^{-2\gamma(l-x)}}{1-\rho_1\rho_2 e^{-2\gamma l}} \tag{4.135}$$

The previous analysis of coupling dealt with a voltage source only. However, an excitation by a current source I_0 can be analyzed using the same formulae, it is just necessary to transform the lumped current source to a lumped voltage source. An example of a source transformation can be seen in Figure 4.32

The open-circuit voltage of the equivalent voltage source in series with the impedance is given as

$$V_{0eq} = Z_1 I_0 \tag{4.136}$$

A simultaneous excitation by a voltage and a current source can be solved by a superimposition of the voltage and the equivalent voltage sources.

Let's treat a transmission line for two limiting cases of the remote end termination, i.e., an open circuit and a short-circuit condition, respectively.

Open circuit	Short circuit
$Z_2 \to \infty \Rightarrow I(l) = 0$	$Z_2 = 0 \Rightarrow V(l) = 0$
$V(0) = V(l)\cosh\gamma l$	$V(0) = Z_C I(l)\sinh\gamma l$
$V(l) = \dfrac{V(0)}{\cosh\gamma l}$	$I(l) = \dfrac{V(0)}{Z_C}\cdot\dfrac{1}{\sinh\gamma l}$

For a lossless line $\gamma = \sqrt{(R' + j\omega L')(G' + j\omega C')} = j\omega\sqrt{L'C'} = j\omega/v$

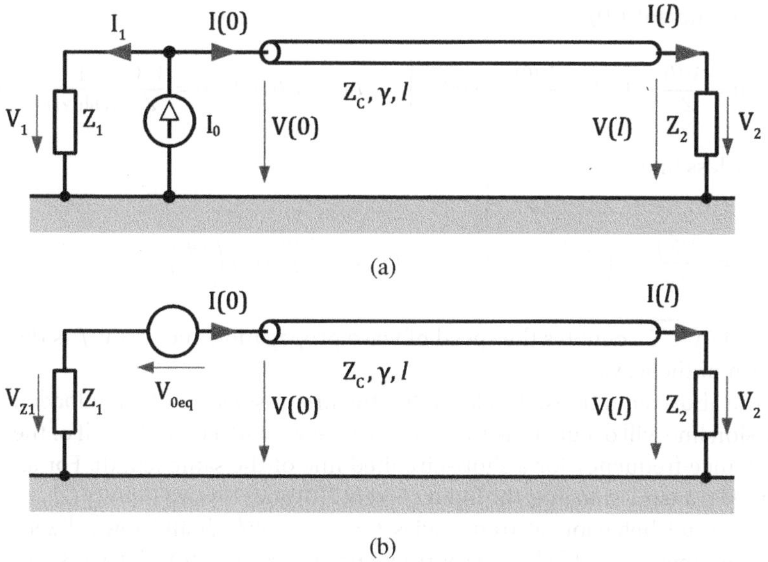

Figure 4.32 Finite line with excitation by an ideal current source at the beginning (a) a current source with short-circuit current I_0; (b) an equivalent voltage source V_{0eq}.

$$V(l) = \frac{V(0)}{\cos\left(\frac{\omega l}{v}\right)}$$

$$V(l) \to \infty \Leftrightarrow \cos\frac{\omega_r l}{v} = 0 \Leftrightarrow$$

$$\Leftrightarrow \frac{\omega_r r}{v} = \frac{\pi}{2}(2n+1)\,\forall n = 0\cdots\infty$$

$$\frac{2\pi f_r l}{v} = \frac{\pi}{2}(2n+1)$$

$$l = (2n+1)\frac{\lambda_r}{4}$$

$$f_r = (2n+1)\frac{v}{4l}$$

$$n = 0 \Rightarrow l = \frac{\lambda_r}{4}$$

$$I(l) = \frac{V(0)}{Z_C}\cdot\frac{1}{j\sin\left(\frac{\omega}{v}l\right)}$$

$$I(l) \to \infty \Leftrightarrow \sin\frac{\omega_r l}{v} = 0 \Leftrightarrow$$

$$\Leftrightarrow \frac{\omega_r l}{v} = \pi(n+1)\,\forall n = 0\cdots\infty$$

$$\frac{2\pi f_r l}{v} = \pi(1+n)$$

$$l = \frac{\lambda_r}{2}(1+n)$$

$$f_r = \frac{v}{2l}(1+n)$$

$$n = 0 \Rightarrow l = \frac{\lambda_r}{2}$$

The input current $I(0)$

$$I(0) = \frac{V(l)}{Z_C}\sinh(\gamma l) = \frac{V(0)}{Z_C}\tanh(\gamma l) \quad \Bigg| \quad I(0) = I(l)\cosh(\gamma l) = \frac{V(0)}{Z_C}\frac{1}{\tanh(\gamma l)}$$

For a lossless line:

$$I(0) = j\frac{V(0)}{Z_C}\tan\left(\frac{\omega l}{v}\right) \quad \Bigg| \quad I(0) = -j\frac{V(0)}{Z_C}\operatorname{cotg}\left(\frac{\omega l}{v}\right)$$

where $v = 1/\sqrt{L'C'}$ denotes the speed of wave propagation and $\lambda = v/f$ is the wavelength of the wave.

From the above equations, it follows that the first resonance of an unloaded transmission line will occur at the frequency where $\lambda_r = 4l$, i.e., at the half of the first resonance frequency for a short-circuited line of the same length. For an open-ended transmission line, the input current $I(0)$ goes toward infinity ($Z_{in} = 0$, a short-circuit behavior) at frequencies $f_r = (1+n)v/(2l)$ and toward zero ($Z_{in} = \infty$, an open-circuit behavior) at frequencies $f_r = (1+2n)\,v/(4l)$, respectively (see Figure 4.33a). For a short-circuited line, this behavior is swapped, i.e., the open-circuit behavior at $f_r = (1+n)v/(2l)$ and the short-circuit behavior at $f_r = (1+2n)\,v/(4l)$, respectively, as depicted in Figure 4.33(b).

4.2.4.4 General Solution for a Terminated Line

The case of a finite line with a voltage excitation V_0 at the beginning represents a limiting case of general problem for a finite line with excitation at arbitrary position x_s along the line as shown in Figure 4.34.

The lumped voltage source divides the transmission line in the left-hand and the right-hand part. The left-hand transmission line can be replaced by its input impedance:

$$Z_{in1} = Z_C\frac{1+\rho_2 e^{-2\gamma x_s}}{1-\rho_2 e^{-2\gamma x_s}} \tag{4.137}$$

This reduced the problem to a problem with an excitation at the beginning treated in the subsection 4.2.4.3. The left-hand reflection coefficient (4.131) is now expressed as

$$\rho_1' = \frac{Z_{in1} - Z_C}{Z_{in1} + Z_C} = \rho_1 e^{-2\gamma x_s} \tag{4.138}$$

After insertion into equation (4.133), the general solution for the voltage at the end of the line is given as

$$V_2 = V(l) = V_0\frac{(1+\rho_2)\,e^{-\gamma l}\,(e^{\gamma x_s} - \rho_1 e^{-\gamma x_s})}{2\left(1-\rho_1\rho_2 e^{-2\gamma l}\right)} \tag{4.139}$$

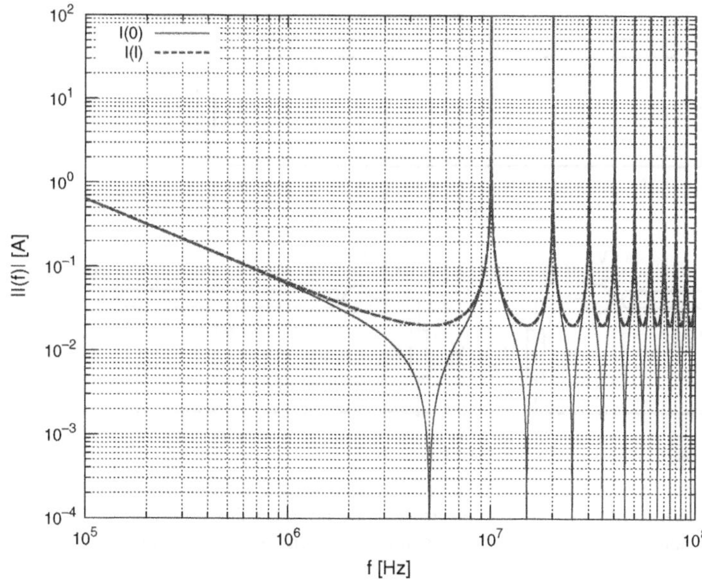

(a)

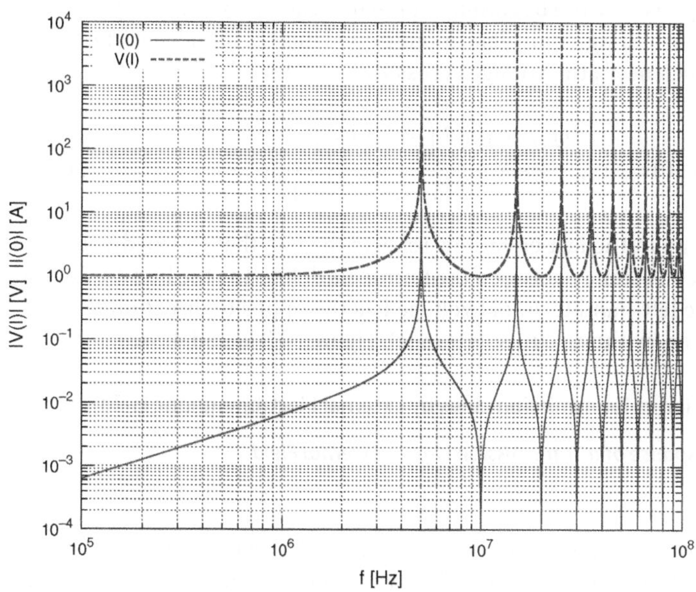

(b)

Figure 4.33 Variation of $I(0)$ and $I(l)$ or $V(l)$ of a transmission line with frequency for $V_0 = 1$ V and $Z_1 = 0\ \Omega$, $Z_C = 50\Omega$, $l = 15$ m and $v = 3 \times 10^8$ m · s (a) remote end short circuited; (b) remote end open.

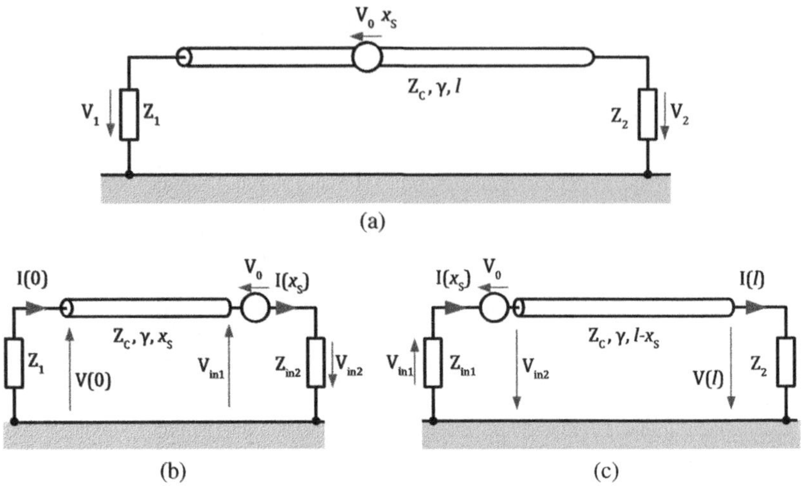

Figure 4.34 Finite Line with excitation at arbitrary point along the line (a) problem definition; (b) approach for the left-hand load solution, (c) approach for the right-hand load solution.

The load response for the beginning of the line can be found using the same procedure. After finding equivalent input impedance Z_{in2} and the equivalent reflection coefficient ρ_2', the voltage across Z_1 is expressed as

$$V_1 = -V(0) = -V_0 \frac{(1+\rho_1)\, e^{-\gamma l}\left(e^{\gamma(l-x_s)} - \rho_2 e^{-\gamma(l-x_s)}\right)}{2\left(1-\rho_1\rho_2 e^{-2\gamma l}\right)} \tag{4.140}$$

The minus sign in the last equations reflects the fact that the load voltage is for the sake of consistency between the terminals defined as the line-to-ground voltage, i.e., the left-hand terminal voltage has the opposite direction than the voltage predicted by the circuit solution (cf. Figures 4.34a and 4.34b).

4.2.4.5 Excitation of Transmission Lines by External Electromagnetic Field

4.2.4.5.1 LF Response of a Transmission Line to External Field

Electric field

The wire-to-ground voltage, as drawn in Figure 4.35, can be estimated through multiplication of the height h of the wire by the vertical component of the field strength E_z at the position of the wire [32].

In the case that both ends of the wire are connected to the ground plane via low impedances, the total current coupled capacitively along the complete length of the line flows out of the line to the left and right terminals split into

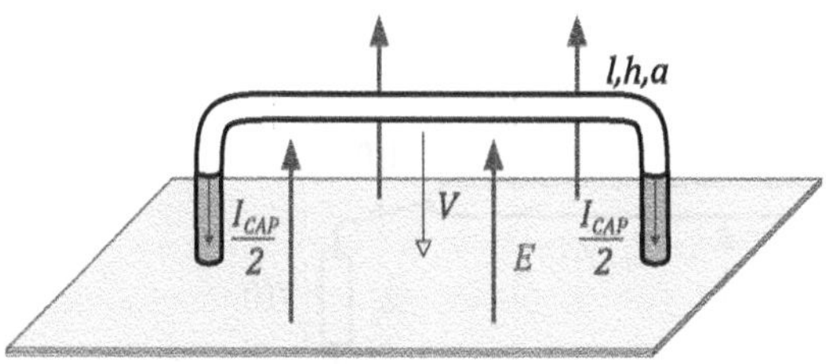

Figure 4.35 Model for the electric coupling.

equal parts. This fact can be approximated by an equivalent ideal current source connected between the middle point of the line and the ground plane. If only one end is connected to the ground plane, the total capacitive current flows in that direction. The short-circuit current of the equivalent current source is given by the equation:

$$I_i(\omega) = j\omega \cdot C' \cdot l \cdot E_z \cdot h \tag{4.141}$$

where C' is the per-unit-length capacitance of the line given by equation (4.122).

Magnetic field

The induced terminal voltage for the loop formed by the cable (see Figure 4.36) above the ground plane is defined by the induction law:

$$V_i(\omega) = j\omega \cdot \mu_0 \cdot H_y \cdot h \cdot l \tag{4.142}$$

This voltage can be represented by a serial lumped ideal voltage source located at the middle point of the line [32]. With exception of an open loop, this induced voltage gives rise to an inductively coupled current, which is limited, by the effective impedance of the loop:

$$I(\omega) = \frac{V_i}{Z} = \frac{j\omega \cdot \mu_0 \cdot H_y \cdot h \cdot l}{R + j\omega L} \tag{4.143}$$

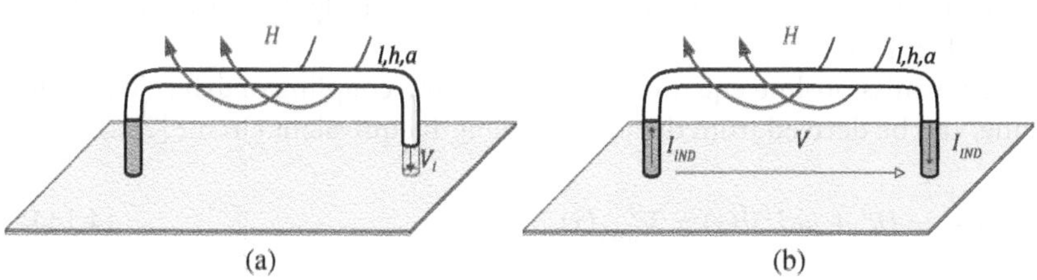

Figure 4.36 Model for the magnetic coupling (a) open loop; (b) loop short-circuited.

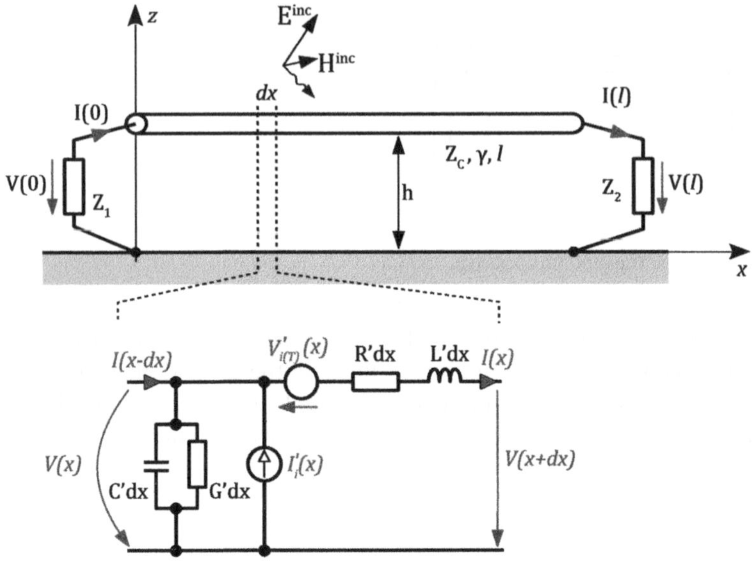

Figure 4.37 Excitation of a conductor over a ground plane by external EM field.

with R denoting the total resistance of the loop comprising the resistance of the line itself and the resistances of the both terminations and L standing for the inductance of the loop given as $L = l \cdot L'$, where L' is the p.u.l. inductance defined by expression (4.121).

In the very low frequency range (where $R \gg \omega L$), the resistance R limits the current. However, in the medium frequency range ($R \ll \omega L$), the inductance L is the governing factor.

4.2.4.5.2 *HF Response of a Transmission Line to External Field* A coupling of external electromagnetic field to a single-conductor over good-conducting structure can be estimated using a model of field coupling to a single-conductor transmission line of length l and wire radius a routed horizontally at height h over a perfectly conducting ground plane, characterized by a characteristic impedance Z_C and terminated by impedances Z_1 and Z_2, respectively. This problem is illustrated in Figure 4.37.

The transmission-line equations, describing the coupling of external field to the line, can be derived from the per-unit-length equivalent circuit [25,31,33]:

$$\frac{dV}{dx} = (R' + j\omega L')I(x) = V'_{i(T)}(x) \tag{4.144}$$

$$\frac{dI}{dx} = (G' + j\omega C')V(x) = I'_i(x) \tag{4.145}$$

where the sources representing influence of external electromagnetic field are expressed as

$$V'_{i(T)}(x) = j\omega \int_0^h \vec{B}_y^{ex}(x, z)\, dz \tag{4.146}$$

$$I'_i = -(G' + j\omega C') \int_0^h \vec{E}^{ex} \cdot d\vec{z} \tag{4.147}$$

with the boundary conditions given by the load voltages and currents

$$V(0) = -Z_1 I(0) \quad \text{and} \quad V(l) = Z_2 I(l) \tag{4.148}$$

The minus sign in the equation for $V(0)$ emerges because the load voltage is once again defined as a line-to-ground voltage and the current flows along the x-axis into the line to be conform with the equivalent circuit in depicted Figure 4.37.

This is so-called the Taylor's formulation (see [25]), comprising both voltage and current sources. An alternate approach, the Agrawal's formulation [25,34], is based on the concept of the scattered voltages, i.e., the voltages only due to the wave processes on the transmission line as the response to the external field. Since the total electric field around the transmission line can be expressed as superposition of the external field and the response (scattered) field of the transmission line:

$$\vec{E} = \vec{E}^{ex} + \vec{E}^{sca} \tag{4.149}$$

and the following equation is valid for a voltage of any point x with respect to the ground:

$$V(x) = -\int_0^h \left(\vec{E}^{ex}(x, z) + \vec{E}^{sca}(x, z)\right) \cdot d\vec{z} \tag{4.150}$$

$$V_S(x) = V(x) + \int_0^h \vec{E}^{ex}(x, z) \cdot d\vec{z} \tag{4.151}$$

the Agrawal's formulation is then

$$\frac{dV_S}{dx} = (R' + j\omega L')I(x) = V'_{i(A)}(x) \tag{4.152}$$

$$\frac{dI}{dx} = (G' + j\omega C')V_S(x) = 0 \tag{4.153}$$

where $V'_{i(A)}$ is the distributed voltage source given as

$$V'_{i(A)}(x) = E_x^{ex}(x,h) \tag{4.154}$$

and the boundary conditions are given by the load voltages and currents and the line integrals of the vertical components of the excitation electrical field, which can be seen as additional voltage sources at each end of the line:

$$\begin{aligned} V_S(0) &= -Z_1 I(0) + \int_0^h \vec{E}^{ex}(0,z)\cdot d\vec{z} = -Z_1 I(0) + \int_0^h E_z^{ex}(0,z)\cdot dz \\ V_S(l) &= Z_2 I(l) + \int_0^h \vec{E}^{ex}(l,z)\cdot d\vec{z} = Z_2 I(l) + \int_0^h E_z^{ex}(l,z)\cdot dz \end{aligned} \tag{4.155}$$

Solution of the Taylor's model equation pairs (4.144) and (4.145) with the boundary conditions (4.148) or the Agrawal's model equation pair (4.152) and (4.153) with the boundary conditions (4.155) can be found, for example, by solving the BLT equations for the load currents and the load voltages. Interestingly, the solution to both pairs has the same matrix form [25]:

$$\begin{bmatrix} I(0) \\ I(l) \end{bmatrix} = \frac{1}{Z_c}\begin{bmatrix} 1-\rho_1 & 0 \\ 0 & 1-\rho_2 \end{bmatrix}\begin{bmatrix} -\rho_1 & e^{\gamma l} \\ e^{\gamma l} & -\rho_2 \end{bmatrix}^{-1}\begin{bmatrix} S_1 \\ S_2 \end{bmatrix} \tag{4.156}$$

$$\begin{bmatrix} V(0) \\ V(l) \end{bmatrix} = \begin{bmatrix} 1+\rho_1 & 0 \\ 0 & 1+\rho_2 \end{bmatrix}\begin{bmatrix} -\rho_1 & e^{\gamma l} \\ e^{\gamma l} & -\rho_2 \end{bmatrix}^{-1}\begin{bmatrix} S_1 \\ S_2 \end{bmatrix} \tag{4.157}$$

The external field excitation is covered by the source vector. For the Taylor formulation, this vector is given as

$$\begin{bmatrix} S_1 \\ S_2 \end{bmatrix} = \begin{bmatrix} \frac{1}{2}\int_0^l e^{\gamma\xi}\left[V'_{i(T)}(\xi) + Z_C I'_i(\xi)\right]d\xi \\ -\frac{1}{2}\int_0^l e^{\gamma(l-\xi)}\left[V'_{i(T)}(\xi) - Z_C I'_i(\xi)\right]d\xi \end{bmatrix} \tag{4.158}$$

and for the Agrawal formulation as:

$$\begin{bmatrix} S_1 \\ S_2 \end{bmatrix} = \begin{bmatrix} \frac{1}{2} & -\frac{e^{\gamma l}}{2} \\ -\frac{e^{\gamma l}}{2} & \frac{1}{2} \end{bmatrix}\begin{bmatrix} \int_0^h E_z^{ex}(0,z)\cdot dz \\ \int_0^h E_z^{ex}(l,z)\cdot dz \end{bmatrix} + \begin{bmatrix} \int_0^l \frac{e^{\gamma\xi}}{2} V'_{i(A)}(\xi)d\xi \\ -\int_0^l \frac{e^{\gamma(l-\xi)}}{2} V'_{i(A)}(\xi)d\xi \end{bmatrix} \tag{4.159}$$

Either of both formulations can be used for determination of terminal responses, since both give the same answer if applied properly [25, 34]. Actually, it is possible to convert one model to another using the integral form of the Faraday's law [31]:

$$\oint_l \vec{E}\cdot\vec{dl} = \frac{\partial}{\partial t}\int_A \vec{B}\cdot\vec{dA} \tag{4.160}$$

Evaluation of the source integrals allows calculation of load responses at the line terminals for arbitrary external electromagnetic fields. However, an analytical integration is possible only for a limited number of cases. Generally, numerical methods have to be applied to find the values of the source components.

4.2.4.5.3 Plane Wave Excitation of a Single-Conductor Above an Ideal Ground Plane

An example of problems for which an analytical solution is possible is the case of illumination of the line by a uniform plane wave [31]. Since the spherical waves radiated from radiating structures locally resemble in the far field uniform plane waves, this approximation can be used to find induced voltages and currents at the line terminals in cases, where the distance from the source of the incident electromagnetic field to the receiving transmission line is much greater than the length of the line.

Although a general solution can be found of e-plane wave at arbitrary polarization and an angle of incidence (see e.g., [25]), there are three special cases of external field incidence for which an analytical solution of the source integrals is straightforward [31]: end-fire, edge-fire, and broadside incidence excitations. These are shown in Figure 4.38.

In the case of vertical polarization (end-fire and broadside excitation), the total field is twice the incident field, that is:

$$E_z^{ex} = 2E_z^{inc} \tag{4.161}$$

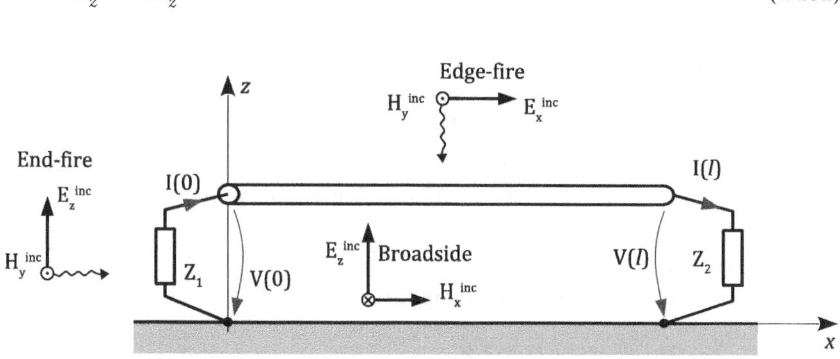

Figure 4.38 Illumination of the line by a uniform plane wave.

For horizontally polarized waves (edge-fire incidence in Figure 4.38), the total field on the conductor is equal to

$$E_x^{ex} = E_x^{inc}\,(j2\sin\beta h) \tag{4.162}$$

where β is the phase constant of line:

$$\beta = \frac{\omega}{c} = \frac{2\pi}{\lambda} \tag{4.163}$$

Case 1—End-Fire incidence
The expressions for the load currents for end-fire incidence are

$$I(0) = \frac{2hE_z^{inc}}{D}\left[\cosh\gamma l + \frac{Z_2}{Z_C}\sinh\gamma l - \cos\beta l + j\sin\beta l\right] \tag{4.164}$$

$$I(l) = \frac{2hE_z^{inc}}{D}\left[1 - \left(\cosh\gamma l + \frac{Z_1}{Z_C}\sinh\gamma l\right)(\cos\beta l - j\sin\beta l)\right] \tag{4.165}$$

where

$$D = (Z_1 + Z_2)\cosh(\gamma l) + \left(Z_C + \frac{Z_1 Z_2}{Z_C}\right)\sinh(\gamma l) \tag{4.166}$$

The near-end responses for the end-fire illumination of a transmission line terminated in different loads are shown in Figure 4.39.

Case 2—Broadside incidence
The load currents for broadside incidence are

$$I(0) = \frac{2hE_z^{inc}}{D}\left[\cosh\gamma l - 1 + \frac{Z_2}{Z_C}\sinh\gamma l\right] \tag{4.167}$$

and

$$I(l) = \frac{2hE_z^{inc}}{D}\left[1 - \cosh\gamma l - \frac{Z_1}{Z_C}\sinh\gamma l\right] \tag{4.168}$$

The near-end load currents for the broadside illumination of a transmission line terminated in different loads are depicted in Figure 4.40.

Case 3—Edge-Fire incidence
The load currents for edge-fire incidence are

$$I(0) = j\beta\frac{2hE_x^{inc}}{\gamma D}\left(\frac{\sin\beta h}{\beta h}\right)\left[\frac{Z_2}{Z_C}(\cosh\gamma l - 1) - \sinh\gamma l\right] \tag{4.169}$$

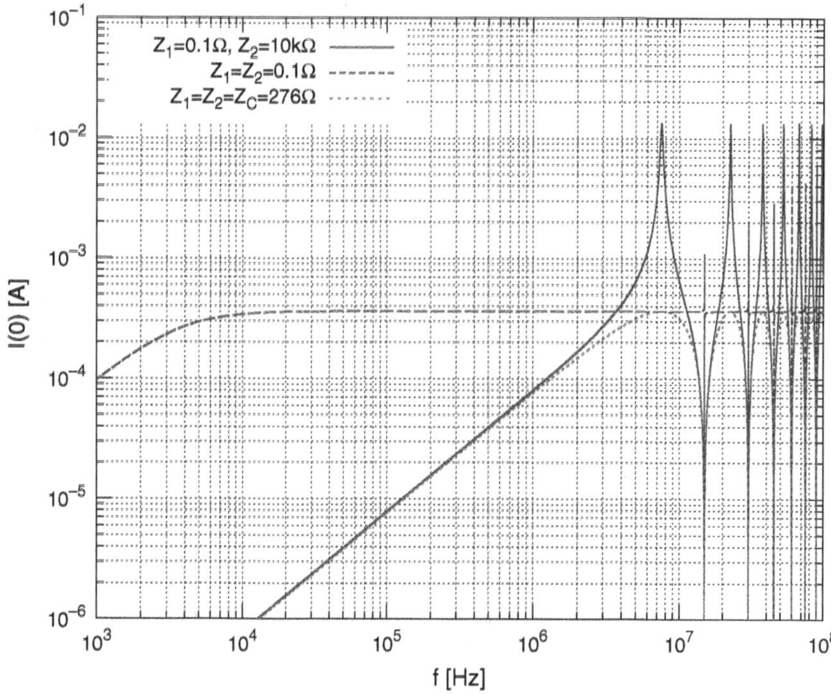

Figure 4.39 Load responses for End-Fire incidence; $a = 1$ mm; $h = 5$ cm; $l = 10$ m; $E_z = 1$ V/m.

and

$$I(l) = j\beta \frac{2hE_x^{inc}}{\gamma D} \left(\frac{\sin \beta h}{\beta h} \right) \left[\frac{Z_1}{Z_C} (\cosh \gamma l - 1) + \sinh \gamma l \right] \tag{4.170}$$

The near-end responses for the edge-fire field incidence on a transmission line terminated in different loads are presented in Figure 4.41.

4.2.4.6 Influence of Shields

4.2.4.6.1 Introduction and Basic Parameters The intent of the shield on the cable is to completely enclose the circuit wires in order to prevent coupling of interference fields and currents from outside to the internal electronic circuitry.

At this point, it may be worth to repeat that significant common-mode (CM) voltages, i.e., voltages of the circuit conductors with respect to the ground plane, can be avoided only when the shield is connected at both ends as a minimum. This statement can be explained as follows:

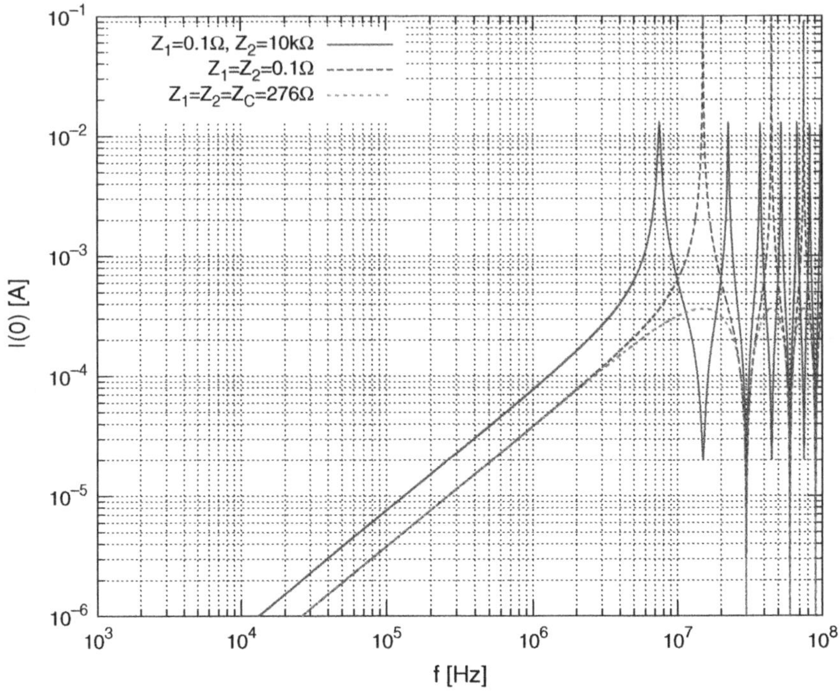

Figure 4.40 Load responses for Broadside Incidence; $a = 1$ mm; $h = 5$ cm; $l = 10$ m; $E_z = 1$ V/m.

- Concerning the electric field (capacitive) coupling, the shield inherently eliminates it when it is "grounded" at least one end and only at the frequencies, at which the shield can be considered electrically short. At higher frequencies, transmission line effects occur.
- Considering magnetic field (inductive) coupling, the shield must be grounded at both ends. Only in this case, the penetrating external magnetic field characterized by magnetic flux Φ_{ex} through the shield-ground loop can induce a secondary current I_S flowing in this loop, as illustrated in Figure 4.42. As predicted by Lenz's law, the magnetic flux of this induced shield current, Φ_S, has the opposite direction to the change in the inducing magnetic field, thus reducing the net magnetic flux through the loop. If the shield is not grounded at both ends, then there is no loop for a current I_S to flow, and consequently, no canceling effects occur. Therefore, if the shield is not grounded at both ends, inductive coupling will not be reduced.

Only a shield made of a solid, perfectly conducting material without any apertures would be capable of preventing any penetration. A real shield can be

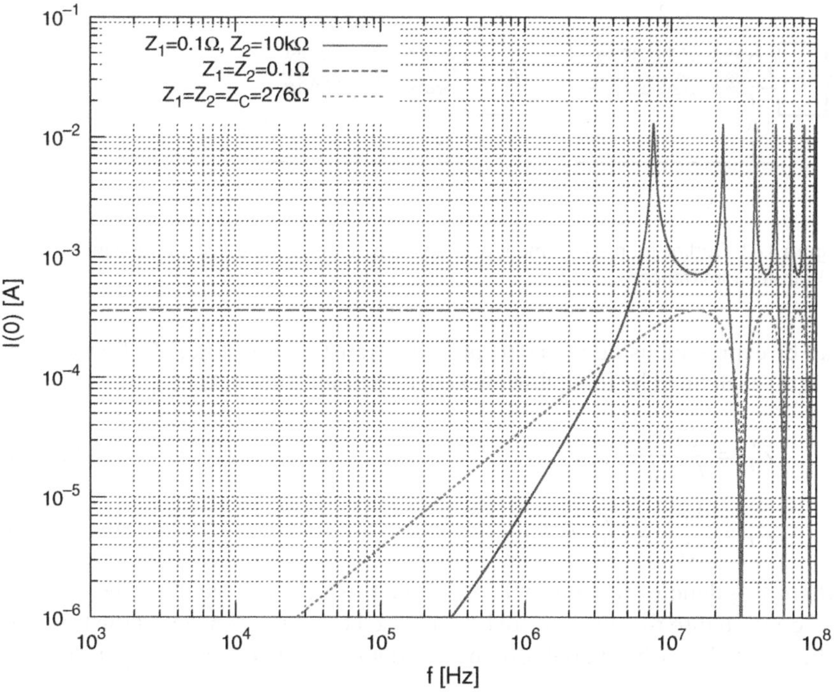

Figure 4.41 Load responses for Edge-fire incidence; $a = 1$ mm; $h = 5$ cm; $l = 10$ m; $E_z = 1$ V/m.

analyzed using the step-by-step coupling process described in section 4.2.2. Following the coupling chain, the external disturbances excite a current on the shield. Then, the shield current and the associated fields at the external shield surface penetrate into the internal volume via conductive or field coupling.

When the exterior shield current I_S is known, a voltage drop on the interior surface of the shield of due to the current diffusion and magnetic field diffusion

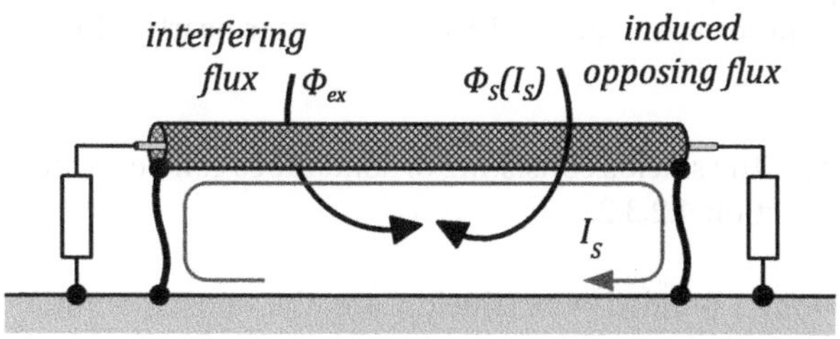

Figure 4.42 Effect of shield around a circuit wire on inductive coupling.

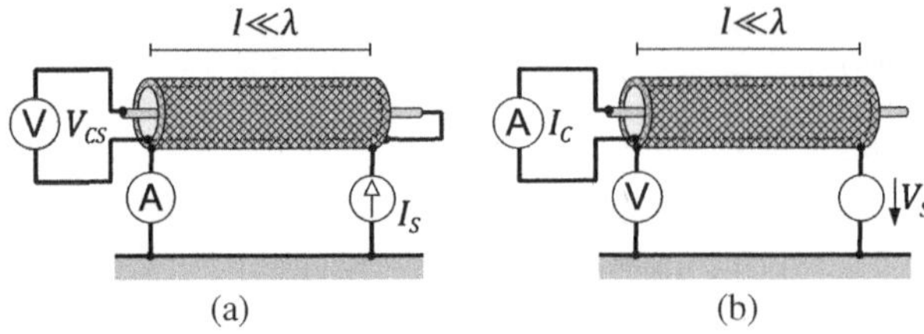

Figure 4.43 Outline for measurement of LF shield parameters (a) transfer impedance; (b)transfer admittance.

can be expressed using the concept of the transfer impedance Z_T' (per unit of length) as defined by the following equation (see Figure 4.43a):

$$Z_T' = \frac{1}{I_S}\frac{dV_{CS}}{dx} \tag{4.171}$$

$$Z_T = \frac{V_{CS}}{I_S} = Z_T' \cdot l \tag{4.172}$$

where V_{CS} denotes the internal conductor-to-shield voltage. The electric field penetration into the interior can be expressed using the concept of the transfer admittance Y_T' as dual variable to the transfer impedance (refer to Figure 4.43b):

$$Y_T' = j\omega C_T', = \frac{1}{V_{SG}}\frac{dI_C}{dx} \tag{4.173}$$

$$Y_T = \frac{I_C}{V_{SG}} = Z_T' \cdot l \tag{4.174}$$

with V_{SG} denoting the shield-to-ground structure voltage and I_C representing the internal conductor current.

The transfer capacitance is a function of the capacitances between and inner conductor and the shield, between the inner conductor and reference, and between the shield and reference, respectively. The capacitance between the inner conductor and reference and between the shield and reference are both a function of the size, shape, and position of the reference. Generally, the closer the reference, the greater these capacitances. Presence of other nearby objects also affects these capacitances.

4.2.4.6.2 ***Solid Tube Shields*** The shields consisting of solid tubes can be treated using the theory given in section 4.2.3.2.

4.2.4.6.3 ***Braided-Wire Shields*** Braided-wire shields are the most common shields used in the aerospace industry. They consist of braids of wire woven in a "herringbone" pattern to give flexibility. Obviously, transfer impedance

Figure 4.44 Illustration of unwrapped braided shield.

Z'_T depends on the weaving characteristics. The per-unit-length transfer impedance of a braided-wire shield is given by the approximate formula [14,25]:

$$Z'_T = Z'_d + j\omega M' \tag{4.175}$$

where Z'_d denotes the diffusion term given as:

$$Z'_d \approx \frac{R'_w}{NC} = \frac{4}{\pi d_w^2 NC\sigma \cos\alpha} \frac{(1+j)\frac{d_w}{\delta}}{\sinh\left[(1+j)\frac{d_w}{\delta}\right]} = R'_{DC} \frac{(1+j)\frac{d_w}{\delta}}{\sinh\left[(1+j)\frac{d_w}{\delta}\right]} \tag{4.176}$$

where (refer to Figure 4.44) d_w is the diameter of an individual braid wire, C denotes the number of carriers (braids); δ stands for the skin depth in the shield (i.e., in the braid wire), α is the weave angle and N denotes the number of strands per carrier (braids) and R'_{DC} denotes the per-unit-length DC resistance of the shield.

The per-unit-length mutual inductance term M' represents the magnetic field penetrating via periodic apertures in the shield due to woven construction. These diamond-like apertures between the individual braids are distributed along the shield circumference and length. They can be modeled as being of an elliptical in shape (see subsection 4.2.3.6). Depending on the weave angle α, the per-unit-length mutual inductance can be expressed as follows [25]:

For $\alpha < 45°$, the eccentricity function $e = \sqrt{1 - (\cot\alpha)^2}$ and:

$$M' = \frac{\pi\mu_o}{6C}\sqrt{(1-\mathbf{K})^3}\frac{e^2}{E(e) - (1-e^2)K(e)} \tag{4.177}$$

For $\alpha > 45°$, $e = \sqrt{1 - (\tan\alpha)^2}$ and:

$$M' = \frac{\pi\mu_o}{6C}\sqrt{(1-\mathbf{K})^3}\frac{1}{K(e) - E(e)}\frac{e}{\sqrt{1-e^2}} \tag{4.178}$$

with $K(e)$ and $E(e)$ representing the complete elliptical integrals of the first and the second kind:

$$E(e) = \int_0^{\pi/2} \left(1 - e^2 \sin^2 \xi\right)^{1/2} d\xi \quad K(e) = \int_0^{\pi/2} \left(1 - e^2 \sin^2 \xi\right)^{-1/2} d\xi$$

Variable **K** is the optical shield coverage defined as

$$\mathbf{K} = 2F - F^2 \tag{4.179}$$

with F denoting the fill of the shield:

$$F = \frac{CNd_w}{4\pi a \cos\alpha} \tag{4.180}$$

where a denotes the radius of the shield. The p.u.l. surface (internal) impedance of the shield Z'_S, i.e., impedance seen by the external circuit is given as:

$$Z'_S = R'_{DC}(1+j)\frac{d_w}{\delta}\coth\left[(1+j)\frac{d_w}{\delta}\right] \tag{4.181}$$

Typical frequency variation of the transfer impedance of a woven cable shield is illustrated in Figure 4.45.

It has to be mentioned that the given formulae gives only rudimentary outline and rough estimation of HF behavior of the braided shields. Several more advanced models have been developed that take into account additional factors influencing the shield behavior [25,35,36]. Nevertheless, the experimental determination of the cable transfer impedance by measurements can be considered as the most reliable method. Measured values of transfer impedances can be found tabulated in [14] or [25]; or in graphical form in [37] or [38].

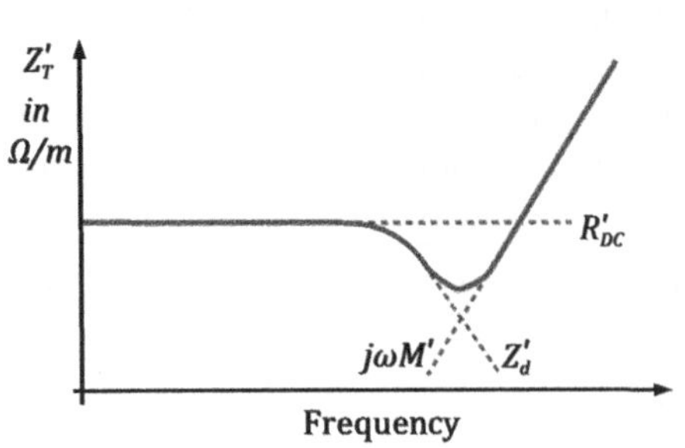

Figure 4.45 Example of frequency variation of the transfer impedance.

4.2.4.6.4 ***Tape-Wound Spiral Shields*** The transfer impedance of a tape-wound non-ferrous shield in which the turns are not overlapped and have no contact between turns is expressed by [39]:

$$Z'_T \approx R'_{DC} \left\langle \frac{(1+j)\frac{\Delta}{\delta}}{\sinh\left[(1+j)\frac{\Delta}{\delta}\right]} + \left\{ \frac{(1+j)\Delta}{\delta} \coth\left[(1+j)\frac{\Delta}{\delta}\right] + \right. \right. \tag{4.182}$$
$$\left. \left. +j\left(\frac{\Delta}{\delta}\right)^2 \frac{a}{\Delta} \right\} (\tan\alpha)^2 \right\rangle$$

where R'_{DC} is the p.u.l. DC resistance of a cylindrical tube of the same wall thickness:

$$R'_{DC} = \frac{1}{2\pi a \Delta \sigma} \tag{4.183}$$

a is the radius of the shield, Δ denotes the thickness of the shield, δ stands for the skin depth in the shield and α is the spiral angle:

$$\tan\alpha = 2\pi a N = 2\pi a \frac{\sin\alpha}{W} \tag{4.184}$$

with W denoting the width of the tape and N standing for the number of turns per unit length. The transfer impedance of a single-layer overlapped shield can be obtained by replacing W by $W - W_{ov}$, where W_{ov} is the overlap width.

4.2.4.6.5 ***Connectors*** Connectors can contain slots through which electromagnetic fields can penetrate to the inside of the cables. Additionally, they have a lumped series resistance associated with the contact resistance of the mating surfaces. The transfer impedance of a connector Z_{TC} is given by

$$Z_{TC} = R_C + j\omega M_C \tag{4.185}$$

where R_C is the resistance measured across the connector; and M_C is the mutual inductance between the external shield and the inner conductors of the cable. These parameters are best determined experimentally. The estimate values of R_C and M_C can lie around 1 $m\Omega$ and 10 pH, respectively [39]. Measured transfer impedances of different types of coaxial and aerospace connectors can be found in [25].

If a pigtail connection of the shield to the connector is used, the influence of pigtail can be included by adding its resistance R_{PT} and inductance L_{PT} to the connector transfer impedance:

$$Z_{TC} = R_C + j\omega M_C + R_{PT} + j\omega L_{PT} \tag{4.186}$$

The pigtail inductance can be estimated using formula (4.95).

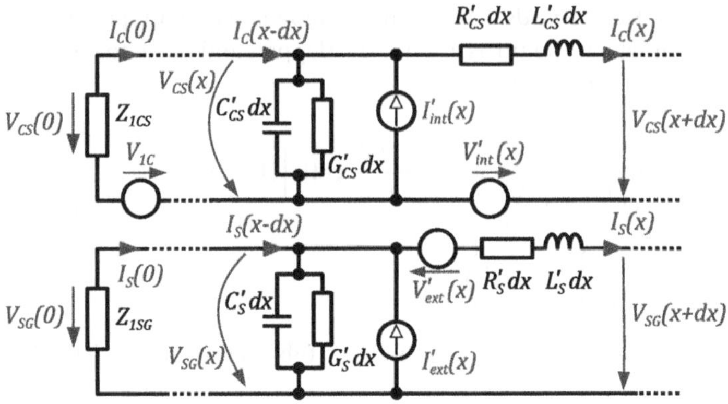

Figure 4.46 Coupled per-unit-length equivalent circuit for evaluation of internal responses with only the left-hand terminations depicted. The right-hand ones are equivalent.

4.2.4.6.6 Evaluation of Internal Voltages and Currents A cable consisting of a wire enclosed by a shield and routed over an infinite ideal ground plane, as shown in Figure 4.42, can be represented by two coupled circuits depicted in Figure 4.46 with the external circuit formed by the shield and the ground plane return and the interior wire-shield circuit.

The current I_S and voltage on the external transmission line V_{SG} can be excited by either a lumped source such as an aperture (see section 4.2.4.2) or by an external electromagnetic field illuminating complete transmission line. The former case is handled in subsection 4.2.4.4. For the latter situation, the analysis presented in subsection 4.2.4.5.2 as the Taylor's formulation given by equations (4.144) and (4.145 can be applied and the following equations can be written for the shield-ground plane circuit:

$$\frac{dV_{SG}}{dx} = (R'_S + j\omega L'_S)I_S(x) = V'_{ext}(x) \tag{4.187}$$

$$\frac{dI_S}{dx} = (G'_S + j\omega C'_S)V_{SG}(x) = I'_{ext}(x) \tag{4.188}$$

where R'_S, L'_S, G'_S, and C'_S are the per-unit-length parameters of the external transmission line. The excitation sources V'_{ext} and I'_{ext} are given by equation pair (4.146) and (4.147).

Applying the Taylor's formulation for the field-to-line coupling, the interior wire-shield circuit is described by

$$\frac{dV_{CS}}{dx} = \left(R'_{CS} + j\omega L'_{CS}\right) I_C(x) = V'_{int}(x) \tag{4.189}$$

$$\frac{dI_C}{dx} = \left(G'_{CS} + j\omega C'_{CS}\right) V_{CS}(x) = I'_{int}(x) \tag{4.190}$$

where R'_{CS}, L'_{CS}, G'_{CS}, and C'_{CS} are the per-unit-length parameters of the internal transmission line circuit. The voltage drop along the interior surface of the shield can be represented by an elementary voltage sources $V'_{int} = Z'_T \cdot I_S(x)$ distributed on the interior surface of the shield. The parasitic electrical field coupling is represented by the distributed current sources $I'_{int} = Y'_T \cdot V_{SG}(x)$. However, a more convenient expression for the internal current source is in terms of the external charge on the shield Q'_S, i.e., independent of the external line voltage V_{SG} and capacitance C'_S [25]:

$$I'_{int}(x) = -j\omega S_S C'_{CS} Q'_S(x) \tag{4.191}$$

with the charge on the cable shield exterior obtained from the continuity equation:

$$\frac{dI_S}{dx} = -j\omega Q'_S(x) \tag{4.192}$$

and electrostatic shield leakage parameter S_S given by the following expressions (refer to Figure 4.44):

For $\alpha < 45°$, the eccentricity function $e = \sqrt{1 - (\cot\alpha)^2}$ and

$$S_S = \frac{\pi}{6C\varepsilon_0}\sqrt{(1 - \mathbf{K})^3}\frac{1}{E(e)} \tag{4.193}$$

For $\alpha > 45°$, $e = \sqrt{1 - (\tan\alpha)^2}$ and

$$S_S = \frac{\pi}{6C\varepsilon_0}\sqrt{(1 - \mathbf{K})^3}\frac{1}{E(e)(1 - e^2)} \tag{4.194}$$

Measured values of electrostatic shield leakage parameter S_S are tabulated in [14] or [25].

With reference to Figure 4.46, the additional lumped source V_{1C} at the beginning (and a similar lumped source at the end) of the internal transmission line represents the excitation due to the transfer impedance of the connector, defined by equation (4.186) in paragraph 4.2.4.6.5. The following expression can be written:

$$V_{1C} = Z_{TC} I_S \tag{4.195}$$

An analytical solution of equation system (4.187), (4.188), (4.189), and (4.190) for a plane wave excitation of the external circuit can be found in [25]. An alternate mixed analytical/numerical approach is presented in [37].

Example Let's consider a cable consisting of a wire enclosed by a shield and routed over an infinite ideal ground plane, as shown in Figure 4.47.

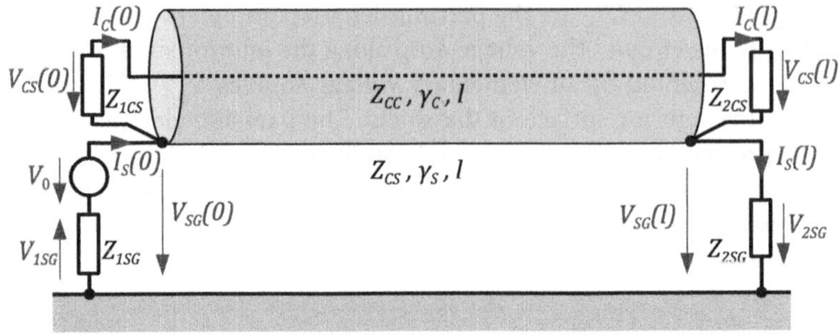

Figure 4.47 Finite shielded line with excitation at the beginning of the shield-to-ground circuit.

This situation can be represented by modifying the model depicted in Figure 4.46 with the external circuit formed by the shield and the ground plane return and the interior wire-shield circuit as illustrated in Figure 4.48.

The estimation of the conductor-to-shield voltages at the terminals of the line can be done by subdividing the internal circuit into N discrete short sections of length Δl, as shown in Figure 4.49. The amplitude of current in each section should be approximately constant, therefore, the maximum length of each section is limited by the condition for electrical shortness (4.8), i.e., $\Delta l < \lambda/10$, thus depending on maximum frequency of interest. In this way, we can convert problem with distributed voltage sources into problem with multiple lumped voltage sources spaced evenly along the length of the shield at discrete points x_k

$$x_k = \frac{2k-1}{2}\Delta l \quad \text{for} \quad k = 1 \ldots N$$

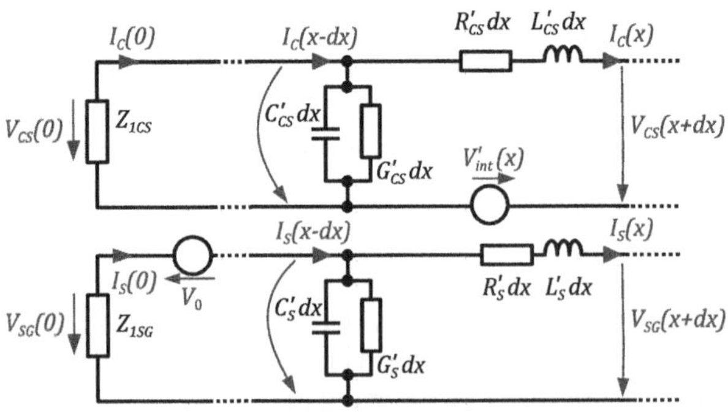

Figure 4.48 Coupled per-unit-length equivalent circuit for evaluation of internal responses to excitation at the beginning with only the left-hand terminations depicted. The right-hand ones are equivalent.

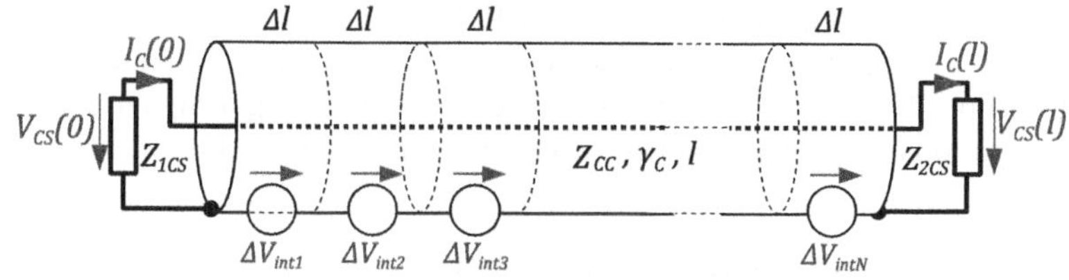

Figure 4.49 Simplified model of the voltage gradient generation along the inner surface shield due to the impressed shield current.

Each discrete voltage source can be defined as

$$\Delta V_{int}(x_k) = \Delta V_{intk} = Z_T' \cdot \Delta l \cdot I_S(x_k) \quad \text{for} \quad k = 1 \ldots N$$

with shield current $I_S(x_k)$ obtained using equation (4.135)

$$I_S(x_k) = \frac{1}{Z_{1SG} + Z_{CS}} V_0 e^{-\gamma_S x_k} \frac{1 - \rho_{2SG} e^{-2\gamma_S (l - x_k)}}{1 - \rho_{1SG} \rho_{2SG} e^{-2\gamma_S l}}$$

The total response across each terminal is then superposition of the responses due to each source. A general problem finite line excitation by a single lumped voltage source at arbitrary position was investigated in subsection 4.2.4.4 with a terminal voltages generated by a single lumped source given by equations (4.139) and (4.140). Consequently, the total terminal voltages can be obtained summing up the contributions due to voltage transferred at each segment:

$$V_{CS}(l) = \sum_{k=1}^{N} \Delta V_{intk} \frac{(1 + \rho_{2CS}) e^{-\gamma_C l} \left(e^{\gamma_C x_k} - \rho_1 e^{-\gamma_C x_k}\right)}{2\left(1 - \rho_{1CS} \rho_{2CS} e^{-2\gamma_C l}\right)}$$

$$V_{CS}(l) = \sum_{k=1}^{N} Z_T' \cdot \Delta l \cdot I_S(x_k) \frac{(1 + \rho_{2CS}) e^{-\gamma_C l} \left(e^{\gamma_C x_k} - \rho_1 e^{-\gamma_C x_k}\right)}{2\left(1 - \rho_{1CS} \rho_{2CS} e^{-2\gamma_C l}\right)}$$

The load response for the beginning of the shielded line can be found equivalently to be

$$V_{CS}(0) = \sum_{k=1}^{N} Z_T' \cdot \Delta l \cdot I_S(x_k) \frac{(1 + \rho_{1CS}) e^{-\gamma_C l} (e^{\gamma_C (l - x_k)} - \rho_{2CS} e^{-\gamma_C (l - x_k)})}{2(1 - \rho_{1CS} \rho_{2CS} e^{-2\gamma_C l})}$$

4.3 HIRF Electromagnetic Environment and Its Effects

The high intensity electromagnetic fields (HIRF) represent the electromagnetic energy radiated from radio, television, radar emitters, and from other sources into free space in the frequency ranges from 10 kHz to 40 GHz. The environment given in [9] is based more than 500,000 emitters in the United States and Western Europe. The global electromagnetic environment might higher since some other nations may operate transmitters at even higher powers.

4.3.1 Standardized External HIRF Environment

The HIRF environments are a composite of transmitters that are airborne, land-based, off-shore platforms, and ship-borne. The emitters cover the entire RF spectrum and their radiated fields vary greatly in energy levels and signal characteristics

The electromagnetic environment has been modeled using the databases that contain parameters pertaining to all known transmitters in the United States and Western Europe [9]. The resulting HIRF envelope is a representation of electromagnetic field strength over a frequency range of 10 kHz to 40 GHz.

The following types of emitters were considered in the calculation of field strengths for the environments [9, 40]:

1. Airport fixed ground emitters, such as marker beacons, ILS (localizer and glideslope), ground controlled approach radars, microwave landing systems, airport and air-route surveillance radars, weather radars, VHF and UHF communications, etc. telemetry.
2. Airport mobile ground emitters, for instance, HF, VHF, and UHF communications, TACAN, radio altimeter, etc.
3. Non-airport ground emitters, for example, commercial MF; HF; VHF AM and FM; and TV broadcast transmitters, radars, satellites, and command and control facilities.
4. Shipboard emitters, including HF, VHF, and UHF communications, navigation and tracking radars, and IFF/ Selective Identification Feature.
5. The off-shore platform emitters, e.g., HF, VHF, and UHF communications; navigation, tracking radars and transponders
6. Air-to-air interceptor and non-interceptor emitters, such as tracking radars; various on-board radars; HF, VHF, and UHF communications, weather radars, etc.

4.3.1.1 Expression of Field Strengths for Modulated Signals

Typical HF communication systems are using of modulation techniques such as Amplitude Modulation (AM), Frequency Modulation (FM), Pulse Modulation (PM), or Continuous Wave (CW).

For all types of modulations, the corresponding worst-case scenarios were used to determine the maximum field strength and so-called peak RMS values are used to express the HIRF environment, i.e., peak of the RMS envelope over the complete modulation period, as depicted in Figure 4.50 [41, 43]. For

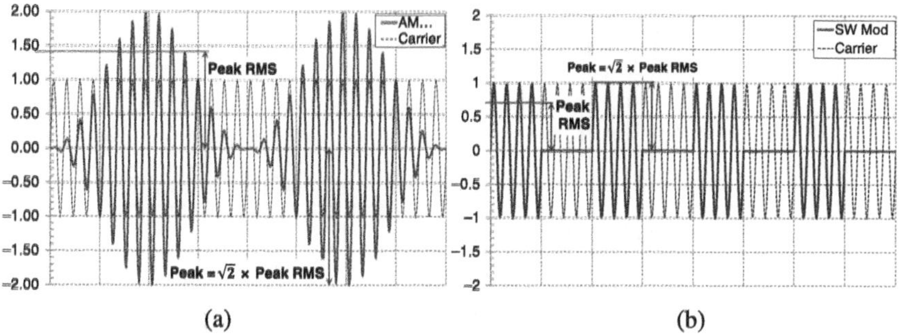

Figure 4.50 Definition of parameters for different amplitude modulation patterns (a) amplitude modulation; (b) pulse/square wave modulation.

FM and PM, the peak field strength is equal to the unmodulated carrier field strength. For 100% modulated AM, the peak field strength is twice the carrier field strength.

4.3.1.2 Peak/Average Power Calculation

Pulse-modulated signals, typically from radar transmitters, have differences between peak and average RMS power. The average power is determined by the ratio of time on to time off over an interval. This time on/off ratio is called the duty cycle dc and is defined as [43]:

$$dc = pw{\cdot}f_r \tag{4.196}$$

where pw is the pulse width and f_r denotes the pulse repetition frequency, as portrayed in Figure 4.51. Then, the average power P_{AV} is the product of peak power P_P and duty cycle dc:

$$P_{AV} = P_P{\cdot}dc \tag{4.197}$$

4.3.1.3 The Civil External HIRF Environment

Four different HIRF environments listed in Table 4.3 have been derived for civil aviation and can be found in civil aviation industry document [9] and its previous issues. Three of them have become civil aviation industry standard environment and they can be found in certification requirements of FAA, EASA, and other national civil aviation authorities [4, 5, 42]:

1. Fixed Wing Severe (not used in HIRF regulations)

   The Fixed Wing Severe HIRF environment is based on the worst-case estimate of electromagnetic field strengths that a civil aeroplane might encounter in the airspace in which fixed wing flight operations are permitted.

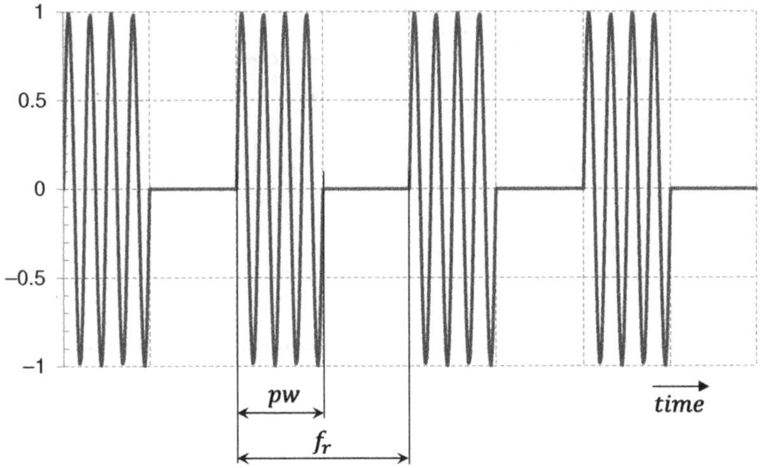

Figure 4.51 Pulse-modulated signal.

2. Certification (HIRF Environment I)
   Certification HIRF Environment was established from the Fixed Wing Severe HIRF Environments by increasing the allowed distance between the aircraft and for non-airfield fixed transmitters, taking into account likelihood of encounter. It is denoted as HIRF Environment I in the FAA/EASA HIRF regulations and advisory materials.
3. Normal (HIRF Environment II)
   The Normal HIRF Environment is the electromagnetic field strength level in the airspace on and about in the vicinity of representative airports in the United States and Europe in which routine departure and arrival operations take place.
4. Rotorcraft Severe (HIRF Environment III)
   The Rotorcraft HIRF Environment III is the worst-case estimate of the electromagnetic field strength levels in the airspace in which rotorcraft flight operations under Visual Flight Rules are permitted, since rotorcraft can fly and hover closer to obstacles and the ground as it was considered for the HIRF Environment I.

4.3.1.4 The Military External HIRF Environment

In case of military aircraft, the situation is different since the unification of the HIRF environment is not present. Different HIRF environments can be found in corresponding national or international defense standards, e.g., [43–45]. Moreover, the levels are evolving between successive issues of the standards and in many cases, there are additional restricted documents where a worst-case operational EM environment is defined [46].

Table 4.3 Summary of different standard HIRF environments

	Field strength (V/m)							
	Fixed wing severe HIRF environment		Rotorcraft severe HIRF environment		Certification HIRF environment		Normal HIRF environment	
Frequency	Peak	Average	Peak	Average	Peak	Average	Peak	Average
10 kHz–100 kHz [1]	50	50	150	150	50	50	20	20
100 kHz–500 kHz	60	60	200	200	50	50	20	20
500 kHz–2 MHz	70	70	200	200	50	50	30	30
2 MHz–30 MHz	200	200	200	200	100	100	100	100
30 MHz–70 MHz	30	30	200	200	50	50	10	10
70 MHz–100 MHz	30	30	200	200	50	50	10	10
100 MHz–200 MHz	90	30	200	200	100	100	30	10
200 MHz–400 MHz	70	70	200	200	100	100	10	10
400 MHz–700 MHz	730	80	730	200	700	50	700	40
700 MHz–1 GHz	1,400	240	1,400	240	700	100	700	40
1 GHz–2 GHz	3,300	160	5,000	250	2,000	200	1,300	160
2 GHz–4 GHz	4,500	490	6,000	490	3,000	200	3,000	120
4 GHz–6 GHz	7,200	300	7,200	400	3,000	200	3,000	160
6 GHz–8 GHz	1,100	170	1,100	170	1,000	200	400	170
8 GHz–12 GHz	2,600	330	5,000	330	3,000	300	1,230	230
12 GHz–18 GHz	2,000	330	2,000	330	2,000	200	730	190
18 GHz–40 GHz	1,000	420	1,000	420	600	200	600	150

4.3.2 Internal Aircraft HIRF Environment

The RF internal electromagnetic environment, in which the given equipment or system is immersed, is different from the external fields present at the surface of the aircraft. Basic coupling processes have been outlined in section. As it can be deduced therefrom, the local levels of attenuation or enhancement achieved for any region are the product of many factors such as materials, bonding, dimensions and geometric form of the region, and the location and size of any apertures allowing penetration into the aircraft. At higher frequencies, cavity resonant conditions can be fulfilled and several hot spots will exist within closed subsections of the aircraft. Additionally, the internal field distributions change significantly with a change in the position and orientation of the aircraft relative to the HIRF source.

The ideal and most accurate assessment of the hardness of an aircraft to HIRF electromagnetic environment is to expose it directly to that environment. However, this "perfect" approach is often technically impractical for several reasons, such as the problems of generating adequate uniform fields over the volume of the aircraft over the complete frequency range of interest.

It is more practical to split the coupling chain into several steps, i.e., first determine the internal electromagnetic environment caused by penetration of external fields and then, as a separate step, expose the internal electrical/electronic installation to the internal fields.

4.3.2.1 Determination of Internal EM Environment Using Numerical Methods

The local levels of internal EM field can be estimated using the formula outlined in Chapter 4.2, especially in frequency ranges where the structures are electrically small (the size of the structure is less than 0.1 λ).

A much more accurate estimation can be made using advanced 3D computational codes for solution of numerical solution of the Maxwell equations (see Table 4.4). Different codes implement or even combine different frequency or time domain methods of finding solution to these equations in differential or integral form. The generally used methods are [47–50]:

- Boundary Element Method (BEM)
- Method of Moments (MoM)
- Finite Difference Time Domain Method (FDTD)
- Finite Elements Method (FEM)
- Partial Element Equivalent Circuit Method (PEM)
- Transmission Line Matrix Method (TLM)

Direct 3D modeling of the complete coupling chain, i.e., from external incident electromagnetic fields to voltages and currents conducted along the cable harnesses to equipment interfaces is very impractical since the physical dimensions of features in wire bundles, such as connectors and shield terminations are

Table 4.4 Maxwell equations

Integral form	Differential form	Designation
$\oint \vec{E}\cdot d\vec{l} = -\frac{\partial}{\partial t}\int_A \vec{B}\cdot d\vec{A}$	$\nabla \times \vec{E} = -\frac{\partial \vec{B}}{\partial t}$	Faraday's law
$\oint \vec{H}\cdot d\vec{l} = \frac{\partial}{\partial t}\int_A \vec{D}\cdot d\vec{A} + \int_A \vec{J}\cdot d\vec{A}$	$\nabla \times \vec{H} = \frac{\partial \vec{D}}{\partial t} + \vec{J}$	Ampere's law
$\int_A \vec{D}\cdot d\vec{A} = \int_V \rho_V dV$	$\nabla\cdot\vec{D} = \rho$	Gauss' law
$\int_A \vec{B}\cdot d\vec{A} = 0$	$\nabla\cdot\vec{B} = 0$	Gauss' magnetic law
	$\vec{B} = \mu\vec{H}$ and $\vec{D} = \varepsilon\vec{E}$	

very small relative to the dimensions of the aircraft. Thus, the discretization of volume and time would involve very small steps yielding immense models and lengthy calculations.

It is more effective to use first the 3D modeling. to calculate local internal electromagnetic fields (see an example in Figure 4.52) and subsequently use a network analysis (outlined roughly in subsection 4.2.4.5) to calculate coupling of these local fields to cables producing the terminal (interface) currents and voltages [51].

However, the simulation of the HIRF hazards is very complex and involves a detailed understanding of the topics and problematics involving

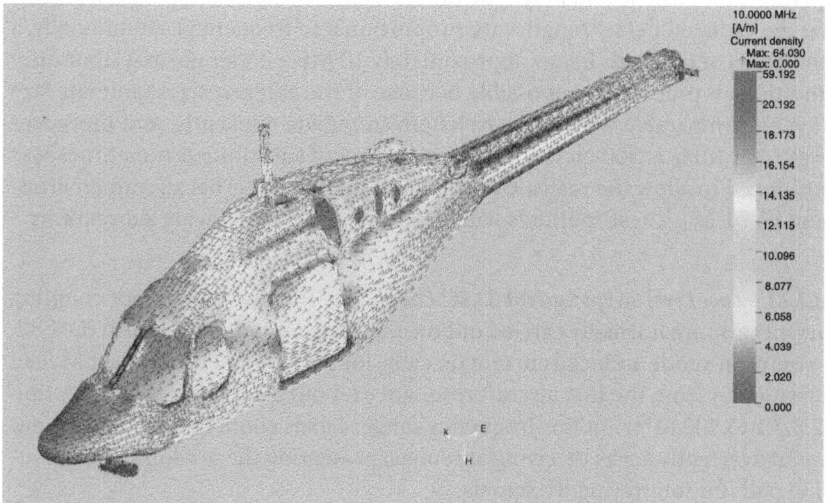

Figure 4.52 Simulated surface current distribution on a helicopter at 10 MHz.

electromagnetic interactions and limitations of the method used. Especially with increase in frequencies of interested, a detailed simulation becomes more and more difficult and error prone.

All test methods require transformation of geometrical (CAD) model to electro-geometrical model. This step involves simplifications (model reduction, omission of some parts, idealization/homogenisation of material properties, etc.). This is key step for simulation and a source of aspect major errors. Therefore, it is necessary to validate the results by comparison with data obtained experimentally by measurements on given aircraft or its parts (fuselage, wing) [50, 52]. Due to these facts, a pure numerical simulation without aircraft measurements is not allowed for aircraft certification for HIRF effects according to [4]. However, it is valuable tool for estimations of internal HIRF environment in early stages of the aircraft design where no real aircraft is available, throughout the design phase to judge the influence of modifications to aircraft structure or during the preparations of final aircraft "certification" tests to identify the worst cases and thus reducing the amount of necessary testing.

4.3.2.2 Experimental Determination of Internal EM Environment

The most straightforward way to determine internal EM environment is to expose the aircraft to external HIRF environment and to measure the internal fields and currents induced on the cable bundles. This approach is so-called high-level testing described in subsection 4.3.3.2.

However, a generation of high-level EM fields at frequencies below 100 MHz would require immense power levels to be fed into antennas (a quick look at the equations for electric and magnetic dipole in subsection section 4.2.3.6 reveals that the radiated field strengths are proportional to frequency). Additionally, a generation of the fields below approximately 10 MHz under normal laboratory conditions is practically impossible because of the antenna sizes (antenna size must be comparable with the wavelength to radiate efficiently, and the wavelength at 1 MHz is 300 meters). Therefore, a set of substitute test methods was introduced to allow the realistic determination of the internal aircraft environment [9, 54, 55]. These methods will be outlined in the following subsections.

4.3.2.2.1 Low-Level Swept Current (LLSC) Coupling The low-level swept coupling current tests are normally carried out on complete aircraft to establish the bulk or common-mode-induced current on cable looms for a given external EM field in frequency from the first aircraft resonance (about 5-10 MHz, see subsection 4.2.3.7.1 to 400 MHz. In this frequency range, cables connecting to equipment can inadvertently act as receiving antennas, presenting the predominating coupling path for interfering RF signals.

However, the measurements are not performed at full HIRF threat levels, but at lower field strengths. These induced bulk currents can then be extrapolated

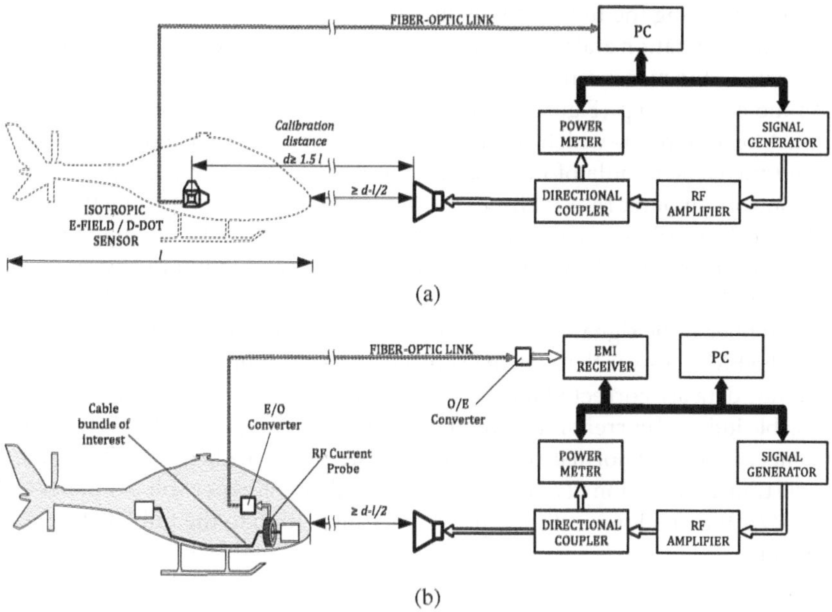

Figure 4.53 Test set-up for LLSC measurements (a) field calibration; (b) measurement.

to derive the cable loom currents expected during the exposition of the aircraft to full HIRF environment, i.e., the currents to which the systems must be immune.

The test consists of two steps. First, depicted in Figure 4.53(a), prior to the installation of the aircraft, the field calibration is performed, i.e., the field is generated at the intended location of the aircraft over the complete frequency range. Both the reached electric field strength E_{cal} and the forward power fed into the radiating antenna P_{FWD} are recorded. The minimum distance between the illuminating antenna and the aircraft is given the 3-dB beamwidth of the antenna. The intention is to have the full length of the aircraft illuminated uniformly with minimum field variation, i.e., the distance of the antenna from the aircraft should be as big as possible. On the other hand, the greater distances yield lower field strengths.

The minimum field strength is limited by the sensitivity of the system for measurement of induced currents. An increase of distance could be compensated by an increase in the power fed into the antenna. This is not always feasible, since often the tests are not performed in shielded chambers but at open-area test sites, so excessive powers fed into antenna can cause RF interference. Generally, a compromise has to be met and the minimum distance of 1.5 times the length of the aircraft is advised [9]. The intention is to have a variation of

3 dB to 4 dB along the full length of aircraft. However, for large aircraft, more localized illumination and more than 4 antenna positions are allowed, when the sensitivity of the measurement system is not sufficient [9]

As the second step (Figure 4.53b), the aircraft is placed at the calibration point. A current probe which is connected to a fiber-optic transmitter is placed around the cable bundle of interest close to the connector of the equipment. The control computer re-establishes the pre-calibrated field by feeding the pre-calibrated forward power P_{FWD} to antenna as the frequency is scanned from the start frequency to 400 MHz. The induced bulk current measured by the current probe I_T is converted to light and transmitted of the fiber-optic link. The received fiber-optic signal is converted back to an analogue electrical form and fed to the EMI receiver that is synchronized with the stimulus signal. The measured raw data are corrected for the current probe transfer impedance and the true cable induced current is recorded as a function of frequency. The measured currents are usually normalized to unit field strength, yielding so-called cable bundle transfer function (TF) curve (given in milliamperes per volt per meter—$mA/(V/m)$) that allows a straightforward extrapolation to the full HIRF environment:

$$TF = \frac{I_T}{E_{cal}} \tag{4.198}$$

The resulting transfer function is then multiplied by the prescribed HIRF environment electric field strengths E_{HIRF} to obtain the expected full-threat in-flight cable bundle currents I_{Efull}:

$$I_{Efull} = TF \times E_{HIRF} \tag{4.199}$$

As an additional parameter, surface current densities at the aircraft skin J_{ST} can be measured using the B-dot sensors. These results can be then compared with the values obtained using the 3D numerical modeling codes to determine the influence of the test setup components that are not present under flight conditions, for example, the ground plane.

The test should be carried out at four different positions of radiating antennas in both horizontal and vertical field polarizations to reach a uniform illumination of the aircraft from all four sides, as illustrated in Figure 4.54.

An example of results obtained during the LLSC testing is shown in Figure 4.55, where the final test results are normalized to 1 V/m. The current levels are an envelope of the worst-case results obtained for all illumination angles and polarizations.

With decreasing frequency, it is becoming more and more difficult to generate and radiate the electromagnetic field. For example, a dipole antenna radiates

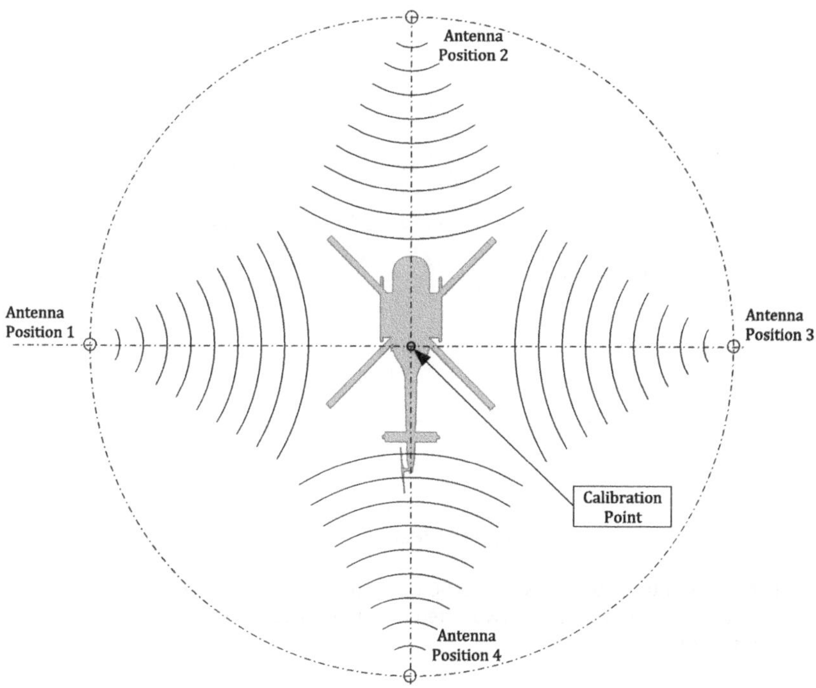

Figure 4.54 Antenna positioning for LLSC measurements.

at the frequency, where its length is equal to a quarter of the electromagnetic wavelength $l = \lambda/4$. At a frequency of 1 MHz, a wavelength in air is

$$\lambda = \frac{c}{f} = \frac{3 \times 10^8 m/s}{1 \times 10^6 Hz} = 300m \tag{4.200}$$

This means that an effectively radiating dipole should have a length l around 75 meters. Electrically shorter antennas (i.e., $l \ll \lambda$) are poor radiators requiring very high-input power to produce field strengths sufficient to measure internal electromagnetic environment.

One option to avoid the need for immense input powers is to use so-called bound wave antennas such as strip lines or multiple-wire antennas that are basically open waveguides (two-conductor transmission lines) guiding an electromagnetic wave across a test object from a generator to the other end loaded by a termination, as depicted in Figure 4.56. The transmission line consists of a top wire plate above a bottom plate, generally the ground plane. The field between plates is a travelling TEM wave with a vertical electric field and a

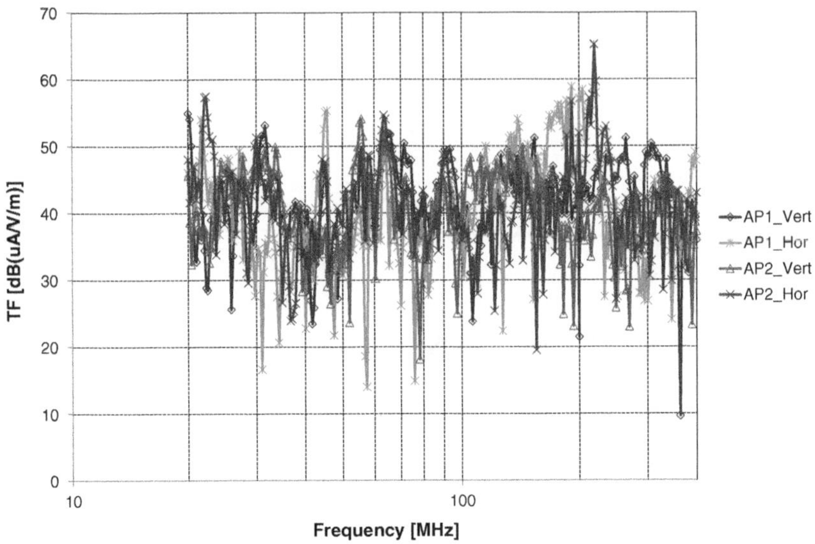

Figure 4.55 Transfer function as result of LLSC test in the frequency range 20 MHz–400 MHz measured in the centre of cockpit for two different antenna positions (AP1/AP2) and two polarizations (vertical/horizontal).

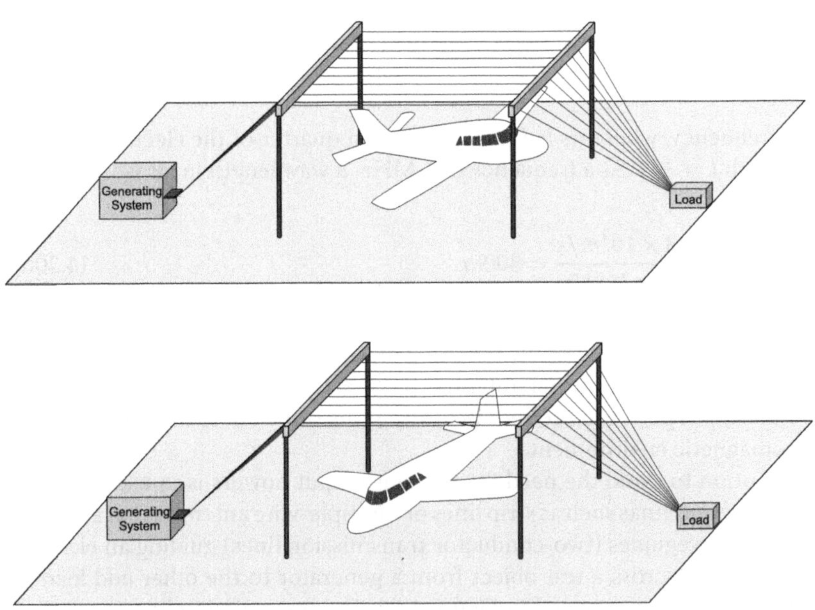

Figure 4.56 A small aircraft placed in the multiple-wire antenna.

horizontal magnetic field. This means that vectors $\vec{E}$ and $\vec{H}$ are both perpendicular to the direction of propagation and their amplitudes are related by $|E|/|H| = 120\pi\Omega \approx 376.7\Omega$.

The measurement again follows in two steps, first calibration and then the actual harness current measurement. Typically, the test object should not be greater than one-third to half the height of the working volume of the antenna, i.e., the separation between the top wire plate and the bottom ground plate. That means that this method allows exposure of relatively small aircraft to vertically polarized (electric field component) fields.

4.3.2.2.2 Tests at Low Frequencies and Low-Level Direct Drive (LLDD) Method In order to overcome the above-mentioned limitations ruled by the size of antennas, a substitution direct current injection method outlined in Figure 4.57 has been developed. This method is based on the following two assumptions [9]:

1. Skin currents on exterior aeroplane surfaces induce currents on cable harnesses only by magnetic field coupling, i.e., electric field coupling is neglected and
2. The internal magnetic field and thus the induced cable current is proportional to the exterior local skin current density; i.e., surface current density at the location of the aperture.

It can be deduced from comparison of the formulae for direct injection of external current on cylinder presented in subsection 4.2.3.2 and the formulas valid for a cylinder immersed in external electromagnetic field given in subsection 4.2.3.4, the surface current distributions are not the same. Additionally, direction of the surface currents with respect to the aircraft apertures depends both on the illumination angle and polarization of the electromagnetic field [56].

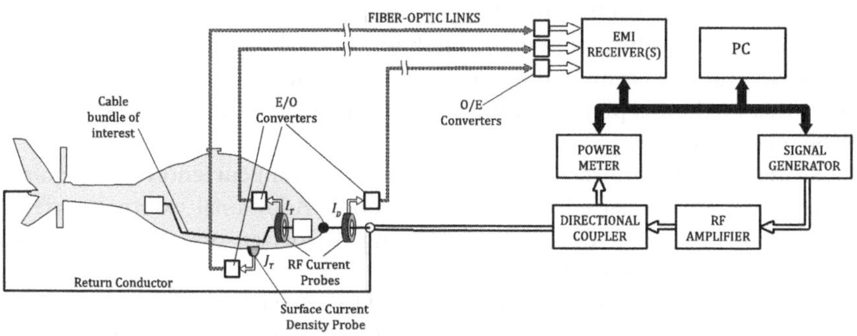

Figure 4.57 Outline of LLDD test set-up.

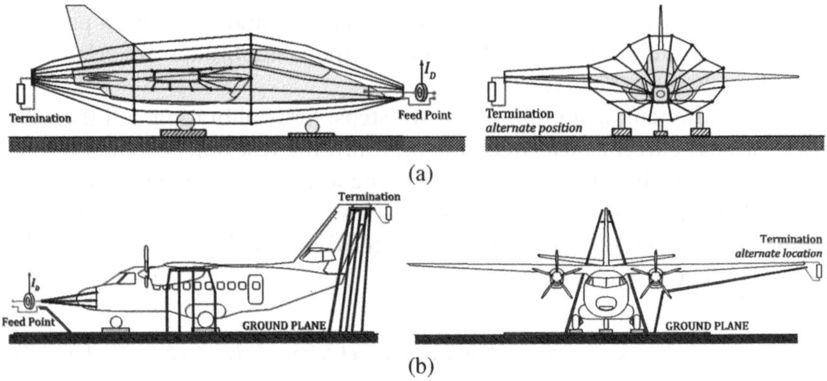

Figure 4.58 Arrangement of return conductors for LLDD test (a) coaxial arrangement, (b) ground plane.

Therefore, a 3D mathematical modeling using numerical codes (refer to subsection 4.3.2.1) is performed to determine the relationship between free field external radiation and skin current for all illumination angles and polarizations. Additionally, a reduced set of skin current measurement results collected at several frequencies up to the aircraft first resonance should be compared with the simulation results to verify the modeling accuracy.

The Low-Level Direct Drive (LLDD) test is used to determine the transfer function between the aircraft skin currents and the currents induced on individual cable bundles applying relatively low signal levels. Typically, the currents are injected directly between various combinations of entry and exit points on the aircraft—wing, engine, nose, or tail. Figure 4.58 depicts a simplified LLDD test arrangement with the return current either via the ground plane or using a coaxial rig, forming a transmission line structure. The coaxial rig is actually the preferred option since it partially eliminates the influence of the ground plane and thus allows more realistic distribution of surface currents, but its feasibility is limited by the size of the aircraft.

Irrespective of return conductor method, the real aircraft is not homogeneous structure and cannot, therefore, be ideally matched to avoid wave reflections, this method is practically limited to frequencies below the first aircraft resonance frequency.

The injected (drive) aircraft skin current and resultant currents on the equipment cables are measured with current probes. Additional external surface magnetic field strengths are measured using B-Dot sensors. An example of measured quantities is shown in Figure 4.59.

The resultant currents are then normalized to external unit field strength by a comparison of the measured skin current densities with the theoretical ones.

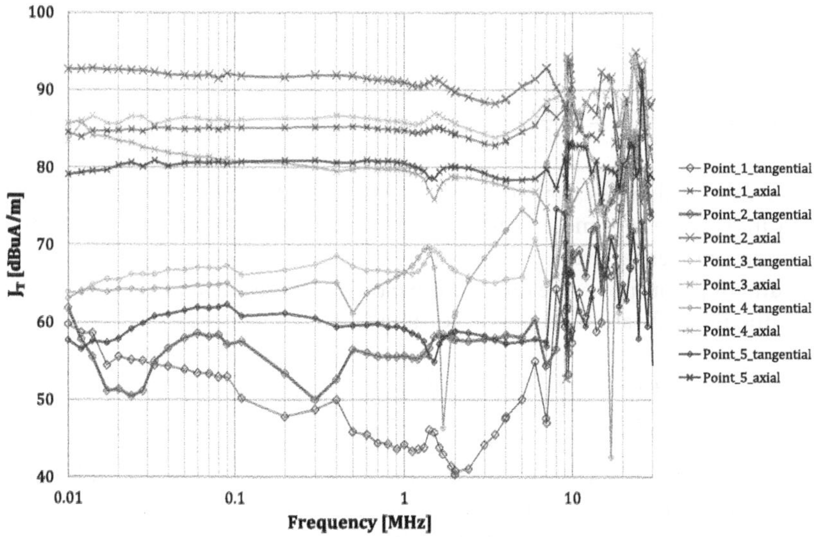

(a)

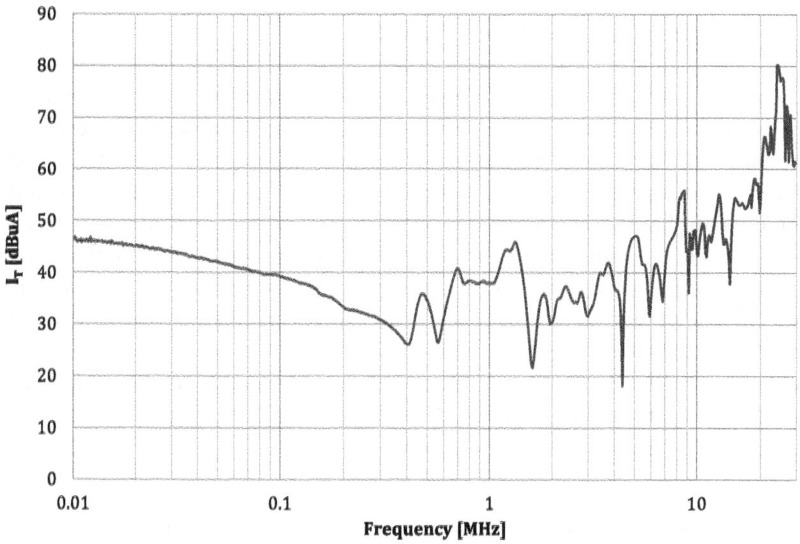

(b)

Figure 4.59 Results of LLDD tests (a) axial and tangential surface current densities at different points along the fuselage; (b) cable bundle current.

This normalization process is as follows: Assumption b) above can be mathematically expressed as:

$$\frac{I_F}{J_{SF}} = \frac{I_T}{J_{ST}} \Rightarrow I_F = I_T \frac{J_{SF}}{J_{ST}} \quad (4.201)$$

where symbol I denotes cable current in Amperes, J_S is skin surface current density in Amperes per meter, and the subscripts F and T refer to in-flight and test conditions, respectively.

Document [9] proposes the following expansion of the formula 4.201 in order to split the normalization process according to individual steps:

$$I_F(H_{inc}) = \left(\frac{I_T}{I_D}\right) \times \left(\frac{J_{ST}}{I_D}\right)^{-1} \times \left(\frac{J_{SF}}{2H_{inc}}\right) \times 2H_{inc} \quad (4.202)$$

where the dimensionless ratio in the first brackets represents the test cable current normalized to the injected (drive) aircraft skin current I_D. The ratio in the second brackets relates the measured surface current density to the drive current and for circular and quasi-circular structures it should be approximately equal to the circumference of the structure.

The dimensionless ratio in the third brackets normalizes the local in-flight (i.e., calculated) surface current density J_F, and hence the in-flight magnetic field strength at the skin surface to the theoretical value obtained by the low-frequency cylinder model immersed in transversal magnetic field defined by arbitrary field strength H_{inc} (see section 4.2.3.4.3). Values of $J_F/2H_{inc}$ exceeding one mean that the aeroplane skin surface current density is higher than the predicted one, possibly due to aeroplane resonance. The in-flight skin surface current density J_F is obtained using a 3D numerical analysis of aircraft exposed to specified incident electromagnetic plane wave specified by the same arbitrary magnetic field strength H_{inc}.

Since the HIRF environment is specified by the electric field strength, the magnetic field strength H_{inc} can be determined as $H_{inc} = E_{inc}/377\Omega$. Then, by inserting into equation (4.202), one obtains:

$$I_F(E_{inc}) = \left[\left(\frac{I_T}{I_D}\right)\left(\frac{J_{ST}}{I_D}\right)^{-1}\left(\frac{J_{SF}}{2H_{inc}}\right)\frac{2}{377\Omega}\right] E_{inc} = TF \times E_{inc} \quad (4.203)$$

where TF denotes the cable bundle transfer function.

A multiplication of TF by the prescribed HIRF environment electric field strengths, E_{HIRF} allows obtaining the expected full-threat in-flight cable bundle currents I_{Ffull}:

$$I_{Ffull} = TF \times E_{HIRF}$$

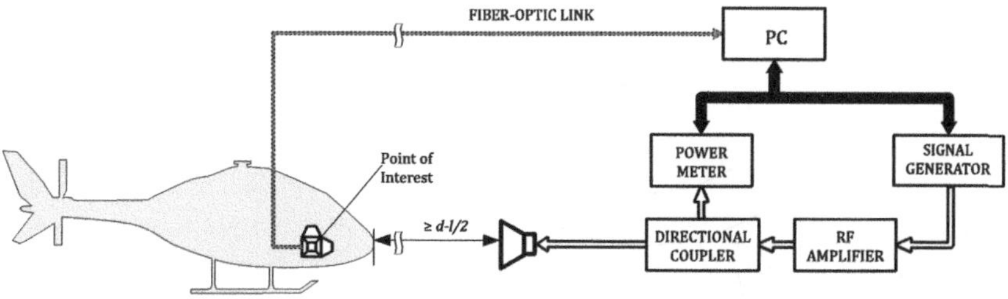

Figure 4.60 Test set-up for LLSF measurements.

4.3.2.2.3 Low-Level Swept Field Coupling The Low-Level Swept Field (LLSF) Coupling method is used to determine internal fields at the location of equipment installations in the frequency range from 100 MHz to 18 GHz. This method is similar to LLSC method. The main difference is that a fiber-optically coupled field sensor is placed at the test location and the electric fields in the internal areas are measured instead of the cable bundle currents (see Figure 4.60).

The output of the test is the aircraft structure attenuation ATT—a ratio between the internal field E_{int} and the external incident electromagnetic field level E_{inc}, expressed in decibels:

$$ATT = 20 \log \frac{E_{inc}}{E_{int}} \tag{4.204}$$

For example, 12 dB or a 4:1 attenuation means the test level is the applicable external HIRF environment electric field strength reduced by a factor of 4.

The measurement follows again in two steps, first calibration and then the actual field measurement.

1. Below 400 MHz, where the field-to-cable coupling is still a significant contributor, the same considerations concerning the mutual aircraft to radiating antenna placement and the field calibration apply (refer to Figures 4.53 and 4.54).
2. Above 400 MHz, only the last $\lambda/2$ length of the EUT connector and wiring, plus direct case penetration, are the dominant coupling paths. Therefore, reduced antenna distances and hence multiple more localized illumination of the aircraft, as sketched in Figure 4.61, is permitted and often more practical. It has just to be ensured that the complete area where the equipment is located is uniformly illuminated and realistic field penetration through all possible points of entry (such as access panels, doors, seams, windows, etc.) into the area is guaranteed.

The minimum antenna distance from the aircraft surface is calculated from the beamwidth of the antenna used and the minimum size of area to be

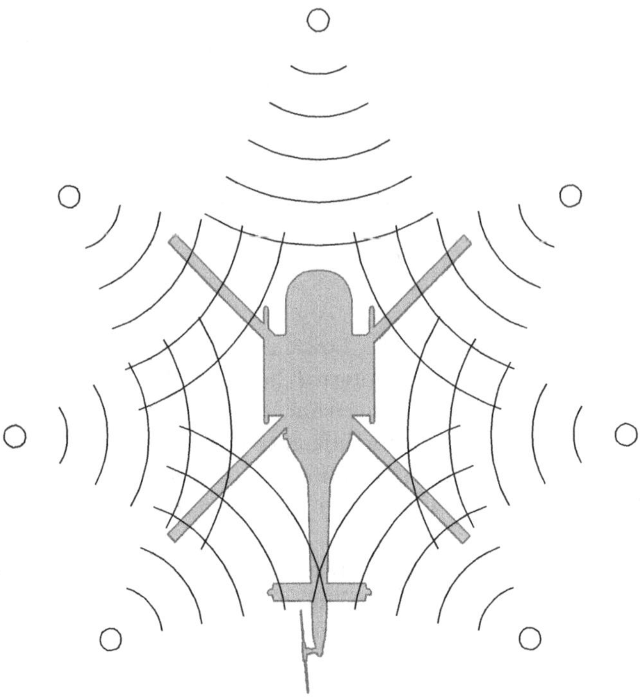

Figure 4.61 An example of Antenna positioning for LLSF measurements above 400 MHz.

uniformly illuminated. At this distance, both the calibration without aircraft and the measurements with the aircraft in place are performed.

The internal field is measured for all illumination angles and for both vertical and horizontal polarizations of the transmitting antenna. Multi-point measurements (at least 3) or mode stirring should be used in order to ensure that the maximum internal field in the vicinity of the equipment is measured.

The attenuation is derived from the difference between the internal fields measured and the calibration file. A worst-case attenuation plot for all illumination angles and polarizations is produced for use during the radiated susceptibility test.

An example of attenuation curves obtained via the LLSF method is presented in Figure 4.62. In the figure it can be seen, a positive attenuation, i.e., amplification of the field intensities occurs at the external and internal resonance frequencies (refer to section 4.2.3.7).

4.3.2.3 Generic Aircraft Structure Transfer Functions and Attenuations

Although measured transfer functions or attenuations provide a more accurate estimation of the real internal HIRF environment in aircraft, it is possible to

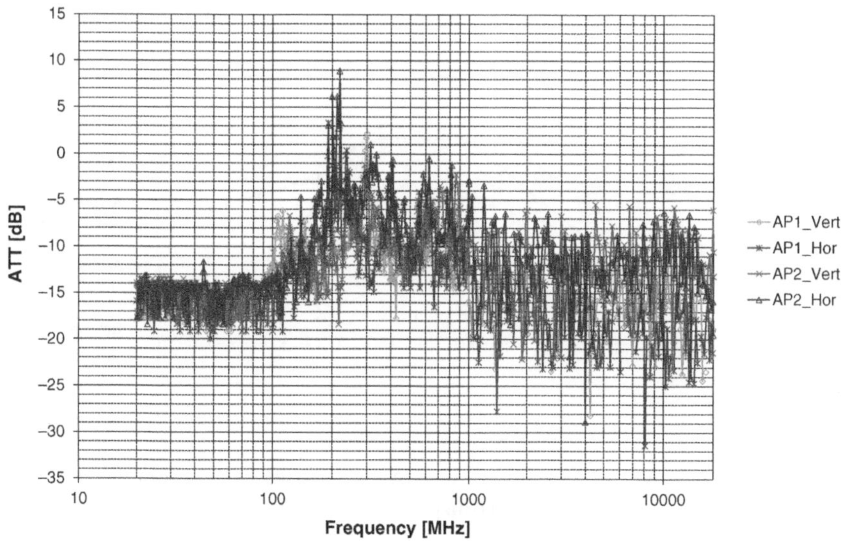

Figure 4.62 Electric field attenuation in the frequency range 20 MHz–18 GHz measured in the centre of cockpit for two different antenna positions (AP1/AP2) and two polarizations (vertical/horizontal).

use so-called generic transfer functions and attenuation given in documents [4] or [9]. Especially in early stages of the design program, the information on internal HIRF environment may not be available and, therefore, test levels can be developed using the generic transfer function and attenuation curves.

4.3.2.3.1 ***Generic Transfer Functions*** Generic transfer functions provided in [4] are the envelope of the maximum currents (in mA) that might be expected to be induced on aircraft wiring bundles in an external HIRF environment of 1 V/m in the frequency range from 10 kHz to 400 MHz. Depending on the type and the size of aircraft, there are four different transfer function curves (see Figure 4.63):

1. for aeroplane with a fuselage length of ≤25 m
2. for an aeroplane with a fuselage length of >25 m and ≤50 m
3. for an aeroplane with a fuselage length of >50 m
4. for a rotorcraft (helicopter)

Break points in the envelope were set to frequencies corresponding to resonant frequencies of the fuselage having the maximum and minimum dimensions for each category. For example, for the 25 m to 50 m category, the low-frequency break point is defined by $l = \lambda/4 = 50$ m, i.e., 1.5 MHz and the high-frequency break point is when $l = \lambda/2 = 25$ m, i.e., 6 MHz.

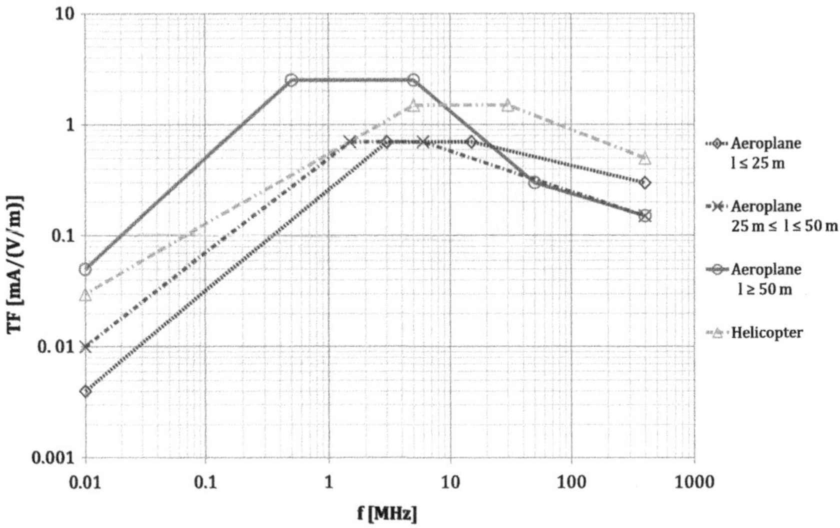

Figure 4.63 Generic transfer function normalized to 1 V/m [4].

It is important to remember that the generic curves are not the worst-case values. They have been derived by a statistical analysis of some several hundreds of spectra from 16 civil aircraft, and induced current levels covering 95 percent population's probability were used as the basis for the transfer function curves [4, 9].

The transfer functions are normalized to a 1 V/m and shall be multiplied linearly by the applicable external HIRF environment to determine the expected bulk currents, as illustrated in Figure 4.64. The User's Guide [9] thoroughly describes the use of generic transfer functions.

4.3.2.3.2 Generic Attenuations As it was mentioned in subsection 4.3.2.2.3, the relation between the internal and external HIRF environment can be expressed by means of attenuation. Document [4] proposes attenuation values for determination of the internal electromagnetic field levels in the frequency range 100 MHz to 18 GHz. Following the guidance of this document, and depending on the location of both the equipment of interest and its associated wiring, the following generic attenuation values can be selected:

No Attenuation (0 dB) for aircraft areas with no HIRF shielding

6 dB Attenuation for aircraft areas with minimal HIRF shielding

12 dB Attenuation for aircraft areas with some HIRF shielding

20 dB Attenuation for aircraft areas with moderate HIRF shielding

32 dB Attenuation for areas with very effective HIRF shielding to form an electromagnetic enclosure

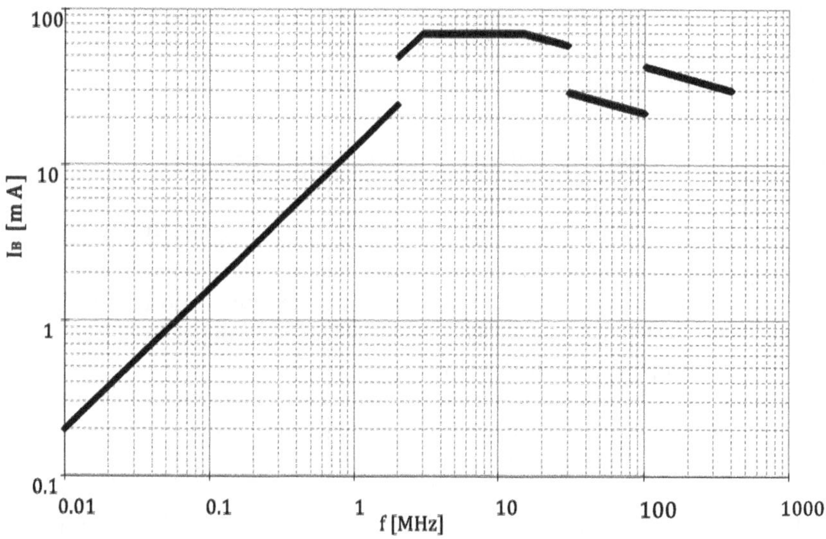

Figure 4.64 Cable bundle current obtained by application of generic TF for small aeroplane ($l \leq 25$ *m*) to HIRF Environment I.

Once again, a thorough understanding of different aspects of electromagnetic interactions and corresponding experience from aircraft measurement is a key point when selecting the attenuation levels and the selection shall be supported by analysis.

The measurements have shown that it is often necessary to select different attenuation values for different frequency ranges, and sometimes even negative attenuation values (i.e., internal field strengths higher than the external HIRF environment) were observed in cockpit areas around the frequencies of the cavity resonances.

The example given in [4] suggesting 0 dB attenuation for the frequency range of 100 MHz to 400 MHz, 6 dB attenuation for the frequency range of 400 MHz to 1 GHz, and 12 dB attenuation for the frequency range of 1 GHz to 18 GHz could be applicable to cockpit areas.

4.3.2.4 Determination of Internal EM Environment by Similarity

The internal electromagnetic environment can be estimated using the results from HIRF attenuation and transfer functions measurements previously performed on a similar aircraft. The extent of reliability of such estimation depends on the degree of similarity between the old and the new aircraft.

The similarity assessment should take into account all differences between aircraft that could have an impact on the internal HIRF environment. The comparison should consider location of equipment and routing of cable harnesses, airframe materials and construction, and size and location of apertures.

When the assessment finds only minimal differences, the similarity may be used to determine aircraft attenuation and transfer functions without the need for additional aircraft tests.

4.3.3 Avionic Equipment/System Functional Verification

4.3.3.1 Integration Bench Approach

There are several options how to verify the correct functioning of the given system/equipment when exposed to the radio-frequency electromagnetic fields. The most common one is so-called test rig testing in a test laboratory, where the equipment under test and its associated wiring is arranged over a ground plane representing the aircraft structure.

This approach is thoroughly described in civil HIRF document [57] with additional information given in [41] and [9]; or corresponding defense specifications [58], therefore, it will be just outlined in the following subsections.

4.3.3.1.1 Set-up In order to provide a reference plane provided in aircraft by the aircraft conductive structure, the system equipment under test shall be mounted on a solid, preferably copper conducting bench—the ground plane. The size of the ground plane must be sufficient to accommodate the system under test and its associated wiring arranged in specified manner; therefore, especially for complex avionic systems the size of the ground plane may be extensive.

The all units to be tested shall be positioned approximately 100 mm from and parallel to the front edge of the ground plane, allowing adequate room for cable bending radii and arrangements.

Cables should be arranged at a distance of 100 mm back from the front edge of the ground conducting bench for as much of their length as possible to guarantee a defined exposure to the electromagnetic fields. The cables shall be supported above the ground conducting bench on 50 mm insulated stand-offs in order to simulate a typical ground current loop area. If greater distances are more representative of the real aircraft installation, these may be used.

When an interconnecting harness between two items of the system is longer than the exposed length along the front edge of the ground plane, the excessive length shall be zigzagged at the back of the test bench on 50 mm supports. Coiling of the cable should be avoided since it significantly increases the cable inductance, thus reducing the induced conducted interference and modifying the natural resonance frequencies.

The bonding provisions specified in the installation instruction shall be used. Bonding jumpers shall have their cross-sections, lengths, routing, and the method of connection as similar as possible to those specified for the real installation.

In order to provide defined impedance relatively independent of the impedance of different power sources, Line Impedance Stabilizing Networks

(LISNs) shall be used. The LISN presents defined impedance between its EUT terminal and the ground plane.

4.3.3.1.2 Bulk Current Injection (BCI) Test The basic concept of conducted susceptibility—BCI test is to simulate currents that will be induced on aircraft cabling by internal and external RF fields but avoiding the need for immense powers fed into radiating antennas of significant size (see subsection 4.3.2.2).

Due to the standing wave effects on the cable under investigation, this method is frequency-limited to 200 MHz or 400 MHz [57,58] (although already above 100 MHz the method can be questioned), since the induced currents are heavily influenced by the size, type, and location of the current injection probe.

The test comprises injection of the specified bulk current levels (determined by LLDD, LLSC, using generic transfer functions or in similar way) into the cable harness under test via a high-power current injection probe. The level of injected current is monitored by a current measuring probe while observing the EUT for signs of malfunction.

The test consists of two steps. First, the probe calibration is performed. The injection probe is placed in the calibration fixture terminated at both sides by 50 Ω terminations.

The control computer sets the frequency and the output signal level of the signal generator, thus adjusting the power level to the injection current probe fed via a directional coupler from a power amplifier. The unmodulated power is increased until the current measured on one of the 50 Ω terminations reaches the prescribed level at given frequency. The necessary forward power to the injection probe is recorded. This forward power plot will be used during the actual test. The frequency range is scanned at logarithmically spaced steps with the above-described process repeated at each step. The minimum requirement is 100 frequencies per decade. Additionally, it should be dwelt at known internal equipment frequencies, such as the local oscillator frequency, clock frequencies, etc.

During the actual test, the injection and monitor probes are placed on the harness to be tested as shown in Figure 4.65. Attention must be paid that any short grounding wires bundled in the cable harnesses are excluded from the test, i.e., taken out of the bundle and routed outside both the monitoring and the injection probes.

The control computer sets the frequency and the output signal level of the signal generator, thus adjusting the power level to the injection current probe fed via a directional coupler from a power amplifier. The injected current is increased until the EUT malfunctions occur or the prescribed test level is reached.

Remark *With the evolution of the test standards, the way of defining the test level has been changing. At the moment [57], the valid approach is as follows:*

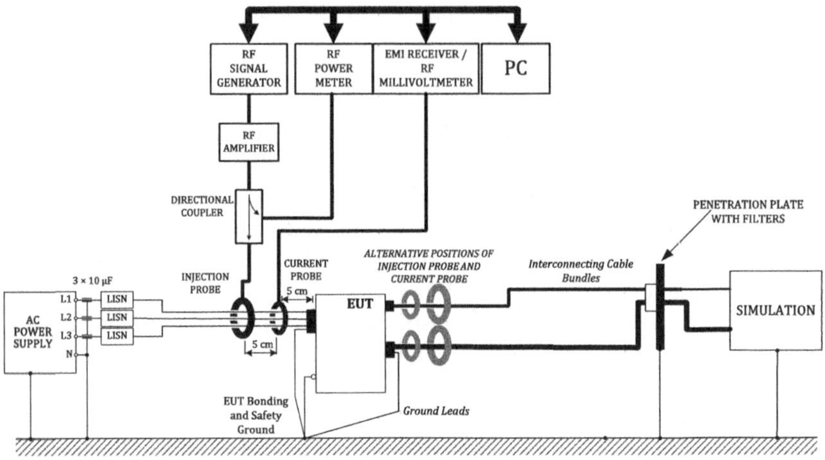

Figure 4.65 Example of the test arrangement for BCI testing.

"The forward power shall be adjusted to achieve the prescribed induced current (the calibration current) on the cable bundle while limiting the forward power to not more than 6 dB above the value of the forward power determined during the probe calibration procedure at the given frequency."

This means following the 20 dB per decade roll-off below the first break point in the calibration current limit (one example is shown in Figure 4.64). Theoretical analysis of the field-to-wire coupling presented in sections 4.2.4.2 and 4.2.4.5 or in [59–61] as well as LLDD/LLSC measurements on aircraft suggest that this approach sufficiently well simulates the field coupling to high-impedance cable loops but it can be questioned for low-impedance (e.g., well-shielded) wiring, at least at low frequencies. Therefore, other approaches (described for example in [55, 58] or [62]) can be defined by the airframer test specifications. For the flight critical systems, the envelope of the maximum currents induced during the BCI test should be compared with the envelope of the transfer function extrapolated to the appropriate environment and demonstrated to be at least equal to it.

The control computer records the current flowing in the cable as measured by the current measuring probe via the EMI receiver, spectrum analyzer, or other suitable device. With the desired test level set, all prescribed modulations (see section 4.3.3.3) are applied for defined dwell time. The dwell time at should be set to sufficiently long value, taking into account EUT, auxiliary and test equipment response times and modulations to be applied. A minimum value given in [57] is 1 second, however, this is not a rule and some other documents require longer dwell times.

While monitoring the operation of the tested system, the process is repeated as each new frequency is tested until the whole band of interest has been

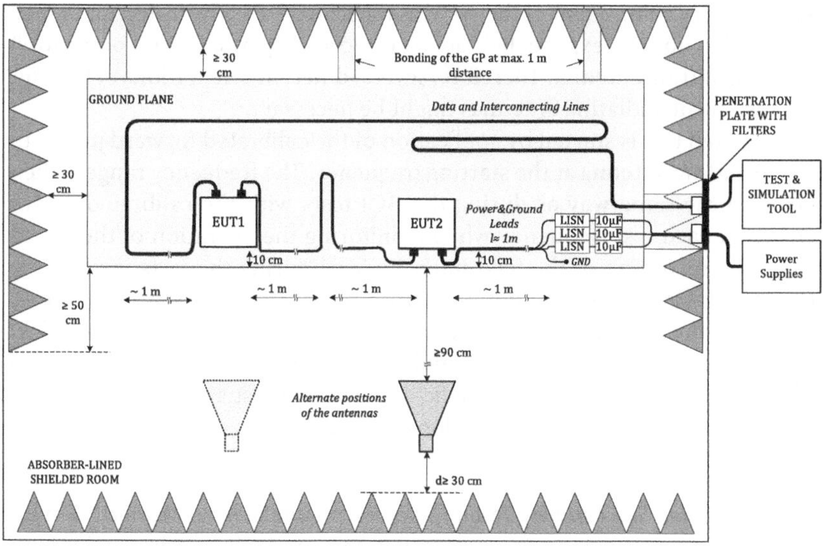

Figure 4.66 Example of a test set-up for RS testing in (semi)anechoic chamber.

covered. The frequency range is scanned at logarithmically spaced steps. The minimum requirement is 100 frequencies per decade plus at known internal equipment frequencies.

4.3.3.1.3 Radiated Susceptibility Testing—Anechoic Chamber Method For showing compliance with requirements for immunity of avionic systems and equipment to radiated electromagnetic field, the test specifications (e.g., [57, 58]) describe a so-called substitution method.

The substitution method allows the field level to be "recalibrated" without the EUT to establish a given field. This includes measuring the forward power fed to the transmitting antenna that is necessary to reach the prescribed field strength measured by the electric field probe. The calibration is performed for the complete frequency range of interest. The rationale is that this pre-calibrated field simulates an incident electromagnetic wave impinging onto the equipment, i.e., field that would exist without presence of EUT and distortions of the field due to the EUT will be similar to ones occurring in a real-world environment.

After the calibration step, the EUT is placed at the test location (see Figure 4.66). The antennas are positioned and aimed to expose the EUT and interconnecting wiring to the EM fields.

The beamwidths of the antenna shall totally cover the unit under test and the first half-wavelength of its wiring. If this is not the case, multiple area scans shall be performed. The multiple area scan is also required for complex systems where the every component of the systems must be tested. The tested units shall be oriented to directly expose their apertures (e.g., displays, CRTs, connectors) to the transmitting antenna. Therefore, several different orientations of the unit under test or the radiating antennas might be necessary.

The radiated test is started by application of the calibrated forward power to the transmitting antenna at the starting frequency. The frequency range is then scanned in the same way as during the BCI tests with the calibrated power reapplied at each test frequency while monitoring the operation of the tested system. The recording of the field intensity level during the tests is generally not required, since it will be different from the pre-calibrated values due to reflection of the equipment under test. It is, however, advised to place a field measuring probe close to the EUT and monitor the presence of the field to verify the correct functioning of the field-generating test system. Vertical and horizontal transmit antenna orientations are required.

4.3.3.1.4 Radiated Susceptibility Testing—Reverberation Chamber Method Reverberation chambers are basically cavity resonators producing desired fields with minimum input power. They are shielded enclosures with the smallest dimension being large with respect to the wavelength at the lowest usable frequency [9, 23, 43, 57].

The chamber is equipped with a mechanical tuning/stirring device with dimensions comparable with the chamber dimensions (see Figure 4.67). When the chamber is excited with RF energy by an antenna, the standing wave resonant patterns are "stirred" by the mechanical tuner/stirrer. The resulting environment varies substantially as the tuner is rotated. Any physical location in the chamber will achieve same peak field strength at some position of the paddle wheel. However, the environment is averaged over a sufficient number of positions of the mechanical tuner/stirrer. The resulting field distribution is the considered both isotropic and uniform.

Much has been written about the advantages and technical justification of this technique. Similarly, the complete validation process using the statistical methods is quite extensive and it is not possible to review it here. Therefore, only explanation of the key points is given here.

The testing process consists of again three steps:

1. Prior to the fitting of the test bench with the system under test, a chamber calibration technique is carried out to demonstrate the chamber meets the field uniformity requirements [57]. This field uniformity calibration is carried out over a complete test/working volume, i.e., the location of the ground plane with the tested equipment within the reverberation chamber. The complete frequency range is scanned in the same way as during the absorber chamber tests.

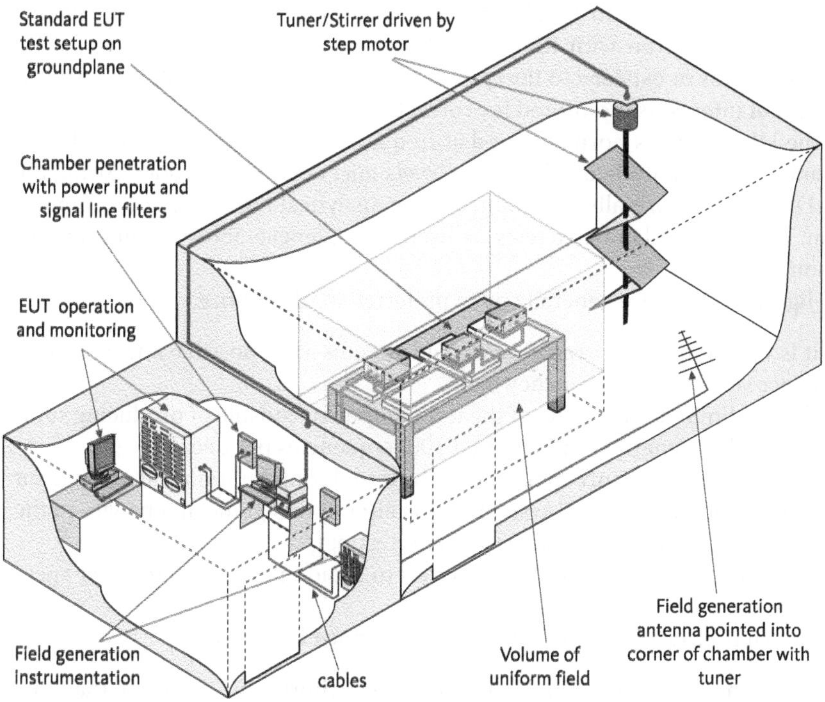

Figure 4.67 Example of a test set-up for RS testing in reverberation chamber.

2. With the test bench and EUT installed inside the working volume of the chamber, the chamber loading measurement is performed, i.e., determination of the decrease in the field strength due to loading of the chamber by the test rig presence.
3. Actual testing with the forward power to the radiating antenna increased by the loading factor determined during the step 2. The tuner paddle is rotated at a maximum rate of 4 revolutions per minute below 1 GHz and 2 revolutions per minute above 1 GHz, while continuously applying power for one tuner revolution. For low frequency modulations, the tuner rotation rate must be reduced. The complete frequency range is scanned from the lower to the upper frequency limit using the appropriate modulations while monitoring the operation of the tested system.

4.3.3.2 High-Level Full Aircraft Tests

4.3.3.2.1 High-Level Direct Drive Tests and High-Level Radiated Tests Aircraft high-level testing includes high-level direct drive (HLDD) tests and high-level radiated tests. These methods are—concerning the test setup and the test procedure—identical to LLDD and LLSF methods, respectively; but the aircraft is being driven with RF current expected for the specified external RF field strengths or exposed to the specified external RF fields, while the installed system of interest is monitored for correct operation.

The HLDD test is normally used at frequencies below the first aircraft resonance. The amplitude of RF current levels injected on the aircraft during the HLDD test are normally determined using analytical modeling and computation. The high-level radiated tests are used at frequencies above the first aircraft resonance.

High-level tests are generally not a preferred method since:

1. it is very difficult to irradiate installed systems at all possible angles of incidence over the full frequency band;
2. there is limited availability of the test equipment capable of producing sufficient field strength at the required 100 frequencies per decade;
3. the test results are only applicable to the tested configuration and subsequent installation of additional systems in the aircraft or significant changes in the configuration would possibly require retesting;
4. there is a need for special test facilities to avoid interference to other spectrum user;
5. special precautions concerning fuel safety must be taken.

While the high-level full aircraft tests are shorter in duration than the approach combining the low-level aircraft measurements and the laboratory test rig testing, the overall costs might me higher due to higher equipment and facility costs

The high-level aircraft testing is usually employed when there is only one system physically available, and this is already installed in aircraft. Even in this case, the high-level aircraft testing are combined with other methods. For example:

1. The generic transfer curves are used to determine the levels the induced RF cable bundle currents up to 400 MHz. Subsequently, the conducted susceptibility—BCI testing is carried out on the system installed in the aircraft, see the next subsection.
2. Above 100 MHz, the high-level radiated susceptibility tests are carried by irradiating the aircraft containing the system of interest.

4.3.3.2.2 BCI Testing on Systems Installed on Aircraft Although testing on a system rig in a laboratory is the preferred, conducted susceptibility testing can be performed on systems installed in the aircraft.

The installed system is tested using the procedures based on those described for the system rig laboratory testing (refer to subsection 4.3.3.1.2). However, a thorough analysis of the installation and careful preparation must be done prior to testing, focusing on (but not only on) the following points:

- Every bundle in the system shall be completely tested by injection. If a bundle branches, each branch is also to be tested.
- The problem area of short grounding wires bundled in the cable harnesses. These must be taken excluded from the test.
- shields bonded at multiple locations. All segments separated by the shield bonding points must be tested. If possible, the shields connections may be interrupted at the bulkheads or injection with the shields interrupted and separated at all points from the ground structure (i.e., direct injection on the core wiring) might be an option.

Additional challenge is the accessibility of cable bundles and provision of a sufficient space to place both the injection and the monitor probes on the cable under test.

4.3.3.3 Modulations

For certain types of circuits and components, a modulated RF signal can be a more severe test than an unmodulated signal, although the average delivered power is lower. In addition, peak modulation electromagnetic field strengths are significantly higher than a pure CW signal.

The amplitudes of the modulated signals are expressed as "peak rms values", defined in subsection 4.3.1.1. When performing radiated susceptibility testing on civil avionic equipment and systems, there are three commonly used types of signals (see subsection 4.3.1.1):

- Continuous wave (CW)—unmodulated signal,
- Square-wave modulation (SW)—modulation of signal amplitude by 1-kHz square wave with at least 90% modulation depth—a special type of pulse modulation with a duty cycle *dc* of 50% and a pulse repetition frequency f_r of 1 kHz as described in subsection 4.3.1.1
- Pulse-modulated signals (PM)—a pulse-modulated signal commonly specified by test standards is defined by pulse widths *pw* between 1 and 4 microseconds and a pulse repetition frequency f_r of 1 kHz, as defined in subsection 4.3.1.2.
- The PM and SW modulations can be additionally gated (switched on and off) at a 1–3 Hz rate with 50% duty cycle, i.e., an additional very low frequency SW modulation is superimposed on the modulated signal. This approach should simulate the effect of rotational radars [57].

For definition of modulation patterns applicable to avionic systems intended for military applications, the corresponding documents (e.g., [44, 45, 58]) shall be consulted.

4.3.4 Showing Compliance with HIRF Certification Requirements

As it was mentioned in the section 4.1.3, the systems installed in aircraft are divided into three groups according to their criticality. Logically, the most stringent requirements apply to Level A function systems, with the severity consecutively reduced for Level B and C function systems.

4.3.4.1 Level A Systems

According to the current requirements of FAA and EASA, the level A systems must be designed and installed so that they are not adversely affected during and after the time the aeroplane is exposed to specified HIRF environments (for definition of HIRF environments see Table 4.3). The exact requirements on functionality versus the HIRF environment can be found in [4] and [5].

The main task for showing compliance is to prove that the system in the real aircraft configuration was tested and proven immune to:

1. either the external electromagnetic HIRF environment I, II, and III, as applicable; when it was installed in the aircraft exposed to external threat using methods described in subsection 4.3.3.2;
2. or the internal electromagnetic fields and the corresponding cable bundle currents, when tested using the integration test rig approach described in subsection 4.3.3.1 and the applicable internal electromagnetic fields and cable bundle currents were determined:
   - experimentally using methods described in subsection 4.3.2.2
   - using similarity with a previously certified aircraft and installation, (subsection 4.3.2.4)
   - by numerical simulation supported and validated by experimental results
   - using generic transfer functions and attenuations (see subsection 4.3.2.3). However, current issues of [4] and [9] allow application of this method to Level A Display system only.

4.3.4.2 Level B Systems

Any level B system must be designed and installed so that the system is not adversely affected when the equipment providing the function is exposed to equipment HIRF test level 1 or 2 defined as:

4.3.4.2.1 Test Level 1

1. Conducted susceptibility tests from 10 kHz to 400 MHz Continuous wave (CW) and 1 kHz square wave modulated signals at minimum levels defined by the following curve:
   (a) Increasing from 0.6 mA at 10 kHz to 30 mA at 500 kHz at a rate of 20 dB per frequency decade
   (b) Constant level of 30 mA from 500 kHz to 40 MHz

   (c) Decreasing by 20 dB per frequency decade from 30 mA at 40 MHz to 3 mA at 400 MHz
2. Radiated susceptibility tests
   (a) From 100 MHz to 400 MHz, CW and 1 kHz square wave modulated fields at minimum level of 20 V/m.
   (b) From 400 MHz to 8 GHz, pulse-modulated electromagnetic fields at minimum of 150 V/m. The modulation parameters shall be as follows: duty cycle of 4%, pulse repetition frequency of 1 kHz, the signal must be switched on and off at a rate of 1 Hz with a duty cycle of 50 per cent.

This definition corresponds to Category R defined in Section 20 of [57].

4.3.4.2.2 Test Level 2 Test level 2 is obtained by reducing HIRF environment II using acceptable aircraft transfer function for BCI testing and attenuation curves for radiated susceptibility testing (see subsection 4.3.2.3). Testing must cover the frequency band of 10 kHz to 8 GHz.

If a level B function system consists of equipment qualified to sufficient levels using laboratory tests [57] and wire types, wire bundle composition, connectors, shields, and shield terminations in the intended aircraft installation are same as specified in the equipment installation instructions, and the lengths defined in [57] were used, a separate full system test is not required.

4.3.4.3 Level C Systems

Level C systems must be designed and installed so that the system is not adversely affected when the equipment providing the function is exposed to equipment HIRF test level 3 defined as:

1. Conducted susceptibility test levels from 10 kHz to 400 MHz with minimum levels defined by the following curve:
   (a) Starting with 0.15 mA at 10 kHz, increasing 20 dB per frequency decade to 7.5 mA at 500 kHz.
   (b) Constant level 7.5 mA from 500 kHz to 40 MHz
   (c) Decreasing at a rate of 20 dB per frequency decade from 7.5 mA at 40 MHz to 0.75 mA at 400 MHz.
2. Radiated susceptibility tests from 100 MHz to 8 GHz at a minimum of 5 V/m.

These definitions correspond to Category T defined in Section 20 of [57]. Again, if a level C function system consists of equipment qualified to sufficient levels using laboratory tests [57] and wire types, wire bundle composition, connectors, shields, and shield terminations in the intended aircraft installation are same as specified in the equipment installation instructions, and the lengths defined in [57] were used, a separate full system test is not required.

4.4 Electromagnetic Effects of Lightning

4.4.1 Lightning Flash

4.4.1.1 Formation of Thunderclouds

Approximately 40 to 100 lightning flashes take place each second around the globe with most of the lightning flashes taking place over land regions. Over land regions, most lightning storms occur in the tropics with 150–200 thunderstorm days in certain regions near the equator with the lightning activity decreasing toward the poles. The lightning activity over oceans is an order of magnitude lower than that over land due to lower convective intensity in oceanic areas.

Generally, it is possible to divide storms into two main classes, heat storms and thermal storms. The heat or convective storms are caused by direct heating of moist air at ground level and predominate in the tropic regions, in summer months in temperate climate regions (summer thunderstorms) and in the mountainous areas (orographic thunderstorms), where the formation of clouds is aided by the lifting of incoming air along the slopes.

The other group involves called frontal thunderstorms. The frontal storms, which prevail in temperate regions, are caused by the impact of a front of cold air on a mass of arm wet air or vice versa, lifting the warm humid air to higher levels of the atmosphere.

A cumulonimbus, or the thundercloud, is formed when a parcel of lighter warm humid air rises from ground and is replaced by heavier colder air drifting down. While rising up, the hot air is cooled down, condenses and forms tiny water droplets that collide to form one rain drop. As the temperature of the air falls below 0°C, some of these small water droplets freeze and form minuscule ice crystals. However, a part the tiny droplets become supercooled water droplets liquid even at temperatures below the freezing point. Some supercooled water droplets collide with tiny ice crystals and freeze directly on them, yielding a growth of the tiny ice crystals, producing so-called graupel particles. As the graupel particles grow, they cannot float in the rising air; they begin to fall creating a downdraught. In this stage, the thundercloud reaches its mature stage in which it is capable of generating lightning flashes. When the downdraught due to falling precipitation causes blocks completely the updraught, the droplet formation stops and thundercloud dissipates [63].

The charging mechanism is not yet fully understood and several theories were introduced. According to the ice-graupel collision mechanism based on collisions between graupel particles and ice crystals or supercooled water droplets [63], at higher cloud layers with temperature below −10°C to −15°C (the reversal temperature), heavier particles charge negatively and due to the gravitational force move downwards while lighter positively charged ice crystals are carried by air currents higher, resulting in the positive net charge layer

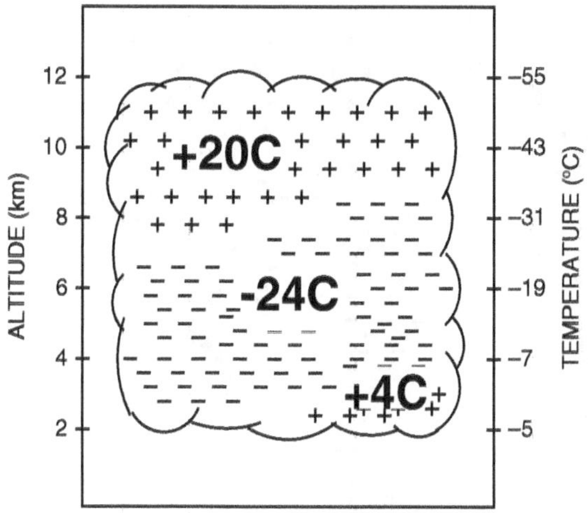

Figure 4.68 Charge distribution in thundercloud.

above the negative net charge of the falling graupel particles. Above reversal temperature (i.e., lower altitudes), the lighter ice crystals charge negatively while carried upward by the updraught and heavier particles charge positively producing a positive net charge pocket located below the negative charge centre. The accumulated charge inside the main charge centres can reach several hundreds of 100°C [64].

Summing up, the main negative charge centre is located in the region around the −10°C to −15°C isotherm, the main positive charge centre is located above the negative one and a small positive charge pocket can be below the negative charge centre, as outlined roughly in Figure 4.68. This is so-called tripolar structure of cloud charge, consistent with some experimental observations showing that a typical cloud contains two main charge centres.

The recent research has shown that the charge distribution can be much more complex with several interleaved oppositely charged layers. Nevertheless, the above-described oversimplified outline of the charge structure suffices to explain many storm-related electrical phenomena at the ground level.

4.4.1.2 Types of Lightning Discharges

The separation of charge outlined in the previous section produces enormous electrical potential differences within the cloud and between the cloud and ground, as well as between neighboring clouds. This can amount to millions of volts, and eventually the electrical resistance in the air breaks down and a flash begins. Lightning, then, is an electrical discharge between positive and negative regions of a thunderstorm, which may be of two basic types, namely (see Figure 4.69):

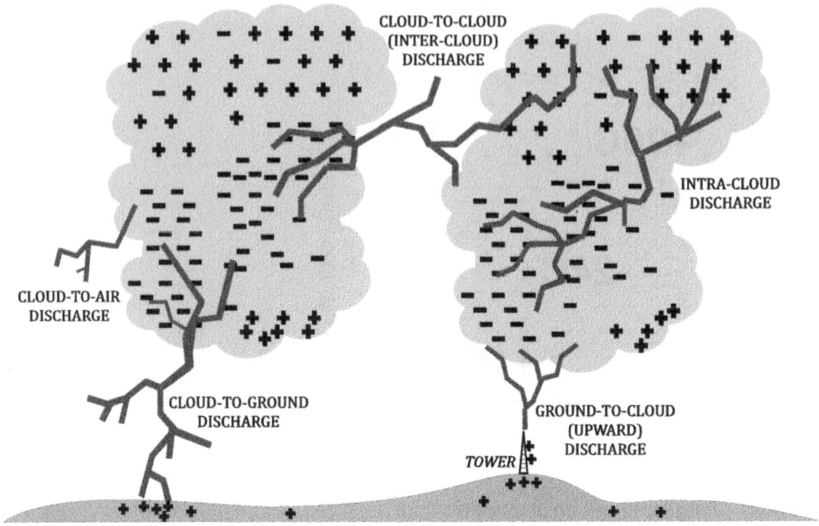

Figure 4.69 Types of lightning discharges.

1. Cloud flashes, comprising
   (a) Flashes between regions of opposite polarity within a cloud (intracloud discharges)
   (b) Flashes between regions of opposite polarity in different clouds (inter-cloud charges)
2. Ground flashes that include:
   (a) Cloud-to-ground discharge, i.e., a discharge originating in a thunder-cloud
   (b) Ground-to-cloud discharge, originating at the tips of taller objects (e.g., towers and mountains)

The ground discharges together with the cloud discharges yield the sum of all flashes. The ratio of the ground flashes to the cloud lightning flashes depends on the latitude with the number of cloud discharges generally prevailing. The following estimation can be made [64]:

$$\frac{N_G}{N_T} = 0.1 + 0.25 \cdot \sin(\lambda) \tag{4.205}$$

where N_G denotes number of ground flashes, N_T is the number of all lightning discharges (ground and cloud flashes together) and λ denotes latitude in degrees. Other similar estimation formulas can be found in [65].

The lower amount of ground flashes in the tropics ($\lambda \approx 0°$) lies in the fact that there the charge centres are due to the higher temperatures at higher altitudes, thus the electric field between the thundercloud and earth due is reduced is often not sufficient to develop a breakdown down to the ground. Therefore, the probability increases for a cloud-to-cloud lightning. The guideline is based on a long-term average, so that the actual proportion the ground flashes varies from a thunderstorm to a thunderstorm. In Central Europe, the proportion of ground flashes ranges between a few percent to about 50% [64].

4.4.1.3 Negative Cloud-to-Ground Flash

4.4.1.3.1 First Return Stroke Once an electrical breakdown is initiated inside a thundercloud, it gives rise to streamer discharges, and these streamer discharges in turn give rise to a leader discharge. If the process is initiated in the negative charge region and the leader lowers negative charge toward the ground, the flash is considered to be negative. A positive flash lowers positive charge to earth.

The above-mentioned leader discharge propagates toward the ground as faintly glowing discharge splitting into several branches along its path, as shown in Figure 4.70. The leader channel and its branches are extended toward earth not continuously but in discrete steps; therefore, it is termed "stepped leader". Each step is about 10 to 100 m long but the steps are lengthened with increase in the electric field as the leader approaches the ground. A development of each individual step takes approximately 1 μs and is associated with a current pulse characterized by the peak value up to few kA, the rise time of in the order of 100 ns and the decay time exceeding 1 μs, causing the step illumination for a

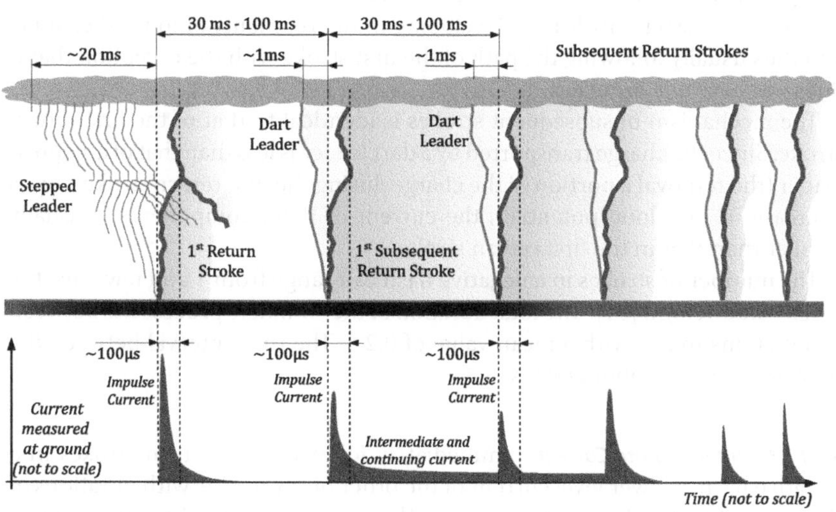

Figure 4.70 Development of lightning discharge.

few microseconds [66]. Additionally, a continuous current of about 100 A flows in the leader channel from the tip to the cloud. The time interval between steps is approximately 10–100 μs. The velocity of leader tip propagation varies from 10^5 to 2×10^5 m/s.

When the leader tip is in height of 50 to 100 meters above the ground, the electric field strength is high enough to initiate upward connecting leaders originating from the tallest objects in the vicinity of leader tip. The length of the connecting leader can be several tens of meters.

At the instant of connection of leaders, a strongly luminous current pulse called return stroke is initiated that propagates along pre-ionized path of the stepped leader from the ground to the cloud. This discharge represents short circuit between the negative charge distributed along the channel and the electrostatically induced positive charge on the ground surface. The return stroke is approximately a hundred times faster than the leader stroke. The current flowing in the channel can exceed 100 kA with rise times of the order of several to tens of microseconds and durations of tens to hundreds of microseconds. The rapid generation of heat associated due to the return stroke current raises the temperature of the discharge plasma channel to approximately 30,000 K in a few microseconds. This almost instantaneous heating causes sudden expansion of the channel, generating a shock wave in the air experienced as thunder.

4.4.1.3.2 Subsequent Strokes Typical negative flash discharges several charge centres in succession, with the result that flash contains several individual discharge processes called subsequent strokes. Each subsequent stroke is initiated by the much faster dart leader developing in the remnants of ionized channel and thus usually following the path of the first stroke with the exception that it does not have any branches.

The mechanism of subsequent strokes is identical to that of the first return stroke. Since the charge transported by a dart leader is less than that of a stepped leader (the removal a portion of the charge during the first return stroke causes decrease in the cloud potential), the current of these components is usually smaller than that in the first return stroke.

The number of strokes in a negative flash can range from 1 and few tens, the mean value being 3 [64]. The total duration of the complete process lies between about 20 ms to 2 s, with a mean value of 0.2 s. The time interval between the strokes is typically about 60 ms.

4.4.1.3.3 Intermediate Current Immediately following the return strokes in a negative flash, a lower level current of the order of several kA with durations of the order of few milliseconds may flow. This portion of the lightning current is known as an *intermediate current component* [10].

4.4.1.3.4 Continuing Currents For some subsequent strokes, the high current pulse is followed by a so-called continuing current with magnitude of tens to hundreds of amperes and duration of up to 1 second, so that there is substantial charge transfer in this phase. Between 30 and 50% of all negative cloud-to-ground flashes contain long continuing currents [63, 64].

4.4.1.4 Positive Cloud-to-Ground Flash

Positive flashes to ground generally occur less frequently than negative flashes; however, in certain geographic locations, there may be more positive flashes to ground. Present standards assume an average of around 10% positive flashes to ground.

Positive flashes normally consist of one stroke only. They have slower rise times than negative flashes, with high peak current and charge transfer; the duration is longer than a single stroke of a negative flash but usually shorter than a complete negative flash. The stroke may be followed by a continuous current.

4.4.1.5 Inter- and Intra-Cloud Flashes

In comparison to ground flashes, much less information is available on cloud discharges, as there was less interest in them since they did not pose any danger to people, structures and livestock at the ground level. Additionally, there was no possibility to measure directly lightning discharge currents and charge transfers, as it was done at the ground level during direct strikes to towers and poles.

However, they have become of great practical interest with the growth of air and space transport. After intensive research using VHF-UHF imaging, the following scenario describing the several stages during a cloud flash has been proposed [63, 67].

1. A development of a vertical channel within the first 10 to 20 ms, initiated by negative discharges from the negative charge centre toward the positive one in a more or less vertical direction. This channel is a few kilometers in length and growing at a speed of approximately 1.5×10^5 m/s from the beginning of the flash.
2. The main activity—The horizontal extension of channels in the upper positive charge centre, followed by discharges spreading from the lower level to the upper level along the vertical channel with repeated breakdowns take place between the lower and upper levels along the vertical channel for a period between approximately 20–140 ms of a cloud flash.
3. Extension of the channels in the lower-level negative charge centre, spreading out in successive steps from the flash origin toward more remote parts of the negative charge centre, occurring between approximately 140 to 200 ms.
4. A decrease in the conductivity of the vertical channel and separation of the upper-level channels from the low-level channels.

The present knowledge of electrical parameters characterizing inter- and intra-cloud flashes follows from research projects carried out in USA and France. During these projects, instrumented aircraft was flown in vicinity and through the thunderclouds in order to be hit by lightning and consequently record the characteristics of cloud flashes. The collected results show that cloud flashes are less severe than flashes to the ground regarding to peak current, charge transfer, and specific energy but may be more severe with regard to the rate of rise of current pulses [29, 67].

4.4.2 Lightning Interactions with Aircraft

4.4.2.1 Introduction

Frequency of lightning strikes to aircraft depends on various parameters (aircraft geometry, size of aircraft, the local climate, flight profile, and route, etc.). For commercial transport aircraft, different sources give different statistical probabilities falling somewhere between one strike per 1000 and 20,000 flight hours. A generally accepted estimate is that lightning strikes to aircraft occur on average, roughly once to every commercial transport aircraft per year in service [10].

A lightning strike effects in aircraft can be subdivided into two groups [10]:

1. Direct effects of lightning—effects directly due to the attachment or passage of the lightning channel currents such as metal burn through, hot spots and ohmic heating, mechanical forces, voltage and thermal sparking, dielectric breakdown of non-conducting surfaces, welding, etc. Extensive information on direct effect of lightning can be found in [29, 68].
2. Indirect effects of lightning—effects due to coupling of the lightning induced electromagnetic fields to the electrical and electronic installation in aircraft (transient voltages and currents induced in aircraft wiring)

4.4.2.2 Lightning Strike Initiation

When the aircraft flies through the quasi-static electric field present in the atmosphere, the electric field around aircraft will be deformed with significant field enhancement around the aircraft extremities such as the nose, wingtips, empennage, etc. [10, 29].

There are two basic scenarios considering a lightning strike to aircraft. The first one, outlined in Figure 4.71, is so-called an aircraft-intercepted lightning strike, where the aircraft is flying close to the path of a naturally developing stepped leader. If the tip of the leader advances within the approximately 50 meters of the aircraft, the electric field around extremities of the aircraft can be sufficiently intensified to initiate connecting leaders from the aircraft extremities. The connecting leader will propagate toward and subsequently join the

Figure 4.71 Development of an aircraft-intercepted lightning strike.

approaching stepped leader. In this way, the aircraft will become a part of the stepped leader path. Simultaneously, the leader development will be initiated at some of aircraft's extremities, from which the stepped leader will proceed further, until it reaches the ground (a ground discharge) or an opposite charge pocket elsewhere in a cloud (cloud discharge). In this way, the aircraft becomes a part of the lightning discharge channel.

In the second scenario, denoted "an aircraft-triggered lightning strike" the aircraft itself triggers the lightning flash. When an aircraft flies in a region with an intense electrostatic field created by cloud electric charges, it can distort and enhance the ambient electric field to the extent that the leader development will start at opposite aircraft extremities, as depicted in Figure 4.72. These leaders will propagate in the direction of the ambient electric field toward the opposite polarity charge reservoirs (in the clouds or on the ground). When these charge centres are reached, a lightning return stroke (i.e., a high current discharge) will be initiated, that would not otherwise occur without presence of the aircraft.

Points at which lightning current enters or leaves the aircraft surface are known as attachment points. Since the aircraft is a part in the lightning development path almost from the initiation, it experiences current pulses involved in all phases of lightning flash lifetime. A possible lightning current pattern experienced by aircraft in flight is depicted in Figure 4.73.

The results from in-flight lightning research projects hint that almost 90% of recorded lightning strikes were triggered by aircraft themselves. Since the research aircraft flew intentionally through or close to thunderclouds, this number is not fully representative for aircraft in normal operation, i.e., usually trying to avoid thunderstorms. Therefore, the number of aircraft-triggered events can be lower.

4.4.2.3 Swept Channel Process

As it was mentioned before in section 4.4.1.3, a complete lightning flash including subsequent strokes and continuing currents can persist for more than

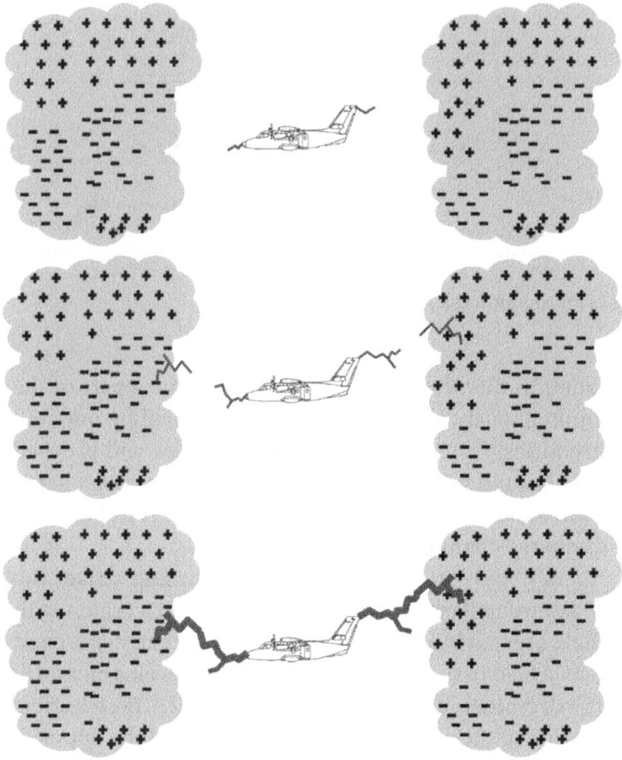

Figure 4.72 Development of an aircraft-intercepted lightning strike.

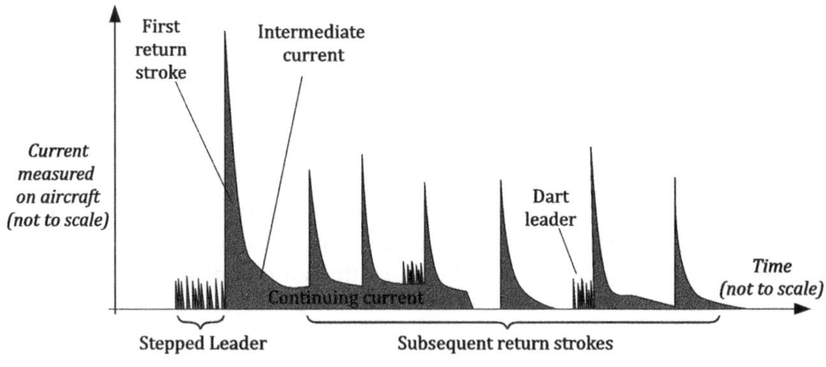

Figure 4.73 Pictorial representation of possible current components in an aircraft lightning strike.

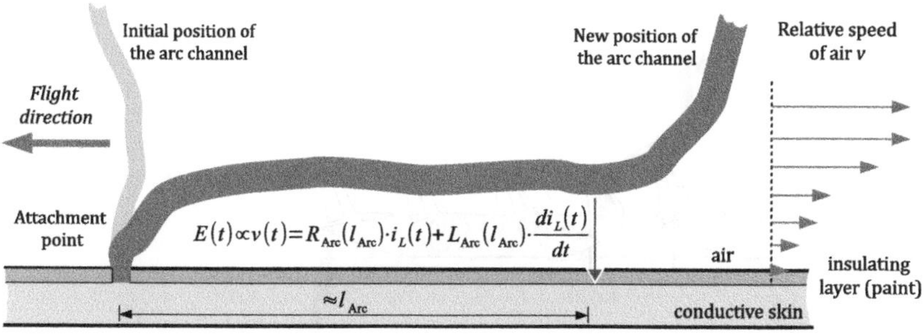

Figure 4.74 Outline of channel elongation and restrike process.

1 second. While the relatively stationary channel remains in its original location, the aircraft will move some distance during the duration of the flash [10, 29].

Apart from the initial attachment points are determined by the mechanisms described in the previous section, there will be also other lightning attachment points on the airframe due to the motion of the aircraft through the channel.

For example, the nose becomes an initial attachment point. As the aircraft flies further, its surface moves forward through the lightning channel. The lightning arc channel is continuously bent and elongated (see Figure 4.74). The elongated channel is insulated from the aircraft structure by the surface paint and a layer of air dragged along the surface due to air viscosity. The increase in the length yields an increase in both the plasma channel inductance and resistance, resulting in the increase in the voltage drop along the channel. When the local field intensity between the plasma channel and the aircraft structure exceeds combined dielectric strength of the surrounding air and the surface painting, then a breakdown of the insulation occurs and the arch channel reattaches at a new point.

This process can be repeated several times throughout the flash duration, with the channel attaching and dwelling at various surface locations in discrete steps for different periods of time. The size of steps depends on the local geometric profile, the paint dielectric properties and thickness and the lightning current parameters.

The lightning channel entry and exit points move along the surface of aircraft channel and the channel appears to be swept backward along the surface, as depicted in Figure 4.75. Therefore, this process is called the swept channel phenomenon. However, when the lightning channel has attached or has been swept to a trailing edge, it cannot progress any further and will remain there, or *hang on*, for the remainder of the flash. Under some condition, the entry and exit points of the lightning channel may be swept to the same or adjacent

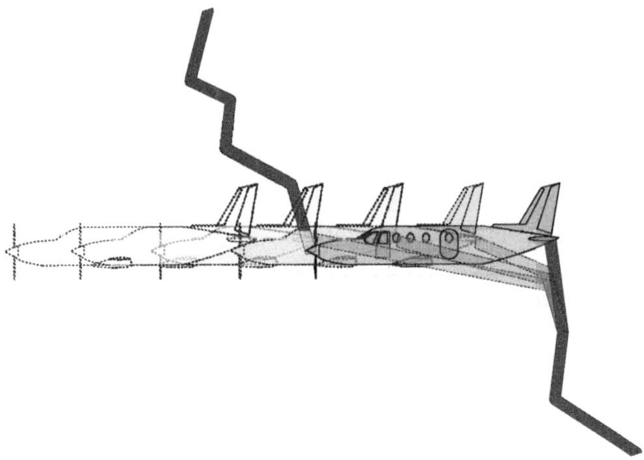

Figure 4.75 Channel swept along the aircraft surface.

trailing edges and the channel will rejoin behind the aircraft and the aircraft will leave the lightning current path.

4.4.2.4 Lightning Strike Zones

Based on the swept channel effect, the following conclusions can be made:

1. With exception of trailing edges, the effects of the flash are spread out over a multiple number of points with no single point receiving the full energy of the flash.
2. The proportion of the flash experienced by any particular point depends on its location on the vehicle surface.
3. Thus, the surface can be divided into lightning strike zones depending on the probability of initial attachment, sweeping, and hang-on.

Therefore, in order to optimize lightning protection, the aircraft is divided into different lightning strike zones and all structural parts and equipment located in these zones shall be designed to withstand their applicable components of the lightning environment.

In general, an aircraft can be divided into the following zones [11]:

- Zone 1A: First Return Stroke Zone; assigned to all areas of the aircraft surfaces where a first return stroke is likely during lightning channel attachment with a low expectation of flash hang on.
- Zone 1B: First Return Stroke Zone with Long Hang-On; denoting all areas of the aircraft surfaces where a first return stroke is likely during lightning channel attachment with a high expectation of flash hang on.

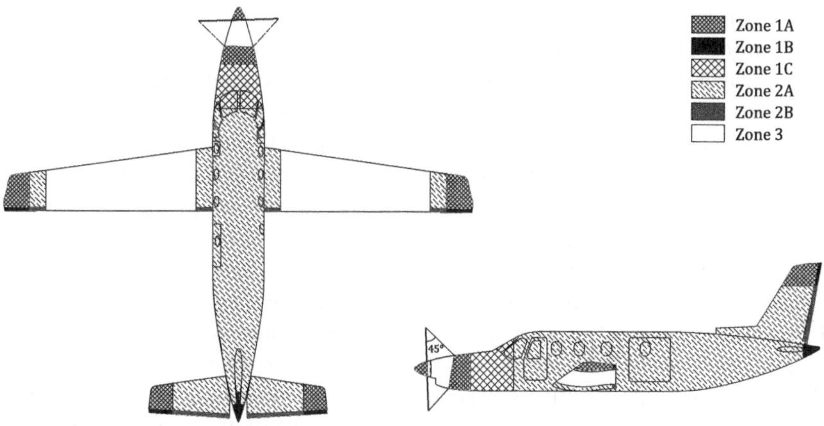

Figure 4.76 Example of lightning strike zoning for a small single-engine aeroplane.

- Zone 1C: Transition Zone for First Return Stroke; assigned to the aircraft surfaces where a first return stroke of reduced amplitude (i.e., at higher altitudes) is likely during lightning channel attachment with a low expectation of flash hang on.
- Zone 2A: Swept Stroke Zone; denoting all aircraft surfaces where subsequent return stroke is likely to be swept with a low expectation of flash hang on.
- Zone 2B: Swept Stroke Zone with Long Hang-On; i.e., the aircraft surfaces into which a lightning channel carrying a subsequent return stroke is likely to be swept with a high expectation of flash hang on.
- Zone 3: Current Conduction Zone; i.e., the surfaces not in Zones 1A, 1B, 1C, 2A, or 2B, where any attachment of the lightning channel is unlikely, and those portions of the aircraft lying beneath or between the other zones and/or conduct substantial amount of electrical current between direct or swept stroke attachment points

Zone definitions and methods of locating them on particular aircraft are given in [11] and analyzed in [69]. An example of the aircraft zoning is shown in Figure 4.76.

4.4.3 Idealized Standard Lightning Environment

Based on from the information about mechanisms of ground and cloud flash plus the process of lightning attachment to aircraft (both triggered and intercepted), it can be assumed that the aircraft can experience all lightning currents as measured on the ground plus the currents involved in leader development during attachment and stepped leader process.

As it can be deduced from content given in section 4.4.1.2, the lightning strike parameters vary from flash to flash. Therefore, standardized voltage and current waveforms have been derived to represent the lightning environment external to an aircraft. As the next step, the standardized external current waveforms have been used to derive standardized transient voltage and current test waveforms that can be expected to appear on cable bundles and at equipment interfaces within an aircraft. These generic waveforms are given in [10] and they are accepted as being adequate for the demonstration of compliance for the protection of an aircraft and its systems against the lightning environment.

These waveforms do not replicate a specific lightning event, but they blend all parameters relevant to effects of lightning into composite waveforms. For example, the waveform representing the first return stroke combines the rise time of a negative first return stroke and negative subsequent stroke currents (i.e., fast rising currents of lower energetic contents) with the current peak current and the duration of characterizing a positive return stroke (high energy, long duration but a slow rise).

4.4.3.1 Idealized Standard External Voltage Waveforms

The idealized standard external voltage waveforms represent the portion of the electric field important for assessment of different discharge processes associated with a lightning strike to aircraft. The tests involve an application the voltage waveforms to a plate-rod or plate-plate electrode arrangement, thus simulating electric fields characterizing pre-breakdown conditions. The test object is then exposed to the generated electric fields by its placing between the electrodes.

Voltage Waveform A rises at a rate of 1000 kV/μs (±50%) until its increase is interrupted by voltage breakdown of the air gap between the high voltage electrode and the object under test. This waveform represents fast rate-of-rise electric fields associated with lightning re-attachment to aircraft surfaces during swept channel processes described in subsection 4.4.2.3.

Voltage Waveform D is a slow front waveform with a rise time between 50 μs and 250 μs. This relatively long rise time is intended to provide sufficient time for streamers from an object to develop. The fall time to 50% of 2500 μs is just approximate value, since the breakdown shall occur during the rising slope of the pulse. This waveform is used to determine locations of aircraft initiated leaders through simulation of slow rate-of-rise electric fields occurring at aircraft extremities during an initial leader attachment stage. Further voltage waveforms specified are:

The current aerospace lighting testing standards require use of voltage waveforms A and/or D for the majority of tests [12,70]. Their parameters are shown in Figure 4.77. Further voltage waveforms specified are:

Voltage Waveform B is the standard 1.2/50 μs waveform which is used in the electrical industry for lightning impulse dielectric tests. This waveform is used

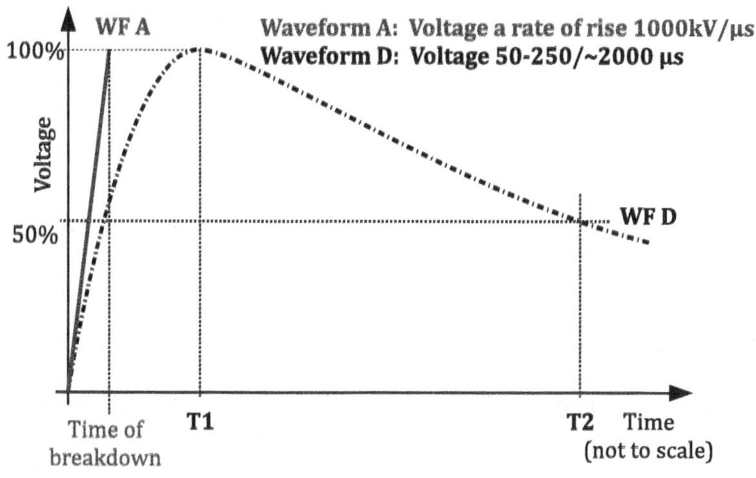

Figure 4.77 Voltage test waveforms.

to simulate effects of strong electric fields in order to evaluate possible hazards from corona and streamers but without a breakdown or flashover.

Voltage Waveform C is a chopped voltage waveform in which breakdown of the gap between an object under test and the test electrodes occurs at 2 μs (±50%). It is used to determine naturally occurring strike locations.

4.4.3.2 Idealized Standard External Current Components

Records of lightning current pulses presented in the literature confirm that the lightning currents differ from stroke to stroke [67, 71]. Therefore, the general approach is to standardize their temporal variation using suitable mathematical functions.

The idealized standard external lightning environment, as defined in [10] or [45], is comprised of the following current components:

- Component A simulates the first return stroke
- Component A_H represents the first return strokes at higher altitudes
- Component B represents the intermediate current
- Components C and C* represent a long and a short continuing current, respectively
- Component D simulates the first of the subsequent strokes
- Component D/2 simulates the one of the remaining subsequent strokes
- Component H simulates the current pulses in leader development, attachment and detachment process and similar pulses occurring throughout the flash duration.

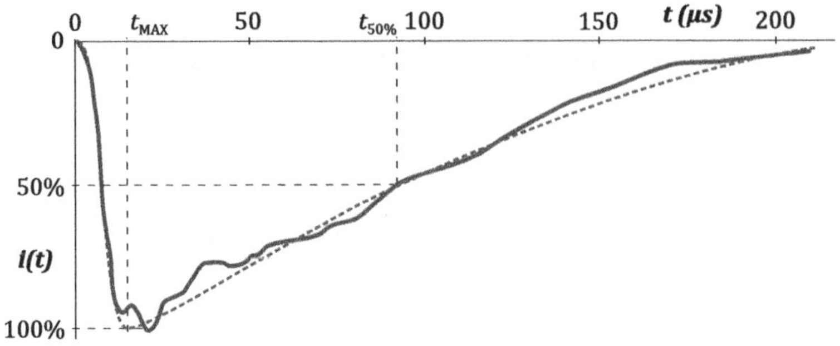

Figure 4.78 Lightning current waveform approximated by a unipolar aperiodic pulse.

With exception of Components C and C*, which are given as rectangular pulses, all lightning current components are represented mathematically as aperiodic pulses with specified rise and decay times, as depicted in Figure 4.78.

The multiple stroke events (see subsection 4.4.1.3.2) are represented by Multiple Stroke (MS) Waveform Set comprising 14 random-spaced pulses of Components D and D/2, as depicted in Figure 4.79.

Multiple Burst Waveform Set consists of three random-spaced bursts, where each burst of 20 randomly spaced Component H pulses sequences (see Figure 4.80). It simulates the bursts of steep pulses during the leader attachment to aircraft and stepped leader development, when the aircraft is part of the stepped leader process. Similar pulses were also observed randomly occurring throughout the lightning flash duration. These pulses are not likely to cause any structural damage to the aircraft. However, due to their random and repetitive spacing, they are capable of inducing transients that may cause interference or upset to certain systems [10].

4.4.3.2.1 ***Waveforms for Indirect Effects Evaluation*** Especially in case of the indirect effects of lightning is the exact mathematical waveform of great importance. Therefore, in order to replicate the temporal variation of the lightning current pulses, several different mathematical functions have been proposed

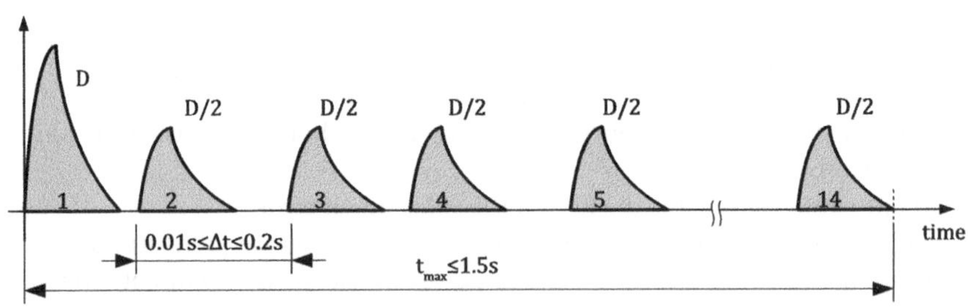

Figure 4.79 Multiple stroke pulse pattern.

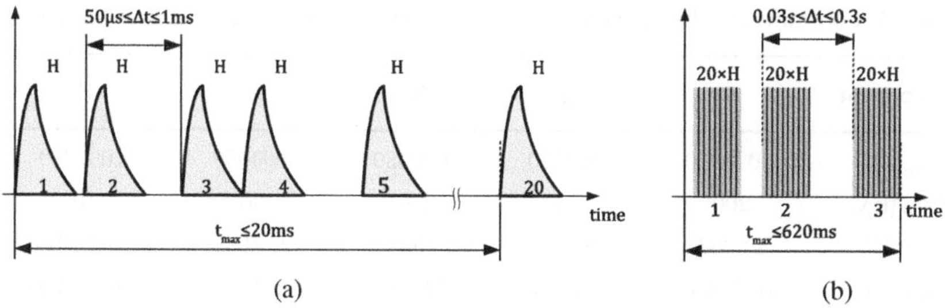

Figure 4.80 Multiple burst pulse pattern (a) one burst of 20 pulses; (b) one multiple burst sequence.

[72]. For simulation of effects of lightning on aircraft, the latest current waveform definitions use the quadruple exponential (QE) description of the lightning current waveform [10]. The older versions used the double exponential definition (DE) of current waveforms. The time domain expression covering both types is:

$$i(t) = \begin{cases} 0 & \text{for } t < 0 \\ k(t) \cdot I_{peak}(e^{-\alpha t} - e^{-\beta t}) & \text{for } t \geq 0 \end{cases} \quad (4.206)$$

where β represents the rise-time constant, α denotes the decay constant, and I_{peak} is the peak value of the lightning current. The peak multiplier k_p ensures that the maximum of $i(t)$ equals I_{peak}. Depending on the used form, the peak multiplier can be expressed as:

$$k(t) = \begin{cases} k_p(1 - e^{-\gamma t})^2 & \text{for } QE \text{ waveform} \\ k_p & \text{for } DE \text{ waveform} \end{cases} \quad (4.207)$$

The additional $(1 - e^{-\gamma t})^2$ term influences only the rising slope of the waveform, ensuring that the maximum rate of change (the first derivative of the function) does not occur at the time $t = 0$, as it is the case for the DE waveform. The maximum rate of change at times $t > 0$ more realistically represents the real lightning current and contributes to the stability of numerical simulations. Mathematically, setting $\gamma = \infty$ reduces the QE form to DE form, both for time domain and the frequency domain expressions.

The parameters to be inserted into the above-given expressions are tabulated in Table 4.5.

Parameters resulting from insertion are listed in Table 4.6. Additional lightning parameters occurring in the table are:

- the pulse rise time T1 (the time to peak) measured from 0 to 100% and the pulse decay time T2 (the time to half value) measured from 0 to 50% of the falling slope

Table 4.5 Constants of lightning current waveforms relevant to indirect effects of lightning

Component	A	A_H	B	D	H
k_p [-]	1.094050	1.099350	2.690480	1.094050	1.057200
I_{peak} [kA]	200	150	4.2	100	10
α [s^{-1}]	11,354	16,605	700	22,708	187,191
β [s^{-1}]	647,265	858,888	2,000	1,294,530	19,105,100
γ [s^{-1}]	5,423,540	7,253,750	22,000	10,847,100	153,306,000

- the pulse duration t_D for the rectangular pulses measured from 50% of the rising slope to 50% of the falling slope
- Action Integral (the specific energy):

$$AI = \int_0^\infty i^2(t)dt \tag{4.208}$$

- Charge transfer Q_T:

$$Q_T = \int_0^\infty i(t)dt \tag{4.209}$$

- the average current I_{avg}, for rectangular pulses specified as

$$I_{avg} = \frac{Q_T}{t_D} = \frac{1}{t_D}\int_0^\infty i(t)dt \tag{4.210}$$

The lightning current waveform described by equations (4.206) and (4.207) can be transformed analytically in frequency domain using the forward Fourier

Table 4.6 Parameters of lightning current components relevant to lightning indirect effects

	A	A_H	B	D	H	C	C*
I_{peak} [kA]	200	150	4.2	100	10	400	400
$\max\frac{di}{dt}$ [kA/μs]	140	140	0.0147	140	200	-	-
T1 [μs]	6.4	4.72	813	3.18	0.245	-	-
T2 [μs]	69	49	2340	34.5	4	-	-
Q_T [Q]	18.9	10.1	10.5	4.8	0.056	200	18
AI [A^2s]	2×10^6	8×10^5	2.85×10^4	2.5×10^5	2.9×10^2	-	-
t_D [ms]	-	-	-	-	-	500	45

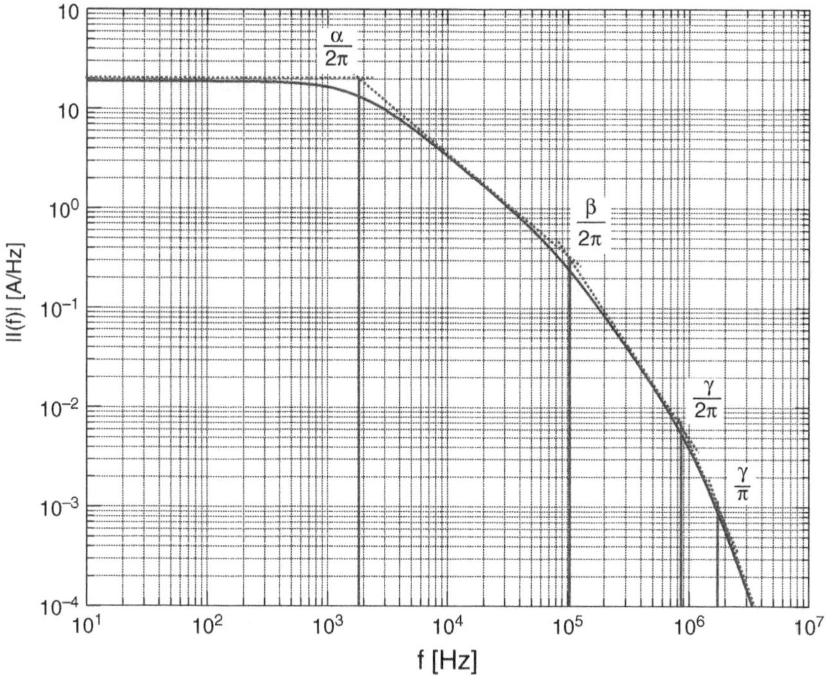

Figure 4.81 Frequency spectrum of lightning current component A.

transform given by equation (4.3). The corresponding frequency domain expression for QE waveform is

$$I(f) = \left[\left(\frac{1}{\alpha + j2\pi f} - \frac{1}{\beta + j2\pi f}\right) - \left(\frac{2}{\alpha + \gamma + j2\pi f} - \frac{2}{\beta + \gamma + j2\pi f}\right) + \right. \\ \left. + \left(\frac{2}{\alpha + 2\gamma + j2\pi f} - \frac{2}{\beta + 2\gamma + j2\pi f}\right)\right] k_p I_{\text{peak}} \tag{4.211}$$

The corresponding frequency spectrum of the lightning first return stroke current, i.e., lightning current component A, is depicted in Figure 4.81.

The first frequency break point is at $f = \alpha/(2\pi)$ Hz and the second frequency break point is at $f = \beta/(2\pi)$ Hz. Between the first and the second break points, the magnitude falls at a rate of 20 dB/decade. At frequencies over the second break point, the magnitude falls at 40 dB/decade. The comparison of the current spectra of lightning current components A, D, and H is shown in Figure 4.82.

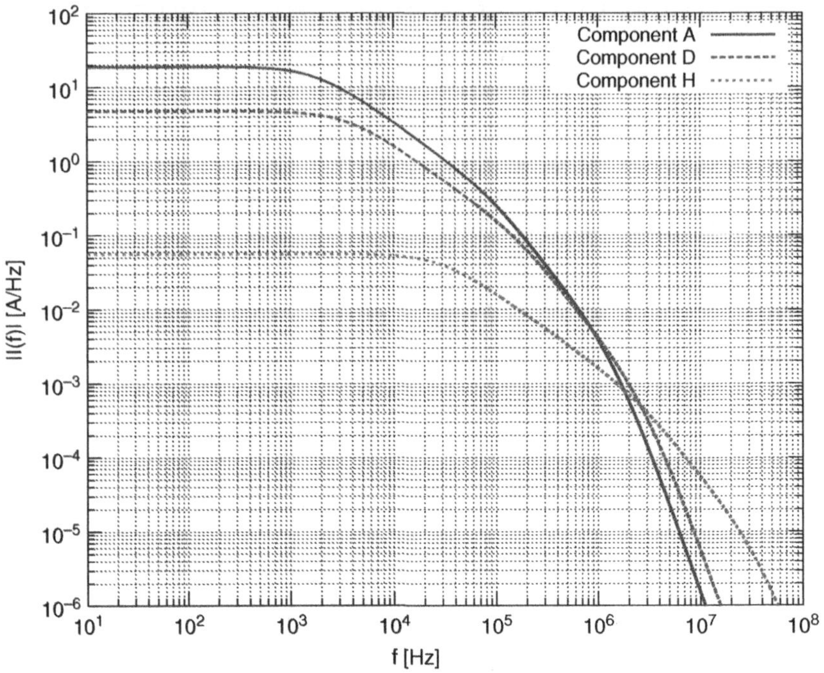

Figure 4.82 Frequency spectra of the lightning current components relevant to IEL.

4.4.3.2.2 Relaxed Waveforms for Direct Effect Evaluation In the definition of idealized lightning current waveforms given in the previous section, the emphasis was put on parameters of importance for both the direct or indirect effects. However, when only the direct effects of lightning are of interest, requirements on some parameters can be relaxed. This is done because of practical reasons, e.g., testing, since generation of high current pulses of high amplitudes with steep fronts is very difficult.

For high current components (components A, A_H, and D), the peak current amplitude and the action integral (the specific energy) are decisive parameters influencing extent of damages, and thus of primary importance. Therefore, they are defined precisely with allowed tolerances. The requirements on the rise and decay times are relaxed.

In case of low-level longer duration current components B, C, and C*, the effects depend predominately on average current value as well as charge transferred, therefore the values of average current, the charge transfer and the duration are specified.

The exact waveforms are not defined. Components A, A_H, and D can be unidirectional or oscillatory. The components B, C*, and C shall be unidirectional, e.g., rectangular, exponential, or linearly decaying. No matter

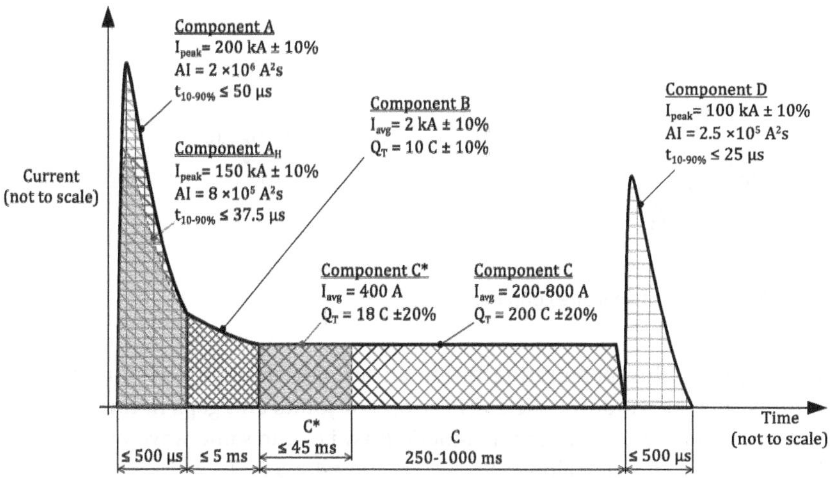

Figure 4.83 Lightning Current Components for direct effect evaluation.

the form, the parameters of individual waveforms given in Figure 4.83 must be reached within allowed tolerances.

4.4.3.3 Idealized Standard Induced Transient Waveforms

Real-world amplitudes and waveforms of transient induced by a lightning strike to aircraft will differ from aircraft to aircraft, in the same aircraft from system to system, and even in the same system from flash to flash. There are several mechanisms by which the external environment induces transients. These can be broadly divided into aperture flux coupling, diffusion flux coupling, and resistive coupling. Most actual induced transients are complex waveforms that result from superimposition of two or more coupling mechanisms.

In order to guarantee a unified approach to design and verification of adequate lightning indirect effects protection of systems and equipment, a set of idealized induced transient waveforms has been derived. These waveforms represent transient responses to external lightning environment defined in section 4.4.3.2.1, induced in the aircraft electrical wiring by basic coupling mechanisms under simplified coupling conditions with each effect treated separately.

For analysis, only lightning current components A, D, and H are considered, since due to their high amplitudes and rates of change, they can induce the significant induced transients.

4.4.3.3.1 Aperture Coupling Lightning current flowing along aircraft produces magnetic field of the same waveform as the lightning current. Aperture flux

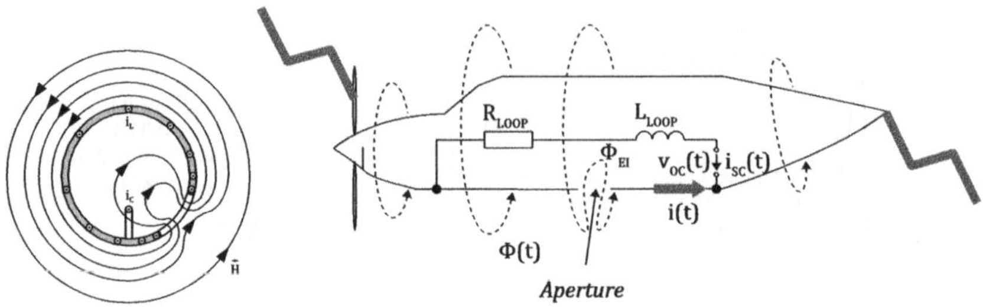

Figure 4.84 Outline of aperture coupling.

is the magnetic flux penetrating inside an aircraft through electromagnetic apertures such as windows, dielectric covers or joints and gaps in the skin. Its waveform is that of the external magnetic flux, i.e., the same waveform as the lightning current flowing along the skin (section 4.4.3.2.1).

A situation, where a wire loop formed by cables and the structure and characterized by a loop resistance R_{LOOP} and inductance L_{LOOP}, is exposed to time-varying aperture flux is shown in Figure 4.84. The magnetic field penetrating through apertures and through the loop will induce:

1. Voltage $v_i(t)$ that follows approximately the rate of change of the driving current and it, therefore, has a steep rise at near initial time and pass through zero at the time of peak current. The limit case is the open-circuit voltage of a loop $v_{OC}(t)$, i.e., the case where one terminal point is separated from the structure:

$$v_{OC}(t) = v_i(t) = \frac{d\Phi_{EI}}{dt} = M\frac{di(t)}{dt} \tag{4.212}$$

   where Φ_{EI} denotes the aperture flux penetrating coupled to the loop, M stands for the mutual inductance between aperture and the loop (depending on the aperture size, its shape and the aperture-cable separation, see subsection 4.2.4.2) and $i(t)$ is the external lightning current in the structure.
2. Current $i_{SC}(t)$, whose waveshape is similar to that of the driving external current, when the loop has R_{LOOP}/L_{LOOP} ratio. This can be deduced using the following circuit equation for the loop:

$$v_i(t) = M\frac{di(t)}{dt} = R_{LOOP}{\cdot}i_{SC}(t) + L_{LOOP}\frac{di_{SC}(t)}{dt} \tag{4.213}$$

   For a limit case of $R_{LOOP}{\approx}0$:

$$M\frac{di(t)}{dt} = L_{LOOP}\frac{di_{SC}(t)}{dt} \Rightarrow i_{SC}(t) = \frac{M}{L}i(t) \tag{4.214}$$

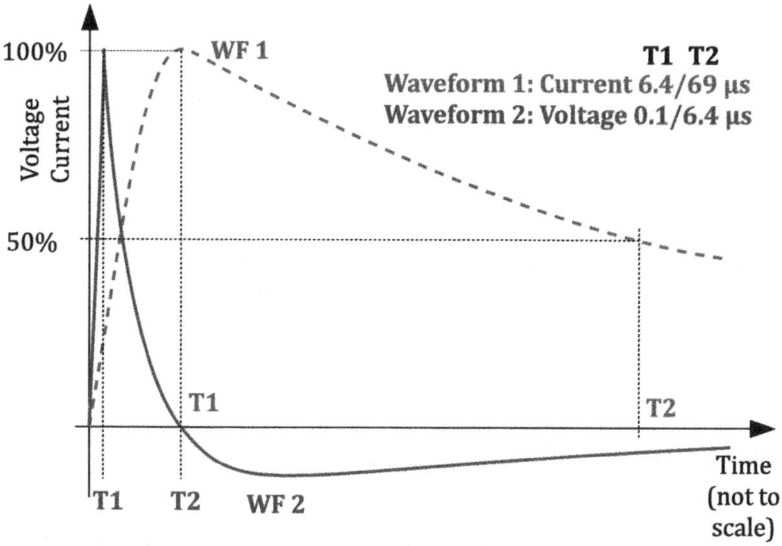

Figure 4.85 Time variation of waveforms 1 and 2.

From the points 1 and 2, it follows that (see also Figure 4.85):

1. The idealized induced current waveform, in [10] and [73] denoted as Waveform 1 (WF1) and applicable to low-impedance loops will have exactly the waveform of the external lightning current component A (see subsection 4.4.3.2.1), but at reduced amplitude.
2. The idealized induced voltage waveform, in [10] and [73] denoted as Waveform 2 (WF2), that is applicable to high impedance loops will have exactly the waveform of the rate of change of the external lightning current component A.

4.4.3.3.2 Resonances Magnetic fields penetrating through apertures will drive or excite resonances on cables yielding oscillatory currents and voltages. The amplitude, decay time, and frequency will be dependent on the aircraft structure length (see section 4.2.3.7.1) as well as on cable lengths, cable impedances, and terminating impedances (refer to section 4.2.4.3). The superimposition of several resonant processes at different frequencies is also possible. The oscillations can occur either individually, i.e., decaying to zero; or superimposed on top of the much longer transients.

These transients are represented by voltage and current pulses that have the form of damped sinusoids or cosinusoids, as portrayed in Figure 4.86. Documents [10] and [73] denote their shape as Waveform 3 (WF3) with standardized

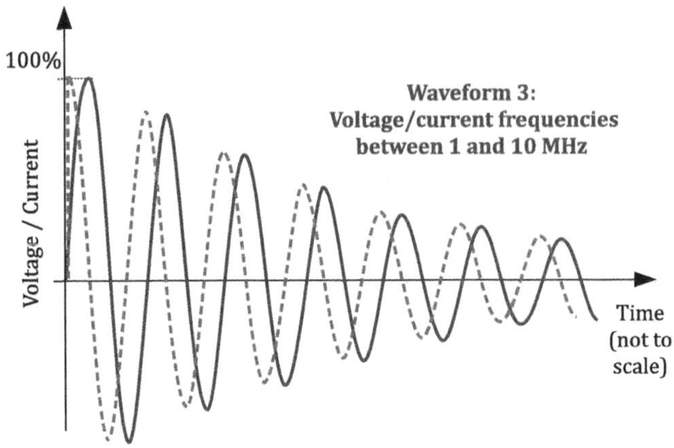

Figure 4.86 Idealized waveform simulating resonances.

frequencies ranging between 1 and 10 MHz. However, other aircraft and/or specified frequencies outside this range can also be applicable.

4.4.3.3.3 Resistive Coupling and Current Redistribution Resistive coupling denotes all processes leading to potential differences between any two structural points, e.g., voltage gradients along the inner surface of the aircraft skin or voltage drop on structural joints. These potential differences are proportional to the current flowing through the structure and the time-dependent resistance of the aircraft skin or structure. In this section it is shown that at lower frequencies, relevant to the lightning effects, the time-dependent resistance of the structure $Z(t)$ can be replaced by the DC resistance R. The voltage drop is then defined as

$$\Delta v(t) = Z(t) \cdot i(t) \approx R \cdot i(t) \tag{4.215}$$

A situation, where a wire loop formed by cables and the structure and characterized by a loop resistance R_{LOOP} and inductance L_{LOOP}, bridges a part of aircraft structure characterized by a resistance R is shown in Figure 4.87. The following considerations apply:

1. Voltage $\Delta v(t)$ follows approximately the variation of the driving current. For the limit case of the open-circuit voltage of a loop defined, i.e., one terminal point separated from the structure:

$$v_{OC}(t) = \Delta v(t) = Z(t) \cdot i(t) \approx R \cdot i(t) \tag{4.216}$$

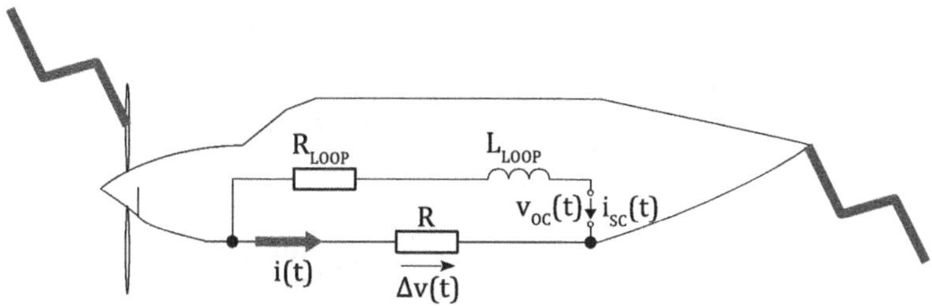

Figure 4.87 Outline of resistive coupling.

2. The amplitude and waveshape of a current $i_{SC}(t)$ in low resistance cables connected to a resistive structure at both ends is governed by the redistribution mechanism resulting from the relatively high inductance of the cable with respect to the structure. The circuit equation can be written as

$$\Delta v(t) = R \cdot i(t) = R_{LOOP} \cdot i_{SC}(t) + L_{LOOP} \frac{di_{SC}(t)}{dt} \tag{4.217}$$

The analytic solution for an external lightning current described by a double exponential waveform

$$i(t) = k_p I_{peak} (e^{-\alpha t} - e^{-\beta t}) \tag{4.218}$$

is

$$i_{SC} = k_p I_{peak} \frac{R}{R_{LOOP} + R} \left[(e^{-\alpha t} - e^{-\sigma t}) \frac{\sigma}{\sigma - \alpha} - \right. \\ \left. - (e^{-\beta t} - e^{-\sigma t}) \frac{\sigma}{\sigma - \beta} \right] \tag{4.219}$$

where

$$\sigma = \frac{R_{LOOP} + R}{L_{LOOP}} \tag{4.220}$$

Depending on time constant σ, the current will generally have significant amplitude but a lengthened waveshape characterized by longer both rise and decay times than the external lightning current. The fraction $R/(R_{LOOP} + R)$ in front of the square brackets hints toward the fact that at the late times the current will be distributed between the structure and the wire according to the resistances (a current divider).

From the points 1 and 2, it follows that (see Figure 4.88):

1. The idealized resistive drop voltage waveform, in [10] and [73] denoted as Waveform 4 (WF4) and applicable to high impedance loops will have exactly

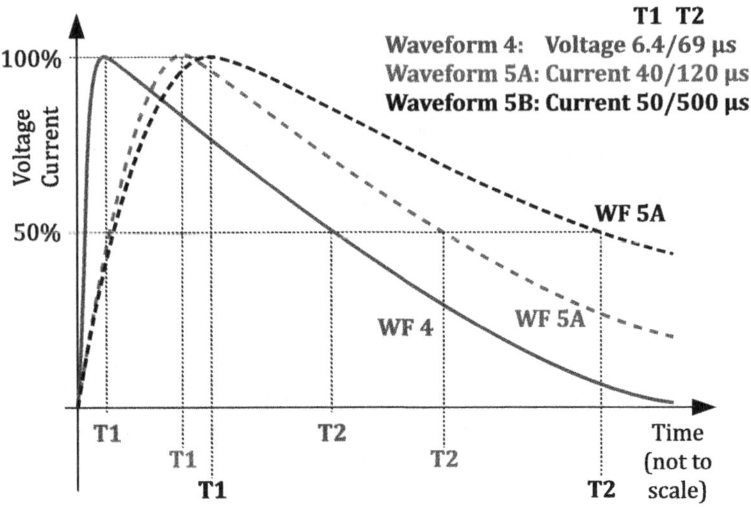

Figure 4.88 Waveforms 4 and 5 simulating resistive coupling / diffusion and redistribution.

the waveform of the external lightning current component A defined in subsection 4.4.3.2.1, but at reduced amplitude.

2. The idealized current waveform in low resistance cables connected to a resistive structure at both ends, in [10] and [73] denoted as Waveform 5A or 5B (WF5A, WF5B), have significant amplitude but a lengthened waveshape characterized by longer rise and decay times than the external lightning current. Two different waveforms allow selection of the more suitable one.

4.4.3.3.4 Diffusion Flux Coupling Diffusion flux is the magnetic flux that has penetrated by diffusion through a conducting portion of the skin. The diffusion process governed by the eddy currents induced in the aircraft structure that distort and attenuate the external flux (see section 4.2.3.4). The amplitude depends on the resistivity of the material through which it has diffused, being negligible for good conductors such as metal and higher for high resistivity materials such as CFC. An example of field diffusion into elliptical structure simulating a wing profile is depicted in Figure 4.18. Another example of the diffusion coupling process through an aperture in well-conducting cylinder covered by a highly resistive cover is illustrated in Figure 4.89

As it is shown in section 4.2.3.4, the aircraft structure behaves as a low-pass filter. Therefore, the induced voltages and currents will have the waveshapes similar to the ones of voltages and currents due to the resistive coupling, hence the same waveforms (i.e., voltage WF4 and current WF5A/WF5B) are used to simulate the diffusion flux coupling.

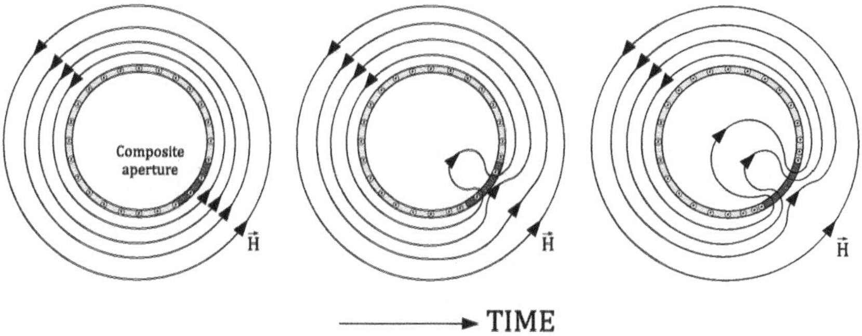

Figure 4.89 Magnetic field diffusing through an aperture with a resistive cover.

4.4.4 Determination of Internal Lightning Electromagnetic Environment

The internal lightning EME can be represented by the resulting lightning-induced voltage and current transient waveforms and their levels that can appear at the electrical and electronic equipment interface circuits. The applicable advisory circular [6] uses the following definitions of the lightning-induced transients:

Actual Transient Levels (ATL)—The voltage and current amplitudes and waveforms actually induced on the given aircraft wiring and appearing at the equipment interfaces when the aircraft is exposed to lightning. They are real transients determined by aircraft test, analysis, or similarity.

ETDLs—A set of the qualification test voltage and current amplitudes and waveforms to which is the system or equipment qualified, i.e., exposed to during the qualification testing without damage or functional upset.

During the certification process, it must be shown that there is acceptable margin between the "worst-case" ATLs collected for all possible external different lightning current paths (refer to Figure 4.90) and the corresponding ETDLs. The margins account for uncertainties in the determination and verification techniques. The circular proposes 6 dB margin (factor of two) for level A systems when the ATL are determined by aircraft tests or by analysis supported by aircraft tests [6].

Especially in the cases of new aircraft designs, the ETDLs must be specified before the ATL are available. Therefore, it is necessary to set a design target values that should not exceed under any circumstances. These target values are denoted as:

Transient Control Levels (TCL)—The expected maximum allowable level of transients that appear at the equipment interface circuits because of the defined external environment. They are often determined before measurements on real installation in real aircraft. Therefore, they are based upon results of lightning tests on existing aircraft, generic values, engineering analysis, etc.

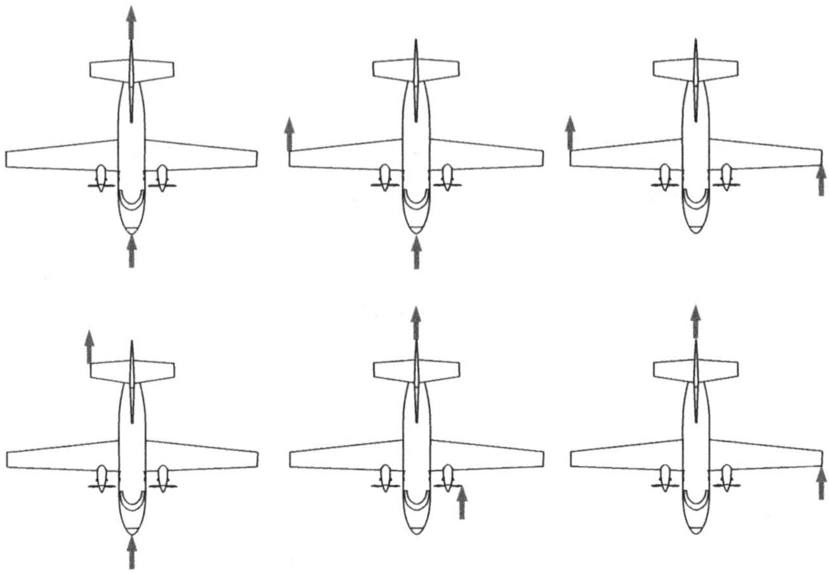

Figure 4.90 Several examples of possible airframe lightning current paths, all other potential combinations of attachment points shall be also taken into consideration.

During the design process, all the lightning protection measures (shielding, equipment location, cable routing, grounding, etc.) shall be designed to guarantee that the TCLs are not exceeded. Therefore, a sufficient margin should be added to TCLs to define ETDLs. The value of margin depends on the reliability and precision of the TCL determination method. The more imprecise the method, the higher safety margin shall be used.

Vice versa, when commercial off-the-shelf (COTS) equipment with previously verified ETDLs is to be installed in aircraft, the stated ETDLs shall be reduced by given margin to define TCLs, and consequently, the aircraft lightning protection shall be designed to meet TCLs.

Since the ATLs and TCLs should be directly comparable to ETDLs. Since ETDLs and corresponding equipment qualification tests are based on DO-160/ED-14, Section 22 [73], all above-mentioned transient levels are defined in the same manner as combination of

- cable bundle currents
- individual wire open-circuit voltages
- individual wire short-circuit currents

The following methods can be used to determine ATLs on real aircraft installation, as well as for estimation of TCLs when the real installation is not yet known or available:

4.4.4.1 Numerical Methods

The numerical methods can be used for determination of lightning transient levels, if the reliability and precision of the method can be proven, e.g., by measurements on models or structural parts. More information and general considerations are given in subsection 4.3.2.1 and in [69, 74, 75].

The main aim of the modeling is to define transfer function. Parameters to be considered should be

- lightning entry and exit points
- structural characteristics, such as materials and apertures
- system characteristics including position of equipment, routing, and shielding of cables, etc.
- non-linear effects influencing linear behavior

Generally, 3D modeling in combination with the network modeling analysis can be used to compute internal waveforms for the different waveforms of the external threat defined in subsection 4.4.3.3 [76]. 2D analysis can also be used as a complement to the 3D analysis for some simple installations.

4.4.4.2 Similarity

Under some conditions, it is possible to perform ATL verification using similarity to previously certified aircraft without performing more tests. It can be done when [6]:

1. There are only minor differences between the previously certified system and installation, and the system and installation to be certified;
2. There are no unresolved in-service system problems related to lightning strikes on the previously certified system; and
3. The previously certified system ATLs were verified by aircraft tests.

To use similarity, the differences to the previously certified aircraft and the system installation shall be identified and analyzed for possible effect on the system susceptibility. The assessment should cover:

1. aircraft type, equipment locations, airframe construction, structural materials, and apertures that could affect attenuation of the external lightning environment;
2. system wiring size, length, and routing; wire types, twisting, connectors, wire shields, and shield terminations;
3. lightning protection devices such as transient suppressors and lightning arrestors; and
4. grounding and bonding.

Use of similarity for ATL verification in a new aircraft design with new systems is not recommended.

4.4.4.3 Generic Levels

It is possible to determine or estimate transient levels using the generic values and guidance given in [73]. The test levels and test categories are based on collection of values and transient waveform obtained during measurements on different types of aircraft.

The selection of applicable waveforms can be based on table for selection of test categories. Following this guidance and knowing the properties of the aircraft structure, it is possible to assign to each cable bundle a set of the waveforms.

1. For metallic aircraft structures, where the resistive coupling can be neglected and only the aperture coupling must be considered, the waveforms can be assigned to cables as follows:
   - to unshielded cables, WF2 and WF3 for single and multiple stroke testing and WF3 for multiple burst testing
   - to shielded cables, WF1 and WF3 for single and multiple stroke testing. Concerning the burst, WF6 is applicable to relatively short low-impedance cables. For long shielded cables, WF3 is applicable waveform for multiple burst testing
2. For aircraft structures, where the resistance cannot be neglected so both the resistive coupling and the aperture coupling must be considered, the waveforms can be assigned to cables as follows:
   - to unshielded cables, WF2, WF3, and WF4 for single and multiple stroke testing and WF3 for multiple burst testing
   - to shielded cables, WF3 and WFA for single and multiple stroke testing. Concerning the multiple burst tests, WF6 is applicable to relatively short low-impedance cables. For long shielded cables, WF3 is applicable waveform for multiple burst testing

The amplitudes of transients can be selected from 5 different test levels according to the equipment installation location and according to the routing of the associated wiring:

- Level 5 is applicable when the equipment and its associated wire bundles are located in aircraft areas with materials providing no or poor shielding, i.e., areas with very severe electromagnetic environment. The equipment is thus exposed to very high lightning transients.

  Analysis might show that in high current density regions such as wing tips, landing gear, and so on) there is need for ETDLs higher than Level 5.
- Level 4 should be assigned when the equipment and its associated wire bundles are in aircraft areas exposed to severe lightning transients, such as wings, fairings, wheel wells, pylons, control surfaces, etc.
- Level 3 can be selected when the equipment and its associated wire bundles are entirely in aircraft areas with moderate lightning transients. Examples of

such areas are avionics bays not enclosed by bulkheads, cockpit areas and further locations with large apertures (doors, windows, access panels, etc.); all inside metal aircraft structure or composite aircraft structure with shielding properties as effective as metal aircraft structure.

- Level 2 is applicable when the equipment and its associated wire bundles are inside metallic or composite aircraft structure with effective shielding. Wire bundles are installed close to the ground plane and have shields terminated at the bulkhead connectors. Hydraulic tubing, metallic cable trays, etc. are electrically grounded at all bulkheads.
- Level 1 applies to the equipment with its associated wire bundles are completely enclosed in well-protected aircraft areas.

When some wires exit the area toward an area with more extreme environment, a higher level applies to these and an analysis shall be done to estimate possible cross-coupling to the co-routed wiring.

It is important to remember that the levels given in [73] are already EDTLs, therefore, no additional margins must be added. The selection of the transient must be supported by analysis of the installation.

4.4.4.4 Lightning Transient Analysis

The lightning transient analysis is used to determine the levels and waveforms of the transients induced into aircraft electrical/electronic systems wiring by injecting pulse currents onto the aircraft structure and measuring the induced transient current and voltage amplitudes and waveforms on installed wire bundle shields and individual wires [12, 29]. For the certification, the test shall be performed on a complete and functioning production aircraft. For the estimation or validation purposes, an empty aircraft, wing or fuselage parts or mock-up can be used.

In this method, the aircraft is successively subjected to low amplitude current pulses with waveshapes as similar as possible to the waveforms of lightning current components A, D, and H, respectively. The amplitude of the low level pulses is generally agreed between the test house, the manufacturer and the authority. Higher the amplitude, more reliable representative the results, but on the other hand, more complicated the test and the test equipment and higher the probability of latent damages to aircraft. Usually, the component A/D test amplitudes are within the range of 1000 to 5000 A and current Component H test current is usually applied at amplitudes up to 1000 A.

4.4.4.4.1 Test Set-up The levels of induced transients depend on the lightning attachment points. The lightning attachment points are represented by the current generator and return conductor attachments to the aircraft. Therefore, the currents are injected directly between various combinations of entry and exit points on the aircraft—wing, engine, nose, or tail.

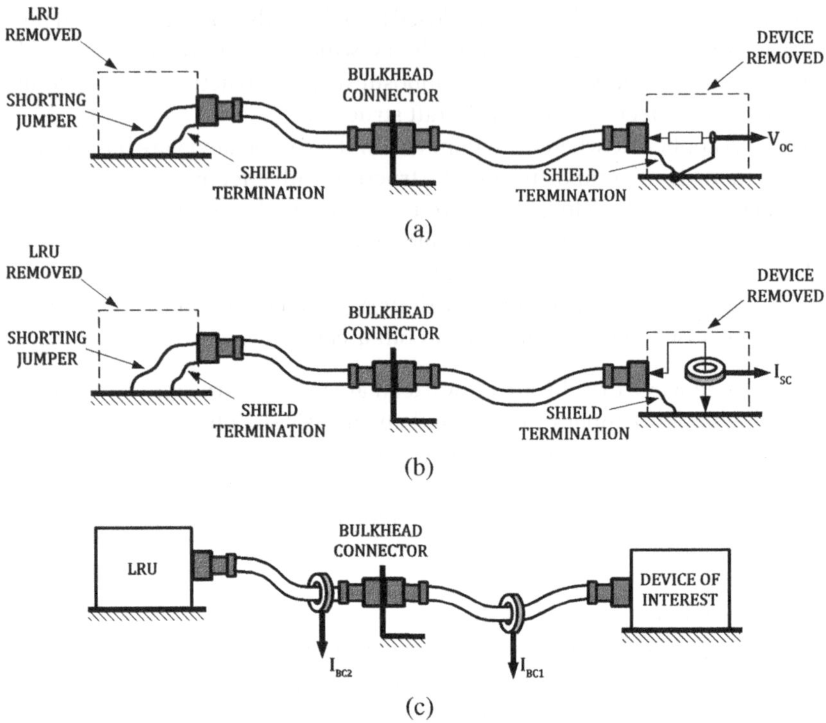

Figure 4.91 Schematic representation of measurement types (a) the wire open-circuit voltage, (b) the wire short-circuit current, (c) bulk cable current.

The test setup is similar or almost identical to that used for LLDD testing depicted in Figure 4.58. Also in this case, a coaxial setup is preferred to ensure representative current distribution on the airframe, but it is limited to smaller aircraft. The configuration with the ground plane as return conductor is generally used for bigger aircraft.

The tires must be isolated from the ground plane return conductors using insulating pads or stands. The insulating pads or stands must withstand the voltages developed between the aircraft and return conductor. The return conductors shall be grounded at a single point to facility ground near the test generator ground point to meet health and safety requirements.

4.4.4.4.2 Measured Quantities Several types of measurements shall be made as shown in Figure 4.91. These include:

1. Open-circuit voltages (V_{OC}), which are induced voltages measured between an individual open-ended wire and adjacent aircraft grounded structure,

with the other end of the wire grounded at the remote equipment location using a low-impedance grounding jumper. Equipment at either end of the measurement wire is disconnected from the wire bundle, but shields of the measured wire, (if present) and any other shields in the same wire bundle should be grounded in the normal way.

   High input impedance voltage probes are used for these open-circuit measurements

2. Short-circuit currents (I_{SC}), which are induced currents measured on individual wires with both ends of the wire grounded using low-impedance grounding jumpers. Other conditions are as described in paragraph 1 above.

   An isolated current probe should be used for this measurement.

3. Wire bundle currents (I_{BC}), which are induced currents, measured in a wire bundle, with the aircraft equipment that use the wire bundle installed in their normal manner and the wire bundles connected to the equipment at each end, in the normal manner. If the wire bundle is crossing grounded bulkhead, the currents all individual segments shall be measured.

The test probes and the complete measuring chain shall have bandwidth appropriate for the anticipated response. A total bandwidth of 50 MHz is advised by [12].

Prior to tests at general measurement location of the aircraft, transient noise check shall be made by recording the transient outputs measured without test wiring present. Noise responses for current measuring system shall be measured with the current measuring probe removed from the shield or wires, and placed close to the wires in that aircraft location. The current transformer should be isolated from the aircraft structure. Noise responses of the voltage acquisition chain shall be measured with the voltage probe disconnected from the test wire and grounded to at the same location as the voltage probe shield.

4.4.4.4.3 Data Processing and Extrapolation The measured pulse transients should be extrapolated to obtain the full-threat external environment. When the exact temporal waveforms A/D and H were used for the measurements, the extrapolation can be carried out using ratios of full threat to test currents.

If waveforms with time parameters different to those of Components A and H will be used for the ATL determination, in order to scale the induced transient response to component A, D, or H responses, the dominant coupling mechanism must be determined. This will determine whether the induced transient response should be scaled by peak current (i.e., resistively coupled voltage) or peak current rate of rise (i.e., voltage due to magnetic field that has penetrated through apertures). Some induced transient responses may have complicated waveforms with contribution of more than one parameter. In such cases, a more exact method such as Fourier analysis should be used (see subsection 4.4.4.5).

4.4.4.5 CW Injection—Swept Frequency Aircraft Tests

The swept frequency aircraft test is used to determine transfer functions relating the external injected current to the internal induced voltages and currents. This test is practically identical with the LLDD test used for HIRF environment determination described in subsection 4.3.2.2.2. The differences to LLDD are:

1. Not only the cable bundle currents, but individual wire currents and voltages have to be acquired and related to the external currents.
2. Not only amplitude but also the phase angle relation between the injected and induced quantities has to be determined, that means use of vector network analyzers.
3. The frequency range of acquisition should be approximately from 100 Hz to 50 MHz.
4. Need for system reference calibration of the complete test system (measurement probes without presence of aircraft, cables and fiber-optic links, etc.) yielding the calibration transfer function $TF_C(f)$ in the vector form (amplitude and phase). In this way, any distortion, loss (or gain), and line length effects can be compensated.

In order to determine the time domain responses for each test point, the transfer function is multiplied by the lightning current component frequency spectrum. This product is then transformed into the time domain using inverse Fourier transform.

The excitation signal $x(t)$ and the response signal $y(t)$ are transformed by Fast-Fourier-transformation (FFT) from the time into the frequency domain:

$$X(\omega) = F\left[x(t)\right] \quad (4.221)$$

$$Y(\omega) = F\left[y(t)\right] \quad (4.222)$$

The complex transfer function of the system under test $TF(\omega)$ for each measurement point and configuration is then defined as:

$$TF(\omega) = \frac{Y(\omega)}{X(\omega)} \quad (4.223)$$

However, actual quantities acquired during experimental determination of the transfer function are the outputs of the measuring sensors transferred over transmission paths to the measurement equipment, i.e., injected signal $a(t)$, the response signal $b(t)$, or their frequency domain equivalents $A(\omega)$ and $B(\omega)$.

As it is shown in Figure 4.92, the real signals at the inputs of the sensors can be reconstructed from the measured signals $A(\omega)$ and $B(\omega)$ as follows:

$$A(\omega) = TF_{S1}(\omega)TF_{P1}(\omega)X(\omega) \Rightarrow X(\omega) = \frac{A(\omega)}{TF_{S1}(\omega)TF_{P1}(\omega)} \quad (4.224)$$

$$B(\omega) = TF_{S2}(\omega)TF_{P2}(\omega)Y(\omega) \Rightarrow Y(\omega) = \frac{B(\omega)}{TF_{S2}(\omega)TF_{P2}(\omega)} \quad (4.225)$$

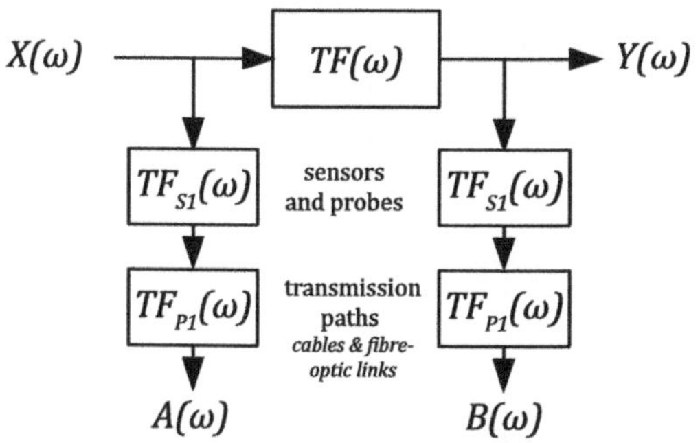

Figure 4.92 Signal definitions for determination of transfer function.

Knowing the complex transfer functions of the cables and sensors from calibration using, e.g., vector impedance analyzer, the transfer function of the given system can be calculated as:

$$TF(\omega) = \frac{Y(\omega)}{X(\omega)} = \frac{TF_{S1}(\omega)TF_{P1}(\omega)}{TF_{S2}(\omega)TF_{P2}(\omega)}\frac{B(\omega)}{A(\omega)} \tag{4.226}$$

This transfer function can then be used to calculate time domain transient response $z(t)$ for any ideal lightning current waveform using inverse Fourier transform method as follows:

$$z(t) = F^{-1}\left[TF(\omega)\cdot I(\omega)\right] \tag{4.227}$$

where $I(\omega)$ represents the frequency domain spectrum of the given lightning current waveform $i(t)$:

$$I(\omega) = F\left[i(t)\right] \tag{4.228}$$

This approach can also be used for analysis of the data acquired during lightning transient analysis, especially when the test current waveshapes were not exactly the same as those defined for components A, D, and H.

4.4.4.6 Identification of Standard Waveforms

For the sake of compatibility with aircraft lightning protection documents and for feasibility of subsequent testing, the ATLs should be directly comparable with the standardized induced transient waveforms introduced in subsection 4.4.3.3. In order to fulfill this requirement, it is practicable to assign to the ATLs the waveshapes of the ideal waveforms. For this purpose, the shapes and the temporal parameters (e.g., the rise and decay times) of the ATLs should be investigated to find the best-matching idealized waveform. An example of this process is depicted in Figure 4.93

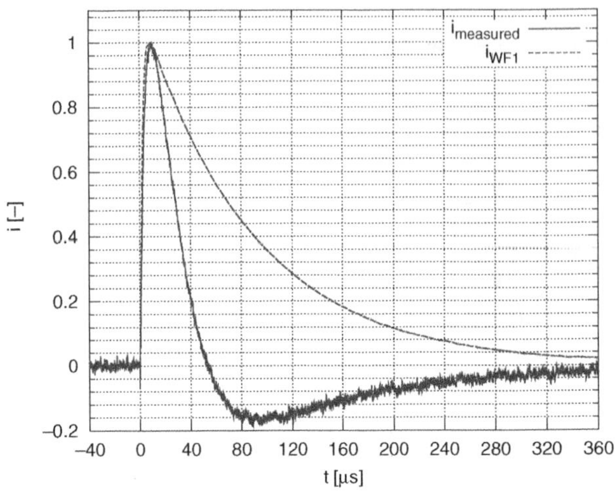

Figure 4.93 Assignment of Standard Waveforms to Measured Transients.

Some induced transient responses may have significant contributions from more than one parameter. For example, in a case of composite aircraft structures transient voltages contain significant structural resistance and aperture contributions or resonances, yielding complex waveforms. In this case, the ATLs can be determined as shown in Figure 4.94.

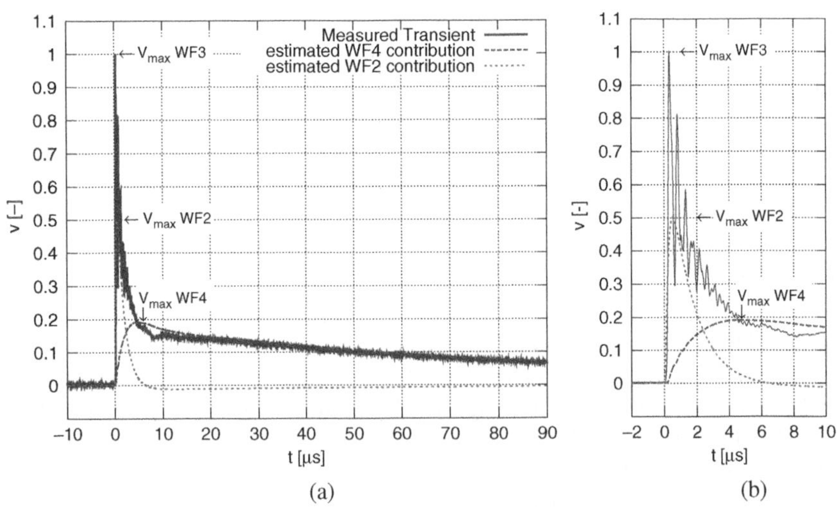

Figure 4.94 Example for Establishing Transient Levels of Standard Waveforms (a) full wave, (b) a detail of wavefront.

4.4.5 Avionic Equipment/System Functional Verification

4.4.5.1 Introduction

In order to verify the capability of equipment to withstand effects of lightning-induced electrical transients, the idealized standard induced waveforms 1 to 6 as defined in section 4.4.3.3 are generally applied using test methods defined in Section 22 of consecutive issues of ED-14/DO-160 [73]. Although this document is primarily intended for testing of avionic equipment, document [12] gives an additional guidance and expands the test methods to the complete systems.

In these documents, two groups of tests are used for equipment qualification, the damage tolerance tests and functional upset tests. Both types of testing are thoroughly described and explained in the above-mentioned documents; therefore, they will be just roughly outlined in the following subsections.

4.4.5.2 Damage Tolerance Tests

Damage tolerance testing, also known as pin injection testing, is a technique whereby the applicable transient waveforms are applied from a generator with defined both the internal impedance and the output transient level directly to individual pins of the EUT connector, a pin after pin, usually between each pin and case ground. This method is intended to assess the dielectric withstand voltage or damage tolerance of equipment interface circuits.

Since the EUT shall be generally tested in a powered state, with a power supply connected the power pins, all other interconnecting wiring should be disconnected. The powered state shall enable setting and testing of input/output circuits in defined states. Therefore, for some pins, it might be necessary to repeat tests for several I/O states.

The test levels are defined as an open-circuit voltage V_{OC} and a short-circuit current I_{SC} at the injection point, thus specifying a specified source impedance $Z_i = V_{OC}/I_{SC}$. When testing an EUT a powered state, some form of blocking device may be necessary to ensure that the test generator does not produce excessive loading of power supply or signal lines. In the case that a power supply is a part of the injection loop, a by-pass circuit might be necessary to guarantee constant low-impedance behavior and protection of the power supply. Examples of the test setup for damage tolerance tests are shown in Figure 4.95.

When the remote load impedance characteristics (including dielectric strength characteristics) are specified and the load does not employ a protective device that would bypass the load impedance, this impedance may be inserted in series with the generator and EUT.

To account for the high-frequency response of arbitrary terminated cable (see section 4.2.4.3) governed by cable characteristic impedance, the maximum inserted series impedance shall be limited to 75 ohms for high for HF part of testing, e.g., application of Waveform 3 transients.

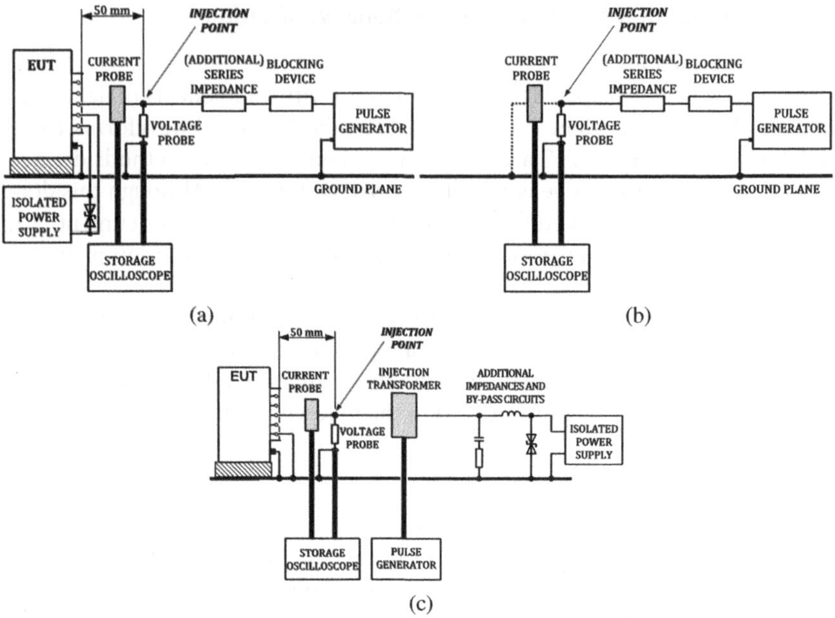

Figure 4.95 Pin injection test arrangement (a) direct injection, (b) calibration setup, (c) test using transformer injection.

When the local signal and power grounds are connected to the grounded structure outside the equipment in the real installation, they shall be also connected to the ground plane during laboratory tests.

For every applied type of transient (waveform), the test consists of two steps:

1. Calibration/performance verification

   Without the EUT present, the generator setting is adjusted to produce required waveform at the required open-circuit voltage V_{OC} at the injection point. Then, the injection point is shorted to ground and at the same generator setting, the current through injection point is measured and verified, if the level I_{SC} is reached (thus the internal impedance) and the waveform temporal parameters comply with requirements.

   All additional power supplies and by-pass and blocking components influence the source internal impedance and, therefore, they shall be included in the calibration setup. However, the output voltage of the AC/DC supplies shall set 0V during calibration.
2. Actual testing

   The injection point is connected to the designated pin of the EUT. At the setting established during calibration, 10 transients of each polarity are applied directly to the interface pins while monitoring the waveforms for any unexpected changes hinting at possible degradation or damage.

In order to reduce a testing time, groups (four or more) of EUT circuits (pins) with the same circuit design for both protection and operation may be qualified by testing three representative pins of each group. The remaining pins in the group are qualified by similarity.

4.4.5.3 Functional Upset Tests on Integration Bench Test Rig

The functional upset tests are performed on the equipment/systems to verify its immunity to transients induced in the associated wiring by the single stroke, multiple stroke and multiple burst lightning environments during a lightning strike to aircraft (refer to 4.4.3.2.1). The most common way is to perform the test on the test rig in a test laboratory, actually the same approach as one described in section 4.3.3.1 used for RF susceptibility testing, but without need for an anechoic chamber. The setup arrangement and requirements are basically the same.

During these tests, the pulse voltages and currents directly excited in cable harnesses. One method is the so-called cable induction (CI) testing, i.e., induction of WF1, WF2, and WF3 transients by means of transformer placed around the cable harness of interest (see Figure 4.96). In this way, the cable under test is the secondary winding of the transients. The injection probe is fitted with a short, low reactance, single turn monitor loop (around the probe) which is connected to an oscilloscope probe to measure the voltage induced into the cables. The levels of the induced cable current are measured using a monitor probe placed approximately 5 cm from the EUT connector.

Injection of the relatively long WF4 and WF5 pulses via transformer would require an employment of transformers with relatively huge magnetic cores to avoid saturation effects. Therefore, an alternate method, called the ground injection (GI) method. For this test, the EUT housing is separated from the ground plane and the test pulses are applied between the EUT housing and the

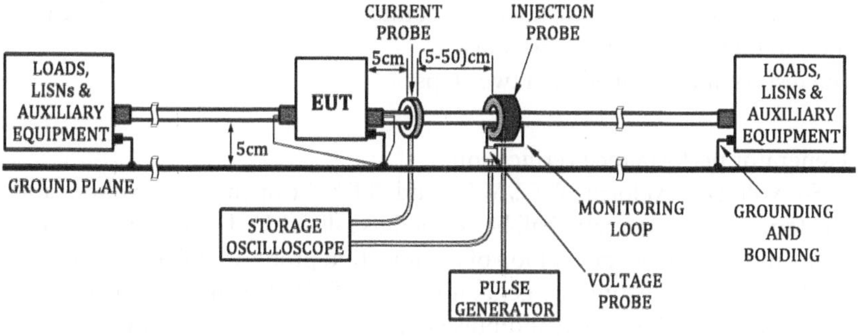

Figure 4.96 Test set-up for cable induction test.

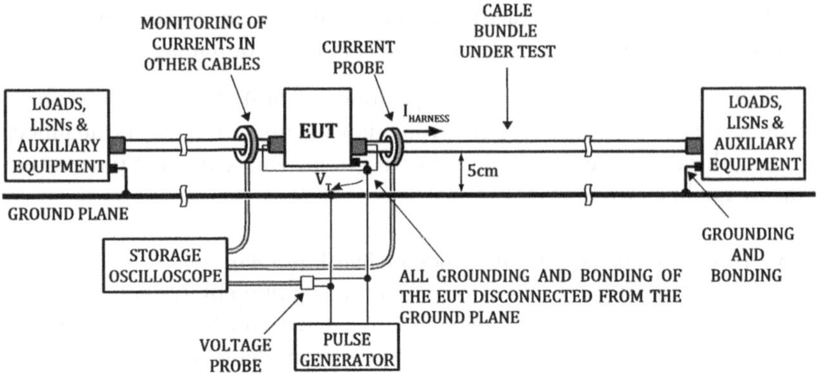

Figure 4.97 Test set-up for ground injection test.

ground plane, as shown in Figure 4.97. All bonding straps safety earths, local power and signal grounds, etc., which are intended to be grounded to the same part of the aircraft structure within 1 m of the EUT shall be also disconnected from the ground plane and connected together to form an isolated grounding point. If the EUT case is grounded directly via a mounting point, the isolated grounding point shall be created at this point on the housing. The pulse generator is then connected between the ground plane and the newly created isolated grounding point. In this way are the transients applied to all cable harness of the EUT simultaneously. The voltage levels of applied transients are measured between the isolated grounding point and the ground plane by means of an oscilloscope probe. The levels of the induced cable current are measured using a monitor probes placed approximately 5 cm from the EUT connectors.

The intention of the tests is to reach prescribed voltage or current levels on all cable bundles connected to the EUT. However, if the individual cable harnesses have significantly different impedances, it might be impossible to reach the required test level in some cable without overstressing other ones. In such cases, a selection of alternate injection points or even a deviation to the cable induction method might be necessary.

The test itself is subdivided in two steps:

1. Generator performance verification

   For voltage waveforms WF2, WF3, and WF4, the open-circuit voltage and the temporal parameters shall be verified usually at V_T level, for CI method measured in the secondary loop of the injection probe, for GI across the end of the cable connecting the generator to the EUT. For multiple stroke and multiple burst, the timing and pulse counts shall be verified as well. Then, the secondary loop of the injection probe for CI or the end of the generator cable for GI shall be shorted and the short-circuit current waveforms shall be recorded.

For current waveforms WF1 and WF5, the short-circuit currents and the temporal parameters shall be verified usually at I_T level, for CI method measured in the secondary loop of the injection probe, for GI at the end of the cable connecting the generator to the EUT. For multiple stroke and multiple burst, the timing and pulse counts shall be verified as well. Then, the secondary loop of the injection probe for CI or the end of the generator cable for GI shall be opened and the open-circuit voltage waveforms shall be recorded.

2. Actual test procedure

The test equipment shall be arranged according to Figure 4.96 for CI or Figure 4.97 for GI. The system/equipment under shall be switched on and set to an operation mode as defined by the test specification. While applying transients, the generator setting is to be increased until the either test or limit level is reached.

During single/multiple stroke tests, 10 single strokes/multiple stroke sequences of each polarity are applied while monitoring the EUT performance.

During multiple burst tests, one multiple burst sequence is applied every 3 seconds for a period of at least 5 minutes at each polarity while monitoring the EUT performance.

This procedure is repeated for every required waveform, every cable bundle, and all modes of operations.

4.4.5.4 Functional Upset Testing on Systems Installed on Aircraft

Functional upset testing on a system rig in a laboratory is the preferred option. However, when there is only one system physically available, and this is already installed in aircraft, the functional upset testing can be performed on systems installed in the aircraft.

The test is carried out using the procedures that have been introduced for the system rig laboratory testing (refer to section 4.4.5.3). However, prior to testing, a thorough analysis of the installation and careful preparation must be again conducted, with a focus on (but not only on) the following points:

- every bundle in the system shall be completely tested by injection. If a bundle branches, each branch is also to be tested.
- the problem area of short grounding wires bundled in the cable harnesses. These must be taken excluded from the test.
- shields bonded at multiple locations. All segments separated by the shield bonding points must be tested. If possible, the shields connections may be interrupted at the bulkheads. The test current level is then obtained by

evaluation of contributions of the individual segments to the total core-wire voltage (see section 4.2.4.6):

$$V_{in} = Z'_{T1} l_1 I_{ex1} + Z'_{T2} l_2 I_{ex2} + \cdots + Z'_{Tn} l_n I_{exn} \tag{4.229}$$

Under assumption that $Z'_{T1} = \cdots = Z'_{Tn} = Z'_T$, the test current level can be obtained as:

$$I_T = \frac{V_{in}}{Z'_T l} = \frac{l_1 I_{ex1} + l_2 I_{ex2} + \cdots + l_n I_{exn}}{(l_1 + l_2 + \cdots + l_n)} \tag{4.230}$$

where l is the total length of the cable bundle, $l = l_1 + l_2 + \cdots + l_n$

Alternatively, injection with the shields interrupted and separated at all points from the ground structure (i.e., direct injection on the core wiring) might be an option. In this case, the core-wire test levels can be determined using the transfer impedance measurements.

Additional challenge is the accessibility of cable bundles and provision of a sufficient space to place both the injection and the monitor probes on the cable under test. This is generally a major issue, especially for waveforms WF4 and WF5, where the injection transformers employ huge magnetic cores.

4.4.6 Showing Compliance with IEL Certification Requirements

Similarly to the requirements for HIRF compliance, also when dealing with the indirect effects of lightning the most stringent requirements apply to Level A function systems, with the severity consecutively relieved for Level B and C function systems.

4.4.6.1 Level A Systems

According to the current requirements of FAA and EASA, the level A systems, i.e., systems performing functions whose failure would contribute to or cause a condition that would prevent the continued safe flight and landing, must be designed and installed so that the operation and operational capabilities of the systems to perform these functions are not adversely affected when the aeroplane is exposed to lightning.

Therefore, the main task for showing compliance is to prove that:

1. the system in the real aircraft configuration was tested and proven immune to transients voltages and currents defined by ETDLs applied to cable bundles using the integration test rig approach described in subsection 4.4.5.3 and
2. There is sufficient margin between ETDLs and ATLs, i.e., transients induced in cabling during lightning strike to the given aircraft, when the ATLs were determined:

- experimentally using methods described in subsections 4.4.4.4 and 4.4.4.5
- using similarity with a previously certified aircraft and installation (subsection 4.4.4.2)
- by numerical simulation supported and validated by experimental results
- using generic transfer functions and attenuations (see subsection 4.4.4.3). However, the current issue of [6] allows application of this method to Level A Display systems only.

4.4.6.2 Level B and C Systems

A level B or C system must be designed and installed so that the functions performed by this system can be recovered in a timely manner after the aircraft is exposed to lightning. The compliance shall be shown using the equipment test approach defined in [73].

The EDTL for testing may be determined using aircraft tests or analysis performed for Level A systems. Alternatively, the following levels can be chosen:

1. Generic Level 3 [73] for most Level B systems. For Level B systems and associated wiring installed in aircraft areas with more severe lightning transients, Level 4 or 5 shall be considered.
2. Generic Level 2 [73] for most Level C systems. For Level C systems installed in aircraft areas with more severe lightning transients, Level 3 is applicable.

The selection of the generic ETDL for Level B and C systems shall be supported by an analysis of the aircraft and system installation confirming the selected levels.

Again, if a Level B or C function system consists of equipment qualified to sufficient levels using laboratory tests [73] and wire types, wire bundle composition, connectors, shields, and shield terminations in the intended aircraft installation are same as specified in the equipment installation instructions, and the lengths defined in [73] were used, a separate full system test is not required.

4.5 Precipitation Static (P-Static)

4.5.1 Introduction

During the early days of aircraft operations under all-weather conditions, it was observed that when flying in precipitation, significant radio frequency interference with the operation of communication and navigation systems was observed. This concurrence of electromagnetic interference ("static noise") and precipitation resulted in the name "precipitation static" or p-static [77].

However, Precipitation Static (P-Static) is term denoting both the causes and the effects, i.e., both electromagnetic phenomenon (disturbance) as well as interference effects primarily, but not only on antenna-connected receiving

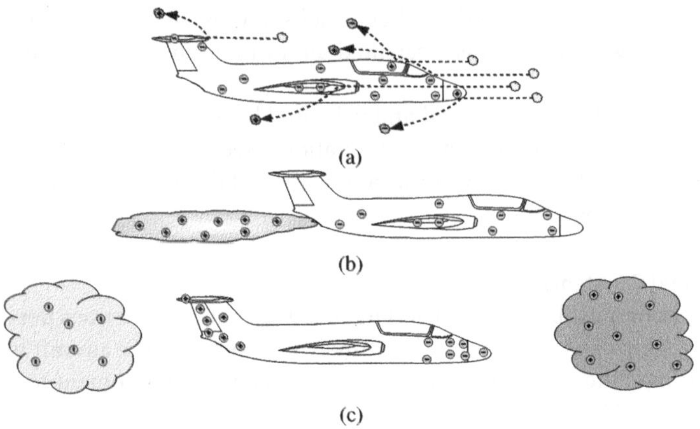

Figure 4.98 Aircraft charging processes (a) triboelectric charging; (b) engine exhaust charging (c) exogenous charging.

equipment, associated with all electrostatic charging processes of flying aircraft [78].

4.5.2 The charging processes

The aircraft surfaces can gain and accumulated charge a result of several processes described below and depicted in Figure 4.98 [77,79,80]:

1. Frictional (triboelectric) charging during collisions with the particles suspended in the air, Figure 4.98(a)

   As uncharged precipitation particles strike the aircraft, the two dissimilar materials come into contact. Charge carriers move from one material to the other to equalize their electrochemical potentials. As the particles reflect from the surface and materials separate, one material keeps extra charge carriers, and the other gives them away, yielding the charge imbalance. The colliding particles can be precipitation particles, such as ice crystals, rain and snow, as well as sand, dust, ash, etc.
2. Engine exhaust charging, Figure 4.98(b)

   This type of charging occurs when the hot exhaust gases carry away charge, leaving a net opposite charge on the aircraft. Depending on the type and the operating characteristics of the engine, absolute currents from 0 to 400 μA were observed. In this case, the charging process is independent of meteorological conditions.

   This phenomenon negligible in the case of propeller driven aircraft is very important in aircraft having engines with after-burn.
3. Exogenous charging, Figure 4.98(c)

It occurs when the vehicle flies through an external electric field, such as that existing between oppositely charged regions of clouds. External electric field causes in conductors separation of charges, thus creating an internal electric field compensating the external one. Net charge of the aircraft remains zero, only extremities become charged oppositely.

4.5.2.1 Aeroplane Charging Process

4.5.2.1.1 Charging Treating the aircraft can be treated as isolated conducted body that is charged and discharged by processes described in the previous paragraph. Following the charge conservation law, the charging process can be described by a continuity equation:

$$Q_A\,(t) = Q_A(t_1) + Q_{in} - Q_{OUT} \tag{4.231}$$

where $Q_A\,(t)$ is the electric charge stored on aircraft at time t, Q_{in} is the amount of charge deposited on the aircraft between time instants t_1 and t, and Q_{OUT} is the amount of charge taken away from the aircraft during the same time period.

The total net charge Q_A accumulated on the aircraft is also given by

$$Q_A(t) = C_A \cdot V_A(t) \tag{4.232}$$

where C_A is the capacitance of aircraft in flight and V_A denotes aircraft potential difference with respect to infinity. The capacitance of an aeroplane in flight cannot be calculated analytically. The paper [82] gives the capacitances of the two aircraft, 600 pF for the CV-580 (wingspan 32.10 m, length 24.84 m) and 1 nF for the C-160 (wingspan 40 m, length 32.40 m), respectively; obtained as numerical solution of the Laplace equations for aircraft surface. Alternatively, the aircraft capacitance can be estimated using formulas for oblate equation (4.233) or prolate equation (4.234) spheroid of the major axis with length $2a$ and the minor axis with the length $2b$ (see Figure 4.99), respectively [83]:

$$C_{\text{obl}} = 4\pi\varepsilon_0 a \frac{\sqrt{1 - \frac{b^2}{a^2}}}{\arcsin\sqrt{1 - \frac{b^2}{a^2}}} \tag{4.233}$$

$$C_{\text{prol}} = 4\pi\varepsilon_0 a \frac{\sqrt{1 - \frac{b^2}{a^2}}}{\ln\left[\frac{a}{b}\left(1 + \sqrt{1 - \frac{b^2}{a^2}}\right)\right]} \tag{4.234}$$

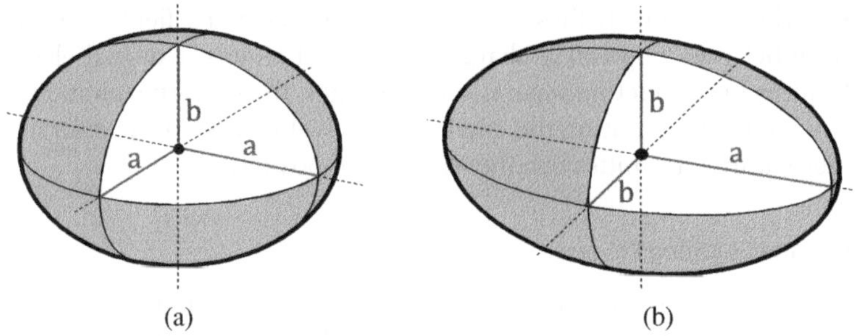

Figure 4.99 Oblate (a) and prolate (b) spheroids.

Differentiating equation (4.231) with respect to time, and keeping in mind that $i(t) = dQ/dt$, the electrostatic state of an in-flight aircraft is described by the following equation [81,84]:

$$I_C(t) = C_A \frac{dV_A(t)}{dt} + I_{Ion}(t) + I_{Dis}(t) \tag{4.235}$$

where I_C is total charging current $I_C = dQ_{in}/dt$. Charge Q_{OUT} carried away from aircraft is represented by two currents. I_{Ion} denotes total discharge current to the surrounding air caused by ionization processes, such as corona or streamers and I_{Dis} denotes total dissipative current to the surrounding air due to non-zero conductivity of the surrounding dielectric given as

$$I_{Dis}(t) = G_A \cdot V_A(t) \tag{4.236}$$

with G_A denoting aeroplane conductance to infinity depending on the conductivity of surrounding dielectric.

If the total charging current is higher than the sum of discharge and dissipative currents, i.e., the deposited charge is higher than the charge bled away from the aeroplane, the potential of an aeroplane increases.

4.5.2.1.2 Precipitation-Charging Currents

The triboelectric charging current I_{TC} is dependent on weather conditions, the aircraft frontal surface area and the speed of the aircraft v. It can be estimated using following equation [79]:

$$I_{TC} = q_p \times c \times v \times A_e \tag{4.237}$$

where q_p is the charge transfer per particle, c is the density of particles, v is aircraft velocity, and A_e is the effective charging area of aircraft defined as [79]:

$$A_e = A_p \times k(v, A_p, \ldots) \tag{4.238}$$

Table 4.7 Charging parameters for different flight regimes

Flight regime	Cloud type	Charge deposited per particle	Particle concentration
		q_p pC	c $1/m^3$
Cruise	Cirrus	1 to 10	2×10^4
Approach or landing	Thunderstorm anvil	1 to 35	5×10^4

where A_p is actual projected frontal area of aircraft and k is effective area factor ($k \leq 1$). This factor is usually much smaller than one, and increases with velocity and particle mass but decreases with increase in the size of aircraft [81]. The reported values of effective area factor vary between 5% to 40% [79, 84].

Typical charge densities and charges deposited by single particle during different stages of flight are given in Table 4.7 [79].

4.5.2.1.3 Engine Exhaust Charging Currents Engine exhaust charging can be significant contributor to overall charging process, however, the engine-charging currents are generally lower than the currents generated by precipitation charging [81, 84]. Measurements presented in document [84] hint that the engine charging of aircraft can vary between 50 and 800 μA, depending upon the aircraft and type of engine.

4.5.2.1.4 Discharging by Corona During the charging, the potential of an aircraft and thus the intensity of electric field surrounding the aircraft increase with the increase in the total net charge stored on the aircraft. The extremities of the aircraft with small radii of curvature, such as wing tips, vertical and horizontal stabilizers and other similar protrusions, enhance the local electric field strength. When the local field intensity exceeds the threshold value—the on-set corona value, the corona discharge occurs.

Corona discharge is a repetitive discharge process by which a current flows from a conductor with a high potential into surrounding air by ionizing it. Depending on polarity, the generated ions and electrons are either attracted and collected by the conductor or repelled and propagate away where recombine to form neutral gas molecules.

These short repetitive pulses with fast rise times and short durations radiate radio frequency energy with spectral content extending into the hundreds of megahertz. The energy from these pulses can couple via antennas, producing the noise signals. These can be strong enough to interfere with the associated receivers resulting in loss of, or interference with, the communication, navigation, etc.

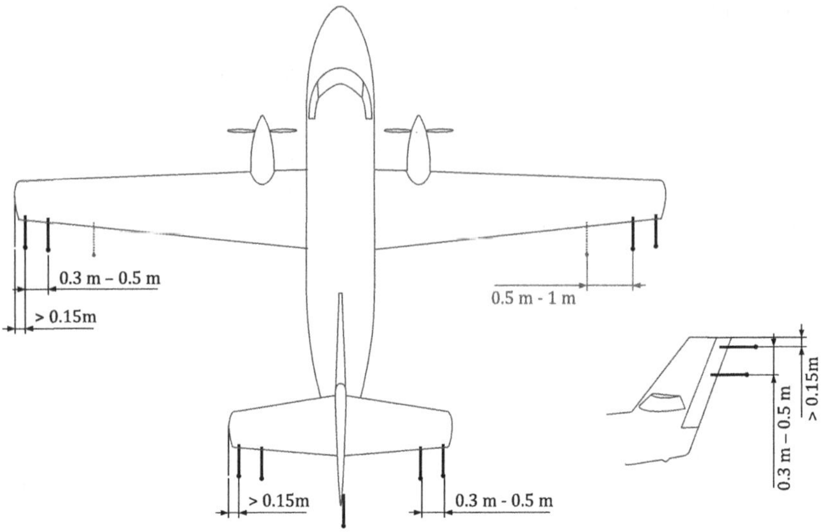

Figure 4.100 Example of placement of static dischargers according to [79].

Solution to corona noise is place static dischargers at locations where the corona discharges are most likely to develop. These are the locations on aircraft with the smallest radii of curvature, i.e., wing tips or vertical and horizontal stabilizer, see Figure 4.100.

Static dischargers (also known as static wicks) provide controlled discharge via high resistance (lower currents generate less noise) at lower levels than the corona inception voltages for antennas. An example of a static discharger design is shown in Figure 4.101.

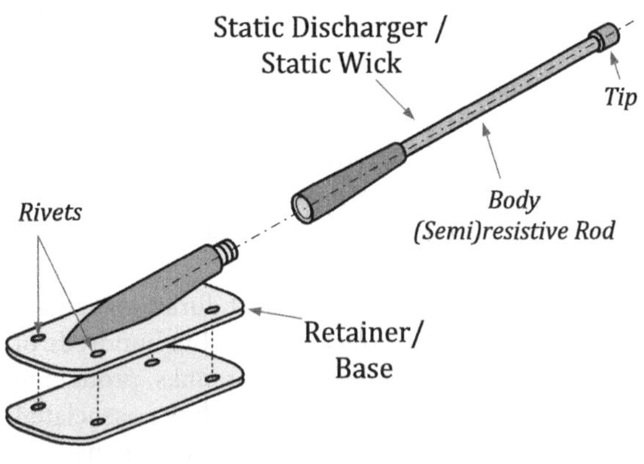

Figure 4.101 Example of static discharger design.

With the limited current carrying capabilities of individual dischargers, sufficient number of dischargers must be installed to prevent of increase in airframe potential above the corona threshold levels. The number of necessary static dischargers N_D can be estimated using the following equation:

$$N_D = \frac{I_{Ion}}{I_D} \approx \frac{I_{TC}}{I_D} \tag{4.239}$$

where I_D denotes the nominal current of one discharger and I_{Ion} is the total discharge current due to corona. The latter can be approximated by the triboelectric charging current I_{TC} defined by equation (4.237).

Example The number of static discharges for a small acrobatic aeroplane with projected frontal area A_p of 4.1 m^2 can be estimated using the worst-case scenario, where the effective charging area is set equal to the projected frontal area, $A_e = A_p$; and inserting the values given in Table 4.7 into equation (4.237):

In cruise, assuming the maximum speed of plane v to be 377 kn $\equiv$ 685 km/h $\equiv$ 190 m/s:

$$I_{TC} = q_p \times c \times A_p \times v = 10 \times 10^{-12}\text{C} \times 2 \times 10^4 \frac{1}{\text{m}^3} \times 4.1\text{m}^2 \times 190\frac{\text{m}}{s} = 156.8\ \mu\text{A}$$

During landing, assuming $v = 130$ kn $\equiv$ 241 km/h $\equiv$ 67 m/s and then

$$I_{TC} = q_p \times c \times A_p \times v = 35 \times 10^{-12}\text{C} \times 5 \times 10^4 \frac{1}{\text{m}^3} \times 4.1\text{m}^2 \times 67\frac{\text{m}}{s} = 480.7\ \mu\text{A}$$

Assuming the nominal current of one discharger to be 50 μA according to the requirements given in [85], the number of necessary static dischargers N_D is then

$$N_D = \frac{i_{TC}}{i_D} = \frac{481\ \mu\text{A}}{50\ \mu\text{A}} = 10 \tag{4.240}$$

The obtained number of dischargers is quite conservative since the effective charging area is usually less than one half of the projected frontal area, as mentioned above. The recommended numbers of static dischargers according to SAE ARP5672 [79] is tabulated in Table 4.8.

4.5.2.2 Localized Charging and Discharging

4.5.2.2.1 Streamering and Insulation Breakdown at Dielectric Surfaces Dielectric surfaces such as windows, radomes, winglets, etc. on frontal impact areas of aircraft have, unless properly treated, an extremely high electrical resistance

Table 4.8 Recommended minimum number of trailing-type static dischargers

Wing span (meters)	Total minimum number Design/(CDL)	Placement per area (minimum)	
		Wing tips	Stabilizer tips
10	5 (4)	1	1
20	10 (8)	2	2
30	15 (12)	3	3
40	24 (20)	6	4
50	31 (25)	8	5
60	44 (36)	14	5 Hor / 6 Ver

CDL—Configuration Deviation List—allows 20% to be missing.
Configuration Deviation List condition should be used during flight tests and becomes the minimum discharger requirement, allowing aircraft dispatch with one or two dischargers missing.

and thus prevent or limit the flow of electrons across its surface or through material. Thus, a considerable amount of charge generated on their surface by triboelectric effect can remain trapped thereon for very long time. With more particles impacting the area, the voltage increases until either

1. a flashover along the surface to the adjacent conductive objects (window frame, fastener, skin, etc.) occurs
2. a dielectric breakdown through the insulating material to the conductive parts underneath.

Both types of discharge generate broadband radio frequency noise. Additionally, these discharges can induce transient pulses in the adjacent wiring.

This phenomenon is also observed at metal surfaces painted with high dielectric strength paint or covered by dielectric decals. If a thin layer of insulating decorative paint is applied over conductive part, the accumulated charge can pinhole the insulating layer or flow via microscopic cracks to the conducting surface underneath.

However, as paint thickness increases, the voltage required to pinhole to the underlying conductive layer (the breakdown voltage of the paint) also increases, with more charge trapped on the paint surface. This process can lead to the effects already described above, i.e., the surface discharge to the nearest conducting component, an insulation breakdown with significant visible holes burnt in the paint.

4.5.2.2.2 Protection Conductive coatings such as expanded metal foils, embedded conductive wires, or resistive coatings, and paints can be applied to the exposed surfaces of these parts to prevent flashover or puncture. Surface treatments applied to non-conducting structure or parts should have a surface resistance that is less than 300,000 ohms per square. For radomes and antennas, the surface coating should not significantly reduce the transmissivity of the radomes or degrade antenna performance.

Therefore, the insulating paint thicker than 100μm should be verified to ensure that charge dissipation is still effective [79].

4.5.2.2.3 Arcing Arcing is sparking between two isolated conductors charged to different potentials. It can be avoided by bonding of all conductive components that can be potentially charged, to the primary structure of the aircraft.

4.5.2.3 Charging Effects on Rotorcraft

P-static charging on rotorcraft is significantly different from that on aeroplane. Helicopters become electrostatically charged by ion emission from the engines and by the triboelectric charging on airfoils, where the rotation of the main rotor blades considered as the main source [45, 87, 88]. Consequently, P-static interference can occur in an hovering helicopter.The charge accumulation effects can be divided into two groups:

1. P-static, i.e., corona discharges due to charge build-up can interfere with the performance of radio communication systems. In this case, all precautions valid for fixed wing aircraft also apply. However, the use of static dischargers might be limited by the lack of suitable locations for static discharges.
2. In the case of external load operations, the accumulated charge can cause an arc between the hook and the cargo during pick-up or between the suspended cargo and the earth during delivery. Therefore, the helicopter should be discharged prior to contact in order to prevent electrical shocks to personnel on the ground or damage/malfunction of delivered cargo.

The accumulated charge (refer to equations (4.231) and (4.232)) can vary over a wide range depending on the helicopter size and its proximity to the earth. The values of helicopter capacitances obtained by measurement on helicopters of various sizes at different altitudes lie in the range 300 to 1000 picofarad [91,92]. These values can be confirmed by calculations of capacitance assuming that the helicopter is a finite cylinder of length l and radius a at height h above ground. For this case, the capacitance C_A can be estimated as [20]:

$$C_A = \frac{2\pi\varepsilon_0 l}{\ln\left(\frac{l^2}{4ha}\left[\sqrt{1+\left(\frac{4h}{l}\right)^2}-1\right]\right)} \tag{4.241}$$

Inserting the values for $a = 1.5$ m and $l = 20$ m into the above equation, a variation of the capacitance with the flight altitude can be obtained, as depicted in Figure 4.102.

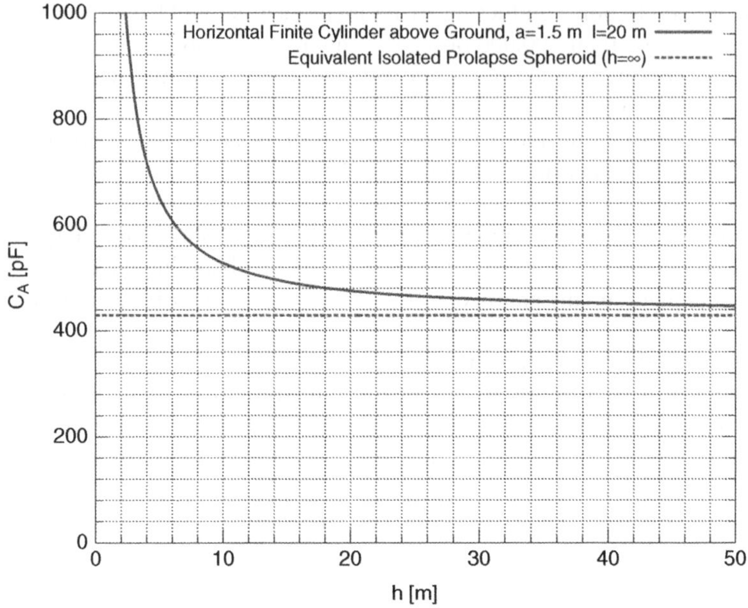

Figure 4.102 Dependence of the cylinder capacitance on the height above ground.

For testing purposes, a widely used upper bound is represented by a 1000-picofarad capacitor charged to 300 kilovolts [45, 87, 88].

4.6 Lightning Effects and Protection in Aerospace

4.6.1 Introduction

Since both the natural and artificially initiated lightning (vehicle-intercepted versus vehicle-triggered, see section 4.4.2.2) are demonstrated hazards to the launch of space vehicles, atmospheric electricity must be considered an important factor not only in the design but also in all phases of operation of space vehicles. Considering the lightning effects, an inadequately protected space vehicle can be upset, damaged, or even destroyed by a direct lightning strike to the vehicle or launch support equipment before or after launch. Damages and upsets can also result from transient voltages and currents induced in the vehicle internal electrical wiring and/or the umbilical connections to launch platform by a nearby lightning strike or a strike to launch platform itself.

Natural lightning occurring during thunderstorms was recognized as a threat to spacecraft since early years of the space mission programs. Therefore, the launch rules prohibited flight through thunderstorms. However, until

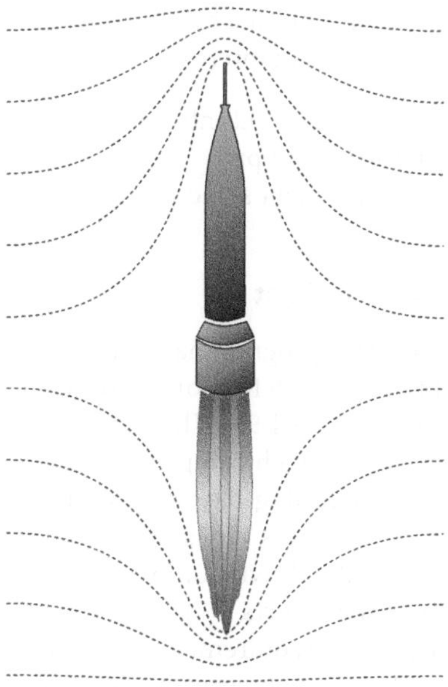

Figure 4.103 Enhancement of electric field at extremities of a rocket extended by plumes.

Apollo XII lightning accident happened on November 14, 1969, it was not considered that a climbing-up rocket could initiate lightning if flying into or close to highly electrified clouds [93].

Apollo XII was struck twice by vehicle triggered lightning within a minute after take-off. Severe electromagnetic disturbances occurred resulting in nine non-essential instrumentation sensors permanently damaged and temporary upsets of equipment accompanied by loss of communications and navigation, interference with instruments, initiation of warning lights and alarms, etc. However, all errors in mission-critical systems were cleared and the mission was successfully completed.

Similar accident occurred on March 23, 1987, when an Atlas/Centaur rocket and its payload were lost when the unmanned NASA vehicle was struck by a triggered lightning. The lightning-induced disturbances caused the rocket's Digital Computer Unit (DCU) upset resulting in an issue of an extreme yaw command yielding a severe vehicle rotation. The stresses associated with this motion led to the vehicle breakup in flight [95, 96].

For both incidents, the lightning was initiated by the launch vehicle itself in situations where natural lightning otherwise would not occur. Launch vehicles are extended by the exhaust plumes and the combination of both significantly enhances the ambient electric field, as depicted in Figure 4.103. This effect can initiate lightning in even electric fields that are tens to hundreds of times smaller than what would be required to initiate natural lightning.

Based on the aforesaid possibilities, the consideration on the space vehicle lightning protection can be subdivided into three areas of interest:

- Direct lightning strike to a space vehicle during take-off, climb, and landing
- Direct lightning strike to a vehicle docked in the launch pad
- Effect of lightning strikes to the launch pad and adjacent structures.

4.6.2 Direct Lightning Strike to a Space Vehicle in Flight

As a result of the early accidents, the American space program has established a set of Lightning Launch Commit Criteria (LLCC) and Definitions to mitigate the risk of a direct lightning strike to a flying spacecraft [94,96]. The constraints given by LCCC are based on the known cloud types which can produce lightning discharges, the distances to charge regions, lightning-flash free periods, and monitoring the atmospheric electrical field strengths. These LLCC have been further developed and later adopted by the Federal Aviation Administration for application at state-operated and private spaceports [97].

Based on the LLCC, the standard defining EMC requirements for space equipment and systems standards issued by the American Institute of Aeronautics and Astronautics (AIAA) [98] or European Space Agency (ESA) [99, 100] exclude a possibility of a direct strike to a spacecraft in flight.

On the other hand, the specifications prepared by NASA [93, 101] or the International Organization for Standardization (ISO) [102] require that the spacecraft shall be designed to withstand the direct and indirect effects of a lightning strike to the vehicle itself, or to nearby objects, before or during launch and ascent operations.

The protection designs for physical effects of direct lightning arc attachment as well as the indirect effects on avionics subsystems and components shall be based on considerations, procedures, and documents identical with those applicable to aircraft lightning protection of described in chapters 4.2 and 4.4. Additional information on the lightning effects and protection of rockets in flight can be found in [103, 113].

4.6.3 On-ground Lightning EM Effects on Spacecraft

In order to prevent a direct lightning strike to spacecraft installed on launch pad and to safely divert the prospective lightning current away from the launch pad/complex, a lightning protection structure should be installed. An initial straightforward idea was to use the umbilical tower itself to intercept lightning flashes and ensure that no flash would strike the vehicle directly [104]. Following at those times common approach to lightning protection, the "protection angle" and the associated "cone of protection" this was considered feasible since the umbilical tower was sufficiently massive, thus a lightning flash would not

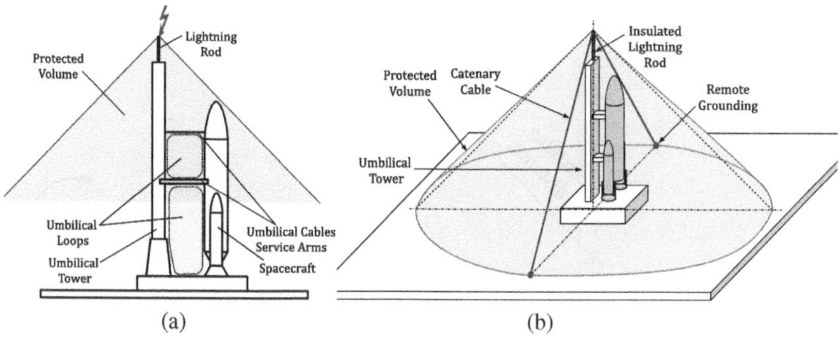

Figure 4.104 Spacecraft lightning protection using the umbilical tower (a) protection by umbilical tower; (b) protection by catenary/downconductors.

damage the tower itself. With the tower considerably taller than the airborne vehicle, a 45° cone of protection (the value of 45° was commonly recommended in contemporary lightning protection standards) was provided for the vehicle (as illustrated in Figure 4.104a) and the tower itself acted as a low-impedance down conductor. However, this solution was associated with the following critical points:

- The current entering the soil at the foot of the tower will raise the ground potential. In order to achieve equipotentialization, all ground support structures (umbilical tower, service tower, cable raceways, pipelines, etc.) must be connected to the tower earthing system and provide the appropriate conductive paths for partial lightning currents.
- In this configuration, not all the current flows purely down the tower. Since the umbilical arms and connected umbilical cables provide electrical paths between the tower and the spacecraft, a portion of the lightning current can flow to ground through the vehicle.
- Current concentrated in the umbilical tower will be associated with high intensity, fast changing magnetic fields surrounding the tower. The umbilical cables connected to the spacecraft create loops, in which excessive transient voltages and currents can be induced (see Figure 4.104a) and simultaneously carry these transients into the vehicle. Open access hatches and doors will further compromise and degrade the vehicle's own lightning protection measures.

In order to minimize the problems pinpointed above, an alternate scheme, depicted in Figure 4.104(b), was introduced. A long insulating mast at the top of the umbilical/service tower supports the lightning rod that was connected to remote ground points via long massive catenary downconductors [105]. In this way, the lightning currents were diverted as far away as possible from the

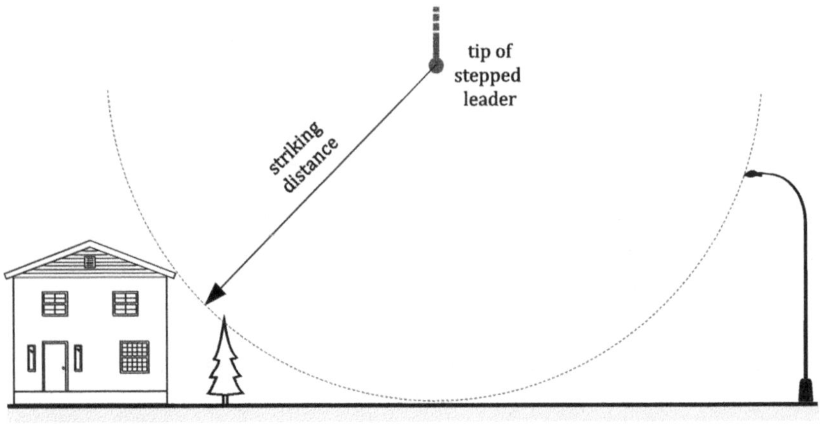

Figure 4.105 Outline of striking distance concept.

launching zone, minimizing both the ground-potential rise and the magnetic fields at the vehicle location. The airborne vehicle again lied within the cone of protection.

However, experiments and experience on lightning strikes to tall structures showed that attachment can occur to parts of the structure lower than the tips of the air-termination rods (i.e., top of the structure) or to the metallic structures that lie within the protective angle. The method was modified and a fixed value of the protection angle was replaced by a variable one, whose value decreases with the increase in the height of protected structure [106]. For tall structures, a protective angle method was proven as unsuitable and the rolling sphere method was proposed as a more suitable method since it was able to explain occurrences of side strikes and failures of the cone of protection [107].

The rolling sphere method is the simplest version of the electrogeometric model based on assumption, that for an expected peak return stroke current the striking distances from the leader tip to any structure (a sharp tip of a rod or tower, an edge, a flat surface) or the Earth are equal. Then, as illustrated in Figure 4.105, the lightning is predicted to attach to that conducting object or the Earth, which the tip of the leader first approaches to within that striking distance.

However, this assumption clearly ignores the fact that the attachment of lightning flashes to grounded structures depends not only on the prospective return stroke peak current, but also on the geometry of the structures considered, since their shapes influence the magnitude of the local electric field and thus subsequent development of connecting leaders.

Additionally, the concepts of the cone of protection, or later the protected volume or space hint that everything within their volume is protected from

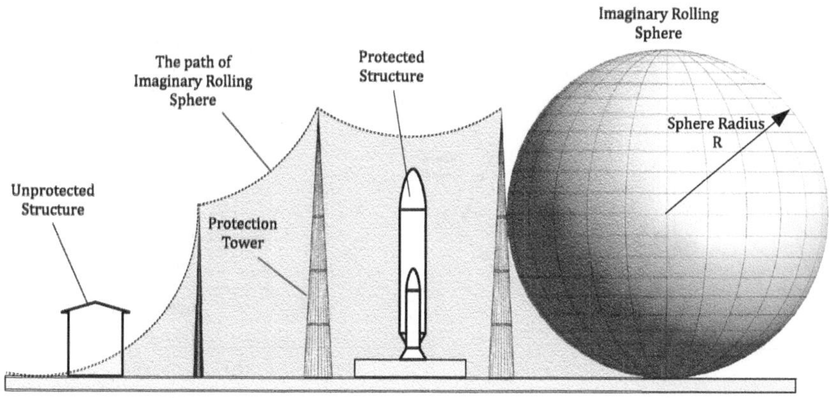

Figure 4.106 Sketch of Rolling Sphere Method application.

the direct lightning attachment. This assumption would only be valid, if this volume would be empty. In reality, depending on their position and shape, the objects inside the "protected" space can be exposed to direct strikes. Therefore, the space is not 100% protected from penetration of lightning [109].

Therefore, more advanced models, such as improved electrogeometric model, several leader inception, and several leader progression models were formulated by different authors [107]. However, despite several shortcomings and need for further development, the rolling sphere method allows quick visualization the lightning attachment points.

Practical application involves rolling of an imaginary sphere over the complete structure and its surroundings. The radius of the sphere is equal to the striking distance associated with the minimum current level for the chosen protection level. Any point of the scene coming into contact with the sphere is a possible point of lightning attachment, as illustrated in Figure 4.106. Therefore, the intentional lightning protection components shall be located in such a way that the rolling sphere touches only the external lightning protection systems. It can be clearly seen that when the height of the structure is more than the radius of the sphere, the lightning can attach also to the sides. The exact guidance on using the rolling sphere method is given in [106, 108].

Application of the rolling sphere method to the single tower structure shows that the lower parts of the structure are not sufficiently protected from the lightning strikes of lower amplitudes, even with the catenary downconductors [110]. Therefore, structures depicted in Figure 4.107 with multiple lightning protection towers were proposed. The towers with the lightning rods interconnected by shield wires (air-termination network) are suitably placed around the launch-pad complex. The difference in both approaches is the way of conducting of the lightning current to ground. Both previously mentioned ways of installation types of the lightning rods and downconductors are implemented.

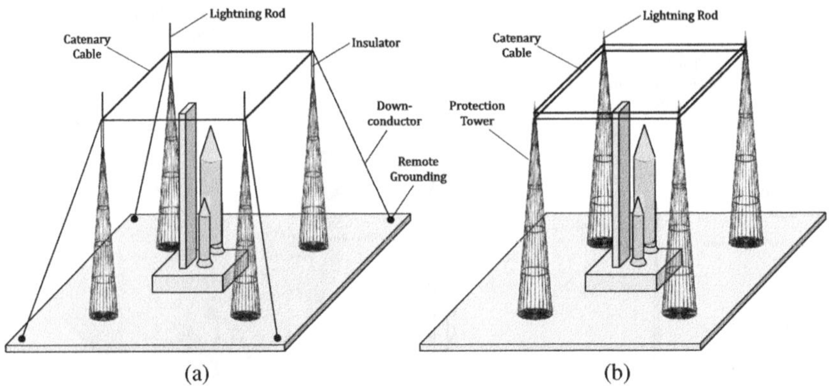

Figure 4.107 Lightning protection by multiple protective towers (a) lightning rods and downconductors isolated from towers; (b) towers functioning as the downconductors.

Figure 4.107(a) depicts an approach with the air-termination network insulated from the supporting towers by insulating supports. The downconductors start at the lightning rods and terminate to the ground end at remote earth points [110]. On the other hand, the approach shown in Figure 4.107(b) involves lightning rods installed directly on the towers, with towers acting as the downconductors [111].

4.6.4 Electromagnetic Effects at the Launch-pad

Whichever method of the launch-pad external lightning protection is applied, an assessment of possible indirect lightning effects introduced into vehicle avionic systems via the umbilical connections or diffusion through composite skins before lift-off shall be performed. Configuration- and site-dependent ATLs coupled to equipment should be evaluated to determine threat levels and test criteria. Correspondingly, both terminals of umbilical circuits should be hardened against lightning transients.

In order to determine the current redistribution within the mutually interconnected components of the lightning protection system, the associated electromagnetic fields and corresponding induced transients, a complex simulations using 3D numerical electromagnetic codes, or combination of a 3D modeling and transmission line network modeling can be used for solving the Maxwell equations given in Table 4.4 [100, 110]. Even in these cases, numerical simulations of the frequency-dependent components, such as the tower grounding impedance or the non-linear behavior, such as sparking, pose difficulties.

Therefore, just to sketch the problem-solving approach, a low-frequency approach using lumped elements will be outlined in the following subsections.

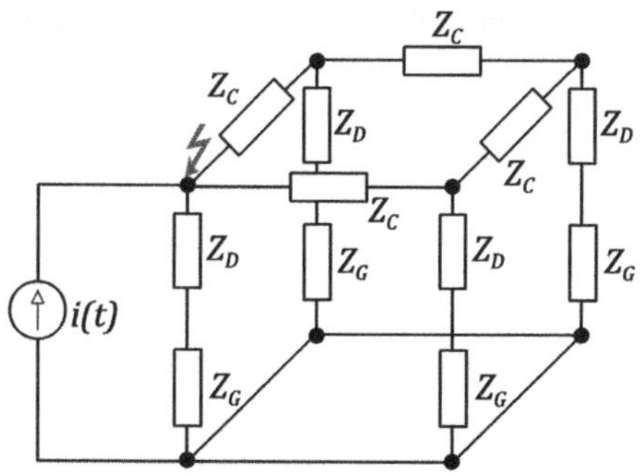

Figure 4.108 Equivalent circuit of the lightning protection system.

4.6.4.1 Current Distribution in the LPS

The lightning protection systems depicted in Figure 4.107 can be simulated by a lumped-element equivalent circuit illustrated in Figure 4.108, where Z_C stands for the impedance of the corresponding segment of the catenary wire, Z_D is the impedance of a downconductor (or the tower), and Z_G represents the grounding impedance of the down-conductor (tower).

When building the model, it is important to consider not only the resistance and self-inductances of the model, but also the mutual inductances between the individual segments. The process outlined in subsection 4.2.4.1 can be applied. Additionally, in the presence of a well-conducting ground plane, such as soil of high conductivity, extensive equipotentialization by means of a meshed earth-termination system or an idealized case with an ideal infinite ground plane, the method of images shall be also applied [27]. The model can be further expanded by including lumped models of all incoming conductors, such as cables and pipelines.

The final current distribution can be then calculated in the frequency domain and inversely transformed into the time domain, or computed directly in the time domain numerically using circuit solvers, such as SPICE.

4.6.4.2 Magnetic Field of Vertical Downconductors and Loop Voltage

The currents flowing in the downconductors and the associated magnetic fields have high time derivatives and, therefore, induced voltages can reach high values and cause damage to equipment. Time-varying electromagnetic field causes the induction of voltages in either closed or open loops. The loops can be formed by conducting structures, e.g., a data or communication networks, a power network, cryogenic plumbing, etc.

Magnitude of the induced voltage depends on the number of downconductors forming the lightning protection system (LPS). In order to outline the influence of the downconductors on induced voltage, a simplified system of Figure 4.109 was studied. The current distribution between individual downconductors can be obtained by network analysis of the network depicted

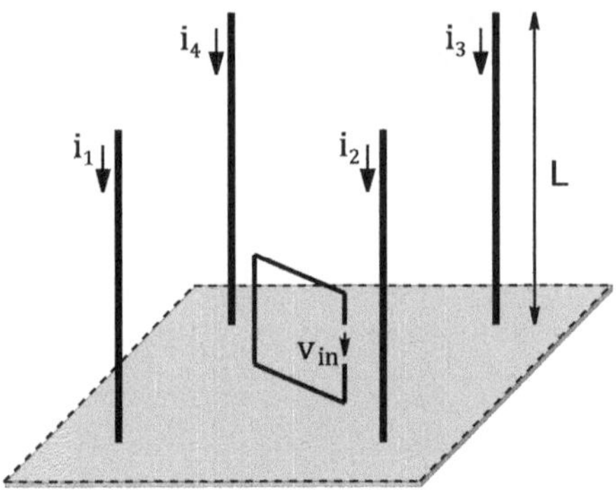

Figure 4.109 Simplified model of a loop between the downconductors.

in Figure 4.108. Then, the magnetic field at the loop location can be determined as follows.

Induced terminal voltage in an open or closed loop has the opposite polarity as the electromotive force *emf* and can be determined using the Faraday's law:

$$v_{in} = -emf = \frac{d\phi}{dt} = \frac{d}{dt}\oint_S \vec{B} \cdot d\vec{S} \tag{4.242}$$

where ϕ denotes the magnetic flux flowing through a closed loop with a surface S and $\vec{B}$ is the magnetic flux density. The vector of the magnetic flux density can be determined employing a magnetic vector potential $\vec{A}$ of a downconductor. Since the current flows in vertical direction, the vector potential has only a z-component which is given by

$$A_z = \int_{-L}^{L} \frac{\mu_0 I}{4\pi r} dz' \tag{4.243}$$

where I is the current flowing through the conductor and r is the distance between a current element dz' and a point of interest P(x,y,z) $r = \sqrt{\rho^2 + (z' - z)^2}$, ρ represents the perpendicular distance of point P from the conductor and L is the length of the conductor.

From the definition of the vector magnetic potential, the following is valid for a vector of magnetic flux density:

$$\vec{B} = \nabla \times \vec{A} \tag{4.244}$$

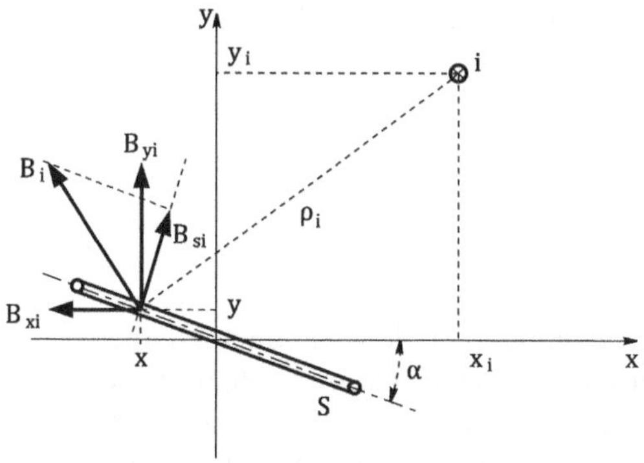

Figure 4.110 Geometry for evaluation of coupling between the loop and a conductor.

After integration of equation (4.243), insertion into the expression for a vector magnetic potential (4.244) and after mathematical modification:

$$B = \frac{\mu_0 I}{4\pi\rho}\left\{\frac{L-z}{\sqrt{\rho^2+(L-z)^2}} + \frac{L+z}{\sqrt{\rho^2+(L+z)^2}}\right\} \tag{4.245}$$

Total magnetic flux density of n vertical downconductors through a loop can be determined by following algorithm. For a conductor i, it is possible to decompose the vector of flux density due to the downward flowing current into x and y components (Figure 4.110).

$$B_{xi} = B_i \frac{y-y_i}{\rho_i} \tag{4.246}$$

$$B_{yi} = -B_i \frac{x-x_i}{\rho_i} \tag{4.247}$$

where B_i denotes a magnitude of a vector $\vec{B}_i$, x_i, y_i are the coordinates of conductor i, x, y are coordinates of a point within the loop and

$$\rho_i = \sqrt{(x-x_i)^2 + (y-y_i)^2}.$$

Then, the total flux is expressed as the sums of contributions of all downconductors:

$$B_x = \sum_{i=1}^{n} B_{xi} \qquad B_y = \sum_{i=1}^{n} B_{yi} \tag{4.248}$$

$$\vec{B} = B_x\hat{x} + B_y\hat{y} = \vec{B}_x + \vec{B}_y \tag{4.249}$$

Finally, the induced voltage in the loop can be determined by numerical evaluation of the integral (4.242).

$$v_{in} = \frac{d}{dt} \oint_S (\vec{B}_x + \vec{B}_y) \cdot d\vec{S} \tag{4.250}$$

$$v_{in} = \frac{d}{dt} \oint_S (-B_x \sin\alpha + B_y \cos\alpha) dS \tag{4.251}$$

Generally saying the higher the number of downconductors, the lower induced voltage in loop is. It is caused by both the division of a lightning current and the mutual cancellation of magnetic fields of individual downconductors. The above-described analysis can be expanded to arbitrary-inclined downconductors and horizontal catenary wires.

References

1 "Requirements on Equipment, Systems, and Installations," § 23.1309(e) of Code of Federal Regulations 14CFR Part 23.
2 "High-intensity Radiated Fields (HIRF) Protection," § 23.1308 of Code of Federal Regulations 14CFR Part 23.
3 "Electrical and Electronic System Lightning Protection," § 23.1306 of Code of Federal Regulations 14CFR Part 23.
4 "The Certification of Aircraft Electrical and Electronic Systems for Operation in the High-intensity Radiated Fields (HIRF) Environment," Advisory Circular AC 20-158A, Federal Aviation Administration, 2014.
5 "Aircraft Electrical and Electronic System High-Intensity Radiated Fields (HIRF) Protection," Acceptable Means of Compliance AMC 20-158, European Aviation Safety Agency, 2015.
6 "Protection of Aircraft Electrical / Electronic Systems against the Indirect Effects of Lightning," Advisory Circular AC 20-136B, Federal Aviation Administration, 2011.
7 "Aircraft Electrical and Electronic System Lightning Protection," Acceptable Means of Compliance AMC 20-136, European Aviation Safety Agency, 2015.
8 "Protection Of Aircraft Fuel Systems Against Fuel Vapor Ignition Caused By Lightning," Advisory Circular AC 20-53B, Federal Aviation Administration, 2006.
9 "Guide To Certification Of Aircraft In A High-Intensity Radiated Field (HIRF) Environment" EUROCAE ED-107A (SAE ARP5583A), EUROCAE, 2010.
10 "Aircraft Lightning Environment and Related Test Waveforms," SAE ARP5412B (EUROCAE ED-84A), SAE International, 2013.
11 "Aircraft Lightning Zoning" SAE ARP5414A (ED-91 including amendments 1 and 2), SAE International, 2005.

12 "Aircraft Lightning Test Methods" SAE ARP5416A (ED-105A), SAE international, 2013.
13 S. A. Schelkunoff, "The electromagnetic theory of coaxial transmission lines and cylindrical shields," *Bell System Technical Journal*, vol. 13, Oct. 1934.
14 E. F. Vance, *Coupling to Shielded Cables*, John Wiley & Sons, 1979.
15 H. Kaden, *Wirbelströme und Schirmung in der Nachrichtentechnik*, Springer Verlag, 1959.
16 A. J. Schwab and W. Kürner, *Elektromagnetische Verträglichkeit*, Springer-Verlag, 2011.
17 C. Christopoulos, *Principles and Techniques of Electromagnetic Compatibility*, CRC Press, Taylor & Francis Group, 2007.
18 H. E. Knoepfel, *Magnetic Fields*, John Wiley & Sons, 2000.
19 H. A. Haus and J. R. Melcher, *Electromagnetic Fields and Energy*, Englewood Cliffs, NJ: Prentice-Hall, 1989.
20 K. L. Kaiser, *Electromagnetic Compatibility Handbook*, CRC Press, 2004.
21 A. K. Jain, *Basic Theory of Magnets*, RHIC Project, Brookhaven National Laboratory, Upton, New York 11973-5000, USA.
22 K. S. H. Lee et al., *EMP Interaction: Principles, Techniques and Reference Data*, Air Force Weapons Laboratory, Albuquerque, New Mexico, 1979.
23 D. A. Hill, *Electromagnetic Fields in Cavities*, Institute of Electrical and Electronics Engineers, John Wiley and Sons, 2009.
24 J. D. Jackson, *Classical Electrodynamics*, John Wiley & Sons, 1999.
25 F. M. Tesche, M. V. Ianoz, and T. Karlsson, *EMC Analysis Methods and Computational Models*, New York: John Wiley & Sons Inc. 1996.
26 A. A. Smith, *Radio Frequency Principles and Applications: The Generation, Propagation, and Reception of Signals and Noise*, John Wiley & Sons-IEEE Press, June 1998.
27 C. R. Paul, *Inductance—Loop and Partial*, John Wiley & Sons-IEEE Press, 2010.
28 E. F. Vance, "Nuclear electromagnetic pulse" in *Handbook of Electromagnetic Compatibility*, Academic Press, 1994.
29 F. A. Fisher, J. A. Plumer and R. A. Perala, *Lightning Protection of Aircraft*, LTI, 2004.
30 K. S. H. Lee and F. C. Yang, *A Wire Passing by a Circular Aperture in an Infinite Ground Plane*, Interaction Note IN317, ADA055583, AFWL-TR-77-52, 1977.
31 C. R. Paul, *Analysis of Multiconductor Transmission Lines*, John Wiley & Sons-IEEE Press, 2008.
32 C. R. Paul, *Introduction to electromagnetic compatibility*, 2nd ed., Wiley-Interscience, 2006.
33 F. M. Tesche, "Plane wave coupling to cables" in *Handbook of Electromagnetic Compatibility*, Academic Press, 1994.

34 C. A. Nucci and F. Rachidi, "On the contribution of the electromagnetic field components in field-to-transmission lines interaction," *IEEE Transactions on EMC*, vol. 37, No. 4, Nov. 4, 1995.

35 A. Tsaliovich,*Cable Shielding for Electromagnetic Compatibility*, Chapman & Hall, 1995.

36 T. Kley, "Optimized single-braided cable shields," *IEEE Transactions on EMC*, vol. 35, No. 1, Feb. 1993.

37 K.-H. Gonschorek and R. Vick, *Electromagnetic Compatibility for Device Design and System Integration*, Springer Science & Business Media.

38 D. A. Weston, *Electromagnetic Compatibility: Principles and Applications*, CRC Press, 2001.

39 "Electromagnetic compatibility (EMC)—Part 5: Installation and mitigation guidelines—Section 4: Immunity to HEMP—Specifications for protective devices against HEMP radiated disturbance. Basic EMC Publication," IEC/TS 61000-5-4 ed1.0, International Electrotechnical Commission, 1996-08-13.

40 Naval Air Warfare Center Aircraft Division (NAWCAD) Technical Memorandum, Report No. NAWCADPAX-98-156-TM, High-intensity Radiated Field External Environments for Civil Aircraft Operating in the United States of America (Unclassified), dated November 12, 1998.

41 "Radio Frequency Susceptibility Test Procedures" ED-90B, EUROCAE, 2010.

42 "Electromagnetic compatibility," Advisory Circular AC 21-53, Version 1.0, Civil Aviation Safety Authority of Australia, Airworthiness and Engineering Standards Branch, Dec. 2015.

43 Allied Environmental Conditions and Tests Publication AECTP 250 "Electrical and Electromagnetic Environmental Conditions," NATO International Staff—Defence Investment Division, Jan. 2011.

44 Defence Standard 59-411 Part 2 "Electromagnetic Compatibility—Part 2 -The Electric, Magnetic and Electromagnetic Environment" Issue 2, Defence Equipment and Support—UK Defence Standardization, Mar 31, 2014.

45 Department of Defense Interface Standard MIL-STD-464C "Electromagnetic Environmental Effects Requirements for Systems" U.S. Department of Defense, Dec. 1, 2010.

46 "Electromagnetic Compatibility in the Defense Systems of Future Years," RTO Technical Report RTO-TR-059, The Research and Technology Organisation of NATO, June 2002.

47 T. Hubing at al., *Survey of Current Computational Electromagnetics Techniques and Software* Technical Report CVEL-08-011.2, Clemson University, 2008.

48 A. Schroder et al., "Analysis of high intensity radiated field coupling into aircraft using the method of moments" in *IEEE Transactions on EMC*, vol. 56, Issue: 1, Aug. 2013.

49 J. Alvarez et al., "HIRF interaction with metallic aircrafts. A comparison between TD and FD methods" in *2012 International Symposium on Electromagnetic Compatibility (EMC EUROPE)*, Rome, Italy.

50 Z. Reznicek et al., "TD and FD simulations of internal EM environment in small aircraft and experimental test comparison," *6th European Conference on Antennas and Propagation (EUCAP)*, 2012.

51 C. Weber et al.,"Evaluation of complexity of wire harness models in a HIRF environment" in *2013 IEEE International Symposium on Electromagnetic Compatibility*, Denver, USA.

52 G. A. Rasek at al.,"HIRF transfer functions of a fuselage model: Measurements and simulations" in *IEEE Transactions on EMC*, vol.: 56, Issue: 2, Nov. 2013.

53 C. A. Balanis et al., *Penetration of High Intensity Radiated Fields (HIRF) Into General Aviation Aircraft*, Telecommunications Research Center, Arizona State University, USA, 2004.

54 Allied Environmental Conditions and Tests Publication AECTP 500 "Electromagnetic environmental effects test and verification," NATO International Staff—Defence Investment Division, Jan. 2011.

55 Defence Standard 59-411 Part 4 "Electromagnetic compatibility—Part 4: platform and system test and trials" Issue 2, Defence Equipment and Support—UK Defence Standardization, Mar 31, 2014.

56 G. A. Rasek and S. E. Loos, "Correlation of direct current injection (DCI) and free-field illumination for HIRF certification" in *IEEE Transactions on EMC*, vol.: 50, Issue: 3, Aug. 2008.

57 "Radio frequency susceptibility," Section 20 of Eurocae ED-14/ RTCA DO-160, Revision G.

58 Defence Standard 59-411 Part 3 "Electromagnetic compatibility—Part 3 -Test methods and limits for equipment and sub systems" Issue 2, Defence Equipment and Support—UK Defence Standardization, Mar. 31, 2014.

59 K. Javor, "On field-to-wire coupling versus conducted injection techniques" in *IEEE 1997 International Symposium on Electromagnetic Compatibility*, Austin, USA.

60 G. A. Rasek, M. Gabrišák "Wire bundle currents for High Intensity Radiated Fields (HIRF) and Indirect Effects of Lightning (IEL) with focus on Bulk Current Injection (BCI) test," *21st International Conference Radioelektronika*, Brno, Czech Republic, 2011.

61 K. Javor, "(More) On field-to-wire coupling versus conducted injection techniques" in *IN Compliance magazine*, Oct. 2014.

62 Department of Defense Interface Standard MIL-STD-461G "Requirements for the control of electromagnetic interference characteristics of subsystems and equipment" U.S. Department of Defense, Dec. 11, 2015.

63 V. Cooray, *An Introduction to Lightning*, Springer Science+Business Media, 2015.

64 F. Heidler and K. Stimper, *Blitz und Blitzschutz*, VDE VERLAG GMBH, 2009.

65 S. Prentice and D. Mackerras, "The ratio of cloud to cloud-ground lightning flashes in thunderstorms," Journal of Applied Meteorology, vol. 16, No.5, 1977.

66 E. Defer and P. Laroche, "Observation and interpretation of lightning flashes with electromagnetic lightning mapper" in *Lightning: Principles, Instruments and Applications*, Springer Science+Business Media B.V., 2009.

67 V. A. Rakov and M. A. Uman, *Lightning—Physics and Effects*, Cambridge University Press, 2007.

68 E. Rupke, *Lightning Direct Effects Handbook*, Report Reference Number: AGATE-WP3.1-031027-043-Design Guideline, Lightning Technologies Inc., 2002.

69 *Analysis of Experimental Data and Models for Upgraded Lightning Protection Requirements (FULMEN)*, Fourth Framework Program, Transport, RTD Programme, 1996.

70 *Lightning Direct Effects*, Section 23 of Eurocae ED-14/ RTCA DO-160, Revision G.

71 V. Cooray et al., *The Lightning Flash*, The Institution of Electrical Engineers, 2003.

72 K. Elrodesly, *Comparison Between Heidler Function And The Pulse Function For Modeling The Lightning Return-Stroke Current*, Ryerson University, Toronto, Ontario, Canada, 2010.

73 *Lightning Induced Transient Susceptibility*, Section 22 of Eurocae ED-14/ RTCA DO-160, Revision G.

74 R. A. Perala,T. Rudolph, and F. Erolsen, "electromagnetic interaction of lightning with aircraft" in *IEEE Transactions on EMC*, Vol. 24, Issue: 2, May 1982.

75 J.-P. Parmantier, F. Issac, and V. Gobin, "Indirect effects of lightning on aircraft and rotorcraft" in *Lightning Hazards to Aircraft and Launchers*, Aerospace Lab Journal, Issue 5, Dec. 2012.

76 M. Apra et al., *Lightning Indirect Effects Certification of a Transport Aircraft by Numerical Simulation* in *IEEE Transactions on EMC*, vol. 50, Issue: 3, Aug. 2008.

77 J. E. Nanevicz, *Static Charging and Its Effects on Avionic Systems*, IEEE Transactions on Electromagnetic Compatibility, vol. EMC-24, no. 2, pp. 203–209, May 1982.

78 *American National Standard Dictionary of Electromagnetic Compatibility (EMC) including Electromagnetic Environmental Effects (E3)*, ANSI C63.14-2009, Oct. 23, 2009.

79 "Aircraft precipitation static certification," SAE ARP5672 (ED-152), SAE international, 2009.

80 British Standard Aerospace Series BS G 257: Part 1 : 1998 *Design of Electromagnetic Hazard Protection of Civil Aircraft, Part 1. Guide to Theory and Threats.*

81 R. A. Perala, *A Critical Review of Precipitation Static Research Since the 1930s and Comparison to Aircraft Charging by Dust*, Electro Magnetic Applications, Inc., Denver, Colorado, USA.

82 J.-P. Moreau, J.-C. Alliot, and V. Mazur, *Aircraft Lightning Initiation and Interception From In Situ Electric Measurements and Fast Video Observations* in Journal of Geophysical Research Atmospheres 97(D14):15903, Oct. 1992?

83 T. H. Shumpert, "Capacitance calculations for satellites, Part 1. isolated capacitances of ellipsoidal shapes with comparisons to some other simple bodies," Sensor and Simulation Notes, Note 157, Sep. 1972.

84 R.L. Tanner and J. E. Nanevicz, *Precipitation Charging and Corona Generated Interference in Aircraft*, Stanford Research Institute, USAFCRL Contract AF 19(604)-3458, Apr. 1961.

85 Department of Defense Detail Specification MIL-DTL-9129G, *Dischargers, Electrostatic, General Specification for*, Jun. 19, 2014.

86 G. C. HUANG et al., *Interference Characteristics of Streamer Discharges*, IEEE Transactions on Electromagnetic Compatibility, vol. EMC-12, no. 2, pp. 54–63, May 1970.

87 Department of Defense Interface Standard MIL-STD-331C "Fuze and Fuze components, environmental and performance tests for" U.S. Department of Defense, Jan. 5, 2005.

88 Allied Environmental Conditions and Tests Publication AECTP 250 "Electrical and electromagnetic environmental conditions," NATO International Staff—Defence Investment Division, Jan. 2011.

89 H. Z. Fu et al., *Analysis of Corona Discharge Interference on Antennas on Composite Airplanes*, IEEE Transactions on Electromagnetic Compatibility, vol. 50, no. 4, pp. 822–827, Nov. 2008.

90 CAP 426 *Helicopter External Load Operations*, Safety Regulation Group, UK Civil Aviation Authority, 2006.

91 TCREC Technical Report 62-33 *Helicopter Static Electricity Discharging Device*, Willow Grove, Pennsylvania: Kellett Aircraft Corporation, 1962.

92 NRL Memorandum Report 5676 *Electrostatic Charging of the CH-53E Helicopter*, Washington, D.C.: Naval Research Laboratory, Nov. 29, 1985.

93 NASA/TM-1999-209734 *Lightning Protection Guidelines for Aerospace Vehicles*, National Aeronautics and Space Administration, 1999.

94 B. D. Fisher, "Effects of lightning on operations of aerospace vehicles" in *AGARD Flight Mechanics Panel Symposium on Flight in Adverse Environmental Conditions*, Gøl, Norway, May 8–11, 1989.

95 H. J. Christian et al., "The Atlas/centaur lightning strike incident" in *Journal of Geophysical Research*, vol. 94, Issue D11, Sep. 30, 1989.

96 NASA/SP-2010-216283 *A History of the Lightning Launch Commit Criteria and the Lightning Advisory Panel for America's Space Program*, NASA, Aug. 2010.

97 14CFR Part 417 *Launch Safety–Appendix G to Part 417—Natural and Triggered Lightning Flight Commit Criteria*, Electronic Code of Federal Regulations, 2015.

98 AIAA Standard S-121-2009 *Electromagnetic Compatibility Requirements for Space Equipment and Systems*, American Institute of Aeronautics and Astronautics, Sep. 2009.

99 ECSS-E-ST-20-07C *Space Engineering—Electromagnetic Compatibility Standard*, ESA Requirements and Standards Division, European Space Agency, Rev. 1, Sep. 5, 2012.

100 ECSS-E-HB-20-07A *Space Engineering—Electromagnetic Compatibility Handbook*, ESA Requirements and Standards Division, European Space Agency, Rev., 1 7 Feb. 2012.

101 M.B McCollum, S. R. Jones, and J. D. Mack, "NASA Manned Launch Vehicle Lightning Protection Development" in *2009 International Conference on Lightning and Static Electricity*, Pittsfield, USA.

102 ISO 14302, *Space systems Electromagnetic Compatibility Requirements*, International Organization for Standardization, 2002.

103 B.C. Gabrielson *Lightning Protection for Rockets*, 1982.

104 F. A. Fisher *Lightning Protection of Launch Facilities at Kennedy Space Center*, Environmental Electromagnetics Unit, Corporate Research and Development, General Electric Company, 1973.

105 NASAfacts FS-2005-10-031-KSC *Lightning and the Space Program*, National Aeronautics and Space Administration, 2006.

106 IEC 62305-3:2010 "Protection against lightning—Part 3: Physical damage to structures and life hazard," International Electrotechnical Commission, 2010.

107 V. Cooray and M. Becerra "Attachment of Lightning Flashes to Grounded Structures" in *Lightning Protection*, The Institution of Engineering and Technology, 2010.

108 IEC 62305-1:2010 "Protection against lightning—Part 1: General principles," International Electrotechnical Commission, 2010.

109 T. Horvath "The protected space proved to be an undefined term" in *2012 International Conference on Lightning Protection (ICLP)* Vienna, Austria, 2012.

110 F. Issac et al. "Space Launching Site Protection against Lightning Hazards" in *Aerospace Lab Journal*, Issue 5, Dec. 2012.

111 U. Kumar, "Lightning protection of satellite launch pads" in *Lightning Protection*, The Institution of Engineering and Technology, 2010.

112 E.Bachelier et al, *Lightning Protection of SOYUZ and VEGA Launching Pads*, International Conference on Lightning Protection (ICLP), Vienna, Austria, 2012.

113 R. C. Scully, *Lightning Protection for the Orion Space Vehicle*, 7th AIAA Atmospheric and Space Environments Conference. Dallas, TX., 2015.

5

Techniques to Design Robust Lightning Protection Circuits for Avionics Equipment

Dr. Clay McCreary

5.1 Introduction

This chapter presents techniques for the design of robust lightning protection circuits for electronic equipment on aircraft (avionics) provided in the context of a handbook.

In addition to providing techniques for the initial design of lightning protection circuits, methods for evaluating existing circuits against new requirements and designing additional protection if needed to meet the new requirements are provided.

Tools exist to assist in the design process. These tools are free of charge and facilitate the design and/or the evaluation of an existing suppression circuit. These tools and directions for their use are provided.

5.2 Clean Sheet Design

When starting a *clean sheet* design, the following steps should be followed to create a robust design:

- The transients to which the interface will be tested must be defined
- Component values are then selected
  - Define the clamping voltage of the protection circuit
  - Define the series impedance of the protection circuit
- Determine potential components by simulating the test with a circuit simulator (i.e., PSPICE)
- Use the waveform transformation algorithm to select the final components from the potential components

Handbook of Aerospace Electromagnetic Compatibility, First Edition. Edited by Reinaldo J. Perez.

- Calculate the minimum trace width
- If no components can be found that will tolerate the test transient and provide sufficient protection, design a multiple-stage protection scheme

5.2.1 Test Waveforms

The first step in designing a protection circuit is to determine the transient against which the circuit will be protecting. This is usually provided during requirements capture.

In avionics, DO-160 [1] or the equivalent test standard for the applicable airworthiness authority requires the indirect effects of lightning be tested using two methods: cable induction and pin injection. Cable induction uses transformers to couple the transient into cable bundles connected to the avionics. This is a functional test, meaning that the avionics must operate during the application of the transient or recover without human intervention. Pin injection applies the transient directly to each signal line. This is a damage test, meaning that the avionics must not have physical damage as a result of the test nor may its functionality be degraded. Pin injection is the more severe of the two types of signal injection. This chapter concentrates on designing for pin injection testing because experience has shown that if the protection circuit is robust to pin injection, it will be sufficient for the cable induction.

For DO-160 [1] testing, there are two types of transients used for pin injection testing: double exponential given by equation (5.2) and damped sinusoid. The double-exponential waveform is referred to as either waveform 4 (WF4), which has a 6.4 μs rise time and pulse width of 69 μs, or waveform 5A (WF5A), which has a rise time of 40 μs and a pulse width of 120 μs. The damped sinusoid is waveform 3 (WF3). The levels for these transients are expressed in the form of the open circuit voltage (V_{OC}) and short circuit current (I_{SC}). Defining the level in this manner provides both the peak voltage and the source impedance of the transients. Typical source impedances are:

- WF3 – 25 Ω
- WF4 – 5 Ω
- WF5A – 1 Ω

The testing or simulation the aircraft manufacturer uses to derive the transients and levels to which each signal line is to be tested is defined in references [2] and [3]. This is beyond the scope of this handbook, but reference [4] provides a great deal of background, theoretical information concerning this testing should the reader be interested.

Due to the source impedance and the shapes of the waveforms, the double-exponential waveforms shown in Figure 5.1 and Figure 5.2 impart more energy in the form of current to the equipment under test (EUT). The peak voltage level of WF3 shown in Figure 5.3 is typically double that of WF4 or WF5A, so the

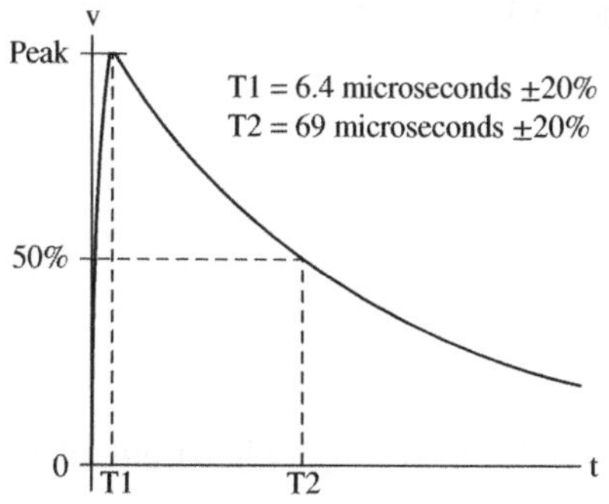

Figure 5.1 WF4 from Section 22 of [1].

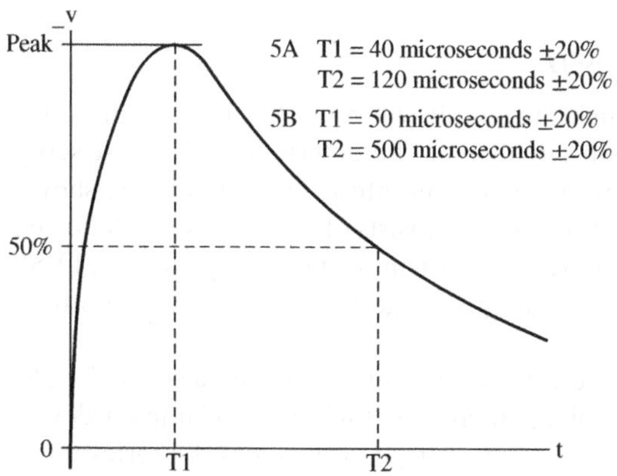

Figure 5.2 WF5A from Section 22 of [1].

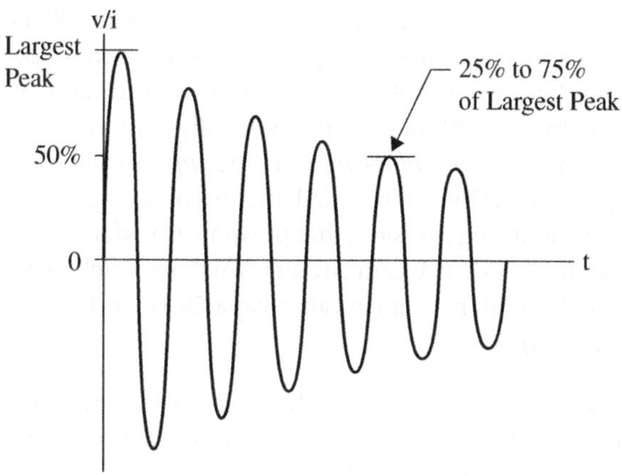

Figure 5.3 WF3 from Section 22 of [1].

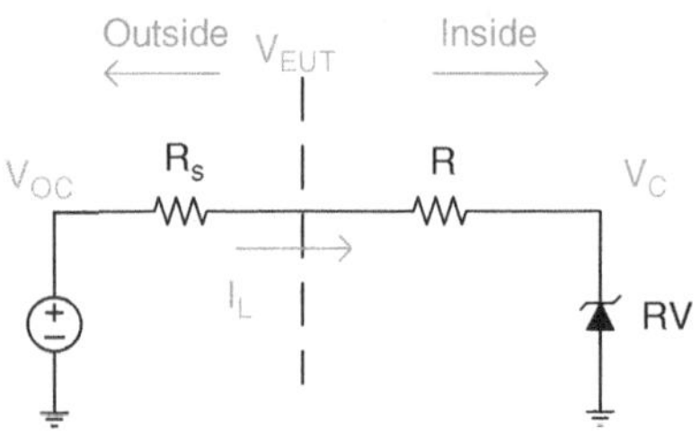

Figure 5.4 Basic protection circuit.

energy imparted to the EUT in the form of voltage, creating a higher potential for electrical overstress to the components.

5.2.2 Protection Circuit Topology

After the transient against which the circuit will protect is known, the component values of the protection circuit must be determined. The test setup external to the EUT and the protection circuit internal to the EUT are shown in Figure 5.4. The typical protection circuit consists of a series resistor followed by a shunt voltage-clamping device. Transient voltage suppressors (TVS), metal oxide varistors (MOV), or gas discharge tubes (GDT) are common voltage-clamping devices.

The voltage-clamping device adjusts the current to maintain a constant voltage. As the applied transient voltage increases, the voltage-clamping device draws more current, resulting in a larger voltage drop across the series resistor keeping the voltage across the voltage-clamping device to a predetermined level. This clamping action limits the voltage to which the downstream circuits are exposed during the lightning transient.

In Figure 5.4, the *R* is a generic symbol for the series impedance and the *RV* is generic for the voltage-clamping device. These components do not have to be resistors or TVSs; nor do both have to be included in the circuit. The value of *R* and the current sinking capability of *RV* determine the configuration of the circuit between the extreme ranges of *pure blocking* and *pure suppression.*

For a *pure blocking* configuration, *RV* is not included, the reason being that *R* presents high impedance to the lightning transient that prevents transfer of sufficient energy to cause damage to the downstream circuits. This can be achieved through the use of resistors with a high resistance value, capacitors, and transformers (not connected to ground):

- If neither end of the transformer coil connected to the interface is connected to ground, the pin-injected lightning transient does not have a return path, so there is high impedance preventing current flow.
- The highest frequency of the lightning test transients is 10 MHz. The double exponentials are <2 kHz. Therefore, the capacitive reactance of a series

capacitor will be a high impedance to the lightning transients, the worst case being 10 MHz.

For a *pure suppression* configuration, *R* is not included. *RV* is capable of sinking enough current that R_S drops enough voltage to allow for effective voltage clamping. This is not an optimum design practice because the actual impedance on the aircraft may be different than that used for testing, so it should be avoided if possible.

Most lightning protection circuits are going to be somewhere in between *pure blocking* and *pure suppression.* Thus, they will include both *R* and *RV.* Therefore, methods for selecting both components are provided.

5.2.3 Component Selection Overview

The challenge of designing protection circuits that can provide sufficient protection in a more compact form factor can be broken down into two parts: selecting the optimum components and determining the minimum trace width from the connector to the protection circuit required to tolerate the transient without fusing. The details about the development of the techniques presented in this chapter can be found in references [5–7].

When designing the protection circuit, the components that are to be protected must be evaluated to determine the voltage to which the voltage-clamping device must clamp in order to provide sufficient protection. Components for providing the required clamping are then selected. Lastly, the components selected for the protection circuit must be evaluated to determine if they can tolerate the required lightning test transients.

After the required clamping voltage is determined and potential components are selected, simulate the lightning transients applied to the protection circuit and the circuit being protected using a circuit simulator. Then, using data from the simulation, the transients to which each component in the protection circuit are exposed are transformed into all the reference transients typically used by the component datasheets.

The last step of this process is not obvious. Intuitively, the clamping voltage should simply be provided on the datasheet along with the power rating. This is true, but the transients used to derive these ratings must be considered. These transients are very different than the transients to which the components will be exposed during the application of the lightning test transients. References [8–13] are examples of datasheets for the components used in lightning protection circuits.

The simulation verifies that the protection circuit provides adequate protection. The transformation allows components to be selected based on direct comparison (aka apples to apples) to the ratings on the datasheets.

5.2.4 Select Component Values

5.2.4.1 Voltage-Clamping Device (*RV*)

The first step in designing a lightning protection circuit is to look at the datasheets of the components being protected to determine the amount of clamping that is required. Typically, it is desirable to clamp the voltage to less than the *absolute maximum value* listed on the datasheet. *Absolute maximum values* are provided as steady-state values in most cases. Therefore, exceeding them by a small amount may be allowable since the lightning transients are short pulses. The amount by which they can be exceeded is determined either through experience or experimentation. However, it is easier and more desirable to simply clamp the voltage lower than the *absolute maximum value.*

The clamping voltage of the voltage-clamping device typically varies with current. There are graphs of this dependence available on the datasheets, but they are difficult to use. Therefore, the use of a circuit simulation tool covered in Section 5.2.5 is the preferred method for determining the actual clamping voltage during the application of the lightning test transients.

5.2.4.2 Series Impedance (*R*)

The series impedance is typically driven by other requirements such as controlled impedance lines, AC coupling, and current limiting. As mentioned above, *blocking* configurations are preferred. Therefore, the maximum allowable series impedance should be selected:

- If the signal is a differential signal that can use a transformer, and there is physical space for it, the transformer should be used.
- If the signal for the interface being protected can be AC coupled with a series capacitor that presents high impedance to signals 10 MHz and lower, the capacitor should be used.
- If the signal is DC or other requirements drive the use of a series resistor, use the maximum allowable resistance.

5.2.5 Simulation

The circuit simulation tool preferred by the reader should be used. Models for the voltage-clamping devices are typically available on vendor websites in a variety of configurations (SPICE models appear to be the most prevalent). The simulation should include the circuit in Figure 5.4 followed by the circuit being protected or a resistor representing the impedance of the circuit being protected. There are instances in which the impedance of the circuit being protected is low enough to preclude the voltage-clamping device from activating (i.e., the protected circuit clamps the voltage below the clamping voltage of the device). In those instances, some series impedance between the protection circuit and the circuit being protected may have to be added. In addition, including the circuit being protected provides parameters of the resulting transient applied to that circuit to verify that the voltage is clamped to the appropriate value.

If VSRC is used as the voltage source for the circuit in PSPICE, the transient parameters to model the waveforms are as follows:

- WF3
  - sin(V_{offset} V_{Amp} Freq TD θ)
    - V_{offset} is a DC offset and is set to 0
    - V_{Amp} is the peak value of the waveform and is set to V_{OC}
    - Freq is the frequency of the sinusoid and is typically 1e^6 (1 MHz)
    - TD is the time delay before application of the transient and is usually set to 0
    - θ is the damping factor and is set to 250,000 for a 1 MHz sinusoid and 2,500,000 for a 10 MHz sinusoid
- WF4
  - exp(V_1 V_2 TD$_1$ τ_1 TD$_2$ τ_2)
    - V_1 is the starting voltage, essentially a DC offset and is set to 0
    - V_2 is the peak voltage and is set to V_{OC}
    - TD$_1$ is the time delay before the rise time and is set to 0
    - τ_1 is the time constant for the rise time and is 1.28e^{-6} or 1.28 u
    - TD$_2$ is the time delay until the fall time and is set to 6.4e^{-6} or 6.4 u
    - τ_2 is the fall time constant and is set to 90e^{-6} or 90 u
- WF5A
  - exp(V_1 V_2 TD$_1$ τ_1 TD$_2$ τ_2)
    - V_1 is the starting voltage, essentially a DC offset and is set to 0
    - V_2 is the peak voltage and is set to V_{OC}
    - TD$_1$ is the time delay before the rise time and is set to 0
    - τ_1 is the time constant for the rise time and is 8e^{-6} or 8 u
    - TD$_2$ is the time delay until the fall time and is set to 40e^{-6} or 40 u
    - τ_2 is the fall time constant and is set to 115e^{-6} or 115 u

τ_1 is calculated by dividing T1 from Figure 5.1 or Figure 5.2 by 5. τ_2 is found using equation (5.1):

$$\tau_2 = \frac{T2 - T1}{\ln(0.5)} \tag{5.1}$$

Equation (5.1) and dividing the rise time by 5 allows for the parameters of double-exponential waveforms other than WF4 or WF5A to be determined by the designer.

5.2.6 Waveform Transformation

In addition to the voltage ratings required of the components based on the voltage drops determined by the simulation, the component current/power rating must be determined. Larger components can dissipate more power, but they occupy more physical space. Therefore, the smallest component that will tolerate both the voltage and current/power to which it will be exposed is desirable. This section outlines the waveform transformation techniques that convert the current/power transients to which each of the components are exposed into those used on datasheets to rate the amount of current/power each size component can tolerate.

The parameters determined in the previous sections are needed for the waveform transformations. To review, they are:

- Series resistance of R
- Clamping voltage of RV (determined by simulation)
- Return line impedance (optional)
- Waveform type/number
- V_{OC}
- I_{SC}

Inputting these parameters into the waveform transformations yields the following parameters:

- Resistor rectangular pulse width
- Resistor peak power
- Resistor energy
- 8×20 μs (8 μs rise time, 20 μs pulse width) current for protection device
- TVS double-exponential pulse width
- TVS double-exponential peak power

There is a tool to perform all the calculations for component size selection that is shown in Figure 5.5 and may be downloaded at the links provided in reference [14] or [15]. Reference [14] requires the user to have MATLAB installed, and reference [15] is a standalone executable file that does not require MATLAB be installed.

To use the tool, input the values on the left side and press the "Calculate" button. The values on the right side are then calculated and displayed. Sections 5.2.7 through 5.2.10 explain how to use the displayed values to select components. Section 5.2.11 explains how to use the trace width that is output, and it also explains the "copper weight" input.

Impedance of the test cables (return line impedance) may also be included, but is not required. The return line impedance is a characteristic of the test setup. A typical method of compensating for return line impedance is to increase V_{OC}, within the $\pm 10\%$ tolerance allowed by DO-160 [1], during the short circuit component of the calibration procedure until I_{SC} is achieved. This method of compensating for the return line impedance results in testing above the nominal requirement for the avionics, so including this parameter as 1/10 of the source impedance will provide a *worst-case* scenario.

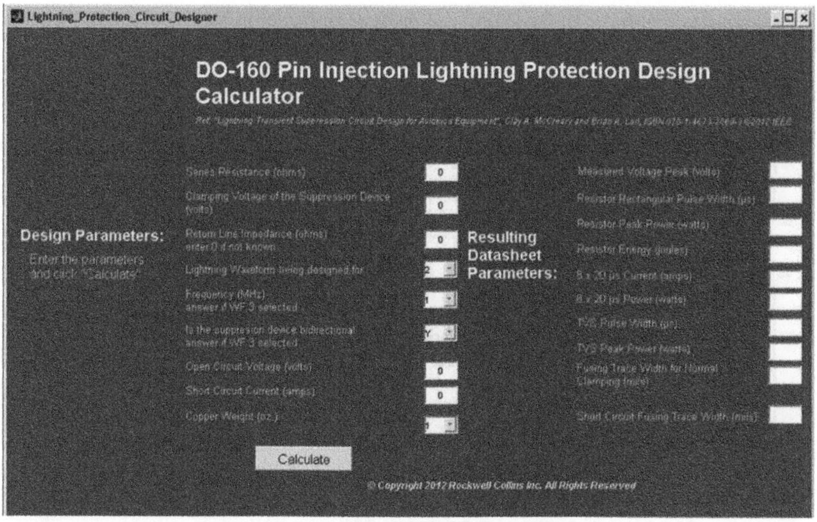

Figure 5.5 Lightning protection circuit design tool GUI [14, 15].

5.2.6.1 Installation and Setup of the Design Tool

Reference [14] requires that MATLAB version R2009b or later be installed on the end users' computer. Two files are contained in a .zip file download: Lightning_Protection.m and Lightning_Protection.fig. Extract and save these files in the working directory being used in MATLAB as shown in Figure 5.6.

After the files are placed in the working directory, run the GUI either by right clicking on Lightning_Protection.m and selecting "Run File" or by typing "Lightning_Protection" on the command line. The Design Tool shown in Figure 5.5 will appear on the screen.

If MATLAB is not installed or the engineer would just like the convenience of obtaining the tool as a standalone executable file, this tool may also be found at the link provided in reference [15]. The package that may be downloaded is a self-extracting .zip file that will automatically install the MATLAB compiler runtime (MCR) if it is not installed on the engineer's computer. A prompt is received to let the installer run, so if MCR or MATLAB is already installed, do not allow the installer to run. In addition to installing MCR, the self-extracting zip file will extract LightningProtection.exe. Simply run the file, and the design tool will appear in the desired folder ready for use.

5.2.6.2 Transient Current Hand Calculation

This handbook also explains the calculations performed by the tool in case the designer does not have access to the tool and must perform the calculations by hand.

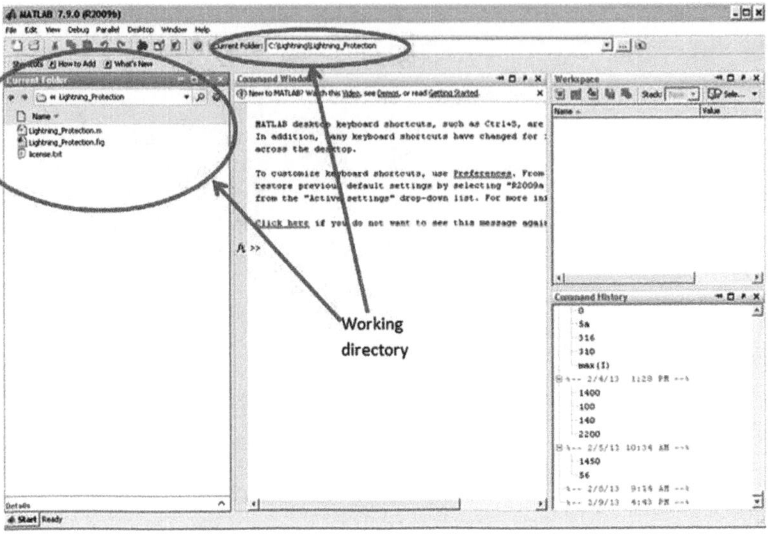

Figure 5.6 MATLAB screen showing working directory.

The first step in the calculations performed by the tool is to determine the current transient that results from the lightning voltage waveform being applied to the protection circuit. The tool assumes an open circuit following the protection circuit. This is the worst-case condition as all of the energy is dissipated by the protection circuit. The current transient is based on the source impedance, the resistance of the series resistor, the clamping voltage (V_{clamp}) of the voltage-clamping device, and the return line impedance (if included). After the current transient is calculated, it is transformed into the reference waveforms used on datasheets. The transformations for each reference transient are covered in the respective components section.

Equation (5.2) is the equation for the double-exponential voltage or current waveform and will be referred to for either throughout the remainder of this chapter. The voltage waveform is applied to the circuit. Equations (5.3) and (5.4) provide the coefficients for WF4 and WF5A that were determined empirically using MATLAB. The result of equation (5.2) is used in equation (5.5) to create $V_{clamp}(t)$. Then, the results of both equations (5.2) and (5.5) are used in equation (5.6) to calculate the resultant $i(t)$.

$$\begin{aligned} v(t) &= V_0(e^{-\alpha t} - e^{-\beta t}) \\ i(t) &= I_0(e^{-\alpha t} - e^{-\beta t}) \end{aligned} \tag{5.2}$$

$$WF4: \ \alpha = 11{,}300\text{s}^{-1} \ \beta = 670{,}000s^{-1} \ V_0 = \frac{V_{OC}}{0.9166} \tag{5.3}$$

$$WF5A: \ \alpha = 13{,}000\text{s}^{-1} \ \beta = 41{,}000s^{-1} \ V_0 = \frac{V_{OC}}{0.4006} \tag{5.4}$$

where

$v(t)$ = resultant waveform as a function of time
V_0 = scaling coefficient (not the peak value)
α = coefficient most closely related to the decay portion of the waveform
β = coefficient most closely related to the rise portion of the waveform
t = time in seconds

$$V_{Clamp}(t) = \begin{cases} v(t) \text{ for } v(t) < V_{Clamp} \\ V_{Clamp} \text{ for } v(t) \geq V_{Clamp} \end{cases} \tag{5.5}$$

$$i(t) = \frac{v(t) - V_{Clamp}(t)}{Z_{SRC} + R_{series} + Z_{ret}} \tag{5.6}$$

where

$i(t)$ = current transient as a function of time
Z_{SRC} = source impedance
R_{series} = resistance of the series resistor in the protection circuit
Z_{ret} = return line impedance

5.2.7 Series Resistor

The outputs of the design tool shown in Figure 5.5 used for resistor selection are:

- Resistor rectangular pulse width
- Resistor peak power
- Resistor energy

If using a metal film resistor, select one in which the pulsed power handling capabilities are specified. The pulsed power handling capability is typically provided with a Wunsch–Bell curve shown in Figure 5.7. This is a graph of peak power (P_{PK}) versus *t*. Here, *t* is the pulse width of a rectangular pulse. P_{PK} is the peak power dissipated by the resistor. To determine if the resistor can tolerate the specified lightning transient applied to the protection circuit, plot the point (t, P_{PK}) on the Wunsch–Bell curve. Each limit line on the curve represents the peak power limit for a given size resistor. Therefore, any size resistor having limit lines greater than the point (t, P_{PK}) may be used in the circuit.

If wirewound resistors are used, the resistor energy output of the tool is used to select the appropriate size resistor. The following thumb rule should be used: $E_{resistor} < 0.72\,\text{J} \times$ *resistor power rating*. For instance, a 1 W wirewound resistor can be used for resistor energies up to 0.72 J, 2 W resistor can be used up to 1.44 J, ½ W up to 0.36 J, and so on. There are no documented studies to substantiate this thumb rule. However, it is being presented in the context of

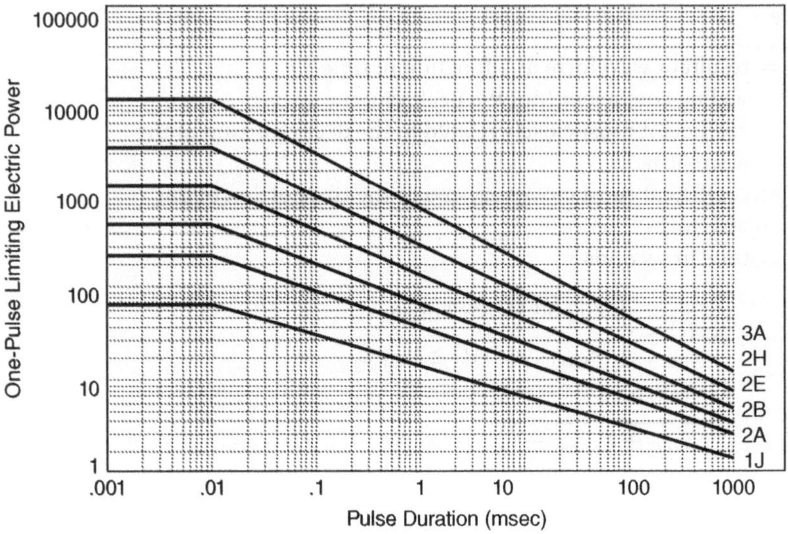

Figure 5.7 Wunsch–Bell curve from KOA SPEER SG73 resistor family [9].

this handbook because it has been shown to produce satisfactory results. This thumb rule has been used for over 10 years in thousands of protection circuits without a failure.

5.2.7.1 Calculations Being Performed by the Design Tool

The resistor energy is calculated using equation (5.7) and the coordinates (t, P_{PK}) are calculated using equations (5.8) and (5.9).

$$E_{\text{resistor}} = \int i^2(t) R dt \tag{5.7}$$

$$P_{PK} = \max(i^2(t) R) \tag{5.8}$$

$$t = \frac{E_{\text{resistor}}}{P_{PK}} \tag{5.9}$$

5.2.7.2 Example

The following is an example for a typical lightning protection circuit. All of the parameters in Table 5.1 must be known prior to design.

Inputting the parameters from Table 5.1, the outputs of the design tool in Figure 5.5 [also the results of the algorithm using equations (5.2) through (5.9)] are:

$$t_p = 82\ \mu s$$
$$P_{pk} = 2612\ W$$
$$E_{\text{resistor}} = 0.21\ J$$

Table 5.1 Series resistor example parameters

Parameter	Value
Series resistance	36.5 Ω
Clamping voltage	12 V
Return line impedance	0.1 Ω
Waveform	5A
Open circuit voltage	300 V
Short circuit current	300 A

Using a basic Kirchoff voltage loop at the peak of the transient or the results from the circuit simulator to find the peak voltage across the resistor reveals:

$$V_{\text{PKresistor}} = 288\ \text{V}$$

From these results, a ½ W wirewound resistor may be used based on the thumb rule: 0.72 J per 1 W power rating (i.e., 0.36 J for a ½ W wirewound resistor) provided the voltage rating is greater than 288 V. A 1 W SG73 resistor is also suitable since the coordinates resulting from the algorithm plotted on Figure 5.8 are below the 1 W rating curve identified by the alphanumeric designator 3A, and the *absolute maximum overload voltage* for a 1 W SG73 resistor is 400 V.

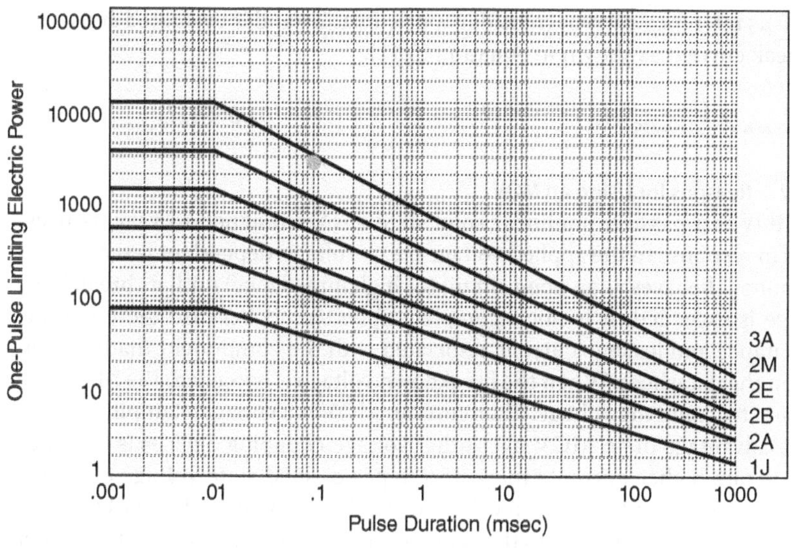

Figure 5.8 Plot of example coordinates.

5.2.8 MOV

The output of the design tool shown in Figure 5.5 used for MOV selection is the 8 × 20 μs current.

The peak current rating for an MOV is typically provided on the datasheet using an 8 × 20 μs reference transient. The designer should verify that the datasheet in question uses the 8 × 20 μs reference transient to rate peak current. After its use is verified, the 8 × 20 μs current output from the design tool shown in Figure 5.5 can be directly compared with the values for peak current shown in the datasheet of the MOV.

5.2.8.1 Calculations Being Performed by the Design Tool

Equations (5.10) and (5.11) provide the calculations required to determine the peak 8 × 20 μs equivalent to the current transient experienced by the MOV during the application of the lightning test transient. Using $i(t)$ calculated by equations (5.2) through (5.6), equation (5.10) calculates the scaling coefficient of equation (5.2) to approximate the 8 × 20 μs reference transient:

$$I_{0\_8\times 20} = \frac{\int i(t)dt}{\int (e^{-120000t} - e^{-140000t})dt} \tag{5.10}$$

where $I_{0\_8\times 20}$ is the scaling coefficient of the 8 × 20 μs reference transient similar to the transient experienced by the MOV during the application of the lightning test transient.

After the scaling coefficient is found, it is used in equation (5.2) for the 8 × 20 μs reference current transient approximation and is evaluated at 8 μs for the peak current as shown in equation (5.11).

$$I_{\max\_8\times 20} = I_{0\_8\times 20}[e^{-120000(8\ \mu s)} - e^{-140000(8\ \mu s)}] = 0.056613 I_{0\_8\times 20} \tag{5.11}$$

5.2.8.2 Reasons for Using an MOV

An MOV is a passive, nonlinear solid state device designed to vary its resistance in response to the applied voltage (i.e., a voltage-dependent resistor). The component behaves as an open circuit (or high impedance) until its breakdown voltage is exceeded. It is not a complete open circuit because there is some capacitance and leakage current associated with the component that is specified on the datasheet. When the breakdown voltage is exceeded, the MOV will conduct to clamp the voltage. Therefore, an MOV is a voltage-clamping device.

The advantage of an MOV over other voltage-clamping devices is that they can tolerate higher currents than other types of solid state voltage-clamping devices with similar breakdown voltages. The disadvantage is that the clamping voltage of an MOV varies with the applied current to a greater extent than other voltage-clamping devices. This makes MOVs preferable for applications that

Figure 5.9 Size comparison of traditional and new MOVs with similar performance characteristics.

can tolerate a higher variance in clamping voltage and must have a low value of series resistor.

In the past, MOVs typically were physically large with a relatively large capacitance. The large capacitance and the inductance of the leads made the MOVs slow to react to the transient. These characteristics made them undesirable for many applications. Recent advances in MOV technology have reduced the size of the components greatly as shown in Figure 5.9. In addition, the removal of the leads has lowered inductance, and new configurations of the component have reduced capacitance. This has reduced the reaction time of the component to nanoseconds. For lightning protection, this can be considered instantaneous.

5.2.8.3 Example

Using the parameters from Table 5.1, the peak 8 × 20 μs current output from the design tool shown in Figure 5.5 is 49 A. Equations (5.2)–(5.11) yield the same result.

This value may be used to select a suitable MOV. The MOV selected must have an 8 × 20 μs current rating that is greater than 49 A. V5.5MLA1206N from reference [12] may be a suitable selection. However, the actual clamping voltage using this component should be found using a circuit simulator. PSPICE simulation using this component reveals that the peak clamping voltage is 13.2 V. Therefore, the 8 × 20 μs current must be reevaluated by replacing 12 V with 13.2 V as the clamping voltage for the input to the design tool shown in Figure 5.5. This has little effect on the 8 × 20 μs current; it changes to 48 A. Thus, the component is suitable for this application.

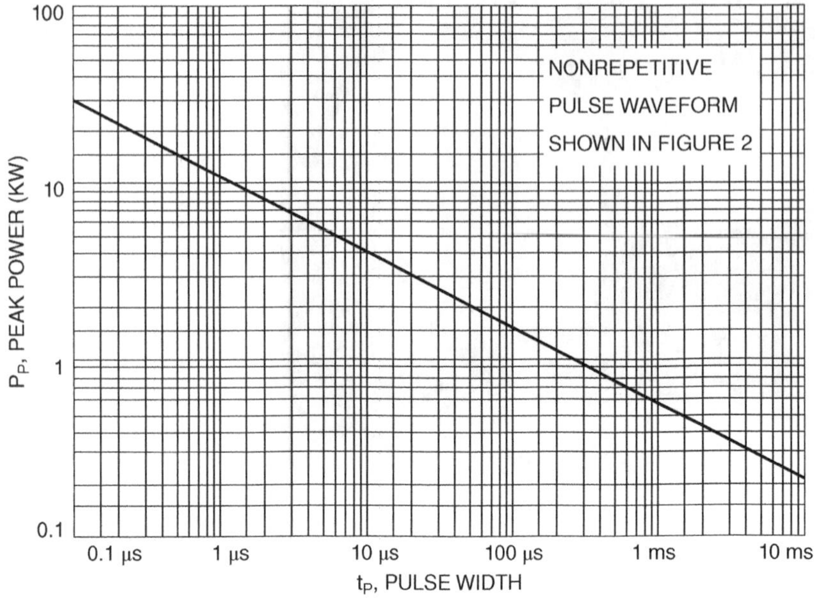

Figure 5.10 TVS Wunsch–Bell curve taken from reference [8].

If the circuit can tolerate this higher peak voltage, then the component is suitable, but if not, a TVS may have to be used.

5.2.9 TVS

The outputs of the design tool shown in Figure 5.5 used for TVS selection are:

- TVS pulse width (t_p)
- TVS peak power (P_P)

TVS datasheets, like reference [8], typically contain a peak power rating in the title or in the "Features" section. A power of 400 W is used in reference [8]. This rating is not considered for lightning protection circuit design. Instead, as with the resistor, the Wunsch–Bell curve, shown in Figure 5.10, is used. This curve is typically developed for the TVS using a double-exponential *current* transient with 10 μs rise time and variable pulse width. The designer should verify that this is how the curve was developed for the potential part prior to continuing.

Once the designer has verified the manner in which the Wunsch–Bell curve was developed, they should plot the point (t_p, P_P) on the curve. If the point is below the curve, the component is suitable.

5.2.9.1 Calculations Being Performed by the Design Tool

The Wunsch–Bell curve is in terms of peak power (P_P) versus pulse width (t_p). The peak power values are the peak powers resulting from the current waveforms with the corresponding pulse widths. Basically, t_p is in terms of the current waveform, not the power. Pulse width of a double-exponential waveform represents the point of the decaying potion of the waveform that corresponds to ½ the peak value. Since the ¼ peak point on the power waveform will correspond to the ½ peak point on the current waveform, τ can be calculated such that t_p will correlate to the ¼ peak point of an exponential decay. Since the rise time of this transient will be small compared with the fall time (typically, the fall time is > 7 × the rise time), $1/\tau$ is a good approximation of α in equation (5.2). This approximation is used in equations (5.13) and (5.14).

The technique for selecting a TVS is a recursive process in which various t_P is tried until the power waveform, $p(t)_{\text{original}}$, to which the TVS will be exposed upon the application of the lightning transient is converted to the transient used in the Wunsch–Bell curve. For the following algorithm, the subscript "*original*" refers to parameters resulting from the lightning transient to which the component is exposed. "*Desired*" is the double-exponential waveform being tried. The recursive process continues until the "*desired*" is approximately equal to the "*original*."

1. Select a t_p and use equation (5.13) to calculate the scaling coefficient, $P_{0\text{desired}}$
2. Use equation (5.14) to calculate the power transient, $p(t)_{\text{desired}}$, for the given t_p
3. Use equation (5.15) to calculate the resultant energy of equation (5.14)
4. If the results of equation (5.12) are greater than those of equation (5.15), repeat with a wider t_p and vice versa for the results of equation (5.15) greater than equation (5.12) until the desired convergence error is achieved (the design tool selects t_P in 1 μs steps)

$$\text{energy}_{\text{original}} = \int p(t)_{\text{original}} dt \tag{5.12}$$

$$P_{0\text{desired}} = \frac{P_P}{\max\left(e^{\frac{\ln(0.25)}{t_p}t} - e^{-500000t}\right)} \tag{5.13}$$

$$p(t)_{\text{desired}} = P_{0\text{desired}}\left(e^{\frac{\ln(0.25)}{t_p}t} - e^{-500000t}\right) \tag{5.14}$$

$$\text{energy}_{\text{desired}} = \int p(t)_{\text{desired}} dt \tag{5.15}$$

where

$\text{energy}_{\text{original}}$ = the energy of the transient to which the TVS is exposed
$p(t)_{\text{original}}$ = power of the transient to which the TVS is exposed as a function of time
$P_{0\text{desired}}$ = scaling coefficient for the $10 \times t_p$ waveform
P_P = peak power of the transient to which the TVS is exposed and the $10 \times t_p$ waveform
$p(t)_{\text{desired}}$ = power of the $10 \times t_p$ waveform as a function of time
$\text{energy}_{\text{desired}}$ = the energy of the $10 \times t_p$ waveform

5.2.9.2 Reasons for Using a TVS

A TVS (sometimes referred to as an *avalanche diode*) is a passive, nonlinear device designed to operate in the avalanche region of the diode curve to clamp the voltage to a fixed value over a large range of current. The TVS will be essentially an open circuit until the breakdown voltage is applied. However, since it is a reverse biased PN junction, there is a finite amount of capacitance and leakage current associated with the component that is specified on the datasheet. When the breakdown voltage is applied, the TVS will conduct to clamp the voltage to that value. As the current is increased, the clamping voltage is nearly constant, rising at a much slower rate for higher currents than an MOV.

TVSs are preferable for applications protecting components that are sensitive to overvoltage conditions, so they cannot tolerate much variance in the clamping voltage. However, TVSs cannot tolerate as much current as an MOV. As a result, a TVS is a larger component, requires the series resistor to be a larger value, or an additional stage of protection must be added to reduce the transient to a level to which the TVS circuit can tolerate. Methods for designing additional stages of protection are detailed in later in this chapter.

5.2.9.3 Example

SMAJ6.0CA is a good candidate for the example in Table 5.1 because the clamping voltage found by simulation for the transient is 12 V as required in Table 5.1, and the breakdown voltage is higher than the peak voltage of the normal signal on this interface.

Entering the parameters from Table 5.1 into the design tool shown in Figure 5.5, $t_p = 156\ \mu s$ and $P_P = 102$ W. This can also be found using equations (5.12)–(5.15) recursively as explained at the end of Section 5.2.9.1. These results are plotted on the Wunsch–Bell curve from reference [8] in Figure 5.11. Since the point is below the Wunsch–Bell curve for SMAJ6.0CA, the component can tolerate the test transient and is suitable for this application.

5.2.10 GDT

The output of the design tool shown in Figure 5.5 used for GDT selection is the $8 \times 20\ \mu s$ current.

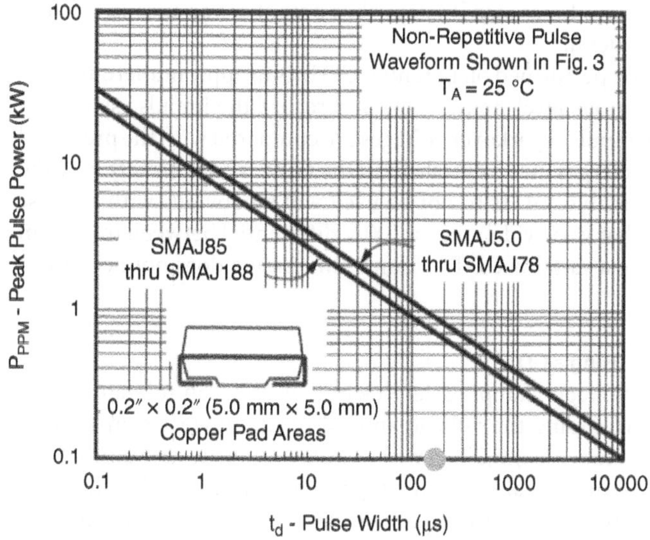

Figure 5.11 Plot of the example coordinates.

GDTs are selected using the $8 \times 20\,\mu s$ current rating and the impulse sparkover voltage. The $8 \times 20\,\mu s$ current rating is used to determine if the GDT will tolerate the lightning transient, and the impulse sparkover voltage is used to determine the maximum voltage to which the protected circuit will be exposed.

GDTs **may not be used on interfaces with a constant DC voltage applied**. Although there are certain instances if the current is limited sufficiently, a GDT can be used with a constant DC voltage applied; lightning protection circuit applications for GDTs have either no or very little current limiting, so they do not fall into this category.

All of the calculations used to find the $8 \times 20\,\mu s$ current for an MOV can be used to find the $8 \times 20\,\mu s$ current for a GDT. The voltage used as the clamping voltage is called the "arc voltage" on the GDT datasheet.

The GDT datasheet provides groups of $8 \times 20\,\mu s$ current ratings along with a "life span" for each. Therefore, the $8 \times 20\,\mu s$ current rating used to select the GDT must have a "life span" of more than 20 applications.

The impulse sparkover voltage is specified on the datasheet with the 100 V/s (DC) and the 100 V/μs values. The impulse sparkover voltage varies linearly between these rates. However, the GDT must be pretested to determine if the particular GDT will comply reliably with the specification. The reason for this is that some GDTs are designed to be activated many times throughout their life. The tradeoff for this greater life time is reduced accuracy for impulse sparkover voltage. This is *not* specified in the datasheet. These GDTs will be specified in the same manner, according to reference [17], as GDTs suitable for lightning protection. Reference [6] describes potential variance in impulse sparkover voltage in detail.

The pretesting to be performed on potential GDTs is to apply numerous lightning transients of various levels to the GDT and record the impulse sparkover voltages. Then, the impulse sparkover voltages are compared with the predicted impulse sparkover voltages in the datasheet. If the GDT is consistently near the specification, it is suitable. GDTs to not be used will be obvious with impulse sparkover voltages that vary widely, i.e., > 100 V differential between minimum and maximum. Once a suitable GDT is found, this testing does not need to be performed again.

5.2.10.1 Reasons for Using a GDT

GDTs are excellent voltage-clamping devices in that they can tolerate a large amount of current clamping the voltage applied to the downstream components to a very low value. Also, GDTs have a very low capacitance values (typically < 1 pf), making them ideal for protecting radio frequency (RF) circuits.

Therefore, if the circuit being protected is capable of tolerating the impulse sparkover voltage for a couple of microseconds, GDTs are a suitable option. In fact, GDTs are a good choice for multistage protection designs. For instance, when a series capacitor is used to AC couple an RF signal and provide a *blocking* design presenting high impedance to lightning transients, using a GDT at the interface allows for a capacitor with a lower working voltage to be used. Another example would be hardening an existing design to tolerate higher levels.

5.2.10.2 Example

Using the values from Table 5.1, it is desired to place a GDT at the terminal before the series resistor. As a result, the GDT will be exposed to the full lightning transient with only the source impedance limiting the current.

Figure 5.12 shows that the $8 \times 20\ \mu s$ current is 1692 A. The GDT that was tested and found to have a predictable impulse sparkover voltage in reference [6] is rated for 10 operations at 5 kA or 1 operation at 10 kA. Since the current to which it will be exposed is less than half the severity of the 10 operation rating, it is safe to deem this part suitable for the application.

5.2.11 PCB Trace Width Determination

The outputs of the design tool shown in Figure 5.5 used for printed circuit board (PCB) trace width selection are:

- Fusing trace width for normal clamping
- Short circuit fusing trace width

The recommended trace width for the traces that will experience the current flowing through the protection circuit during the transient (the path from the

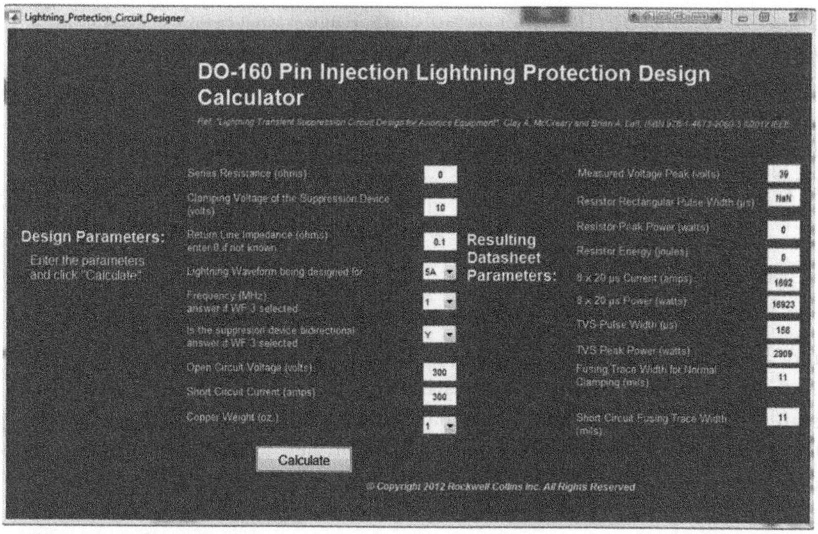

Figure 5.12 Design tool output for GDT.

interface terminal and the protection circuit, the traces connecting the components in the protection circuit, and the path from the protection circuit to ground) is 2 × short circuit fusing trace width. If constraints prevent this trace width, 2 × fusing trace width for normal clamping should be used. The absolute minimum trace width would be 1.25 × fusing trace width for normal clamping. However, this will not be a robust design.

The term *fusing* refers to the PCB trace dimensions in which the trace will melt and open under the specified conditions. The short circuit fusing trace width is the trace width that will fuse should *RV* from Figure 5.4 fail shorted. The trace width for normal clamping is the trace width that will fuse when the protection circuit functions normally during the lightning transient. Reference [7] provides a detailed study used to develop the methods used to predict these dimensions. However, that is beyond the scope of this handbook.

The reason that 2 × is recommended is that doubling the fusing trace width has resulted in thousands of successful designs with no trace fusing. The reason that it is recommended to use the short circuit fusing trace width when selecting the trace width is that *RV* components typically fail shorted if they should fail. Thus, if 2 × short circuit fusing trace width is used for the trace width, and *RV* fails, the fault is not compounded by the trace fusing also.

5.2.11.1 Calculations Being Performed by the Design Tool

The trace fusing calculations using equation (5.16) are functions of the cross-sectional area of the trace. Once the thickness is input in the design tool shown in Figure 5.5 along with the other parameters, trace width is calculated in terms of mils (0.001 in or 25.4 μm). Trace thickness is input in terms of *copper weight.* This is an industry standard. The term "1 oz. copper" refers to the thickness of 1 oz. of copper rolled out over an area of one square foot. This thickness is 1.4 mils (35.56 μm). The thickness varies linearly with the weight, so ½ oz. copper is 0.7 mils and so on.

$$W = \frac{\left[\int i^2(t)dt\right]^{\frac{1}{2}}}{0.187400T_h} \tag{5.16}$$

where W is trace width (mils) and T_h is the trace thickness (mils).

The dielectric material also affects the fusing trace width calculation. The design tool shown in Figure 5.5 calculates the fusing trace width without consideration of the dielectric. The study in reference [7] determined that the results are similar to those if the dielectric used is FR-4 and it is considered in the calculation. Since FR-4 is commonly used for lightning protection circuits, this technique was used. For specialized applications, such as RF designs, using different dielectric materials; the outputs of the design tool shown in Figure 5.5 cannot be used. One of the models from reference [7] will have to be used. The simplest model that may be used to change the dielectric is the closed-form calculation.

5.2.11.2 Example

Figure 5.12 provides the output of the design tool shown in Figure 5.5 for the high current application of using a GDT for lightning protection. Since the series resistance is 0 Ω and the clamping voltage is very low (nearly a short) the short circuit fusing trace width and the fusing trace width for normal clamping are both 11 mils. Therefore, the trace from the interface terminal to the GDT should be 22 mils wide of 1 oz. copper.

5.3 Evaluating and Hardening Existing Protection

There may be occasions in which there is a request for an engineer to perform an evaluation to predict the performance of existing WF4 lightning protection to WF5A transients. The process for this evaluation can be broken down into three steps. Each step determines if the next step is needed. This "weeding out" process facilitates the effort when there are a great number of interfaces requiring the evaluation.

1. Transform the WF4 transient test data from previous testing into an equivalent WF5A level and compare to the required level.

2. If the required WF5A level is greater than the equivalent WF5A level, determine if the existing lightning protection can tolerate the required WF5A level.
3. If the existing lightning protection can tolerate the required WF5A level, determine if the existing lightning protection provides sufficient protection for the downstream circuit.

If the first step determines the signal line does not need additional protection, that signal line requires no further evaluation. If the second step determines that the signal line needs additional protection, that signal line requires no further evaluation. The remaining lines are subjected to the final step of evaluation.

Finally, when dealing with legacy equipment, there is the desire to not modify the signal paths. Additional lightning protection is going to have an effect on signal lines, so this chapter presents techniques that minimize this effect.

5.3.1 Description of Evaluation

5.3.1.1 First Step

The tool shown in Figure 5.13 to perform this step may be found in ref. [16]. All of the inputs needed for this tool can be found on the datasheets for the previous WF4 testing. If there is no previous WF4 testing, proceed to step 2.

If V_{OC} of the equivalent WF5A transient is greater than the required V_{OC} for WF5A, then the signal line has already demonstrated that it can tolerate

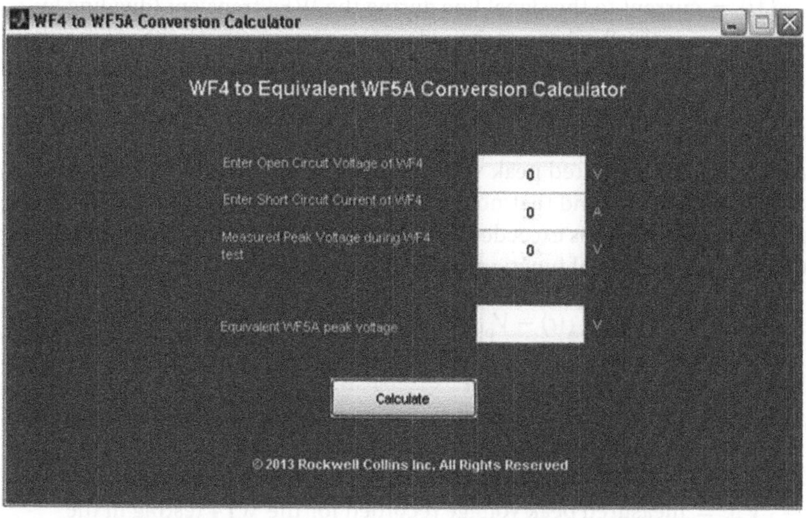

Figure 5.13 Tool that may be used to perform the first step of the evaluation.

the energy that will be imparted to it when the required WF5A level is applied. Therefore, no additional protection is required, and the evaluation is complete for that signal line.

If the first step indicates additional protection is required, proceed to the second step of the evaluation.

5.3.1.1.1 Calculations Being Performed by the Design Tool The method used for this step of the evaluation is to transform the previous WF4 test results into equivalent WF5A levels using the following assumptions: the clamping voltage is the same for WF5A as WF4 and no current is drawn by the unit until the clamping voltage is exceeded. This results in a very conservative *equivalent* WF5A level.

Using the data in the test report for the previous testing to determine the variables in equation (5.17), the energy imparted to the signal line during the application of the WF4 transient during that testing is calculated with equation (5.18).

$$I_l(t) = \frac{V_{WF4}(t) - V_l(t)}{5} \tag{5.17}$$

$$E_{WF4} = \int V_l(t)\, I_l(t)\, dt \tag{5.18}$$

where

$V_{\mathrm{WF4}}(t)$ = Equation (5.2) using the parameters of equation (5.3)
$V_{\mathrm{l}}(t)$ = voltage transient measured at the interface to the avionics during the WF4 transient for the certification testing
$I_{\mathrm{l}}(t)$ = current to the signal line during the WF4 transient (dividing by 5 is the WF4 source impedance)
E_{WF4} = energy absorbed by the signal line during the WF4 transient for certification testing

Assuming the measured peak voltage remains the same for the application of a WF5A transient, and that no current is drawn by the signal line until the measured peak voltage is exceeded by the applied voltage, the current transient is the positive portion of function created in equation (5.19).

$$I_{\mathrm{lWF5A}}(t) = V_{\mathrm{WF5A}}(t) - V_{\mathrm{pk}} \tag{5.19}$$

where

$I_{\mathrm{lWF5A}}(t)$ = the calculated current transient resulting from the application of WF5A, only the positive values are used
$V_{\mathrm{WF5A}}(t)$ = equation (5.2) using the parameters of equation (5.4)
V_{pk} = measured peak voltage recorded for the WF4 testing in the certification test report

The source impedance is omitted from equation (5.19) because it is 1 Ω for WF5A. Equivalent WF5A V_{OC} is determined recursively starting at 1 V and being increased in 1 V increments until $E_{IWF5A} \geq E_{WF4}$. E_{IWF5A} is calculated using (20).

$$E_{IWF5A} = \int I_{IWF5A}^{2}(t)dt \tag{5.20}$$

5.3.1.2 Second Step

The second step of this evaluation determines whether the existing lightning protection can tolerate the new WF5A requirements. This step of the evaluation only determines whether the existing protection components can tolerate the new transient, it does not determine whether they provide adequate protection, which is determined in the third step.

The reason for this step of the evaluation is that most of the time, the additional lightning protection for the new requirements is needed to protect the existing lightning protection components.

The techniques presented in Section 5.2 are used to determine if the components used in the existing lightning protection can tolerate the new requirement.

5.3.1.3 Third Step

The third step of the evaluation determines if the protection is adequate for the downstream circuits. This is accomplished by using the results of the simulation described in Section 5.2.5 to determine the resultant transient to which the circuit being protected will be subjected with the application of the required level of WF5A. Then, the process described in Section 5.2.4 using *absolute maximum* ratings and experience is used to determine if the circuit can tolerate the transient.

5.3.2 Lightning Protection Hardening Techniques

For every signal line that requires additional protection, there is the desire for the additional protection to have the least effect on that line. This is due to the fact that some signal lines have specific current requirements, controlled characteristic impedance, or RF characteristics. Although not an exhaustive list of lightning protection hardening techniques, the following techniques are very effective and minimize the effect on the existing signal lines:

- Add a GDT
- Add small series resistor
- For signals that may be AC coupled, add a series capacitor
- Add another stage of solid state voltage-clamping device

5.4 Design Examples

5.4.1 Design of Multiple-Stage Protection of a DC Power Line [18]

Consider the design of lightning protection for a 28VDC power input to tolerate DO-160 level 4 requirements in Table 5.2. To protect the downstream components, the transient voltage level must be clamped below 80 V.

At first, the engineer will try to design a single-stage protection circuit to provide the desired protection using the techniques described earlier in this chapter. However, the engineer is unable to find components capable of tolerating the transient produced when subjected to Level 4 Lightning testing, so the engineer considers using multiple stages of protection circuits. The following sections detail the design of a two-stage protection circuit.

5.4.1.1 Simulation

The engineer designing this circuit starts with simulations in PSPICE. Figures 5.14–5.16 show the resulting circuit and the simulation results for WF5A, WF4, and WF3. R1 is the source impedance, U8 and U12 are the clamping devices, R2 is a resistor between the stages, and R12 simulates the high impedance load.

The TVS used for U12 is selected for its clamping voltage and was the component used in the initial single-stage design.

The component used for U8 is an MOV because it must be capable of tolerating a large amount of current since there is no series resistance between it and the terminal. A GDT could not be used because this line has a constant DC voltage applied.

A clamping voltage is selected for U8, and the results of the simulation input into the design tool as described in subsequent sections. If suitable components can be found, the design is complete. If not, a three-stage design may be required. For this design, the resultant U8 is an MOV with a 130 V working voltage.

The resistor between U8 and U12 is *very important.* When the transient is applied, the clamping votlage of U12 will be exceeded first, so it will start conducting. Without the resistor drop, the clamping voltage of U8 may never be exceeded resulting in the MOV not activating. This will cause U12 to become

Table 5.2 DO-160 level 4 requirements [1]

Waveform	Voltage open circuit (V_{OC}) (V)	Current short circuit (I_{SC}) (A)
3	1500	60
4	750	150
5A	750	750

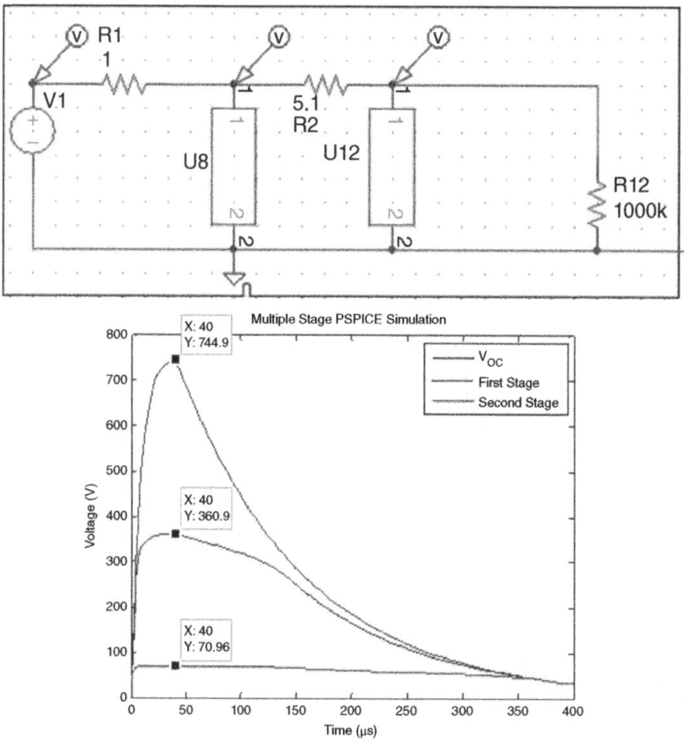

Figure 5.14 PSPICE simulation of the WF5A transient.

damaged. Thus, whenever multiple stages are used, resistors must be placed between the stages to prevent one stage precluding another from acitvating.

5.4.1.2 First Stage

The object of the first stage is to clamp the transient to a level that can be tolerated by the second stage using minimal series resistance for the second stage. Ideally, the first stage does not use series resistance in this design. In most designs, it is undesirable to place a suppression device on the input without series resistance because the series impedance is not controlled and known. However, the desire to minimize power dissipated during normal operation takes priority over the operation during a random transient condition. Thus, to minimize series resistance on the 28VDC power input, the first stage does not use series resistance.

PSPICE simulation using MOV models provided by the vendors revealed that a particular 130 V working voltage MOV will clamp the WF5A transient to 360 V.

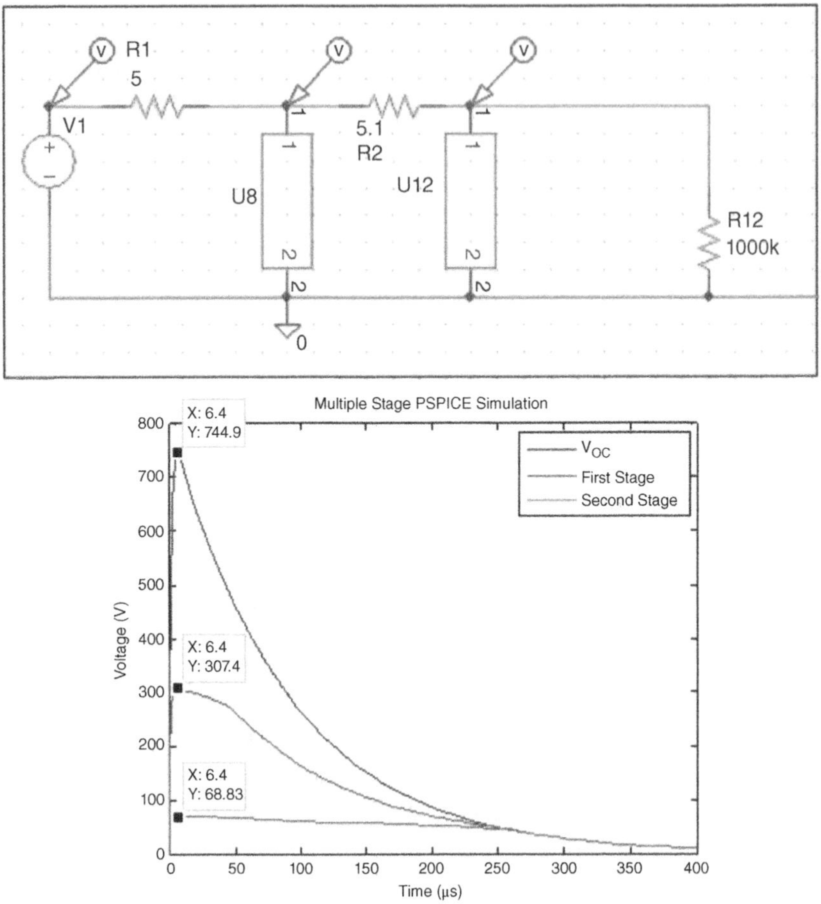

Figure 5.15 PSPICE simulation of the WF4 transient.

Inputting the series resistance of 0 Ω, clamping voltage of 360 V, and the level 4 transient conditions into the design tool as described in previous sections and pressing "Calculate" yields the results shown in Figure 5.17.

These results indicate the equivalent 8 × 20 μs transient peaks at 1271 A. The 8 × 20 μs current rating for this MOV is 2500 A. Thus, it is suitable for this design.

The short circuit fusing trace width indicated in Figure 5.17 for 1 oz. copper is 27 mils, so the recommended trace width for the first stage is 54 mils to create a robust design.

If the resulting fusing trace width is too wide for the application, 1.5 or 2 oz. copper may be selected. Selecting 1.5 oz. copper from the "copper weight" drop-down box and clicking "Calculate" outputs 18 mils for the "short circuit fusing

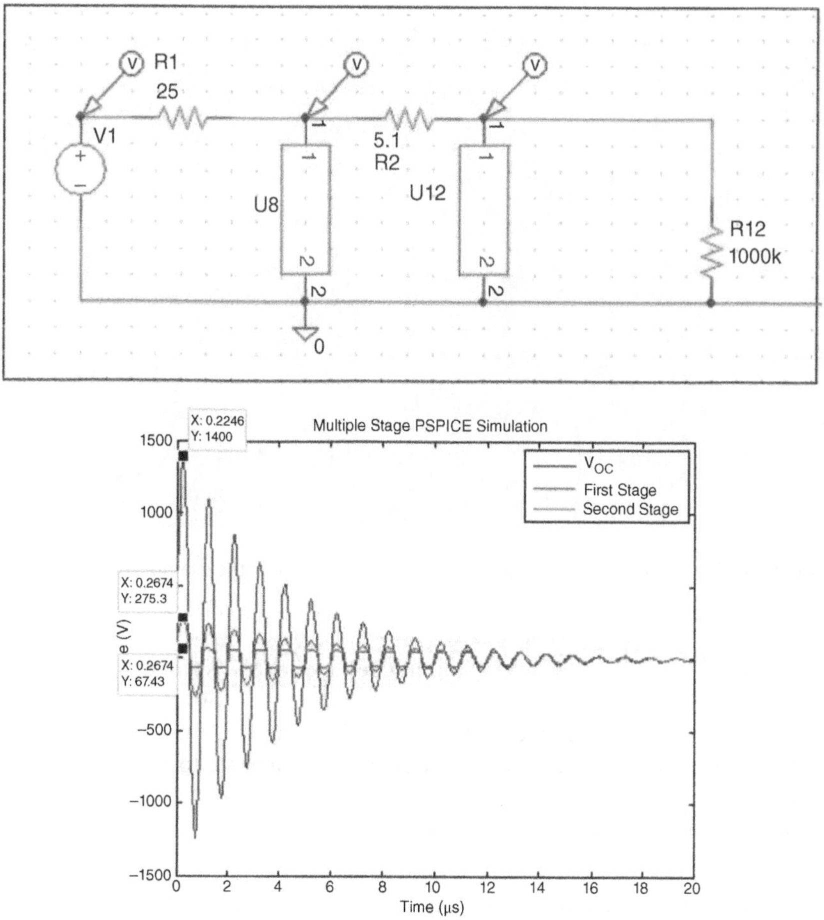

Figure 5.16 PSPICE simulation of the WF3 transient.

trace width," so 36 mils would be the recommended trace width for 1.5 oz. copper. The "short circuit fusing trace width" is 14 mils for 2 oz. copper, resulting in 28 mils for the recommended trace width.

5.4.1.3 Second Stage

To design the second stage of the protection circuit:

1. The clamped voltage transient on the first stage is exported from the PSPICE simulation into a .txt file.
2. Select .txt for the lightning waveform being designed for.

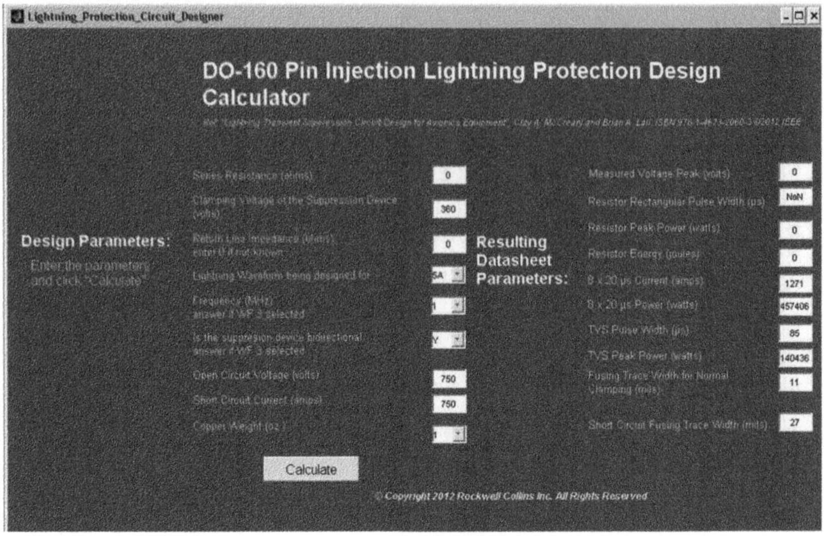

Figure 5.17 Results for the design parameters of the first stage of the example design.

3. Input the series resistor and clamping voltage for the second stage of the protection circuit design.
4. There are no values needed for the rest of the inputs.
5. When "Calculate" is pressed, the engineer will navigate to the .txt file and select it.
6. At that point, the outputs will be populated.

A suitable TVS could not be found, but a 30 VDC working voltage MOV is found that will clamp the 360 V transient to less than 80 V. This is determined by first using a very low series resistance and determining if the MOV can tolerate the transient and if the clamping voltage is near the required voltage. Then, the series resistance is increased until the clamping voltage is less than 80 V. The value of the resistor resulting in a clamping voltage less than 80 V is 5.1 Ω. As long as the transient is less than 80 V, the downstream circuit is protected.

The 8 × 20 μs current for the MOV is 452 A, and a surface mount MOV is selected.

The peak power of the resistor, 16.5 kW, is too high for a thick film resistor to be used, so wirewound resistors were considered using the thumb rule of 0.72 J/rated W. The resistor energy is 2.125 J. The thumb rule would be violated by 35 mJ using a 3 W wirewound resistor. Instead, multiple resistors were installed in series to avoid violating the thumb rule for any of them. The energy dissipated by each is equivalent to the ratio of the individual resistance to the total resistance, like a voltage divider. The total resistance is 5.1 Ω. The resultant power dissipated during normal operation is 0.46 W. This does not cause the maximum power rating of the equipment to be violated.

Although the power dissipated by the series resistance does not result in the violation of the maximum power rating of the equipment, it is not optimum. The series resistance can be reduced by adding more stages; however, other design constraints may prevent this, most notably, the amount of physical area allowed for the circuit to occupy. In this instance, there is not enough room to add another stage. At this point, the decision must be made whether to change the mechanical structure to allow the addition of another stage, or just accept the less than optimal power dissipation under normal operation. Since the maximum power requirement is not violated, it is acceptable to use this circuit.

5.4.2 Evaluation of Existing Designs and Hardening Circuits Where Necessary [19]

Table 5.3 is an example of the evaluation results for several signal lines. These results were extracted from a spreadsheet containing over 400 signal lines. Spreadsheets are a very effective tool to manage the bookkeeping associated with evaluations of this size. The spreadsheet allows the use of "If" statements. Thus, for signal lines with similar test results and different WF5A level requirements, the equivalent WF5A can be copied and pasted for all the appropriate signal lines and the "If" statement will determine the results of the first step by comparing to the new WF5A level requirement. Carefully setting up the spreadsheet can shorten the time required for this evaluation significantly.

The first signal is a discrete input with an MOV on the interface with no series resistance. The clamping voltage during the original certification testing to WF4 level 3 (300/60) [1] requirements is 145 V. The tool from Figure 5.13 [16] is used to calculate the equivalent WF5A level of 175 V.

Table 5.3 Example evaluation of lightning protection to new WF5A requirements

Signal type	Test level WF5A	Equivalent WF5A from WF4 in cert. test report	Requires second step	Second step results	Requires third step	Third step results	Additional protection required
Discrete input	442	175	Yes	Damaged	No		Yes
Discrete input	116	No Data in Test Report	Yes	On	Yes	Pass	No
Differential	288	No Data in Test Report	Yes	Damaged	No		Yes
Differential	614	179	Yes	Damaged			Yes
Analog	206	119	Yes	On	Yes	Fail	Yes
RF	964	Never Tested	Yes	Damaged			Yes

Test data for the second, third, and sixth rows could not be found. The tool from Figure 5.13 was used to fill in the cells for the equivalent WF5A of the fourth and fifth rows. Using an "If" statement, the "Requires second step" column is populated automatically. The last column, "Additional protection needed," monitors the results of each step. It fills in automatically when a step definitively determines the need or lack of need for additional protection.

Using simulation and the design tool from Figure 5.5, the second step is performed on the applicable signals in Table 5.3. The 8 × 20 μs current for the MOV in row 1 exceeds its rating. Therefore, this signal line requires additional protection. The third fourth and sixth rows have similar results.

The components in the second and fifth row signals are shown to be capable of tolerating the transient produced by the new WF5A levels.

Using the results of the simulation, the resultant transient to which the circuits being protected will be exposed is compared with the absolute maximum ratings on the datasheets for the components in these circuits. This is the third step. The results indicate that the signal line in the second row requires no additional protection. The existing protection circuit can tolerate the new levels and provides adequate protection. The protection circuit for row 5 does not provide adequate protection, so additional protection is required.

5.5 Conclusion

The guidance provided in this chapter allows for careful design of lightning protection circuits using the smallest components and trace dimensions required to tolerate the lightning test transient for which they are being designed.

This chapter introduced tools to assist in the design of these circuits along with design examples for using them. In addition, the equations used by these tools to perform the calculations needed for these designs are presented allowing for hand calculation if desired.

References

1 "Environmental conditions and test procedures for airborne equipment," RTCA/DO-160E, RTCA Inc., Dec. 9,2004.
2 "Aircraft lightning environment and related test waveforms," SAE ARP5412, Nov. 1999.
3 SAE-ARP5414, "Aircraft lightning zoning," SAE, 1999.
4 F. A. Fisher and J. A. Plumer, "Lightning protection of aircraft," NASA Reference Publication 1008, 1977.
5 C. A. McCreary and B. A. Lail, "Lightning transient suppression circuit design for avionics equipment," in IEEE 2012 International Symposium on Electromagnetic Compatibility (EMC), pp. 93–98, 2012.

6 C. A. McCreary and B. A. Lail, “Choosing the appropriate gas discharge tubes (GDT) for avionics lightning protection,” in IEEE ElectroMagnetic Compatibility Magazine, vol. 2, no. 3, pp. 57–59, 2013.
7 C. A. McCreary and B. A. Lail, “Determining minimum printed circuit board (PCB) trace dimensions for transient suppression circuits,” *IEEE Trans Compon Packag Manufact Technol*, vol. 3, Issue 11, pp. 1876–1888, 2013.
8 “SMAJ5.0 thru SMAJ188CA,” Datasheet, Vishay General Semiconductor Inc. April 2011.
9 “SG73,” Datasheet, KOA SPEER Inc., Jan. 23, 2008.
10 “CMA0204,” Datasheet, Vishay Inc., Jul. 18, 2006.
11 “WSC, WSN,” Datasheet, Vishay Inc., Aug. 3, 2007.
12 “MLA Varistor Series,” Datasheet, Littelfuse Inc., 2010.
13 “M50-C90XSMD Surge Arrester,” Datasheet, Epcos Inc., Feb. 14, 2007.
14 C. A. McCreary, “Lightning protection circuit design tool for avionics,” MATLAB File Exchange, http://www.mathworks.com/matlabcentral/fileexchange/39650-lightning-protection-circuit-design-tool-for-avionics, Dec. 26, 2012.
15 C. A. McCreary, “Executable lightning protection circuit design tool for avionics,” *EDN Magazine*, Design Tools Section, http://www.edn.com/uploads/tools/LightningProtection_pkg.exe
16 C. A. McCreary, “Tested WF4 to equivalent WF5A transformation calculator,” MATLAB File Exchange, http://www.mathworks.com/matlabcentral/fileexchange/39813-tested-wf4-to-equivalent-wf5a-transformation-calculator, Jan. 11, 2013.
17 International Telecommunications Union, “Characteristics of gas discharge tubes for the protection of telecommunications installations,” Recommendation ITU-T K.12, May 2010.
18 C. A. McCreary and B. A. Lail, “Design of multiple stage avionics lightning protection for DC power input lines using a graphical user interface (GUI),” in IEEE 2013 International Symposium on Electromagnetic Compatibility (EMC), pp. 177–181, 2013.
19 C. A. McCreary and B. A. Lail, “Hardening lightning protection for avioincs on composite aircraft,” in IEEE 2013 International Symposium on Electromagnetic Compatibility (EMC), pp. 171–176, 2013.

6

Pyrotechnic Systems in Aerospace Applications

Karen Burnham

6.1 Introduction

The pyrotechnic system on an aerospace vehicle is always subject to intense scrutiny. By nature of the literally explosive potential represented by each pyrotechnic device, the consequences of failure are often extreme. Most pyros are both "must work" and "must not work" devices: they must not go off accidentally, but they must work at exactly the moment they are needed. By far the safest pyro device on any vehicle is the one that's not there—one that is not used because the same task can be accomplished in a different way. Unfortunately, when mass is a design driver, pyro devices are often the best choice to accomplish a given task with the minimum possible mass.

Pyrotechnics are always one-shot devices. They can be used for applications as diverse as gas-generating mortars that deploy parachutes, very small pyro valves that can stop or start fluids flowing, frangible nuts and joints that help separate spacecraft stages, or even the t-handles that blow off emergency access hatches. There is also a wide variety of pyrotechnic system architectures, many of which do not have any electromagnetic interference (EMI) concerns. Items that are initiated by a percussive force (a pull handle in an emergency exit) will not be sensitive to EMI, nor will those initiated by a chemical fuse, such as reefing line cutters in a parachute system. However, one of the most common pyro architectures has significant potential vulnerability to EMI. In this design, the primary explosive charge is relatively inert. It can be frozen, dropped, shot, set on fire, or even hit with an electric spark and it will not detonate. The only thing that can cause the primary explosive to detonate is a smaller explosion, and this explosion is provided by an initiator. When those initiators are activated electrically, which is common, it is the element most vulnerable to EMI. In these

Handbook of Aerospace Electromagnetic Compatibility, First Edition. Edited by Reinaldo J. Perez.

electrically initiated explosive devices (EIEDs), if the initiator is accidentally activated, the entire system will trigger, causing potentially catastrophic failures. These systems will be the subject of this chapter.

Cases of pyro failures are all too real. Perhaps the most tragic case of an explosive failure due to electrical interference is the fire aboard the naval vessel USS *Forrestal* in 1967. The *Forrestal* was an aircraft carrier carrying combat aircraft during the Vietnam war. On July 29, 1967, a ZUNI rocket carried on an aircraft inadvertently fired into another aircraft, rupturing its fuel tank and setting it on fire. Then, the ordnance on the victim aircraft exploded from the heat, causing the destruction to spread across the carrier deck. In the end, 134 men died and another 161 were injured, many fighter planes had to be pushed into the sea to prevent them from exploding and damaging the ship further, and the carrier had to limp back to port for temporary and then extensive permanent repairs, at a cost of $72 million to the US Navy ($538 million in 2016 dollars).

The cause of the initial ZUNI rocket firing remains uncertain, but it is believed that either stray voltage or a power surge during aircraft power up caused the initial firing. A safety pin that inhibited these systems was known to sometimes dislodge during windy conditions and may not have been in place. Another safety precaution, an electrical pigtail not meant to be connected until the aircraft was ready to launch, was sometimes bypassed during times of heightened combat readiness, since there was concern that a faulty pigtail connection could delay a launch. So an electrical connection may have been in place, allowing a power surge to flow into the rocket initiation system, and a safety pin may have blown loose. From such small failures catastrophes are born. In response to this and other ordnance failures, the US Navy has developed a rigorous HERO (hazards of electromagnetic radiation to ordnance) research program. This chapter heavily leverages the work that has been done to address HERO concerns.

Design, analysis, and testing of pyrotechnic devices must be approached with caution and attention to detail. There are many aspects of pyrotechnic safety that do not fall under the umbrella of EMC—lot-acceptance testing, redundant architectures, chemical composition, low temperature testing, etc. This chapter will shed some light on those aspects of pyro system design and safety that EMC engineers may become responsible for in their careers. It will lay out various issues that should be considered during the design phase of any pyro system designed for inclusion in an aerospace program. From component-level selection and testing to subsystem design and system-level testing, the more thought that is put into the design upfront will save countless dollars and hours—and perhaps lives—later on.

6.2 Component-Level Concerns

6.2.1 Different EIED Types

The first thing the EMC engineer will want to determine is what types of EIEDs are being considered for use on the vehicle. There are many kinds of pyrotechnic initiators available. One of the more common systems is one that is also particularly sensitive to EMI, the low-voltage single bridgewire (or "hot bridgewire") device. In this case, the EIED is fired using an initiator that uses a single low-resistance (around 1 Ω) bridgewire. The initiator is fired when a DC current exceeding a certain threshold flows through the bridgewire. At that point the bridgewire rapidly heats, setting off an explosive mixture and initiating the pyrotechnic chain of events. The amount of power needed to initiate the device may be quite small. Dual bridgewire devices are based on similar principles but use two bridgewires per initiator to provide built-in redundancy. Instead of two pins per initiator, these devices will have four. From an EMI point of view, these devices do not provide much added safety, since two firing lines inside the same harness will be equally susceptible to interference. Single bridgewires can be made even more sensitive (able to be fired using very low power sources) if the bridge is made with a deposited carbon film. These carbon bridge EIEDs become so sensitive that their use is often banned.

Exploding bridgewire (EBW) devices have a similar architecture but are usually powered by a high-voltage source, often greater than 500 Vdc, delivering hundreds of amperes. The higher power delivered to the bridgewire allows it to heat and vaporize much more rapidly, allowing for more precise timing of the detonation. Although a single bridgewire device might fire in hundreds of microseconds to a few milliseconds, an EBW device might fire in a few microseconds. Exploding foil initiators (EFIs) similarly use high voltage for actuation, delivering energy to a thin membrane which explodes and propels fragments into the explosive material. In general, these systems are safer from an EMI perspective; the higher voltages and faster rise times needed to trigger the device are harder to achieve with only stray voltage or radio frequency interference. However, the large power sources needed to fire these devices mean that they are less commonly found on aerospace vehicles where weight is of primary importance. These devices may also be more likely to dud if their foil or bridgewires become damaged through exposure to RF, heating, or electrostatic discharge (ESD).

In addition to systems that use simple low-resistance wires for their bridgewires, there are also semiconductor bridge (SCB) systems and thin film bridge initiators (TFB). In a TFB device, a thin film conductor applied to a substrate acts as the bridgewire for the device. By varying the composition and geometry of the thin film, the device can be tailored for reaction time, all-fire and no-fire energies, and robustness to ESD and EMI. These devices are particularly popular in automotive airbag systems. NAVSEA-OD-30393 [14] describes SCBs as "essentially a polysilicon resistor on a silicon or sapphire substrate." The polysilicon resistor forms a bridge which can be designed with specific resistance and good consistency between devices. Because of their very small size, transient suppression and other protective features can be integrated into the same EIED as the bridge. They are fired with high power, short duration pulses that achieve firing in a few microseconds.

The remainder of this chapter will focus on those EIEDs most susceptible to EMI, the low-voltage single bridgewire devices. As this presents the worst case (and also a very common case in aerospace applications), the various protection techniques described here can be used for other EIED types as needed.

6.2.2 Device Sensitivity

The next most important piece of information for the EMC engineer to discover is the sensitivity of the EIED to DC power. This may be presented as the "All Fire" and "No Fire" limits, with another common term being the "Mean No Fire Stimulus (MNFS)." NAVSEA-OD-30393 [14] defines MNFS as "the greatest firing stimulus which does not cause initiation, within five minutes, of more than 0.1 percent of all electric initiators of a given design, at a 95 percent confidence level." One of the most common EIED types is the single bridgewire device with a 1 A/1 W no-fire level. The "all-fire" level is similarly defined as the level at which 99.9% of devices will fire within the specified operating time (usually on the order of milliseconds), with 95% confidence. The system is then designed to deliver something more than this stimulus to the device when it will be operated. Usually these numbers for DC activation power/current are available from the manufacturer, as well as the resistance of the bridgewire if one is being used.

What is generally *not* available from the manufacturer is something the EMC engineer needs to be concerned with: the device's sensitivity to RF power. Research projects done by organizations such as the Naval Surface Warfare Center, the Franklin Institute of Applied Physics, and NASA [2] can provide some information on the sensitivity of specific devices. Technically, determining a device's sensitivity to a specific frequency of RF power is no different from determining its pin-to-pin DC all-fire and no-fire levels—the problem is the limited number of frequencies that any organization can realistically test to.

In the past, the Bruceton test was the standard way of experimentally determining these numbers (the test is named after Bruceton, Pennsylvania, home of the US Bureau of Mines) [4]. In the Bruceton test method, at least 40 devices are tested to obtain information about device sensitivity at a single frequency.

Estimates are made of the expected mean and standard deviation of the MNFS prior to testing. Based on those estimates, set intervals of firing stimulus are chosen. A device is installed and exposed to an initial firing stimulus for 5 minutes. If it fires, the next unit will be exposed to a level one interval lower than the original. If the device fails to fire, the next device is exposed to a level one interval higher than the previous unit. By stepping through 40 devices in this manner, a reasonable estimate of the MNFS and all-fire levels can be made.

Another sensitivity test is the Langlie method [7], developed in 1965. In this case, the test director initially specifies the upper and lower bounds, and the starting point is the average between them. Steps are calculated based on the average between certain steps or end points. This test is not as dependent on starting guesses as the Bruceton method is, but is still relatively easy to implement. Another method that can be computed by hand if necessary is the Probit analysis [1], a graphical method in which plots of stimulus levels that produce mixed results (some fires and some no fires) are analyzed to yield information about the mean and standard deviation of the MNFS.

Today, an alternate statistical approach known as the Neyer d-optimal test is often used [15]. This method allows for equal confidence in determining sensitivity thresholds using fewer test samples—an important point when test cost is a factor. The Neyer test requires the use of a computer algorithm to determine test levels from one test to the next based on the history of fire/no-fire levels in the test sequence. Unfortunately, using any of the methods discussed so far, a test using anywhere from 10 to 50 EIEDs will still only yield information about the MNFS and all-fire levels of a device at a single frequency of power.

The testing becomes even more burdensome when different firing modes are taken into account, since some devices are equally or more sensitive to RF voltages that develop pin-to-case (common mode) than they are to the intended pin-to-pin (differential) firing signal path. These two firing modes are illustrated in representative circuit schematics in Figures 6.1 and 6.2. Pin-to-case voltages are likely to develop in common mode interference situations on the cable harness bringing the firing lines to the EIED. Depending on the device under consideration, there may be some history or other data indicating which mode is more sensitive, and whether it is more sensitive to continuous wave (CW) power or pulsed power. In the case of pulsed power, it can happen that each pulse, containing higher levels of energy than the CW average, heats up the bridgewire or explosive mix, and the element does not have time to fully cool off before the next pulse hits, heating it further. Eventually, the heat builds up to the point that the device either fires or duds (Figure 6.3). This is one of the reasons for the 5-minute exposure time that is standard in EIED testing, to fully gauge if this effect (known as thermal stacking) is affecting the device.

Mil-Hdbk 240A [9, p. 26] contains a useful formula for estimating the vulnerability of a given EIED to the thermal effect from the peak power contained in narrow pulses for a given threat transmitter. It calculates a multiplying factor

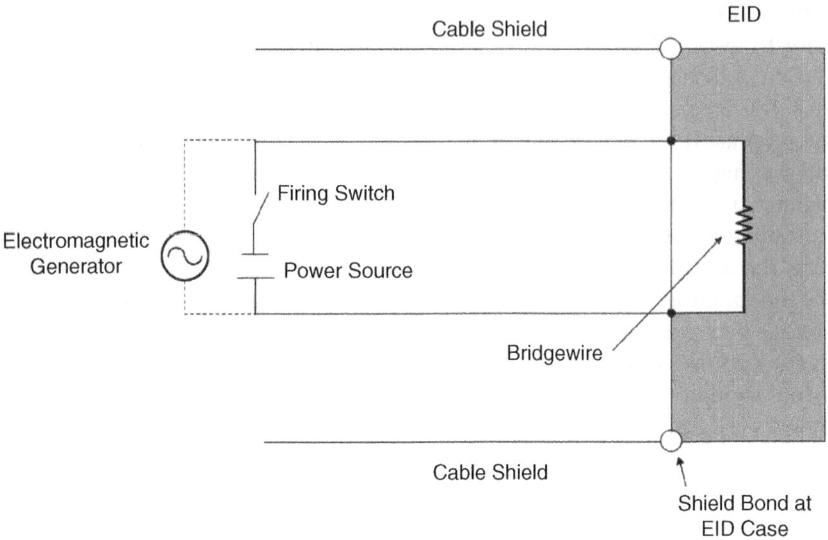

Figure 6.1 Differential firing scenario of a typical two-wire firing system. *Source*: [9, Figure 10].

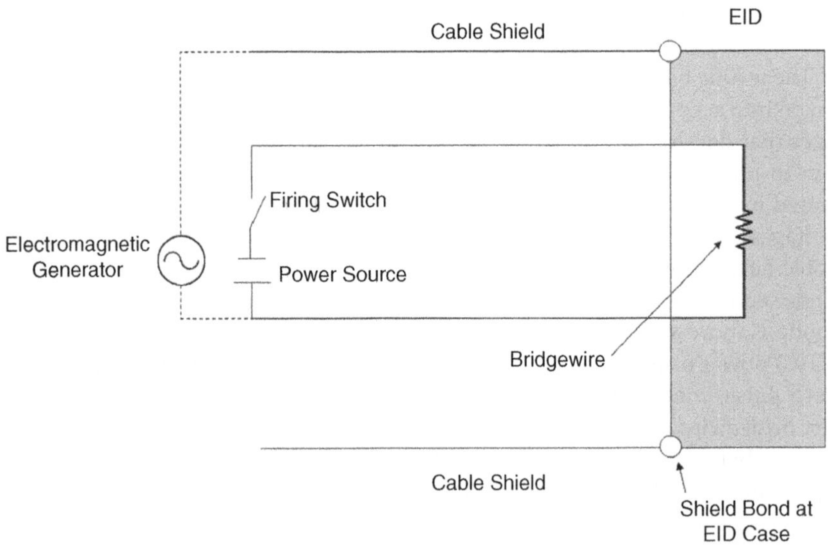

Figure 6.2 Common mode firing scenario of a typical two-wire system. *Source*: [9, Figure 12].

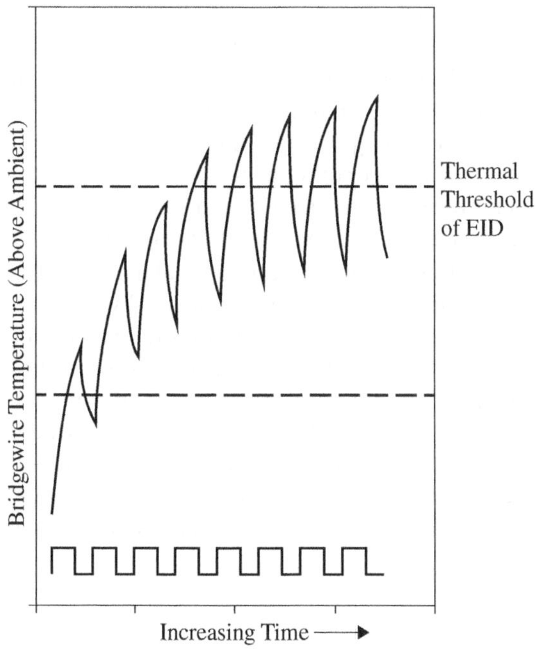

Figure 6.3 Thermal stacking effect. *Source*: [14, Figure 3-1].

(MF) that, given knowledge of the device's vulnerability to continuous wave or average power (CW), can predict an MNFS for peak pulsed power:

$$\text{MF} = (1 - \text{e}^{(-t2/\text{tau})})/(1 - \text{e}^{(-t1/\text{tau})})$$

where $t1$ is the transmitter pulse width (seconds), $t2$ is the transmitter pulse interval (1/PRF, the pulse repetition frequency) in seconds, and tau is the EIED time constant in seconds. The MNFS of the EIED at CW is multiplied by the MF to get an MNFS for peak pulsed power. If tau is less than or equal to $t1$ and $t2$ (i.e., the duration of the pulse at peak power lasts longer than the firing time of the device), the MF will be 1, meaning that the device is just as susceptible to the pulsed power as to CW power.

Mil-Std 1576 [10] provides procedures for a research program to characterize the RF sensitivity of a device, expending 230 different units for testing (although that document was last updated in 1984 and bases its testing on the Bruceton method, so this program could likely be improved with the introduction of the more recently developed sensitivity test methods and analysis). Even if that many units can be spared (along with the time to test them), at the end of that testing campaign the devices will only have been exposed to 10 specific frequencies of RF power. The concern then is that the device happens to resonate with

RF power at some frequency in between the frequencies actually tested and will be much more sensitive than the measurements show. According to research conducted for HERO purposes, "Calculations indicate that the error at a given resonant frequency could produce a current 2.6 times that at two test frequencies on either side of the resonant frequency" [17, p. 5-2]. Another source of variability is the presence of a crew member or other personnel; the addition of human capacitance can routinely change the induced currents by a factor of 2. Combining these two factors (assuming that we could have a device that fires at $1/2.6 \times \frac{1}{2}$, the predicted MNFS) and then adding some additional margin yields a guideline that under no circumstances should currents caused by the electromagnetic environment (EME) or stray voltages be allowed to exceed 15% (−16.5 dB) of the MNFS for safety critical devices. This number is sometimes changed to 10% (−20 dB) in organizations with a preference for round numbers. For EIEDs which are being used in nonsafety critical applications (and where their initiation would not cause harm to personnel) the margin can be relaxed. HERO recommends restricting currents in the firing line for nonsafety critical EIEDs to no more than 45% of the MNFS (−6.9 dB), which again sometimes is changed to 50% (−6 dB) [17].

6.2.3 Preinstallation

There are two main methods of protecting an EIED from RFI during storage and handling: a shorting plug and a Faraday cap (Figure 6.4). A shorting plug is a metal device that fits between the two pins of the EIED, keeping any stray voltage from developing between them and activating the bridgewire. These are not always the best protection from RFI since as a worst case they can act as an antenna and allow RF energy to couple directly into the device. Also, they do not always protect against voltages developing pin-to-case, which can be another cause of inadvertent activation. The preferred method of protecting an EIED is a Faraday cap, a metal cap that firmly connects to the case of the EIED with 360° contact and an electrical bond sufficient for RF exposure, preferably less than 2.5 milliohms at DC. This creates a complete shield around the entire device, preventing RF energy from coupling into the bridgewire or the explosive mix.

The most vulnerable moment in the life of an EIED is when it is being installed. At this point it is being handled, its safety device has been removed (shorting plug or Faraday cap), and it does not yet have the benefit of the shielding and filtering from the overall system. A primary concern is the threat from an ESD event. A spark can easily occur in cases where personnel are involved, or in a region of high EME such as the deck of an aircraft carrier. The main prevention for ESD events is to ensure that the handler, EIED, and installation site are at the same electrical potential at all times, usually through the use

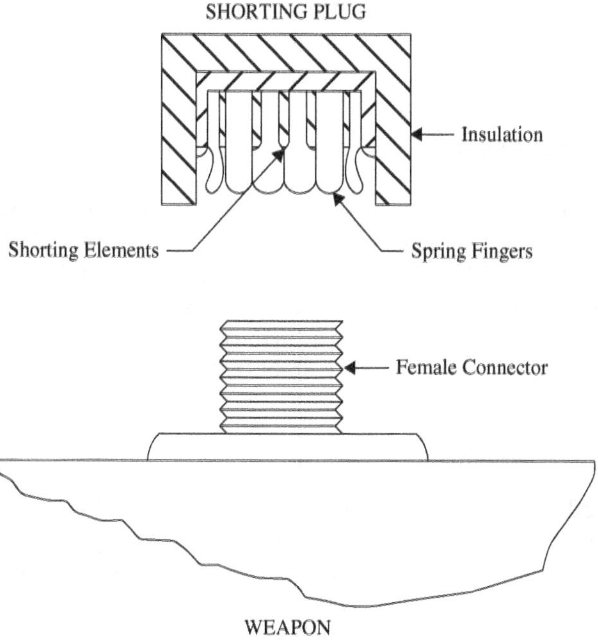

Figure 6.4 Shorting plug with Faraday shielding. *Source:* [14, Figure 4-7].

of ground straps which are tested before the EIED is unpackaged. EIEDs can also be designed to be insensitive to ESD events, such as those designed with a spark gap to divert ESD energy away from the explosive mix. Regardless, all EIEDs should be tested with a human body capacitance model for generating the spark. An example of this kind of test is outlined in NASA's JSC 62809 [6], in which the EIED is "subjected to an electrostatic discharge of 25,000 volts from a 500 picofarad capacitor applied in the pin-to-case mode with no series resistor and in the pin-to-pin mode with a 5 kilohm resistor in series."

Another concern is the threat from lightning. During installation, if there is any lightning or lightning potential reported in the vicinity (often within 5 miles), all pyro handling operations should cease and the EIEDs should be packed away in their safe conditions (shorted and/or capped). Once installed in a system and provided with proper shielding and grounding (about which more later), most pyro devices should be adequately protected from the worst effects of lightning. As NASA research conducted by the author showed, 1 A/1 W no-fire single bridgewire devices suffer few affects from lighting pulses as attenuated by system shielding; the majority of the energy in a lightning strike is delivered in the first 70 μs, and these single bridgewire devices take milliseconds to fire [3]. Devices with extremely fast firing times may be more at risk, but they tend to also require higher firing voltages which an attenuated lightning pulse should not reach in a properly designed system.

6.3 Vehicle-Level Concerns

There are a number of important concerns when considering the overall vehicle system as opposed to the EIED independently. The first is knowledge of the electromagnetic environment (EME) in which it will be operating, and the second is the overall architecture of the firing system and how EMI protection will be included at each level or interface. A basic pyro system architecture will contain a power source, safe/arm/fire circuitry, and EIED (Figure 6.5).

6.3.1 Electromagnetic Environment

In order to analyze the design of the firing system, the EME must first be characterized. EIEDs that need to be installed on an aircraft carrier deck may have different requirements than ones that are installed inside an operations and checkout facility. Items for use on a submarine may be concerned with very different frequency lists than those destined for space. Items that will be used on space-based platforms may need to deal with plasma charging or ionizing radiation environments. Ideally, the EMC engineer for a project will be provided with a list of frequencies of interest and the expected average and/or peak electric field strength (in Volts/meter) or power density (in Watts/square meter) at each frequency. The equation to convert from field strength (E) to power density (PD) in free space [10] and assuming the far field of a transmitting antenna is:

$$\mathrm{PD} = E^{2/(120\pi)}$$

Where exactly the far field (where the EM waves propagating from an antenna are plane waves) begins is a matter of some debate, but a commonly accepted rule of thumb is that it is $2D^{2/\lambda}$ meters away from the antenna, where D is the length in meters of the largest dimension of the antenna and λ is the wavelength in meters. If the pyrotechnic system is going to be used in the near field of powerful transmitting antennas, a more complex analysis of the expected field strengths may be needed.

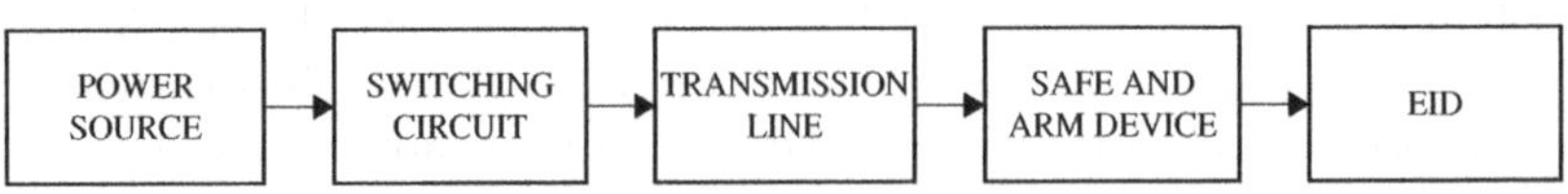

Figure 6.5 Firing system basic architecture. *Source*: [14, Figure 4-1].

The engineer should also find out information about all onboard transmitters which will be sharing the platform with the pyro devices; both intentional transmitters such as communication antennas and unintentional transmitters such as the frequencies of the onboard clock oscillators. If no information about the external EME is forthcoming, there are different sources which can provide default levels for use in testing and flowing requirements to subsystem vendors. Mil-Std-1576 [10] offers a "worst-case" EME suggestion of 2 W/m^2 from 1 to 50 MHz, and 100 W/m^2 from 50 MHz to 32 GHz. This corresponds to a field intensity of 200 V/m from 50 MHz to 32 GHz, which is also commonly found in the aircraft and shipboard requirements of Mil-Std-461 [11] (going up to 40 GHz). NAVSEA-OD-30393 [14] also provides a more up-to-date default list of EME threats by frequency and field strength, but it is always better to characterize an EME tailored for the particular operating environment of the specific system. Military standards which define worst-case EMEs usually do so in terms of peak and average field strengths. As discussed in Section 6.2.2, knowledge of the pulse widths and PRFs of the peak transmitters compared with the time constant of a given EIED can allow for estimation of vulnerability to peak power once vulnerability to average powers is established.

6.3.2 Shielding

Armed with the knowledge of the internal and external threats facing the pyrotechnic subsystem, the EMC engineer can proceed with designing the protection for the EIEDs. Perhaps the most important factor to consider is shielding. In an ideal world the signals, firing mechanism, power supply, power lines, and the EIED itself would all be contained in one box that was made of metal (Faraday cage) with no apertures or seams through which stray electromagnetic energy could couple (Figure 6.6). Obviously, this is not possible: if nothing else, common sense dictates that the source of the firing command (often a person) wishes to be conveniently far away from the resulting explosion. Instead, the shielding efforts should be an effort to get as close to that continuous reality as practically possible (Figure 6.7).

One particular area of concern is cable shielding. Again, perfect shields (solid metal conduit electrically bonded to metal boxes or structure at each end) are unlikely to be allowed. A more flexible option is metal braiding of some sort. The usual guideline is that this braid should have greater than 90% optical coverage (from MIL-C-27500G [8]), defined as the ratio of metal to potential total surface area. Again, the braid should be terminated with 360° circumference to the connector or box it is connecting to, with an RF termination bonding value of < 2.5 mΩ (measured at DC). (This bonding guideline comes from NASA 4003 [13] and Mil-Std-464C [12], which define a Class R bond to prevent RF interference from propagating on a system.) This kind of termination is illustrated in Figures 6.8 and 6.9. Proper shielding termination is crucial—shields

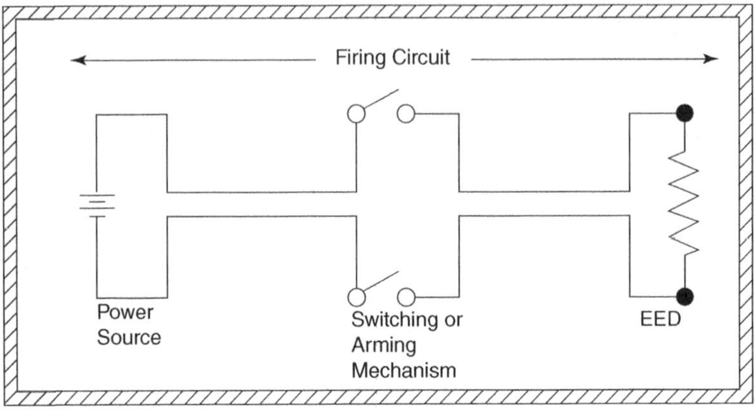

Figure 6.6 Ideal shielding situation. (*Source*: [14, Figure 1-2].

should be terminated each time the wires pass through a bulkhead connector, never tied to a pin and brought through the bulkhead with the signal lines. It should also never be terminated with a pigtail or jumper, which concentrates all noise currents along a narrow path and can cause radiated emissions problems. Figure 6.10 shows the effect of different shield termination (pigtail) lengths as RFI increases in frequency.

Another critical point is shielding the box that houses the controlling electronics (and often the power source as well, either battery or capacitor bank) for

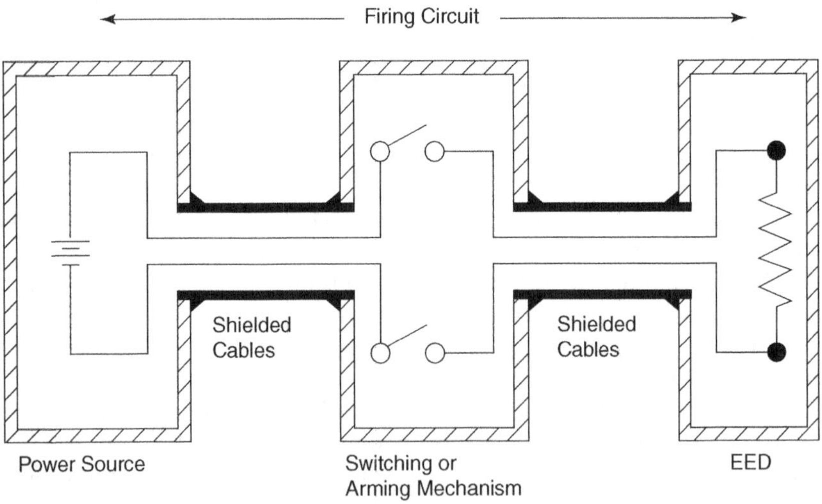

Figure 6.7 Shielding situation with interconnects. *Source*: [14, Figure 1-3].

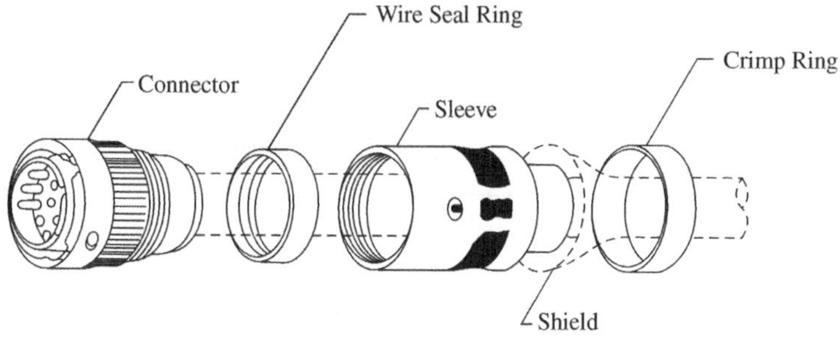

Figure 6.8 Shield termination using crimping. *Source*: [14, Figure 5-15].

the pyrotechnic system. It is uncontroversial that these boxes should be made of metal. However, these metal boxes are rarely if ever completely continuous—they are not welded shut after installation. There are usually seams from the panels of the enclosure, and sometimes apertures for thermal regulation as well, depending on what other systems may be housed in the same box (ideally as few as possible). Remember that apertures can act as antennas themselves—a slot cut in a sheet of solid metal will radiate in resonance with a particular frequency of RF just as a rod antenna of the same length would. It is important to control these potential sources of RFI that can couple into the control circuitry of the pyro system.

The best way to handle a seam is to have overlapping portions of flat metal, bonded together by screws or other fasteners spaced at relatively close distances (closer is better). Overlapping flat metal, with some mechanical force to keep them in intimate contact, if they have been properly cleaned and have no non-conductive coatings in the joint, is the best possible kind of electrical bond, and essentially it ensures electrical continuity of the joint. Be aware that some kinds of fasteners (e.g., screws, bolts, and rivets) have a tendency to "relax" over time, slightly releasing pressure on the bond and degrading its effectiveness. Also consider the potential of corrosion when dissimilar metals are in contact in a

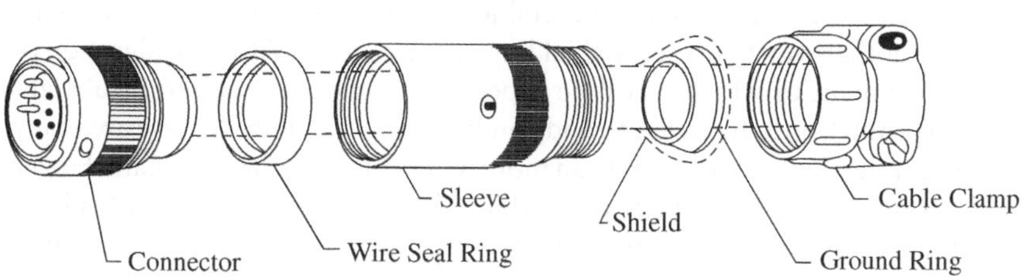

Figure 6.9 Shield termination using a threaded assembly. *Source*: [14, Figure 5-16].

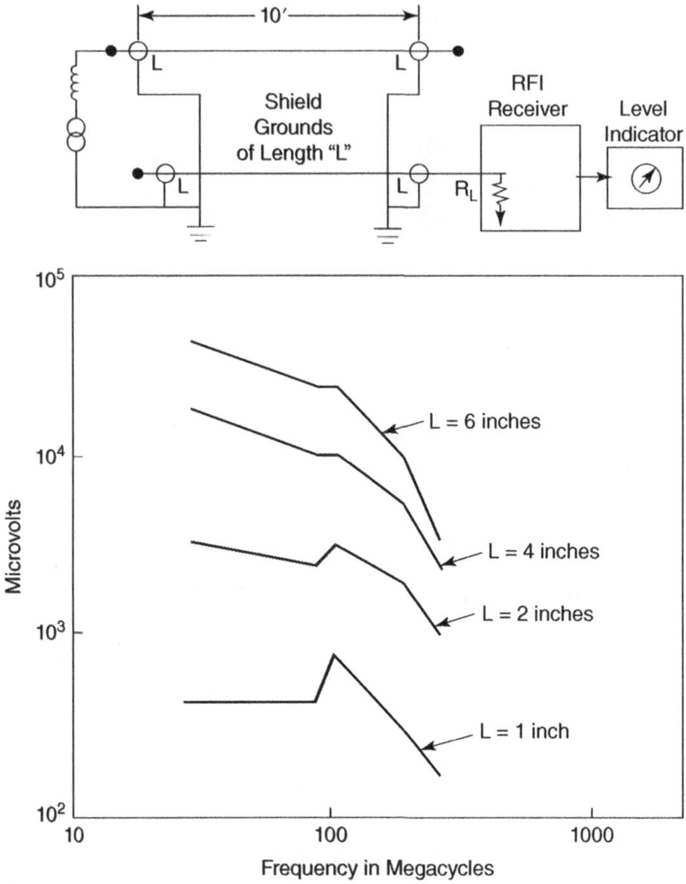

Figure 6.10 Coupling as a function of termination length of shield.
Source: [14, Figure 7-6].

corrosive environment (such as the salt air and water near ocean coastlines or on ocean vessels) and make sure to choose compatible metals for the join.

In some cases, a flush metal-to-metal join is not practical. Conductive gaskets are another way to preserve shielding effectiveness. With consideration to the environment to which it will be exposed (e.g., does it need to be airtight or waterproof?), and the strength and conductiveness required, there are many options for flexible materials that can more or less fill a gap in a box seam. Finger stock is one option, often made of beryllium copper. Different kinds of rubber combined with metal can provide gaskets that are both flexible and conductive (although vendor claims of conductivity should be verified in each specific application, as properties can vary dramatically between lab conditions and the

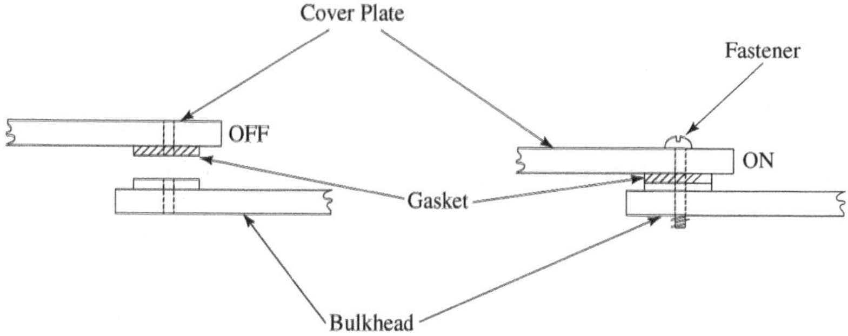

Figure 6.11 Flat cover plates with gaskets. *Source:* [14, Figure 5-25].

"real world"). Figures 6.11 and 6.12 illustrate different ways to install a conductive gasket.

Other apertures in a pyro-related box should be avoided if possible. But in the event that they may be needed for thermal regulation, there are a few ways to mitigate their provision of an entry point for RFI. Depending on the requirements for attenuation of RFI and maximum airflow, metal screen meshes or honeycomb meshes might be used (honeycomb is the best for maximizing airflow and RF attenuation, but is subject to cost, mass, and volume constraints). Apertures can be sized if the maximum frequency that is a potential RFI threat has been identified. Each aperture can be considered to be a waveguide, allowing frequencies higher than its cutoff to propagate and attenuating frequencies below that. For rectangular apertures, the cutoff frequency (f_c) is:

$$f_c = (5.9 \times 10^9)/W$$

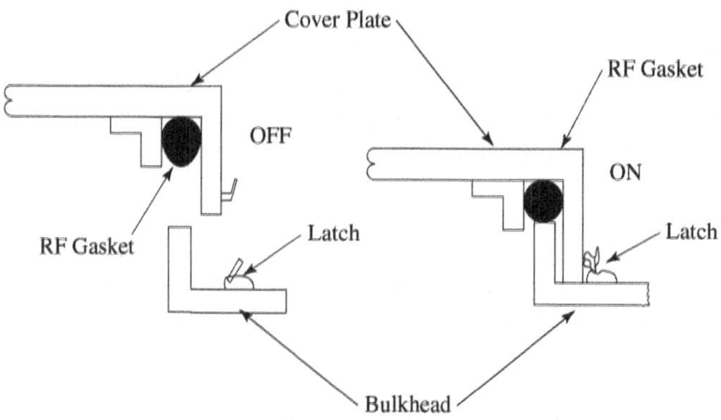

Figure 6.12 Angled cover plates with gaskets. *Source:* [14, Figure 5-26].

where W is the longest dimension of the aperture in inches, and f_c is in Hertz. For circular apertures the relationship is:

$$f_c = (6.9 \times 10^9)/d$$

where d is the radius of the circular aperture in inches [16, Chapter 6].

6.3.3 Cabling

The shields over cable harnesses have already been discussed. However, it is important to consider this shielding in light of the expected EME and the worst-case possible coupling between the EME and the cable itself. In the very worst case, the cable shield acts as a tuned half-wave dipole antenna at the frequency of interest (when doing a safety analysis, usually the frequency with the highest expected power and exposure with the lowest frequency). We can use the following equation to calculate the power received by this unintentional antenna (from [14, pp. 2–18]:

$$P_r = (G_r \times \lambda^2 \times \mathrm{PD})/(4\pi)$$

where P_r is in Watts, λ is the wavelength of interest in meters, PD is the power density in Watts/meter^2, and G_r is 1.64, the gain of a tuned half-wave dipole. The current induced in the cable is then approximated by the DC Ohms law equation:

$$I = \sqrt{(\mathrm{P}/\mathrm{R})}$$

where P is power in Watts, R is resistance in Ohms, and I is current in amperes.

As stated in NAVSEA-OD-30393 [14], "The current in the bridgewire has never been found to exceed the value calculated by this method." The idea in this worst-case analysis is to discover the worst possible induced currents on the cable shield caused by the EME, and then to calculate the shielding effectiveness needed to prevent any current greater than 16.5 dB (or 20 dB) below the MNFS from developing in the EIED bridgewire or firing lines. If the current calculated above is converted to dBA (dB compared with 1 Ampere):

$$I_{dBA} = 20\log(I_A)$$

and the mean no-fire current is converted using the same formula, then the necessary attenuation that must be provided by the shielding and cable design is simply:

$$\text{Cable attenuation} = I_{dBa}\,(\text{induced by EME}) - I_{dBa}\,(\text{mean no-fire current})$$

If the EME is not well defined by the specifying authority, 200 V/m is equivalent to PD = 100 W/m^2, and the military standards often impose this environment from 50 MHz up to 40 GHz (0.075 m). However, the equation above shows that lower frequencies of longer wavelength can impose the more severe threat, and Mil-Std 1576 [10] suggests 2 W/m^2 at 1 MHz (300 m). Thus, the worst-case points will likely be 2 W/m^2 at 1 MHz and 100 W/m^2 at 50 MHz. Remember that at a certain point, the physical limitations of the vehicle can come into play: no possible firing line harness on a 3-m long satellite will be a half-wave dipole at 1 MHz, so common sense can rule out that particular threat, leading the engineer to focus instead on frequencies that might more readily couple into the harnessing.

Although shielding is critical to achieving this attenuation, another critical feature of the cable is that the wires of each firing circuit should be *twisted shielded pairs.* A real risk in any pyrotechnic system is crosstalk between firing wires co-routed in the same harness—the risk that a firing signal sent down one wire pair might induce currents in a neighboring pair of sufficient magnitude that another device, meant to be safed, ends up firing. One way to avoid that potential is to avoid co-routing pyro leads in the same harness, but that is often impossible to avoid. The simplest way to minimize the risk is to twist the wires for each firing circuit together. This minimizes any loop area that can cause current-induced magnetic fields to couple from one circuit to a neighboring circuit. Twisting also ensures that the firing leads will be the same length, another easy way to avoid induced voltages. The number of turns per foot is debatable, with guidance ranging anywhere from 6 to 20 turns per foot [14,16]. Figure 6.13 shows several different shielding and twisting configurations, with shielding effectiveness numbers from experimental data.

In addition, each individual pair should be shielded as well as the overall cable harness shielding. These individual pair shields should never be brought through a connector, such as a connector that might route a cable harness through a bulkhead. Individual shields within a harness should be terminated at each connector by removing them from their conductors as close to the connector as possible and tying them to the outer cable shield. At that point, assuming the outer shield is terminated properly, the inner shields will also be well terminated. Through bulkhead connectors are the place where crosstalk between different cables within a harness is most likely, since when all conductors in a harness are brought through on pins they are neither shielded or twisted. This environment should be considered when assessing risk of routing different firing, signal, and power lines together.

For these reasons, pyro lines should never be routed in the same harness as high current AC power. They should also be separated from sensitive communications or low voltage differential signaling lines, as the fast rise-time transients caused by a firing pulse can interrupt signals on those lines during a firing

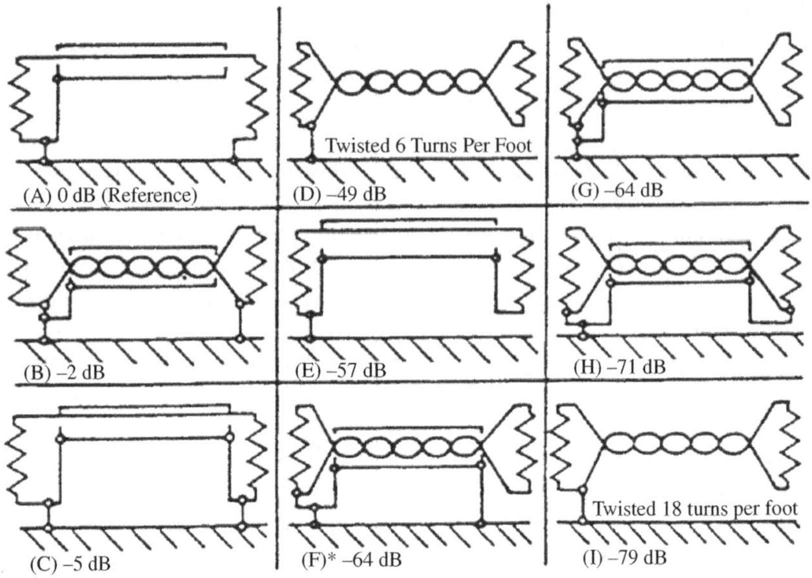

*Preferred circuit for high frequencies

Values given are for circuits 1 inch above ground plane but are about the same for other distances from ground plane

Figure 6.13 Shielding effectiveness provided by different twisting and shielding configurations. *Source*: [14, Figure 7-10].

event. Pyro lines should also be kept separate from lines that are carrying RF power or signals for communication or radar antennas.

Cables should be as short as feasible and, whenever possible, routed close to a ground plane or vehicle structure—in this way avoiding large loop areas that could encourage magnetic coupling. Long conductors can be efficient antennas when their length approaches $\lambda/4$. If there are particular frequencies of interest (nearby AC power lines or communication transmitters in the vicinity), cable shields should be grounded at intervals of no more than $\lambda/6$ to prevent this coupling from occurring. Multiple point grounding of cable shields is another way to prevent large loop areas from allowing EM fields to couple into the harness.

6.3.4 Grounding

Shielding without proper grounding may prove to be less effective than hoped. The fundamental goal of grounding is to prevent potential differences between any two points in a system that could allow unintended currents to flow. The ideal is a perfectly conductive ground reference system that has zero impedance over the entire range of RF energy and that is easily referenced by every electrical/electronic device in the system. However, due to the nature of realistic materials and the challenges of locating the myriad different electronic units within a vessel, the goal is to get as close to this ideal as possible within design constraints.

In most aerospace platforms, we refer to a "ground plane" in the vehicle. This is a system of elements that are all intended to be held at the same electrical potential, more accurately termed a "reference plane." This may be the chassis of a vehicle, or an actual metal sheet upon which the electronics boxes are mounted, or a system of structural elements, mounting racks, and cable shielding that are all joined together with the best possible bonding. Historically, the metal chassis or skin of an airplane, spaceship, or other vehicle could be used as a ground plane for the internal electronics. However, in the age of composite materials, this is often no longer possible. Composite materials may be electrically conductive to some extent (varying dramatically based on the materials involved and the method of construction), but are rarely conductive enough to be used as a ground plane for RF-sensitive electronics.

For the pyro system specifically, the housing of the boxes that include the firing control circuitry must be grounded by the most direct means possible. If bonding straps are necessary to make the connection, they should be as short as possible and ideally should follow the 5:1 length-to-width ratio rule of thumb that reduces the inductance that might present high impedance at higher frequencies. Bond straps can turn out to be ineffective at RF frequencies, and should be avoided if possible.

The shields of the cables emanating from those boxes should be grounded at the box end and again at their terminations at the EIEDs themselves, provided that the EIEDs are mounted so as to be connected to the same ground system. Figure 6.14 shows an example of bonding a connector to prevent RFI. The reference potential must be confirmed: if the pyro controlling box and the EIED housing are not referenced to the same ground, then induced currents are almost guaranteed to flow on the outer surface of the firing cables, causing enormous potential risk. At that point the shielding is a path for interference instead of protection from it. Remember that some low-voltage firing devices, such as the typical single bridgewire 1 A/1 W no-fire device, can be more susceptible to RF potentials developing between the pins of the EIED and the case of the unit than they are to RF potentials developing pin-to-pin across the bridgewire. Keeping the shields and housing of the pyro system from developing unintended potentials is critical for system safety and functionality.

Grounding is also important for protection from ESD hazards. Continuous low impedance ground paths prevent static charge from building up at any one spot in the system and potentially causing a spark or other ESD event [18]. Keep in mind that ground paths that may not be suitable for protection from RF energy can be more than adequate for bleeding off static charges. Many

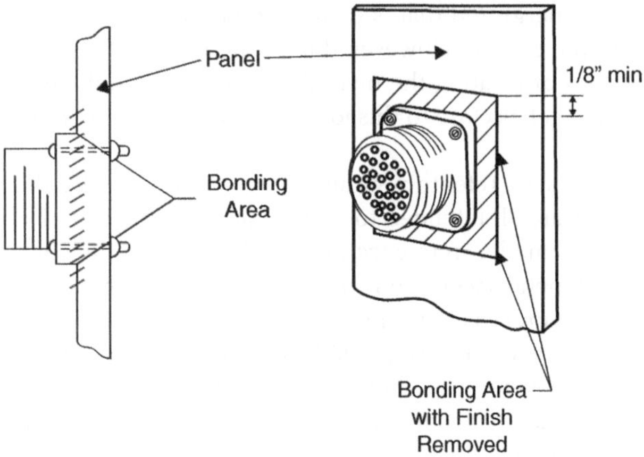

Figure 6.14 Bonding a connector. *Source*: [14, Figure 6-4].

composites, especially those made with graphite, have enough conductivity to allow for static bleed off, even if they would never be considered part of the primary ground plane.

6.3.5 Protection of Firing Circuitry

If we imagine a continuous shielded volume that includes the box housing the firing control circuitry and continues through the firing line cables all the way to the EIED itself, then we must be aware of the different signals that penetrate that shielded volume. At the very least, there will be a signal line that carries the firing commands into the pyro controller circuitry. Quite possibly, there are many lines penetrating the volume: sensors, power supply, and different kinds of signaling. Protecting the firing circuits inside their boxes is very important.

The first thing to consider is filtering the signals as they come into the pyro controller box. The design of EMI filters is a topic worthy of a book all on its own, and there are many sources available that can offer guidance. The requirements will differ depending on the type of line that is being filtered, the source and load impedances, and the amount of expected current and power. Generally speaking, low-pass filters will be needed to allow intended power and signals to pass through while blocking high-frequency noise from entering the shielded volume. Shunt capacitors placed close to the signal/power inputs to the box can be effective in this regard, although more complicated filter types may be needed depending on the application.

Transient suppression should also be considered as part of circuit protection. The firing system itself can produce significant transients when the firing signal is finally switched on and then ceases (when the EIED is fired and becomes an open circuit, no longer drawing power). Even if there are no other sources of transients in the box, the firing system should be designed to be self-compatible, which may mean transient suppression elements on the firing lines. Resistors can be used to slow rise times, but at the cost of dissipating more power. Other options include RC circuits and various combinations of passive elements, diodes, and transistors.

Another consideration should be what other systems are packed into the same box. Ideally, a pyro controller box would contain no circuitry or printed circuit boards unrelated to the pyro system. In reality, space and weight constraints make that isolation impractical, and the pyro firing system may be controlled by a board or set of boards included as one of many inside an electronics box. In this case, it is critical that the cards dealing with pyros should be isolated from the other boards, at a minimum by a ground plane between the two sets of cards. Each pyro card should be protected by ground cards on either side of it, offering shielding from any potential sources of interference. Physical separation between the different kinds of cards will help (but again may be impractical). As much as possible, cards handing power switching or RF signals should be kept well away from the firing circuit cards.

Sometimes a monitoring circuit is required for the overall system: a monitor that can tell the safe/arm/fire status of each EIED circuit and the fired/unfired status of each EIED. These monitors should use the lowest amount of current possible while still being effective; at the very least, they should draw no more than 10% of the firing current of the EIED in use.

6.3.6 Platform-Level Testing

Once the entire system is assembled, it must be tested. The EIEDs will be tested separately as discussed above, but pyro subsystem and then vehicle-level testing will likely be required. Testing pyrotechnic systems is uniquely challenging, as the consequences of inadvertent ignition (one likely consequence of a failed test) are more catastrophic than informative. Live EIEDs should always be removed from the vehicle, system, or subsystem prior to testing, and the vehicle should be designed in such a way that EIEDs can be installed and removed at whichever level of vehicle integration testing is planned, all the way up to final assembly if needed. Keep in mind that conditions that exist during ground handling will need to be tested as well as the fully integrated pyrotechnic systems.

6.3.7 Worst-Case Scenarios

Test definition is extremely important, and thorough knowledge of the design can help immensely. Specifically, knowing particular points of entry for RF power (apertures, discontinuities, lid seams, etc.) and their dimensions can emphasize certain frequencies and orientations to test; ground handling procedures can point toward specific ESD test levels, cable lengths can suggest potential resonant frequencies, etc. When it comes to choosing specific testing scenarios, Mil-Hdbk 240A [9] suggests pretest characterizations that can be done relatively quickly, such as placing a current probe on a ground strap between the ordnance and ground and quickly trying a number of different test configurations (polarizations, illumination angles, etc.) to see which ones induce the largest ground currents.

The concept of "worst case" also allows us to contend with the fact that while the average EME is specified in terms of far-field values (200 V/m or some other well-behaved value), almost all platform- and device-level testing will occur in the near field of a given transmitting antenna. Although this is not ideal, it is generally the only practical way to proceed. In this case, a consistent calibration method is crucial, with field probes being placed in the test area and monitored to ensure that the realized field strengths match the prescribed EME. It is important to keep in mind, however, that near-field power is not nearly as well behaved as far-field power and may have unpredictable or counterintuitive results; in some aircraft testing, it has been seen that more intense fields develop near the wing *farther away* from the radiating antenna than on the closer wing, which may undermine naïve determinations of "worst-case" orientations [9, p. 65].

6.3.8 EIED Surrogates

Since live EIEDs must be removed during testing, there are a few different techniques available to determine how much energy is being coupled into the firing system and whether the EIEDs are at risk of inadvertent operation or dudding. (It should be noted that EBW and EFI devices, being initiated by high voltages, are usually robust in terms of inadvertent operation but damage from heat dissipation due to RF energy in the foil/bridgewire can cause dudding. In these cases, a mean no damage stimulus will be defined and should be tested to at the system level.) One option is to place an RF-sensing device near the location of each EIED bridgewire, to accurately measure the RF voltages developed in the area of the EIED during operations.

Another approach is instrumented surrogate EIEDs. For several different types of single bridgewire EIEDs, such as the NASA Standard Initiator (NSI), devices have been made which create surrogate EIEDs, which have the same form and fit as the originals, but with thermocouples attached to the bridgewire and no explosive material. A generic example is shown in Figure 6.15. The measurement from the thermocouple is sent to a data-acquisition device via fiber optic cable. These are sometimes better used in subsystem-level testing than in system-level testing due to the challenges of installation, but can provide detailed measurement data that yield solid information about margins (how

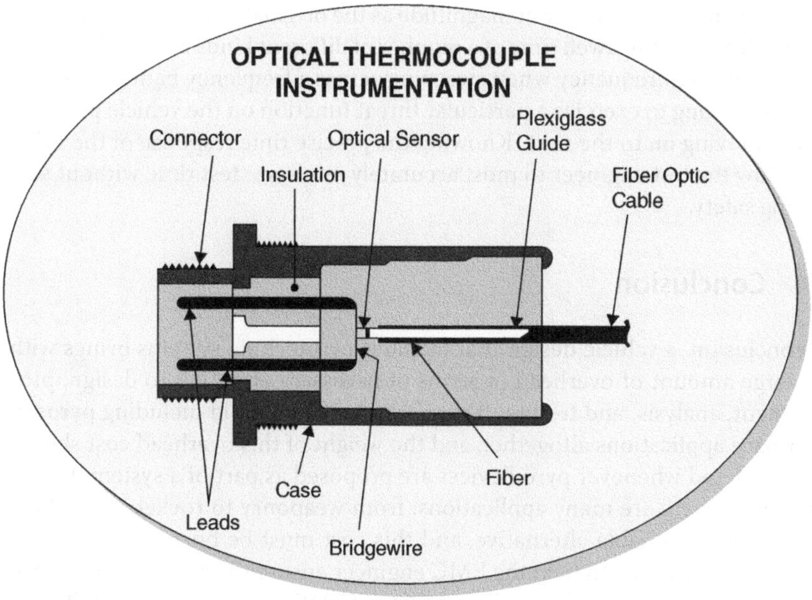

Figure 6.15 Optical thermocouple instrumentation of a hot bridgewire EIED.
Source: [9, Figure 17].

close the device was to the MNFS). This is especially helpful in those cases where very high average power levels are specified in the EME definition document but cannot be replicated in the test lab. Devices such as these provide enough information for a reasonable extrapolation from device response to test levels to expected response to true EME levels. Any such device should have the ability to measure energy at least as low as 5% of the MNFS of the device, and 1% is better. It should be noted that whenever expended EIEDs are refurbished and instrumented to serve as test surrogates, their impedance may be significantly different from the original live devices and this difference should be characterized if possible.

A less desirable but still feasible option is to replace EIEDs in the vehicle with fuse devices in which the fuses will blow if the MNFS is reached. Then, the system can be monitored for the timing of any fuses blowing and the timing correlated to the stress that caused the failure. These devices are relatively simple and, if any access at all to the EIEDs is possible, can be installed simply in place of the device. However, they yield no information about safety margins and cannot be used to extrapolate from lower field strength performance to higher.

An important note about any EIED surrogate being used: it should have a response time (tau, or the time constant as defined in the discussion of thermal stacking) on the same order of magnitude as the original device. This time constant will dictate the dwell time of a number of different kinds of tests: how long to dwell at each frequency when sweeping across a frequency band in discrete steps; how long to exercise a particular threat function on the vehicle platform before moving on to the next. Knowing the precise time response of the EIED can allow the test engineer to most accurately minimize test time without sacrificing safety.

6.4 Conclusion

In conclusion, a vehicle design that includes pyrotechnic systems brings with it a large amount of overhead in terms of necessary attention to design, procurement, analysis, and testing. It would be better to avoid including pyros in aerospace applications altogether, and the weight of this overhead cost should be considered whenever pyro devices are proposed as part of a system design. Obviously, there are many applications, from weaponry to rocketry, in which there is no reasonable alternative, and this cost must be borne. In that case, many resources exist to help the EMC engineer ensure the safe and functional operation of the vehicle. Generally speaking, if EMC best practices are followed meticulously, the only difference between a pyro subsystem and any other critical system on a vehicle is the design margin of safety—15 to 20 dB of margin compared with 3 to 6 dB for many other applications.

The power and danger of explosive devices should be intuitively obvious to everyone involved in designing and managing aerospace projects, and it may be easier to get the needed design allowances and resources needed than for some other EMC concerns—if these concerns are raised early and often. "Fixing" a pyro system that shows vulnerability to EMC late in vehicle validation is (very) difficult and expensive, so full attention to EMC principles as early as possible in the program process will pay enormous dividends later on. Although this chapter has focused on "worst-case" situations, such as single bridgewires and Navy HERO procedures, the general principles described herein should assist those tasked with designing pyro systems in any aerospace application. When it comes to subsystems with this level of potential catastrophic impact, "overkill" in terms of system design, protection, and testing is much preferable to the alternative.

References

1 Bliss, C. I. "The Method of Probits." *Science*, vol. 79, 1934, pp. 38–39.
2 Burnham, Karen, et. al. "NASA Standard Initiator Susceptibility to UHF and S-Band Radio Frequency Power and Lightning Strikes." *Proceedings of the EMC Symposium in Denver, CO*, 2013.

3 Burnham, Karen. "Induced Currents and Voltages from Indirect Lightning Strikes in Cable Harness Suspended Above Composite Panel." JSC Technical Report, JSC-66640, October 2013.

4 Dixon, W. J. and A. M. Mood. "A Method for Obtaining and Analyzing Sensitivity Data." *Journal of the American Statistical Association*, vol. 43, March 1948, pp. 109–122.

5 Joint Ordnance Test Procedure (JOTP)-061: Hazards of Electromagnetic Radiation to Ordnance (HERO) Safety Test. Naval Ordnance Safety and Security Activity, 2013.

6 JSC 62809: Human Rated Spacecraft Pyrotechnic Specification. Rev. D, NASA, 2010.

7 Langlie, H. J. "A Reliability Test Method for 'One-Shot' Items." *Proceedings of the Eighth Conference on the Design of Experiments in Army Research Development and Testing*, 1965.

8 Mil-C-27500G: Cable, Power, Electrical and Cable Special Purpose, Electrical Shielded and Unshielded, General Specification. US Department of Defense, 1988.

9 Mil-Hdbk-240A: Hazards of Electromagnetic Radiation to Ordnance Test Guide. US Department of Defense, 2011.

10 Mil-Std-1576: Electroexplosive Subsystem Safety Requirements and Test Methods for Space Systems. United States Air Force, 1984.

11 Mil-Std-461G: Requirements for the Control of Electromagnetic Interference Characteristics of Subsystems and Equipment. US Department of Defense, 2015.

12 Mil-Std-464C: Electromagnetic Environmental Effects Requirements for Systems. US Department of Defense, 2010.

13 NASA-STD-4003A: Electrical Bonding for NASA Launch Vehicles, Spacecraft, Payloads, and Flight Equipment. NASA, 2013.

14 NAVSEA OD 30393: Design Principles and Practices for Controlling Hazards of Electromagnetic Radiation to Ordnance (HERO Design Guide). Second Rev., Naval Sea Systems Command, 2001.

15 Neyer, B. T. "A d-Optimality-Based Sensitivity Test." *Technometrics*, vol. 36, 1994, pp. 61–70.

16 Ott, Henry. *Electromagnetic Compatibility Engineering*. New York: Wiley, 2009.

17 Philosophy and Methodology of the Hazards of Electromagnetic Radiation to Ordnance (HERO) Program. Revised. Naval Surface Warfare Center, 1988.

18 Soriano, J. Francisco. "Reducing/Eliminating ESD Hazards During Pyro Operations." NASA TM-2001-210256, June 2001.

7

Assembly-Level EMC Testing of Space Components/Subsystems

Leslie R. Warboys

The purpose of this chapter is to discuss the testing of spacecraft components and subsystems. The discussion will include requirements and clarification, derived requirements, testing approaches and types, potential component to system impacts and vice versa. The engineer must keep in mind that the success of any mission is highly dependent upon the proper operation of subsystems and their electromagnetic interplay. For ease of reading, the chapter will refer to component or subsystem as just the word subsystem. If an individual component must be tested to ensure the entire subsystem is functioning, then test it following all the recommendations of the subsystem level within the chapter.

7.1 Preliminary Steps

The following four items should be considered prior to contract award as much as possible. These items will impact the type of testing executed and, of course, the cost of the project. However, in real-world contracting, it may not always be feasible. It is best practice to plan ahead for the subsystem-level testing. The steps an engineer should execute to clarify the electromagnetic requirements are as follows:

1. **Review/discuss the contractual E3 requirements set forth by the customer.**
   This will clarify the meaning; within the customers mind eye, and rationale for the EMC requirements (MIL-STD, Aerospace, AIAA, IEEE, IEC, etc.) or any specific preferential approach or combination.

Handbook of Aerospace Electromagnetic Compatibility, First Edition. Edited by Reinaldo J. Perez.

2. **Ascertain what launch facility will be used.**
   This will allow the engineer to assure that the subsystem will function properly and not impacted by the environment exhibited at the launch facility. Subsystems are usually tested at only one facility. However, note that there are always multiple facilities: factory, transport, assembly area, transport to pad, pad, etc. Variations to the basic requirements that may impact the operation of the subsystem should be tailored (altering the requirement to meet the real requirement) into the test requirements and agreed upon with the system engineer and or the customer.
3. **Determine what launch vehicle type will be employed through orbit insertion.**
   Different launch vehicles exhibit differences in grounding schemes, conducted and radiated emissions will vary throughout the prelaunch, launch phase and other mission phases, particularly if the subsystem is sharing launch vehicle power.
4. **Identify unique subsystem requirements.**
   The engineer should inform the customer of any unique subsystem requirements. These requirements could be as simplistic as receiver frequency, bandwidth, and threshold levels as well as transmitter frequencies, keying, encryption. Some components may require DC/DC converters or filters, to provide isolation from the spacecraft power bus to preclude any known bus noise issues, or for isolation purposes. Addressing these needs early on in the integration process avoids costly modifications later in the spacecraft build.

7.2 Basic Testing Concepts

There are four basic must types of testing performed at the subsystems level. It is important to note that some of these tests cannot be performed at the system/spacecraft level. Once the subsystem is integrated into the entire system or spacecraft, it is extremely difficult to isolate from all the other radiated or conducted components on the vehicle. The extremely difficult test would be conducted emissions, conducted susceptibility, radiated emissions, and radiated susceptibility. The aforementioned tests, or variations thereof, regardless of the contractually specified documentation, are required. Basically, the tests tell the engineers the following with respect to spacecraft and the subsystems:

- **Conducted Emissions**
  Typically, conducted emissions tests are performed at the subsystem level. These are not reliable at a system level test due to the numerous contributions on to the system power bus from multiple other subsystems. There is inherent danger to other subsystems at the spacecraft level, if test limits are

significantly exceeded. The measured results will inform the system engineer as to what to expect, in the way of bus noise contribution, from the subsystem.

- **Conducted Susceptibility**
  Typically, these tests are performed at the subsystems level but not at the system level, because there is a risk of damage to hardware. These tests cannot be executed at the system level due to the numerous contributions to the system power bus, from other subsystems, and the inherent danger of potential damage to any of the other subsystems at the spacecraft level. The measured, threshold levels of the injected signal, if below the test limits, will inform the system engineer as to what to expect, in the way of performance, from the subsystem once installed on the spacecraft bus.
- **Radiated Emissions**
  Some radiated emissions tests can be performed, if *very carefully* executed, at the system level. It requires a highly experienced EMC engineer and specialized test equipment. However, it would not be advisable if there are any extremely RF-sensitive units on the spacecraft, due to damage potential. It is not uncommon to see highly sensitive receivers or optical units on a spacecraft and if the radiated levels at the subsystem level are excessive damage will occur. The tailored limits at the subsystem level should be representative of the sensitive frequencies elsewhere on the spacecraft or launch vehicle. The measured results will inform the system engineer as to what to expect, in the way of radiated emission levels, from the subsystem.
- **Radiated Susceptibility**
  Radiated susceptibility tests are performed at the subsystem level. The tailored levels, which the customer provides, would represent those radiated emissions on the spacecraft as well as the launch vehicle and launch facilities. The results, particularly any threshold levels recorded, will inform the system engineer as to how well the subsystem will integrate into the spacecraft from an RF susceptibility point of view.

7.3 Commonly Performed Tests

This segment will describe the most commonly used tests and explain their rationale. Within the basic concepts previously described, there are a number of executable subtests. MIL-STD-461G constitutes the structural basis for the majority of tests currently performed for subsystems. Hence, we will review the tests within this document that are most commonly called for, since the inception of the 461 series, and break down the aforementioned testing concepts above.

The variations of the MIL-STD-461 document (Basic through G) have offered a suggested table of tests to be performed on subsystems. (It has also been used on ground-based weapons, ships, aircraft, and spacecraft systems throughout the years.) MIL-STD 461 is referenced in many other documents to include the latest AIAA Standard, S-121A-2017 "Electromagnetic Compatibility Requirements for Space Equipment and Systems." The current document of MIL-STD 461G has continued that tradition. Keep in mind that the limits can be, and are, tailored to the needs of the system by the systems engineer and the needs of the customer. The latest table, "Table 7.1," currently depicts MIL-STD-461G.

The recommended tests for spacecraft are in bold italic red print. Since problems with component parts, sensitive to frequencies as low as 50 and 60 Hz, have been experienced in the past, tailoring CE01 or CE101 in the current standard may be required in conjunction with agreement of the system engineer or customer. Depending on the experience of the spacecraft systems engineer or the customer, expect variations from Table 7.1 to occur. This chapter will follow the table with respect to testing regiment depicted with the customer specifying all the legend S categories.

7.4 Test Plan

Now that we have covered available test options and sources, let us look where testing really starts. The first thing the electromagnetic compatibility engineer (EMC) must do, after receiving the finalized test and or tailored requirements from the spacecraft systems engineer or customer, is prepare a test plan for the subsystems to be tested. The following tests would be candidates for inclusion in the plan depending on applicability and customer approval. The engineer should keep the test plan execution flexible, to allow for schedule changes or test equipment failures.

A test plan consists of the requirements as agreed upon by the contracting bodies. A list of the tests and how they are to be performed is given in the documents. There is a general sketch or block diagram of the item to be tested. Look for the item's location within the whole system if possible. Because of classified information, the test engineer may not know where the testable item is located or how it will interplay within the entire spacecraft.

7.4.1 Pass or Fail Criteria

There is no room for misinterpretation of test results as to whether the equipment under test (EUT) has passed or failed. According to the requirements, an item either passes or fails, no in-betweens. The EMC engineer then needs to have clear definitions of what constitutes a pass and what constitutes a fail and how to monitor for such. Failures should be documented in as much detail as possible for the design engineers and system engineers so they know what to do next.

Table 7.1 Requirement matrix

Equipment and subsystems installed in, on, or launched from the following platforms or installations	Requirement applicability																		
	CE101	***CE102***	***CE106***	***CS101***	***CS103***	***CS104***	***CS105***	CS109	***CS114***	***CS115***	***CS116***	CS117	CS118	RE101	***RE102***	RE103	RS101	***RS103***	RS105
Surface ships	A	A	L	A	S	L	S	L	A	S	A	L	S	A	A	L	L	A	L
Submarines	A	A	L	A	S	L	S	L	A	S	L	S	S	A	A	L	L	A	L
Aircraft, Army, including flight line	A	A	L	A	S	S	S		A	A	A	L	A	A	A	L	A	A	L
Aircraft, Navy	L	A	L	A	S	S	S		A	A	A	L	A	L	A	L	L	A	L
Aircraft, Air Force		A	L	A	S	S	S		A	A	A	L	A		A	L		A	
Space systems, including launch vehicles		***A***	***L***	***A***	***S***	***S***	***S***		***A***	***A***	***A***	L			***A***	L		***A***	
Ground, Army		A	L	A	S	S	S		A	A	A	S	A		A	L	L	A	
Ground, Navy		A	L	A	S	S	S		A	A	A	S	A		A	L	L	A	L
Ground, Air Force		A	L	A	S	S	S		A	A	A		A		A	L		A	

Legend:
A: Applicable
L: Limited as specified in the individual sections of this standard.
S: Procuring activity must specify in procurement documentation.
Source: [1, p. 26].

In the case of susceptibility testing, the EUT needs to have thresholds for the susceptibility signal levels. One must determine at what level the EUT functions improperly. These levels are then recorded and documented in the final testing reports. Even if the engineer rectifies the issue, it must be documented in the report. The reason is if the EUT later fails in a like manner at the next level of a system, the engineers can look back and learn what caused the failure and what was done to fix or address the failure.

In the case of emissions testing, any emission outages and levels should also be noted in the report. If the threshold or emission level cannot be rectified and retested successfully, it is documented and reported to the spacecraft systems engineer to allow for evaluation as to its system compatibility/acceptability.

The system engineer will evaluate susceptibility and emissions outages against the other spacecraft systems to determine acceptability. Those items not acceptable, of the subsystem in conjunction with the system, will have to be resolved prior to flight. The system engineer and customer are consulted if any adjudication requires significant power consumption (DC/DC converter, significant filtering change, etc.) or mass change, due to power, footprint, or mass properties impacts.

Typically, the spacecraft bus impedance will be lower than that of the defined line impedance simulation network (LISN) in MIL-STD-461 G. Hence, some of the perceived outages may be found to be acceptable. It would be in the interest of the subsystem engineer to ask what anticipated bus impedance is expected.

7.4.2 Grounding

The subsystem/component ground configuration must be reflective of the grounding configuration on the spacecraft. This will allow the testing to better simulate the responses of the EUT in its operational configuration.

7.4.3 Cabling

The cabling used during the subsystem testing should reflect the cabling used, in the spacecraft interface, as closely as practicable, to include shields, twists, length, etc., for data as well as power cables. Testing needs to as close as possible to simulate spacecraft configuration.

7.4.4 Line Impedance Stabilization Network

In the past, consensus was to use a LISN; it represents the actual spacecraft bus. It is more cost effective to use a standardized LISN approach; however, this requires conversion, on the part of the system engineer. The system engineer must translate the resultant data to the system bus for all the conducted

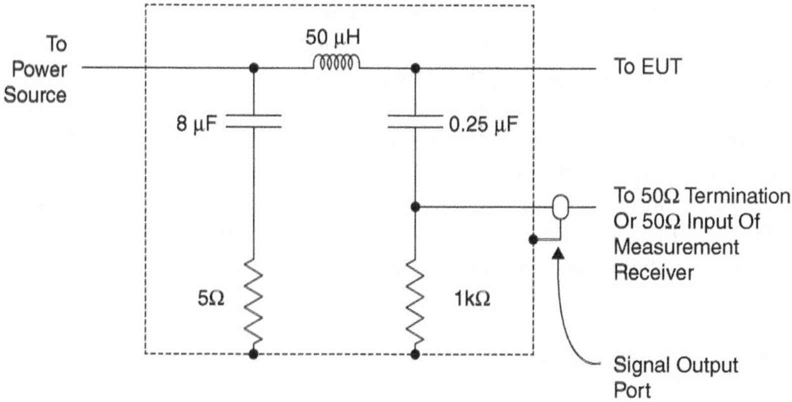

Figure 7.1 LISN schematic. *Source*: [1, p. 22].

test resultant data. The test engineer should ensure the systems engineer is well aware of the standardized LISN approach used in testing. The schematic diagram of the LISN used in 461G testing is depicted in Figure 7.1, as well as its impedance by frequency in Figure 7.2.

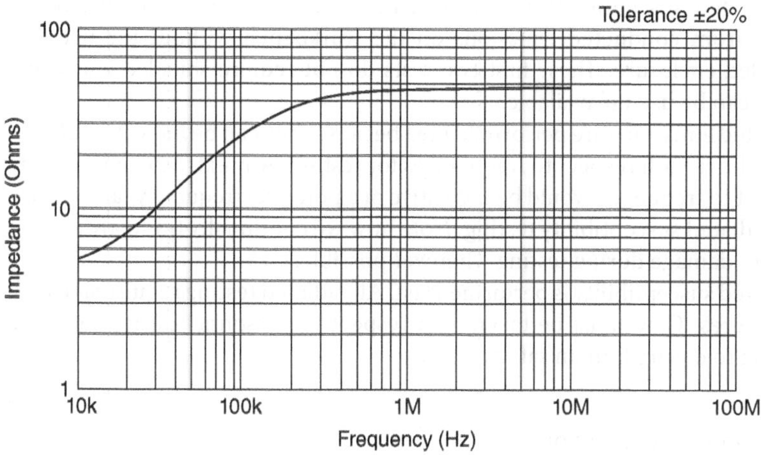

Figure 7.2 LISN impedance. *Source*: [1, p. 23].

The AIAA document S-121A-2017 has modified the LISN figure (Figure 7.1) in the following manner, the 5-ohm resistor on the power source side is removed and the 8 uF capacitor replaced by a 1 uF capacitor tied to the case ground on the power source side. The 50 uH coil has been replaced by a 5 uH coil, the 0.25 uF capacitor replaced by a 1 uF capacitor, and the 1 k ohm resistor by a 5 k ohm resistor to case. Copyright forbids including the 5 uH LISN figure. If the reader needs to visualize it, please refer to the AIAA document. This will increase the impedance rise time slightly with respect to frequency depicted in the impedance graph (Figure 7.2) from MIL-STD-461G.

The engineer should always remember that a subsystem is his/her responsibility. From an EMC point of view, any failure on the ground, that is addressed, is preferable to a failure on orbit or deep space.

7.5 Testing Sequence

Typically testing sequence is performed in a logical manner to protect the EUT and preclude any subsystem damage.

- Conducted emissions are performed first because the emissions seen on the power lines can be indicative of what to expect to see on the radiated emissions. Problems seen on conducted emissions if addressed quickly tend to reduce radiation emissions issues. Additionally, conducted emissions can be indicators of significance of pending conductibility issue, which can cause damage to the EUT.
- Conducted susceptibility is performed second because the setup already exists from conducted emissions on the workbench without the need of shielded enclosure. These first two tests do not require shielded enclosure but it can be used if available.
- Radiated emissions are performed next because they can be indicative of vulnerabilities with respect to frequencies for testing. Any extreme outage with respect to frequency would be indicative of a potential frequency failure during radiated susceptibility testing. Note how we continually try to avoid subsystem damage during testing wherever possible.
- Radiated susceptibility is performed last. All radiated testing require shielded enclosures. Care must be taken with respect to personnel during radiated testing to ensure human safety.

7.5.1 Conducted Emissions

CE102-Conducted Emissions Radio Frequency Potential Power Leads: This test, which was previously known as CE03, is used to monitor power line emissions from the subsystem onto the spacecraft bus that could additively or on its own merit, disrupt or potentially damage other spacecraft subsystems. The frequency range has been reduced from 50 to 10 kHz. The untailored limits, shown Figure 7.3, from the 461G document.

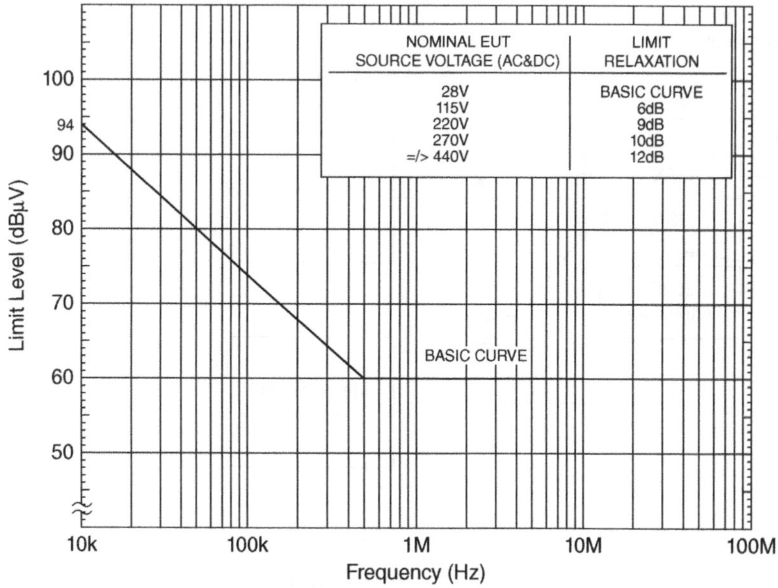

Figure 7.3 CE102 limit (EUT power leads, AC and DC) for all applications. *Source*: [1, p 37].

The above limit is dependent on the spacecraft bus impedance. In the past, every effort was taken to carefully match impedance of the system bus; this is a costly and time-consuming process. The LISN standardized the bus impedance network, requiring additional analytical effort on the part of the engineer to translate any outage results, is much more cost effective as described previously. The engineer (spacecraft or subsystem) must ensure any test outages, to the spacecraft system bus, are acceptable. Typically, the spacecraft bus impedance is lower than that of the LISN across the frequency spectrum.

Figure 7.4 is a schematic picture of the CE102 test setup from MIL-STD-461G.

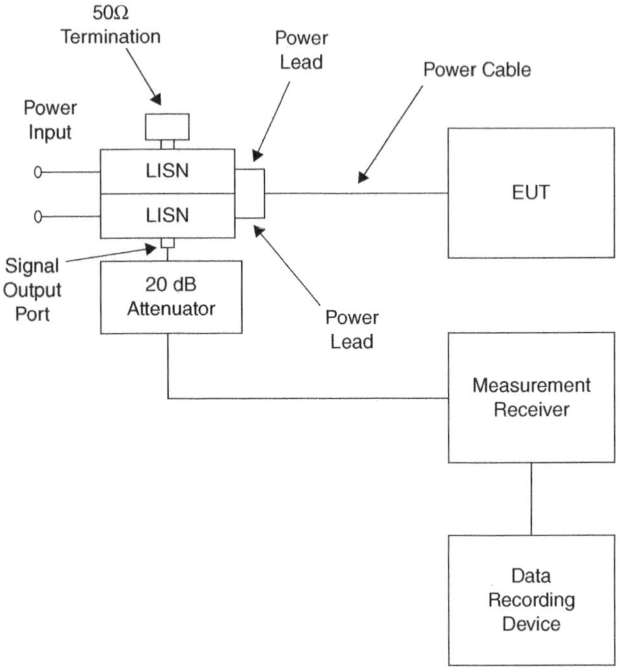

Figure 7.4 Measurement setup. *Source*: [1, p. 39].

CE106: This test was previously known as CE06, again the frequency range has been expanded from 26 to 40 GHz. Obviously, this requirement would not apply if the subsystem/component has no antenna, transmitter, amplifier, or receiver; this test primarily looks for out-of-band emissions such as harmonics and intermodulation products outside the primary operational frequency band.

"EUT operating frequency range	Start frequency of test
10 kHz to 3 MHz	10 kHz
3 MHz to 300 MHz	100 kHz
300 MHz to 3 GHz	1 MHz
3 GHz to 40 GHz	10 MHz"[1]

1 Requirements for the Control of Electromagnetic Interference Characteristics of Subsystems and Equipment MIL-STD-461G 11 Dec. 2015, EUT operating frequency range, p. 40.

7.5.2 CE106 Limits

Conducted emissions at the EUT antenna terminal shall not exceed the values given below.

(a) Receivers: 34 dBμV
(b) Transmitters and amplifiers (standby mode): 34 dBμV
(c) Transmitters and amplifiers (transmit mode): Harmonics, except the second and third, and all other spurious emissions shall be at least 80 dB down from the level at the fundamental. The second and third harmonics shall be suppressed to a level of – 20 dBm or 80 dB below the fundamental, whichever requires less suppression." [2]

7.5.3 Conducted Susceptibility

7.5.3.1 CS101

This is equivalent to the old CS01, the intent of which is to simulate bus noise. The CS101 test as described in MIL-STD-461G only appears to simulate differential mode noise and does not address common mode noise. Differential mode noise and line-to-line injection can be addressed by twisting the power line or by filtering. Common mode noise line to chassis does not appear to be addressed but exists when there is inadequate cable shielding and ground shifts during operational events through launch and mission certain operational events during missions. Common mode noise is exhibited, if there is insufficient or lack of shielding on the spacecraft. Common mode noise can be addressed by adequate shielding, Baluns, [3] and Ferrites, both of which impact mass and magnetic balance/dipole of the spacecraft. The lack of definitive common mode testing has existed for some time even in the old CS01, but common mode noise tends to exhibit itself frequently on spacecraft as well as launch vehicles.

The CS101 limits are shown in Figure 7.5 as reflected in the current MIL-STD-461G and are reminiscent of those used in MIL-STD-461C.

---

[2] Requirements for the Control of Electromagnetic Interference Characteristics of Subsystems and Equipment MIL-STD-461G 11 Dec. 2015, "CE106 limits," p. 40.

[3] Wikipedia, A **balun** /ˈbælʌn/ is an electrical device that converts between a balanced signal (two signals working against each other where ground is irrelevant) and an unbalanced signal (a single signal working against ground or pseudoground). A balun can take many forms and may include devices that also transform impedances but need not do so. Transformer baluns can also be used to connect lines of differing impedance. The origin of the word balun is **bal**ance + **un**balance.

Baluns can take many forms and their presence is not always obvious. Sometimes, in the case of transformer baluns, they use magnetic coupling but need not do so. Common mode chokes are also used as baluns and work by eliminating, rather than ignoring, common mode signals.

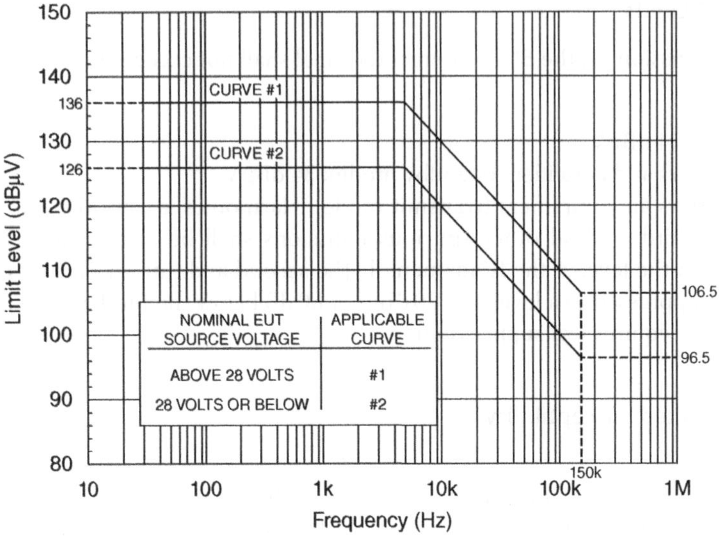

Figure 7.5 CS101 voltage limit for all applications. *Source*: [1, p. 50].

The waveform defined for this test is sinusoidal in nature, which is correct as per the standard. (Many signals contain modulation, and the EUT may pass with a sine wave but fail with a modulated sine wave. The test engineer must always remember that "Mission Success" is what is strived for, and failure of any subsystem may constitute a mission failure. You are only as good as your last success, and a failure will reflect poorly on the engineer as well as his company and could reflect poorly on future project awards. Modulation of the sine wave, as well as common mode test injection, can make the difference between a successful mission and a failure. The modulation used for the test should be reflective of that used within the spacecraft system; if that is not known, then a square wave will usually address worse case for most modulation types.)

CS101 typical setup schematic is shown in Figures 7.6 and 7.7, as depicted in MIL-STD-461G.

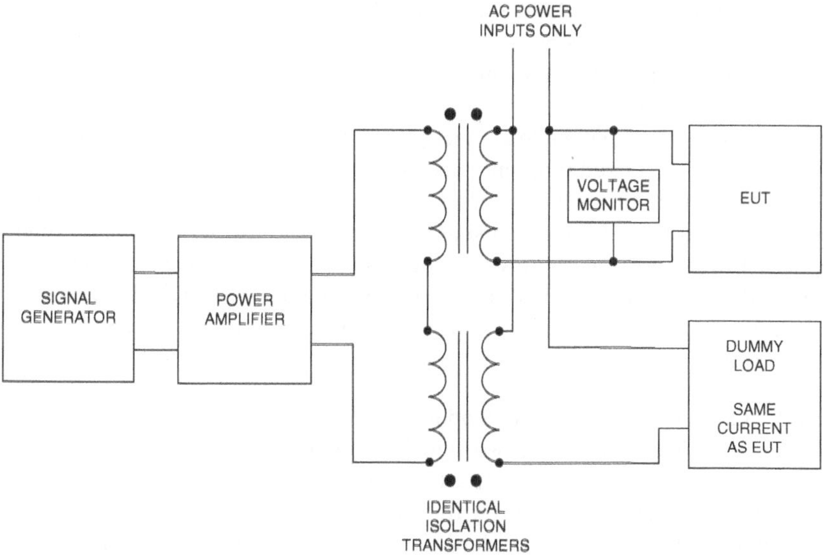

Figure 7.6 CS101 power amplifier protection. *Source*: [1, Appendix A, p. 220].

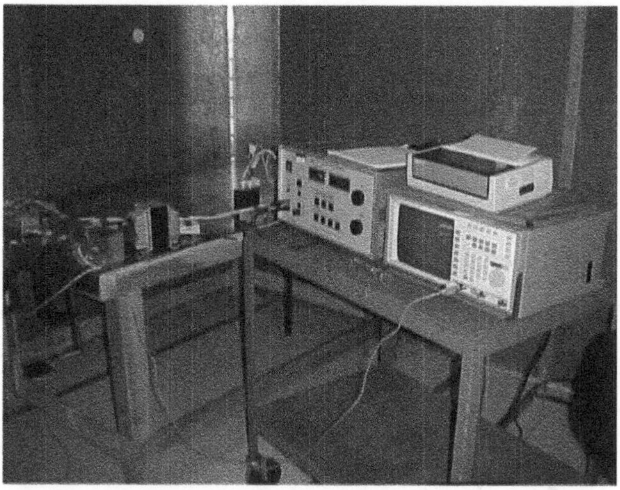

Figure 7.7 Photograph of a CS101 setup with monitor and printer. *Source*: L. Warboys photograph in test application of CS 101 New York 2000.

7.5.3.2 CS103, CS104, and CS105 F_0 Pass Band Measurement

The next three CS tests require the test engineer to find the pass band for the front end of any receiver, transmitter, or amplifier on the EUT. The specifications for the EUT define the theoretical pass band around the F_0. The upper and lower frequencies, of signal loss, labeled mF_0 and nF_0, respectively, constitute the pass band for a fixed frequency receiver, for each test.

Note that some of the test generators throw spurs out of the selected frequency, particularly if they are using frequency synthesizers to develop their signals. The monitoring of the signal sources is imperative, to ensure that they are not the source of spurs within the EUT pass band creating a false failure.

In the absence of a receiver, transmitter, or amplifier, CS103, CS104, and CS105 would not apply.

7.5.3.3 CS103, Antenna Port, Intermodulation, 15 kHz to 10 GHz

This test consists of injecting two signals simultaneously at locations up and down the band from 15 kHz to 10 GHz, outside of the receiver pass band, as previously measured/swept above. The out-of-band signals are at mF_0 and nF_0, one signal modulated using standard receiver modulation, the second signal is CW. Each of the two signals injected is as defined in MIL STD-461G paragraph A 5.8 for the receiver. The receiver shall not respond to any intermodulation products generated. The difficulty here is ensuring that neither of the generators is throwing a spur, by itself, into the receiver pass band.

The setup for this procedure as shown in Figure 7.8, as depicted in MIL-STD-461G.

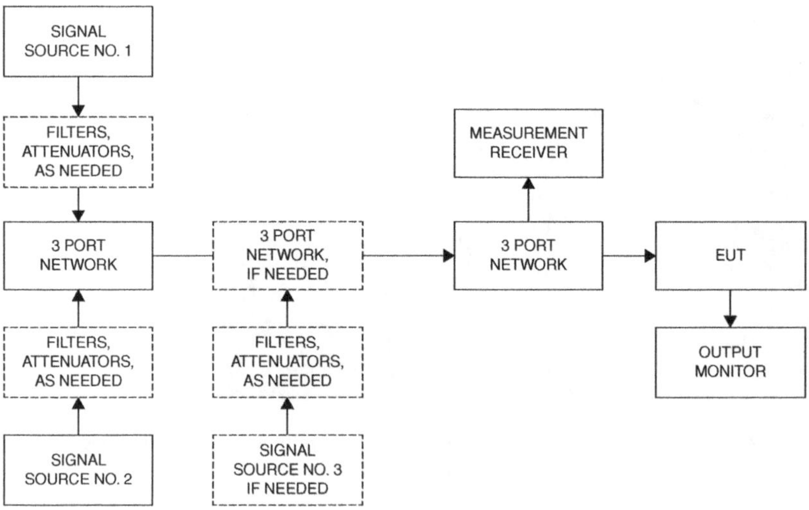

Figure 7.8 CS103 General test setup. *Source*: [1, Appendix A, p. 222].

7.5.3.4 CS104, Antenna Port, Rejection of Undesired Signals, 30 Hz to 20 GHz

The pretest pass band sweep is the same as described in paragraph A5.9 of MIL-STD-461G.

This writer has preferred to use the single signal method using a modulated signal as referenced in paragraph A5.9 (modulation waveform as normally seen by the EUT receiver) sweeping up from the upper pass band frequency to 20 GHz and down in frequency from the lower pass band frequency to 30 Hz. The receiver is to be monitored for any response during the execution of the out-of-band frequency sweeps. The alternative method of this test is to place a modulated standard reference signal defined in A5.9 of the MIL STD above the standard reference signal modulated sweeping up and down in frequency out of band as previously described, monitoring the receiver for any response.

The setup for this procedure is shown in Figure 7.9, as depicted in MIL-STD-461G.

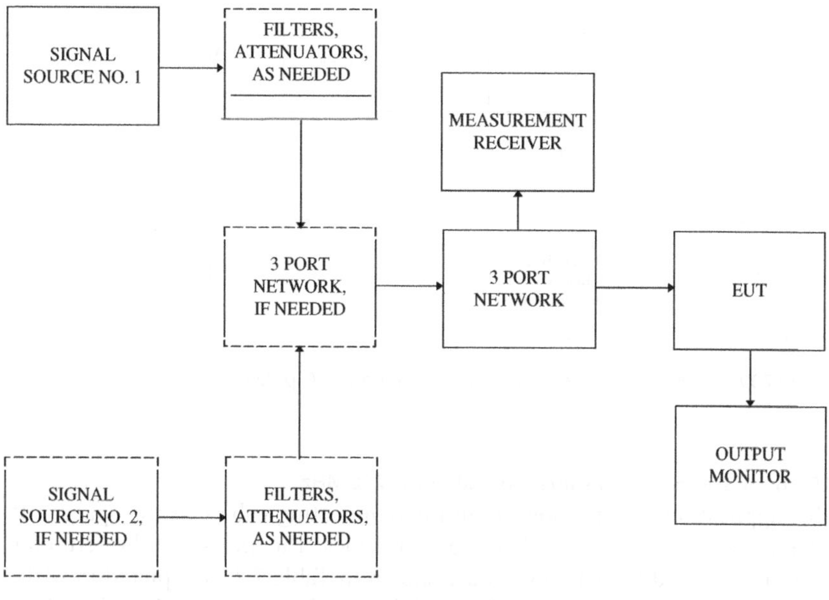

Figure 7.9 CS104 general test setup. *Source*: [1, Appendix A, p. 225].

7.5.3.5 CS105, Antenna Port, Cross-Modulation, 30 Hz to 20 GHz

This test is similar to the previous two tests, in setup and pass-band measurement. In this test, an unmodulated signal is injected at the receiver F_0, at the maximum tolerable level, and a modulated signal, using the receiver-referenced modulation, is swept up and down the band in accordance with MIL-STD-461G paragraph A.5.50 above the receiver standard referenced level, plus or minus the receiver intermediate frequency (IF) band width. The receiver is monitored for any response change. This test would only be applicable to receivers that would process information from amplitude modulated (AM) carrier.

This test would not be applicable to any amplifier or receiver not using AM modulation.

The typical setup for this test is shown in Figure 7.10, as depicted from MIL-STD-461G.

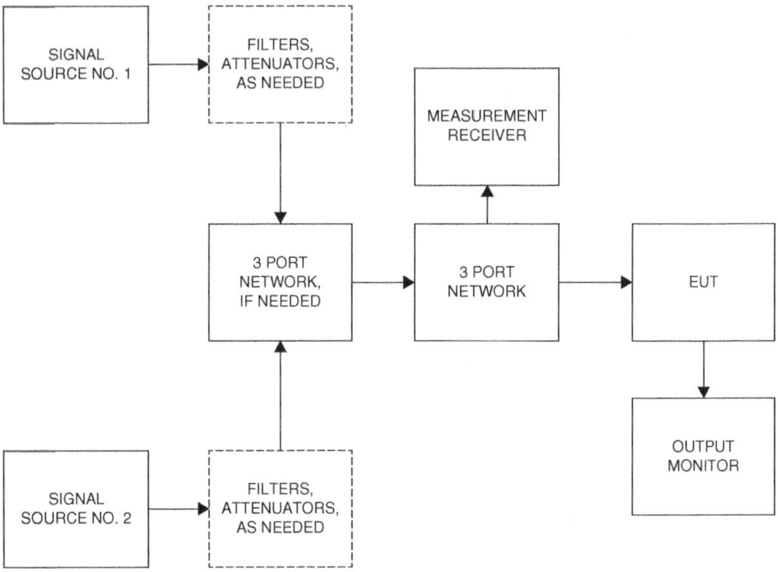

Figure 7.10 CS105 general test setup. *Source*: [1, Appendix A, p. 228].

7.5.3.6 CS114, Bulk Cable Injection, 10 kHz to 200 MHz

This application is applicable to all interconnecting cables to the spacecraft from the subsystem to include the power cables. The limits and the setup for both the test and the calibration are shown in Table 7.2 as depicted in MIL-STD-461G. This test does specify modulation, defined as 1 kHz pulsed 50% duty cycle. As can be seen in the table, this test begins at 1 MHz for spacecraft applications and is intended to emulate induced currents from other spacecraft components as well as emitters. The setup and the test can be quite time consuming depending on the number of interfaces the subsystem has with the spacecraft. The cables tested should be of flight cable length and configuration to include any shields and twisted pairs. Failure of this test would lead to any change in the normal operation of the subsystem to include significant changes in waveforms on cables interfacing with the spacecraft.

Table 7.2 CS114 limit curves

Platform / Frequency range		Limit curve numbers shown in figure CS-114-1 and limits							
		Aircrfaft (external or safety critical)	Aircraft internal	All ships (above decks) and submarines (external)*	Ships (metallic) (below decks)	Ships (non-metallic) (below deck)**	Submarine (internal)	Ground	Space
4 kHz to 1 MHz	N	-	-	77 dBμA	77 dBμA	77 dBμA	77 dBμA	-	-
10 kHz to 2 MHz	A	5	5	2	2	2	1	3	3
	N	5	3	2	2	2	1	2	3
	AF	5	3	-	-	-	-	2	3
2 MHz to 30 MHz	A	5	5	5	2	4	1	4	3
	N	5	5	5	2	4	1	2	3
	AF	5	3	-	-	-	-	2	3
30 MHz to 200 MHZ	A	5	5	5	2	2	2	4	3
	N	5	5	5	2	2	2	2	3
	AF	5	3	-	-	-	-	2	3

KEY: A = Army
N = Navy
AF = Air Force
* For equipment located external to the pressure hull of a submarine but within the superstructure, use SHIPS (METALLIC) (BELOW DECKS)
** For equipment located in the hanger deck of Aircraft Carriers
Source: [1, p. 66]

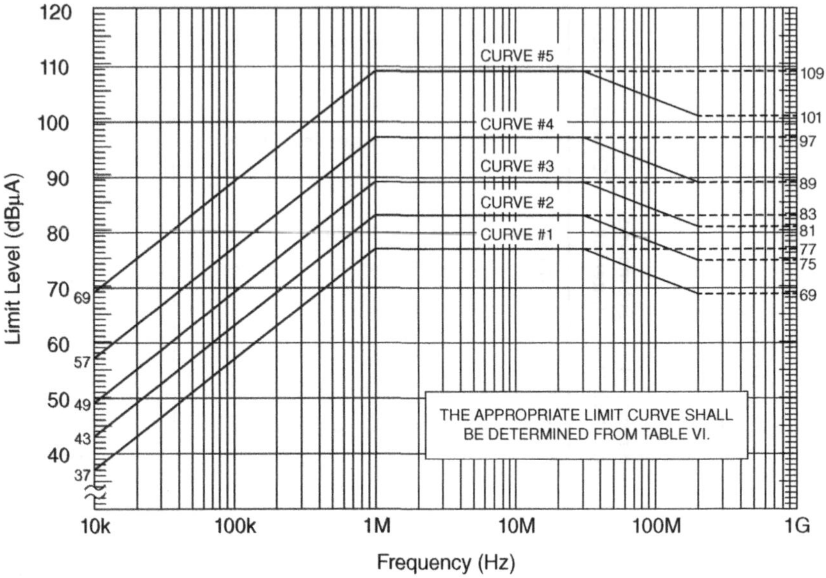

Figure 7.11 CS114-1 calibration limits. *Source*: [1, p. 67].

Table 7.2 indicates that spacecraft will test to curve 3 of the graph shown in Figure 7.11 (CS114-1). Figure 7.11 (CS114-1) indicates the test levels (Curve 3 as indicated in TABLE 7.2) , Figure 7.12 indicates the parameters for insertion loss on the probes, and Figure 7.13 is the setup for calibration and test.

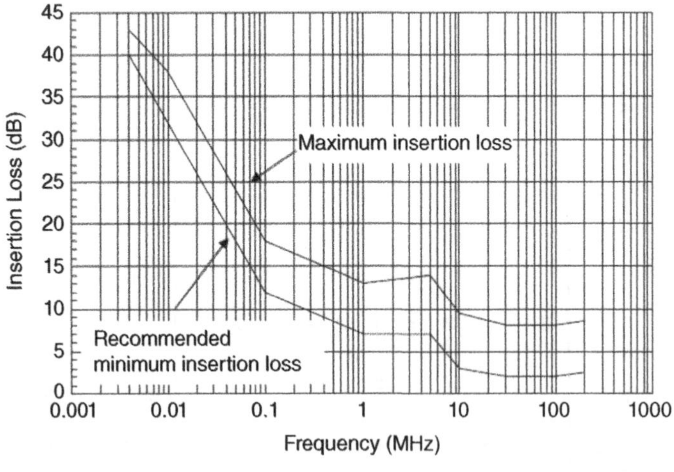

Figure 7.12 CS114-2 maximum insertion loss for injection probes. *Source*: [1, p. 68].

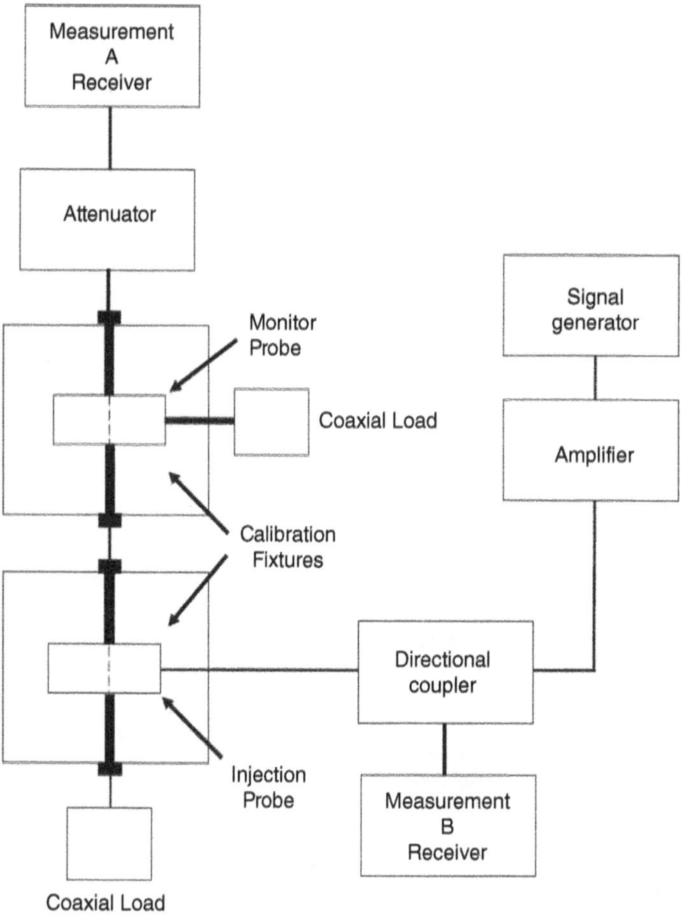

Figure 7.13 CS114-3 calibration setup. *Source:* [1, p. 69].

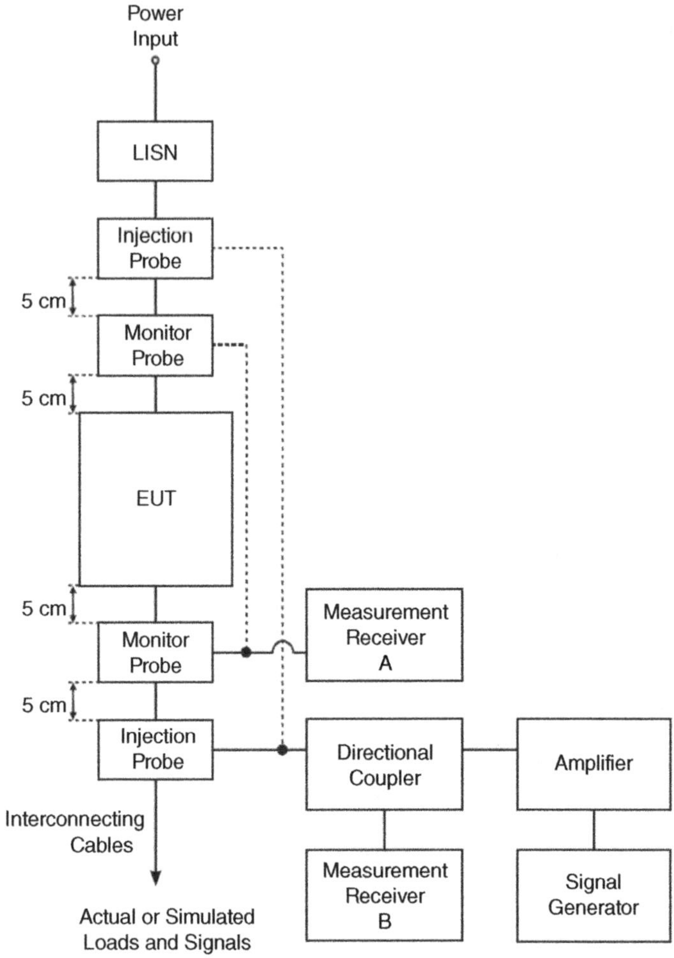

Figure 7.14 CS 114-5 bulk cable injection evaluation. *Source*: [1, p. 71].

7.5.3.7 CS115: Bulk Cable Injection, Impulse Excitation

The intent of this test is to demonstrate no degradation in the performance of the EUT when the pulsed waveform (shown in Figure 7.15 from MIL-STD-461G) is coupled onto the cables interfacing with the spacecraft, to include the power cables.

Figure 7.16 shows the calibration and test configurations as schematically depicted in MIL-STD-461G for CS115.

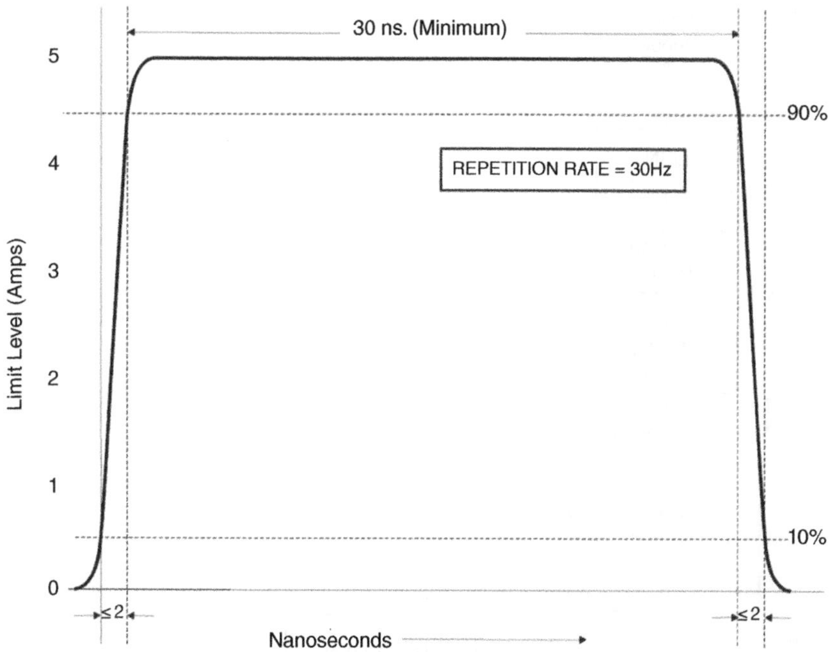

Figure 7.15 CS115 signal characteristics for all applications. *Source*: [1, p. 74].

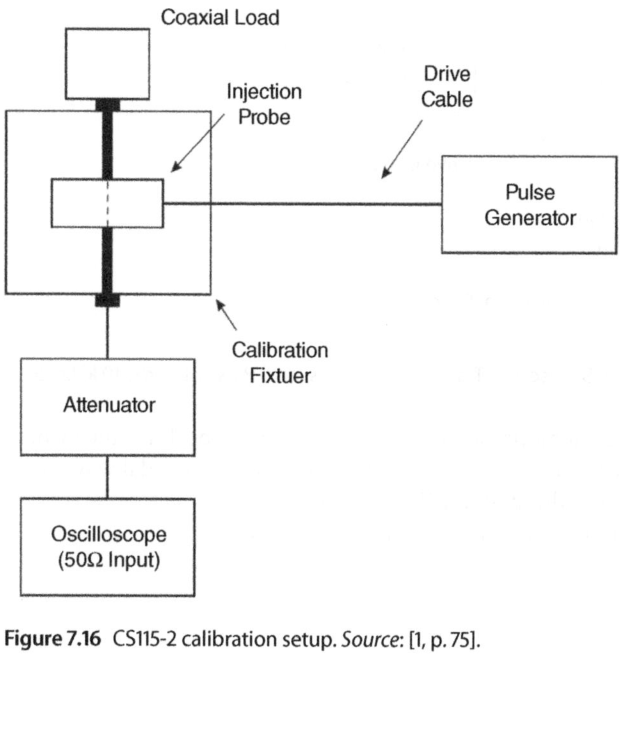

Figure 7.16 CS115-2 calibration setup. *Source*: [1, p. 75].

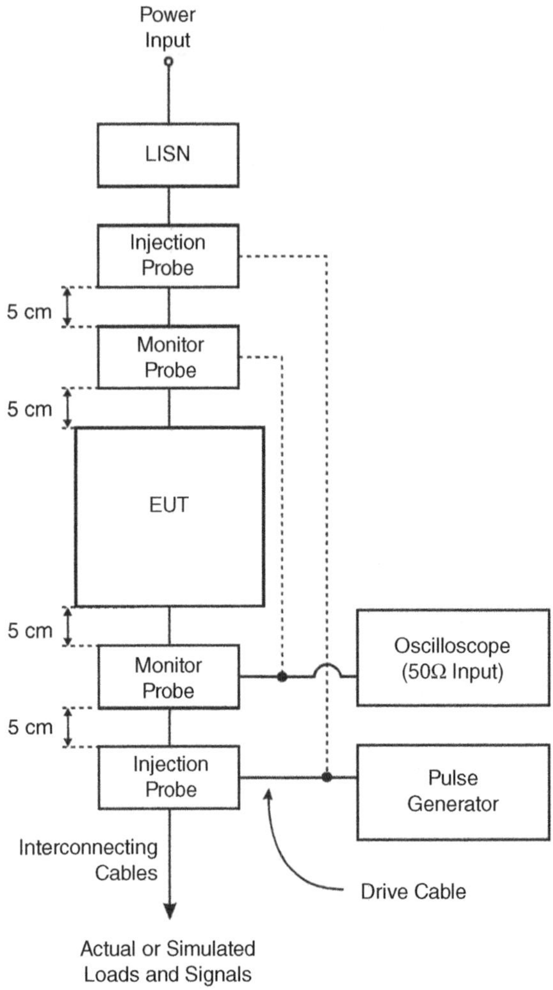

Figure 7.17 CS115-3 bulk cable injection. *Source*: [1, p. 76].

7.5.3.8 CS116: Damped Sinusoidal Transients, Cables, and Power Leads, 10 kHz to 100 MHz

This test is intended to demonstrate no degradation in the performance when the power and interface cabling is excited by the damped sinusoidal waveform shown in Figure 7.18 as exhibited in MIL-STD-461G:

The current is limited as shown in Figure 7.19.

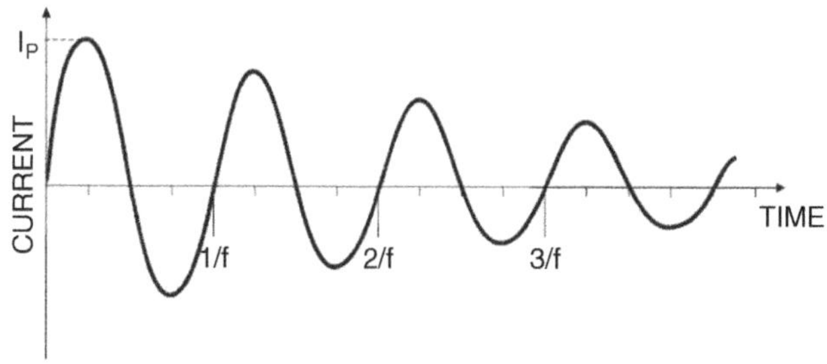

NOTES: 1. Normalized waveform: $e^{-(\pi f\, t)/Q}\sin(2\pi ft)$

Where:
f = Frequency (Hz)
t = Time (sec)
Q = Damping factor, 15±5

2. Damping factor (Q) shall be determined as follows:

$$Q = \frac{\pi(N-1)}{\ln(I_P/I_N)}$$

Where:
Q = Damping factor
N = Cycle number (i.e. N = 2, 3, 4, 5,...)
I_P = Peak current at 1st cycle
I_N = Peak current at cycle closest to 50% decay
ln = Natural log

3. I_P as specified in Figure CS116-2

Figure 7.18 Typical CS116 damped sinusoidal waveform. *Source*: [1, p. 80].

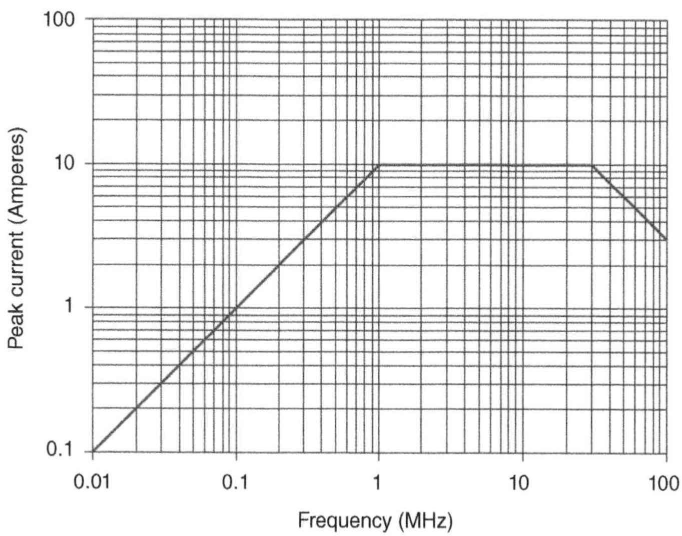

Figure 7.19 CS116-2 limit for all applications. *Source*: [1, p. 81].

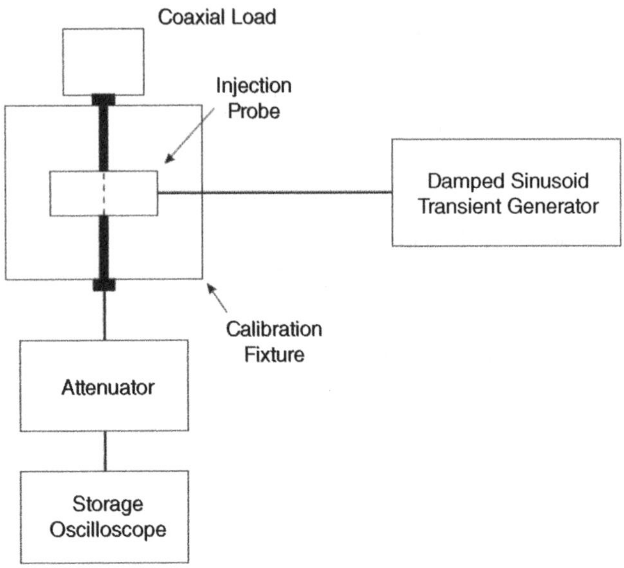

Figure 7.20 CS116-3 Typical setup for calibration of test waveform. *Source*: [1, p. 82].

The setup and test configuration for this test differ very slightly as can be seen in Figures 7.20 and 7.21; the difference would be the suggested equipment as listed in the Military Standard.

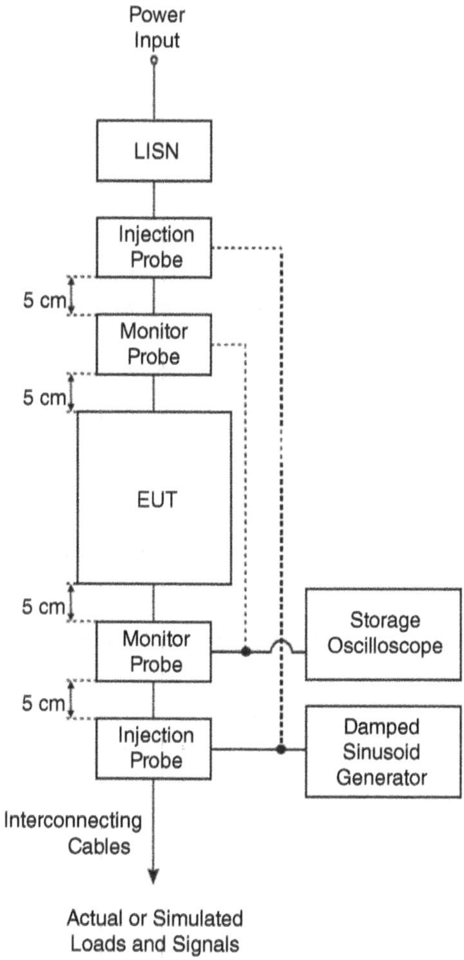

Figure 7.21 CS116-4 Typical test setup for bulk cable injection of damped sinusoidal transients. *Source*: [1, p. 83].

7.5.3.9 RE102, Radiated Emissions, Electric Field, 10 kHz to 18 GHz.

This test ensures that, via a standardized measurement procedure, the subsystem and its associated cabling will not interfere with the other spacecraft systems. The exemptions would be the frequencies and associated bandwidths for intentional emitters on board the subsystem. The test limit curves are shown in Figure 7.22.

Typically, there will be notches inserted in the table to accommodate any spacecraft receivers as well as receivers on board (usually X-band or UHF), launch vehicle receivers, such as range safety command destruct, and GPS. A typical notional RE102 limit is shown below (Figure 7.23) with respect a

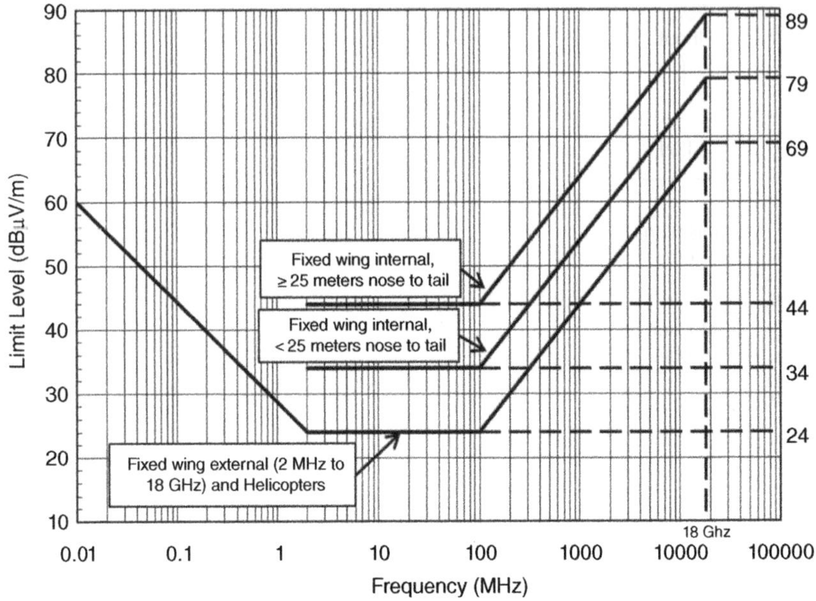

Figure 7.22 RE102-3 limit for aircraft and space system applications. *Source*: [1, p. 116].

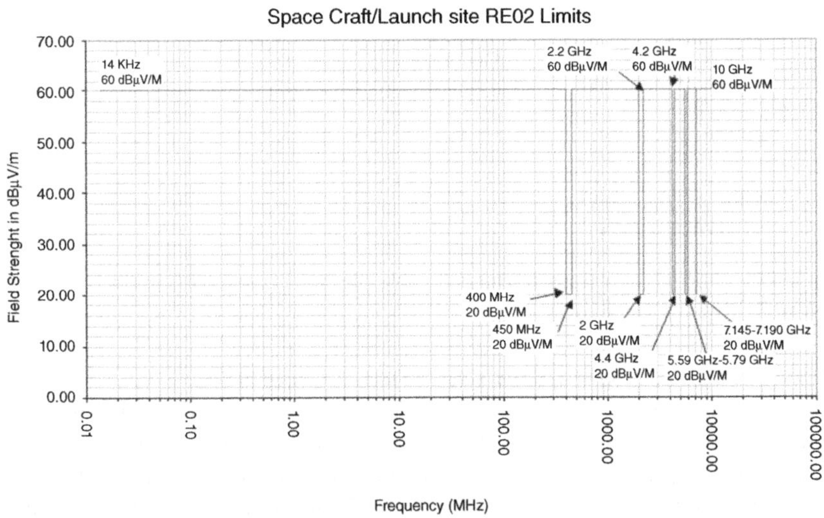

Figure 7.23 Typical tailored RE 102 limit for launch vehicle radiated emission limits. *Source*:: L. Warboys, Denver. CO. 2000.

subsystem and the rest of the spacecraft and launch vehicle radiated emission limits. It should be noted that the notches for the receivers on board the spacecraft and the launch vehicle are difficult to measure. Typically, the best level that can be measured efficiently should be discussed with the spacecraft systems EMC engineer, the final evaluation would be made at the system level. The aforementioned can be best accomplished at the spacecraft level using the flight antennas in what is known as a "Look Back" test.

A typical test setup shown in Figure 7.24 as depicted in MIL-STD-461G.

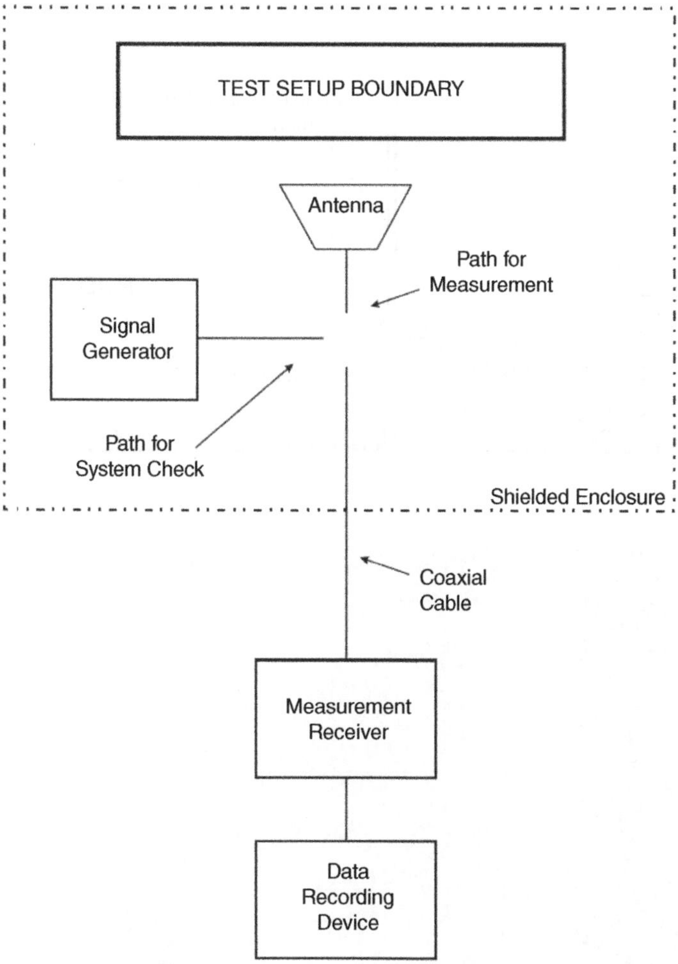

Figure 7.24 Basic test setup. *Source*: [1, p. 118].

Typically, there are three to four types of antennas to cover the frequency range required for spacecraft operation, those antennas are as follows (depending on the antenna type and frequency range):

Rod antenna 10 kHz–30 MHz Vertical

Figure 7.25 shows rod antenna positioning RE 102-6 pictorially and photographically.

Biconical 30 MHz–300 MHz some to 1 GHz Vertical/horizontal

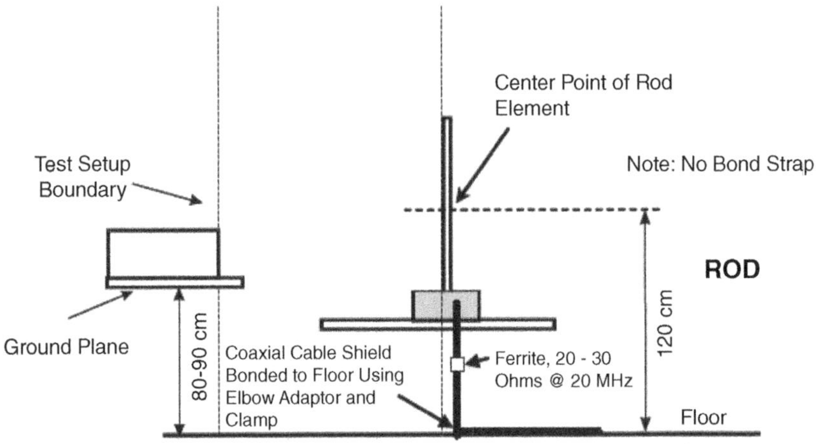

(a)

(b)

Figure 7.25 Antenna positioning. RE 102-6. *Source*: [1, p. 119].

Figure 7.26 is biconical antenna positioning RE 102-6 pictorially and photographically.

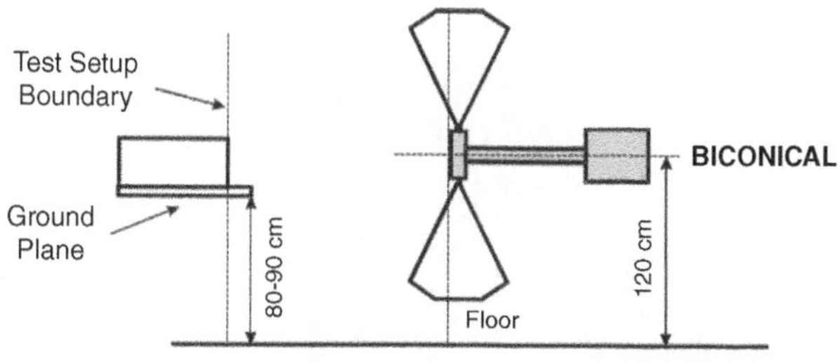

(a)

(b) Biconical antenna Vertical

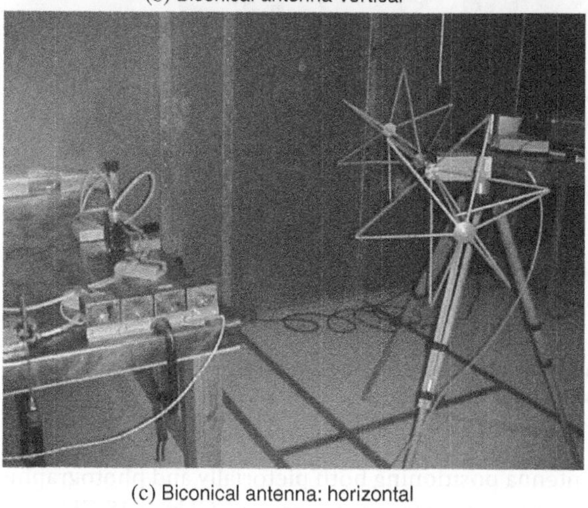

(c) Biconical antenna: horizontal

Figure 7.26 RE102-6, Antenna positioning. *Source*: [1, p. 119].

Figure 7.27 is for ridge horn antenna positioning RE 102 -6 pictorially and photographically.

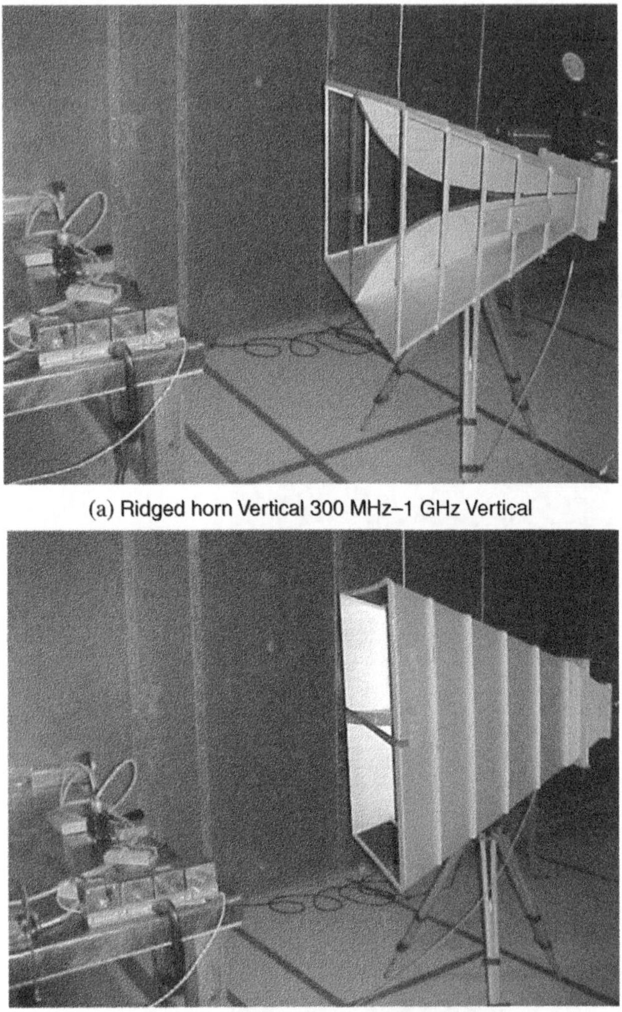

(a) Ridged horn Vertical 300 MHz–1 GHz Vertical

(b) Ridged Horn 300 MHz--1 GHz Horizontal

Figure 7.27 Ridge horn antenna positioning RE 102-6 pictorially and photographically. *Source*: [1. p. 119; L. Warboys Denver. CO. 2000].

Figure 7.28 shows antenna positioning both pictorially and photographically is a Double Ridge Horn antenna for frequency range 1 GHz to 18 GHz.

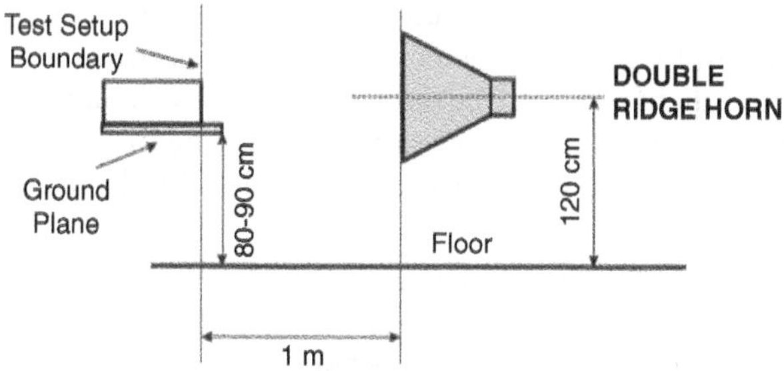

(a)

(b) Ridge horn antenna vertical

(c) Ridge horn antenna horizontal

Figure 7.28 Double ridged antenna positioning. *Source*: [1, p. 119; 4].

7.5.3.10 RE103, Antenna Spurious and Harmonic Outputs, 10 kHz to 40 GHz

To monitor emissions, this test would be in lieu on CE106 and performed on transmitters using the actual flight antenna. The test would use the same antennas but would focus on the area of the transmission antenna. The frequency ranges and setups for the calibration and test are shown in Table 7.3 as depicted in MIL-STD-461G.

Table 7.3 Frequency ranges and setups for the calibration and test

Operating Frequency Range (EUT)	Start Frequency of Test
10 kHz to 3 MHz	10 kHz
3 MHz to 300 MHz	100 kHz
300 MHz to 3 GHz	1 MHz
3 GHz to 40 GHz	10 MHz

Source: [1, p. 122]

Depending on the operating frequency range of the EUT, the start frequency of the test is as follows:

The reader should be aware that the F0 of the operational frequency and its bandwidth is notched out for this test.

The calibration setups for RE103 as depicted in MIL-STD-461G are shown in Figures 7.29 and 7.30.

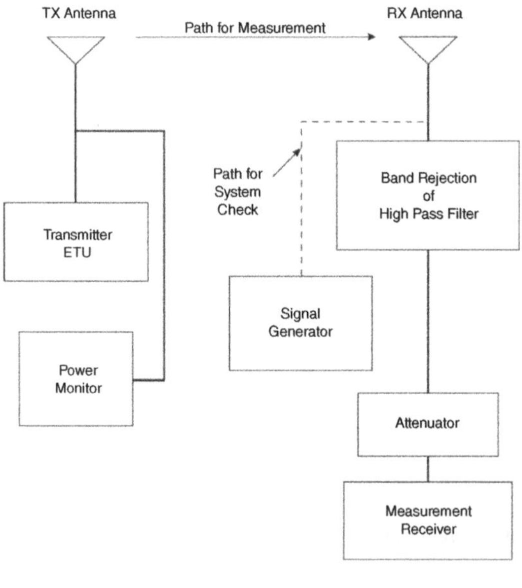

Figure 7.29 Calibration and test setup for radiated harmonics and spurious emissions, 10 kHz-1 GHz. *Source*: [1, p. 126].

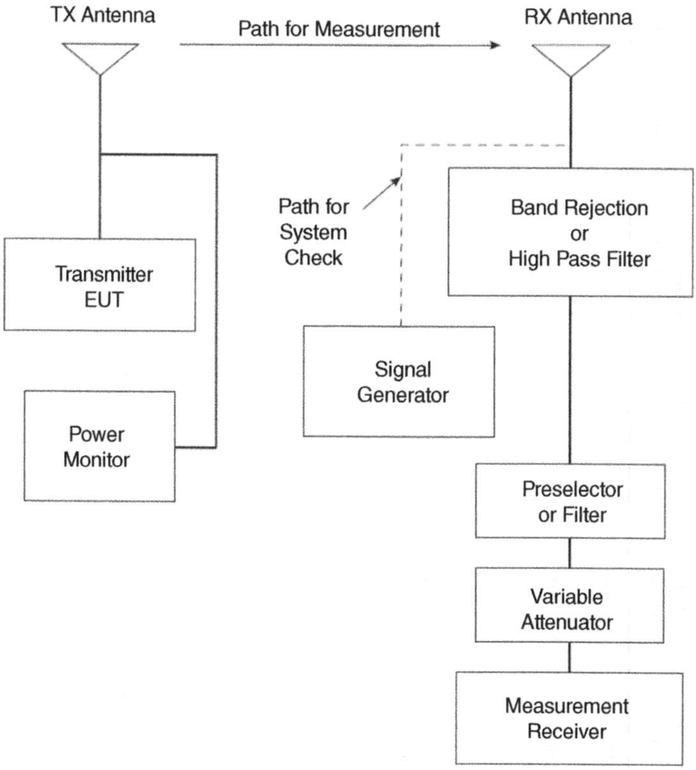

Figure 7.30 Calibration and test setup for for RE 103-2 radiated harmonics and spurious emissions, 1 GHz to 40 GHz. *Source*: [1, p. 127].

7.5.3.11 RS103, Radiated Susceptibility, Electric Field, 2 MHz to 40 GHz

These tests are intended to emulate the exposure of the EUT to the fields of exposure it will experience throughout its operational mission. The limits will vary depending on what exposure the subsystem may see at the launch facilities, during build up at the launch pad and throughout the launch phase. The limits depicted in MIL-STD-461G, shown in Table 7.4.

As shown in Figure 7.31, the notional graphical depiction of a launch facility RF environment and Table 7.4 differ somewhat from the spacecraft environment presented in MIL-STD-461G. In this testing, if limited to the standard, the probability of mission failure would be high.

Table 7.4 RS103 limits

Platform / Frequency range		Limit levels (volts/meter)							
		Aircrfaft (external or safety critical)	Aircraft internal	All ships (above deck & exposed below deck) and submarines (external)*	Ships (metallic) (below decks)	Ships (non-metallic) (below deck)**	Submarine (internal)	Ground	Space
2 MHz to 30 MHz	A	200	200	200	10	50	5	50	20
	N	200	200	200	10	50	5	10	20
	AF	200	20	-	-	-	-	10	20
30 MHz to 1 GHz	A	200	200	200	10	10	10	50	20
	N	200	200	200	10	10	10	10	20
	AF	200	20	-	-	-	-	10	20
1 GHz to 18 GHz	A	200	200	200	10	10	10	50	20
	N	200	200	200	10	10	10	50	20
	AF	200	60	-	-	-	-	50	20
18 GHz to 40 GHZ	A	200	200	200	10	10	10	50	20
	N	200	60	200	10	10	10	50	20
	AF	200	60	-	-	-	-	50	20

KEY: A = Army
N = Navy
AF = Air Force
* For equipment located external to the pressure hull of a submarine but within the superstructure, use SHIPS (METALLIC) (BELOW DECK)
** For equipment located in the hanger deck of Aircraft Carriers
Source: [1, p. 145]

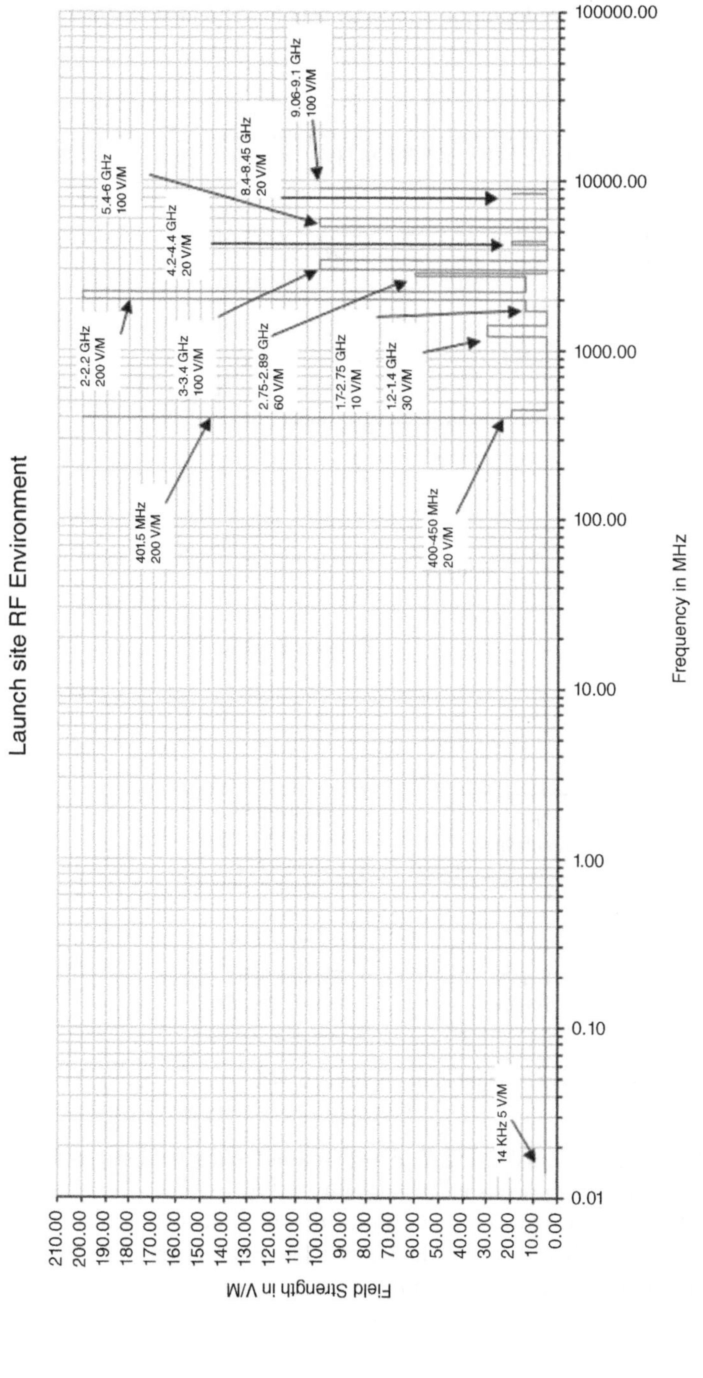
Launch site RF Environment
Field Strength in V/M
210.00
200.00
190.00
180.00
170.00
160.00
150.00
140.00
130.00
120.00
110.00
100.00
90.00
80.00
70.00
60.00
50.00
40.00
30.00
20.00
10.00
0.00
Frequency in MHz
0.01
0.10
1.00
10.00
100.00
1000.00
10000.00
100000.00
14 KHz 5 V/M
401.5 MHz 200 V/M
400-450 MHz 20 V/M
2-2.2 GHz 200 V/M
3-3.4 GHz 100 V/M
2.75-2.89 GHz 60 V/M
1.7-2.75 GHz 10 V/M
1.2-1.4 GHz 30 V/M
5.4-6 GHz 100 V/M
4.2-4.4 GHz 20 V/M
8.4-8.45 GHz 20 V/M
9.06-9.1 GHz 100 V/M

Figure 7.31

The setups for this testing are shown in Figure 7.32 as depicted in MIL = STD-461G. This depiction shows testing in a shielded enclosure or anechoic chamber.

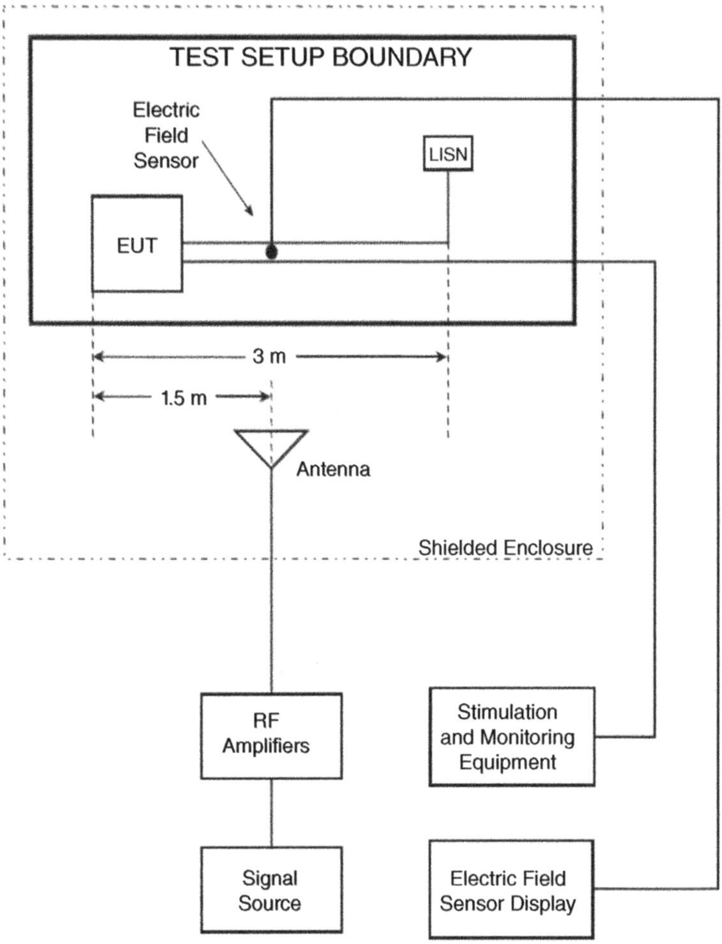

Figure 7.32 Test equipment configuration. *Source*: [1, p. 147].

RS103 testing at JPL using log Yagi antenna notes the field strength sensor, the antenna polarity shown is vertical (Figure 7.33).

The MIL-STD also shows configurations for stir mode testing, also known as reverberation chamber testing as depicted in Figures 7.34 and 7.35. Because all the test methods have their pros and cons, the author prefers the first method for control and troubleshooting purposes.

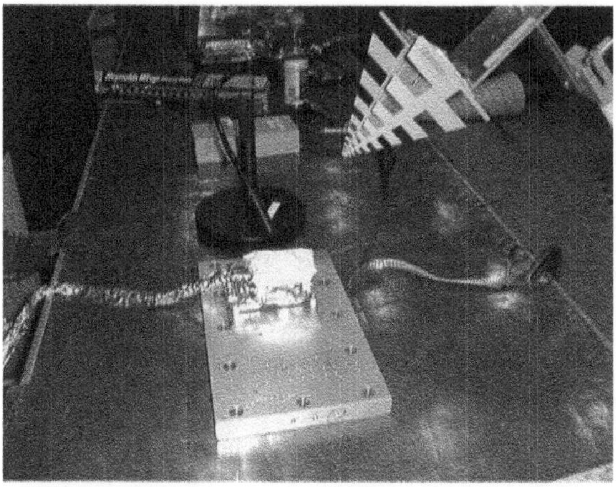

Figure 7.33 RS103 testing at JPL using log Yagi antenna.

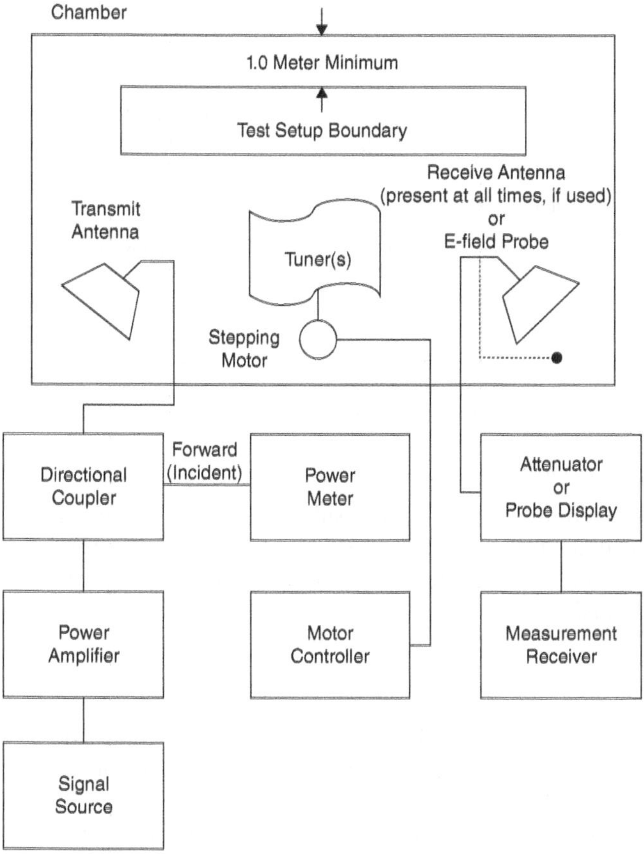

Figure 7.34 Reverberation chamber setup. *Source*: [1, p. 150].

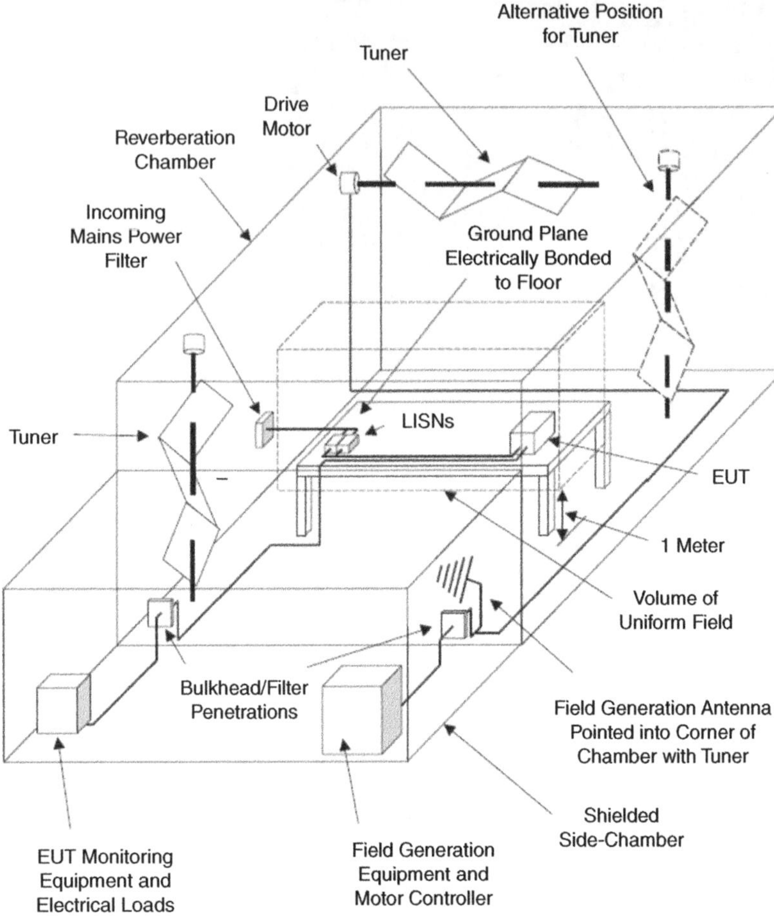

Figure 7.35 Reverberation chamber overview. *Source*: [1, p. 151].

The above tests are required on the part of subsystems, if called for by the system engineer/customer, from MIL-STD-461G as well as from the AIAA S-121A-2017 document "Electromagnetic Compatibility Requirements for Space Equipment and Systems."

References

1. Department of Defense Interface Standard: Requirements for the control of electromagnetic interference characteristics of subsystems and equipment. MIL-STD-461G, Dec. 2015.
2. AIAA Standard, S-121-"Electromagnetic Compatibility Requirements for Space Equipment and Systems." 2017.

8

System-Level Testing of Spacecraft

Johannes Wolf

This chapter discusses system-level testing of spacecraft, meaning at the level of the entire integrated spacecraft, and aims to share some of the experience gained in electromagnetic compatibility (EMC) support for spacecraft missions over more than 15 years. First, a classification of system-level testing is provided, starting with a description of the EMC verification approach and discussing the purpose of testing at the system level. Then, system-level test requirements are defined and divided into four different groups: launcher compatibility, autocompatibility or self-compatibility, RF compatibility, and compatibility with instrument-driven scientific requirements. Finally, the execution of the system-level tests is discussed, detailing the specifics on several practical examples of satellite projects, where the author and his team have been or still are involved for EMC support.

8.1 Classification of System-Level Testing

In order to put the topic of system-level testing into context, this section provides an overview of the entire EMC verification approach, discusses the model philosophy and the EMC verification throughout the various project phases at the equipment and the subsystem levels. The purpose of testing on system level is then derived and discussed.

8.1.1 EMC Verification Approach

A typical spacecraft system consists of several subsystems, e.g., power, attitude control, command and data handling, and the payload or instrument(s), where each of these subsystems is composed of one or more equipment boxes. The entire EMC verification process is performed in several steps on different

Handbook of Aerospace Electromagnetic Compatibility, First Edition. Edited by Reinaldo J. Perez.

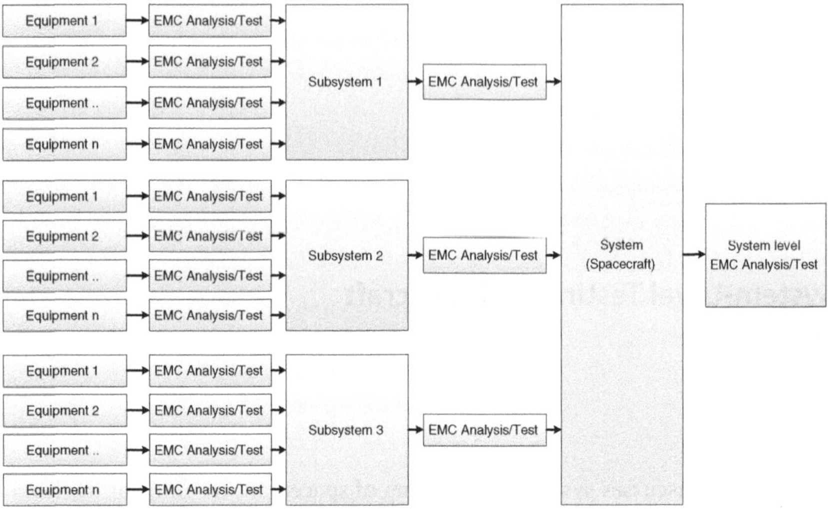

Figure 8.1 EMC verification approach.

levels, starting with equipment- or box-level tests, continuing to subsystem-level tests, and ending with system (integrated spacecraft)-level tests, as illustrated in Figure 8.1. The verification methods used are analysis and test. This approach requires dedicated EMC requirements that must be derived from system-level EMC requirements by tailoring and breaking them down for each level. During this process, margins can be implemented in the subsystem and equipment-level requirements with respect to the system-level requirements.

EMC analysis forms a very important part of the EMC verification process [1]; it starts with the definition of mission specific EMC requirements for the spacecraft (at the system level) and is then breaking the requirements down to subsystem- and equipment-level, implementing EMC safety margins as required. The EMC analysis report is considered a living document that will be updated throughout the project phases with more detailed information and measurement data becoming available as the project progresses. In the analysis procedure, measured emissions can be compared with predicted emissions (from modeling and simulation) and can be evaluated against defined emission limits and margins. Finally, EMC analysis is a very useful tool and provides all the information to properly evaluate nonconformances raised in the course of the entire EMC verification process.

EMC tests are very important to verify predictions and models from the analysis and to verify EMC requirements [2]. Whereas emission measurements do not impose any stress on the hardware other than operating it, susceptibility tests do have the potential to stress or, in the worst case, to damage the equipment under test. This potential can be reduced to a large extent by a proper analysis performed in parallel to the electrical design, taking into account the required susceptibility test levels. This is an important consideration when the model philosophy is discussed.

The approach to EMC verification is also dependent on the model philosophy. The model philosophy is chosen taking into account the development status, the integration and test program, and programmatic constraints. An overview of model definitions and verification strategies can be found in ECSS-E-HB-10-02A [3]. The classical approach is to have a qualification model (QM) and a flight model (FM), whereas, today, the usual practice is to follow a protoflight approach, using one and the same model for qualification and flight, the so-called protoflight model (PFM). Both approaches have several advantages and disadvantages. The major disadvantage of the QM/FM approach is that, basically, two flight models must be built, which can require serious efforts in terms of parts procurement, time for manufacturing, and cost. The major disadvantage of the PFM approach is that the FM is exposed to a certain stress due to the qualification tests, which are performed at levels above the ones to be expected during launch and flight. Even in the case where no potential damage to the hardware is expected, the tests may cause performance degradation, which can have an impact on the mission later on. Nevertheless, the industry prefers the PFM approach since the cost aspect predominates over the risk of having damage or degradation. The economic aspects of system-level testing are discussed in [4].

In terms of EMC verification during the project phases, a guideline is given in ECSS-E-HB-20-07A [5], providing the rationale for unit test requirements, describing system-level EMC activities and EMC design techniques, discussing EMC test methods, as well as analysis methods and computational models, and presenting troubleshooting and retrofit techniques. In the Annex, cables, ground planes, and common mode voltage reduction are discussed based on fundamental concepts and models.

As regards the overall EMC verification process, there are two extremes in the approach and, in practice, one can find all possible variations between those extremes. One extreme (minimalistic case) is, basically, to perform only one single EMC test campaign on the QM or the PFM; the other extreme (mainly for new developments) is to analyze and test for EMI throughout all development phases, starting from the breadboard level via engineering models until the QM/FM or PFM. The effort to be spent should be adapted to the needs of the project and is dependent on various factors like heritage, experience, and severity of the EMC requirements for the particular mission.

In general, it is considered a less risky approach to take care of potential EMC issues starting at an early stage of the design and to continue throughout the project phases. Although it is possible to achieve a successful EMC verification following the minimalistic approach and, thus, saving some costs and effort, experience tells us that this will only be successful as long as the EMC requirements are not demanding and the system neither employs strong sources of interference nor has very sensitive equipment that will suffer from (being the victim of) interference.

Concerning EMC verification throughout the integration levels, different approaches can be followed as well. Two typical approaches (also representing two extremes) are as follows:

1. Minimalistic approach:
   - Equipment-level verification is performed by analysis (of previous test results from other projects),
   - subsystem-level verification is performed by test,
   - system-level verification is performed by analysis (based on subsystem-level test results)
2. Full verification approach:
   - Equipment-level verification is performed by analysis and test for all equipment
   - subsystem-level verification is performed by analysis and test,
   - system-level verification is performed by analysis and test for a subset of requirements

Figure 8.2 provides an overview of EMC requirement levels and their verification.

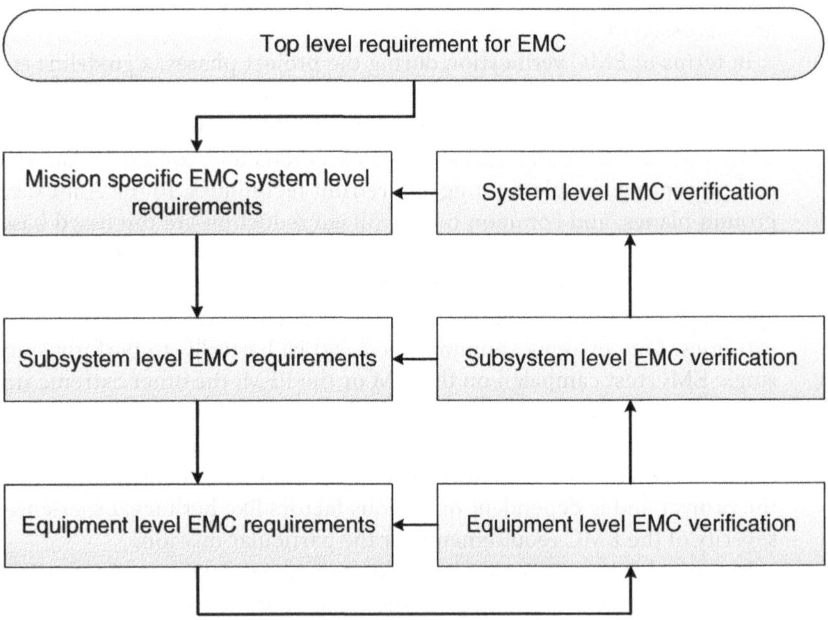

Figure 8.2 EMC requirements and verification flow.

In [6], the importance of incorporating EMC engineering principles in system-level testing for a complex project is discussed. The principal process described therein can be useful for space projects as well.

8.1.1.1 Equipment-Level Tests

On the equipment level, it is often the case that equipment is heritage from other projects and EMC test results are available. These test results can be used as valuable inputs to the EMC analysis and, in case the requirements are not too much different between the actual project and the one for which the equipment has been qualified, it is a possible way forward not to repeat EMC testing on that equipment or to perform only a subset of EMC tests (delta qualification). An argument against this usual practice is that considering electronic equipment for space applications, we typically manufacture a single unique item, as opposed to large series of items for mass production. This implies that each piece of equipment is unique, may incorporate different workmanship errors, and, therefore, can exhibit different EMC behaviors. One typical example are power input filters, which clearly have a direct influence on conducted emissions and susceptibility. In such filters, often inductors are employed that are custom designed and handmade, which is the source for quite large parameter tolerances. Thus, especially if there are stringent EMC requirements, the better approach is to test each individual equipment for EMC.

8.1.1.2 Subsystem-Level Tests

On the subsystem level, several pieces of equipment are integrated close to each other and have to interact in order to fulfill a certain function. The aspects of being integrated and the need to interact, to share resources (power), and being interconnected, are not covered by tests on the equipment level. It can also be the case that equipment is able to fulfill its function only in combination with other equipment and, therefore, can only be properly tested at the subsystem level. In terms of EMC requirements, the integrated configuration on the subsystem level can create new coupling paths between the equipment for disturbances, which can change the situation for emissions and susceptibility as compared with the test configuration at the equipment level. This means that, on the subsystem level, even in case all individual pieces of equipment have successfully passed the EMI tests, it is not ensured that the subsystem has no issues in terms of emissions and/or susceptibility, thus, tests on the subsystem level are necessary and useful.

8.1.2 Purpose of Testing at the System Level

After successfully testing all subsystems (this includes mechanical, thermal, and electrical tests as well), the next step in the manufacturing, assembly, integration, and test (MAIT) flow is the integration of the subsystems into the spacecraft, which is then considered as the system. At the system or spacecraft level, it is the first time that all subsystems are located in their final positions, electrically connected, and operated together in a way that is representative of the mission/flight. At the system level, the scenario is very much similar to the one discussed on the subsystem level before: although the subsystems are successfully tested before, after integrating them into a system, it cannot be taken for granted that there are no issues with emissions and/or susceptibility. The reason is that new (multiple) coupling paths are created and the interactions between subsystems and their equipment become more and more complex. Another aspect is that, in this configuration, all of the wiring and cabling between the equipment of several subsystems is in the final flight configuration and provides the correct terminations (the right equipment at both ends, no simulators). It is well known that, especially in the lower frequency range (<1 GHz), EMC at the system level is very much dependent on cable effects [7], rather than on leakage through seams and other box openings. This underlines once again the need for tests at the system level.

With respect to grounding and bonding, only at the system level is the final and representative configuration achieved. Also, software plays an important role here because this is typically the first time that the flight software will be used to operate the equipment on board the spacecraft.

The main purpose of the system-level test is, thus, to ensure that all integrated subsystems are able to properly work together without degradation of performance in the presence of noise created from other equipment/subsystems in the system (designated as intrasystem compatibility, self-compatibility, or autocompatibility) and in the presence of noise created from sources outside the system (intersystem compatibility). A dedicated part of the system-level test is the verification of the radio frequency compatibility (RFC). This test is performed to verify that the transmitters on the spacecraft do not interfere with other subsystems and is, thus, a particular case of the definition: " … spacecraft subsystems and equipment have to properly work together without degradation of performance … in the presence of noise created from other equipment/subsystems in the system …," in this case, the RF subsystem.

When the system employs electro-explosive devices (EED), a dedicated safety margin is required and, typically, will be verified by test during the system-level test campaign.

In some cases, tests at the system level are also used to cover for missing EMC tests or to repeat failed tests on the subsystem or equipment level [8]. Table 8.1 provides a short summary of what are considered advantages of system-level testing versus disadvantages.

Some of the disadvantages listed in Table 8.1 do not play a role if a proper EMC program has been followed, starting with analysis and test on the

Table 8.1 Advantages/disadvantages of system-level testing

Advantages	Disadvantages
Hardware is most representative for flight	Difficult to access interfaces
Representative cable routing and shielding	Too late to make design changes
Representative RF interactions/coupling	Hard to fix problems that occur
Real flight software is loaded and operated	EMC solutions often forced to "Band-Aids"
	Very time consuming to bring the system into a defined mode

equipment and subsystem levels, but can be detrimental in case the system-level test is the only EMC test performed on the spacecraft.

In [9], a good overview is provided about EMC performance verification and what can be done to make EMC verification at the spacecraft level more efficient based on the example of the METOP satellite. Propositions are made to replace the more complex and time-consuming susceptibility tests with emission tests and to limit testing to critical areas to reduce testing time, effort, and, therefore, cost. It is also pointed out that efficient EMC verification can only be achieved on the basis of a detailed EMC analysis during the early project phases.

Due to the high complexity that a spacecraft may have, the EMC test at the system level is not purely an EMC test. Because the flight software is involved during the system-level EMC test, a lot of software testing is performed, which, from experience, can cause delays in the test campaign. Typically, functional and electrical tests are also performed during a system-level EMC testing campaign. In some cases, it is also possible/necessary to carry out performance tests in the presence of electromagnetic disturbances. Also, very often, tests on the telemetry and data links are performed at the system level, since the environment (e.g., in an anechoic chamber) allows these tests to be performed (to ensure compliance to regulations for intentional RF emissions).

It is important to understand that, typically, for spacecraft testing at the system level, there are no conducted interfaces to other systems except power and data connections for test purposes, thus, at the system level, no conducted susceptibility or conducted emission tests will be performed. One exception is a lightning test, where the lightning current is injected into the spacecraft skin/structure. Up to frequencies of a few hundred MHz, bulk cable current injection can be considered as an alternative to radiated susceptibility tests [10]. Sometimes, also, conducted emission measurements are performed at the system level by opening a power connection from one equipment to the power control and distribution unit (PCDU) and inserting a breakout box in order to measure and verify the conducted noise level on the power bus with respect to the values assumed in an analysis/simulation.

8.2 System-Level Requirements Definition

As already mentioned in Section 8.1, the system-level EMC requirements must be defined taking into account the specifics of the mission, the launcher to be used, communication and telemetry frequency bands, and whether or not there are sensitive instruments or particularly noisy equipment on board the spacecraft. An overview of EMC activities performed on space missions is provided in [11], addressing the impact of EMC on the spacecraft procurement, the impact on the requirements definition, and on the entire EMC control program. This is illustrated by a number of example projects where future trends and needs are discussed.

It can be stated that it is neither practical nor would it make sense to define a set of unique susceptibility test levels and emission limits applicable to all spacecraft/missions. The top-level EMC requirement "the spacecraft subsystems and equipment have to work without damage or degradation of performance with each other and in the presence of external interference sources" remains valid. As a logical consequence in EMC standards for space, e.g., ECSS-E-ST-20-07 [12], one does not find such values in the requirements section, but in the appendix to be used as an example. Often, it is usual practice to refer to standards like ECSS-E-ST-20-07 in order to copy values for susceptibility requirements and emission limits. Without performing the EMC engineering work to tailor these values and limits for the particular mission, this approach might either lead to requirements that do not ensure that the top-level requirement can be fulfilled or to requirements that are too stringent, thus wasting resources. From experience, it is rather the first case that can be observed in many projects.

For many spacecraft, e.g., for telecom spacecraft, a so-called platform is used. The platform is providing the basic and common subsystems like propulsion, attitude control, power, command and telemetry, and the launch adaptor. The so-called payload, containing the subsystems that determine the mission—for the chosen example this is the telecom equipment (antennas and transponders)—is integrated on top of the platform, thus forming the satellite/system. In this case, there are also interface requirements in terms of EMC and power quality defined by the platform provider. With regard to the integration level, the platform, as well as the payload, can be considered individual subsystems and, after integration of the payload, they form the satellite/system. In terms of the system requirements definition, the situation may become difficult, because, on the platform side, requirements are imposed on the payload, whereas, on payload side, the mission might impose requirements from the payload (sensitive instruments or very noisy equipment) toward the platform. This, then, is becoming a serious issue considering that the platform is, in most cases, already built, tested, and qualified. The solution has then to be found on a case-by-case basis, depending on what are the mitigation options, whether the sensitivity or the emissions on the payload side can be reduced rather than imposing a more stringent requirement toward the platform. However, typically, platforms are built for a series of similar satellites (e.g., platforms for telecom satellites or platforms for navigation satellites) such that the platform requirements already take into account the typical and expected requirements from the payloads.

Since the process in the system requirements definition is rather complex and based on many different inputs, one could also consider it as an optimization process with multiple criteria. The main inputs for the system requirements definition are:

- top-level EMC requirements
- frequencies and power levels from transmitters (telemetry, data downlink)
- notch requirements for receivers on board (command, GPS)
- emissions from the launcher and the range as well as launcher receiver notches (launcher requirements)
- particular requirements from the payload (instruments) in terms of sensitivity, required notches
- particular sources of noise in the payload (e.g., high-power lasers)
- platform requirements (if a platform is used)
- known emission and susceptibility levels from equipment (heritage)
- standard emission and susceptibility levels (conducted)
- required EMI safety margins for EEDs
- required overall EMI safety margins

All these data have to be considered in order to define the radiated susceptibility levels, the radiated emissions limit as well as the conducted emissions limit and conducted susceptibility levels. This is part of the EMC engineering work to be performed and is usually documented in the EMC analysis document. The outcome, being detailed conducted and radiated EMC requirements on the subsystem and the equipment level, will be reflected in the EMC specifications or in a general interface requirements document.

In Figure 8.3, the scenario at the system level is shown, indicating the EMC aspects for launcher compatibility, autocompatibility, and RF compatibility based on a disturbance source-to-victim relationship. The basic principle to achieve compatibility is to establish and verify limits for the maximum emissions from potential sources and to define and verify susceptibility levels for the potential victims, thus ensuring an EMC safety margin with respect to the maximum emissions.

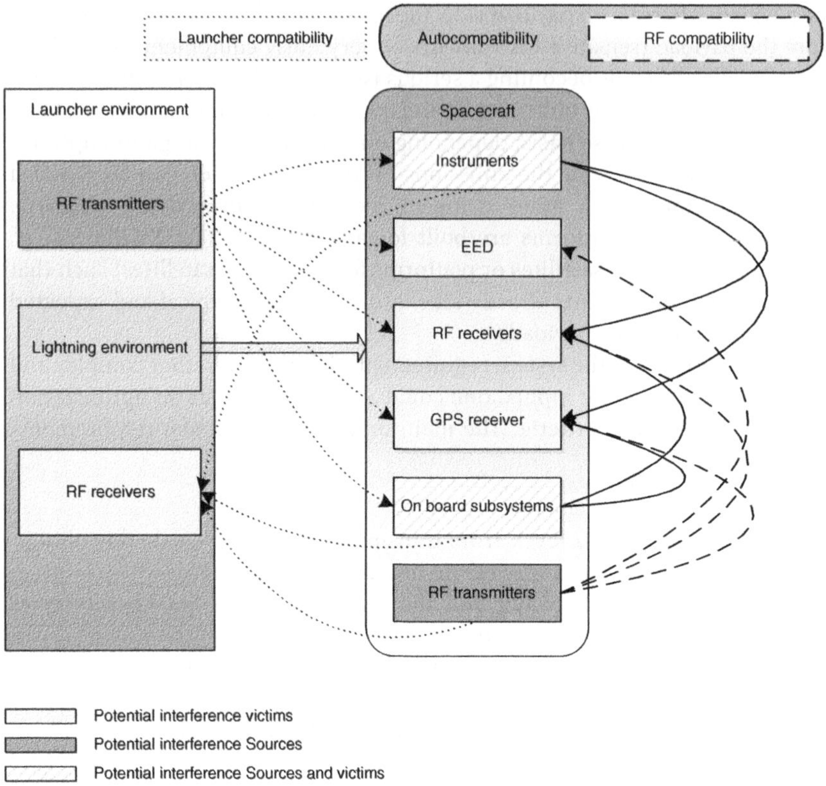

Figure 8.3 EMC on satellite level: sources and victims.

Looking at system-level testing, due to the absence of a wired interface (once in orbit), only the launcher compatibility and the autocompatibility (especially the radiated aspects) are of concern and will be discussed in the following section. The scientifically driven requirements play a special role and will be addressed in a dedicated section.

8.2.1 Launcher Compatibility

In order to bring the spacecraft into its desired orbit, a launcher is required. The launcher and the range are creating a certain electromagnetic environment that the spacecraft has to be compatible with. Especially if the spacecraft is powered on during launch, it must be ensured not only that the spacecraft is not disturbed by the launcher environment, but also, and even more important in that phase, that the spacecraft is not disturbing the launcher operations. This mainly concerns the RF links between the launcher and the ground for

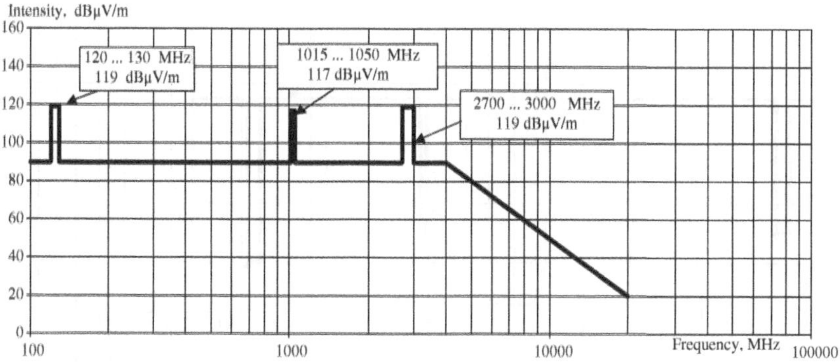

Figure 8.4 Rockot launcher-radiated emissions (from Rockot user's guide [13]).

commanding, tracking, and telemetry, as well as the launcher abort system, which under no circumstances shall be disturbed.

Regarding the emissions from the launcher and the launch pad, the values are usually provided in the launcher user manual. Figure 8.4 provides an example of the Rockot launcher emissions profile as defined in the Rockot user's guide [13]. Characteristics are the positive notches for the frequency bands of transmitters used onboard the launcher or on the range. This emissions profile has to be considered as a radiated susceptibility profile for the spacecraft. The envelope of the launcher emissions is fairly low, except for the positive notches, where high levels are present. Although being protected to a certain degree by the fairings, the spacecraft multilayer thermal insulation, the metallic spacecraft structure elements, and the box enclosure, the spacecraft electronics will be exposed to emissions generated by the launcher and/or the range, including nearby sources (radars). When comparing the launcher emissions with the susceptibility requirements of the spacecraft, the spacecraft susceptibility levels typically cover the launcher emissions over a large frequency band, except for the transmit frequencies of the launcher. Thus, the major areas to be verified for susceptibility are the transmit frequency bands. In case the spacecraft is completely powered off, receiver and/or instrument antennas can be of concern, picking up RF energy and, therefore, creating a potential risk to damage the very sensitive receiver front-end or low noise amplifier (LNA).

The pyrotechnic actuators (if present) in the form of EED are also of concern. The EMI safety margin of the EEDs with respect to launcher emissions (typically 20 dB) must be ensured and verified.

Figure 8.5 provides the Rockot launcher susceptibility profile as per the Rockot user's guide [13], which is the maximum emissions limit from the spacecraft. Here, the requirement is typically rather benign (in the sense that high emission levels from the spacecraft are tolerated), except in the launcher receive

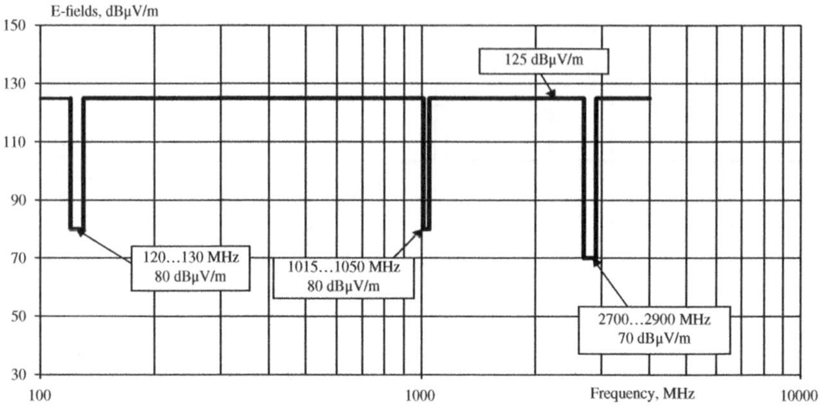

Figure 8.5 Rockot launcher-radiated susceptibility (from Rockot user's guide [13]).

frequency bands, one of them being the frequency band used for the launch abort system (for the Rockot launcher this is 120–130 MHz). As long as the spacecraft is launched in the powered off state, this is not of concern. However, the trend today is more and more to keep at least some of the spacecraft subsystems powered on during launch. Since this can include transmitters, it is very important to ensure that the launcher receive frequency bands are not disturbed by the spacecraft. These frequency bands and their related levels are taken into account by implementing notches that usually go down to very low levels.

Regarding the notches, there are different philosophies; one is to state that notches are not necessary, because the radiated emissions limit for the spacecraft is already so low, that, in combination with additional attenuation, the receivers on the launcher will not be affected. The additional attenuation is justified due to the shielding of the spacecraft in combination with the fairings (inside to outside) and the fact that the spacecraft emitted noise will not directly enter the antennas, since they are directed to earth and, therefore, only side lobes would cover the spacecraft. The other philosophy is to introduce notches in the requirements, which, at some point, might be too conservative due to the attenuation effects described above and might also cause additional effort, in case the limit is exceeded and further analysis is required to justify the exceedances. In Europe, the approach to introduce notches is usually followed, but, typically, the notch levels are not as low as the receiver-sensitivity levels.

In case spacecraft transmitters are active on launch, the EMC verification is usually done by analysis, because the spacecraft would then be covered by the fairings, which form a cavity and, therefore, attenuate the emitted RF signals.

When on the pad, the launcher, including the spacecraft (payload) integrated in the fairings, may be exposed to lightning. The launch pad usually provides a lightning protection system (lightning rods around the pad) to protect the launcher and its payload from direct lightning strikes. This is not a 100% protection, but the probability of getting a direct lightning strike onto the launcher is fairly low. Nevertheless, there are indirect lightning effects, with which the launcher, as well as the payload, have to be compatible. The related requirements are defined in the launcher user manual. In [14], the EMC verification of the launch pad protection system for the SOYUZ and VEGA launchers is primarily addressed, but it also provides a good description of the lightning protection system.

To summarize, depending on the status of the spacecraft during launch, the main concern is the compatibility of antenna ports to the launcher-generated emission levels and/or to ensure there is no spacecraft-generated disturbance in the launcher receiver notch frequency bands.

8.2.2 Self-Compatibility: Autocompatibility

Due to the absence of external noise sources and victims for the spacecraft, the requirement for self-compatibility or autocompatibility is the most important EMC requirement next to the launcher compatibility. Although the formulation of this requirement on the top (system) level is very simple, it needs to be broken down for subsystem- and equipment-level tests into dedicated requirements for conducted emissions and susceptibility and for radiated emissions and susceptibility. As already mentioned in Section 8.2, this is a complex optimization process.

Concerning the compatibility to radiated noise, this aspect is covered by the launcher compatibility (see Section 8.2.1), the RF compatibility, and the compatibility with scientifically driven requirements, which will be discussed in the following paragraphs.

As far as the conducted emissions and susceptibility requirements are concerned, they have to be defined for subsystem- and equipment-level tests, implementing an adequate EMI safety margin to finally ensure the compatibility of the system with itself. A usual approach to determine the conducted susceptibility levels is to superimpose the expected conducted emissions (the defined emissions limit) from all equipment and calculate the power bus ripple voltage based on the expected power source impedance. Adding a margin to that voltage one can derive a figure for the susceptibility test level to be required at the subsystem and the equipment levels. As long as there are no particular sensitive instruments, as well as no particularly noisy equipment on board, it is rather straightforward to achieve autocompatibility. It might become a real issue once there are dedicated and stringent requirements imposed by the scientific instrument(s), which will be discussed more in detail in Section 8.2.4.

Regarding compatibility to radiated noise, this can be split into two cases:

1. Nonintentional RF emissions and
2. Intentional RF emissions.

The second case concerns the onboard RF transmitters for telemetry and data downlink and is discussed in the following section. Depending on the type of equipment used onboard the spacecraft, nonintentional RF emissions (first case) are usually not an issue, as long as the launcher receiver notch requirements, as well as the onboard receiver notch requirements, are not violated. The radiated susceptibility requirements can be chosen such that there is sufficient margin with respect to the emissions generated from onboard equipment. A proper grounding and bonding concept plays an important role here, as well as adequate cable routing, shielding, and shield termination.

8.2.3 Radio Frequency Compatibility

RF compatibility is a special case of autocompatibility, because it concerns intentional RF emissions. It is treated separately from autocompatibility, because the RF compatibility test must be performed in a shielded anechoic room in order not to violate the applicable regulations for intentional RF emissions on the ground. RF compatibility basically means that in case all the RF transmitters are powered on and working nominal, the other spacecraft subsystems shall be able to work as intended without performance degradation. In the first instance, RF compatibility is an issue between RF transmitters and receivers, using frequency bands that can be close to each other. RF transmitters also emit RF energy outside the intended frequency band of operation. These out-of-band emissions can fall into the receive band of an onboard receiver, e.g. the telemetry/command receiver or the GPS receiver. Since these receivers are designed to provide high sensitivity, the out-of-band noise from an onboard transmitter can reduce the receiver's performance or, in the worst case, block the receiver function. The compatibility is ensured by establishing requirements for out-of-band maximum emission levels on the transmitter side and requirements for out-of-band susceptibility for receivers in combination with choosing the right position for the antennas, taking into account their antenna patterns, to minimize RF coupling between the antennas. While out-of-band emissions and immunity to them can be measured at the subsystem/equipment level, the interaction between the RF subsystems, including antennas with the spacecraft surface and structure, can only be tested at the spacecraft level.

Especially for telecom satellites that typically host a number of powerful RF transmitters (transponders), it is of high importance to verify the workmanship of the RF connections between transmitters and antennas in terms of RF leakage (unintentional RF emissions from the RF transmission path) and secondary effects such as passive intermodulation (PIM) that can lead to interference, resulting in a lack of performance.

8.2.4 Compatibility with Scientifically Driven Requirements

As far as scientific missions are concerned, the requirements imposed by the scientific payload may become the main driver for the EMC requirements. Scientific instruments are often operated close to their physical limits in order to achieve the high sensitivity required to meet their scientific targets. The high sensitivity, in turn, imposes requirements on the spacecraft and other payloads/instruments onboard to provide an electromagnetic environment, which actually allows to exploit the sensitivity of the instrument. Such requirements can address various parameters of the EM environment such as DC magnetic fields (magnetic cleanliness) for magnetic missions, low-frequency electric or magnetic fields, requiring very low levels that can even come close to the limits of what can be measured. To derive the relevant EMC requirements, it is very important to understand the operating principle of the instruments and, thus, to understand what can act as a potential disturbance source and what can be the coupling mechanism. For scientific missions, the theoretical requirements can become so stringent that no feasible technical solution can be found and even measuring those levels would be beyond today's capabilities. In such cases, compromises must be found between the scientifically driven requirements and what is technically feasible. To illustrate this, the following examples are provided for satellite projects where the author and his team were (or still are) involved supporting EMC engineering throughout the project phases from requirements definition to qualification.

8.2.4.1 Swarm

The three identical Swarm satellites [15] have the mission to precisely measure the Earth's magnetic field and its fluctuations. The main instruments onboard each of the Swarm spacecraft are the vector field magnetometer (VFM), the absolute scalar magnetometer (ASM), and the electric field instrument (EFI). The magnetometers are mounted on the satellite boom and impose a requirement on the spacecraft for generating a maximum DC magnetic field of <0.3 nT at the ASM position (end of the boom) and <1 nT at the VFM position (middle of the boom). In comparison, the Earth's magnetic field is in the order of 40–60 μT. Concerning the EFI, the requirement imposed on the spacecraft is to keep the generated DC and low-frequency magnetic field at levels <10 μT. In order to achieve the required DC magnetic field levels, a magnetic cleanliness program was set up to keep the magnetic field generated by equipment and the structure under control. In this process, requirements have been derived and were verified for structural parts, equipment, and subsystems. At the system level, a simulation was performed based on the inputs from all subsystems and equipment and compared with DC magnetic field measurements that have been performed in a special magnetic test facility to verify the magnetic cleanliness of the spacecraft.

8.2.4.2 Solar Orbiter

The Solar Orbiter mission [16] is to perform close-up, high-resolution studies of the sun and the heliosphere. The Solar Orbiter comprises a set of 10 in situ and remote-sensing instruments, among them the radio and plasma wave analyzer (RPW) and the magnetometer (MAG) have been identified to be the most demanding instruments in terms of EMC requirements. The task of the RPW instrument is to measure electromagnetic and electrostatic waves in the solar wind from DC to 20 MHz. This imposes on the satellite a requirement to limit the electric field emissions in the nonstandard frequency range from DC to 20 MHz to a very low profile. The challenge is twofold: (a) to measure such very low emissions in the low-frequency range and (b) to ensure that the emissions from all other equipment are kept as low as required. The MAG instrument imposes additional stringent requirements for magnetic cleanliness. Figure 8.6 illustrates the relationship between the Solar Orbiter magnetic field limits and the standard ones for RE101 from MIL-STD-462F.

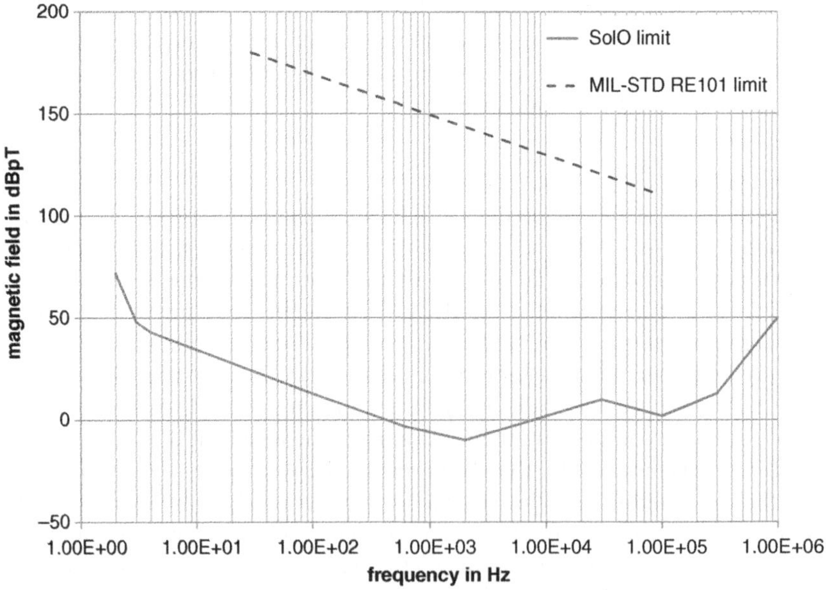

Figure 8.6 Solar orbiter magnetic field requirements versus MIL-STD RE101.

8.2.4.3 BepiColombo

BepiColombo is a joint mission between the European Space Agency and the Japanese Aerospace Exploration Agency to explore Mercury. The spacecraft consists of a transfer module (MTM) and two individual orbiters: the Mercury planetary orbiter (MPO) with instruments to map the planet and the Mercury magnetospheric orbiter to measure the magnetosphere. EMC requirements for BepiColombo [17] are mainly driven by the high-power electric propulsion system, which can generate very high current transients during operation and by the fact that the steerable high-gain antenna for communication to the ground can partially illuminate the spacecraft and, therefore, parts of the spacecraft are exposed to high RF field strengths. Further, on the MPO, a differential magnetometer (gradiometer) instrument is employed that requires stringent limits for DC magnetic fields on the spacecraft in order to ensure the instrument's accuracy.

In the cases of the Swarm and the Solar Orbiter mentioned earlier, the EMC requirements imposed by the instruments became the main driver for the spacecraft system-level EMC and required a rigorous EMC approach to analyze and define the requirements properly, to find the appropriate test and measurement methods to verify them, and to closely follow the verification throughout the project phases. The example of BepiColombo shows that there can be combinations of driving EMC requirements, like conducted emissions from the electric propulsion system in combination with RF compatibility with the RF fields from the steerable high-gain antenna, as well as magnetic cleanliness requirements for AC and DC magnetic fields driven by the magnetic instrument.

8.3 Test Execution at the System Level

For the execution of the tests at the system level, a number of constraints must be dealt with:

- The spacecraft must be kept in cleanroom conditions
- Mechanical ground support equipment (MGSE) is required
- Extensive electrical ground support equipment (EGSE) is required
- The spacecraft is not powered via the solar arrays, but from a power EGSE
- The spacecraft is connected via dedicated cable to perform command uplink and data downlink
- The spacecraft is operated under atmospheric conditions on the ground
- For cooling the spacecraft electronics, special cold plates and fluid ground support equipment (FGSE) may be required
- Some instruments may have constraints for operation on the ground
- The operation of the spacecraft can be very time consuming

- Additional sensors from other environmental tests and their related wiring might still be present (thermocouples, accelerometers, etc.)
- Great care must be taken to avoid damage of the flight hardware

The requirement to avoid contamination of the spacecraft basically means that all the testing on the spacecraft must be performed under cleanroom conditions with the required level (class) of cleanliness. Thus, typically, the tests are performed within integration areas or in test facilities that can provide the required cleanliness level. This might become a problem for transport, as well as for the use of special test facilities that cannot provide the required cleanliness level. In such cases, the spacecraft could be kept under a cleanroom tent to improve the cleanliness level.

In order to keep the spacecraft in a certain position and to transport it, usually a dedicated MGSE is used. Depending on the type and size of the MGSE, the electromagnetic properties of the spacecraft can be influenced in terms of reflections for RF tests. From a practical point of view, the MGSE might also cover or hinder the access to antennas or ports on the spacecraft. Since the MGSE usually has a galvanic connection to the spacecraft, it must be included in the grounding of the spacecraft. Especially for spacecraft that require a certain magnetic cleanliness, the MGSE must be chosen/designed such that the magnetic cleanliness is not compromised. All these aspects need to be assessed and taken into account when planning the system-level tests.

For satellite tests, the amount of EGSE can become quite large. Since the solar arrays cannot be used to generate electric power, a solar array simulator usually provides the power on the ground or the spacecraft has a dedicated interface for the power from EGSE. As an alternative, the batteries can be used as a power source for a limited amount of time. In order to command the spacecraft and to visualize, check, and record the telemetry data, a dedicated data EGSE is used. In many cases, these data EGSE can communicate with the spacecraft via a wired data link or in combination with RF special checkout equipment (SCOE) or a wireless link to the spacecraft can be established and used. The problem linked to the use of EGSE is that wired connections and, therefore, potential ground loops are generated that can have a negative influence on spacecraft performance and/or test results. Therefore, it is of high importance to have a good grounding concept for the spacecraft and the EGSE, including measuring instrumentation and the facility ground, in order to minimize the influence of ground loops and related disturbances.

Testing spacecraft on the system level is normally performed under atmospheric conditions at ambient temperature. This is not the environment for which the spacecraft is designed and might cause limitations for the operation of the spacecraft. For instance, there is the thermal aspect, if subsystems require cooling; this problem can be solved using cold plates in combination with FGSE. But, there are also instruments that can only operate in vacuum and/or at cryogenic temperatures. In this case, alternative solutions must be found.

Another aspect to be dealt with is the time required to bring the spacecraft into a dedicated mode to be used for testing. Since, in many cases, flight software is already installed on the spacecraft, the entire sequence of procedural steps must be followed as if the spacecraft was already in space. Thus, it can take a long time (in the order of 30 minutes up to several hours) from power up until the spacecraft has reached the desired operational state. This can be considered a good opportunity to test the flight software, but, unfortunately, many procedural steps must be performed that are time consuming and not necessary for the purpose of the test. In [18], a test execution control tool is described to automate the testing and [19] reports on a standardized, repeatable test approach that can be used for communication satellites that are very similar and would require only minimum adaptations to the test suite.

EMC is not the only test performed at the system level, spacecraft usually undergo a certain sequence of environmental and functional tests. Thus, it can happen that additional sensors used for other tests are still attached to the spacecraft. These can be, for example, thermocouples used for thermal testing or accelerometers used for vibration testing. The problem are the wires connected to these sensors that might be present in numbers ranging from a few up to a couple of hundred. The wires can act as antennas and, thus, support coupling of RF energy into the spacecraft during RF tests.

Last, but not least, it has to be taken into account that during testing at the spacecraft level, very expensive, sensitive, and fragile test hardware is involved. This means that proper measures for product assurance and safety must be put in place and special care must be taken in order not to damage the hardware during handling and, in particular, during the tests. RF susceptibility tests, for example, can have a potential to degrade or damage the hardware, especially if (by accident) too high levels are applied. This is, in particular, true for sensitive receiver front ends and sensitive instruments. For some equipment or subsystems, it is also required to adhere to a dedicated sequence when changing the operational modes. There can also be constraints regarding the maximum time of operation or the selection of operating modes, which, if not adhered to, can lead to degradation or damage.

8.3.1 Launcher Compatibility Tests

Since launcher compatibility tests include the verification of radiated susceptibility, these tests must be performed in an environment that allows generating intentional RF signals, but ensures that neither personnel nor the external environment is exposed to nonpermissible high RF field levels. Thus, these tests are usually performed in a shielded anechoic chamber (EMC facility). As an alternative for RS tests, in [20], the use of a transportable reverberation chamber is proposed. Not all EMC facilities are suitable because either they do not provide sufficient size (test volume) or the cleanliness constraints cannot be met. Launcher compatibility is typically verified in two steps: the susceptibility test with respect to radiated emissions from the launcher and the launch site and the radiated emissions measurement of the spacecraft. As already detailed in Section 2.1, the radiated susceptibility test is limited to frequency bands determined by the launcher and the launch site, whereas the radiated emissions measurement is focusing on the notches required by the launcher receivers. In case the spacecraft transmitters are powered on during the launch, the related transmit frequency bands must be considered as well. If the spacecraft is switched off during launch, no launcher compatibility test is normally carried out. The EMC analysis is used to verify whether sensitive antenna front ends are safe with respect to the launcher-induced electromagnetic field levels to be expected under the fairings. A special case is, for example, the Rockot launcher, where the fairings are ejected at a certain point after the launch and the spacecraft will pass by the launcher and, therefore, can be illuminated by the launcher transmit antennas. In the launcher compatibility tests, the EMI safety margin for EEDs can also be tested. To avoid the risk of igniting the EEDs and to enable the measurement of the igniter current, instead of the real EEDs, dedicated EED simulators can be used. By measuring the currents during the susceptibility tests, one can derive the safety margins. This is dealt with in more detail in Chapter 6 on pyrotechnic systems.

In the following, an example is provided for the launcher compatibility testing of the Aeolus Satellite. Aeolus [21] is a satellite with the mission to acquire profiles of the wind on a global scale. It was launched in August 2018 using a Vega launcher, but was qualified to be launched with a Rockot launcher as well. The option to have a second launcher means, for launcher compatibility, that the requirements for both launchers must be enveloped. This is illustrated in Figure 8.7 up to 1 GHz, the maximum emissions are determined by Vega requirements to 100 dBμV/m, the first notch from 120 to 130 MHz is for the Rockot launch abort system, the second notch from 440 to 460 MHz is for the Vega destruct receiver, the third notch from 1015 to 1050 MHz is for Rockot, as well as the notch from 2700 to 2900 MHz, whereas the last notch and the upper limit are taken from Vega requirements. It can be seen that, in the Vega destruct receiver notch, the Aeolus emission peaks are exceeding the limit. Therefore, an additional measurement has been taken as depicted in Figure 8.8. These emissions have been investigated in depth, and the outcome was that the skin connectors could be identified to be the root cause for the leakage.

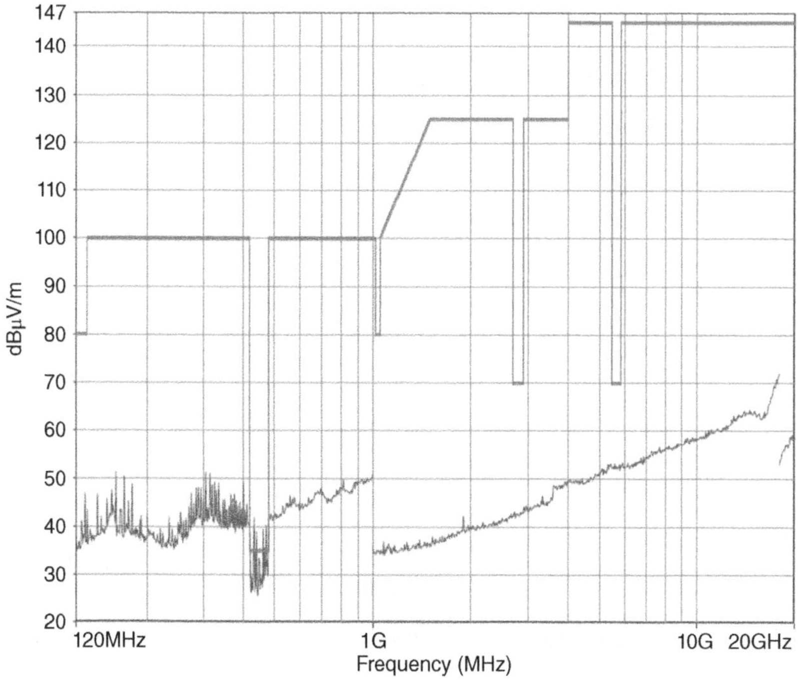

Figure 8.7 Aeolus satellite emissions (courtesy Airbus D&S Ltd).

In Figure 8.9 the panel with skin connectors used for testing on the ground is shown. There is an MIL-Bus connection and two connectors with jumpers. Two features basically supported the leakage in the frequency range from 420 to 460 MHz. One problem is that the jumper connections are implemented as long wire loops. Taking into account that they are connected to digital electronics inside the spacecraft, clocked with high clock rates and featuring high slew rates, the loops act as antennas and radiate. The other problem was the use of connector savers. Such connector savers serve to limit the number of mating/demating cycles for the flight connector, but, unfortunately, in this case they electrically isolated the connector bodies and, therefore, interrupted the bonding between the spacecraft and the connector backshell/cable shield. In this configuration, the shield of the attached cable cannot work properly as a shield and will radiate noise. In the case of Aeolus, this problem was solved easily because these skin connectors are only used on the ground and had to be covered with metal tape as already implemented for the outer left connector in Figure 8.9.

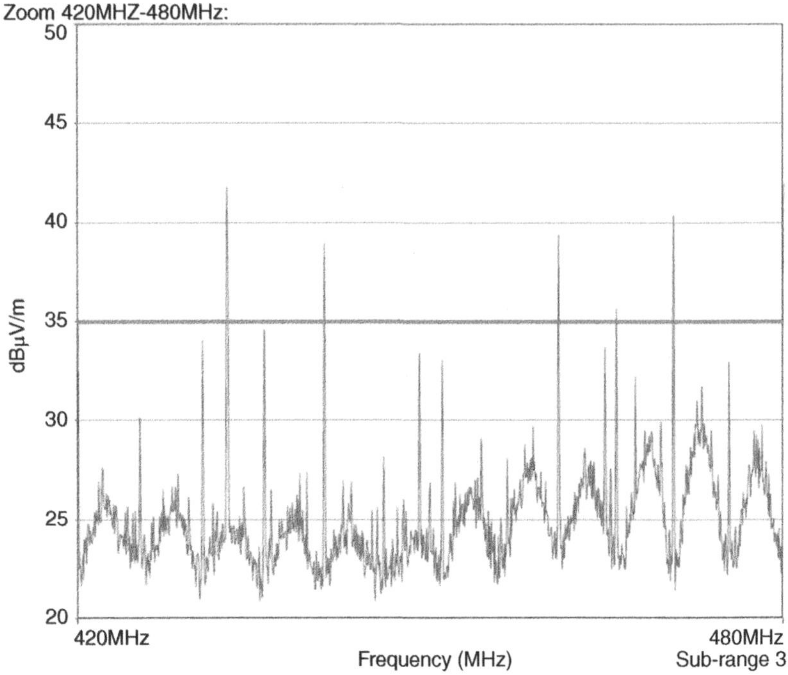

Figure 8.8 Aeolus satellite notch emissions 420–480 MHz (courtesy Airbus D&S Ltd).

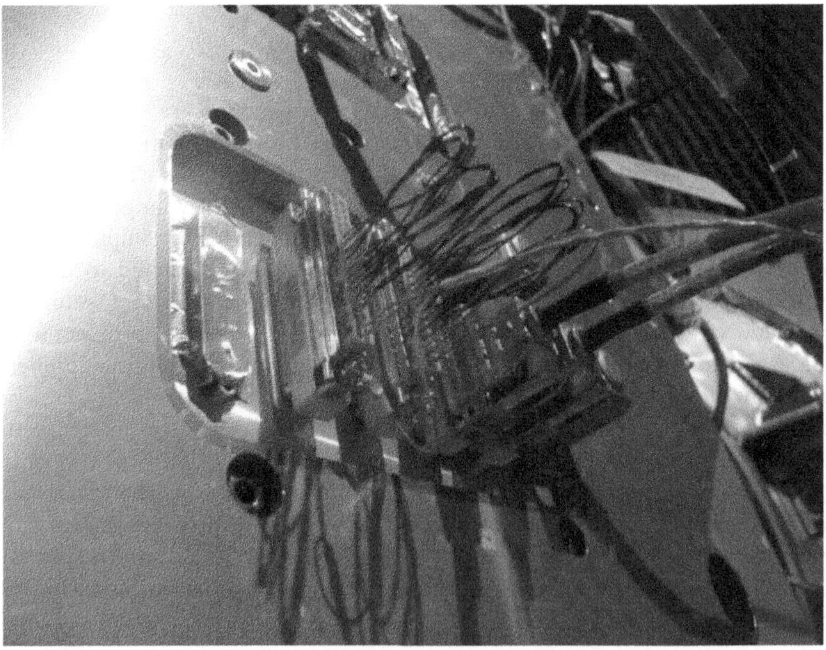

Figure 8.9 Aeolus skin connectors with connector savers (courtesy Airbus D&S Ltd).

8.3.2 Testing Autocompatibility

As already pointed out in Section 8.2.2, the autocompatibility test focuses on the operation and performance of the spacecraft onboard equipment to test whether one equipment or subsystem is causing disturbances and, therefore, malfunctions or a lack of performance to any other equipment or subsystem being part of the same system (spacecraft). It should be noted that the autocompatibility test is a qualitative test and does not verify EMI safety margins.

For the autocompatibility test, it is required to have all the equipment not only switched on, but operating in a mode that is (a) considered representative for flight and (b) acquires sensitive data in order to detect any susceptibility. Susceptibility can be detected by comparing expected housekeeping, measurement, or science data (e.g., when running the equipment standalone) with acquired data during the autocompatibility test. Thus, it is important to plan and prepare for this test carefully in terms of:

- Operational modes that can be used
- Software required to run these modes
- Simulators to provide required input to sensors or instruments
- Criteria to check susceptibility
- Means to evaluate the criteria/data
- Ground support equipment, e.g., for cooling

In theory, every possible combination of equipment/subsystems including their modes of operation must be tested. For a system like a spacecraft composed of several subsystems, the effort can easily reach dimensions that are not feasible any longer. Thus, it is very important to find an optimum between the coverage of test cases and the required test time. One usual approach is first to identify potential subsystems or equipment that can act as potential sources of interference and such equipment that is very sensitive to noise and, therefore, a potential victim. The required information for this assessment can be derived from the equipment or subsystem-level EMC tests and the EMC analysis. Based on the results of this assessment, the number of test cases can be reduced by considering only the source-to-victim relations, which is illustrated in Figure 8.10. The color code represents the categorization of the equipment into potential sources, victims, and a third category, where equipment can do both, act as a disturbance source and being sensitive. In Figure 8.10, three types of equipment can be identified to be potential victims only: the receivers (with very high sensitivity in the receive frequency bands) and the EEDs (being susceptible, of course, on a much higher level). As pure potential disturbance sources, only the transmitters (mainly for RF) have been identified. Onboard systems as well as instruments play a special role because they can be both source and victim. The example of Figure 8.10 also gives rise to the question what to do if the disturbance sources are transmitters: in case there is a

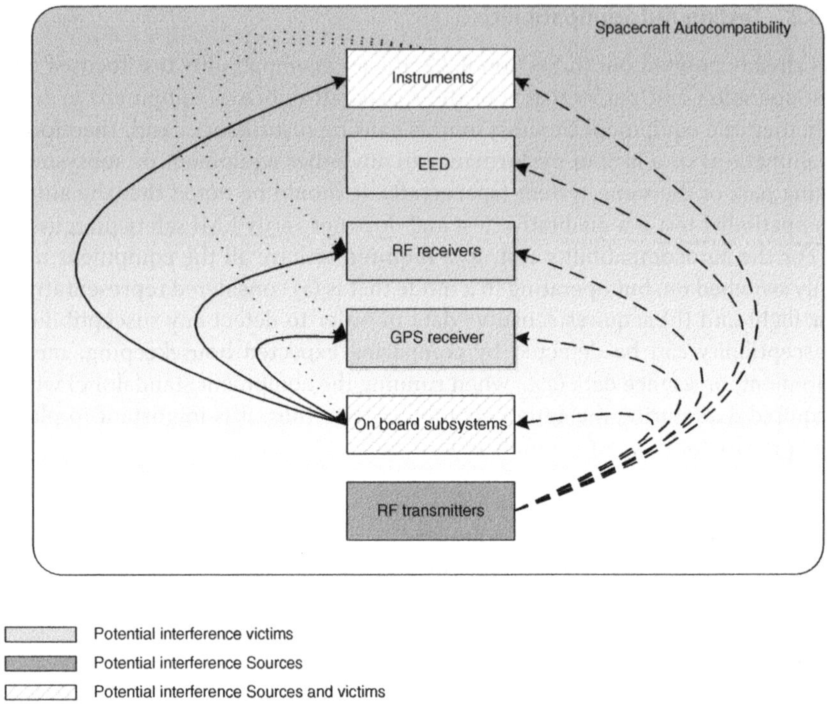

Figure 8.10 Spacecraft autocompatibility.

separate RF compatibility (RFC) test foreseen (then the autocompatibility test does not require a shielded anechoic test facility), the transmitter antennas will be equipped with antenna caps, preventing RF energy from being transmitted to the environment. Otherwise, the autocompatibility test can also be performed inside a shielded anechoic facility and can be combined with the RFC test.

For autocompatibility tests, it is of utmost importance to define and plan the modes of operation to be tested as well as the pass/fail criteria. Regarding the modes of operation, the common practice is to use all the modes defined for the spacecraft. In case there are too many modes or highly sensitive instruments onboard, different approaches can be followed. One is to start with a minimum operational configuration (PCDU and the onboard computer in the ON state), since these subsystems are always required for operation, and then to switch ON the sensitive instrument(s). During the following steps, the equipment and subsystems that are expected to present potential disturbance sources for the sensitive equipment are switched on. Equipment and subsystems that are considered to generate no interference with the sensitive instrument can be grouped for switching on or combined in one step. In particular, for sensitive instruments, but also in general, it is key to define a set of sensitive parameters and/or data to be checked in order to find out potential interference.

A further step is to keep all potential disturbance sources active at the same time and check the potential victims one by one. In case susceptibility is observed, the sources can be switched off one by one in order to determine the equipment causing the interference.

A completely different approach starts from the full operational configuration. This implies the risk that, in the case of an incompatibility, the source of the interference is not known and has to be detected by a sequential switching OFF of equipment and subsystems. It is to be noted that, in particular for sensitive instruments, mode changes or transient events on the spacecraft also need to be exercised and tested for compatibility. For all these modes and mode changes, the spacecraft operation may impose a number of constraints that must be respected in order not to damage the spacecraft hardware. One can easily imagine that this is a rather complex task, and good planning prior to the test is key to have a complete and efficient autocompatibility test.

8.3.2.1 Input Simulation

In general, the quality of an autocompatibility test—and, therefore, the confidence that the spacecraft will work without malfunctions and performance degradation caused by EMI from onboard equipment when in orbit—is very much dependent on how representative for flight are the conditions and modes in which the equipment can be tested. To achieve this representativeness on the ground, the simulation/stimulation of inputs plays an important role. In the following, a number of interfaces and their related simulation of inputs are discussed. As far as antennas and RF links are concerned, there are basically two possibilities: one is to establish an antenna-to-antenna link using a support antenna that is suited for the frequency band and the other is to use antenna caps provided with an RF connection. For the antenna-to-antenna link, a shielded anechoic chamber is required in order not to violate local regulations for intentional RF generation and to provide a sufficiently clean RF environment. Figure 8.11 shows the test setup for the Aeolus Satellite [21] for the autocompatibility test in the EMC facility. RF links are established using the spiral antenna for the S-band link and the horn antenna for the X-band downlink (left side of the picture). In both cases, RF equipment (receivers/transmitters) are required in the signal chain downstream of the counterpart antennas that provide the required RF signals and levels. For command uplink and data downlink, the RF-SCOE is used. It serves to establish a link, to change the RF parameters (e.g., attenuation and transmit levels), to measure RF parameters of the carrier, and it can be linked to the main EGSE. If a GPS signal is required, either a GPS simulator can be used or often a GPS signal received by an antenna outside on the roof of the test facility is linked via coaxial cable and preamplifier to a transmit antenna inside the chamber, placed in view of the spacecraft GPS antenna.

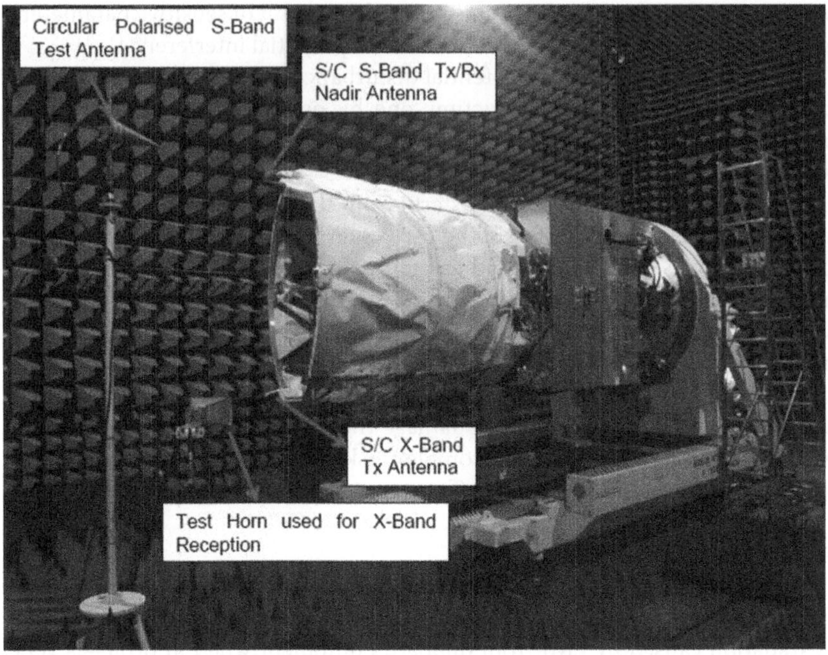

Figure 8.11 Aeolus setup for autocompatibility testing with RF links (courtesy Airbus D&S Ltd).

As far as power is concerned, spacecraft are usually powered from solar generators in combination with batteries to provide a backup for eclipse phases. The spacecraft often provides as well a power interface for operation on the ground. Thus, there are three options to power the spacecraft during the autocompatibility test:

1. Using the power EGSE connected via cable to the spacecraft power interface
2. Using solar array simulators connected via cable to the power interface of the solar wings
3. Using the spacecraft battery

In most cases, either option (1) or (2) is used. In case a real clean configuration must be achieved without introducing any ground loop, option (3) can be used for a limited amount of time. Comparing options (1) and (2), the second option offers the possibility to include the solar array regulators in the test, which is the best simulation of powering the spacecraft via solar generators.

Regarding optical sensors such as star trackers and sun sensors, optical ground support equipment (OGSE) is used to simulate the input. This can range from simple optical hoods that fit on sun sensors, equipped with a single light source (lamp or LED), to more sophisticated OGSE for star trackers that can simulate one or more star configurations (based on LEDs).

8.3.3 RF Compatibility Tests

As mentioned in Section 8.3.2, RFC tests are a subset of the autocompatibility tests with focus on RF subsystems and their compatibility among each other and with the entire spacecraft. RFC tests are often separated from the autocompatibility tests because they need to be performed in an environment that allows intentional RF radiation from the transmitters without disturbing the environment and the EGSE. This is usually fulfilled in an anechoic faraday cage equipped with absorbers (typically an EMC chamber). RFC tests typically have two aspects:

1. Antenna-to-antenna coupling from transmitters into receivers and
2. Internal coupling due to the operation of powerful RF equipment generating disturbances to other equipment and instruments.

The testing of antenna-to-antenna coupling is mainly to test the effects of out-of-band emissions falling into receiver bands used onboard. In some cases, the antenna gains and patterns and the distance and orientation to the receiving antenna are well defined, such that this part of the RFC test can also be performed at the receiver input port (receiver antenna disconnected) by injecting the test levels and frequencies directly from a signal generator. Usually, this kind of direct testing at the receiver input port is already performed at the subsystem or the equipment level.

Internal RF coupling effects can also be tested in a mode of operation where the RF equipment is operating while the transmitter antennas are covered with antenna caps that are suitable to handle the transmitted power. If the nominal RF power is too high to be handled by the antenna caps, the test must be performed in a shielded anechoic chamber.

As soon as there are payloads or instruments embarked that can be susceptible in the transmit frequency ranges and to the transmitted levels, it is very important to include test steps in the RFC procedure that allow the verification of the proper function of that equipment in the presence of nominal RF levels. In particular, this is true for scientific instruments, which can even be illuminated from the transmitted electromagnetic field. In this case, it is unavoidable to perform the test in an anechoic chamber with the antennas connected, uncapped, and transmitting the nominal levels.

To perform the RFC tests in a representative way, the signal received from the ground must be simulated. This can be achieved using an RF link, placing a transmit antenna in the chamber in the field of view of the receiver antenna, and feeding the signal with the appropriate level from the RF SCOE to achieve the targeted level at the spacecraft receiver input. Figure 8.12 shows the CryoSat-2

Figure 8.12 Cryosat-2 setup for RFC testing with RF links (courtesy Airbus D&S Ltd).

[22] spacecraft setup for RFC tests. The spiral cone antenna in the foreground and the two horn antennas on the right serve to establish the RF links and to provide the GPS signal.

In particular for telecom spacecraft in terms of RF compatibility, it is very important to test RF leakage from the junctions and connections of RF waveguides connecting the outputs of the transmitter stages to other RF equipment and the antenna feeds. Since telecom spacecraft have multiple RF payloads combined with multibeam networks and redundant paths, the structure of RF connections becomes very complex and, thus, locating potential RF leakages becomes difficult.

Once an RF incompatibility has been detected, the usual way to further narrow down the RF leak is to perform RF sniff tests to locate the source or to perform spray tests to locate the potential weak point where the RF can enter. Depending on the frequency range, different types of sniff antennas can be used: for frequencies below 1–3 GHz, small rod antennas or loops are suitable, whereas, for higher frequencies, often waveguide to coax adapters are used.

In order to perform the sniff test, the test operator must come close with the sniffer antenna to the suspect hardware or cables on the spacecraft and check on a spectrum analyzer the level of the signal caused by the leaking RF. The signal amplitude serves as an indication of how close the sniffer antenna is located with respect to the RF leakage. In case the potential leak where RF can couple into the equipment is to be found, a spray test can be useful. An RF generator is used as a source for the signal in the frequency range of interest and its output is connected to the spray test antenna. Once the antenna is close to the weak point in the shielding, interference can be observed and, therefore, the weak point is found. These methods are very practical to locate and fix potential weak points in the shielding of cables and boxes.

8.3.4 Testing Compatibility with Scientifically Driven Requirements

As mentioned in Section 2.4, scientifically driven requirements can be very demanding and often they cannot be verified with the usual standard EMC test methods and setups. The reason being that either the instrument cannot be operated in ambient conditions on ground (e.g., requires vacuum or cryogenic temperatures) and/or required emission limits for these instruments are so extremely low that standard test facilities together with the EGSE cannot provide the required environment in terms of ambient noise. Also, it can be the case that the scientifically driven requirements are outside the frequency range of the standard EMC tests or addressing phenomena that are not covered by the standard EMC methods and requirements. In such cases, dedicated solutions need to be found that meet the particular needs of the instrument(s) and allow to evaluate compliance with the scientifically driven requirements. In the following, some exemplary approaches and solutions are discussed.

8.3.4.1 Example of Herschel: Testing Susceptibility against Transient Magnetic Fields.

The Herschel Satellite [23] employs a very large and powerful infrared telescope with three scientific instruments, one of them being the Photoconductor Array Camera and Spectrometer (PACS). The PACS photometer is based on highly sensitive bolometer detectors, which transform the absorbed radiant energy into temperature. The electrical output is derived from the temperature elevation that is caused by absorbing energy. In order to reach the desired sensitivity, the bolometers (grouped in detector arrays) have to be operated at a cryogenic temperature of 300 mK. A dedicated feature of PACS is that the bolometers are assembled in matrices and are multiplexed in order to reduce wiring and to meet the very demanding thermal constraints (power is limited to 10 μW for detector and readout). The important characteristic of the bolometers in PACS (as for any other bolometer) is that they are not dependent on the wavelengths and are sensitive to different types of energy. This makes them in particular vulnerable to noise generated on the spacecraft [24]. Thus, from an EMC point of view, the cryogenic subsystem with the detectors and cold electronics is the most critical part.

Due to thermal reasons, the electrical connection between the cold electronics and the warm electronics in the service module cannot be shielded and does not provide good electrical (and thus thermal) conductivity. This harness is routed outside the spacecraft and in close proximity to the solar panels, which has been identified in the course of the EMC analysis as a potential problem: the solar panel-generated magnetic field can couple into the harness.

In order to understand the potential impact, it is important to know the specifics of the solar power generator used on Herschel. This is based on a sequential switching shunt regulator (S3R), which typically has to switch currents in the order of 14 Amps (for Herschel) between the solar cell strings and the spacecraft power bus, with rise times in the order of a few μs and repetition frequencies ranging from a few Hz up to 10 kHz and higher. This means the nominal operation of the solar arrays produces pulsed magnetic fields that can couple into the harness between cold electronics for the bolometers and warm readout electronics, disturb the measurements and therefore have a direct impact on scientific output and performance.

Analysis and tests have been performed during the development on equipment level and the susceptibility was confirmed. The test also showed that the observed susceptibility was strongly dependent on the loop area determined by the harness between cold and warm electronics. As a mitigation to reduce the risk, it was decided to reroute the harness.

In order to verify that the rerouting had the desired effect and to verify the operation of the solar power generator has no impact on the bolometer readout, a test on system level was planned and performed. A transient magnetic field susceptibility test was conducted at fully integrated satellite level using a mockup to simulate the solar array and related magnetic field coupling. Figure 8.13 shows the mockup for the solar panel from the front side. This panel was then mounted close to the spacecraft in the position of the original solar panel as depicted in Figure 8.14 (the black panel is the mockup and the wiring can be seen). The waveform expected to flow through the solar cell strings was generated with an arbitrary waveform generator, amplified by means of a power

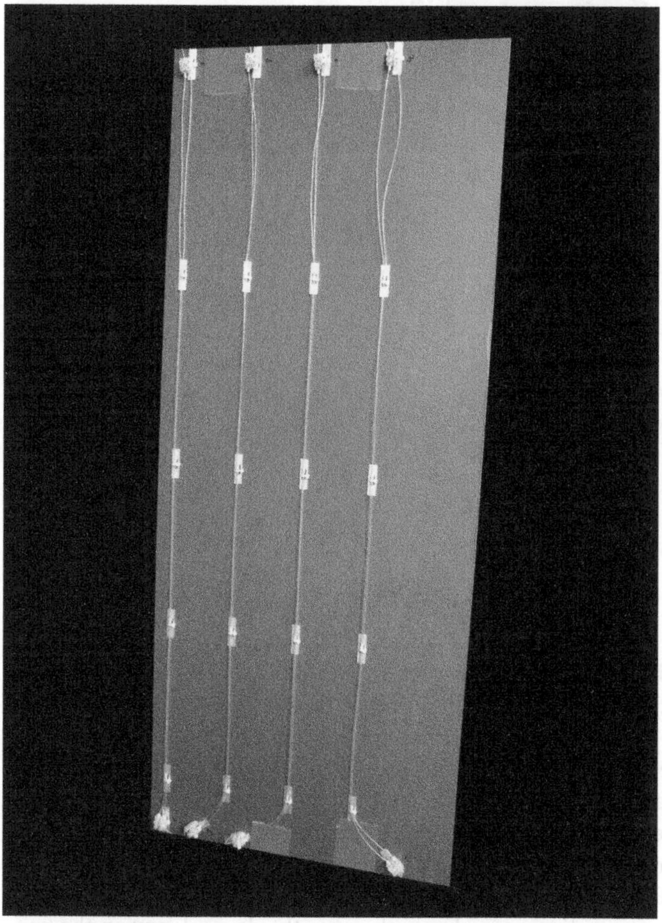

Figure 8.13 Herschel solar panel mockup front side.

Figure 8.14 Herschel solar panel mockup mounted (backside and cabling visible).

amplifier to achieve the targeted current amplitudes of 14 A with a rise time of ~1 μs and injected to the coupling circuit of the mockup. Bursts of current pulses have been injected with 100 different pulse repetition frequencies. The results were assessed by comparing the RMS values derived for quiescent (no pulses applied) and noisy mode (pulse bursts applied). The measurements have been further characterized by a power spectrum density (PSD) with 1/f power spectrum. When the pulse bursts were applied, this resulted in a noise increase of 250%, which is a clear indication of strong interference. Investigations on the harness routing were clearly showing the dependency on the harness routing and an optimum routing path ensuring minimum coupling was determined.

Table 8.2 MetOp search and rescue instruments

Instrument	Frequency (MHz)	Receiver level (dBm)
SARR1	121.5	−145
SARR2	243	−145
ADCS and SARP	401.65 and 406.05	−147
SARR3	401.65 and 406.05	−147

Conducting this special test on spacecraft level was the only way to assess the severity of the transient magnetic field disturbances from the solar array generator and to draw conclusions in order to make decisions on the spacecraft harness routing.

8.3.4.2 Example of MetOp: Measurements on System Level to Verify the VHF Notch [25]

The MetOp satellites [26] are built in the frame of a large European meteorological satellite program to ensure meteorological observations from a polar orbit. Besides the scientific instruments employed on the satellites, there are search and rescue (S&R) instruments: S&R repeater (SARR), S&R processor (SARP), and the advanced data collection system ADCS working in the VHF frequency band.

As suggested by the very low receiver levels listed in Table 8.2, the derived radiated emission requirements for both satellite platform and payload are very stringent at these frequency notches. In the course of the EMC analysis, the emission requirements have been broken down, taking into account cable shielding, module shielding, and filtering. A particular challenge was the service module, being a recurrent hardware and thus not leaving freedom to implement additional measures to meet the emission requirement. This is an example for payloads imposing requirements to the service module (platform), as discussed earlier in Section 8.2.

A number of special measures were implemented in order to achieve the required emission levels:

- Structure cutouts have been closed
- Baffles for structure venting have been implemented
- Structure gaps have been closed with aluminum tape with conductive glue
- Harness has been carefully shielded, especially the external harness
- Solar array cables have been filtered
- Special harness feed-through connections have been designed and implemented
- Shielding has been implemented between service module and payload module

Figure 8.15 MetOp satellite during RFC measurements in the CPTR (courtesy Airbus D&S Ltd).

During the development and at the equipment level, measurements have been performed to verify the harness shielding, to verify the filtering and to measure the shielding efficiency of the structure.

In the frame of the satellite EMC test campaign, the verification of the radiated emission in the sensitive S&R receiver notches constituted the major part. The EMC test of the MetOp payload module engineering model was performed in a compact payload test range (CPTR) as shown in Figure 8.15, which is normally used for antenna measurements. This chamber provides a good shielding in a frequency range from 14 kHz to 40 GHz, but the installed absorbers are effective only in a frequency range from 1 to 40 GHz. To compensate for that and to attenuate reflections from the CPTR reflectors, mobile absorber walls with special low-frequency absorbers have been placed in front of the facilities of RF reflectors. When trying to measure such low levels in the notch frequency bands, it becomes a real challenge to bring down the noise in the chamber and of the measuring equipment to these levels. The EGSE and auxiliary equipment is starting to act as a major disturbance source at such low levels. In the particular case of MetOp, a number of countermeasures were necessary in order to enable measurements in the notches of the S&R instruments:

- Checkout computers were located as far as possible away from the satellite and were connected via a single LAN interface
- Harness had to be reduced to the absolute minimum necessary
- Feed-through connections with built-in filters were used between chamber and control room
- All cables were equipped with overall harness shields
- For the power harness, so-called high-loss cables have been used (capacitance of 650 pF/m) to attenuate RF carried from outside the chamber into the CPTR

One particular measure was required to avoid emissions from the telecommand and telemetry connection to the satellite: emissions representing the 232nd harmonic of the clock frequency (524.288 kHz) were clearly visible in the 121.5 MHz notch. The solution to lower these emissions was to insert filter connectors in the chamber wall feed-through and place ferrite beads on the cable toward the satellite.

To lower the noise floor of the measurement instrumentation, the original S&R antenna was used as receiving antenna to measure exactly at the position of the receive antenna for the S&R instruments. Furthermore, a high-gain LNA was required to compensate the cable losses and to reduce the instruments noise floor. High-quality cables were used to keep the cable losses at a minimum.

Finally performing the test campaign on system level, the compliance of the payload and service module to the defined notch frequency bands could be successfully verified, although this required a large effort. This would not have been possible without proper preparation, starting from analysis and tests on equipment level and implementing appropriate measures up to the preparation and planning of the system-level test.

References

1 J. D. M. Osburn, "The role of systems EMC analysis in large system EMC management by equipment EMC assurance, electromagnetic compatibility," in *Symposium Record., IEEE, International Symposium on*, pp. 189–192, 1991.

2 R. J. Perez, "Correlating the EMC analysis and testing methods for space systems in MIL-STD-1541A: Electromagnetic compatibility," in *IEEE International Symposium Record*, pp. 25–32, 1990.

3 ECSS-E-HB-10-02A, *Space Engineering – Verification Guidelines*. Noordwijk: ECSS Secretariat, ESA ESTEC, 2010.
4 D. Farren and T. Ambler, "The economics of system-level testing," *IEEE Des. Test. Comput.*, vol. 14, no. 3, pp. 51–58, Jul.–Sep. 1997.
5 ECSS-E-HB-20-07A. *Space engineering – Space Systems Electromagnetic Compatibility Handbook*. Noordwijk: ECSS Secretariat, ESA ESTEC, 2011.
6 R. Zamir, V. Natan-Bar, and E. Recht, "System level EMC – From theory to practice," in *International Symposium on Electromagnetic Compatibility*, vol. 3, pp. 741–743, 2005.
7 J. R. Moser, "Peripheral cable-shield termination: The system EMC Kernel," in *IEEE Transactions on Electromagnetic Compatibility*, vol. 28, no. 1, pp. 40–45, 1986.
8 D. S. Dixon, "Radiated emission and susceptibility EMI testing: The advantages of conducting unit level tests in a system level environment," in IEEE National Symposium on Electromagnetic Compatibility, pp. 77–82, 1989.
9 K. Kempkens, "An approach to cost-effective EMC performance verification," in *ESA-CNES EMC workshop*, Feb. 12–13, 2004, WPP-221, pp. 151–157.
10 D. C. Pande and M. K. Aravind, "Some aspects of bulk cable current injection (BCCI) test method at system-level testing of an aircraft," in *Proceedings of the International Conference on Electromagnetic Interference and Compatibility*, pp. 29–34, 1999.
11 A. Ciccolella and F. Marliani, "EMC in space systems: Current practices and future needs: The ESA perspective," *Radio Science Bulletin* No 328, pp. 9–19, 2009.
12 ECSS-E-ST-20-07C, *Space engineering—Electromagnetic compatibility*, Noordwijk: ECSS Secretariat, ESA ESTEC, 2012.
13 EHB0003, *Rockot user's guide*. EUROCKOT Launch Services GmbH, no. 4, Revision 0, 2004.
14 E. Bachelier, F. Issac, L. Guibert, W. Quenum, V. Enjablert, and L. Mohedano, "Lightning protection of SOYUZ and VEGA launching pads," in *International Conference on Lightning Protection (ICLP)*, pp. 1–6, 2012.
15 ESA, "Our activities – observing the earth – Swarm – facts and figures." Available at https://www.esa.int/Our_Activities/Observing_the_Earth/Swarm/Facts_and_figures. Accessed Sep. 23, 2018.
16 ESA, "About science and technology – Solar orbiter definition study report (red book)." Available at http://sci.esa.int/solar-orbiter/48985-solar-orbiter-definition-study-report-esa-sre-2011-14/#. Accessed Sep. 23, 2018.
17 K. Kempkens, "EMC issues on BepiColombo spacecraft," in *International Symposium on Electromagnetic Compatibility Tokyo* (EMC'2014 Tokyo), pp. 844–847, 2014.
18 M. Levesque, J. Louie and A. M. Guerrero, "Test execution control tool: Automatic testing in spacecraft integration and test environments," in *IEEE Aerospace Conference Proceedings*, vol. 2, pp. 389–394, 2000.

19 G. Crandall, C. Abbott and S. Reedy, "Evolution of a standardized," in *IEEE Repeatable Satellite Test Approach: Autotestcon*, pp. 435–444, 2007.

20 F. B. J. Leferink, "In-situ testing of aircraft and satellites using a transportable reverberation chamber." in ESA Workshop on Aerospace EMC, Florence, 2009.

21 ESA, "Adm-aeolus – Earth explorers – ESA's wind mission." Available at https://www.esa.int/Our_Activities/Observing_the_Earth/Aeolus. Accessed Sep. 23, 2018.

22 ESA, "Cryosat-2 – earth explorers - ESA's ice mission. Available at https://www.esa.int/Our_Activities/Observing_the_Earth/CryoSat. Accessed Sep. 23, 2018.

23 D. Schmitt, F. Marliani, B. Voss, and A. Poglitsch, "Magnetic field susceptibility of the Herschel /PACS Instrument: Identifying, Assessing and Fixing a Probable Severe Loss of Performances: Aerospace EMC 2009, ESA workshop on: Florence 2009

24 D. K. Griffin, J. Delderfield, J. J. Bock, V. Hristov, C. Cara, D. Schmitt, C. Cunningham, L. Rodriguez, M. Griffin, B. Swinyard, F. Marliani, P. Mauskopf, de Korte, and A. Cross, "Electromagnetic compliance of high sensitivity bolometric detectors for space science," in *ESA workshop on Aerospace EMC*, Florence, 2009.

25 P. Bisognin, A. Dyne, and K. Kempkens, "Accommodation of a sensitive VHF instrument on a large satellite," in *Proceedings WPP-221, ESA-CNES EMC Workshop*, Feb. 12–13, pp. 95–104, 2004.

26 ESA, "MetOp – observing the earth – MetOp overview." Available at https://www.esa.int/Our_Activities/Observing_the_Earth/MetOp. Accessed Sep. 23, 2018.

9

Subsystem EMC for Aircraft

Paul Kay

9.1 Introduction: The Aim of Subsystem-Level Testing

This chapter discusses electromagnetic compatibility testing of electronic components and subsystems that are intended for integration onto aircraft and describes the central aspects of some of the commonly encountered technical tests performed under aviation electromagnetic compatibility (EMC) standards. It is intended that the reader will gain an appreciation of the technical content and significance of subsystem test specifications and methods, without being burdened by detail that can be found in the standards themselves; the readers should find themselves better placed to oversee the EMC aspects of integration of electronics onto aircraft, and to produce contract details, test plans, or other requirements documentation to support certification of electronic subsystems for flight.

Aircraft subsystem EMC exists and is considered in the wider context of aviation systems safety. When new hardware is introduced to an aircraft, the system-level safety of the aircraft must be reviewed in order to certify the "new" aircraft system that incorporates the new hardware. This is typically done by revisiting the aircraft certification baseline and systems safety case and performing a series of risk analyses in order to demonstrate to the aviation regulator (military or civilian) that the "new" aircraft system is safe to fly—even when the "new" system is substantially the same as the "old" system, with notionally one subsystem change. In the military context, there may also be a requirement to demonstrate that the "new" aircraft system can fulfill its mission as well as operating safely.

Systems safety and mission criticality are determined through a risk assessment process. The level of risk associated with any hazard is determined by considering the likelihood of an event occurring and the consequences brought

Handbook of Aerospace Electromagnetic Compatibility, First Edition. Edited by Reinaldo J. Perez.

about as a result of that event. When making estimates of risk where there is an absence of information and dire consequences of failure, we tend to overestimate the risk. When risk is overestimated, design and integration costs can rise or operational restrictions can be applied unnecessarily; system integrators therefore require as much information as possible about new subsystems and hardware prior to making such assessments. This is where component- and subsystem-level EMC testing plays an important part, as it provides information about the component's/subsystem's maximum emissions and susceptibility (think "electromagnetic ruggedness"). Systems engineers can use this information in performing assessments of the likelihood of an interference event occurring as a result of integrating the new hardware. With sufficiently detailed subsystem-level electromagnetic compatibility data, the systems engineer can form judgments about suitable locations in the aircraft for the new subsystem to be installed, mitigations to protect existing aircraft systems, or begin to populate a source–victim test plan to check for interference after a trial installation, in the case that there is an identified risk of interference or insufficient information. EMC test results, obtained in a test laboratory environment for a particular subsystem, can be analyzed when integrating the same subsystem onto various aircraft types; such analyses may show acceptable risks for a component in some aircraft but unacceptable risks for other aircraft. Introducing an electronic item to an aircraft, either as a portable device or as an integrated (installed) subsystem, raises two questions related to EMC:

(i) What could this item do to the aircraft?
(ii) Will this item function correctly on the aircraft?

The first question really derives from the requirement to revisit the systems safety analysis for the aircraft as a result of possible interactions related to the new subsystem. Even if the new subsystem provides a function as trivial as in-flight entertainment, the potential for interaction with safety-critical aircraft systems exists, and some level of analysis is required to provide confidence that there is no harmful interaction between extant systems and the new subsystem.

The second question acknowledges that an aircraft can provide a hostile electromagnetic environment. Aircraft are host to a plethora of radio transmitters and are sometimes exposed to high-intensity ground- and ship-based transmitters; electronic subsystems installed in and on an aircraft can be exposed to intense radio frequency fields as well as severe electrical power transients, from both aircraft power bus switching and lightning.

One way of determining answers to these questions would be to install the item on the aircraft and then test the aircraft as a system. If the aircraft and the new item pass the necessary functional tests while the new subsystem is operating and while a range of electromagnetic disturbances are applied to the whole aircraft, then we could say that electromagnetic compatibility had been achieved.

The process of testing an entire aircraft is complex, time consuming, and expensive; a comprehensive program for whole aircraft testing can occupy many months, and highly specialized test facilities are required. There are times when whole aircraft testing is necessary or desirable, but the expense, logistics, and comparative dearth of aircraft-level test facilities means that many aircraft types receive perhaps only one whole-system EMC test during development; some aircraft types are not subjected to any whole-aircraft-level EMC tests at all—the entire case for system-level compatibility may rest on subsystem testing and analysis.

EMC standards, such as RTCA/DO-160, MIL STD 461, and the UK DEF-STAN 59-411, provide a "building-block" approach to achieving electromagnetic compatibility by specifying methods, test levels, and limits for subsystems that are to be used on aircraft or in other installations. Compliance with laboratory test levels and limits for subsystems forms the basis for expecting that EMC issues are unlikely when the component is integrated into the host aircraft. That said, subsystem-level compliance with technical standards is not a guarantee that problems will not be found in the integrated system, but the risk is lower. The final installation location for the electronic subsystem is an important determinant of system-level compatibility. For example, a subsystem that complies with the applicable emission tests in a standard could still cause interference if it is mounted in close proximity to the antenna of a victim receiver; this can be especially challenging on small aircraft, where space is at a premium and options for installing new items are limited. Conversely, noncompliance with a subsystem standard is not a guarantee of EMC problems at integration, if additional mitigations such as supplementary shielding or adequate separation (space and/or frequency separation) can be provided.

When a new hardware item is considered for use in an aircraft, compliance with component-level standards helps to ensure that the item has a suitable level of compatibility, based on its expected application and installation location. Proper consideration of the latter aspect may require tailoring of the limits applied to a subsystem's specification, to ensure that unwanted interactions do not occur if the planned installation location is close to victim antennas. "Tailoring" refers to the process of adjusting the published limits in a standard to address identified risks with particular procurement programs or to remove restrictions that may not be relevant to a particular situation.

Interference is considered in two ways: front-door and back-door coupling:

- Front-door coupling refers to interactions where unwanted signals intrude into the victim system via a pathway that is intended to bring signals into the victim. Usually, this refers to unwanted signals impinging on the antenna of a victim radio receiver, where the source's electromagnetic emissions appear as in-band or out-of-band interference in the victim. In the case of

front-door coupling, the interference enters the victim system via a pathway that is intended for the reception of signals.
- Back-door coupling refers to interactions where unwanted signals intrude into the victim system via pathways that are not intended to receive signals, for example, power lines.

One could argue that the concept of front-door coupling could be extended to include nonradio receivers, such as analog-sensing lines for temperature or angle of attack. While the frequency range of these slowly changing signals is as low as a few Hertz, the bandwidth of the processing and encoding circuits may be much higher, rendering them susceptible to "out-of-band" interference—conceptually not very different from a radio receiver out-of-band interference scenario. One mitigation for front-door interference problem is filtering, to ensure that only frequencies within the desired (expected) frequency range can enter the system. A key characteristic of subsystems that are candidates for front-end interference is that they include at least one sensitive receptor pathway, almost always a radio receiver/antenna combination.

9.2 Motivations for Testing: Safety of Flight and Success of Mission

Purpose-built avionics equipment may have been tested by the manufacturer to civil aviation or military EMC standards; statements of compliance and test reports provide key information to inform the integration onto any given aircraft.

Avionics can be grouped into three categories for the purposes of EMC considerations:

(i) safety-critical items—equipment that is critical to safe flight, such as flight and engine control systems, navigation systems, and communication systems;
(ii) mission-critical systems—equipment whose functioning is key to the purpose of the flight, such as weapon systems in a military aircraft or scientific payloads in a research aircraft;
(iii) other systems—systems that are not critical for safe flight, are not central to the mission at hand, but whose correct functioning provides noncritical information or makes the flight more comfortable. Such systems are sometimes subjected to a minimal set of emission testing referred to as "safety-of-flight" (SOF) tests—not to be confused with the full-test program that is applied to safety-critical systems. However, such systems must be either installed or tested in such a way that failure does not bring about a safety hazard for any other aircraft systems, for example, by tripping a circuit breaker on failure.

1. Safety-Critical Systems. Equipment in the first category of subsystems is exposed to harsh and stringent EMC tests. Usually, these systems are integral parts of the aircraft, and testing is covered at the time of manufacture, as part of the certification process for the aircraft. The intention is to ensure that the subsystem under test can tolerate the expected worst-case electromagnetic environment at its installation location. Examples of safety-critical equipment include full authority digital engine controllers, flight control computers, servo controllers, flight instruments, and fire control systems. Safety-critical systems are identified by the aircraft system integrator (very often the manufacturer) as part of the systems safety analysis.

2. Mission-Critical Systems. Equipment in the second category is exposed to similarly harsh tests as safety-critical equipment, in order to represent the expected electromagnetic environment, although in some cases, the severity of the immunity test levels may be reduced somewhat.

For mission-critical systems, there is a nonsafety-related consequence to malfunction of the equipment; if a subassembly or component is critical to the success of the mission, then its ability to withstand the electromagnetic environment on the aircraft must be determined. The test documentation should disclose the electromagnetic susceptibility of the equipment, in addition to the electromagnetic emissions the equipment will introduce to the aircraft. The typical susceptibility test scenario is to apply EMC disturbances to the equipment, one at a time, while monitoring the equipment to determine if any performance degradation, malfunction, or damage occurs. Examples of mission equipment include scientific or research payloads on research aircraft and weapons systems on military aircraft. Mission systems are related to the purpose of the flight, but are not related to the airworthiness of the aircraft—loss of functionality, or reduced functionality, of mission-critical systems would not result in an unsafe condition.

While the term "mission critical" has a military flavor and background, the concept can be equally well applied to commercial, scientific, and industrial equipment. Consider a piece of equipment that may be used in performing a geological survey in a locality that is several hours flying time from the aircraft's base: if the equipment malfunctions with a loss of data or calibration settings (for example) when the aircraft's transmitters are keyed on a particular frequency, or due to other RF or transient disturbances that exist on the aircraft, then the very reason for configuring the aircraft and making the flight has been defeated. For equipment of this kind, it is worthwhile to consider susceptibility testing even if it is not mandatory for the purposes of flight clearance or certification.

3. Other Systems. Equipment in the third category is only required to meet the emissions test components of the relevant civil aviation or military standards. The limits are not relaxed compared with the emission limits applied for the first two categories, but the scope of testing is smaller, as there is generally no requirement for susceptibility testing. The reason for this is that there is no interest in the effect that the aircraft environment (power disturbances, RF fields due to transmitters, electrostatic discharge, lightning, etc.) has on the equipment—if the equipment fails to function or suffers operational failures due to the electromagnetic environment it faces onboard the aircraft, there is no safety or mission consequence—so long as it is not powered from a critical circuit on the aircraft. If a new noncritical item of equipment shares a power circuit with aircraft subsystems that perform safety-critical functions, and the new item suffers a failure as a result of an EMC event such as a lightning strike or power bus switching transient, it could still compromise safety. If a subsystem does not have appropriate protection circuits, it could fail in a short circuit mode, which would trip the breaker for the entire circuit, taking out safety-critical subsystems with it. This is perhaps more likely to arise as a systems integration problem on smaller aircraft, where comparatively fewer power circuits may be available for newly introduced subsystems.

Such problems aside, the effect that a new subsystem has on the aircraft is every bit as important for this category as it is for the other two categories—that is, its conducted and radiated emissions could compromise unrelated safety-critical aircraft systems.

This SOF testing refers to a suite of EMC tests that are performed on a component or subassembly to verify that it will not compromise the key avionics systems on the aircraft, but with little regard for the integrity of the component itself. SOF testing is not to be confused with tests for safety-critical equipment. SOF tests determine the frequencies and intensity (amplitude) of unintended radio frequency emissions; information is used to inform a risk analysis against the hazard the equipment presents to the aircraft's communication and navigation systems. Unless a new transmitter is introduced to an aircraft or the new equipment produces high field intensity or large-amplitude RF currents, there is little additional risk of interfering with safety-critical systems in the baseline aircraft besides those that are radio receivers.

In some operational scenarios, the aircrew may be able to order or initiate disconnection of a nonessential piece of equipment that is suspected of causing interference. This approach is sometimes taken with flight test instrumentation systems, which may be only a temporary fit to an aircraft that is flown under a restricted set of conditions (including location, payload, and weather). If the equipment is found to interfere with aircraft systems, it is possible to mitigate the effect for a few flights in a given locality by restricting use of communications systems (for example, by procedurally blocking out affected channels). While this is hardly the preferred approach for an EMC purist, it is a pragmatic way of dealing with a potential problem to ensure that the essential safety systems for the aircraft are available for the duration of the test flights.

9.2.1 Communications and Navigation Systems

Radio receivers are inherently the most sensitive victims of radio interference. Traditional communications and navigation receivers commonly used in aviation are highly sensitive tuned narrow band designs, usually multistage superheterodyne types; such receivers have good ability to reject signals and noise outside of the intended reception band, but are highly sensitive to degradation by any unwanted signals that may exist in the desired reception band.

Careful frequency planning helps ensure that harmonics from transmitters and intermodulation products do not fall into sensitive receiver channels. The frequencies of these kinds of disturbances are predictable, and correct allocation of channels generally keeps operational problems to a minimum. However, disturbances from emitters that were not considered during the frequency planning process—notably, unintentional emissions from electronic subsystems—could fall into sensitive receiver bands, and the performance of affected receivers could be compromised. EMI problems may be apparent as unwanted squelch breaks (squelch circuits enable the audio output when a signal is received—but they cannot differentiate between a valid signal and an unwanted disturbance), audible noise such as buzzing or squealing, or—less obviously—reduction in radio or range due to receiver desensitization.

Increasingly, aerial navigation makes use of satellite navigation technologies, and the susceptibility of these systems to interference is generally lower than a narrow band superheterodyne receiver, but also the failure modes are different. Satellite navigation signals use code-modulated spread spectrum techniques, and susceptibility to narrow band signals (such as unintentional clock harmonic emissions) is generally low, but where susceptibility does occur, the failure mode is highly nonlinear, in contrast with an analog (particularly AM) systems, where noise due to increasing interference grows to eventually swamp the wanted signal. A digital communications system (including satellite navigation systems) that is suffering interference incorporates redundancy in the modulation process and sophisticated error correction algorithms. These mitigations can extract the required data from a partly corrupted signal, with almost no perceptible degradation of performance, until eventually the interference is so strong that either key synchronization markers are lost, or there is insufficient redundancy to allow reconstruction of the baseband signal—and when this point is reached, the entire signal is lost. Furthermore, digital communications systems require a finite amount of time to "lock" to the wanted signal—a process that can take a few seconds to several minutes, depending on the nature of the signals and receiver. For this reason, the limits for some categories of aviation equipment provide additional protection in the satellite navigation band.

9.2.2 A Note on Commercial and Industrial EMC Standards

While these standards are outside the scope of aviation applications, they are included in this chapter for the benefit of system integrators who may be using commercial-off-the-shelf (COTS) products in aviation applications. Most COTS equipment suppliers can provide EMC compliance information of some sort to the system integrator, and it is important that avionics engineers undertaking such integrations understand the EMC compliance information that is presented. It is common to be able to access EMC compliance documentation that supports "CE Marking" or FCC (Federal Communications Commission) compliance of a device. CE Marking, while ubiquitous, only has legal relevance in the European Union (EU), although many jurisdictions outside the EU recognize the technical standards for EMC on which CE declarations of conformity are often based, and even then is only of relevance in general consumer and industrial applications—not aviation applications. The CE Mark indicates that the supplier has made a Declaration of Conformity to all relevant Technical Directives, some of which include EMC requirements. However, retesting of a COTS item to the applicable aviation or military standard is normally required to maximize the likelihood of a successful integration with the aircraft. Civilian standards used for consumer goods and industrial applications cover a smaller frequency range than aviation standards and usually permit larger amplitude emissions. The measurement methods applied in these standards, along with FCC compliance tests, are not directly comparable with MIL STD 461 or RTCA/DO-160 series tests for emissions or susceptibility. Furthermore, the acceptance criteria applied for susceptibility (commercial standards use the term immunity instead) tests in support of CE Marking are sometimes at the discretion of the manufacturer, whereas RTCA/DO-160 and other aviation susceptibility standards require that the test item complies with its own specifications during or after application of susceptibility tests.

The European EMC Directive does not apply to aeronautical products—the only situation where an aircraft systems integrator should have to work with standards and documentation relating to the EMC Directive is when integrating equipment that was not originally designed for aviation applications. An example of this scenario is installation of a conventional computer onboard an aircraft to support a scientific or research application. Successful integration of the computer may require supplementary shielding, filtering, or operational restrictions in order to achieve electromagnetic compatibility, even though it may be supplied with many pages of certifications and attestations regarding its EMC integrity.

9.2.3 A Note on Software

This chapter is presented in a context of adding a new hardware item to an aircraft. However, since almost all modern electronics has a software component embedded in it, and since the software very often governs how the hardware

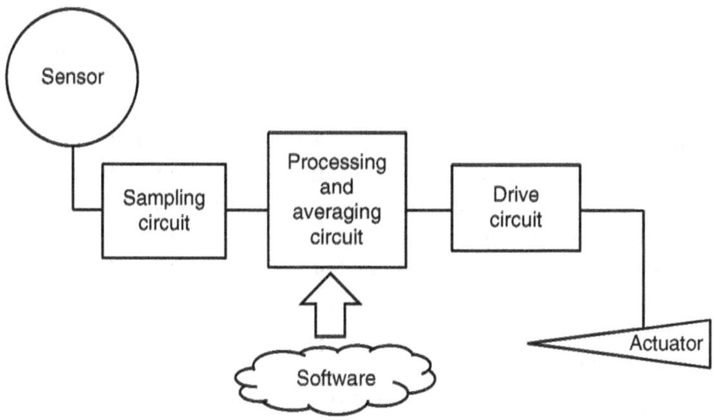

Figure 9.1 Example sensor and programmable averaging circuit.

operates, there is a compelling case for reconsidering electromagnetic compatibility following software upgrades.

Figures 9.1 and 9.2 show a conceptual example. Consider a system that has a programmable averaging function, as shown in Figure 9.1. An analog sensor is sampled, and the samples are passed to a processing and averaging circuit that produces an output to drive an actuator; this is representative of part of a fly-by-wire aircraft control system.

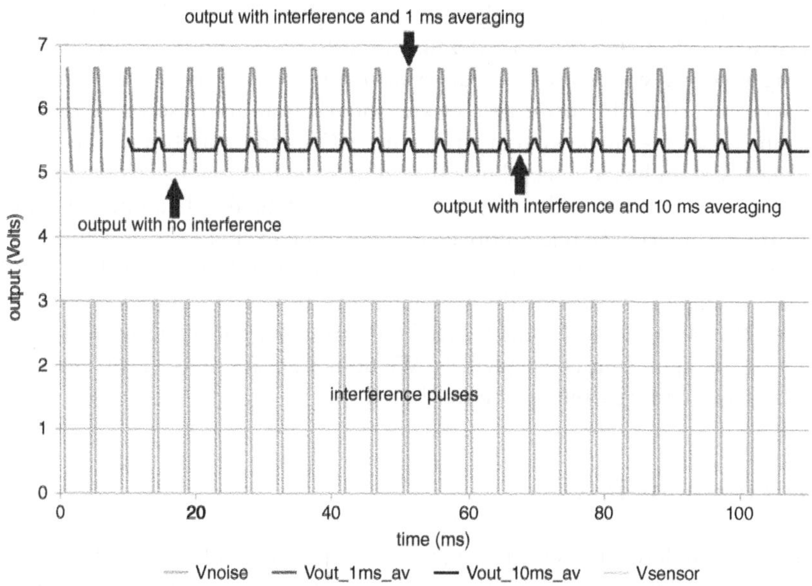

Figure 9.2 The effect of averaging times on outputs in the presence of pulsed interference.

If the sensing circuit and/or wiring is receptive to RF interference from some external source, and the external source is sufficiently strong, then the output for different interference and averaging scenarios changes, as shown in Figure 9.2. A short (1 ms) averaging time gives little smoothing to the pulses, and the output of the averaging circuit is obviously corrupted as shown by the red trace in Figure 9.2; the expected sensor DC output is shown in yellow and the hypothetical induced interference is shown in gray. In contrast, the black trace shows the output (combination of interference and sensor DC output) with an averaging time of 10 ms: the output is somewhat degraded and has an offset error, but the circuit may have passed an immunity test if the deviations were within specifications. In the example, the error of the 10 ms averaging (black trace) output is always less than 10%.

If a software "upgrade" is performed on the processing and averaging circuit, reevaluation of the test results, and possibly retesting, should be considered to provide confidence that the outputs of the circuit are not adversely affected by the change.

For example, software changes could alter the number of cycles over which a slowly varying parameter (such as angle of attack) is averaged. If a software change deliberately reduced the number of averaging cycles (perhaps in order to provide a reduced response time at the output of the circuit), the susceptibility of the circuit to pulsed disturbances would increase. At a subsystem level, this increases the reliance on shielding and filtering in the hardware components of the sensors and circuits, which may have been sufficient for long software averaging times but insufficient at shorter times. It may be possible to perform such evaluations or analysis by software simulations that apply various virtual inputs to the code in order to analyze the code's outputs—this is fine so long as code test scenarios include realistic representations of interference.

It seems worthwhile to at least consider the implications of software modifications to safety-critical systems from an EMC standpoint. To reiterate, if a hardware system has a vulnerability that allows the external electromagnetic environment (EME) to add unwanted noise to a particular data line, but the software has filtering to reject unrealistic data, then overall EMC is achieved. However, if that software was changed in such a way that the filtering effect was lost (for example, by reducing the averaging time from 10 to 1 ms as shown in Figure 9.2), the same EME could bring about erroneous data readings with the potential for significant consequences. We do not always revisit EMC testing after software changes, but maybe we should.

9.3 Emissions Tests

Technical standards for electromagnetic compatibility of subsystems and components define test methods and limits for emissions, or emanations, from subsystems in order to provide protection for the classes of systems that are likely to already exist on the host aircraft. It should be noted that, where no limit is specified, a compliance test is generally not performed even if a method exists, although this should not preclude engineering teams from invoking any test method that they think will help them develop or understand their product. With respect to compliance testing, however, it is uncommon for a subsystem or component supplier to present EMC data other than the legally or contractually required minimum, and this ties the tests that are performed and reported to any limits that are specified. That said, EMC emission tests are usually performed continuously across the frequency range for which limits are specified (rather than only at the frequency bands of interest for a particular aircraft).

There is an intimate connection between the limits applied in any given standard and the test methods that are used in that standard; it is generally not valid to apply the limits of one standard to test results produced using the methods of another. Most test methods have unavoidable compromises built into them, and limits have been established by history and experience, and may have offsets included in them to accommodate properties of the measurement method. In some areas, particularly IEC/CISPR, there have been ongoing efforts to improve the technical aspects of measurements and provide new measurement methods, coupled to efforts to produce or align limits with the new methods in a way that corresponds to the protection requirements of established methods. In general though, IEC/CISPR test methods are considerably different from those used in aviation applications, and extrapolation or application of aviation limits to IEC/CISPR test results is neither helpful nor useful—retesting to the appropriate aviation standard is recommended.

Military and aviation standards have maximum limits for emissions that are established by decades of operational experience; history has shown us that those subsystems whose emissions are under the limits in the standard are unlikely to cause interference problems on the aircraft—so long as appropriate installation practices are followed. Compliance with a standard is not a guarantee of trouble-free operation, but it is a good start. And, of course, the years of experience only apply to those aircraft systems that are years old, such as VHF/UHF communications and navigation. The question as to the suitability of legacy limits to complex wideband digital modulations remains open.

MIL STD 461, DEF-STAN 59-411, and RTCA/DO-160 provide different emission limits for different types of aircraft or different applications. Furthermore, the limits are frequency dependent. The test method is the same within each standard (that is, independent of the limit) for the various emissions tests and limits contained in each standard—and each of these three standards uses similar methods of measurement—but a simple statement of compliance with a standard should be accompanied by some detail relating to what version of the standard, what limit, and under what conditions. For example, a power supply module may comply with the limits for a large aircraft under MIL STD 461, but it may require additional shielding or a filter to comply with the helicopter or external aircraft limits of MIL STD 461. The method of measurement in MIL STD 461 versions F and G is sufficiently different from the radiated emissions test in MIL STD 461E that results in the HF band could be quite different if tested using that method. Stepping back from the somewhat narrow picture of the results from laboratory tests to different versions of a standard, and thinking about the aircraft systems that we seek to protect, we can form a view that a laboratory test result—to any standard—can be used to inform an aircraft source–victim EMI/C test, where the candidate new system is operated on the aircraft and aircraft systems that may be susceptible to emissions identified in the laboratory tests can be checked.

In RTCA/DO-160 G, different equipment types are identified by different categories and limits that are determined by the intended installation location of the equipment under test (EUT). This approach allows the EUT to have higher emissions if it is to be installed in an avionics bay than if it is to be installed close to antennas or in a predominantly nonconductive aircraft. If the EUT is to be installed close to aircraft antennas, the limits have "notches" to apply tighter controls to emissions in bands that are critical for aircraft applications, such as VHF communications, transponder, and GPS bands. As an example of the complexity around limits just for radiated emission tests, RTCA-DO-160 G has six categories of equipment for the purposes of emission limits. The limits embody assumptions about the expected path loss between the test item (the interference source in this example) and the sensitive potential victim systems built into the aircraft.

For radiated emissions for aviation applications, MIL STD 461 and DEF-STAN 59-411 apply limits that embody similar assumptions about RF path loss between the test item (source) and the potential victim equipment on the aircraft:

- Equipment for internal applications on fixed wing aircraft > 25 m nose to tail
- Equipment for internal applications on fixed wing aircraft < 25 m nose to tail
- Equipment for external applications on fixed wing aircraft, and helicopter internal and external applications

The rationale for this is that large aircraft have sufficient space that higher emissions from subsystems can be tolerated, as the likely separation between the source (EUT) and victim (aircraft receivers and sensors) is greater than in small aircraft and helicopters. Recall the EMI/C interference model of source—coupling path—victim. The limits for large aircraft have an inherent assumption that the coupling path from aircraft subsystems has larger losses than in smaller aircraft. Helicopters are subjected to the tightest limits because they are generally smaller than fixed wing aircraft, have larger fuselage apertures, and therefore are likely to have the smallest coupling path losses and correspondingly the largest likelihood of interference. The same limits for emissions from subsystems are applied to external applications on all aircraft because there is the strong possibility of close proximity line-of-sight interference from the source (mounted externally, e.g., in a wingtip pod) to the victim antenna. Recalling our interference mechanisms, the concern here is for front-door interference coupling, where the emissions from the new subsystem impinge on the receptors (antennas) of the victim system.

The smallest difference between the measured emissions (conducted or radiated) from the test item and the applicable limit in the standard is referred to as the *pass margin*. Conventionally, the pass margin is defined as

$$\text{pass margin } (dB) = \min[\text{limit}(f) - \text{emission amplitude}(f)]$$

where *limit* and *emission amplitude* are frequency dependent.

If a test report for a given component or subassembly shows a pass margin of N dB for a given limit, then the margin for an alternative limit can be determined by linearly extrapolating using the difference between the known limit and the alternative limit—*so long as the test method is the same for the different limits.* Care must be taken when comparing results in one version of a standard with limits of another version—a thorough inspection of the methods of the two versions must be made to ensure that no significant changes were introduced.

EMC laboratory tests for cabinet and box-sized equipment faces a challenge dictated by the physics of the measurement antennas. A structure is not a very efficient radiator (or receptor) of electromagnetic energy unless its dimensions are on the order of a quarter wavelength or more. This is not to say that a structure must be tuned to some fraction of a wavelength in order to radiate, but a structure that is much shorter than a quarter wavelength is not a very efficient radiator or receptor.

For cabinet and box-sized equipment tested in a laboratory, structures are inefficient radiators below approximately 80 MHz. The same applies for test and measurement antennas. Nonetheless, there is the potential for interference to occur at these lower frequencies because the cables connected to the EUT when it is installed on the aircraft may be long enough to radiate effectively, or proximity between the EUT (a potential interference source) and a victim system or cable loom on the aircraft may be such that even inefficient radiation exceeds an interference threshold of the victim.

Most EMC standards provide voltage or current measurement methods for lower frequencies and radiated (antenna) test methods are applied above those frequencies. The frequencies for changing from one technique to another are not hard and fast, and there is sometimes considerable overlap between the two notionally complementary approaches. Table 9.1 provides an overview of some standards and their conducted and radiated test frequency ranges. The table is not intended to be an exhaustive list of EMC standards; it is a cross-section of

Table 9.1 Comparison of conducted and radiated emission test frequency ranges

Standard	Application	Conducted emissions test frequency range*	Radiated emissions test frequency range**
MIL STD 461F/G	Military equipment (US standard)	CE101: 30 Hz–10 kHz (I) CE102: 10 kHz–10 MHz (V)	RE101: 30 Hz–100 kHz (H) RE102: 10 kHz–18 GHz (E) RE103: 10 kHz–40 GHz (E)
RTCA/DO-160 G and ED-14G	Civil aviation aeronautical equipment (US and Europe)	Section 21 Para 21.4: 150 kHz–152 MHz (I)***	Section 21 Para 21.5: 100 MHz–6 GHz (E)***
DEF-STAN 59-411	Military equipment (UK standard)	DCE01: 20 Hz–150 MHz (I) DCE02: 20 Hz–150 MHz (I)	DRE01: 88 MHz–18 GHz (E) DRE02: 500 Hz–250 kHz (H) DRE03: 1.6–88 MHz (E)
IEC/CISPR 11	Industrial, Scientific and Medical Equipment (general civilian applications)	9 kHz or 150 kHz–30 MHz (V)	9 kHz–30 MHz (H) 30 MHz–18 GHz (E)
IEC/CISPR 22	Information technology equipment (general civilian applications)	150 kHz–30 MHz (V)	30 MHz–1 GHz (E)
IEC/CISPR 32	Multimedia equipment (general civilian applications)	150 kHz–30 MHz (V)	30 MHz–(max) 6 GHz (E)
IEC/CISPR 25	Vehicles (civilian, protection of onboard receivers)	150 kHz–108 MHz (V) 150 kHz–108 MHz (I)	150 kHz–2.5 GHz (E)

*(I) = current measurement, (V) = voltage measurement.
**(H) = magnetic field measurement, (E) = electric field measurement.
***Earlier versions, e.g., DO-160E, specified conducted current emission limits in the range 150 kHz–30 MHz and radiated E field measurements in the range 2 MHz–6 GHz.

some military, aviation, and commercial/industrial standards that are relevant to aviation systems integrators. Furthermore, susceptibility requirements of the standards are not shown in Table 9.1, but are discussed later and summarized in Table 9.2.

Since conducted and radiated emission tests provide amplitude measurements of emanations from the EUT, it is important for the standards to define measurement instrument parameters that could affect the indicated amplitude, if they were left uncontrolled. The following measurement receiver parameters are defined in the standards listed in Table 9.1:

- Measurement bandwidth (IF bandwidth or resolution bandwidth)
- Measurement bandwidth type (usually 6 dB for EMI measurements; spectrum analyzer resolution bandwidths are usually specified at 3 dB points)
- Dwell or sweep time of the measurement receiver
- Measurement receiver detector type (peak for MIL STD 461, RTCA/DO-160, and DEF-STAN59-411, but quasipeak, average, or rms average for IEC/CISPR)

These factors are yet another point of difference between dedicated aviation equipment standards such as MIL STD 461, RTCA/DO-160, and DEF-STAN 59-411 and general consumer or industrial standards such as those produced by IEC/CISPR. Standards that specifically address aviation equipment and applications require the measurement receiver's peak detector to be used, but commercial and industrial standards allow use of detectors that give lower-amplitude indications when measuring pulsed interference.

9.3.1 Conducted Emission Tests

Conducted emission tests aim to quantify the magnitude of radio frequency emissions that exist on power and other interface wires or cable bundles connected to ports on the equipment. Conducted emission tests are a surrogate for radiated emission tests that are not practicable in a laboratory environment due to the long wavelengths and correspondingly large antennas and large separation distances associated with measurements up to approximately 100 MHz; they also serve to quantify the potential for interference by direct conductive coupling on shared networks (particularly power networks). As illustrated in Table 9.1, there is no hard changeover frequency where conducted testing stops and radiated testing starts, although a general comment is that conducted voltage and current techniques lose effectiveness above 400 MHz, and many standards do not specify conducted emission tests above 30 MHz.

There are two approaches to measuring conducted emissions:

(i) measurement of the RF currents on individual cables or bundles of cables and
(ii) measurement of the RF voltages on individual conductors.

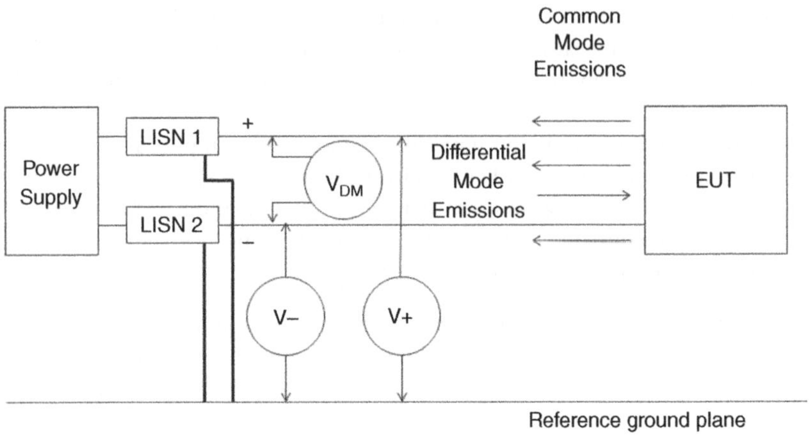

Figure 9.3 Conducted emissions voltage measurement.

Figures 9.3 and 9.4, respectively, show conducted voltage and current power line emissions measurement. The EUT may be powered from a DC source, or a single-phase AC source (multiple phase EUTs are accommodated in a similar manner but the diagram is omitted for clarity). While there is no ground connection shown at the EUT, most standards require that the EUT is grounded in accordance with the manufacturer's instructions or the system integrator's declared installation plan. For a line replaceable unit (LRU) intended to be installed in an avionics bay, the EUT ground is generally provided by a faying (conductive mating) surface, dedicated mechanical earth pin, earthing strap,

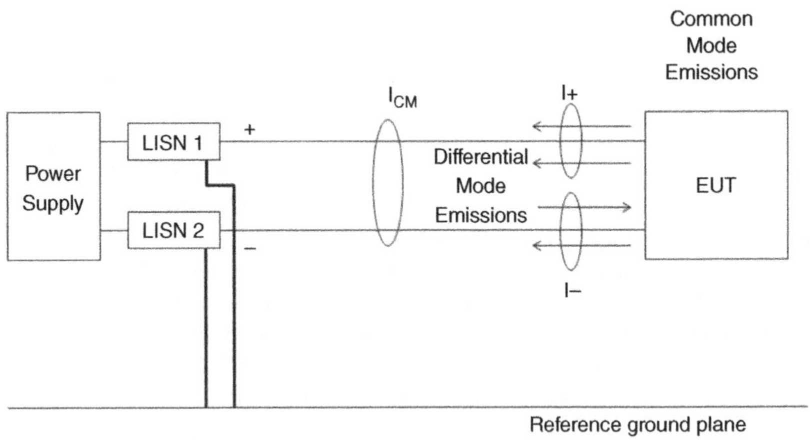

Figure 9.4 Conducted emissions current measurement.

earthing/mounting/locking screw or a combination thereof. In the case of consumer and industrial equipment, such special grounding measures are seldom provided and the EUT reference ground may be provided only by an electrical safety ground (e.g., earth wire in a three-pin mains cable), or perhaps "accidentally" by a cable shield in a data interface (e.g., USB connection). In terms of testing an EUT for EMC, the grounding should be representative of the intended installation. Some standards, for example, MIL STD 461, require a DC milliohm meter measurement of the grounding integrity of the EUT (for grounded EUTs) prior to commencing tests. While the relevance of a DC-bonding check at RF is questionable, the DC check ensures that EUTs with dedicated grounding measures are correctly installed for the test (that is, gross errors such as forgetting the bonding strap, missing spring washers, or dirty mating surfaces are detected by the DC check).

Elementary circuit analysis tells us that the voltages and currents present on power and other cables attached to the EUT are functions of the impedances in the system. For this reason, EMC standards are highly prescriptive with regard to physical arrangement of the EUT, the reference ground plane, and layout of cables, because these factors determine the stray capacitance and coupling of the RF currents and fields on and around the EUT. By specifying the test layout, reproducibility of measurement results between different test facilities is improved. However, one key variable is not addressed by specifying the layout: power supply impedance. One test facility may operate the EUT from a battery while another may use a DC power supply; a third facility may perform conducted emissions tests in a shielded room, with power fed via high RF loss filters. Each of these cases presents the EUT's power input port with a different RF impedance: the RF impedance viewed from the power input of the EUT, looking back toward the power source, is not controlled in any of these scenarios. It follows that measured voltages and currents would vary for the same EUT measured in each different setup. To overcome this problem, conducted power line measurements, whether voltage or current, specify use of a line impedance stabilization network (LISN) in series with the power supply; these are shown in Figures 9.3 and 9.4. Many LISNs provide a coaxial measurement port to allow monitoring of the common mode voltage on each conductor with respect to the reference ground plane; LISNs of this type should have unused coaxial ports terminated with a coaxial 50 Ω termination, as the measurement port forms part of the impedance network.

A safety note on LISNs*:* LISNs incorporate large capacitors between the power lines and the enclosure. This means that they cannot work on circuits that incorporate residual current devices (RCDs), "safety switches," or earth leakage detectors, because the leakage current (more than an Amp) trips the safety device. For this reason, they are usually installed on dedicated circuits in the test facility or operated via an isolation transformer. They should always be securely earthed and nobody should attempt to pick up or relocate a live LISN.

If the *ground connection is not intact* and the LISN input is *connected to an AC power source* (e.g., 50, 60, or 400 Hz), the capacitors will cause the case of the LISN to float up to the mains voltage and a *significant shock hazard* exists if anyone is in contact with the case or any coaxial cable that may be connected to the measurement port. Always follow the LISN manufacturer's installation instructions and avoid having personnel around the LISN when it is in use.

It should be noted that attempts at controlling the test setup and power line impedance aim to provide reproducibility between different facilities working to the same standard: *there is no attempt to provide equivalence of results or reproducibility between different standards.*

Some EMC standards specify limits and tests for conducted emissions on other cable interfaces as well as power interfaces, because any cable attached to the EUT enclosure has the potential to form a radiating structure. In general, these "other" cables are tested by measuring the RF currents with a current probe arranged to measure the common mode current in a similar manner to I_{CM} in Figure 9.4. Current clamps are helpful tools for EMC diagnostics as well, because they can clamp around a cable bundle without having to break the connection.

Radio frequency emissions from the EUT are shown as currents emanating from the EUT on the two conductors; emissions may be common mode (that is, appearing on all conductors equally with respect to earth) or differential mode (existing as an RF potential difference between two conductors). Usually, elements of both are present.

Voltage limits for differential mode emissions are seldom specified. Most of the standards listed in Table 9.1 do not specifically target differential mode emissions using a voltage measurement (V_{DM} in Figure 9.3)—the exception is DEF-STAN 59-411 test DCE03, which captures the amplitude and duration of transient emissions on the power supply lines as a result of manual and automatic switching operations in the EUT. For those tests that require measurements of common mode voltage on power supply lines, the voltages are measured between each conducted in the supply bundle and reference ground. Generally, the LISN includes a voltage measurement port for connection of an EMI receiver. LISNs may be constructed singly, as represented in Figures 9.3 and 9.4, or two or more lines may be combined into a single LISN enclosure. So long as the LISN meets the applicable impedance specification for the required lines, either is acceptable.

Between different standards that require RF current measurements for conducted emissions, including those listed in Table 9.1, there is some variation in the arrangement of the current probes. For power line measurements, placement of the probe may be specified close to the EUT cable back shell, or close to the LISN. "Back shell" refers to the shell or body of the cable bundle connector that connects power and other services to the EUT; since the size of the back shell dictates the distance between the current probe and the EUT enclosure, it is important that the special test cable looms that are generally required for EMC testing are representative of the actual looms that will be used in the aircraft. With reference to Figure 9.4, the loop probe I_{CM} may be placed close to the LISNs (e.g., DEF-STAN 59-411 power line testing) or close to the EUT (e.g., RTCA/DO-160 G). Placement of current probes can be challenging for the test facility: the bulk and weight of some probes, combined with the rigidity, location and orientation of the EUT's interface cables and the need to connect the facility's EMI receiver coaxial cable to the current probe measurement port conspire to make meeting specifications such as "Conducted Emissions Probes to be located 5 cm from the EUT back shell" (RTCA/DO-160 G) more difficult than one would expect.

9.3.2 Radiated Emission Tests

Radiated emission tests are designed to quantify the electromagnetic emissions from the EUT and its associated cabling. The test configuration is specified in order to emulate the expected installation arrangement of the EUT, and to provide consistent results between different test facilities. The testing process should include an exploration of different EUT operation modes and configurations in order to identify the mode and layout that maximizes emissions, consistent with the expected installation and use of the equipment. As indicated in Table 9.1, measurements are based on electric or magnetic field sensing.

9.3.2.1 Radiated Emission Test Sites

Defense and aviation equipment standards such as MIL STD 461, RTCA/DO-160, and DEF-STAN 59-411 provide a pragmatic approach to measuring radiated emissions. The measurement range is specified as 1 m, which allows testing in comparatively compact anechoic chambers. Radiated emission tests are more challenging to perform than conducted emission tests because they are inherently more prone to corruption from unwanted influences. Outdoor tests are complicated because of ambient emissions—signals that already exist at the test site, due to telecommunications, broadcast, radar, and other transmissions as well as unwanted emissions from commercial and industrial sources such as electric motor controllers, welders, spark ignition motors, information technology equipment, and most other electronics, not to mention the support equipment brought to the site by the test team. Furthermore, outdoor testing brings a range of logistical problems—the author has had outdoor test schedules compromised due to bushfire preventing access to the Open Area Test Site (OATS), rain entering cable and turntable pits (damaged cables and non-functioning equipment), livestock straying onto the site, and laboratory vehicles bogged in muddy paddocks; OATS operators in colder climes have to deal with ice and snow. However, the anechoic chamber option is expensive if tests in accordance with IEC/CISPR standards are performed, because the chamber

Figure 9.5 Open air test Site configured for 10 m radiated emission tests. The paved area between the mast and the turntable is covered with a galvanized mesh and is flat to within a few centimeters. The mast and table are predominantly nonmetallic in construction, to avoid unwanted reflections, and the receive antenna carriage on the mast is remotely controlled to adjust the antenna height up to 4 m above the ground plane. The turntable is remotely controlled to allow rotation of the test item in order to identify the maximum emission level at each frequency of interest. The test receiver, operators, and any support equipment are located out of camera shot to the left, away from the receive antenna and the test item.

must be sufficiently large to accommodate a 3 m or 10 m measurement distance and a 4 m antenna height, plus the necessary clearances around the test range, absorber, and antennas; MIL STD 461, RTCA/DO-160, or DEF-STAN 59-411 tests can be performed in a more compact chamber as the test range is 1 m, and there is no requirement to adjust the measurement antenna height. An OATS configured for 10 m testing is shown in Figure 9.5, while Figure 9.6 shows a MIL STD 461, DEF-STAN 59-411, or RTCA/DO-160 chamber.

9.3.2.2 Radiated Magnetic Field Emission Measurements

The reader may recall electromagnetic theory from textbooks or undergraduate class, and probably already knows that a transverse electromagnetic (TEM) wave in free space is composed of oscillating electric and magnetic field components at right angles to each other and to the direction of propagation in free

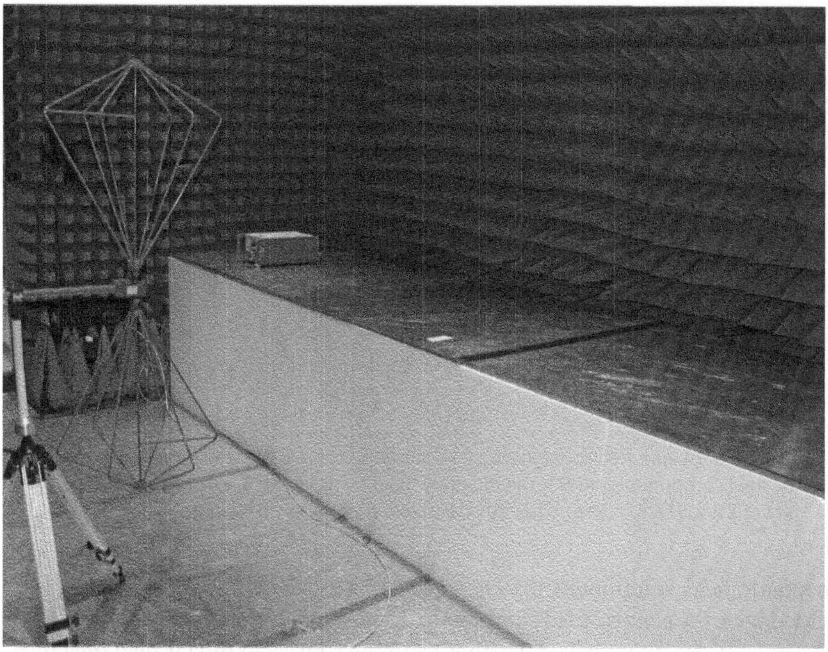

Figure 9.6 Anechoic chamber configured for 1 m testing, as required by MIL STD 461. DEF-STAN 59-411, and RTCA/DO-160 use a similar chamber and setup.

space, as shown in Figure 9.7. From that starting point, if we can determine the magnitude of one component of the field, we can determine the other component of the field and then the Poynting Vector, or power flux density. This is a straightforward matter in the far field of a radiating source, but at closer distances, the relation between the wave components and the structure is complex and not easily predicted.

It seems fair to ask why the standards listed in Table 9.1 sometimes require measurement of the E field and sometimes the H field, and sometimes both. The primary reason for this is that EMC measurements are very often performed in the near field of the EUT and the measurement antenna; plane wave conditions do not exist and either the electric or magnetic component of the field may dominate, depending on the nature of the radiating structures associated with the EUT. The near field for these purposes is considered to be a distance from the source less than $\lambda/2\pi$. In the near field of a radiating source, the conditions shown in Figure 9.7 do not exist, as the wave has not propagated sufficiently far from the source to completely form.

Maxwell's equations tell us that, in the near field of a predominantly high impedance source, the electric field component decays with $1/r^3$, while the

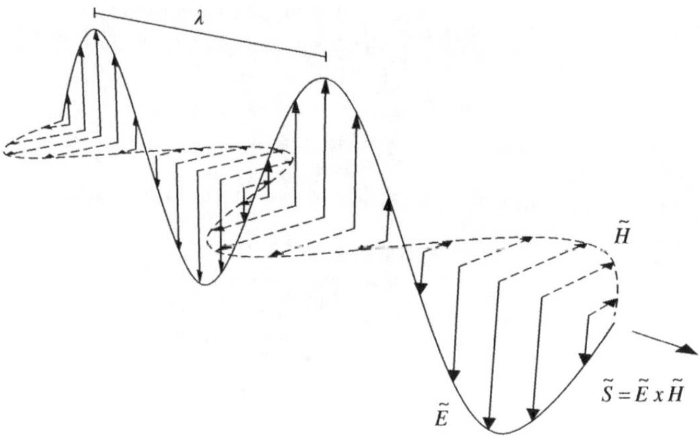

Figure 9.7 Transverse electromagnetic wave in free space. λ is the wavelength, ***E*** is the electric field component and ***H*** is the magnetic field component; ***S*** is the vector cross product of ***E*** and ***H*** and is the power flux density of the wave.

magnetic field component decays with $1/r^2$. Conversely, if the source is low impedance, the electric field component decays with $1/r^2$ and the magnetic field component decays with $1/r^3$. Further from the source, in the far field (distance from the source greater than $\lambda/2\pi$), the electric and magnetic components of the field decay with $1/r$, for either type of source. This means that, depending on the EUT, the electric field may be disproportionately larger than the magnetic field or vice versa.

At typical measurement distances for avionics-sized EUTs in the lower frequency ranges, neither the relationship between E and H nor the nature of the source (high or low impedance) is known. Some aviation systems—potential victims of interference that may exist on the aircraft—are susceptible to magnetic fields, for example, acoustic systems and VLF, LF, or MF communications (very low frequency, 3–30 kHz; low frequency, 30–300 kHz; or medium frequency, 300 kHz–3 MHz). These systems may use loop antennas that are receptive to magnetic field disturbances, but due to shielding and size reject electric fields in those ranges.

Risk of interference to these types of systems is evaluated by EMI tests that are performed using a loop sensor. Inspection of Table 9.1 shows that some standards specify emission measurements in lower frequency ranges twice—once with an electric field sensor and limit and once with a magnetic field sensor and limit.

Test technologies vary somewhat for magnetic field tests, from handheld passive loop sensors to tripod-mounted active loop antennas. MIL STD 461 and DEF-STAN 59-411 specify handheld loop probes, of diameter approximately 13.3 cm. These tests are not applicable to all platforms, only those where magnetically susceptible equipment may be present (for example, antisubmarine warfare equipment on aircraft). Magnetic field measurements are only specified for lower frequency ranges because above approximately 300 MHz, construction of an electrically small loop becomes more difficult and the loop's response to the electric field component becomes excessive.

A final special mention on the subject of magnetic field emissions is RTCA/DO-160 Section 15: Magnetic Effect. This test examines the static (DC) magnetic field produced by the EUT; the output of the test is not a pass/fail result, but a classification for the EUT that indicates the minimum allowable distance between the EUT and any magnetic compass or flux gate on the aircraft. The criterion for the test is a 1° deflection observed on a reference compass at the test site—the separation between the EUT and the compass is reduced until 1° deflection is observed and the maximum distance (for various orientations of the EUT) at which this deflection occurs is recorded. For example, if the EUT produces 1° of deflection at a separation of 0.5 m from the compass, then it falls into category A (0.3 m < 1° deflection distance ≤ 1 m) and the EUT should only be installed 1 m or more away from aircraft compasses or flux gates.

9.3.2.3 Radiated Electric Field Emission Measurements

The majority of EMC measurements for radiated fields use electric field responsive antennas and specify the limits in terms of electric field strength.

Table 9.1 shows the frequency ranges for conducted and radiated emissions tests for aviation, military and some civilian technical standards for EMC. IEC/CISPR standards seldom specify radiated field emissions limits below 30 MHz—a notable exception is CISPR 25, which applies to equipment that is installed in a vehicle and aims to protect radio receivers on the vehicle from interference; CISPR 25 specifies a rod antenna test in the range 150 kHz–30 MHz that is similar (but not identical with) the rod antenna tests in MIL STD 461 and DEF-STAN 59-411. As indicated in the footnotes to the table, earlier versions of RTCA/DO-160 also required rod antenna measurements in the range 2–30 MHz, but these were dropped from revision F onward.

Approaches to radiated emissions testing vary considerably across the suite of standards listed in Table 9.1. IEC/CISPR standards attempt to perform the test in the far field of the EUT and address emission measurement procedures, maximizing emission readings, measurement uncertainty, site validation requirements, antenna calibration methods, antenna calibration site requirements and other sundry matters. The electric field strength results produced from IEC/CISPR tests (with the exception of CISPR 25) are a best estimate of the field strength in the far field of the test item. The intention of this was that the limits could be derived from criteria such as internationally recognized minimum field strengths for service areas of broadcast transmitters—noting that CISPR's historical purpose was to protect broadcast AM radio—a long way from trying to protect avionics! In contrast, MIL STD 461, RTCA/DO-160, and DEF-STAN 59-411 make no attempt to perform a far-field measurement and maximize sensitivity of the measurement system and convenience by placing the measurement antenna approximately 1 m from the EUT. The limits for radiated emission tests in these standards are based on experience with platform installations and are therefore relevant, even though the tests do not produce good free space field strength estimates across much of the frequency range. These differences are a key reason why COTS equipment that complies with commercial or industrial standards must be reevaluated (ideally tested) against the particular avionics-related requirements of specialized aviation standards such as DO-160 and MIL STD 461.

Near-field effects in standards that use 1 m test distances give rise to the potential for poor reproducibility of test results between different facilities. For this reason, MIL STD 461 is more prescriptive of measurement antenna type than IEC/CISPR standards. IEC/CISPR standards allow any antenna that meets basic requirements of dimensions and cross polarization isolation (the ability of the antenna to reject linearly polarized electric fields that are perpendicular to the measurement plane of the antenna) to be used, so long as it has been characterized in accordance with the applicable standard. Since test ranges are usually 3 m or more, the physical form and near-field response of the measurement antenna has generally less influence on the result than it does at 1 m range. MIL STD 461 specifies the type and dimensions of the antennas to be used for radiated emission testing—so any facility that is working in accordance with the standard is using the same equipment and variations due to the difference in response of different antennas capable of doing the same job (for example, a log-period array and a double-ridge guide horn) are removed.

9.3.2.3.1 Antenna Correction Factors Engineers tasked with performing radiated emission tests or interpreting EMC test reports should develop an appreciation of the importance of antenna factors. With few exceptions, EMC antennas have a complicated frequency response; that is, for constant field strength, the voltage that appears across the antenna terminals (i.e., at the coaxial connector) varies with the frequency of the field, across the specified operating band of the antenna. This makes interpretation of emissions spectra more difficult than merely connecting an antenna to a receiver or analyzer and taking amplitude readings.

The transfer function that relates electric field strength to voltage is called the "antenna factor" (AF):

$$\mathrm{AF} = \frac{E}{V}$$

where E is electric field strength (V/m) and is V is the voltage measured at output terminals (V); so the dimensions of the antenna factor are 1/m.

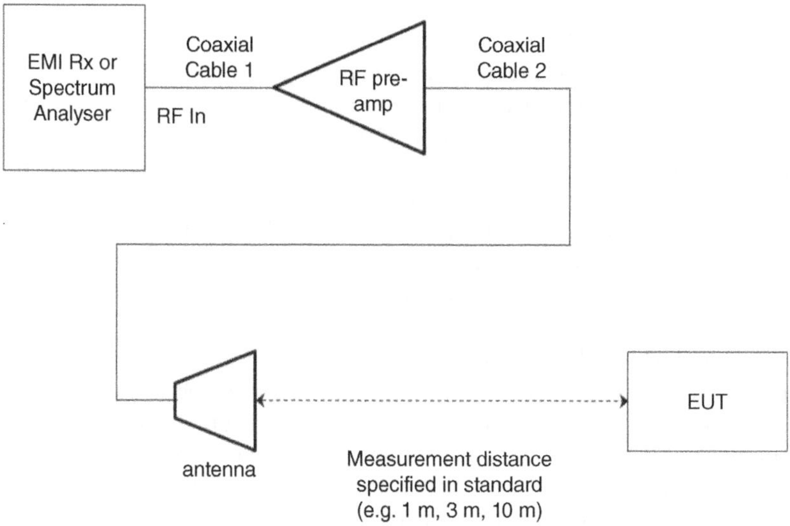

Figure 9.8 Radiated emissions block diagram (typical).

The units of electric field strength are Volts per meter, V/m, and the units of magnetic field strength are Amperes per meter, A/m. In logarithmic terms for emission measurements, the electric or magnetic field strengths are specified as levels relative to 1 μV/m or 1 μA/m, respectively.

Since most EMC measurements and limits are described in logarithmic terms, the estimate of the field strength is obtained:

$$E = V + AF$$

where E is the electric field strength (dBμV/m), V is the voltage measured at antenna terminals (dBμV), and AF is the antenna factor (dB/m).

Most practical EMC emission measurements are more complex than the antenna and the voltage measuring device (EMI receiver or spectrum analyzer); a typical radiated emissions measurement system incorporates more than one coaxial cable and a preamplifier, as shown in Figure 9.8. (This complexity is not often required in conducted emissions measurement systems, as there is usually sufficient sensitivity in the receiver to detect emissions well under the applicable limit, even with lossy devices such as attenuators or transient limiters in the measurement chain.)

In the case illustrated in Figure 9.8, the electric field strength estimate (assuming a 50 Ω measurement system) is formed using:

$$E = V + \text{cable1 loss} - \text{preamp gain} + \text{cable2 loss} + \text{AF}$$

where

E	= electric field strength (dBμV/m)
V	= voltage measured at antenna terminals (dBμV)
cable 1 loss	= the insertion loss of cable 1 when measured in a 50 Ω system (function of frequency)
preamp gain	= the gain of the RF preamplifier when measured in a 50 Ω system (function of frequency)
cable 2 loss	= the insertion loss of cable 2 when measured in a 50 Ω system (function of frequency)
AF	= antenna factor (dB/m) (function of frequency)

Systems of the kind shown in Figure 9.8 are widely used, but care should be taken when strong signals (e.g., transmissions or emissions well above the emission limits for which the system was designed) are present in the measurement environment, as the preamplifier may saturate (be driven out of its linear range of operation). In this case, the assumption built into the estimate equation is no longer valid and the field strength estimate can be wildly inaccurate. Moreover, a saturated preamplifier will produce harmonics and intermodulation products that cannot readily be distinguished from the EUT emissions. A good way to check for saturation problems in an emission measurement system is to temporarily add an RF attenuator to the input side of the preamplifier—if there is no saturation, the measured signals should reduce in amplitude by the value of the attenuator. If saturation is occurring, however, it is likely that many of the apparent signals on the analyzer would disappear (intermodulation products) and the amplitude of the strongest signals remaining may not reduce at all, or may reduce by a smaller amount that the attenuation.

Figure 9.9 shows a typical type of radiated emissions test result. In the chart, the blue trace is background noise, generally referred to as "ambient emissions,"

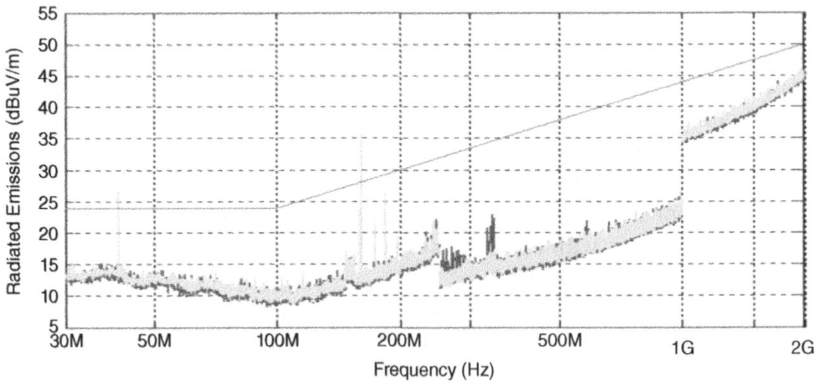

Figure 9.9 Sample radiated emissions test result (Beresford and Li, 2013).

and the green trace is the combination of the background noise and the EUT's emissions. The vertical axis is in field strength units, dBμV/m, indicating that the calculations to account for cables, antenna factor, etc., have been performed prior to display (either in the measurement receiver itself or in postprocessing software). The step in both traces at 300 MHz is due to the difference in antenna factors between the biconical array, used 30–300 MHz in this example, and the double-ridge guide horn, used 300–1000 MHz. The second step, at 1 GHz, is due to another antenna change and the effect of a change in measurement bandwidth (from 100 kHz below 1 GHz to 1 MHz above 1 GHz). The red line is the limit. Although not stated in the chart, the supporting documentation should indicate parameters such as measurement dwell time or sweep rate, measurement antenna polarization, measurement distance, EUT operation mode, and measurement bandwidths. A table of those emissions with the smallest pass margin, or exceedances, is also useful. For these reasons, system integrators should always seek a full copy of the test report for any subsystem under consideration in an aviation integration context—certificates of compliance hold no technical data and even extracts such as emission curves have little meaning outside the full test report. In Figure 9.9, we can see isolated emissions over the limit near approximately 40 and 160 MHz. The cluster of small signals in the blue trace near 300–400 MHz indicates that the ambient (background) emissions were present during the background measurement but not the EUT measurement; usually, background emissions are stable and the EUT trace is not lower in amplitude than the ambient trace. Either way, in the example given in Figure 9.9, ambient emissions are of no concern as they are well below the limit. However, if the background emissions were closer to the limit (stable or otherwise), additional measurements on the EUT would be required to verify that the EUT is really not responsible for any emissions in the obscured part of the band. Techniques for this include measuring at a closer range (if allowed under the standard), near-field scanning, or performing the test under conditions where the ambient is not present (e.g., at a different location, different time, or in a shielded chamber). Postprocessing techniques that arithmetically subtract ambient emissions from the EUT emissions trace are generally not helpful, as the ambient may vary over time, and the combined effect on the measurement result of an ambient emission with an EUT emission is not a simple summation.

For standardized test methods used for aviation subsystems, the test range is 1 m and extrapolation to shorter ranges is not appropriate, because the measurements are performed well in the near field, where the variation in measured field strength with distance is not readily predictable. Fortunately, such tests are usually performed in shielded anechoic chambers, where the ambient emissions are not so much due to background broadcast and telecommunications transmissions (the main source of ambient signals on an outdoor site) as they are due to leakage into the test chamber brought about by

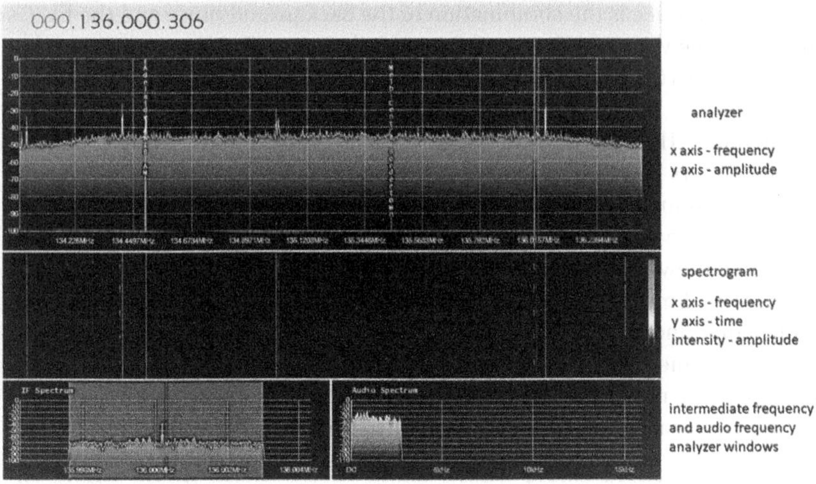

Figure 9.10 SDR GUI showing spectrum analyzer and spectrogram.

the EUT's support equipment. In this case, the best way of reducing ambient emissions is to try to increase the RF isolation between the support equipment and the EUT at the penetration to the test chamber.

Modern RF test and measurement equipment provides a wealth of opportunities for intermittent interference analysis. EMI receivers and spectrum analyzers now commonly have digital intermediate frequency (IF) stages and fast memory, so that scrolling displays of a wide frequency range can be displayed to the operator. Indeed, such capability is now available at modest cost to amateurs and hobbyists through software defined radio (SDR). Figure 9.10 shows the user interface for a spectrum analyzer program that works with an SDR and is analyzing off-air signals; VHF air-band traffic can be seen in the spectrum analyzer display in the top window. Systems of this kind may be helpful design aids for manufacturers who want to understand the EMI performance of their prototypes without the inconvenience and expense of visiting a dedicated EMC test facility.

The spectrogram part of the display in the middle window color codes the amplitude of the signal; the *x*-axis is the same as the spectrum analyzer *x*-axis immediately above it (i.e., frequency), and the *y*-axis (unlabeled) is time—the display continuously scrolls vertically. This type of display is helpful for identifying intermittent emissions: in this example, the spectrogram shows a number of vertical lines. Some of these are nearby airport transmissions received off-air and some are EMI from the host computer (no shielding or grounding precautions were taken in producing the demonstration). The vertical lines in the spectrogram correlate with spikes in the spectrum analyzer display at the top of the window and such vertical lines represent signals that are continuously present. The bright vertical line in the spectrogram display corresponds to the 136.04 MHz spike in the spectrum view immediately above it, which is of stronger amplitude than the spike corresponding to the other continuous signals. However, the spectrogram also shows three artifacts that are not visible in the spectrum analyzer display. Approximately 134.4 MHz, 136 MHz (marker in spectrum analyzer display), and 136.35 MHz have dotted or dashed vertical lines that indicate intermittent disturbances. The spectrogram display mode is a valuable aid in identifying the presence of intermittent emissions—especially when coupled with an analyzer that has a wide IF bandwidth (also referred to as resolution bandwidth or RBW).

9.4 Immunity Tests

A Note on Terminology: Technical standards for aviation, military, industrial, medical, and consumer equipment address the electromagnetic robustness, or susceptibility, of equipment by specifying "limits" for various electromagnetic phenomena. The use of the term "limits" in this context is unfortunate; a better expression would be "test levels," because the limits in electromagnetic susceptibility tests generally describe the intensity of the disturbance that the EUT should withstand. A further layer of semantic confusion is added by the use of the term "susceptibility" in standards such as MIL STD 461 and RTCA/DO-160 to describe tests whose parallels in IEC and other standards are referred to as "immunity" tests. The inference carried by the word "susceptibility" in the context of an EMC test is that there is an attempt to stress the EUT to the point of failure, whereas the term "immunity" carries the connotation of a simpler test that merely seeks to ascertain whether the EUT is immune to a specified level of disturbance. Subtleties of semantics aside, the terms are often somewhat loosely used interchangeably, and tests that increase the electromagnetic stress until the EUT's point of failure are uncommon, unless the EUT demonstrates susceptibility (lack of immunity) at the specified test level ("limit"), in which case the disturbance level is reduced until normal functionality can be restored.

9.4.1 Aims and Interpretation of Immunity Tests

The essential problem for avionics that is exposed to an electromagnetic disturbance is that it is composed of conductors, linear elements, and nonlinear elements. Conductors include

- wires in cable bundles attached to the EUT
- wires internal to the EUT
- circuit board tracks in the EUT
- RF shields, including shields on cables and the EUT's enclosure

Linear elements inside the EUT include amplifiers and detectors; nonlinear elements inside the EUT are mostly logic circuits. Linear elements may be overdriven by unwanted RF signals so that they become nonlinear. This can lead to audio rectification of the interfering RF signal, and this in turn is why RF immunity tests should be performed with a pulsed or amplitude-modulated waveform.

Electromagnetic fields impinging on conductors induce currents on the conductors. These currents flow in loops in the EUT and produce voltages across impedances, some of which can appear at the linear and nonlinear circuits that give the EUT its functionality. When the induced currents and voltages bear a sufficiently close resemblance to the wanted signals that normally exist in those circuits, the analog or digital sequences that comprise the EUT's functioning are disrupted and malfunction results. In terms of the interaction between the EUT and the interfering field, there is no difference between a pulsed telecommunications signal inducing buzzing on an intercom and the same signal inducing errors in a flight control computer. Once an unwanted signal of the right (or perhaps, wrong!) form has reached a sufficiently sensitive part of the EUT, undesired responses at the system level are a distinct possibility.

Technical standards for product- or subsystem-level susceptibility provide limits that subsystems should withstand in order to provide an expectation of compatibility with the electromagnetic environment to which they will be exposed. It should be noted that the baseline levels of EMC compliance for consumer and industrial electronics in most countries outside the EU do not include electromagnetic susceptibility requirements.

The method for testing electromagnetic susceptibility differs somewhat from emissions tests: in emissions testing, a set of measurements are made and the amplitudes of the various emanations from the test item are compared to a limit that is published in the applicable standard. Inherent in this process is the production of a *pass margin*: the amplitude of the emanation from the test item is known and can be compared to the limit in the standards—the smallest difference between the emission amplitude and the limit (noting that both are frequency dependent) is referred to as the pass margin. By convention, a negative pass margin indicates a failure or exceedance of the emission limit.

However, in the case of susceptibility testing, the typical process is to produce a disturbance level at the limit required by the standards and then subject the test item to that disturbance and monitor its performance. If the performance meets the required specification, the test item is deemed to have passed the test. For the system integrator, this is a key point of difference between emissions and susceptibility testing: while emissions tests produce a pass margin as part of the test and evaluation process, no such margin is produced as a result of a successful susceptibility test. A failure during a susceptibility test of this kind may result in exploratory testing at levels lower than the specified limit in order to determine the threshold of failure; such results should be presented in the test report but may not appear on lower fidelity compliance documentation such as certificates and attestations. Some standards, for example, MIL STD 461, require measurement of the EUT's interference threshold in the event that a failure is detected.

Exhaustive susceptibility testing can be considered to be successively testing a component for immunity at increasing levels to determine the point of failure. Since this is generally a function of frequency, full susceptibility characterization is time consuming and is seldom performed. Some immunity test standards require evaluation of a susceptibility threshold for test items that do not meet the specified immunity level (for example, MIL STD 461, tests CS114 and RS103), but such information is not always provided to the system integrator or may not be released by the Original Equipment Manufacturer on the grounds of military security or commercial confidentiality.

For example, if a test specification requires a 50 V/m susceptibility level to radiated field strength, and a test item endures that field strength without deviating from its specified performance level (i.e., passes the test), there is no way of knowing whether the test item would also tolerate 60 V/m field strength, or whether it could also tolerate 200 V/m field strength. For the system integrator, this can be problematic if a subsystem is presented for integration into an aircraft where the expected or required susceptibility level is greater than the level to which the item was originally tested. Such cases can present the system integrator with a choice: do they provide supplementary shielding in their design to guarantee that the item will not be exposed to a higher field strength than its demonstrated compliance level, or do they send the item for supplementary testing to certify it to the desired higher level? Each choice has costs and benefits. In the former, the system integration is more complicated and perhaps delayed, and shielding integrity must be maintained throughout life of type (a maintenance burden), but the system-level design is more protective of subsystems that may be electromagnetically fragile; this can make through-life support easier as it allows more flexibility with subsystem substitution in the future. Alternatively, testing the subsystem to a higher level can lead to a simpler integration (if the subsystem indeed tolerates the desired higher immunity level) because no system-level changes are required, but the test may have to be performed multiple times throughout the life of type if the subsystem is substituted or substantially modified during the system's service life.

Generally, compliance with a conducted or radiated field susceptibility limit provides confidence of compliance at lower levels, but determination of immunity compliance at higher levels can only be determined by further testing at the higher level. One exception to this statement is high voltage and high current transients (e.g., on power and data lines), where transient suppression components have a nonlinear response to the incoming disturbance. Somewhat paradoxically, an input port that is fitted with high-level transient protectors may actually be vulnerable to lower level transients, because such transients may be large enough to cause damage to internal circuits, but too small to trigger the protection components. Selection and rating of power and data inputs for transient protection is not a simple case of opting for the largest component in the suppression parts catalog!

Susceptibility testing mirrors emission testing with respect to the interaction between electromagnetic waves in the environment and the electrical size of the EUT and its attached cables, so conducted susceptibility tests are based around lower frequencies, typically up to 200 MHz, and radiated susceptibility tests extend up to 40 GHz. In common with emission tests, there is considerable frequency overlap; radiated electric field susceptibility tests extend as low as 10 kHz and radiated magnetic field immunity tests as low as 30 Hz. For aviation electronics, there is no static electric or magnetic field immunity test requirement (i.e., there is no susceptibility complement to the static magnetic field test in RTCA/DO-160 Section 15 Magnetic Effects for aviation electronics).

Table 9.2 lists some standards and summarizes their continuous conducted and radiated susceptibility test methods and frequency ranges. Transient susceptibility tests are not included.

Table 9.2 Comparison of conducted and radiated immunity test frequency ranges

Standard	Application	Conducted immunity test frequency range*	Radiated immunity test frequency range**
MIL STD 461F/G	Military equipment (US standard)	CS101: 30 Hz–150 kHz (I) CS114: 10 kHz–200 MHz (I)	RS101: 30 Hz–100 kHz (H) RS103: 2 MHz–40 GHz (E)
RTCA/DO-160 G and ED-14G	Civil aviation aeronautical equipment (US and Europe)	Section 16 DC-800 Hz (V) Section 18 Para 18.3: 10 Hz–150 kHz (I) Section 20 Para 20.4: 10 kHz–400 MHz (I)	Section 19 Para 19.3: 350–800 Hz (I) Section 20 Para 20.5: 100 MHz–18 GHz (E)
DEF-STAN 59-411	Military equipment (UK standard)	DCS01: 20 Hz–50 kHz (V) DCS02: 50 kHz–400 MHz (I) DCS03: 20 Hz–50 kHz (I)	DRS01: 20 Hz–100 kHz (H) DRS02: 10 kHz–18 GHz (E) DRS03: DC (H)
IEC/CISPR 24	Information technology equipment (general civilian applications)	150 kHz–80 MHz (V or I)*****	50–60 Hz (H) 80 MHz–1 GHz (E)

*(I) = current injection (inductive coupling), (V) = voltage injection (capacitive coupling).
**(H) = test by establishing a magnetic field, (E) = test by establishing an electric field.

9.4.2 Conducted RF Immunity Tests

These tests simulate unwanted signals in the avionics subsystem's electromagnetic environment by means other than immersing the test item in an RF electromagnetic field; the test disturbance is introduced to the EUT by either galvanic conduction or capacitive or inductive coupling. We can consider two scenarios for continuous RF immunity tests: front-door and back-door coupling.

9.4.2.1 Front-Door Conducted RF Susceptibility Tests

When simulating front-door interference to radio receivers in a conducted immunity test, the unwanted signals are introduced to the receiver under test by means of coaxial adapters, RF hybrid circuits, couplers and other hardware. The general approach is to establish a reference level of performance in the avionics device (something that incorporates a radio receiver in the case of front-door interference) by providing a sample of the wanted signal at the receiver input, for example, at a level 3 dB above the rated sensitivity of the receiver. Then, the interference signal is introduced at various frequencies and a much stronger level, to evaluate the receiver's ability to reject out-of-band interference. Very often, these test frequencies are selected to target likely harmonic and intermodulation products that could affect the receiver under test, as well as either side of the operating frequency of the system.

The result of this type of test is a set of curves similar to the one shown in Figure 9.11, where the curve indicates the interference level necessary to disrupt the receiver's operation. Tests of this kind should be performed at a few channels in the receiver's range of operation (typically, lowest, mid band, and highest channel), so that a family of curves is produced. Such curves can be used when integrating the subsystem, to help determine suitable locations (or at least, to avoid unsuitable locations) to install the system's antenna on the aircraft, based on avoiding existing transmitters that may produce excessive out-of-band energy at the system's RF receiver input. Tests of this kind are specified in standards such as MIL STD 449D and MIL STD 461.

Out-of-band interference tests applied to digital radio receivers often yield a different kind of result from that shown in Figure 9.11. The details can vary considerably depending on the nature of the digital modulation and coding systems used in the receiver, but the common thread is that digital communications systems spread their information across a wider frequency range than analogue AM/FM systems. The general shape of an out-of-band interference test for a digital communications receiver is shown in Figure 9.12. In Figures 9.11 and 9.12, BW is the receiver's pass bands; it requires only a small amount of interference power in the receiver passband to upset a narrow band AM or FM signal, but comparatively much more is necessary to bring about the same result in a digital modulation receiver, including those that do not employ code-spreading techniques.

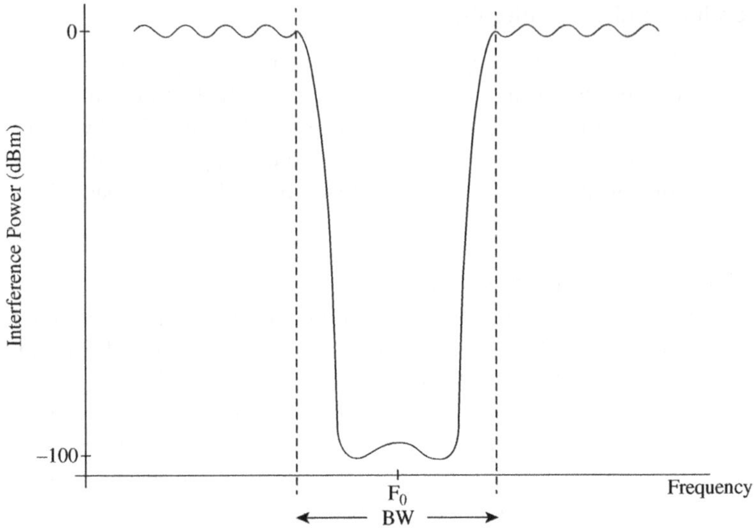

Figure 9.11 Typical out-of-band interference rejection test result for a narrow band AM or FM receiver.

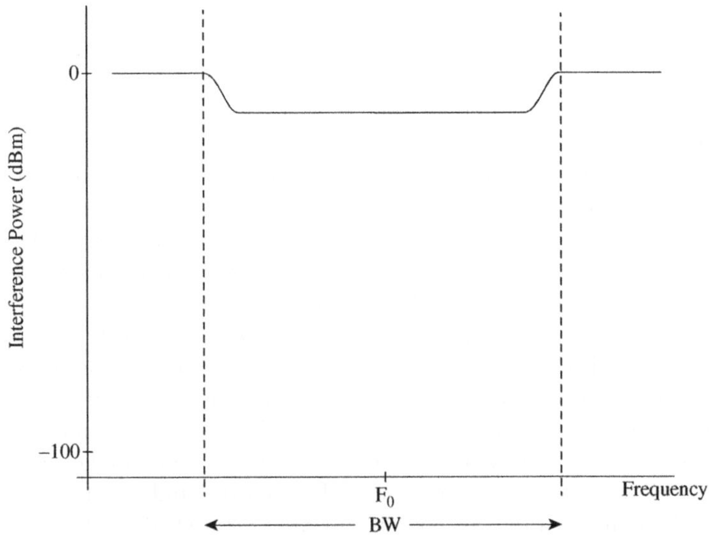

Figure 9.12 Out-of-band interference test result for a digital communications receiver, with a continuous wave (CW) interfering signal tracked across the pass band.

9.4.2.2 Back-Door Conducted RF Susceptibility Tests

Unwanted RF energy in a "back-door" scenario is introduced to an aircraft subsystem either by the interfering RF signal impinging upon the enclosure of the subsystem itself, or by impinging on wiring attached to the subsystem. The victim system's propensity to receive RF energy from an interfering transmitter is a function of the electrical size of its enclosure and its associated wiring. For many aviation related subsystems, the enclosure size ranges from handheld up to cabinet size, while the associated wiring can be many tens of meters long. The interaction between the aircraft wiring and the electromagnetic environment is difficult to simulate in a laboratory environment for the purposes of testing avionics subassemblies, so the preferred approach is to apply the test disturbances by coupling them onto cables that connect to the EUT.

Within this class of test, two further divisions can be identified: power frequency and related disturbances, and radio frequency disturbances. The usual approach (for example, under MIL STD 461 or RTCA/DO-160) is to introduce RF disturbances to the test item by means of a current clamp placed around the cable bundle under test, while a transformer in series with the test item's power line is used for power frequency disturbances.

Power frequency conducted susceptibility, or immunity, tests exist to provide a level of assurance that the avionics subsystem can tolerate an anticipated level of disturbance related primarily to the aircraft power system. Such disturbances may enter the avionics subsystem via its power port or via other cables that may carry power frequency disturbance introduced by inductive coupling, for example due to long closely spaced parallel cable runs. The electromagnetic environment for avionic subsystems includes power supply ripple or distortion that may be present on the aircraft's AC or DC power supply; avionics equipment must be able to tolerate this. The sinusoidal susceptibility test signal in these types of tests is not modulated, and the frequency range is restricted to the lower bands outlined in Table 9.2, usually from the power frequency up to 150 kHz.

Some standards, such as DO-160 section 16, specify rigorous testing of the equipment's ability to withstand a wide range of power-related disturbances, such as power supply brown-outs and interruptions that may occur during normal or abnormal operation of the aircraft's systems. Power supply susceptibility tests can also require that the test item can tolerate voltage, frequency and phase variations in the aircraft's AC power supply. Tests of this kind are commonly performed using highly programmable AC power supplies and sources, to automate the large number of different test scenarios that must be applied. In the case of military equipment, some power susceptibility tests and requirements, such as power supply variations, are presented in MIL STD 704. While this is not always considered to be within the remit of the EMC engineer or integrator, it should not be forgotten.

On the other hand, wiring attached to avionics subassemblies will carry radio frequency currents as a result of electromagnetic fields in that exist in and around the aircraft due to on-board and off-board transmitters and sources. For the purposes of testing to determine if avionics can tolerate this aspect of their electromagnetic environment, radio frequency disturbances are introduced to power and other wiring attached to the test item by means of RF current transformers. The typical frequency range for such tests is within 10 kHz to 400 MHz, and some standards do not apply this technique above 100 MHz. Tests are performed using a stepped frequency generator, with either pulse or amplitude modulation applied to the test signal. If particular frequencies of interest relevant to the test item are known (for example, clock and data bus frequencies), they should be included in the test, as the default logarithmic frequency progression specified in standards such as DO-160 and MIL STD 461 may not include them. If conducted and radiated emission tests have been performed on the test item already, analysis of the results can give an insight into internal clock rates even if detailed design data about the system under test is not available. By targeting the test disturbances to align with critical frequencies, there is a greater likelihood that any susceptibility in the test item will be detectable. A further possible choice for targeted test frequencies is any known threat frequencies on the aircraft that the avionics system will be installed in or in its planned operating environment.

Some challenges of conducted RF susceptibility tests are related to introducing the test disturbance to parts of the test setup that may be more electromagnetically fragile than the test item itself. For example, an avionics subsystem may have been designed with good EMC performance in mind, but the data or sensor simulators used to simulate the aircraft (so-called support equipment) may not be so robust—e.g., plug-in data bus cards in a consumer grade notebook PC used to simulate the aircraft's flight control computer. Experienced practitioners have a range of ways of dealing with such problems, including shielding the support equipment, filtering the support—EUT interface (to the extent that this is possible without preventing the interface from working) or installing optical fiber or other RF isolation in series with the EUT interface. The problem comes when the support equipment fails, or indicates failure of the test item, it is impossible to know whether the EUT itself is functioning correctly.

9.4.2.3 Multiple Interface Testing

Although not a requirement of most standards, if a subsystem provides multiple identical interfaces in order to achieve high reliability, low failure rate performance, it is important to tailor the EMC test program to apply the injection current to each interface simultaneously. The conducted susceptibility tests exist primarily to simulate the currents that would be induced in the presence of an RF field; it follows that all cables connected to the EUT could be immersed in approximately the same field, so currents would be induced in all cables simultaneously. A conducted susceptibility test that only tested one interface at a time may not detect a failure if the EUT can afford to have interference on one interface at a time; an example of such an EUT is a system with duplicate processing and interface circuits—so long as either circuit is functioning, system-level functionality is achieved. The laboratory susceptibility test should endeavor to simulate that by testing all identical interfaces simultaneously, in order to verify the system's ability to reject or otherwise accommodate induced RF currents and voltages. This is especially important with equipment that performs safety-critical functions.

9.4.3 Radiated Susceptibility Tests

Radiated susceptibility tests expose the EUT and its associated cabling to known radiofrequency electromagnetic fields. In common with conducted RF susceptibility tests, the field is modulated to simulate the types of RF threats that may exist in the EUT's intended environment, with some concessions for ease of test and measurement. This process simulates the influence of radio transmitters that may be operating near the EUT when it is installed, either onboard the same aircraft as the EUT or on sources near the aircraft, such as nearby aircraft transmitters or terrestrial transmitters such as telecommunications, airfield or naval radar or broadcast transmitters. Terrestrial transmitters are especially worthy of consideration for helicopters because helicopters frequently fly at low altitudes, in "nap of the earth" flight profiles.

The trend in EMC standards for subsystems and equipment is to reduce the amount of low-frequency-radiated testing that is performed (below 2 MHz in the case of MIL STD 461, or below 100 MHz, in the case of RTCA/DO-160 G) as cable bundles are the dominant pathway for RF energy into the EUT in these ranges, and technical challenges with producing high-intensity fields make performance of a radiated test expensive and difficult. The question of the EUT's susceptibility to continuous RF fields is adequately addressed by conducted injection techniques applied to the cables, for typical box to cabinet sized equipment used in aviation applications. There is no motivation to perform conducted RF susceptibility tests above a few hundred MHz because power, data, and ancillary cables attached to the EUT are poor conductors at those frequencies; the dominant pathway for radiofrequency interference to sensitive circuits then is by fields penetrating the enclosure, for example, through vents, access panels, or poorly bonded seams. For this reason, radiated field susceptibility tests extend to frequencies well above those where conducted immunity tests stop.

Low-frequency-radiated susceptibility tests, up to approximately 100 kHz, use inductive loops or Helmholtz coils to produce a magnetic field at the EUT. Various standards apply this approach as shown in Table 9.2.

- DEF-STAN 59-411 specifies radiated electric field susceptibility testing in the range 10 kHz–40 GHz and radiated magnetic field susceptibility testing in the range 20 Hz–100 kHz.

- MIL STD 461 has a frequency gap between tests RS101 and RS103 from 100 kHz to 2 MHz, where (unless the test plan tailors the frequency ranges) no radiated susceptibility test is required. Former version MIL STD 461D required radiated electric field susceptibility testing as low as 10 kHz, but this requirement was removed with version E in 1999.
- RTCA/DO-160 G does not specify radiated electric field susceptibility tests at frequencies below 100 MHz, which is approximately in line with the long-established 80 MHz minimum radiated field susceptibility frequency used across IEC and CISPR standards. In both cases, EUT exposure to RF threat transmitters in the environment operating at frequencies lower than the lowest radiated field susceptibility test is addressed by conducted RF susceptibility tests.

A block diagram for a typical radiated field susceptibility test is shown in Figure 9.13. This is a generic diagram that encompasses some variations in application between the various standards listed in Table 9.2 and other EMC test methods. Essentially, there are two approaches to determining and therefore controlling the electric field strength at the test item:

(i) Precalibration of the field by a substitution method; this is analogous to the approach described for conducted RF susceptibility tests or
(ii) Closed-loop monitoring and control.

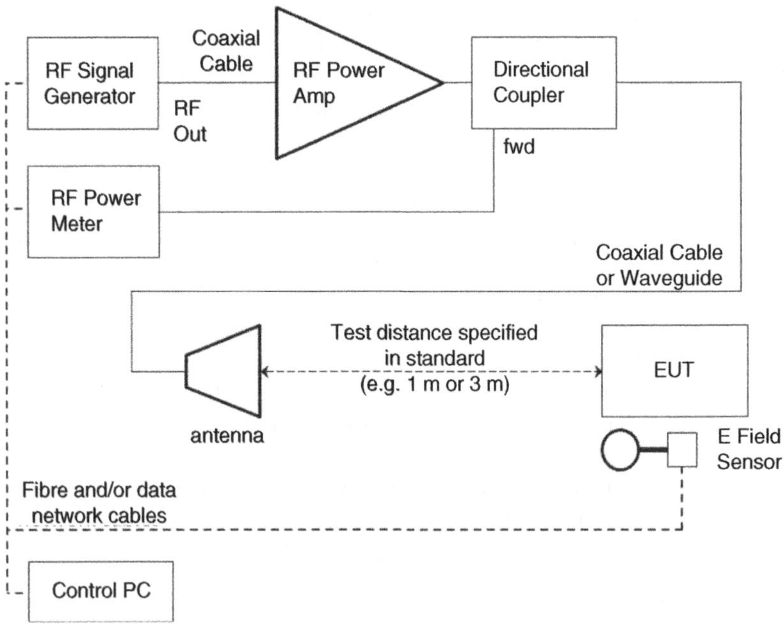

Figure 9.13 Typical radiated susceptibility test system diagram.

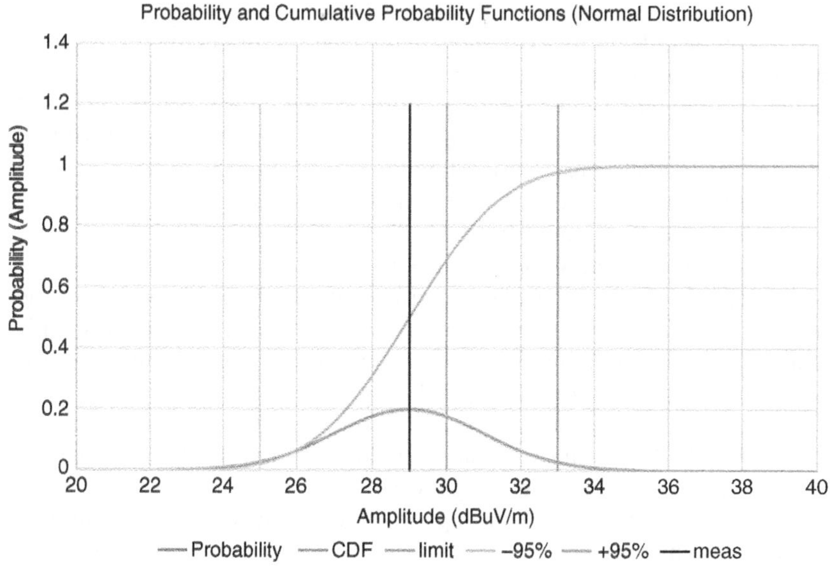

Figure 9.14 Example of measurement uncertainty for a radiated emission test.

The precalibration (substitution) method involves performing the test in two stages: calibration and test. During calibration, the RF signal generator output level is adjusted at each frequency to produce the required field strength at the sensor, with the EUT absent from the test environment. When the desired field has been produced, the forward power monitored at the output of the power amplifier is recorded. This process may include recording the response of the power measurement system with continuous wave (CW) and test modulation enabled, as some E field sensors do not respond well to modulated fields. During the test run, the E field sensor is not required and the signal generator output is adjusted to reproduce the forward power that was recorded during the calibration. RTCA/DO-160 specifies use of such a substitution method; DEF-STAN 59-411 allows a similar method, but requires the EUT to be present in the test environment during the calibration process.

The closed-loop method, as described in MIL STD 461F/G, places the E field sensor in the test environment with the EUT. Version G of MIL STD 461 removed the option to use a forward power leveling technique (the "receive antenna method" from version F was deleted) and requires closed-loop monitoring, making it unique among EMC standards. At each frequency, the signal generator output is adjusted to produce the required electric field strength. This method requires that the sensor not be placed too close to edges or corners of the EUT, as distortions in the field can cause measurement errors. A further problem with this method is that the sensor must either correctly respond to the modulated field, or be characterized to allow a correction to be applied during the test. Furthermore, at higher frequencies where horn antennas are used, it can be difficult to accommodate the field sensor and the EUT within the antenna beamwidth without bringing them closer together than is ideal. Finally, an argument could be made that the interaction between the field and the EUT is part of what the test seeks to investigate (that is, to the extent that the interaction could cause a susceptibility), so adjusting the test generator to produce the field while the EUT is present is not the same thing as producing the required field, then immersing the EUT in it. DEF-STAN 59-411 allows use of a receive antenna procedure above 1 GHz, which is a variation on the substitution method described above and is performed with the EUT absent from the test environment. Reverberation chamber susceptibility tests inherently apply a substitution method to producing the required field strength, although it is common to monitor the field strength during the test.

The discussion so far relates only to establishing the level of the field at a single point, but in the EUT's final installation environment, the field is likely to exist across a wider region encompassing the EUT and its wiring. EMC standards have various approaches to managing the difficult issue of field uniformity in anechoic test environments:

- MIL STD 461 has no requirement for verification of field uniformity over the test volume for anechoic chamber tests (the space occupied by the EUT and associated cables), instead relying on the specification for the absorber to manage field uniformity
- RTCA/DO-160 G has no requirement for verification of field uniformity over the test volume, also relying on the absorber specification to manage field uniformity
- DEF-STAN 59-411 has no requirement for verification of field uniformity over the test volume, although a chamber that can meet the requirements for radiated emissions in this standard is more likely to provide good uniformity than one that did not meet that requirement

However, each of these standards includes a requirement that the EUT be adequately illuminated by the beamwidth of the test antenna, to the extent that multiple antenna locations may be required to adequately expose a large EUT and its associated cabling. DEF-STAN 59-411 provides useful guidance on this point under the procedure for LRUs, where it requires that the test should specifically address displays, apertures, and connectors.

Each of these standards allows use of reverberation chambers for radiated susceptibility testing. The calibration, characterization, and acceptance procedures for reverberation chambers include field uniformity requirements.

9.4.3.1 Radiated Susceptibility Test Sites

In principle, any of the test sites illustrated under the section relating to radiated emission test sites could be used for radiated susceptibility testing. In practice,

OATS are generally not used because the signals produced could cause interference to nearby spectrum users and would almost certainly breach radio communication and interference laws enforced by various government communications regulators (e.g., U.S. Federal Communications Commission, U.K. Office of Communications, and Australian Communications and Media Authority). Standards most relevant to avionic systems (RTCA/DO-160, MIL STD 461, and DEF-STAN 59-411) allow anechoic chamber or reverberation chamber testing. The most recent version of RTCA/DO1-60 (version G) starts radiated immunity testing at 100 MHz, which is above the frequency range where the greatest challenges in producing high-intensity fields lie, and is a suitably high frequency that a modest-sized reverberation chamber (dimensions several meters across) provides good reverberation performance. However, MIL STD 461 and DEF-STAN 59-411 require radiated electric fields tests as low as 2 MHz, and sometimes a tailored test plan may require tests at lower frequencies; in these cases, reverberation chambers and the associated benefits with regard to generating high field intensities cannot be realized.

In contrast with its radiated emission measurements, MIL STD 461 test RS103 (susceptibility) does not specify particular antennas for the test, and any type of radiating structure is allowed. Some radiating structures, such as TEM cells, incorporate the necessary shielding to prevent interference problems otherwise associated with susceptibility testing, so the test site and radiating structure are one and the same thing.

9.4.4 Transient Susceptibility Tests

Conducted and radiated susceptibility tests discussed up to this point seek to determine whether the EUT can tolerate continuous exposure to a radiofrequency threat.

Transient threats exist as a result of various electromagnetic and power bus phenomena, including load switching, personnel electrostatic discharge, precipitation static, and lightning. Most aerospace-related technical standards specify tests for transient phenomena, and there is a plethora of waveform specifications, test generators, and coupling methods. For brevity, a generic process is described here, to provide an insight into the characteristics of a typical transient susceptibility test.

The typical transient susceptibility test waveform generator consists of a high voltage generator, a fast switch, a waveform shaping circuit and a coupling circuit. The coupling circuits are generally similar to those employed for conducted RF susceptibility tests, although this is not always obvious as some transient impulse generators include coupling networks internally. Invariably, the amplitude, rise time, and decay time of the waveform must be verified, usually with a digitizing oscilloscope; it is sometimes a requirement to include such verifications in the test report.

The key difference in performing or specifying transient susceptibility tests compared with continuous immunity tests is that the disturbance itself is generally so brief (usually < 1 ms and sometimes < 1 μs) that meaningful observation of the EUT during the application of a single disturbance is impossible. In recognition of this, and in an effort to produce a suitably harsh test, some standards specify repeated application of the disturbance over a longer time period (e.g., minutes) to allow observation of the EUT's performance. Other standards do not specify EUT performance during the disturbance, but require that correct performance continues after application of the disturbances.

In terms of writing a test plan or monitoring the EUT, key failures to look for in transient susceptibility tests include:

- processor lock-up
- change of operating mode
- loss of function requiring operator intervention
- damage to power protection circuits

Failures of the kinds listed above require some kind of interaction with the EUT to evaluate properly and can be difficult to monitor without a human in the loop.

9.5 Test Plans for Avionics Subsystems

The complexity of running an EMC test means that test facilities are usually managed by EMC specialists; it is not common for the EMC specialists to also be specialists in whatever kind of EUT comes through the facility's door.

The means by which the project sponsors, or test facility customers, and the test facility itself communicate the technicalities related to testing the EUT is the test plan. The overall usefulness of the test program is almost entirely dictated by the test plan, and the attention it pays to simulators and loads, monitoring systems, and pass/fail criteria. It is despairingly common to see test reports describing tests on complex items where inappropriate exercising or monitoring of the EUT renders the test result almost useless. An example of this is a test on a video imaging system where the test object was focused on a fixed scene throughout the test. Since the resultant image is not moving, there is no way of knowing if the test system has locked up; distortion or blanking of a video display is not the only failure mode. Also, a static image gives no dynamic indication that the EUT's internal processors are still active.

9.5.1 Simulating the Aircraft (Excitation/Loads)

If the EUT is one that interfaces with the aircraft's core avionic systems, simulating those systems and providing realistic data and control inputs can grow to be a challenge that can dominate the test program. For example, the MIL STD 1553 avionics data bus is common in aviation applications. If the EUT requires 1553 data in order to function, then testing the EUT without providing 1553 data is not likely to identify the worst-case emissions or the most susceptible configuration. It is not common for a specialist EMC test facility to hold a comprehensive suite of avionics simulation and monitoring equipment—these items are usually provided by the party responsible for the EUT. It sometimes works out that an EUT is so tightly integrated into the host aircraft (e.g., has a large number of disparate modules and data interfaces to the aircraft) that laboratory testing is not a practicable option, and even though the EUT may be physically small enough to fit into a test chamber, simulating its operating environment may be so difficult that testing at the aircraft is the only meaningful way of evaluating its performance.

A methodical approach to simulating the EUT's final operating environment and configuration for the purposes of performing EMC tests in the laboratory warrants consideration of the following factors:

9.5.1.1 Physical Installation Arrangement(s)

Will the device be rack mounted, handheld, used on a nonconductive bench top; is it carried in a bomb bay or on a pylon? The test configuration should be replicated to the extent possible that is consistent with the requirements of the standard. Aviation equipment standards allow tailoring from the default setups variously given therein because it is recognized that obtaining a meaningful and relevant set of test results is more important than blind adherence to a generic set of requirements. When determining the layout for the test item, it may be desirable to deviate from the intended use arrangement of the EUT to a configuration that is perhaps overly pessimistic, but more likely to bring out emissions and susceptibilities. Viewing the aircraft close to the EUT's intended installation or use location as a set of conductive, nonconductive, or partially conductive surfaces can help guide how the EUT should be configured in the chamber.

Cable runs should aim to replicate the grounding arrangements of the aircraft. This can require more intimate knowledge of the EUT and its related aircraft than the test facility is likely to hold. For example, if a shielded cable exits the EUT and then passes a short distance before going through a bulkhead connector, it is reasonable to ask whether the shield is connected to the airframe at the bulkhead. In some aircraft, there may be a connection to the airframe and in others, especially smaller aircraft where a single-point ground strategy may be in place, the bulkhead may only provide mechanical support with no ground connection. If such details are known, it is important to include them in the test plan, so that the laboratory test layout is a good approximation to the planned installation.

Laboratory tests sometimes demand the design and manufacture of special cable looms to facilitate provision of power to the EUT and connection to simulators and monitoring equipment. This can be especially challenging, as it is necessary to ground the shield of shielded cables at the point where they exit the test chamber (otherwise the shielding of the chamber is not merely compromised, but ruined), yet it is not desirable to introduce a ground connection to the cable loom at a distance from the EUT that is not representative of the expected aircraft installation. Test activities are sometimes a compromise and a good test plan and test report should identify those aspects of the test layout that are not electrically representative of the aircraft installation.

Test stands, cable supports, and ancillary equipment that is necessary to make the EUT function, but is not part of the EUT or its cabling (i.e., would not be installed or used in the aircraft) should, to the greatest possible extent, be electromagnetically absent from the environment. Plastic, polystyrene foam, and optical fiber are useful aids to producing EUT support and monitoring systems. Compressed gas actuators can be manufactured to pneumatically operate press buttons and other controls on EUTs where controls must be manipulated to properly excite the system.

9.5.1.2 Provision of AC/DC Power

Most test facilities can provide AC and DC power to the test item, although it is worthwhile to enquire about the availability of a 400-Hz aircraft supply if the EUT requires one. While it is an easy matter to connect a 400-Hz inverter to a chamber, the high insertion loss RF power filters used on most EMC chambers may not always accommodate 400 Hz easily. Leakage currents from a 115 V 400 Hz ac source can be several amps, and power factor correction to compensate for the leakage capacitance inherent in the filters is all but mandatory.

DC power may be provided by conventional laboratory supplies connected into the chamber via high insertion loss RF filters or by a battery located inside the chamber. It is not usually practical to locate a conventional laboratory supply inside the test chamber as it will produce emissions that will pollute the ambient environment and it may be susceptible to the disturbances produced during susceptibility testing. RF isolation between the AC or DC power source and the test environment is essential: if the power supply ambient is too high, there is no way of knowing what emissions are due to the power supply and what are due to the EUT. If the power supply output varies as a result of an susceptibility test disturbance, causing the EUT to falter, there is no way of knowing whether the power supply or the EUT failed.

Power circuits receive special attention under MIL STD 461F/G in that the connection between the LISN and the EUT is not allowed to be by shielded cable, because most aircraft (and other platform) wiring is not shielded. The standard recognizes that while an equipment manufacturer may specify shielded power cable in order to pass an EMC test, the actual installation power wiring is almost certainly unshielded. Since this is documented in the standard, it is likely that the test facility will be reluctant to use shielded power cables, even if the EUT manufacturer specifies them.

9.5.1.3 Antenna Ports

Some EUTs include integral antennas of their own. While intentional transmissions are generally excluded from limits under EMC tests, it is often necessary to provide a realistic radio link to and from the EUT in order to produce a realistic set of operating conditions. Examples include GPS antennas, telemetry antennas, and communication network antennas.

Since most radio receivers are highly sensitive, it is comparatively easy to introduce the necessary signals to the test chamber by means of coupling antennas connected by coaxial cables to the test generator outside the chamber. A simple approach to producing a GPS signal inside the chamber is to install a reradiating system. Due to the low signal levels involved, the ambient performance of the chamber is not compromised unduly by addition of a good-quality GPS reradiator. This approach is fine so long as the EUT does not expect to be moving—if simulated movement is required then a GPS simulating signal generator is really the only option.

Duplex radio links are a little more problematic because they preclude the use of booster amplifiers, unless a complex network of circulators or diplexers is used. If there is sufficient power in the transmitter and sufficient sensitivity in the receiver, it is sometimes possible to use passive reradiating to introduce the test signal to and from the test chamber; this has been used with some success for wireless LAN systems. Good RF filtering to pass only the desired band, directional antennas and other precautions are necessary to prevent undue pollution of the chamber ambient.

Working with antennas on the EUT starts to cross over from EMC testing to radio system testing. Sometimes this is desirable, other times it is not. If the intent of the tests is to evaluate the EUT's ability to function, on the assumption that it has (for example) a sufficiently strong input radio signal, then any susceptibilities that occur because the EUT has inadvertently been provided with too low an input signal are of little interest. Of course, in the process of performing the test, it can be difficult to determine whether a failure is due to insufficient coupled signal or some other failure.

Radio test standards, such as MIL STD 449D, specify a reference wanted signal 3 dB above the minimum detectable signal level for the receiver when testing for out-of-band rejection and other radio-specific performance tests. If the receiver itself is not the primary concern (for example, if an approved receiver module is integrated into a larger system), then it may be valid to apply ample signal strength during EMC tests. There is no right or wrong answer to this point, but it should be considered and specified in the test plan. A more stringent test is produced by specifying a weak RF input signal to the test item as this simulates operation near the edge of radio range, where the emissions of the whole system come into play with the receiver sensitivity and the EUT's electromagnetic environment to produce a set of conditions where failures could occur that may not have been identified during receiver testing at module level.

Coaxial coupling of reference signals to EUTs that have removable antennas is easier for the test facility, but consideration must be given to the impact of the additional ground connection that is made as part of the new coaxial connection.

9.5.1.4 Data Ports

The simple response to the question of how to simulate data on the EUT's ports is to provide a suitable data source, either a simulator or a flight data recorder operating in playback mode. Data links usually operate at rates that mean filtering of the kind applied to power circuits is not possible, because the signals that we want to keep out of the test chamber are in the same band as the data traffic (recall Nyquist and the rule of thumb that transmission of a square wave requires approximately 10 harmonics to give good reproduction).

Since filters cannot be used, other approaches to RF isolation must be applied. The easiest is optical fiber isolation, although the fiber isolators themselves can bring additional EMC baggage of their own to the problem. While fiber isolators for common data interfaces are readily available, many of these are produced to provide extended range rather than good EMC performance. Such an isolator may have to be mounted in a shielded box with a filtered power supply in order to provide adequate EMC performance for test purposes. By placing the support equipment (for example, a PC or data generator) outside the main shielded chamber with a fiber converter, a fiber can be passed through the chamber wall using a waveguide beyond cutoff penetration to another converter inside the chamber; the data interface from the EUT then connects to a comparatively small and more-or-less electromagnetically benign fiber adapter rather than a PC or data generator.

In some cases, data cables may be shielded in the EUT's installation. In this case, it is possible to connect the EUT via a shielded penetration (e.g., shielded LAN bulkhead connector) to the simulation equipment (e.g., network switch) outside the test environment. This is a less effective approach than optical fiber isolation as it degrades the overall shielding performance of the chamber to the level of the cable's shielding, which is generally less than that of the chamber by itself. Also, it introduces a possibly unwanted ground connection at the penetration point. Nonetheless, the technique has been used with some success.

9.5.1.5 Operator Controls and Indicators

Some EUTs require operator interaction to adequately replicate their operational mode during testing. Various methods can be applied to operate user controls such as switches and buttons remotely, including pneumatic systems (which can employ plastic air lines to operate plastic pistons). Other devices can be stimulated by means of a rotating arm mounted to an induction motor—the induction motor does not produce any emissions and is inherently immune to all but the strongest electromagnetic interference. Such systems have been employed to repetitively pass a succession of targets in close proximity to a sensor in order to illicit a known dynamic response from a test item. Such a system could be adapted to trigger a touch screen display.

EMC-hardened test chamber cameras can be used to remotely monitor visual indications on the EUT to confirm functionality; it is reasonable to expect that a test facility would have such a system in place. Coupled with a recording and monitoring system, a chamber camera is an indispensable aid to immunity testing as it allows magnification and remote observation of the EUT's indicators and controls without exposing personnel to the test electromagnetic environment.

9.5.1.6 Other Environmental Factors

With regard to environmental factors such as vibration, shock, altitude, temperature, and humidity, it is generally impractical to excite the EUT realistically while performing an EMC test. If a test item requires, for example, an extreme environmental condition in order to correctly operate, innovative measures will be necessary to devise ways of performing the test. One approach may be to use special test software or firmware to trick the EUT into thinking it is seeing its required environmental inputs. Another approach may be to apply localized heating or cooling to the sensors, preferably with nonmetallic tubes or ducts, to produce the desired effect. Difficult problems of this nature can be resolved during pretest planning meetings between stakeholders.

9.5.2 Monitoring Test Item Performance

Aviation subsystems often require complex support equipment as they are usually integrated into the aircraft systems. Simulators are usually necessary in order to make the EUT "think" it is in the aircraft. This means that items such as MIL STD 1553 bus interfaces, RF transponder signal generators or specialized display interfaces may be required in order to fully monitor and exercise the EUT during laboratory tests.

9.5.2.1 Operation Modes and Exploratory Testing

Many EUTs can operate in different modes and are likely to have different immunity and emissions performance depending on what they are doing. Generally, it is important to capture the worst-case emissions mode of operation for those tests, and the most susceptible mode of operation for susceptibility tests. In the case of emissions tests, it is comparatively straightforward to make rapid preliminary measurements (for example, with near-field probes) to determine the modes that provide the greatest emissions.

Determination of the most susceptible mode of operation is more difficult. One approach is to perform testing in all modes, but the very long run times of aviation related susceptibility tests (particularly radiated susceptibility, which can consume many days of chamber time) often force restrictions on exhaustive testing of this kind. Factors to be considered in determining the most susceptible mode include

- consideration of the modes of operation that are active during different phases of flight, and their impact on overall systems safety or suitability
- RF and transient threats that may be present, or can be ruled out, for different phases of the EUT's operation
- determining a mode of operation that maximizes the likely susceptibility of the EUT

The final case considers cases where an EUT may perform a number of functions, but where a mode of operation can be identified that exercises all functions. For example, a computing device that provides an information display to the air crew of a rotary wing or general aviation aircraft may include a digital display, multiple processors, memory storage, and satellite navigation. It may be possible to configure the EUT so that all of these things can be excited at once for the purposes of EMC tests, by running a software program that interrogates the GPS receiver, processes and displays the result, and creates a log file. In this way, susceptibilities across a range of subsystem internal functions can be investigated in a single test.

Options for EUT monitoring range from operator direct observation and note-taking, through recorded chamber camera files, EUT simulator/monitoring equipment performance logs to EUT internal status logs (for EUT types that record a health log to internal memory). The latter type can provide comprehensive information about the EUT's performance during the test, but are difficult to work with as they cannot be interrogated until the end of a test run.

When preparing a test plan and executing a test, it is worthwhile to make use of the various equipment changes the test facility will make during the process (e.g., amplifier and antenna changes) to perform more in-depth functional tests on the EUT to ensure that all is well. For example, a radiated immunity test in the range of 2 MHz–40 GHz will have at least a dozen pauses in proceedings due to EMC equipment changes. These pauses present opportunities to download internal status data from the EUT, analyze event logs and perform other system health checks. The benefits of this are twofold: first, failures and anomalies are detected early, and second, errors in setup or configuration are detected, minimizing rework.

EMC testing is a time-consuming and sometimes tedious process. As a result, relying solely on the observation of the EUT's operation to determine pass/fail criteria is highly unreliable, as fleeting failures are easily missed. Video recording of the chamber camera system allows subsequent "fast forwarding" to more quickly detect errors or anomalies. Unless there is some synchronization between the video recording system and the test facility's automation systems, it is vital to ensure that the time stamps on the disparate systems are synchronized. Without such synchronization, it is difficult to determine what a given time point on the video recording (e.g., where a failure was observed) corresponds to in terms of polarization, frequency or level—especially at the moment of detecting a failure, when the test sponsor's day rapidly transitions from boring to dreadful, and everyone is a bit frazzled! It is possible for the EMC test systems to embed key technical data from the test in the video record, so the test conditions are always displayed on the screen and in the video archive, reducing the reliance on supporting notes and synchronized time stamps.

9.5.3 Pass/Fail Criteria: Emissions

Setting pass/fail criteria really applies more to immunity testing than emissions testing. For emission tests, there is a published limit in the standard, and the measurement result is either under the limit or over the limit, so compliance can be determined easily. Actually, that is not quite true, because facilities that are accredited (e.g., National Voluntary Laboratory Accreditation Program, National Association of Testing Authorities, and United Kingdom Accreditation Service) are obliged to calculate and disclose their measurement uncertainties—so it is possible to get an emissions pass, fail, or "indeterminate" result.

In the case of avionics for civilian or military applications, irrespective of whether the regulatory or contractual basis is MIL STD 461, RTCA-DO-160, or ED-14, a range of emission test limits are presented. Selection of which limit to apply depends on a range of factors, including the intended aircraft installation, the power supply or the intended location of the EUT on the host aircraft. Repurposing an old test report or result for a new aviation application should include an analysis of the limits applicable to the new scenario, to ensure that overall EMC is still achieved.

9.5.4 Measurement Uncertainty

For the user of EMC test reports and results, and for the regulator, measurement uncertainty is widely regarded as a nuisance as it creates the inconvenient situation where the result of a test is indeterminate, or, more colloquially, too close to call. A test result is considered to be indeterminate when the reported result plus or minus the measurement uncertainty is outside the range allowed by the limit; for example, if the limit for a radiated emission test is 30 dBμV/m and the measurement result is 28 dBμV/m, with a 3 dB measurement uncertainty, then the result is indeterminate, because the entirety of the 95% confidence interval for the result is not below the limit. The special challenge with EMC tests is that the uncertainties are typically large: ± 6 dB is possible for a radiated emissions measurement, and even a really well planned and managed system may struggle to produce better than ± 3 dB. The test facility—particularly accredited ones, which are more or less obliged to report uncertainties if they are accredited under the internationally accepted ISO 17025 standard for accreditation of test and calibration facilities—will report both the result and the measurement uncertainty; it is up to the system integrator to determine if the result is acceptable. In the case of radiated emissions tests to some of the more stringent limits (e.g., rotary wing/external RE102 limit in MIL STD 461), redesign of a product to reduce emissions to fall below the limit by a margin greater than the measurement uncertainty (i.e., to obtain a "clear pass") may be an effort out of proportion with the benefit it would bring to the overall program, unless there is reason to expect interference problems in a given situation.

When dealing with a test result that is closer to the limit than the measurement uncertainty ("indeterminate") in a regulatory or contractual context, options can be limited. If the result is compliant, there is nothing to worry about. If the result is noncompliant, the subsystem should be modified and retested so that it is compliant—usually with cost and schedule implications. Indeterminate results are difficult to deal with as they do not provide firm guidance as to whether or not redesign is warranted. The following considerations may be helpful in coming to a decision:

- If multiple test samples are available, the indeterminate tests could be performed on several samples, to determine if the set of results are more strongly biased on the noncompliant or compliant side of the limit.
- If the subsystem test is not for regulatory or contractual purposes, the potential noncompliance can be flagged for further investigation during system-level integration tests (e.g., aircraft–source–victim testing).
- If the indeterminate result is identified during the test process, the laboratory test setup can be manipulated, within the guidance provided in the applicable technical standard, to determine if there is a sensitivity to some aspect such as EUT operating mode, orientation, or cable arrangement. In doing this, it is important to note that many standards require that the test item be arranged to maximize its propensity to fail the test. This option is not helpful when analyzing data for tests that were performed sometime in the past (e.g., at system integration stage).
- Some regulatory regimes take the simplistic view that if the reported value is compliant, then the result is taken as compliant without any regard for the breadth of the confidence interval. The underlying statistical justification

for this approach assumes that, if the uncertainty band is symmetrical, the odds favor compliance over noncompliance when the reported value is compliant, even if a substantial part (e.g., up to 50%) of the confidence interval is noncompliant. This approach is also valid if measurement uncertainties were considered by the standards makers in setting the limits.
- Many test facilities report a single uncertainty value per test for brevity and ease of reporting, and this would generally be the widest uncertainty applicable for any given test. However, EMC test uncertainties are a function of frequency, and it may be worthwhile enquiring with the test facility what the uncertainty is for a particular measurement result, if a question of regulatory or contractual compliance is clouded by the uncertainty.

Measurement uncertainty in EMC testing is expressed as the extremes above and below the reported result that describe the (by convention) 95% confidence interval of a Gaussian distribution centered on the reported value. Without digressing into the statistical detail, it works out (Central Limit Theorem) that a measurement system with more than about three uncorrelated sources of random error (i.e., pretty much any EMC measurement system) has a resultant probability density function that is approximately Gaussian. The measurement uncertainty—the range around the reported result within which the facility is 95% confident that the result lies—can be estimated by combining the contributions of all the sources of error in the measurement system. CISPR/TR 16.4.1 has a detailed explanation of measurement uncertainty as it applies to EMC measurements that the interested reader can consult.

This example represents the probabilities associated with a radiated emission result at a single frequency where the reported field strength is 29 dBμV/m, with a 4 dB measurement uncertainty, and the limit is 30 dBμV/m. The distribution of a radiated emission test system is expected to be approximately normal. While the reported result is under the limit, the uncertainties for that measurement say that 95% of the values are contained in a range 4 dB above and below the reported value. The cumulative distribution function (CDF) in the figure shows us that approximately 70% of the values are expected to lie below the limit (intersection of CDF and limit lines). Similarly, if the reported result was 1 dB over the limit, part of the probability curve below the limit would be nonzero. In the case that a reported emission result is less than the limit by a margin greater than or equal to the measurement uncertainty, then there is a very small ($\leq 2.5\%$) probability that random errors have conspired to produce an erroneous compliant result.

9.5.4.1 Pass/Fail Criteria—Immunity

In general, the pass/fail criteria for immunity tests can be derived from the manufacturer's specifications for the EUT's performance. For this reason, consideration of the pass/fail criteria really should be one of the first considerations in a test plan, as it can drive the requirements for monitoring and stimulation of the EUT.

Pass/fail criteria fall into two categories:

(i) absolute failure (including damage) and
(ii) acceptable degradation

Absolute failure cases are easy to manage: for example, if the EUT's hypothetical error light comes on, the system stops responding, or there is a loss of the desired output from the system, the failure is easily detected and identified during the test and assessment is not subjective.

Acceptable degradation cases can be more difficult. For example, a data communications system may be warranted to provide no worse than a given bit error rate (BER). Under conditions of optimal signal and no interference, perhaps the system produces no errors at all. A first cut at EMC testing may be to run the tests and see if any errors occur—if none occur, then it has been a good day and the work is done. However, if errors are found, it is then necessary to determine if the error rate (BER) exceeded the specification; this may require specialized equipment. Since, at the time of planning a test, the result is not known, the lowest schedule risk comes at the financial and workload cost of having specialized equipment for detailed performance monitoring available up-front.

Another scenario where gradual degradation may be acceptable is with deterioration of a video or audio output signal. The quality of an audio circuit can be objectively monitored by using a signal-to-noise and distortion measurement if the EUT lends itself to carrying a reference tone. Systems that produce a range of tones to indicate different things to the operator, or video systems, can be more difficult to define criteria for as the effect of the interference can be subjective. For example, how much pixilation in an image is too much? And how can pixilation be measured automatically? These questions must be considered during the test plan stage so that all parties can agree on objective measures of what constitutes a pass and what constitutes a fail.

So far as measurement uncertainty itself is considered in a context of immunity or susceptibility tests, the mathematical estimations usually apply to the ability of the test facility to characterize the test disturbances they create. That is, the immunity measurement uncertainty may be considered to be a "level setting" uncertainty. In the case of RF tests such as conducted and radiated susceptibility under DO-160 or MIL STD 461, the uncertainty could, for example, be managed by adding a margin to the test level so that the lower 95% confidence bound was at the specified limit, to ensure that no disturbance was applied at less than the specified level.

In the case of transient susceptibility tests, the uncertainties are more complex, in the sense that they usually have amplitude and time parameters.

Application of uncertainty margins to rise, fall, and hold times is generally impractical as these parameters are defined within the test generator hardware.

For transient and continuous susceptibility tests, amplitude uncertainty becomes more important when the EUT does not comply with the requirement in the standard, and the disturbance level is adjusted to determine an interference threshold. In such cases, the EUT itself may be a significant contributor to the measurement uncertainty, if it is not consistent in its response to the disturbances.

Further Reading

R. Beresford and L. Li, "Pre-compliance EMC measurements for MRO equipment," in *Asia-Pacific International Symposium on Electromagnetic Compatibility*, Melbourne, Australia, May 2013.

CISPR/TR 16-4-1 Ed. 2.0 (2009), "Specification for radio disturbance and immunity measuring apparatus and methods Part 4.1: Uncertainties, statistics and limit modelling—Uncertainties in standardized EMC tests."

J. Ely, 2005, "Electromagnetic interference to flight navigation and communication systems: new strategies in the age of wireless," American Institute of Aeronautics and Astronautics. Accessed from www.ntrs.nasa.gov/archive/nasa/casi.ntrs.nasa.gov/20050232846.pdf (Oct 24, 2012).

W. Emerson, "Electromagnetic wave absorbers and anechoic chambers through the years," *IEEE Trans. Antennas Propag.*, vol. 21, no. 4, pp. 484–490, Jul. 1973.

F. Grassi, C. Rostamzadeh, D. Bellan, G. Spadacini, and S. Pignari, "Assessment of the bulk current injection test procedure based on the substitution method," in *Asia-Pacific International Symposium and Exhibition on Electromagnetic Compatibility*: APEMC 2013, Melbourne, Australia, pp. 418–421, May 20–23, 2013.

K. Javor, "On field-to-wire coupling versus conducted injection techniques," in *IEEE Symposium on EMC*, Austin, Texas, pp. 479–487, 1997.

P. Kay, "EMC test site validations for 1 m automotive and military emissions tests," in *Electromagnetic Compatibility Symposium*, Perth (EMCSA), 2011, pp. 1–6, Nov. 9–11, 2011.

N. Leveson, *Engineering a Safer World – Systems Engineering Applied to Safety.* Cambridge, MA: Massachusetts Institute of Technology, 2011.

D. Morgan, *A Handbook for EMC Testing and Measurement.* London, UK: Institution of Engineering and Technology, 1994.

J. D. M. Osburn, "EMC antenna parameters and their relationships," *ITEM*, Jun. 1997 (accessed from www.ets-lindgren.com/whitePaper-I0196 (Jan. 28, 2015).

S. H. Voldman, *ESD: RF Technology and Circuits.* West Sussex, England: John Wiley & Sons, 2006.

10

EMI Effects in Flight Control Systems and Their Mitigations

Irfan Majid

"Radio waves suspected in copter falls," *The Philadelphia Inquirer*, November 12, 1987. Anyone who has been associated with the field of electromagnetic interference (EMI)/electromagnetic compatibility (EMC) will surely infer that the headline being referred to relates to the five crashes of US Army Black Hawk helicopters. These crashes were caused by coupling of high-intensity radio frequency (HIRF) energy to helicopters' fly-by-wire (FBW) systems. Does this imply that professionals in the field of aviation are oblivious to the possibility of such an EMI occurrence, indeed not? But this demon has the uncanny nag of surprising even the most prepared.

The subject matter of this chapter is how EMI can affect flight control systems (FCS), what are the peculiarities of EMC testing of a safety-critical system like the flight control system and what are the preventive techniques to keep the FCS meet the requisite EMC requirements, throughout the life cycle of the aircraft. It encompasses military aircraft, commercial airplanes, and unmanned air vehicles (UAVs). The electromagnetic environment (EME) experienced by the airborne aircraft has become considerably harsh, more so for military platforms. Section 10.2 takes stock of the EMI expected from such EME. Section 10.3 narrates some of the aircraft crashes caused by EMI in FCS. Section 10.4 provides an overview of FBW systems as regards functionality and their constituent elements. In numerous modern-day aircraft, the FCS form an integral part of flight management system (FMS). In the context of this chapter, FCS will refer to FBW systems. The final response of FCS is dependent upon a multitude of other sensors and systems. Therefore, introduction to FMS is also included. The next section introduces EMC standards relevant to FCS EMC testing. Section 10.7 illustrates the methodologies used for EMC testing of complete aircraft and FCS. EMC testing of FCS has to follow a system engineering approach; this complete process is explained in this section.

Handbook of Aerospace Electromagnetic Compatibility, First Edition. Edited by Reinaldo J. Perez.

Section 10.8 dwells into details of how EMI affects FCS, and real-life examples are cited to clarify the point. As physical testing and experimentation of complete aircraft is costly, the logical alternative is modeling and simulation; this is the focus of Section 10.9. The UAVs are distinctly different from other aerospace vehicles and are treated separately. It reviews the implementation of FCS in UAVs and their susceptibility to EMI. Urban EME the UAVs can be exposed to is very complex and can have some peculiar EMI effects; this is touched upon in Section 10.10. The focus of the last section is on factors that enhance the probability of malfunction in the presence of EMI. It also annotates some good practices that can ensure better EMC in flight critical systems.

10.1 Introduction

The history of aircraft EMI is as old as the history of aircraft electronics or avionics systems themselves. Installation of every such system meant an addition of a potential EMI victim and at the same time a potential EMI source. Aircraft avionics systems have come a long way since Captain James Doolittle flew the famed blind flight on Sep. 24, 1929 [1]. In the nascent years of aviation, avionics systems mainly catered for the requirements of communications. The observed EMI mainly related to desensitization of receivers or crackling noises in headsets. Navigation systems like nondirectional beacons, radio navigation aids, radio altimeters, and inertial navigation systems started becoming integral part of aircraft. These were supplemented by support systems like ATC transponder/identification friend or foe and instrument landing system (ILS). Radar was used on military aircraft for air surveillance and ground mapping. In commercial and military transport aircraft, weather radars were installed. The EMI effects now observed were degradation in performance or erroneous outputs. These were attributable to lightning/HIRF fields and intersystem EMI [2]. As regards EMC, the focus was making the systems immune to the most ominous EMI threat, namely, lightning.

A milestone transition took place for military aircraft in the 1970s and somewhat later for commercial aircraft (with the exception of Concorde aircraft), when the FCS became predominantly electrical. This type of FCS is generally referred to as FBW systems. Mil-Std-464 defines safety-critical system as a "term applied to a condition, event, operation, process, or item whose proper recognition, control, performance or tolerance is essential to safe system operation or use; for example, safety critical function, safety critical path, or safety critical component. A term also used when a failure or malfunction of a system or subsystem can cause death or serious injury to personnel." Thus, systems such as global navigation satellite system (GNSS) or, more specifically, global position system, whose performance directly influences the aircraft navigation and FMS performance, also become safety critical. GNSS receivers are inherently highly prone to interference. Additionally, engine indication and crew alerting system fall into the category of an electronic flight-critical system. A salient feature of this chapter will be to highlight the essentials of EMC testing of flight-critical systems, more specifically the FBW-type FCS.

The FCS is formed from a multitude of interconnected electronic and electromechanical modules. Small errors of omission and commission in integrating the complete system on the aircraft can allow many a doors ajar for EMI to creep in. Redundancies are always implemented in FBW systems; to minimize the possibility of redundant channels also getting affected by EMI, common practice is to implement these channels from different types of components and to keep them as spatially dispersed as possible. Nonetheless, chances of such an occurrence can never be eliminated. The primary function of FCS is to enable the human pilot or the autopilot to move the control surfaces in a manner so as to execute the intended maneuver in the air. The statement in itself highlights the biggest limitation of conducting any FCS-related testing on ground, EMC testing is no exception. The effect of any EMI occurrence will become more severe when it takes place close to limiting aerodynamic performance condition of a maneuver that is difficult to replicate with aircraft static on ground. Constituent components of FCS are tested for EMC in the chamber, and the complete system undergoes a series of tests but actual flight conditions cannot be emulated in totality.

Furthermore, in the early days of operation, any FBW system, duly qualified for EMC, is highly immune to EMI; however, rigors of operations and routine maintenance are likely to degrade this immunity. The complexities associated with effective EMC testing of FCS and dire consequences of any malfunction of a flight-critical system due to EMI make this subject important and challenging. The inspiration for writing this chapter has been to make the engineers involved in EMC testing or EMI analysis of flight controls and related systems better equipped to shoulder these responsibilities. The material is also useful for aircraft system engineers desirous of enhancing their understanding of EMI and EMC. Aircraft EMI occurrences continue to surprise even the most experienced. A better understanding of systems and their vulnerabilities to EMI can help reduce the probability of such occurrences.

FCS has been selected as an example of flight-critical system as its EMC test requirements are treated as a separate category and are more stringent than the other systems. Historically, it was the implementation of an FBW system in Airbus A-320 aircraft that necessitated complete review of HIRF testing levels of civil aircraft; thus, it merits special consideration. Though the basic methodology for EMC testing of flight-critical systems is similar to that of other aircraft systems, there are certain requirements that are specific to systems whose disruption can be catastrophic and can result in loss of life. These requirements will be elaborated in succeeding sections; however, they are briefly stated here to emphasize that EMC testing of flight-critical systems is to be planned and executed in a special manner.

(i) For high-intensity radiated fields (HIRFs) and lightning-related tests, flight-critical systems are exposed to higher field strengths than the other systems.
(ii) Some level of whole aircraft testing is required for critical functions.
(iii) For complete EMC testing of FCS and usually for other flight-critical systems, a system engineering approach is followed. Close liaison is to be maintained between the EMC engineer and the system-design engineer, and it is executed in the form of a process.
(iv) For all safety-critical systems, sufficient design margins are to be added to cater for EMC test method inadequacies. If all units produced are not tested for EMC but only some selected samples, the margins also compensate for production/integration variations.
(v) How quickly the system recovers from EMI disruption is also to be determined.

EMC testing of FCS is required to be carried out by manufacturers of all classes of commercial and military aircraft and UAVs, during manufacturing and after major modifications or upgrade of aircraft. However, in the latter case, it is usually a subset of complete testing. The EMC testing methodologies covered in this chapter are applicable to both fixed-wing and rotary-wing aircraft; however, the specific details mainly focus on fixed-winged aircraft. The material covered in this chapter will be useful for both EMC engineers and system engineers involved in the planning and conducting EMC tests of FCS.

10.2 Nature of EMI Experienced by Aerospace Vehicles

All electrical/electronic/electromechanical systems are susceptible to EMI. EMI can be defined as the process by which disruptive electromagnetic energy is transmitted from a source-to-the-victim device through radiated and/or conducted paths. When the amplitude of an interfering signal becomes relatively large as compared with the signal of interest, then the interference can upset system performance [3].

The two basic categories of EMI are natural and manmade. The most menacing source of natural EMI are lightning and static electricity. To a considerably lesser extent, 11 years cosmic cycles effecting electronic communications and presence of naturally occurring strong magnetic fields also fall in this category [4]. The manmade sources are all the electrical/electronic/electromechanical

Table 10.1 Natural and manmade sources of EMI with reference to aircraft

Natural sources		Manmade Sources	
Terrestrial	**Celestial**	**Systems**	**Circuits and components**
Natural lightning, precipitation (P-) static	Cosmic noise, solar noise	Communication, radar, navigation, 115 V 400 Hz AC power, 28 and 270 VDC, avionics buses and cables, electronic warfare, entertainment	Local oscillators, switches, motors, filters, relays, strobe lights, actuators, latching contacts, digital circuits, heaters, pumps

devices and equipment developed by human beings. Classes of EMI and its sources are listed in Table 10.1.

The EMI received by victim systems can further be segregated into near-field and far-field interference, which is on the basis of distance between victim and source. This will also determine whether the interference is magnetic (H) field or electric (E) field dominant. Figure 10.1 shows spatial orientation

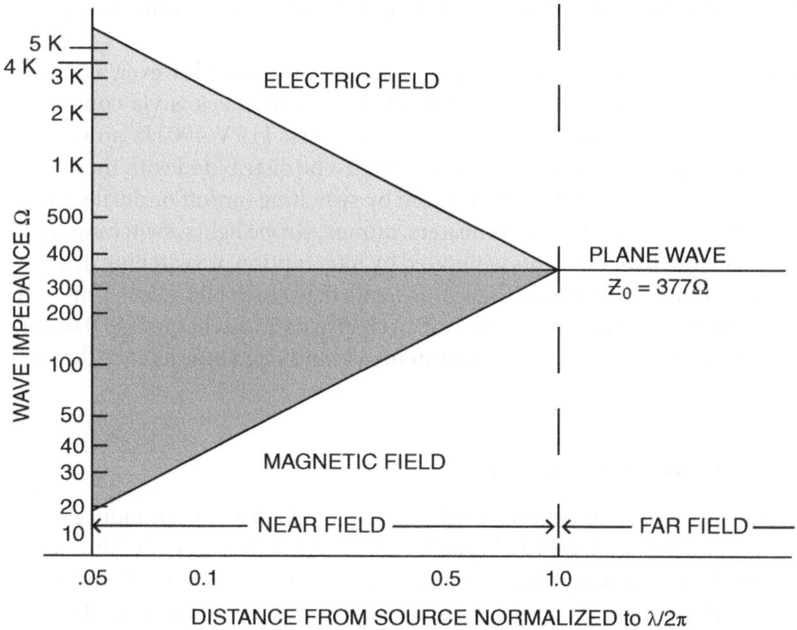

Figure 10.1 Near- and far-field characterization of an electromagnetic wave.

of E and H field as a function of wavelength and impedance. A majority of modern-day military and commercial aircraft comprise a number of electrical/electronic/electromechanical systems. In aircraft that employ FBW FCS, the effect of EMI can be catastrophic [5].

10.2.1 EMI Sources Internal to the Aircraft

All electronic/electrical/electromechanical equipment emit electromagnetic energy through conducted or radiated mechanisms. The degree of radiation will vary with exact type of equipment and how well EMC aspects have been addressed in its design. Contemporary aircraft constitute of a large number of such systems. It is important to realize that all avionics, instrumentation, and electrical systems have subsystems distributed throughout the aircraft. These subsystems in turn will contain receivers, transmitters, high-speed processors, actuators, etc. A large number of antennas associated with transmitting and receiving systems are installed on the aircraft fuselage so that maximum spatial diversity can be achieved. This implies that there are extensive lengths of RF cables running throughout the fuselage. Additionally, in military aircraft, weapon control and electronic warfare equipment, and entertainment systems in commercial aircraft significantly increase the potential EMI sources. It will be discussed in Section 10.8 how EMC tested elements can become susceptible to EMI.

Thus far, radiated interference sources have been stated. However, a considerable portion of interference received by aircraft modules is via conducted means. The voltage supplied by aircraft generators, 115 V 400 Hz power, has associated electric and magnetic fields and must be distributed with these concerns addressed. Transients are produced by switching (on/off or during mode transition) of heavy loads such as heaters, pumps, strobe lights, switching regulators, and inductive transients generated by interruption or switching of 115 V AC supply to inductive loads. These transients may reach 600 V, last 1 ms with rise-time of nanoseconds [6]. In case of an electrostatic discharge (ESD) occurrence, it is ultimately seen by the subsystems as conducted interference through connecting wires [7].

10.2.2 External Sources of EMI

The strongest source of external EMI is lightning. Lightning can induce a few kilovolts per meter of electric fields inside the airframe of an aircraft. The nature of these fields is transient in nature, whose details can be found in [8] and other literature. Below the altitude of about 3000 feet, external EMI is due to multitude of emitters that keep increasing in number to support various types of wireless communications and increase in the number of high-power ground radars. At higher altitudes, transmitters onboard the aircraft have increased considerably, especially for military aircraft. But what has really made matters worse is increasing number of aircraft per unit volume of airspace. Moreover, overlapping of military and civil airspaces has made matters worse [9,10]. A visualization of dense airspace can be accessed through various websites that depict commercial aircraft equipped with ADS-B [11], which does not present all the aircraft flying in the airspace. The present drive for integrating unmanned aerial systems into existing airspace has mainly focused on air traffic management issues. But the effect on overall EME due to a host of command/video link transmitters in multiple frequency bands has not received requisite attention [12, 13]. In case of commercial aircraft, another disturbing source of internal EMI are portable electronic devices, these have been the focus of much attention in last two decades [14, 15]. A physical orientation of internal and external EMI sources affecting the aircraft is depicted in Figure 10.2.

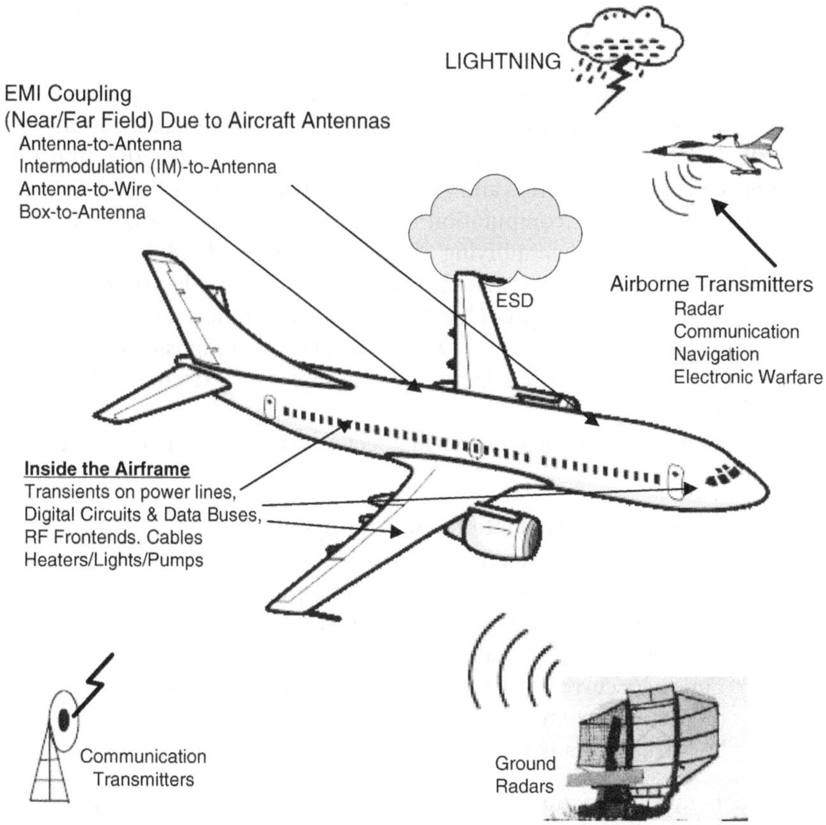

Figure 10.2 Internal and external EMI sources contributing to aircraft EME.

10.2.3 Analysis of the Contemporary Aerial EME

EME refers to the magnitude of electromagnetic power spread over the frequency band of interest, present in the interior and immediate surroundings of the aircraft. EME is represented as amplitude spectral components of electromagnetic signals. A more formal definition is "the measurable electromagnetic field(s) that exist within a definable region(s) where equipment or systems must operate, either individually or in conjunction with other equipment or systems" [16]. Computation of EME is important for the following reasons:

(i) To determine the expected EMI so that the protections can be accordingly designed. In aircraft, EMI protections mean adding cost, weight, and reducing available volume. None of which is desirable. Thus, aircraft EMI protections shall be no more or less than desired.
(ii) EMC standards set their immunity test values according to prevalent EME.
(iii) In certain cases, test standards allow test limits to be adjusted in accordance with EME.

The magnitude of interference is computed using the equations of propagation of electromagnetic energy, taking care of the exact conditions [17–19]. In case of internal EMI, it is to be considered whether the distance between potential sources and victims should be considered as far-field or near-field coupling. The computation of near-field interference is more complex, and generally some simplifying assumptions are made [20]. This also implies that it should be differentiated if the interference is electric or magnetic in nature. The shielding provided by fuselage and other compartments of airframe has to be catered for in calculations. Use of composite airframe in manufacturing of aircraft has increased the impact of external EMI on internal modules.

How much the interference will affect a system is a function of the following [4]:

(i) Frequency bands of interference and system.
(ii) Magnitude of interference.
(iii) Impedance match with victim system input circuitry.
(iv) Electrical dimensions of the victim system vis-à-vis wavelength associated with interference signal.
(v) Time of occurrence of interference in correspondence to the signal level of the victim system.
(vi) Polarization (radiated emission only).

Generally, the points are self-explanatory; possibly, the time of occurrence requires further elaboration. Consider a radar pulse illuminating an aircraft. When the electromagnetic energy of this pulse reaches a data cable carrying digital signals, these data are liable to change if the pulse occurs at a time so that it occupies a substantial portion, generally taken to be 50% or more, of the data bit or symbol time. With clock speeds of digital circuits transcending GHz boundaries, the radar pulses are very likely to corrupt the data stream. The probability that EMI will degrade the performance of the victim system is associated with the EME around the system. Talking in terms of probabilities actually makes sense in EMI occurrences, as usually these are stochastic in nature. This is the reason why EMI occurrences are so difficult to reconstruct [21]. The EME is usually measured in terms of spectral components using some form of a spectrum analyzer. To see these results with reference to a region of space, spatial coordinates are appended with each measurement, and then using plotting features of data analysis software, a contour plot can be displayed on a computer. Figures 10.3(a) and (b) show sample results of these measurements, respectively.

10.3 Reported Catastrophic EMI Occurrences in FCS

In terms of being essential to the flight, aircraft systems are categorized as follows:

Category A, Critical: Prevents safe flight (safety critical)
Category B, Essential: Impaired ability to fly (mission critical)
Category C, Nonessential: No significant degradation in performance (noncritical)

FCS, without doubt, falls in category "A." When the system is characterized as flight critical, its design, installation, and testing requirements are greater than those of other systems; EMI in these systems can have serious consequences. This section covers the instances where EMI crept into FCS and brought down the aircraft.

10.3.1 The Sensitivities of a Flight-Critical System

Any fault induced in FCS can cause the aircraft to crash [6, 23]. Thus, qualification testing requirements of these systems, embedded protection mechanisms, and redundancies are distinctly more than other systems [17]. The realization that EMI can be catastrophic for aircraft has materialized over past few decades. Some of the most infamous occurrences where EMI proved fatal by affecting the FCS are cited in this section [24]. Some of the cases that were detected and corrected in time are analyzed in a later section, where the purpose is to demonstrate how exactly elements of FCS become victims of EMI and what protection measures can be employed. Yet, there may be numerous other incidents that would have been unreported or were not properly diagnosed to be EMI related [25].

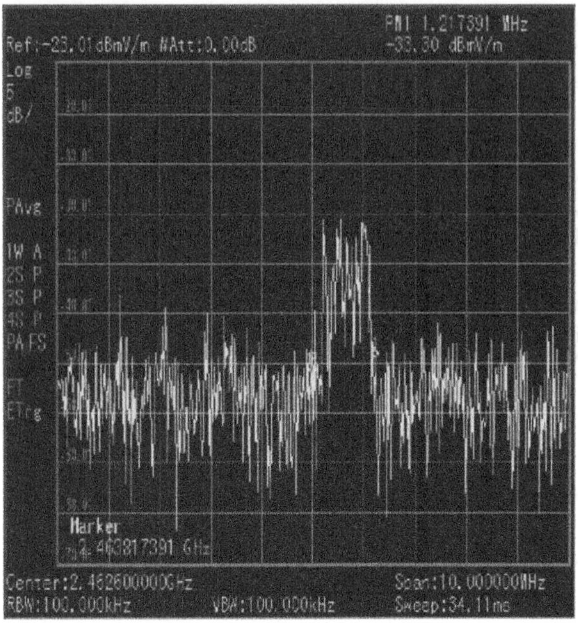

(a)

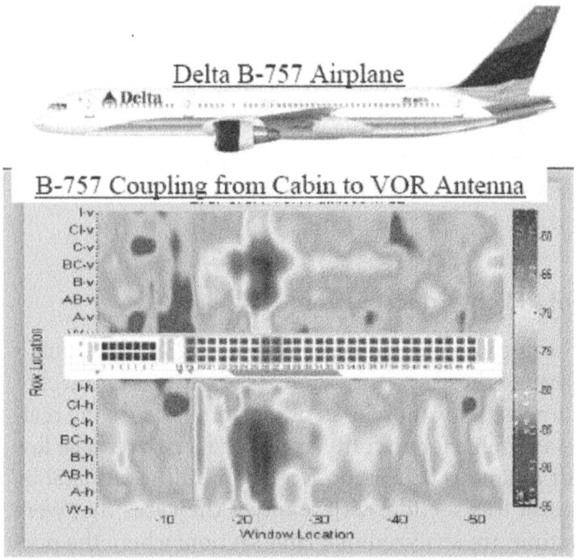

(b)

Figure 10.3 (a) A typical spectrum analyzer display showing the magnitude of spectral components and (b) a contour plot of EM energy in a section inside fuselage due to PEDs [22].

10.3.2 Reported Occurrences

Initially, when electronic FCS were first installed on the B-52 bomber autopilot system, rear flight control surfaces exhibited random movements when the HF radio onboard was operated. The HF radio signals were induced in the wiring system and coupled to the autopilot system. In the time frame from 1981 to 1987, five crashes of US Army Blackhawk helicopters killed or injured a number of personnel onboard. The crashes occurred when these helicopters flew close to radio broadcast transmitters; one of the crashed helicopters is shown in Figure 10.4. Investigations revealed that the flight control electronics including their subsystems had insufficient EMI protection against HIRF. This resulted in random movements of flight control surfaces while flying past high-power radio broadcast towers.

A remotely piloted vehicle, Pioneer, was controlled by using portable remote control box. During its flight testing by US Navy in January 1987 aboard the USS Iowa, the pilot observed that the RPV went into maneuvers that were not demanded; this caused loss of control and consequently resulted in a crash landing. Subsequent investigation found that the remote control boxes received false signals from HF communication transmitting antennas that were located aboard the Iowa. The FCS employed inadequate shielding and improper cable termination.

In 1984 near Munich, Germany, a Luftwaffe Tornado fighter crashed after flying too close to a powerful Radio Free Europe/Radio Liberty transmitter. The suspected cause of occurrence was again coupling of HF energy into FBW flight controls of the aircraft. One crash of F-16 fighter aircraft is attributed to EMI caused by HIRF from a Voice of America transmitter. F-16 is a highly

Figure 10.4 A US Army Black Hawk helicopter that crashed due to EMI in FCS.

agile and maneuverable aircraft employing FBW control systems. The FCS constantly monitor pilot inputs to ensure aerodynamic performance curves are not exceeded. Apparently, an undesired signal was transmitted to the control surface causing the crash. Consequently, many of the aircraft had to be EMI hardened to prevent such occurrences. This incident was a key factor in initiation of the Federal Aviation Administration (FAA) HIRF Certification program.

The occurrences cited here are related to FCS that resulted in crashes. There are numerous instances where interference in some of the sensors or systems affecting the output from flight control computer (FCC) caused inadvertent behavior of the control surface or wrong response from autopilot. Passenger PEDs have been known to cause upset to the navigation systems [26]; if the aircraft is using FMS mode, the FCS can respond in an erroneous manner. Due to inbuilt protection in FCC software, the aircraft does not violate aerodynamic performance curves but it can go off course.

10.3.3 Lessons Learnt

EMI is a harsh reality that poses a very real threat to all aircraft that take to the air. The more "electrical" the aircraft, the more likely it is to be affected. The extent to which an aircraft is affected would depend upon the system that is the victim and the phase of flight at the time of occurrence. All flight-critical systems have to be especially hardened against EMI. Laxity in this aspect can lead to loss of lives and destruction of expensive aircraft. Sharing of data in this regard is of paramount importance. However, at times the incident is not recognized as due to EMI, especially if the crew is not well aware of EMI effects. In case of military aircraft, sometimes facts are not made public due to reasons of confidentiality [25].

10.4 Anatomy of FBW FCS

FCS are essential to the aircraft performing its primary function. In military terminology, it is a flight-critical system and in commercial aviation it will be characterized as a level "A" system. Malfunction in such a system can result in loss of life and aircraft. Thus, their design, fabrication, integration, installation, and embedded protections need to be very robust. Usually, it is understood that only all electronic FBW FCS are susceptible to EMI. However, there are many aircraft in service that do not employ FBW but use some sort of electrical/electronic modules to implement autopilot systems, thus to a varying level all these systems can be disrupted by EMI. This section provides an insight into functioning of FCS. System architecture, including description of modules that constitute FCS, is also covered. The focus remains to be how EMI can affect these modules. Once these modules are installed on the aircraft to form an integrated system, a whole new set of variables comes into play. These peculiarities are the subject matter of Section 10.8. The approach here is to describe the underlying concepts with reference to a generic configuration rather than focusing on a specific type of aircraft. As mentioned in introduction to the chapter, EMC testing of FCS follows a system engineering approach. The knowledge of actual design and implementation of FCS will rest with the system engineer, which will translate into overall EMC test plan in consultation with EMC engineer.

10.4.1 Functional Overview of the FBW System

After an aircraft takes to air, it has to be kept airborne and steered toward its desired destination. It is understood that keeping an aircraft in the air is more demanding than making it take to the air. Flight controls is the system that performs this task, pertinent to mention this will include landing and takeoff roll as well. In scientific terms, flight controls provide the capability to control the forces and moments acting on the aircraft, thus enabling control of accelerations and hence velocities, translations, and rotations. The FCS aims to achieve this via the aircraft's flight control surfaces. The thrust provided by the engines must also be taken into account, since this also produces forces and moments acting on the aircraft [27].

In simplistic terms, the function of flight control system is to move the control surface or surfaces in response to pilot's input, which is also termed as "maneuver demand." This is done through a system of interlinkages or connections between control stick, rudder pedals, and the control surfaces. Initially, these interlinkages comprised of steel wires and pulleys. This was replaced by a hydraulic system. Those were superseded by electrical systems. These systems can be a mix of hydraulic and electrical systems, or it can be an all-electrical system. The all-electrical system is generally referred to as FBW. As regards EMI, basic rules hold for FCS as well, that is, the greater the number of electrical/electronic/electromechanical modules in a system, more susceptible it is to EMI. Many aircraft that do not use an FBW system employ autopilots to ease pilot workload. Though the primary FCS is not electrical but the autopilot may constitute some electrical or electronic systems, their performance can be disrupted by EMI.

The function of FCS is described focusing on a digital FBW (DFBW) system, their exact implementation can vary from aircraft to aircraft, but the basic functions remain the same. The pilot force applied to two primary flight controls in the cockpit, control stick or column, and rudder pedals is converted into electrical signals by means of transducers. In the earlier versions of FBW systems, these electrical signals were in analog format; in later versions, these signals are digitized and distributed to various flight control surfaces through avionics data bus [28], hence the term digital FBW. A generic block diagram of

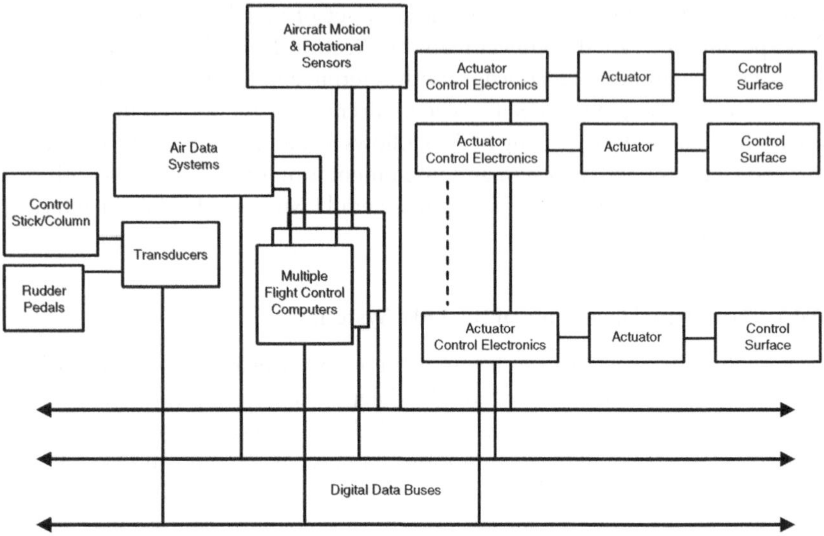

Figure 10.5 Block diagram of the DFBW system.

DFBW system is shown in Figure 10.5 and a pictorial view indicating placement of main modules is shown in Figure 10.6.

These electrical signals, representative of pilot's command, are transmitted to the FCC; in addition to pilot's command, the FCC receives inputs from different sensors to determine aircraft's current velocity vectors, rotational rates, flight envelope, position of control surfaces, and engine parameters. It compares them with aircraft–aerodynamic–performance curves; on the basis of this comparison, it ascertains that the maneuver demand made by the pilot does not upset aircraft stable flight; if that is the case, it alters the pilot's command accordingly. The finally computed command by FCC is transmitted to the electronic circuitry controlling the actuators. The actuators are attached to the required flight control surfaces. The actuator and its control electronics are packaged together as a module. The resultant changes in aircraft motion and attitude is conveyed to FCC as feedback and the control surfaces are thus adjusted until desired change is achieved. Detailed working of FBW flight controls along with associated control laws can be found in [29, 30].

As FCS is absolutely critical to aircraft remaining airborne, multiple levels of redundancies are implemented in it. The computing platforms, control electronics, and the actuators themselves are used in parallel chains. Also, as all the electrical FCS need electrical power to function, power supply has to be ensured to FCS in case of failure of primary or even secondary power supply buses. It is important for the EMC engineer to understand how this fault tolerance is employed, so that he can work out what can be potential EMI upsets and

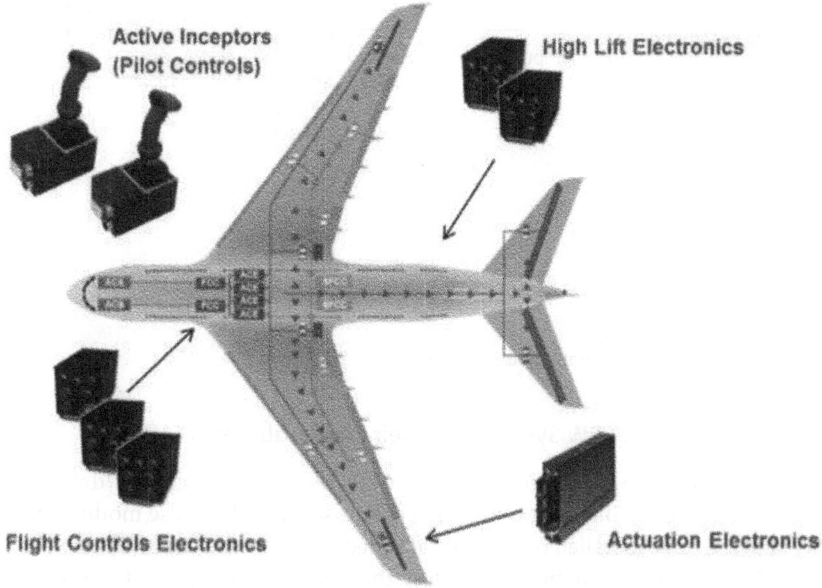

Figure 10.6 Elements of the FBW system and their possible placement in aircraft.

will the system redundancies cater for them or not. In case there is a possible occurrence of EMI that may bypass system fault tolerance, this possibility has to be provided better EMI protection. This will be the general configuration of a full-authority DFBW system. Other aircraft may employ it in analog form with some sort of mechanical backup or some partial functionality of the system may be used. The description above highlights the fact that during EMC testing when it is required to test the complete functionality of the system, testing of FCS needs very elaborate planning and long execution times.

There are aircraft that do not use electrical FCS but do have autopilots that have electrical or electronic modules. Autopilots are basically used to ease pilot's workload [31]. They may be characterized into the following categories.

Stability Augmentation System: Enhances stability of the aircraft. This is achieved by using the control surfaces to make the aircraft more stable. One example is the phugoid damper or the yaw damper. Without it, the aircraft is less stable or possibly even unstable.

Control Augmentation System: The control augmentation system can either be a "longitudinal autopilot" (aircraft motion in vertical plane), a "lateral autopilot" (aircraft motion in horizontal plane) or combination of both.

A generic representation of an autopilot is shown in Figure 10.7. Basic functionality of autopilots is described here because there have been actual instances of autopilots being disrupted by EMI.

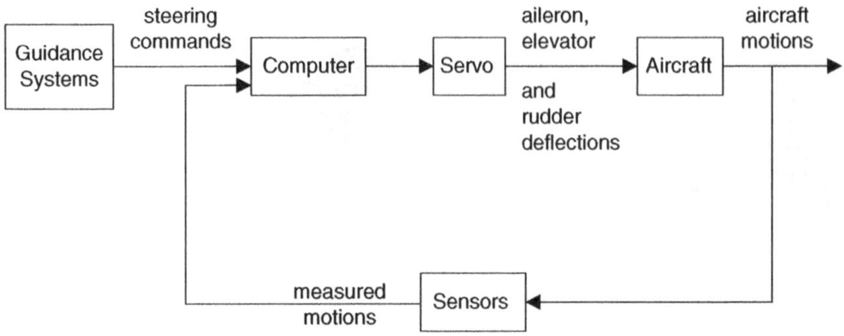

Figure 10.7 Basic block diagram of an autopilot.

10.4.2 Elements of FBW System and Their Susceptibility to EMI

This subsection views the elements of the FBW system in greater detail as regards what are the inherent potential weaknesses to EMI. These modules are only allowed to be installed on the aircraft once they have cleared their applicable EMC testing in the chamber. However, the EMC engineer should, as second line of defense, keep the potential EMI sources as curtailed as possible. Generally, the most effective method but not always possible is to spatially separate source and potential victims. For example, if an electronically controlled actuator is in the close vicinity of a motor-driven pump, it needs to be relocated or sufficiently shielded. When these elements are installed on the aircraft to function as a complete system, many "back-door" channels of EMI come into existence. Except wires and data buses, all elements require electric power to operate; being a flight-critical system even if primary and secondary power bus fails, emergency power has to be provided to all electric FCS. The input power has to be in accordance with Mil-Std-704, DO-160, or equivalent, yet conducted EMI remains a possibility that has to be blocked. A study by Clough et al. indicates that current DFBW systems are not really prone to EMI above 1 GHz but are vulnerable to AM interference in the VHF band [32].

10.4.2.1 Actuators and Actuator Control Electronics

Actuators and actuator control electronics (ACE) are used to move the control surfaces in response to maneuver demand. They can be hydraulic, mechanical, linear variable differential transformer, or electric motor type. As regards EMI, the first type are not of consequence. The actuator always has some control electronics associated with it and is packaged as a unit. The complete actuator module can be susceptible to EMI, e.g., the electric motor type actuator could yield significant displacement errors due to external magnetic field coupling. However, in packaged form, it is well protected. Control electronics of some actuators use pulse width-modulated signals for control of motors; these can be disturbed by signals matching their frequency range and are also a likely source of radiation. EMI through power input lines can disturb the response of an actuator [33].

10.4.2.2 Flight Control Computer

As the name suggests, FCC is a computer that can be in the form of single board housing, multiple processors, multiple PCBs connected via a backplane, or may even be proportional plus integral plus derivative-type controller using analog electronics. Analog electronics was known to be prone to high-intensity amplitude-modulated signals. There have been reported instances where the data stored in memory were found to be corrupted by EMI. Signals traveling on PCB traces, clock circuitries, interrupt lines are the possible weak links for EMI to creep in. EMI filters are generally used on power input lines to guard against conducted EMI.

10.4.2.3 Wires/Data Bus

Data buses carry multiplexed signals in digital format while wires or cables will be used to transmit power, analog sense and control signals, or RF signals. For the purpose of brevity, these have been grouped together. When the complete system is in place in the aircraft, they can be termed as the Achilles' heel of FCS. Chances of EMI affecting a system are directly related to its exposed area. Wires run through the length and breadth of an aircraft. All cables used in aviation meet the desired compliance limits but their connections to various modules and assemblies can create back-door channels of EMI. Though it will be discussed at greater length in later part of the chapter, it will be worth mentioning here that maintenance and aging deterioration can greatly reduce EMI protection and that is most pronounced in the wires.

10.4.2.4 Sensors/Transducers/Displays

Talking of FCS, EMI is to be considered from sensors to display [7]. A whole wide range of sensors are used in an aircraft, whose output is used to determine the actual movement of the control surfaces or display flight parameter's information to the cockpit display. These sensors can be pressure differential type, thermocouples, and displacement transducers. Sensors directly related to FCS include proximity switches, accelerometers, gyros, LVDTs, air data computers, and associated electronics. There have been a number of reported instances where a fuel quantity indicator system (FQIS) has shown erroneous readings due to EMI. This can alter the FMS computations resulting in incorrect response of FCS. The sensors generally are not directly susceptible to EMI, but their associated electronics and interconnections to computing and display modules can be areas of concern. The electronic-type sensors require AC or DC voltage to operate, and DC-excited sensors are normally more immune to EMI.

10.5 Flight Management System

Previous subsections described the function of FBW systems and autopilots. FMS add another layer of automation to these tasks and implement what is termed as 4D navigation. The fourth dimension optimizes the function of time management and that too in a manner so as to economize fuel consumption.

10.5.1 Functional Overview of FMS

Flight plan and trajectory prediction work together to produce the four-dimensional trajectory and fuse this information into a flight-planning algorithm. The navigation function provides guidance, reference, and advisory information relative to the desired trajectory and aircraft current attitude and position. Another important area is the control display function, which also includes aural warnings.

There are typically two databases that support the main flight management functions. These are the navigation database, which must be periodically updated, and the performance database that models aircraft performance characteristics, i.e., engine thrust, aerodynamic, etc. The navigation database contains published data relating to airports, navigational aids, waypoints, airways, and terminal area procedures along with required navigation performance (RNP) values specified for the associated airspace. The purpose of the navigation database is to store information about location, frequency, elevation, and class information for the various ground-based radio navigation aids. This information is necessary to select, tune, and process the data from the navigation radios into an aircraft position. It also provides the flight plan function with airport, landing approaches, departure, and approach procedures, i.e., predefined strings of terminal area waypoints, airways, i.e., predefined enroute waypoint strings, and named waypoint information that allows for swift route creation; this is as per ARINC 424 [34]. The performance database stores aircraft and engine model data containing drag, thrust, fuel flow, speed/altitude envelope, thrust limits, and a variety of efficient speed settings that are unique to the aircraft. Figure 10.8 shows the interrelationships between the core functions and the databases [35].

10.5.2 EMI Susceptibilities of FMS

How does the additional functionality of FMS affect the basic working of FCS and its vulnerability to EMI is addressed here. The basic set of FCS elements remains to be the same. However, by utilizing the FMS, an autopilot mode is set, now the response of the FCS is dependent upon inputs from engine performance and navigational systems as well. Important to note that some navigational systems such as VHF omnidirectional range, distance measuring

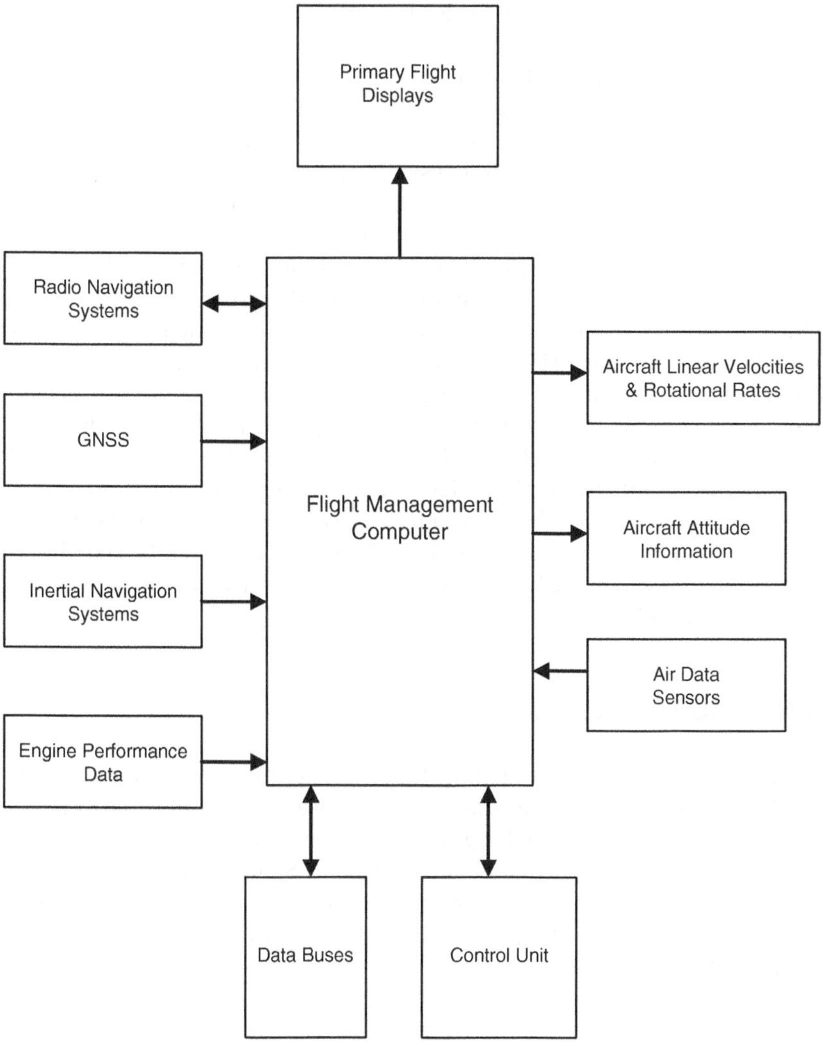

Figure 10.8 Block diagram of data exchange between elements of FMS.

equipment (DME), ILS, and the much relied onGNSS all are prone to external interference, in the sense that the input signal received by antenna is already corrupted by interference in the propagation channel. If EMI causes any error in pick up from engine performance-related sensors or alteration in data being transmitted to cockpit displays, erratic behavior of the FCS can occur. As in other systems, the wires that convey the sensor readouts of engine performance parameters run long distances in areas that are potential sources of strong magnetic and electric fields. Thus, the wiring lying close to engines has to be properly shielded, and special care has to be taken to connect these wires to their terminations at both ends.

One significant difference between EMI susceptibility of basic FCS and FMS is the susceptibility of later to PEDs being used onboard the aircraft. In their basic functionality, the FCS can be considered practically immune to emissions from PEDs. However, the radiation from PEDs can reach the antennas of navigation systems through windows and processed as legitimate input, causing random response from FCS [15,22,26]. Section 8 contains some real-life examples of EMI effects on the aircraft autopilot and FMS system.

EMI occurrence in the navigation system can cause erroneous response of FCS when operating in FMS mode. This can be mitigated through the use of robust software algorithms. These algorithms first determine the suspected errors by correlating the trend of outputs of a particular system with previous outputs. In case of an abrupt change, the second step is to analyze and compare the outputs from other relevant systems and make corrections, e.g., if GPS inputs seem to be effected by interference, they are compared with INS and other land-based navigation systems. Rounding up, it can be stated that when using FMS mode the flight crew has to be aware that aircraft is relatively more prone to EMI.

10.6 EMC Test Standards

Standards are a set of rules and limits for ensuring product quality and operational compatibility. These are established by consensus and approved by a recognized body for common and repeated use. It provides rules, guidelines, and defines requirements for activities or their results. Standards development is a collective work and is created by bringing together the experience and expertise of manufacturers, buyers, users, research organizations, and government departments.

10.6.1 Relevant EMC Standards

In military aviation, the most widely used standard for subsystem-level EMC testing is US Department of Defense Mil-STD-461 [36] and for complete system testing it is MIL-STD-464 [37]. At the time of this writing, the latest revision of former is MIL-STD-461 G and that of later is MIL-STD-464C. In commercial aviation, Radio Technical Committee for Aeronautics (RTCA), an advisory body to FAA of United States, formulated RTCA/DO-160 [38], and what is referred to as its European twin EUROCAE/ED-14 G [39] are mainly followed EMC standards. UK military aircraft manufacturers follow

Def-STD 59-41 part 8 [40]. China has its own GJB standard that is compatible with MIL-STD-461C. For NATO, aircraft guidelines for EMC are available in STANAG 3516 and STANAG 3614 [41, 42]. Being the most widely used standards internationally, the salient features of MIL-STD-461, MIL-STD-464, and RTCA/DO-160 are stated in this section. All aircraft are a composition of equipment from different manufacturers integrated together. This equipment could have been tested for EMC according to a different EMC standard. Thus, it becomes important to be able to compare the conformance of these standards. Even the military aircraft have started to use commercial off the shelf equipment; however, for flight-critical systems such as FCS, the suppliers are asked to abide by a designated standard. The last part of this section provides a sample comparison between military standards and RTCA/DO-160. Major aircraft manufactures such as Boeing have their own guidelines for EMI control such as D6-16050 [43].

The purpose of electromagnetic compliance is to ensure that neither the equipment radiates electromagnetic energy to disrupt the functioning of other electrical equipment in its vicinity nor its operation is susceptible to the radiation by other electrical equipment in the surroundings [44]. The means by which electromagnetic energy can travel from source to victim equipment is either via air, i.e., radiated, or through wires, i.e., conducted means. This creates four distinct classes of EMC tests

(i) Conducted emissions (CE)
(ii) Conducted susceptibility (CS)
(iii) Radiated emissions (RE)
(iv) Radiated susceptibility (RS)

Official documentation of EMC standards provided by the agencies formulating the particular standard and their description is readily available. The endeavor here is to highlight the relevant EMC standards and their specific sections applicable in various stages of EMC testing of FCS. Also, for applicable sections where the EMC or system engineer has to make a selection of the most appropriate test parameters, some guidelines have been provided to make these choices. EMC testing of aircraft can be segregated into two distinct categories: module-level testing and complete system-level testing. EMC testing of FCS modules is carried out as specified in MIL-STD-461 and RTCA DO-160, where the previous is applicable for military aircraft and the later defines test conditions for commercial aircraft. MIL-STD-464 provides guidelines for system-level testing of military aircraft. It is relevant to mention here that some literature on EMC breaks down the system into units that combine to form assemblies, assemblies are integrated to build a subsystem, and subsystems are combined to fabricate a system that implements a complete function. However, in this text, only two divisions are made, i.e., modules and system.

10.6.1.1 MIL-STD-461

The initial version was issued in July 1967 by DoD as a consolidated standard superseding the EMC standards previously being followed by individual defense services. At the time of this writing, MIL-STD-461 G, finalized in December 2015, is the latest revision. It is a very comprehensive standard which is either followed by a majority of countries of the world "as is" or in derived form. It specifies requirements and limits according to the vehicle or system, e.g., surface ships, and aircraft, where equipment will be installed. Physical location (internal or external to structure), and unique platform operational characteristics are also taken into account. Although alteration of the requirements is allowed for individual procurements, MIL-STD-461 comprehensively defines default requirements. It also provides standardized test methodologies, which is consistent among the various requirements. There are setup conditions which are common to all the tests such as groundplane usage, electrical cable construction and routing, and power-line treatment. The standard defines maximum and typical values of test limits, especially for susceptibility testing, according to physical location of the equipment, e.g., as the equipment installed inside the avionics bay will have the shielding from the fuselage, it can be tested at typical values. The applicability of tests is described in Table V of the standard and is titled as the requirement matrix. There are a total of 19 tests, 3 for CE, 10 for CS, and 3 each for RE and susceptibility. As scope of this book is aerospace vehicles extract of "requirement matrix" governing aircraft combined with brief description of each test is presented in Table 10.2. The table indicates what tests are applicable to which type of equipment. The discussion in this section is with reference to MIL-STD-461 G. The major change in this version is addition of two new tests, CS117 and CS118, and removal of test CS106.

All tests applicable for aircraft in the requirement matrix are to be conducted on all the modules of FCS; it is important to mention that the modules are tested along with their associated wiring. The required test limits for test equipment, setup, test procedures, and data presentation are covered in the write up for each test. However, when FCS modules are being tested, there are certain additional factors that have to be taken into account. These are elaborated next.

In tests such as CS114, RS103, the value or limit of the injected current or applied field strength will be what is defined for safety-critical systems. For example, the description of the CS114 test states the test limits for various platforms in Table VI of MIL-STD-461 G. In turn, the table refers to figure for CS114 calibration limits (frequency vs. current plot) to select the appropriate curve. For safety-critical systems, curve 5 with maximum limits is selected.

All modules of FCS that are identified to be directly influencing flight safety will have to demonstrate a 6 dB margin of safety from the limit specified in the standard. The rationale for this will be covered in discussion on MIL-STD-464, and the criterion for selecting the safety-critical modules is covered in test methodology. The tests for which "procurement activity" has to specify the test

Table 10.2 Requirements of MIL-STD-461 G and their applicability to aircraft

Requirement	Description	Army aircraft including flight line	Navy aircraft	Aircraft Air Force
CE101	CE, power leads	A	L	
CE102	CE, power leads	A	A	A
CE106	CE, antenna terminal	L	L	L
CS101	CS, power leads	A	A	A
CS103	CS, antenna port, intermodulation	S	S	S
CS104	CS, antenna port, rejection of undesired signals	S	S	S
CS105	CS, antenna port, cross-modulation	S	S	S
CS109	CS, structure current			
CS114	CS, bulk cable injection	A	A	A
CS115	CS, bulk cable injection, impulse excitation	A	A	A
CS116	CS, damped sinusoidal transients, cables and power leads	A	A	A
CS117	CS, lightning-induced transients, cables and power leads	L	L	L
CS118	CS, personnel borne ESD	A	A	A
RE101	RE, magnetic Fd	A	L	
RE102	RE, electric Fd	A	A	A
RE103	RE, antenna spurious and harmonic outputs	L	L	L
RS101	RS magnetic Fd	A	S	
RS103	RS, electric Fd	A	A	A
RS105	RS, transient electromagnetic Fd	L	L	

A: applicable; L: limited as specified in the individual sections of this standard; S: procuring activity must be specified in procurement documentation.

procedure and limits have to be in accordance with the expected EME with margins added to cater for some specific worst-case situations. But care should be taken so that these limits do not make the design too complex and cause the overall cost to increase significantly.

The definition of some of these tests that are applicable on aircraft equipment, such as CE106 and RE103, refers to antenna or antenna port of the equipment.

Core FCS modules generally will not fall into this category but FMS navigational equipment has to be tested accordingly.

CS117 has introduced lightning-induced transients with limited applicability to all aircraft-safety-critical equipment and interconnecting cables. This can also include non-safety-critical equipment with interconnecting cables/electrical interfaces that are part of or connected to equipment performing safety-critical functions, which in this case will be FCS. Test methodology is similar to RTCA DO160 cable bundle including multiple stroke (MS) waveforms 1 through 5, as well as multiple burst (MB) waveforms 3 at 1 MHz and 10 MHz and waveform 6. CS117 applies the single stroke (SS) and MS waveforms as a combined test. This is a very important test for FCS modules; however, as it is derived from Section 22 of RTCA DO 160, the specifics of test methodology are discussed there.

10.6.1.2 RTCA/DO-160

DO-160 G covers standard procedures and environmental test criteria for testing airborne electrical and electronic equipment (avionics). The tests specified in DO-160 G are typically performed to meet FAA or other international regulations covering electrical or electronic equipment that is installed on commercial aircraft. The tests and test levels/limits (also referred to as "equipment categories") found in DO-160 G are applicable to virtually every type of aircraft in use today, including small general aviation aircraft, business jets, helicopters, regional jets, and large passenger aircraft. The document includes 26 sections and three appendices, but Sections 15 through 23 and Section 25 cover EMC. Examples of tests covered in DO-160 G are temperature, altitude, vibration, sand/dust, power input, radio frequency susceptibility, lightning, and ESD. RTCA/DO-160 G and its European twin, EUROCAE/ED-14 G, are identically worded.

Section 15: Magnetic Effect This "MC" (for "magnetic compatibility," as opposed to "EMC" for electromagnetic compatibility) test is performed to determine how much the equipment under test (EUT) will deflect a compass needle, or effect the indication from a magnetic field sensor, also known as a "flux gate." It does not have any separate category for safety-critical systems.

Section 16: Power Input Although an argument can be made that "power input" (or "power quality" as they are referred to in other standards) tests are not truly EMC tests, they are included here for two reasons:

1. Power input/quality tests are often performed in the EMC lab by the EMC test personnel.
2. In the latest versions of DO-160, the frequency ranges for some of the tests fall well within the domain of typical EMC tests, and the test equipment used is similar to many other "true" EMC tests found elsewhere in DO-160 and other EMC standards.

Section 16 is the only section of DO-160 that contains requirements and tests that cover both the susceptibility and the generation of EMI from the EUT.

Section 17: Voltage Spike This test determines whether the EUT can operate as required during and/or after voltage spikes are applied to the AC and/or DC power input(s). Any method of generating the spike may be used, provided that the pulse produced has a duration of at least 10 ms, a rise-time of less than 2 ms, and a source impedance of 50 ohms. A minimum of 50 voltage spikes are applied within 1 minute. The FCS will be tested as category "A" system. This test is very similar to MIL-STD-461 test method CS106; however, the applicability of CS106 to aircraft was omitted in version "F" of MIL-STD-461, and in the latest revision, i.e., MIL-STD-461 G, this test has totally been deleted. Note that characteristics of both AC and DC power to be fed to the aircraft are specified in detail in MIL-STD-704, at present the latest revision is version "F" [45]; if the supplied power complies with the requirements of MIL-STD-704, EMI due to transients should not be an issue.

Section 18: Audio Frequency Conducted Susceptibility—Power Inputs This test is performed to determine that the EUT will operate as specified when audio frequency interference is applied to the AC and/or DC power input. The test setup and procedure are nearly identical to the MIL-STD 461F test method CS101, with the only difference being the actual test level and frequency range. The audio frequency interference is transformer coupled onto each power input lead, and the peak-to-peak voltage level of the interference signal is measured across the power input and return leads. Test levels are up to 8% of the nominal AC input voltage, and the frequency range is as broad as 10 Hz to 150 kHz. FCS modules use different types of power input for their operation and are to be accordingly tested.

Section 19: Induced Signal Susceptibility The tests in this section are performed to determine that the EUT can operate as required when the equipment and interconnecting cables are subjected to audio frequency electric fields, magnetic fields, and transient voltage spikes. The test levels for the interconnecting cable tests are determined by the length of wire that is exposed to the radiating wire. Test category is defined by a combination of three letters: the first letter defines the test level. For FCS, category "C" is applicable.

Section 20: Radio Frequency Susceptibility These tests are performed to determine that equipment will operate as specified when the EUT and its interconnecting cables are exposed to radio frequency interference. Continuous-wave (CW), square-wave AM (SW), and pulse-modulated (PM) RF signals are required, for both CS and RS. The test-level categories are formed from a combination of two letters: the first for CS and the second for RS. FCS will be tested for HIRF external field requirements; these categories are B, D, F, G, L, M, N, and O grouped into one. The category defines field strength as well as type of modulation, with category "L" being the most severe. The CS test is similar to CS114 and RS test is similar to RS103.

Section 21: Emission of Radio Frequency The tests in this section are performed to determine that the EUT does not emit radio frequency interference either conducted or radiated in excess of the specified limits. Conducted RF emissions appearing on interconnecting cables and power leads are measured. Radiated RF emissions from the EUT, interconnecting cables, and power leads are also measured. Testing of FCS will be carried out as per category "P" or "Q," depending upon the placement of the antenna on aircraft.

Section 22: Lightning-Induced Transient Susceptibility These tests determine whether the EUT can operate as specified during and/or after various lightning-induced transient waveforms are injected into connector pins, interconnecting cables, and power leads using pin injection, and/or cable bundle tests. The pin injection method is normally used to show damage tolerance, while the cable bundle tests are normally used to show upset tolerance. It is an important test for FCS EMC, as modules of FCS are located all over the airframe structure and associated wiring runs through complete aircraft. The waveforms to be injected are characterized by three pairs of characters. The first two letters depict waveform and test level. The second pair indicates single and MS waveform set and level. The last pair indicates MB waveform set and test level. The waveform sets are picked on the basis of whether cable bundles are shielded or not and the type of aircraft structure used. The waveforms can either be current test and voltage limit or voltage test with current limit. The type of structure, i.e., metal, composite, or mixture of both, will determine the type of coupling that is aperture coupling or aperture resistance coupling. This is explained in Section 22.3.1 and Table 22.1 of the standard. FCS test engineer will have to decide about the test waveforms as per the structure of aircraft, layout of FCS, and expected EME the aircraft will encounter during operations.

Section 23: Lightning Direct Effects The tests in this section are performed to determine the ability of externally mounted electrical and electronic equipment to withstand the direct effects of a severe lightning strike. The equipment will not normally be powered "ON" during the test, and these tests usually cause damage to the EUT. High-voltage and/or high-current tests at levels of thousands of kilovolts and/or hundreds of kilo-Amps are required. The test to be conducted are categorized by four characters: the first two indicate high-voltage test and the last two are for high-current test. High-voltage test is carried out to determine the attachment points for high-current test. The high-current test determines the tolerance of the equipment to lightning effect. The actuators of FCS that move the control surfaces can be candidates for this test depending upon how exactly these are mounted on the aircraft. Many of the FMS modules have external antennas, and lightning protection circuitry usually forms the first stage of the RF front end. These also need to be tested for lightning direct effects.

Section 25: Electrostatic Discharge This test determines whether the EUT can operate as specified during and after being subjected to an electrostatic air discharge event. The test procedure and test generator used are similar to most other international ESD standards, except that the EUT is bonded to the ground plane and only air discharge is specified. Test points are chosen based on their accessibility to personnel, with 10 positive and 10 negative polarity discharges at 15 kV applied to each one.

10.6.1.3 Comparison of RTCA/DO-160 and MIL-STD-461 Standards

A broad comparison of RTCA/DO-160 and MIL-STD-461 is summarized in Table 10.3; in line with the scope of this chapter for MIL-STD-461, only aircraft are considered. This indicates that for some of the tests, equipment qualified under some different standard can be considered "pass" in other standard as well. Similar comparisons can be worked out between other standards as well [46].

Table 10.3 Summary of comparison of RTCA/DO-160 and MIL-STD-461 G standards

			Parameters				
S.No	MIL-STD-461 G Test	DO-160 Section	Applicable equipment[1]	Frequency range	Methodology	Limits	Risk[2] Army/Navy/ Air Force
1	CE102[3]	16	Comp	Partial	Comp	Comp	L/M/L
2	CS101	18	Comp	Comp	Similar	Partial	M/M/M
3	CS114	20	Comp	Partial	Similar	Similar	N/L/L
4	RS103	20	Similar	Partial	Similar	Comp	M/M/L
5	RE102	21	Comp	Partial	Similar	Comp	M/M/L
6	CS117	22	Comp	Comp	Similar	Comp	L/L/L

Partial: significant variations; Comp: (Comparable) some differences; Similar: Almost same; L: Low; M: Moderate; H: High; N: Not applicable.

Notes:

1. Generally, MIL-STD-461 is applicable to a wider range of equipment, and DO-160 covers airborne equipment.
2. For risk analysis, only aircraft are considered and not other types of equipment.
3. As explained in the text, Section 16 is not actually an EMC test but covers input power requirements and as such should be seen in conjunction with MIL-STD-704.

Table 10.3 has been presented with the focus that MIL-STD-461-qualified equipment is to be used on commercial aircraft. If RTCA DO-160-qualified equipment is to be installed on military platform, the remaining applicable tests of MIL-STD-461 will have to be performed. This is a qualitative comparison for quick overview, the details of which are comprehensively covered in [46].

Besides these standards, the governing aviation authorities of a country may issue regulations for determining verifying aircraft performance, such as FAA issues, Code of Federal Regulation (CFR), and Advisory Circulars (AC). The purpose of stating the essence of these standards was to highlight the extent of testing airborne equipment has to undergo before being installed on the aircraft.

10.6.1.4 MIL-STD-464

MIL-STD-464 provides guidelines for system-level EMC testing and does not have an equivalent in civil aviation. Unlike the EMC standards used for module-level testing, it does not spell out the exact test procedures but general guidelines and refers to other applicable standards [47]. The user translates these guidelines into system-level EMC test procedures. In this section, the focus is on the requirements that are relevant for EMC testing of FCS.

This standard contains two sections, the main body and an appendix. The main body of the standard specifies a baseline set of requirements. The appendix portion provides the rationale, guidance, and lessons learned for each requirement to enable the procuring activity to tailor the baseline requirements for a particular application. The appendix also permits government and industry personnel to understand the purpose of the requirements and potential verification methodology for a design. This standard establishes EME effects (E3) interface requirements, and verification criteria for airborne, sea, space, and ground systems, including associated ordnance and is applicable for complete systems, both new and modified.

This standard defines the external EME that will be encountered by various types of aircraft, including UAVs, as per their operational role. This is important for all design engineers to be familiar with expected EME at the inception of design work, so that system specifications can be accordingly defined. External EMI protection is then implemented to control the internal environment (including margin considerations) to acceptable levels. The equipment immunity levels must be above the internal environments by necessary margins to account for criticality of the equipment. Although the test requirements in MIL-STD-461 applicable for specific types of external environments, but they may need to be revised in accordance with what will actually be the internal EME induced by external EME in the particular design. The system E3 design must be viable throughout the system life cycle.

From the perspective of EMC testing of FCS, it is important to note that this standard requires that design margins shall be established based on system criticality, hardware tolerances, and uncertainties involved in the verification of system-level design requirements. The term "safety critical" is used for systems or subsystems whose failure or malfunction can cause death or serious injury to personnel. FCS is the most safety-critical system of the aircraft. A minimum margin of 6 dBs is prescribed for safety- and mission-critical systems. The verification of the system-level design process must include normal in-service operation, checkout, storage, transportation, handling, packaging, loading, unloading, launch, and the normal operating procedures associated with each aspect. Margins are not meant to increase the basic defined levels for various EME. The preferred methodology is to verify that electromagnetic fields induced internal to the system by external environments are below equipment immunity by at least the prescribed margin. The size of the assigned margin is dependent upon inherent accuracy of the verification method employed. This can be better understood by considering a test case cited in the standard itself, where lightning protection is to be verified. One method can be to expose an aircraft to lightning like flash (with worst-case attachment points). Such realistic method of verification will require a smaller overall margin. However, if low-level pulsed or CW testing with extrapolation of measured induced levels is used for verification, this approach will typically require an overall margin of 6 dB for EMC qualification of the system. Similar margins may be appropriate for purely analytical approaches that have been proved to yield accurate results for a particular type of system being evaluated. The least accurate verification method is the use of an analysis which has not been previously verified as yielding "accurate" results for the system type of interest. For analysis, techniques supported in appropriate literature but not verified for the system being evaluated margins as large as 30 dB are not unrealistic. However, it is important to note that increasing the level of margins possibly means increase in cost, design complexity, and often weight penalty. Therefore, the requirement of margins has to be very meticulously worked out by the system engineers and the EMC engineer.

The standard requires that margins need to be demonstrated wherever they are applicable. This can be achieved by test, analysis, or a combination thereof. Instrumentation used for testing of margins shall capture the maximum system response and shall not adversely affect the normal operation of the system. If environment simulations below specified levels are used, instrumentation responses may be extrapolated to the full environment where the system has linear responses. Analysis may be acceptable under some conditions; however, the required margins will typically be larger.

It is important to mention that there may be a misconception that if all the modules of a system have been EMC tested according to MIL-STD-461, then there are sufficient margins between emission and susceptibility test levels, thus intrasystem EMI problems shall not occur. This will hold good if the coupling between the modules is one to one, e.g., wire-to-wire coupling. However, if the coupling takes place through an antenna interface, the susceptibility level will be much lower. Thus, margins will have to be established for all possible coupling paths.

10.7 EMC Test Methodologies of FCS

EMC test standards reviewed in the last section provide the basic principles and methodologies for EMC verification of aircraft systems. Safety-critical systems are categorized separately in EMC standards and their qualification criteria are more stringent. FCS being essential to keeping an aircraft airborne has to be tested at levels specified for safety-critical systems. FCS interacts with many other systems, its modules are installed at dispersed locations on the airframe structure, and the associated wiring is spread practically all over the aircraft. It is important to realize that EMC testing of FCS is a complex undertaking and requires very close coordination between the system engineer and the EMC engineer. Proper EMC testing and qualification is the main step toward EMI mitigation in FCS; however, there are other measures that need to be taken care of during system installation, operations, and follow-up maintenance that need to be adopted to avoid EMI effects. The complete EMC test process is explained in this section.

The EMC system-design process for safety-critical systems is quite similar for military and civil aircraft. The process described here is basically that of civil aircraft, mainly described in Society of Automotive Engineers, Aerospace Group ARP 5583 [48], specific differences in EMC test process of military aircraft are elaborated. A simplified process flow chart is depicted in Figure 10.9.

The EMC system-design process will start with categorization of flight critical and non-flight-critical systems. FCS is essential to safe operations of aircraft. Its disruption can be catastrophic and can result in loss of lives, as per definition in Mil-Std-464 and FAA Advisory Circular AC 20-158A [49], it falls in the category of safety-critical or level "A" systems, respectively. As this chapter focuses on FCS, the following steps will be with reference to level "A" systems, more specifically, the FCS. Next FCS modules whose functioning is actually critical to the operation of the system will be identified. It is important to note that all modules of a safety-critical system may not be safety critical, e.g., a monitoring or data recording system, and some particular module of a noncritical system that provides some input to safety-critical system can be categorized as safety critical. The approach of playing safe and declaring a number of modules as safety critical will have cost penalties in terms of increasing design complexity and more stringent test requirements.

The next step in EMC system engineering approach will be to identify the EME an aircraft is likely to experience in its intended role. For military aircraft, MIL-STD-464 provides the requisite data. However, the limits can be altered

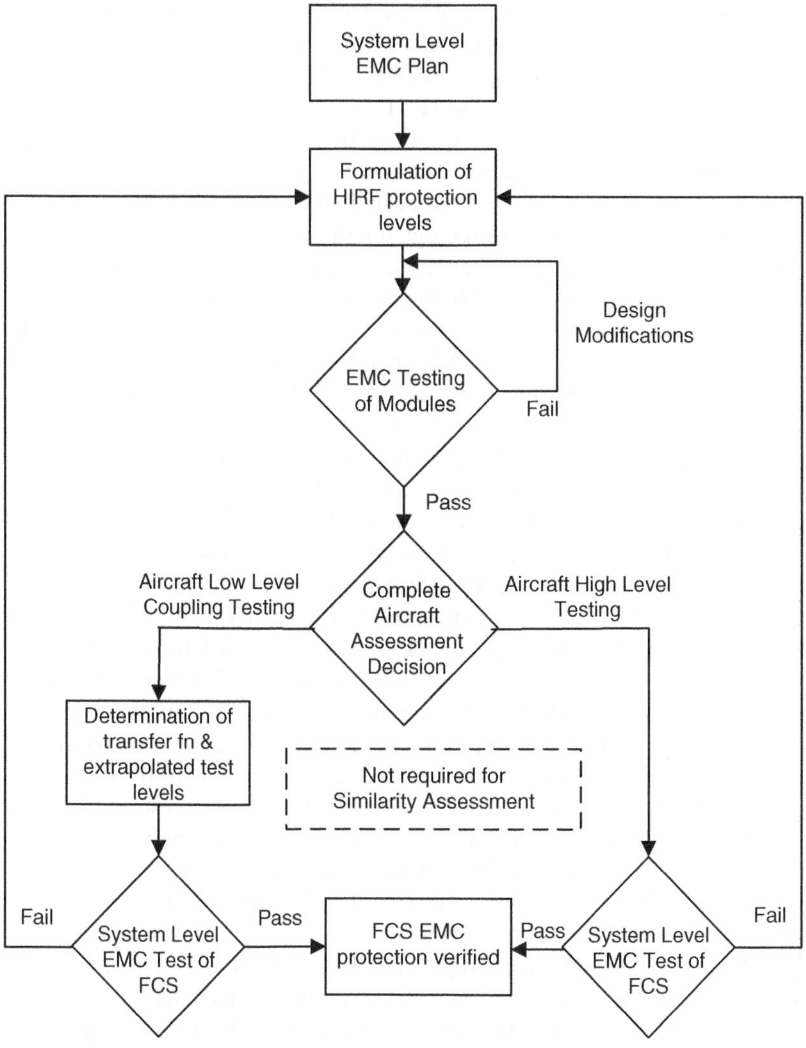

Figure 10.9 Simplified process flow chart for system engineering approach to EMC testing of FCS, derived from test procedures for safety-critical systems.

in accordance with specific requirements. For civilian aircraft, HIRF was recognized as a serious concern when these aircraft started utilizing elements of the FBW system. The EME was empirically studied for thousands of transmitters worldwide and EMC test levels were determined. It is important to note that the HIRF requirements are defined as per the criticality of function, which then translates into the performance levels for system implementing the function. The HIRF requirements are basically defined in CFR 23.1308 Appendix J and CFR 25.1317 Appendices L, D and E, CFR 27.1317 and CFR 29.1317. CFR 27 and 29, derived from [48, 49], are applicable to rotorcraft HIRF and lightning protection. The HIRF environment is tabulated in three categories: general test levels for fixed wing aircraft, test levels for fixed wing aircraft in the proximity of an airfield, and test levels for helicopters; these are higher than fixed-wing aircraft because helicopters have the liberty to fly into any type of area. The requirement for safety-critical systems like FCS, as spelled out in these CFR, is that these systems should not be affected by HIRF and if affected it should be able to automatically recover within a specific time so the functionality of the system is not compromised. The requirements of these CFR are consolidated in Advisory Circular AC 20-158 A. For civil aircraft, the most stringent test limits are provided in lightning direct effects, Section 23, of RTCA DO-160 for modules installed external to the aircraft and for rest of the modules, the limits are defined in Sections 19 and 22.

Safety-critical systems will be required to qualify the EMC tests by considerable margins and not on the edge of the limits. Going by the basic requirement of MIL-STD-464, a margin of at least 6 dBs has to be added to test levels of all safety-critical systems, i.e., doubling the susceptibility limits and reducing the emissions by half. At this stage, the EMC engineer will be required to determine analytically or through simulations that what shall be the acceptable limits, inclusive of margins, for each module and system. This is done by taking into account the susceptibility and emission levels of all the modules, physical location of all the modules, type, and routing of associated wires, coupling paths, uncertainties in EMC test methods etc. An overview of these analytical/simulation methods is presented in Section 10.9. After these EME levels and additional margins have been worked out, the system engineer finalizes and documents the EMI protection requirements that have to be incorporated in each module. When a module is designed and produced, the actual EMC testing phase starts. This is explained next.

The Society of Automotive Engineers (SAE) Aerospace Recommended Practice ARP5883 Rev-A [48] and AC 20-158A [49] distinctly define categories of modules, integrated systems, and complete aircraft. In military EMC standard, the integrated system tests category is not that clearly defined either in MIL-STD-461 or MIL-STD-464. Integrated system testing can either be carried out as a standalone system inside a chamber or while it is installed inside the aircraft, at HIRF levels determined by complete aircraft testing. For the first system, complete testing is carried out; for serial production and minor modifications, similarity analysis for aircraft assessment can be used. This is comprehensively explained in [48]. For the testing of complete aircraft, two methods can be used, namely, high-level testing and low-level coupling tests (LLCT). These are broadly described in the subsection on testing of complete aircraft and references are provided for readers interested in greater details.

10.7.1 EMC Testing of Modules

Early EMC standards for aircraft equipment only catered for emission testing that was basically meant to protect HF communication equipment. The original test limits were derived from the measurement of in-flight emissions on aircraft. Over the years, these limits have been revised, generally upward, as per acquired experience. The frequency range has extended in accordance with the band width of onboard electronics equipment. The immunity testing was introduced when flight-critical systems became more electrical. The immunity test requirements have increased from 1 V/m to 7 kV/m in some frequency bands. At the same time, high-sensitivity receiver front ends have resulted in more stringent emission limits. Consequently, the prevalent limits for EMC standards for airborne equipment are much more demanding than that those of yester years.

The first phase of EMC testing for aircraft systems is at module or subsystem level in test chambers. These are tested against limits set by governing standard. If the test is not cleared, changes are made in the module design or requisite EMI protections are added. Usual failures and remedies are not in the scope of this chapter; however, the subject has been dealt in detail in references [4, 20, 21, 44]. The pass/fail criterion for emission testing is very explicit, and the emission level against each tested frequency is plotted. These are in the form of spectral spikes; the allowable limit for each frequency in the form of a solid line is superimposed on the frequency spectrum plot. For non-safety-critical systems, anywhere the spike crosses this line it is taken as a failure (Figure 10.10). In case of FCS modules, the emission levels have to be below the limit line by prescribed margin. For immunity testing, the "fail" criterion is generally taken to be a fault condition occurring in equipment self-test, e.g.,

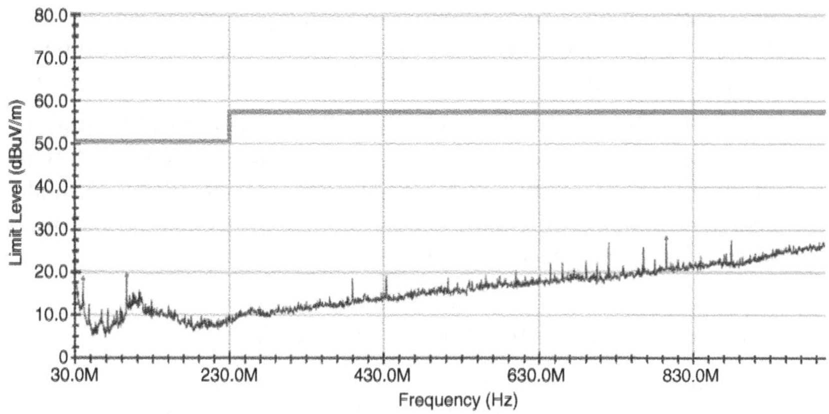

Figure 10.10 Product emissions plotted along with limits of EMC standard (gray line).

in case of data transmit/receive equipment, it can be specified bit error rate (BER) being exceeded. In cases where the output is qualitative, e.g., audio tones or video being discernible, judgment of the intended user is taken as the decision. When the subsystem is being delivered by a contractor, the exact EMC requirements are stated in the initial contract. The contractor is required to provide the EMC test results along with the subsystem itself.

After all the modules/subsystems are installed in the aircraft, the second level of EMC testing takes place. During this phase of testing, all aircraft systems that can generate EMI or are susceptible to conducted and radiated EMI are sequentially operated in tandem. In case any degradation of performance in any subsystem is observed, source and victim systems are analyzed and EMI protections are accordingly added. This will normally comprise of the following steps:

1. **Formulation of Test Plan**. The test plan will list the systems to be tested as victims. Whether the system will be tested with external power supplies or engine running or both, position of switches, systems that will be switched on as sources of potential EMI, fail conditions, and test equipment required are explicitly stated.
2. **Conduct of Test**. The test will be conducted in accordance with the finalized plan. A table similar to one shown in Figure 10.11 will be printed and filled. Status of system status indicators for each system switched on will be recorded. This table usually forms part of the final report. For conduct of certain tests of FCS, some simulated inputs may have to be provided, like that for speed, altitude, etc. Some modules of FCS may not become functional if the aircraft is on ground, thus inputs of landing gear down may have to be disabled by appropriate selection of circuit breakers.
3. **Analysis of Results**. The results recorded in the form of tables are analyzed to check whether some abnormalities were observed in any of the systems. If some system displays susceptibilities to EMI, it is reviewed whether the problem is with design of the system or installation has not been properly done. Thus, EMI immunity is enhanced in the most cost-effective manner.

These tests do not expose the aircraft to external sources of EMI at levels specified by the standards, but are a good check for intrasystem compatibility. It also provides the confidence that during installation and integration, intermodule connections, bonding, grounding, routing of wires, protections for antenna-to-antenna coupling have been implemented satisfactorily.

10.7.2 Methods for EMC Testing of Complete Aircraft

Level "A" systems require both system-level testing and aircraft-level testing [50]. Two methods can be adopted for EMC testing of complete aircraft: high-level tests and LLCTs. In high-level tests, the aircraft is exposed to

ONBOARD EMC TESST SHEET

AIRCRAFT TYPE

AIRCRAFT CONFIGURATION

TEST EQUIPMENT

SUPPORT EQUIPMENT

SYSTEM PARAMETER VALUES

FAIL CRITERIA

S. No	System to be operated	System under check FCS					
		External Power Supply					
		Slats	Flaps	Elevators	Ailerons	Rudder	Cockpit Displays
1	Landing Gear	√	√	√	√	√	√
2	VHF Comm	√	√	√	√	Shuddering	√
:							
:							
7	ATC Transponder	√	√	√	√	√	√

Figure 10.11 Sheet depicting typical information on intrasystem EMC test document.

high-magnitude RF fields and the behavior of the system under test is observed for any disruptions. The LLCTs utilize low power RF fields to find out the aircraft internal EME or induced current in wiring looms. This is then extrapolated to expected HIRF levels and compared with the test levels used in module or system-level testing. This verifies whether testing has been carried out as per actual expected EME or not. Both tests have their advantages and disadvantages, which are well elaborated in [48, 49]. In contemporary times, LLCT are being more widely used. Besides being cost-effective, LLCT generate data that can be computationally extended to scenarios for which actual testing was not carried out. It is pertinent to mention that before the conduct of actual test, system calibration is always carried out.

High-level testing can be further bifurcated into two categories. One is high-level radiated tests (HLRT), where the aircraft is illuminated on all sides so that all aircraft apertures are exposed to RF energy. The aircraft must be in the far field of the antenna, and this may require multiple antenna positions to illuminate the aircraft adequately. The second is high-level direct drive (HLDD) test which injects high-level currents directly onto the aircraft. The HLDD test is usually carried out at frequencies below the first airframe resonance. The amplitude of RF current levels injected on the aircraft during the HLDD test are normally determined using analytical methods and 3D modeling and simulation.

The LLCT can be of three types, namely, low-level swept current (LLSC), low-level swept field (LLSF), and low-level direct drive (LLDD). In the LLSC method, transfer function that relates the external RF fields to currents flowing in wire bundles associated with aircraft equipment is measured. The LLSF method is used to measure the transfer function relating external RF fields to RF fields inside the airframe. These tests are conducted for all possible aircraft orientations, configurations, and polarizations as necessary, to produce transfer functions for the various wire bundles being measured or for the equipment bay receiving RF energy (LLSF). These transfer functions are then optimized to predict the internal wire-bundle current or aircraft internal EME. LLSC is depicted in Figure 10.12.

LLDD relates the wire-bundle current to aircraft skin current and in turn relates aircraft skin current to external EME. Below 400 MHz, LLDD and LLSC tests assess the internal bulk currents on wiring, referenced to a normalized 1 V/m external electric field. Above 400 MHz, the LLSF test assesses the internal electric fields at the location of the equipment, also referenced to a normalized, 1 V/m external electric field. The internal current and electric field data are extrapolated to specified external EME. This is then compared with the test levels, for both currents and field strengths, used for EMC testing of modules or systems. A functional block diagram for conduct of tests is shown in Figure 10.13.

Between 100 and 400 MHz, the significant coupling mechanism is complex. For large aircraft, comparison of laboratory and airplane E-fields is more appropriate. For smaller airplanes and UAVs, with short wire lengths, airplane currents produce dominant coupling mode.

10.7.2.1 System-Level EMC Testing of FCS

For a complex system like FCS, which is spatially dispersed in the fuselage and has long wiring looms attached with majority of its modules, the methodology for system-level testing is very similar to complete aircraft testing. Except for

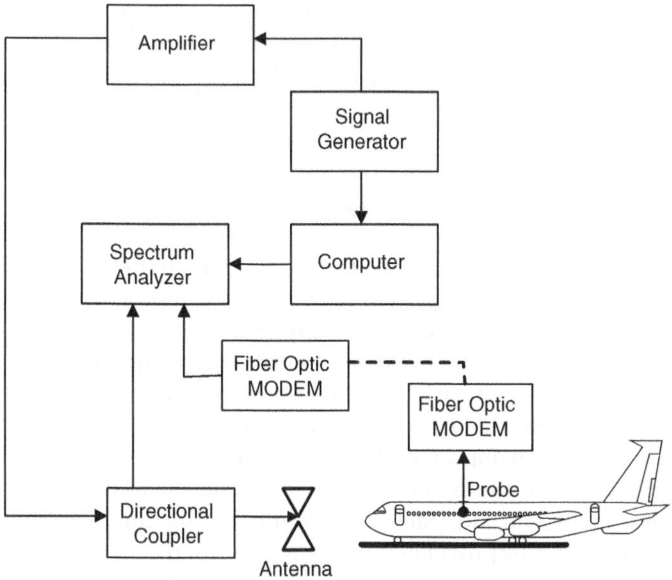

Figure 10.12 Block diagram of LLSC test setup. In the actual setup, a number of probes will be placed to measure currents in aircraft wiring looms. In LLSF, the current probes will be replaced by RF probes that will measure electric field strength.

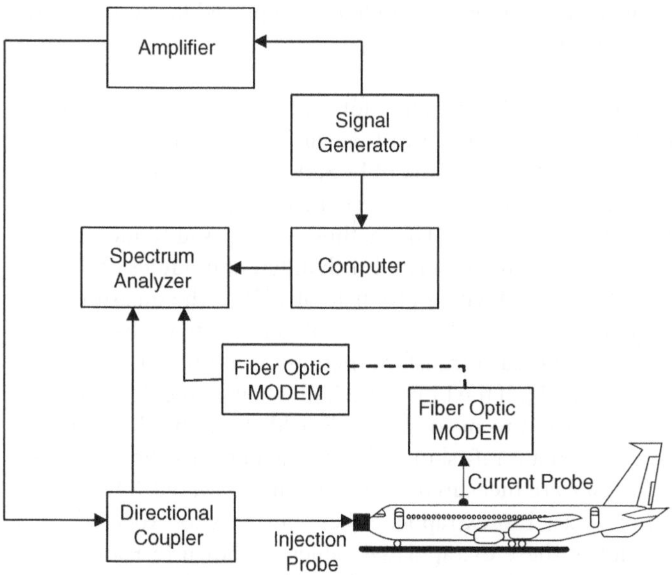

Figure 10.13 Block diagram of the LLDD setup. In the actual setup, a number of probes will be installed on fuselage to measure skin current and its distribution.

some small aircraft such as UAVs, it is generally not possible to set up the FCS in the EMC test chamber exactly as it is installed in the aircraft. The test standards make it mandatory that for system-level EMC testing inside the chamber, the system should be laid out exactly as it is installed in the aircraft. This subsection focuses on the actual conduct of EMC testing of FCS and highlights some aspects of overall EMC plan that influence this testing. How system-level testing is carried out will depend upon which aircraft-level-assessment methodology is used, i.e., LLCT or high-level testing. Regardless of the approach, it is very important that the failure conditions due to exposure to the HIRF test levels be carefully analyzed. For the system to "pass" EMC testing, it has to be ascertained that HIRF testing did not induce any adverse effect (which also include system failure effects) to the aircraft functional performance. For FCS, this will require that disruption in any system that provides any input to FCS does not cause system failure. This aspect is explained in greater detail in testing of autopilot mode. It is also important that the redundant elements that are added to diminish chances of failure should also be tested during EMC testing. Moreover, the time required for the system to recover from a disruption due to HIRF, either automatically or by providing an annunciation for the pilot, is to be analyzed in terms of whether the aircraft is able to maintain safe flight for that long a period of time or not. For FCS, this time is generally very small, and there are instances where a system disruption of fraction of a second can be catastrophic, e.g., where an aircraft is making an automated landing and ILS starts experiencing serious interference. The manner in which system-level testing is carried out vis-à-vis the assessment methodology employed for complete aircraft test is described next.

Case I: Low-Level Coupling Test When some form of LLCT is used for aircraft HIRF assessment, the system-level FCS testing will be performed as described in the text to follow. Due to the criticality of FCS, the testing cannot be carried out on general transfer functions, except for cockpit displays.

Below 400 MHz, either LLSC or LLDD method will be used to measure the currents setup in the wire looms associated with different modules of the FCS when the aircraft is illuminated with an RF field of 1 V/m. The measured currents over the frequency band are then extrapolated to relate to the expected HIRF environment. The standard transfer functions as a function of fuselage length are available in [48,49]. If the LLDD method is being used, first the fuselage skin currents are related to normalized RF field strength. These current levels are then injected on aircraft skin and then the currents they set up on looms inside the fuselage are measured. The FCS is not powered "ON." Again, the transfer function is used to extrapolate these currents to expected HIRF levels. The actual plot of these extrapolated values as shown in Figure 10.14 indicates very sharp peaks and roll offs, which will be difficult to generate. So an envelope fit is carried out for generating the desired signals. The extrapolated

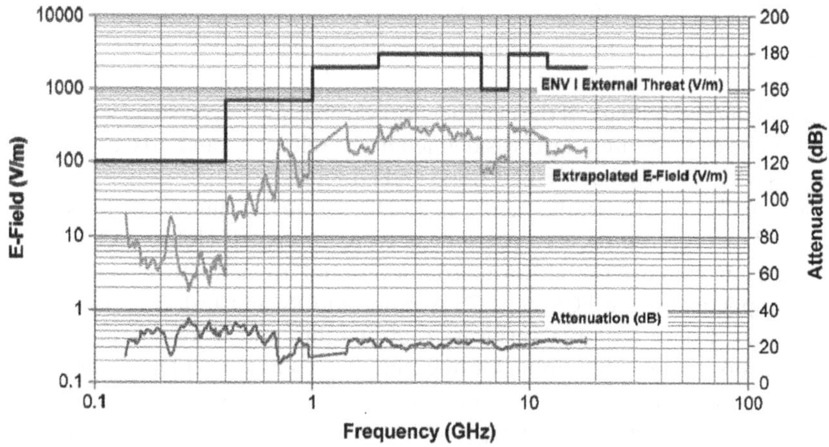

Figure 10.14 An example of transfer function curves for LLSF testing. From the enveloped curve of external HIRF environment, internal RF field strengths are computed as a function of measured fuselage attenuation.

current levels are then injected on to the looms attached to FCS modules and they are monitored for any failure conditions to occur. To determine the margin available between failure condition and currents induced due to expected HIRF, the current levels are increased to a point where failure occurs. Second, when failure condition occurs, EMI is then removed to find out the time required by the module or system to start functioning normally. The most common method of injecting currents on to the wire looms is bulk current injection. If accessibility of looms is a problem, another method to use is direct current injection (DCI), which injects currents on to fuselage skin. However, determining the exact current distributions on the skin requires some complex computations.

Above 400 MHz, the LLSF method is employed to determine the field strength inside the fuselage, with normalized external field. This determines the aircraft attenuation to external fields. The system-RS field strength (E_{RS}) is then computed using the formula

$$E_{RS} = 10^{\left(\frac{-A_{LLSF}(dB)}{20}\right)} \times E_{HIRF}$$

where A_{LLSF} is aircraft LLSF attenuation in dBs and E_{HIRF} is the selected external HIRF environment in V/m.

The FCS modules must have been tested for RS at extrapolated levels, the plot for extrapolated level computed as a function of external HIRF level and fuselage attenuation is shown in Figure 10.14. For the system-level test of FCS inside the aircraft, such field levels have to be generated and applied in various antenna configurations. If such high-power amplifiers and high-gain antennas are not available, the distance can be reduced between the antenna and the aircraft; however, far-field conditions must be maintained.

Case II: High-Level Testing If the facilities are available and FCS system-level testing is to be carried out at levels as defined in the relevant standards, two types of tests will be carried out: HLDD tests and HLRTs.

Principally, HLDD is the same as LLDD; however, due to high power involved, a cage return wire network is to be designed. Unlike LLDD, the injected currents are not at normalized levels and are actually meant to test the compliance of the FCS. So FCS is powered ON and tested in all possible modes. Failure criteria, verification of margins, and recovery time from EMI disruptions are determined in the same manner as in LLDD. For power ON, ground power unit supplies the power and aircraft tires have to be insulated from the ground. HLDD is conducted up to first aircraft resonance frequency.

For RS testing of FCS, the HLRT method is employed; the aircraft is suspended in the air with its landing gears retracted and it is exposed to emissions as per selected HIRF levels. It is important to note that in HLRT, testing is carried out in discrete frequency steps. The minimum number of frequency steps in a particular range of frequencies, type of modulation to be used, and dwell times are defined in [48]. This is not explicitly available in MIL-STD-464. The aircraft must be illuminated from positions all around the aircraft so that no section of fuselage is left out. The number of positions will vary with the size of the aircraft. Complete FCS performance is checked against all positions. During this high-power testing, safety measures related to adding filters to protect avionics equipment other than FCS from high-power EM energy, personnel health, and eliminating fire hazard by ensuring that no fuel is present in the aircraft are to be adopted.

10.7.2.2 Monitoring of EMC System During EMC Testing

The monitoring of FCS during EMC testing is described here, which is applicable to any of the test methodologies explained earlier. Testing of FCS involves inputs from a number of other systems, and its failure criteria are best defined by the aerodynamics experts. Thus, system-level test plan and failure conditions of FCS are to be worked out at an initial stage of the project by the system engineer. To carry out the EMC assessment, the aircraft must be configured so that the operation of FCS represents actual flight conditions as much as possible. The control loop gain can be set to values where they are most sensitive to interference. In most cases, this is accomplished by means of providing simulated inputs for critical flight parameters. The highest loop gain is at a relative low speed and high altitude. Generally, interference in feedback loop can be most disruptive [51]. On the other hand, in terms of flight safety the most critical time is when aircraft is at a lower altitude, like final phases of landing, and close to peripheries of aerodynamic performance curves. In these instances, the aircraft should be able to recover from any disruptive effects of EMI within fraction of seconds.

To check the operation of the FCS, it should be in power ON condition. For reasons of personnel safety, all switching and monitoring actions are to be remotely carried out. In digital avionics systems, all modules are controlled through avionics data buses, and the same are used to convey the status of the modules to flight control and maintenance computers. During EMC testing, primarily data bus monitors are used to analyze the behavior of different modules. However, all these remote devices should be connected through fiber optic links and positioned well outside the region of influence of RF fields. If remote monitoring of all the systems is difficult to accomplish flight data recorders (FDR), a regulatory essential requirement on all aircraft, and maintenance computers can be used for the purpose; however, the EMC qualification of these systems must have been carried out prior to be made part of FCS testing. The built-in self-tests of the modules can also be remotely initiated to monitor its performance. In addition to monitoring of digital data, the physical movement of the control surfaces is also recorded. This can be measured by angular displacement sensors installed alongside flight control surfaces, and their movement is visually checked by CCTV cameras [52]. Consultation with aerodynamics and system engineers is required to determine the acceptable surface movements.

10.7.2.3 EMC Testing of Autopilots

EMC testing of autopilot modes can be considered as an extension to system-level testing of FCS. It is important that systems that provide input to the FCS are assessed sequentially, so that their resistance to EMI can be individually assessed. For example, if the autopilot mode is for autolanding, radar altimeter, inertial navigation system, ILS, etc., will have to be integrated with the system. In all these modes, some of the inputs will have to be simulated, e.g., in autolanding, ILS inputs will be simulated and fed to FCC, bypassing ILS RF front end. When providing external stimulus for EMC testing, it is to be ascertained that these external systems used are either outside of the area where high field strengths are present or qualified for same EMC standard as the aircraft. Monitoring of system responses is done as described above. In autopilot modes, the FMS system is also checked where the output is also monitored to check drifts in navigational data.

The final step in EMC testing of aircraft and FCS is the actual flight test. This is monitored by the flight test engineer [53], who observes the response of the system in varying flight profiles and checks for any fault indication coming "ON" when other systems are being operated.

10.7.2.4 Limitation of Ground Testing

The fact remains that ground testing of aircraft can only be partially realistic. As the aircraft is not moving in any of the axes, there is no aerodynamic feedback;

thus, no precession of any of the rate gyros takes place. However, straight and level flight can be well modeled, and any disturbance to the control surfaces is a measure of the FCS susceptibly. Moreover, greater part of the tests cannot be performed with engines running. Running engines are a significant factor in aircraft internal EME, especially with regard to magnetic coupling.

10.7.3 Test Facilities for Complete Aircraft

The complete aircraft may be tested in either of the following types of facilities, that are applicable for EMC testing of other equipment as well but for aircraft the size has to be very large:

(i) **Open-Area Test Sites (OATS)**: An OATS in the true sense may not be available for aircraft testing, as in today's world of wireless communications an area accessible to aircraft yet free of electromagnetic signals is hard to manage. Radiating high fields in open area over a wide range of frequencies can be disruptive for other spectrum users and may lead to legal issues. Exposing under belly area to required field strength may not be possible. Below 200 MHz, one solution to this problem is use of DCI techniques. Where instead of radiating very high power levels, equivalent current is injected into aircraft airframe; the aircraft is positioned on a conducting surface to provide a return path. Though not suitable for qualification testing as per standards, primary testing for diagnostics and compatibility testing required after aircraft modifications is carried out in relatively remote areas of airfield (Figure 10.15a). Also, this is the available option for engine running tests.

(ii) **Anechoic Chambers**: Other than the size, characteristics of the anechoic chamber are the same as that of EMC testing of other electronic equipment. There also has to be a mechanism for hoisting the aircraft few feet above the ground. This allows illumination of underbelly areas and retraction of landing gear. Testing can be carried out as per requirements of the EMC test standards. However, engine-running tests cannot be performed. Figure 10.15(b) shows Eurofighter being tested in a chamber.

(iii) **Reverberation Chamber**: A reverberation chamber is a specialized EMC test chamber generating very high field strength using modest amplifier power. The reflections between the RF-shielded walls are optimized to add up to maximum of as many standing waves as practically possible. A rotating paddle or other means of altering the geometry of the room is almost always used in a reverberation chamber for "mode-stirring" or "mode-tuning." Sometimes a frequency-modulated source is used to achieve "electronic mode-stirring." The chamber is mainly used for testing electrical immunity of test objects. Here, the aircraft is tested over almost all illumination angles and polarizations during one rotation of the tuner.

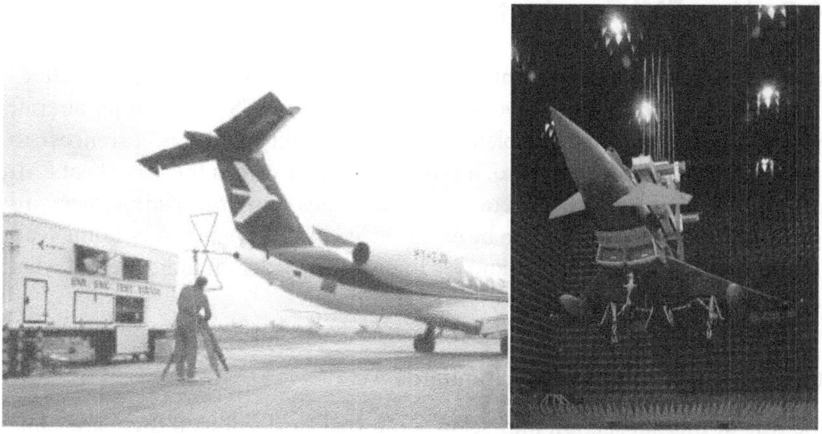

(a) (b)

Figure 10.15 (a) EMC testing of aircraft in open area. (b) EMC testing of eurofighter 2000 in an anechoic chamber. These are high-level radiated test methodologies.

Moreover, the size of the chamber proportionally varies with wavelength of test signal. So if minimum frequencies are low, size becomes large. However, larger chambers incur greater losses and yield less efficiency. Nonetheless, it is considered a viable option for generating high-intensity fields and research is ongoing to provide solutions for its shortcomings.

(iv) **Transverse Electromagnetic (TEM) Cells**: A TEM cell may be thought of as the expansion of a coaxial transmission line, the outer conductor is connected to a shielded cavity and the inner conductor to the end of a septum that is suspended in the cavity. The other end of the septum is terminated with a load. The space between the septum and outer cavity is made large enough to house a test object. Theoretically, it is a feasible option for low-frequency EMC testing of complete aircraft. However, as per known sources these are not being used in practice.

(v) **Gigahertz TEM (GTEM)**: The GTEM cell is a self-contained shielded enclosure designed for making both RS and RE EMC measurements. The GTEM's control of resonances and reflections allows precise signal levels, and the elimination of antennas provides faster test throughput. Electrically, the GTEM is an extension of a 50-ohm transmission line. The transmission line's outer-conductor flares and expands into a pyramidal form, becoming the GTEM. It is useful as both RE and RS measurements can be made. Relatively high fields can be created by applying moderate power. It is broadband, and generated fields are very uniform. It is transportable and only requires external power for operation. GTEM can be used to test modules of FCS.

10.8 How EMI Couples to FCS

After reviewing the requirements of EMC test standards and FCS test methodologies, it seems hard to believe that EMI can affect FCS or any other aircraft system. This section aims to explain why the probability of EMI occurrences can never be totally eliminated. Also, it is described that why in the process of EMI mitigation EMC standards and test methodologies cannot provide a safeguard against all types of EMI that can be experienced in actual situations. Some real-life occurrences are cited as examples to give the reader an indication of what can be expected during the life cycle of an aircraft.

10.8.1 Shortcomings in EMC Standards

The title of this subsection may seem somewhat misleading; if there are some shortcomings in the standards, why are they being followed and why are not the regulatory bodies taking some actions to remove these inadequacies? The shortcomings are with regard to what can be termed as engineering compromises, inability to enact all possible real-life situations, production timelines, and considerations of cost. These issues are elucidated in this subsection.

10.8.1.1 Immunity Testing Does Not Cover Simultaneous Multiple EM Disturbances

During immunity testing, the excitations are generated one at a time. Where as in real life, conducted and radiated disturbance may occur in tandem, a strong RF field and a transient burst can take place at the same time [54]. Although the probability of such an occurrence is low but it can occur. If the scenario has to be accommodated during EMC testing, the susceptibility test levels and conditions have to be further enhanced, which will cause massive increase in time and cost of EMC testing.

10.8.1.2 Real-Life Scenario Is Not Truly Replicated

In an anechoic chamber, there are no reflected waves. In real life, reflecting surfaces in proximity of the aircraft, like aircraft in a formation or in near earth flight profiles, can cause the EM waves to strike the airframe at multiple angles. Another aspect in this regard is that it is generally known that in addition to magnitude of the EMI its type of modulation can be important in causing disruption of signals. However, generally EMC testing only caters for amplitude modulation (AM). As already mentioned in the case of FCS, ground testing cannot be representative of all flight profiles; in high "g" maneuvers and final phases of landing etc. even a small deflection of control surface can have catastrophic effects.

The EMC tests related to transient voltages are carried out against ESD/P-Static, lightning or power-line inductive switching transients that are usually of very large magnitude (hundreds to thousands of volts). The amplitudes are kept at reduced levels so that permanent damage to circuit does not occur. Some inductive switching transients have ringing frequencies of 10 MHz reoccurring at 1 MHz repetition rate and lasting for around 1000 ps. During test, the transient is repeated every two seconds. (RTCA/DO-160 specifies 8–10 pulses per second for 10 seconds.) The overall duration of transient and its recurrence rate in comparison to the digital data bits and words being transmitted will determine the extent of data loss. Lightning transients have low frequencies and high ringing frequencies, for instance, at 1, 3, and 7 MHz, these are set up by the electrical resonant lengths of an aircraft and damped out in a few microseconds. But then they may reoccur again and again. Thus, it is not possible to test the system against every possible type of transient noise. A representative lightning curve is shown in Figure 10.17.

10.8.1.3 Representative Sample Testing

Although not recommended for flight-critical system like FCS there is a possibility that all the aircraft in a production lot may not undergo the same rigorous level as the first aircraft.

10.8.1.4 EMC Conformance during Life Cycle

In normal practice, the EMC testing of the aircraft is only carried out prior to initial roll out and after some major modifications/upgrades. In the latter case, partial EMC testing may be conducted, whereas functional performance of components and interconnects is perpetually degrading. This is covered in greater detail in last subsection.

After the aircraft is commissioned, other factor that comes into play is maintenance. If there is one clause of standards that is pushed to the background, it is section 5.10 of MIL-STDtd-464. It is about ensuring good electromagnetic compliance practices throughout the life cycle of the aircraft. During first-line maintenance, the maintenance crews are generally under pressure to get the aircraft back in the air as early as possible. In commercial aviation downtimes mean financial losses and in case of military aircraft delay of seconds can be the deciding factor between success and failure of a mission. Thus, in time-compressed situations, chaffing of wires and shielding enclosures may occur, and the connections of wires with connectors can be done in a manner not exactly conforming to good EMC practices. The effect of sustained vibrations and harsh environmental conditions also takes their toll. It is only at the time of periodic maintenance that these problems may be addressed. It may be noted that flight-critical systems such as FCS are generally required to meet the reliability levels of one fault in one billion flying hours.

10.8.2 Aircraft Design Philosophy

Though it cannot be denied that over the years EMI has caused many anomalies in functioning of the aircraft systems, some of which were catastrophic.

Yet, the fact of the matter is that EMI protection is one of the important but not the most important concern in aircraft design. Equipment location on the aircraft is usually settled on the basis of temperature characteristics, form factor, accessibility for maintenance, and shift in the center of gravity issues rather than EMC. Avionics equipment units are tested to requisite EMC standards and then along with their interconnecting wiring, installed directly in the aircraft. They generally perform properly. This method has been cost effective and acceptable in terms of flight safety. Thus, most of the time EMI protection has been reactionary, not anticipatory [7].

The crux of all this is that if it were possible to shield every electronic module, route every wire sufficiently away from possible sources of EMI and use of aircraft structure as ground could be omitted EMI problems could be made almost nonexistent. However, weight, volume, and cost constraints do not permit this. Thus, the modules installed in the aircraft are protected against EMI to the extent that they qualify as EMC standards and levels of interference expected in normal operations and no more. Nonetheless, there always exists a probability that the level or type of interference will penetrate the protections to disrupt the performance of electrical or electronic modules.

10.8.3 Specific EMI Protection Measures for FCS

Subsection 10.4 stated some of the elements of FCS; perhaps seemingly, the most insignificant element is most critical in compromising EMI protection and causing other faults [55], i.e., interconnecting wires. If the shielding of a wire is weakened, it acts as an antenna, either receiving or transmitting. The EMI susceptibility of any element is directly related to its exposed area. As the control surfaces are located away from the main fuselage, data buses of FCS and other wiring required for supplying power to modules run throughout the length and breadth of the aircraft. The specified type of wiring for avionics data buses usually has good EMI immunity, like MIL-STD-1553 and ARINC-429 [56] utilize twisted shielded pair. The data receiver input resistance and data transmitter output resistances are sufficiently high and low, respectively. As data being transmitted are digital, the bit low and high levels should have a sufficient range and a clear margin in between [28]. During EMC testing, bus monitors are used to check if data on the bus are corrupted; however, it needs to be verified that the change in data is due to EMI or other system fault [57].

At certain frequencies, the internal structure of fuselage may act as a resonant cavity, enhancing the magnitude of interference. The wires are also resonant at integer multiples of wavelengths associated with frequency of signal they are carrying. In other cases, radiation inside the fuselage can couple to antennas installed on the external surface of airframe through apertures and windows. This is one of the reasons for criticality of using PEDs inside the aircraft.

10.8.4 Real-Life Examples

Some examples of EMI effecting elements of FCS are presented here. However, unlike the occurrences mentioned in Section 10.3 these incidents were detected and corrected or occurrence of EMI was probable. The mitigation procedures adopted to overcome these problems are also mentioned. This provides an insight into the options available to the EMC engineer to enhance EMI protections after the FCS has been integrated in the aircraft. It is relevant to add that for reasons of confidentiality the exact type of aircraft of some of the occurrences is not made public.

10.8.4.1 Coupling between HF Radio and Autopilot

On September 8, 1994, US Air flight 427, a Boeing 737-300, entered an uncontrolled descent and impacted terrain near Aliquippa, Pennsylvania. Extract of the NTSB report NTSB/AAR-99/01 (PB99-910401) stated: "although the Boeing 737 flight controls are primarily hydromechanical (not electrical), there are electrical links to the flight control systems through the autopilot (which interfaces with the ailerons, spoilers, elevators, and stabilizer trim but not the rudder), the rudder trim, and the rudder system's yaw damper. Therefore, EMI and/or HIRF could theoretically affect the autopilot, rudder trim, and/or yaw damper systems within the limits of those systems." The actual cause of the crash could not be pinned on EMI, but the system was considered susceptible to transmissions from HF radio, and wire shielding was improved to make it more immune to EMI. Boeing Service Bulletin 737-22-1112 was recommended by FAA.

10.8.4.2 Coupling between VHF Communication System and Flap Controller

In a business jet aircraft, electronic flap position controller malfunctioned when it was simultaneously operated with VHF radio set. In such cases, flaps would stop at random positions or move in opposite direction. Flap drive, controller, and wiring were installed below the rear seats of the aircraft. The VHF communication antenna was installed on bottom centerline of fuselage, just below the flap drive circuit. Flap controller wiring to flap drive unit used improperly shielded wires. The airframe was made from fiber glass; aluminum foil was additionally used for shielding and grounding. Flap actuator was electrical motor-driven-type jackscrew, which had three distinct positions and a magnetic proximity sensor provided positional feedback. The flap controller used switch selection and magnetic sensors to move the jackscrew motor. Electronic flap controller employed TTL circuit logic to control and command flap position. VHF transmitter induced RF currents on flap control wires that were conducted into flap controller and rectified by TTL circuit elements. Rectified RF currents were interpreted as a logic change by the TTL circuit elements. It was discovered that during primary EMC testing of aircraft, simultaneous functioning of

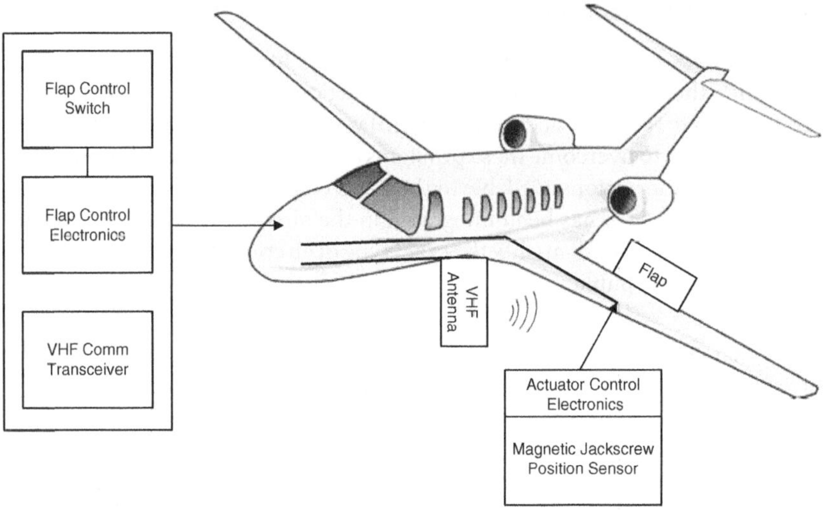

Figure 10.16 EMI coupling from VHF antenna to ACE wiring.

flap and VHF communication was not tested. A service bulletin was issued to add EMI suppression to flap system wiring and install ferrite cores to flap control electronics wiring. The lesson learnt is that the wires can act as emitters and receptors. The affected circuitry, which in this case was ACE, may not be the basic victim of EMI [58]. Figure 10.16 represents the layout of this system.

10.8.4.3 Coupling between HF Set and Control Surfaces/Fuel Indication System

In a military FBW aircraft carrying an HF transmitter, the operation of the HF set caused random movement of control surface by about 8°. Also, the fuel quantity indicator displayed erroneous readings. Analytic trouble shooting revealed that at some frequencies, resonance effect was created, inducing currents in FCC wiring. The problem was solved procedurally by prohibiting HF transmitter to use the frequencies that caused EMI in FCC. One aspect to be taken care of in such cases is to determine how the built-in redundancies are handling the interference. If the FCS has multiple channels where voting is being employed to compute the correct maneuver demand, it is important to determine that whether all the channels are being affected. Ideally, the redundant channels should be on different data bus and physically separated [59].

10.8.4.4 Disruption in ILS Localizer Receiver due to Weather Radar

A transport airplane used autopilot for Cat-I landing. During landing the aircraft drifted left of runway center line, and no warning message was generated for the pilot. The aircraft used an autopilot with mode control panel, whose wire looms were routed in the near vicinity of weather radar wiring. Weather radar wires ran through nose bulkhead to weather radar, where localizer antennas were installed. Mode control panel generated RF CE from internal processor clock, these conducted currents coupled to collocated localizer coaxial cable that were conducted to localizer antenna. Localizer receiver detected these emissions as a valid localizer signal, causing the aircraft to drift. It may be mentioned here that ILS localizer is highly prone to VHF interference.

10.8.4.5 Susceptibility of INS to High-Power Transmitters

In the mid-1980s, flights from the UK to Barbados reported problems with INS, which were attributed to shipborne and airborne high-power transmitters deployed by the USA to monitor drug trafficking in the Caribbean Sea. The net result was that INS memory was corrupted. The problem was solved by altering operating procedures and that was to keep the INS powered ON during turnarounds [25].

10.8.4.6 Mach Probe Heater Causing Erroneous Fuel-Flow Reading

Another recorded occurrence was erratic reading of fuel-flow reading when Mach probe heater was switched "ON." The heater utilized 115 V 400 Hz power for operation. It was found that unshielded fuel-flow circuit wires ran in the same 100 ft loom as heater return line. Shielding the fuel-flow circuit solved the problem. Interestingly, the system had cleared RTCA/DO-160 testing. If the aircraft is using FMS mode, FCS makes corrections to cater for fuel consumption and could alter desired flight profile. Highly automated systems automatically correct for noise or unbalance. When such systems are switched off, large and disconcerting imbalances may be hard for pilot to handle as he disengages autopilot and the aircraft may go out of control [60]. One element of FMS that is known to be prone to EMI is FQIS. When autopilots are using FMS mode and such an occurrence takes place, an erroneous response from FCS can be the result [61].

10.8.5 Software

An important element of FCS is the software; effects of some potential faults caused by EMI can be predicted to a reasonable level of accuracy. Normally, this will be the interference caused by 400 Hz line voltage and faults in digital circuits. Thus, the software can have the algorithms to cater for these faults [62]. Faults due to other elements like damaged wire etc. are difficult to model and predict. In such cases, remedial actions cannot be embedded into software algorithms. Sometimes parallel chains of FCS use software from different programmers to eliminate the possibility of software bugs.

10.9 Modeling and Simulation

This section is a review of modeling and simulation options available to predict EMI effects, compute EME, and evaluate protections designed to mitigate potential interference. In addition, applicability and limitations of commonly used techniques are also stated. Modeling is an essential ingredient of simulation. The material in this section is primarily meant for EMC and aircraft system engineers. Nonetheless, some useful references are available for modeling and simulation engineering desirous of focusing on aircraft EMC.

Analytical and simulation results cannot be used to qualify level "A" systems like the FCS, even by incorporating prescribed margins. However, in EMC testing of FCS these methods are used for

(i) Predicting the HIRF levels each module will be exposed to and coupling of external RF energy to looms of the FCS system at the start of the project. As no measured data are available at this stage, these results provide the initial EMI protection requirements.
(ii) Predicting the transfer function of LLSC testing for noncritical elements of FCS, e.g., displays, or, computing the EME inside the fuselage and currents induced in wire looms due to external RF fields.
(iii) Predicting the current magnitudes and distributions of skin currents when exposed to a given HIRF environment; this is used to determine current levels and placement of injection probes for direct drive tests.
(iv) The testing of a complete aircraft for all HIRF environments is very time consuming. It is not possible for a complete aircraft to be made available for such extended time for ground testing before production certification. As a compromise, testing may be carried out for some expected environment and for other, extrapolation may be carried out using simulation and analytical methods.

10.9.1 Computational Electromagnetics (CEM) Solutions

Going by the classifications of physics, EMI falls in the domain of applied electromagnetics. The solution to problems of electromagnetics is based on solutions of Maxwell equations. However, working out exact analytic solution of Maxwell equations even for slightly complex problems is a very elaborate exercise. Thus, numerical or computational techniques are used to find an approximate solution for relatively complex scenarios. With proper application of correct computational technique, results are very close to those found analytically [63]. However, for complex or electrically large structures, the computational effort required is colossal. For some of the repeatedly required calculations, researchers, through tedious analytic derivations, previously run simulations or empirical data have formed closed-form expressions that can be readily employed to arrive at the results. A brief review of methods that can be used to find out the solutions to equations representing EMI problems is presented here.

10.9.1.1 Analytic Solutions

Analytic solutions provide closed-form solutions of Maxwell equations by well-grounded mathematical techniques. This implies solving the system of partial differential equations. However, their applicability is limited in case of complex problems. The most commonly used methods are [64] as follows:

(i) Separation of variables
(ii) Series expansion
(iii) Conformal mapping
(iv) Integral methods

Besides these methods, there are other methods such as "transfer admittance method" that have been used to derive mathematical expressions for establishing interaction of external HIRF waves with aircraft fuselage [65]. The coupling between EM waves with aircraft wiring and other avionics modules after penetrating the fuselage has also been derived on similar lines [66]. Moreover, cross coupling among 2–3 wires can be computed using analytical methods. It is to be remembered that some simplifications are made in the formulating solutions, e.g., aircraft body may be considered as a perfect cylinder.

10.9.1.2 Numerical Techniques

CEM is a modeling process and therefore a study in acceptable approximation and numerical solutions. Central to all numerical techniques is the idea of discretizing some unknown electromagnetic property. For example, in method of moments, it is surface current, in finite element method it is the electric field, in finite difference time domain, electric or magnetic field is discretized. Meshing is used to subdivide a large geometry into a number of nonoverlapping subregions or elements.

Within each element a simple functional dependence (basis functions) is assumed for spatial variations of the unknown parameter. Numerical techniques replace the problem with an approximate, one which causes physical (geometric) and numerical limitations that have to be taken into account. During the creation of approximate model, assumptions and simplifications are generally introduced; consequently, limitations are introduced on the solution accuracy, e.g., assuming an infinite ground plane or substrate in antenna structure or simplifying a thin wire by a current filament. The prominent numerical methods are:

(i) finite difference time domain
(ii) finite element method
(iii) method of moments
(iv) transmission line method

(v) optics based methods
(vi) multilevel fast multipole method
(vii) finite integral method
(viii) transfer impedance method

All these methods have their strengths and limitations and must be carefully employed as per the geometry of space and desired resultant parameters [67]. Their solution is computation extensive and if electrically large structures are involved or high level of precision is desired, simulation times can be enormous. CEM techniques are useful in determining field-to-wire interaction in limited space and coupling between wires in a harness, interference between collocated antennas, shielding effectiveness, ESD modeling, etc. The limitation of numerical methods in dealing with spatially large or complex problems either require very elaborate processing hardware or analytically breaking down the problem into smaller segments [68]. Creating exact model of the problem is very important in getting precise answers. The EM properties of materials, e.g., permeability and permittivity, exact dimensions, boundary conditions, and application of ports representing signal sources must be done very carefully.

10.9.2 Other Methods

There are some problems routinely encountered for which researchers through rigorous mathematical derivations or by analysis of empirical data have formulated mathematical relationships, in which values of independent parameters are provided to calculate the results. Another category is the use of circuit simulators like SPICE to predict parameters such as wire-to-wire coupling [20]. This is a simpler and much faster way of getting the results, but the prerequisite is that accurate circuit models of the elements have to be available. Computation of antenna-to-antenna coupling on aircrafts is a useful application of CEM, as this is a commonly occurring requirement; the level of expected coupling taking into account antenna characteristics, type of fuselage surface, frequency bands, distance in terms of wavelengths, radiated power, and polarization relatively simple expressions have been worked out [69]. In a vast majority of practical situations, these expressions can be relied upon. As an additional margin of safety, the distance between antennas may be kept slightly more than what the calculations indicate safe. Where the available space is too limited or very high levels of accuracies are required, complete CEM-based simulation can be run.

It was mentioned in Section 10.7 that during EMC testing of FCS on ground, all flight conditions cannot be generated. Simulated data can be provided for elements such as rotational feedbacks that do not become operational when aircraft is static. These data are not accumulated from EM simulations but other software that can simulate aircraft flight controls and aerodynamic performance. These simulations will primarily fall in the category of control systems engineering. These simulations are also used to predict aircraft recovery time from EMI disturbances [70].

10.9.3 Current Status of Modeling and Simulation Work

Ideally, simulation should be able to replace physical testing in totality; however, this may not be possible due to technological restrictions and certification requirements, i.e., no regulatory body accepts simulated results to clear the equipment for EMC for use on aircraft, especially for flight-critical system such as FCS. What an EMC or aircraft system engineer needs to know are the scenarios where modeling and simulation can be a useful tool. Some of the conditions where modeling and simulation work can be beneficial are as follows:

(i) Study the effects of lightning and other external HIRF on different subsystems and cable bundles inside aircraft fuselage (Figure 10.17).
(ii) Determining coupling between antennas installed on aircraft.
(iii) Finding cross-coupling between aircraft wires.

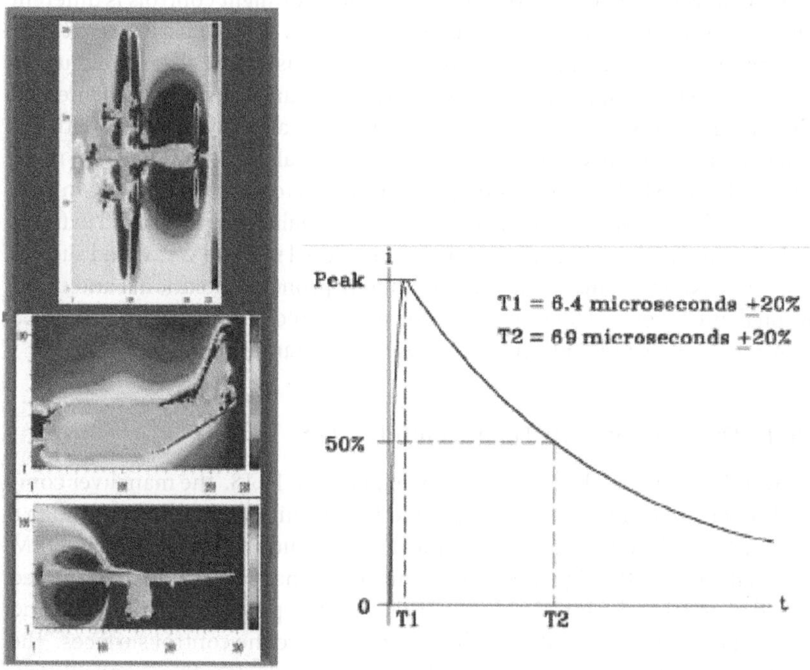

(a) (b)

Figure 10.17 (a) Simulation of lightning effects on aircraft. (b) Equivalent transient waveform to compute lightning effects on aircraft.

(iv) Approximating radiation from damaged cables.
(v) Analyzing resonance effects in fuselage and compartments such as avionics bay.
(vi) Predicting EMC compliance of PCBs.

Though it would have been a great help if the option of using simulations was available for estimating compliance of complete aircraft, but it is not perceived as a likely possibility in the near future. In conclusion, it can be stated that modeling and simulation is a useful tool in the arsenal of an aircraft EMC engineer. Nonetheless, he or she has to be well aware of what method is best suited in a situation and what are its limitations.

10.10 FCS of UAVs

UAVs or unmanned air systems are pilotless aircraft that are remotely flown via a wireless link. Thus, it is imperative that FCS of UAVs is all electronic type, as mechanical force cannot be distantly exerted. In comparison with the FCS of a manned aircraft, the physical orientation of the UAV flight controls is different from conventional aircraft. Unlike manned aircraft, UAVs have lesser weight-carrying capacity, smaller available space, and, because of the absence of human beings to react intelligently in case of system malfunctions, the implemented redundancies and safety measures are not as elaborate. This, coupled with the existence of a radio link in flight-critical system, makes the FCS of UAV more susceptible to EMI. Some UAVs employ autopilots to keep them flying in case of loss of radio link on a predetermined path. The main navigational aid in flying the designated path is generally a global navigational satellite system aid such as GPS. The GNSS systems themselves are highly prone to intentional and unintentional interference. However, the basic methodology of EMC testing of FCS of UAV will follow the same guidelines as that of manned aircraft.

10.10.1 Overview of a UAV Flight Control System

The overall system block diagram is shown in Figure 10.18. The maneuver command generated by the pilot on the ground is transmitted to the UAV through a pair of MODEMs. One can consider the decoded output of the UAV MODEM to be analogous to the signal from the transducer that senses the force applied to rudder pedal and control column. This signal is processed by an FCC and conveyed to ACE and resultantly the actuators move the control surfaces. The number of control surfaces in case of UAVs is generally less than conventional aircraft. The cable lengths are comparatively shorter but shielding protections are restricted due to weight limitations. Also, spatial limitation between EMI sources and potential victims is relatively small. For elements of FCS onboard the UAV, the basic principles of EMI susceptibility and protection remain the

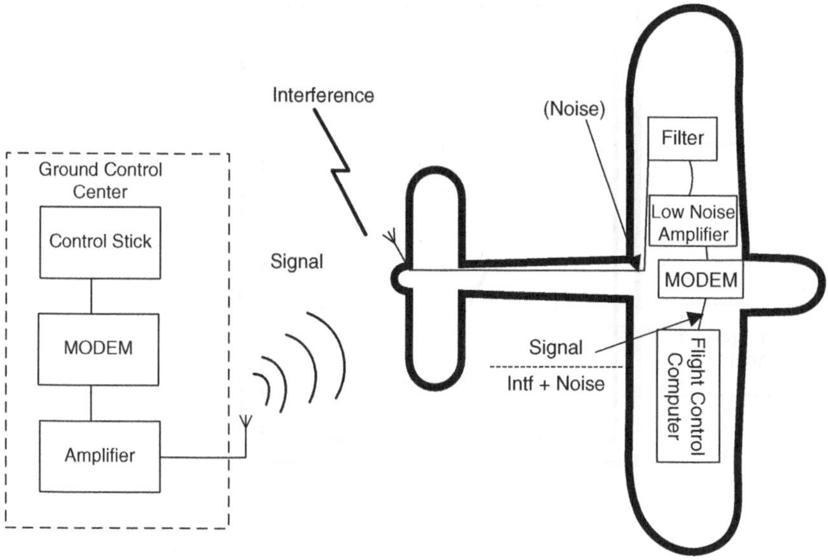

Figure 10.18 Basic block diagram of UAV flight control.

same as that for manned aircraft. The main difference is in the wireless communication link.

10.10.2 EMI Susceptibilities of UAV Flight Control System

The modules of the communication link are tested for EMC as per relevant EMC standard. In case of military purpose UAVs, the modules will be tested in accordance with MIL-STD-461 or equivalent standard, and the complete UAV will be expected to conform to MIL-STD-464 or comparable EMC test standard. UAVs meant for commercial use can be tested against commercial standards, though the limits may have to be altered in accordance with expected EME. It is relevant to add here that the rationale that prescribed HIRF EME is higher for rotary-wing aircraft is that their flight paths are more flexible and fly at relatively low altitudes over high transmission zones holds good for UAVs as well. One of the comprehensive contemporary EMC test standards for communication equipment is ETSI EN 301 489-1 EMC standard for radio equipment and services [71]. However, the UAV manufacturer may decide to use any other relevant standard. The main criterion for qualification of communication link will be signal-to-noise ratio or BER.

Except for military use high-altitude long endurance and medium-altitude long endurance category of UAVs, other types are not likely to be exposed to HIRF associated with lightning. Unlike manned aircraft that have been in service for over a century, UAVs have become regular part of the airspace for about

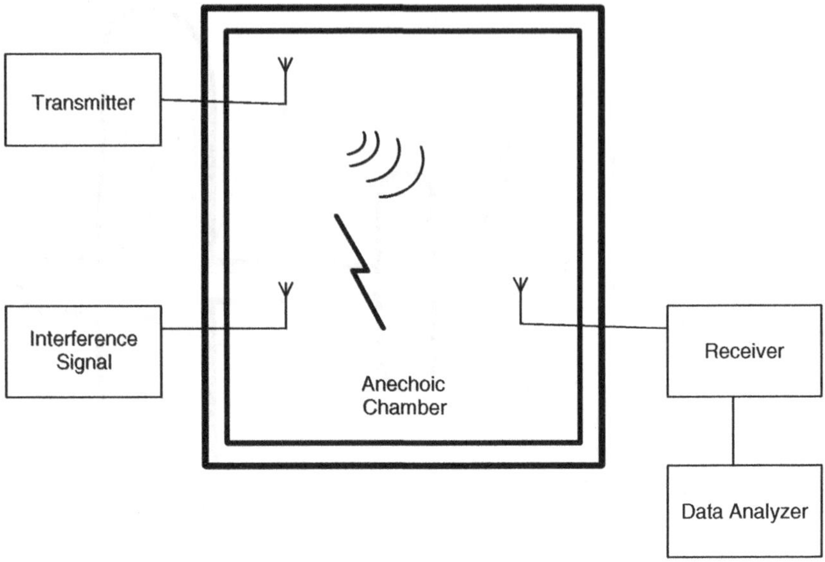

Figure 10.19 EMC test methodology for UAV flight control wireless link.

last two decades. Thus, the regulations and test standards are not as mature as those of conventional aircraft [72]. Some proposals for complete UAV EMC testing have been presented by various researchers [73]. The use of wireless link forms a front-door coupling channel for interference to be accepted by the receiver chain of the UAV. Anytime the signal to interference plus noise ratio drops below what is required for data extraction, the receiver is desensitized and no information is received. The general procedure for testing of communication link is shown in Figure 10.19. Inside the chamber an interference source representative of actual channel noise is introduced and induced error in data transmission is monitored. A vast majority of UAVs, except very short range small UAVs, use digital modulation schemes for transmitting error-free information in available bandwidth, higher data security, and better quality. Through injected interference signal, it is determined what magnitude of in-band and out-of-band signal level is required to desensitize the receiver for a particular type of modulation.

Additionally, frequency management of other systems, especially those employing any transceiver, e.g., video or telemetry links, is to be carefully done. It is to be ascertained that not only the main operating frequencies but their harmonics and mixing products do not interfere.

So far, radiated interference has been discussed, but the FCS of UAV can be disrupted by conducted EMI as well. Generally, UAVs have nonconducting airframes, and special care needs to be exercised in grounding and bonding.

Although there are weight limitations, strong sources of EMI such as clock circuits should be adequately shielded. UAVs DC power sources generally use DC–DC converters that are a strong source of radiated and CE. These should be shielded and line filters must be used at the power line inputs of modules receiving this DC power. Engine ignition coils are also a strong source of EMI. Cables and sensitive circuits, especially elements of FCS such as actuators and FCC, should be as far from this source as possible and protected against radiated and conducted EMI, by using appropriate shielding and line filters.

When the UAV is flying in autopilot mode or a predetermined flight profile, the most susceptible element is the GPS. The first thing to check is that the radio link being used for flying the UAV is not interfering with the GPS. The GPS antenna is mounted on the top of fuselage that gives it a degree of immunity from ground-based transmitters. But when flying at low altitudes or during high bank angles, it becomes more susceptible to interference from ground-based transmitters. Like other receivers, not only the basic operating frequency of a system but also the harmonics or intermodulation products due to nonlinear mixing of frequencies can have some components in the GPS C/A frequency, i.e., 1575.42 MHz. Therefore, the immunity of GNSS receiver to noise has to be addressed in the systems engineer's EMC plan. Some useful information about EMI mitigation in UAVs is well addressed in [74].

10.10.3 Some Peculiarities of EME Experienced by UAVs

Another factor to be taken into account in case of UAVs is that they fly over urban areas at relatively low altitudes. Thus, there are some typical characteristics about the EME UAVs are exposed to. The multitude of transmitters including ultra wide band, industrial scientific medical bands being used even by unlicensed products create a very harsh EME. It may not generate high-magnitude electric fields but definitely in terms of waveforms and number of simultaneous transmissions the EME is very harsh and difficult to predict. Some experimentation has been carried out to study these effects [75]. For hi-tech military UAV flying at altitudes above 10,000 feet, the effect is not very pronounced, but for other UAVs not employing very robust EMI protections and flying at lower altitudes, the situation can be very different. However, as UAVs are nascent in the realm of aviation, reported information about EMI problems of UAV is very limited. EMC testing of UAVs, their EMI susceptibilities, and EMI protections are active topics of current research.

10.11 Some Special Considerations for EMI Mitigation

This concluding section highlights some factors that need to be taken care of to reduce the probability of EMI occurrences in aircraft systems, especially FCS.

The final subsection views some future trends in aviation vis-à-vis the effect of EMI in FCS.

10.11.1 Mitigation of EMI in FCS

In the previous sections, some "front-door" and "back-door" channels for EMI to creep into aircraft systems have been mentioned. Also, some protective measures taken to negate the effect of some specific EMI problems have been discussed. The standard recommended practices for EMC hold good for aircraft systems and FCS. It remains a reality that systems like FCS are designated as flight critical due to their functionality. However, one garbled radio call in an emergency situation or suppression of an aural warning regarding a crucial flight parameter, due to interference, can also become safety critical. In modern-day aircraft, numerous systems work in an interconnected manner. Thus, no relaxation can be accorded in implementing EMI protections in any of the systems. A large majority of EMI-related problems can be nipped in the bud if it can be ensured that the magnetic fields due to currents in signal wire and return wire cancel out each other and fuselage structure is not used for return path, e.g., a balanced, isolated circuit using shielded twisted pair wires. A brief description of factors that affect EMI protection follows, details can be found in [4,7,20,21,44]. It may be surprising, but almost 99% of protection from HIRF can be achieved by proper bonding, shielding, and routing.

10.11.1.1 Shielding

Aircraft shielding has to be considered in layers. The first layer of shielding is provided by the metallic aircraft structure. This layer is extremely useful in providing protection against lightning, ESD, and high-power RF signals. For flight-critical systems, two layers of shielding is a necessity. If the aircraft is not metallic, conductive linings can be added. The second layer is provided by the metallic box enclosing the circuit itself. Generally, aluminum is used as the material for shielding enclosures in aircraft. The thickness of the material corresponds to the level of attenuation required in dBs. The total level of attenuation provided by the combined effect of these shields should be sufficient to attenuate the external EMI to a level that will not disrupt the functioning of internal circuitry. The wires used to connect different enclosures together shall also be shielded. The shielding of the wires and the enclosures has to be properly grounded. Wire shielding can be single braid, double braid, braid and foil, solid conduit, etc. [76,77]. Shielding is required to be carried out on cables and bundles. Magnetic field shielding is not possible without circulating current wire wrapped around bundle. Below 400 MHz, mostly interference couples to the modules through wire bundles. These cables can act like antennas; they become resonant at certain frequencies and impart a gain to interference energy in some frequency bands. Shielding provides protection against this phenomenon. The shielding has to be connected to the ground, usually through a backshell. For FBW systems, the recommended backshell is 38999 but it is expensive and heavy. For other wires, an over braid for additional protection may be used, then a pigtail will be used for grounding the shield. For better EMI protection, the shielding should be terminated at both ends. The shielding effectiveness has to be maintained throughout the life cycle of the aircraft.

10.11.1.2 Grounding and Bonding

The basic purpose of good grounding and bonding is to ensure a very low impedance path between circuit return and zero volt potential. Grounds are required for AC power, DC power, shielding, low-frequency analog circuits, digital circuits, and RF circuits. Isolation must be maintained between circuits and aircraft structure. The wire shields are terminated, typically at both ends [78]. Shielding enclosure on skin panels and trays unless electrically bonded will develop harmful voltages or electrostatic charge centers. Backshell connectors are used to protect leakage of EM energy from apertures that are used to allow passage of wires in shielding enclosures; these backshell connectors are to be bonded to the shielding surface. The generally accepted resistance level for bonding is less than 2.5 milliohms [79]. The MIL-STD-464C section describing why good bonding is a requirement cites an example where omission of bonding in aileron rudder interconnect system resulted in shaking of rudder surface when exposed to external EMI [37]. The best bonding methodology is metal-to-metal contact with the maximum surface area in contact. Or, the structures can be joined together using rivets. This is also a good practice but the number of rivets should be as large as possible. In cases where these methodologies cannot be used bonding jumpers or straps are used.

10.11.1.3 Routing

To reduce the flux density around the wires, it is necessary to route the wires as close to the ground plane as possible. Wherever possible, the wires of safety-critical systems may be routed away from high emission areas and radiating antennas. Just by using correct shield terminations and routing the wires close to ground interference can be reduced by an order of 40 dBs. However, in case of FCS, routing flexibility of wire looms is relatively less due to accessibility problems and long wire lengths.

10.11.1.4 Lightning Protection of FCS

Safety-critical systems such as FCS are to be protected from catastrophic failures due to lightning. Lightning protection guidelines are provided in CFR 25.581. Metal structures have better protection against lightning strikes and lightning can be diverted from critical systems. Protection has to be provided at entry, exit points, and current paths. For metal structures, the entry and exit points are not much critical. However, the dwell time of currents due to lightning strikes is relatively large. This means that bonding straps and other conductive paths provided for the flow of currents due to lightning must be able to sustain high currents for about 1–2 seconds. If not properly sized, they can burn out. For nonmetallic structures, current flow paths have to be designed using metal straps. The addition of these metal straps is liable to increase overall weight of the aircraft; therefore, the system engineer will like it to keep it to a minimum. Nonetheless, all important components of the aircraft have to be protected from lightning-induced currents by diverting them away from critical modules.

In FCS, the control surfaces are located on external surfaces; thus, these have to be protected against lightning. The control surfaces are connected to their controlling actuators through some type of linkage mechanisms. If proper protection mechanism is not designed, the control surface and its attachments can be severely damaged. All control surfaces must have multiple bonding straps attached across their hinges. This provides redundancy in protection and reduces current density in each bonding strap. The size of these straps, in terms of diameter, has to carefully calculated; they should be able to handle the heating factor due to flow of current for relatively long time. Also, the contact area where it is attached to the surface should be sufficient, and it should make a perfect contact to make the resistance between two surfaces close to zero. If there is resistance, flow of current will generate voltage across it that can cause damage to the surface. Conductive sealant should be used to keep moisture out of this contact. One important aspect is to keep these bonding contacts corrosion free during aircraft life cycle. Figure 10.20(a) shows a bonding jumper across an elevator and Figure 10.20(b) is a picture of a strap burnt out due to lightning [80].

10.11.1.5 Wiring and Packaging

For FCS, the return wires are preferred to be twisted with signal wire to have ideally zero net current flow. It also keeps electric and magnetic fields from interfering with signals being carried by the wire. Wires must not be

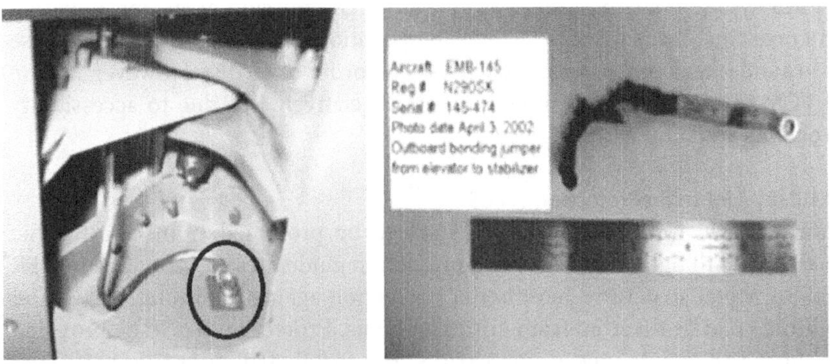

(a) (b)

Figure 10.20 (a) A bonding jumper across elevator hinge; surface contact is encircled. (b) An improperly designed jumper to stabilizer strap was burnt out like a fuse when aircraft experienced lightning strike.

installed across apertures. The apertures for passage of cables need to be properly designed and generally must be placed in one localized area. Line drivers/receivers must be packaged close to the connector aperture area.

10.11.2 Better Management for EMI Mitigation

Besides the factors mentioned above that deal with physical design and installation of elements, there are some other factors for better EMC that relate to management issues. These are as follows.

10.11.2.1 Care during Maintenance

As stated in previous sections, the major reason for EMC-qualified aircraft to become susceptible to EMI is degradation during routine and corrective maintenance. Normally, EMI protections rank very low on the priority list of maintenance crews. The primary concern is to make the aircraft airworthy in shortest possible time. The elements that need to be addressed are:

(i) All EMI protective shielding must remain damage free during the life cycle of the aircraft.
(ii) The routing of wires must conform to that of original design. No changes are to be made.
(iii) Shield terminations must be kept as short as possible.
(iv) Grounding and bonding should remain good enough to keep the conductivity within prescribed limits.
(v) Maintenance practices must ensure avoidance of ESD.
(vi) Wiring insulations are to be closely monitored.
(vii) Corrosion control must be effective to avoid deterioration of grounding and bonding, besides causing other structural problems.

10.11.2.2 Awareness

Though the engineers involved in design of aircraft are well aware of EMI issues, the same cannot be said about aircraft maintenance and flying crews [25]. The trainings conducted in this regard generally inculcate an understanding of EME effects (E3) but not practical implementation of EMC concepts. Awareness programs must be conducted for maintenance and flying crews that should highlight the effects of EMI and how it can affect the system performance. The flying crews are also to be made aware that areas in which probability of EMI is high, relying on autopilots may be avoided. Another aspect is to educate the crews to recognize EMI occurrences and maintain a record of these occurrences and their remedial measures. During the service life of the aircraft, EMC testing is generally carried out when avionics upgrades take place. Some Air Forces conduct limited EMC testing on some aircraft of the fleet after completion of designated flying hours [81] but the same is not practiced in civil aviation. The main impediment in this regard is the aircraft downtime that ultimately translates into cost.

10.11.3 Future Trends and EMC Requirements

Some of the present-day aircraft employ fiber optic cables to transmit critical signals. Also, in regions where wiring has to pass through areas of high levels of radiations, fiber optic cables may be preferred. Even the concept of "Fly by Light" is in place [29]; this is to eliminate the EMI susceptibility of interconnecting wires, which are the cause of a majority of EMI problems. However, there is always an associated electronics with fiber optic cables for converting the signals from electrical to optical and vice versa. The signal levels are generally low, and unless properly configured for EMC, this portion becomes a weak link in EMI protection [32].

The future requirements in air traffic management aim to accommodate a greater number of aircraft in the same airspace [82]. The airspace will also be riddled by the presence of UAVs. One major derivative of these requirements is in the form of RNP [82]. In straightforward terms, this will necessitate the aircraft to follow the designated flight paths very strictly. Consequently, not even slight deviation in flight path can be tolerated. Thus, EMI effects in FCS will have to be absolutely minimized. At the same time, EME has become very complex for UAVs and military aircraft supporting urban warfare. With international terrorism being a stark reality, the chances of intentional EMI cannot be ruled out. This will require even further improvement in EMI protections and perhaps modifications of EMC procedures for aircraft and more specifically for FCS testing. The focus shall not only be to increase the magnitude of power for susceptibility testing but also to cater for different modulation schemes and multiple EMI occurrences. That is to say that test environment should be more representative of actual EME; however, this has to be achieved in a manner so as to keep the EMC testing time and cost as low as possible.

References

1 Albert Helfrick, *Principle of Avionics*, 4th ed. Leesburg, VA: Avionics Communication Inc., 2007.

2 J. Carter Nigel, *Aircraft EMC: The Past, the Present and the Future: An Overview*. Defense Evaluation and Research Agency, HMSO, Nov. 1996.

3 D. R. J. White, *A Handbook Series on Electromagnetic Interference Compatibility, Vol. 5, EMI Prediction and Analysis*. Germantown, MD: Don White Consultants, 1988.

4 Mark I. Montrose, and Edward M. Nakauchi, *Testing for EMC Compliance Approaches and Techniques*. Hoboken, NJ: IEEE Press, 2004.

5 J. Carter Nigel and D. A. Bull, *EMC Testing of Whole Aircraft*. IEE, HMSO, Savoy Place, London WC2R OBL, UK, 1994.

6 Clifton A. Clarke and William E. Larsen, *Aircraft Electromagnetic Compatibility*. Washington, DC: U.S. Department of Transportation, Federal Aviation Administration, Jun. 1987.

7 *IEEE Guide on Electrostatic Discharge Characterization of the ESD Environment.* IEEE C62-47-1991. New York: IEEE, 1992.

8 J. Molan, *The Physics of Lightning.* London: The English Universities Press, 1963.

9 Elaine Scarry, "The fall of TWA 800: The possibility of electromagnetic interference," Special Supplement to AFI 51-103 Crash Investigations, 1998.

10 James. Burrel, "Disruptive effects of electromagnetic interference on communication and electronic systems," Master's Thesis, George Mason University, School of Information Tech & Engg, Apr. 2003.

11 "International Civil Aviation Organization Automatic Dependent Surveillance – Broadcast (ADS-B)," Study and Implementation Task Force, ADS-B/SITF/WP/9, Mar. 2003.

12 UAV in Civil Airspace – OUTCAST or Friend, National Aerospace Laboratory NLR, Rpt # NLR-TP-2008-246, Jun. 2008.

13 Ashish Sharma, "UAVs flying in the NAS," *Avionics Magazine*, Aug. 2012.

14 Peter B. Ladkin, "Electromagnetic Interference with Aircraft Systems Why Worry?" University of Bielefeld, Article RVS-J-97-03, October 1997.

15 "In-Flight Radio Frequency Spectrum Measurements of Commercial Aircraft Cabins," Office of Aviation Research and Development Washington, DC 20591, DOT/FAA/AR-06/41, Sept 2006.

16 Randy J. A. Jost, "Primer on EME," in *IEEE International Symposium on Electromagnetic Compatibility Fort Lauderdale*, Florida, USA, 2010.

17 Richard Hess, *Chapter 25, The Avionics Handbook.* Washington, DC: CRC Press, 2001.

18 John S. Seybold, *Introduction to RF Propagation.* Hoboken, NJ: John Wiley & Sons, 2005.

19 Keith G. Balmain and Edward C. Jordan, *Electromagnetic Waves and Radiating Systems*, 2nd ed. Englewood Cliffs, NJ: Prentice Hall Inc., 1968.

20 C. R. Paul, *Introduction to Electromagnetic Compatibility*, 2nd ed. Hoboken, NJ: John Wiley & Sons, Inc., 2006.

21 Christos Christopoulos, *Principles and Techniques of Electromagnetic Compatibility*, 2nd ed. Boca Raton: CRC Press, 2007.

22 Jay J. Ely, *Electromagnetic Interference to Flight Navigation and Communication Systems: New Strategies in the Age of Wireless.* Hampton, VA: NASA Langley Research Center, AIAA 2006.

23 F.N. Stoliker, "Introduction to flight test engineering," RTO AGARDograph 300. RTO/NATO 2005.

24 *Electronic Systems Failures and Anomalies Attributed to Electromagnetic Interference.* Washington, DC: NASA Publication 1374, 1995.

25 Martin L. Shooman, *A Study of Occurrence Rates of Electromagnetic Interference (EMI) to Aircraft with a Focus on HIRF (External) High Intensity Radiated Fields.* Washington, DC: NASA Contractor Report 194895, 1994.

26 "Portable electronic device generated electromagnetic fields on board a large transport UK civil aviation authority," *Aeroplane*, 2005.

27 D. F. Anderson and S. Eberhardt, *Understanding Flight*. 2nd ed. New York, NY: McGraw Hill, 2010.

28 Cary R. Spitzer, *Digital Avionics Systems*. Caldwell, NJ: Black Burn Press, 2000.

29 R. P. G. Collinson, *Introduction to Avionics*, 3rd ed. New York: Springer Press, 2011.

30 Greg F. Bartley, Chapter 25, *The Avionics Handbook*. Washington, DC: CRC Press, 2001.

31 J. Blakelock, *Automatic Control of Aircraft and Missiles*, 2nd ed. New York: John Wiley & Sons, Inc., 1991.

32 Bruce T. Clough, Brad Cope, and Shawn Donley, "Microwave induced upset of digital flight control systems," in *IEEE Digital Avionics System Conference*, 1994.

33 Hobart Patrick, *Minimizing Electrical Noise in Actuator Drive Systems for Maximum Reliability and Performance*. Tolomatic Inc., 2013.

34 ARINC 424, Navigation System Database Standard, Aeronautical Radio Inc. 35. Walter Randy, Chapter 25, *The Avionics Handbook*. Washington, DC: CRC Press, 2001.

35 Department of Defense Interface Standard Requirements for the Control of Electromagnetic Interference Characteristics of Subsystems and Equipment, MIL-STD-461 G.

36 Department of Defense Interface Standard Electromagnetic Environmental Effects Requirements for Systems, MIL-STD-464C.

37 RTCA DO-160 G, Environmental Conditions and Test Procedures for Airborne Equipment.

38 EUROCAE ED-14C, Conditions D'Environment et Procedures D'Essais Pour Les Equipments de Bord.

39 Ministry of Defense, Defense Standard 59-41, EMC Part 8.

40 STANAG 3516: *Electromagnetic Interference and Test Methods for Aircraft. Electrical and Electronic Equipment*. 3rd ed., May 10, 1993.

41 STANAG 3614, *Electromagnetic Compatibility of Aircraft Systems*. 3rd ed., Jun. 8, 1989.

42 D6-16050-2, "Electromagnetic Interference Control Requirements," The Boeing Company.

43 T. L. Clark, M. B. McCollum, D. H. Trout, and K. Javor, *Marshall Space Flight Centre Electromagnetic Compatibility Design and Interference Control (MEDIC) Handbook*, NASA Reference Publication 1368, 1995.

44 Department of Defense Aircraft Electric Power Characteristics, MIL-STD-704F.

45 Results of Detailed Comparisons of Individual EMC Requirements and Test Procedures Delineated in Major National and International Commercial Standards with Military Standard MIL-STD-461E, 2001.

46 MIL-STD-464 Revision C: Electromagnetic Environmental Effects Requirements for Systems, www.document-center.com.

47 Society of Automotive Engineers (SAE), Aerospace Group, Aerospace Recommended Practice ARP 5583 Rev A, *Guide to Certification of Aircraft in a High-Intensity Radiated Field (HIRF) Environment*, 2010.

48 US Department of Transportation Federal Aviation Administration (FAA), Advisory Circular AC 20-158 A, The Certification of Aircraft Electrical and Electronic Systems for Operation in the High-intensity Radiated Fields (HIRF) Environment, 2014.

49 Nigel J. Carter, "Aircraft EMC: The Past Present and Future Challenges of Aircraft EMC," *IEEE EMC Magazine*, Quarter 1, vol. 1, 2012.

50 Brian L. Stevens and Frank L. Lewis, *Aircraft Control Simulation*, 2nd ed. New York: John Wiley & Sons, Inc., 2003.

51 M. B. Redman, *Aircraft Test Conditions for Electromagnetic Compatibility Assessment*. New York: IEE Press, 1996.

52 E. McShea Robert, *Test and Evaluation of Aircraft Avionics and Weapon Systems*. Edison, NJ: SciTech Publications, 2010.

53 Armstrong Keith, "Why EMC testing is insufficient, and what is necessary," *EE Times Online*, Sep. 2008.

54 Cynthia Furse and Randy Haupt, "Down to the Wire," *IEEE Spectrum*, Feb. 2001.

55 ARINC 429, "Digital Information Transfer System," Aeronautical Radio Inc.

56 RTO NATO, Electromagnetic Compatibility in the Defense Systems of Future Years, 2002.

57 Dave Walen, Aircraft Electromagnetic Compatibility, FAA IVT Course, 2002.

58 Elya B. Joffe, "Some EMI Anomalies on Aircraft and Their Solutions," EMC in Aircraft Systems Workshop, Technical University Hamburg-Harburg, 2005.

59 L. Lee, *The Day the Phones Stopped – The Computer Crisis – The What and Why of It, and How We Can Beat It*. New York: Donald I. Fine, Inc., 1991.

60 D. Dowling and A. Washler. "Modernization and Standardization of Aircraft Fuel Quantity Indicating Systems," *IEEE/AIAA 11th Digital Avionics Systems Conference Proceedings*, 1992.

61 Electromagnetic Compatibility and Functional Safety, Institute of Engineering and Technology Fact file Annex H – Software, 2000.

62 David B. Davidson, *Computational Electromagnetics for RF and Microwave Engineering*. Cambridge: Cambridge University Press, 2005.

63 Matthew N. O. Sadiku, *Numerical Techniques in Electromagnetics*, 2nd ed. Boca Raton: CRC Press, 2000.

64 Elya B. Joffe and Jacob Gavan, "External Interaction of High Frequency Electromagnetic Waves with Aircraft Fuselage," in *Proceedings of the 1992 IEEE International Symposium on Electromagnetic Compatibility*, Anaheim, CA, 1992.

65 Elya B. Joffe and Jacob Gavan, *Coupling of external High Frequency Electromagnetic Waves to Aircraft Wiring – Theoretical Background*,

Proceedings of the IEEE International Symposium on Electromagnetic Compatibility, Dallas, U.S.A., August 1993.
66 Thomas Weiland, Martin Timm, and Irina Munteanu, "A practical guide to 3-D simulation," *IEEE Microwave Magazine*, Dec. 2006.
67 Sevgi Levent, *Complex Electromagnetic Problems and Numerical Simulation Approaches*. Piscataway, NJ: IEEE Press, 2003.
68 Thereza M. Macnamara, *Introduction to Antenna Placement and Installation.* Chichester, West Sussex, UK: John Wiley and Sons, 2010.
69 Oscar R. Gonzalez et al, "Direct effects of lightning: Requirements and design aspects," in *Proceedings of the 40th IEEE Conference on Decision and Control,* vol. 5, Feb, 2001.
70 ETSI EN 301 489-1, Electromagnetic Compatibility, European Telecommunications Standards Institute, 2005.
71 NATO UAV Systems Airworthiness Requirements (USAR) 1431, STANAG 4671.
72 Li Bo, Huang Daqing, Riurui Nie, "A novel method of EMC test at the system level of UAV, SPIE," in *6th International Symposium on Instrumentation and Control*, China, 2006.
73 Joshua Tristancho, Marcos Quilezy, Enric Pastor, "An electromagnetic interference reduction check list for unmanned aircraft system," in *48th AIAA Aerospace Sciences Meeting*, Ontario, 2010.
74 Bragg Heather, C-Band and Ku-Band UAV Line-of-sight data link. EMC Analysis for Two Operational Scenarios, Project Report for OSAD, 2004.
75 L. O. Hoeft, "A simple theory for predicting the electromagnetic performance of enclosures," in *IEEE EMC Symposium*, 1985.
76 L. O. Hoeft, "Current division and shielded conductive cables," *IEEE International Symposium on Electromagnetic Compatibility*, Arlington, VA, USA, August 1983.
77 H. W. Denny, *Grounding for the Control of EMI*. Gainesville, VA: Interference Control Technologies, Inc., 1983.
78 J. Brettle, "Electrical bonding in aircraft," *IEEE International Symposium on Electromagnetic Compatibility*, San Diego, CA, USA, October 1979.
79 Billy M. Martin, PE EMC Cessna Aircraft Co, Video Lecture, Direct Effects of Lightning: Requirements and Design Aspects, Apr. 2013.
80 Furnell Tac and Harris Michael, *Management of EMC in Ageing Aircraft in the Australian Defense Force*, University of South Australia Research Paper, 2003.
81 ICAO, Defining the Future of Air Traffic Management, 2012.
82 FAA, United States Standard for Required Navigation Performance (RNP) Approach Procedures with Special Aircraft and Aircrew Authorization Required 2005.

11

EMC Considerations for Unmanned Aerial Vehicles

Paul Kay

11.1 Introduction

It is sometimes reported that unmanned aerial vehicles (UAVs) suffer from loss of control and unpredictable performance due to interference. Radio interference is a subset of a wider range of electromagnetic interference problems that are referred to by industry professionals as "electromagnetic compatibility" (EMC), and EMC is usually considered in two distinct sections, namely, emissions and immunity.

"Emissions" refers to the unwanted electromagnetic emanations from an electronic system or subsystem. These are radiofrequency energies that are present on the conductive surfaces of the subsystem or its cables; due to their uncontrolled nature (that is, the unplanned frequencies and amplitudes with which they occur), even comparatively small amplitude emissions can compromise the performance of nearby equipment, especially nearby radio receivers or other electromagnetic sensors. For example, an unshielded sensor cable acts as a transmitting antenna for any RF energy that is present on it, such as high-frequency currents associated with a data acquisition circuit.

"Immunity" refers to the ability of a system or subsystem to tolerate radio frequency electromagnetic fields that may be incident upon its enclosure, sensors, or cables. Maxwell's equations tell us that a time-varying electromagnetic field incident on a conductor will produce time-varying currents on the conductor; these can present as "noise" on electronic circuits, which can compromise operation of the circuit. That is, the ability of an electronic system or subsystem to accept radio frequency energy is a property of its physical structure—its size, composition, and disposition. So, in our earlier example, the unshielded sensor cable acts as a receiving antenna as well and can convert energy in radio

Handbook of Aerospace Electromagnetic Compatibility, First Edition. Edited by Reinaldo J. Perez.

frequency fields around the cable to currents and voltages on the cable, which may upset the functionality of the associated data acquisition circuit.

Two further concepts are useful when discussing EMC: "front-door" and "back-door" interference. "Front-door" interference occurs when the victim system suffers interference from unwanted RF sources that enter its circuits via a pathway that is designed to receive RF energy, such as an antenna/receiver subsystem. Colocated radio systems operate harmoniously when careful frequency planning and management ensure that transmissions, harmonics, and possible intermodulation products do not fall into frequency bands that are required for reception of radio signals. These mitigations contribute to the avoidance of front-door interference as a result of transmissions, but sensitive receivers remain vulnerable to interference that falls within their operating bandwidth. An example of this is an automotive broadcast radio receiver suffering interference due to electric fans or wiper motors. "Back-door" interference occurs when the interference enters the victim system via a pathway that is not designed to receive RF energy; an example of this is a telephone extension buzzing due to a cellular telephone handset located close by. In this case, frequency planning is not a mitigating factor, as the victim system is not operating in any particular frequency band, its tendency to admit RF energy is a property of its composition and form, and its tendency to respond to any admitted RF energy is a property of its internal operation (analog and digital processing, for example).

The reliance of UAVs on radio communications, both for their survival and their ability to complete their missions, means that radio communication and navigation subsystems warrant special attention in the context of overall EMC for the system. In addition to radio communication aspects of EMC, the generally nonmetallic construction of UAVs and tight weight restrictions means that data and control circuits and wires and circuit assemblies often have poor shielding. This makes them more vulnerable to "back-door" interference from transmitters in the operating environment or onboard the UAV.

Aviation regulators generally classify UAVs according to their weight, with some international variation in categorization, registration, and licensing requirements; commercial operation of a UAV can be a trigger for regulatory requirements in some jurisdictions, depending on the size of the UAV. However, for the purposes of EMC discussion and analysis in this chapter, UAVs are considered as two types: "large" and "small." This distinction is made because of the different types of EMC problems that exist in the two cases, driven by the types of electronic and radio systems that are used in the two applications, and the relative amounts of space available for design features that mitigate EMC problems.

Small UAVs are characterized by a mass of typically less than 5 kg, are battery operated, employ flight control and payload technology that is available to the general public and use radio communication technology that operates in spectrum allocations available to the general public. Radio technologies used in these vehicles do not require special licensing and transmitter powers are sufficiently low that there is no RF exposure ("RADHAZ") regulatory burden on the operator (RF exposure is managed by the manufacturer of the product, in the same way that cellular telephone handset RF exposure compliance is managed by the manufacturer). Such systems are themselves readily available to the public and are operated generally within line of sight of the ground-based pilot, who may not necessarily require any formal training or be subject to any regulatory oversight. Nonetheless, the wide range of applications and growing popularity of such systems warrants consideration of the EMC aspects, as their small size does not preclude them from causing injury or property damage if an interference event causes a loss of control. UAVs of this kind may be used by hobbyists, farmers, building consultants, surveyors, etc., in fields as diverse as agriculture, mining, and real estate.

In contrast, large UAVs incorporate complex avionics that are similar to those used on piloted aircraft. Indeed, a large UAV physically and electrically resembles a full-sized aircraft except there are no seats or windows! UAVs of this scale, cost, and complexity are generally not available to the public; designers, operators, and maintainers are qualified, skilled, and generally subject to regulatory oversight in each of their respective domains. UAVs of this type are subject to EMC considerations that are more in line with conventional aircraft and radio systems for control and payload are more likely to operate in parts of the spectrum where licensing is required. Transmitter powers at the ground station and the UAV itself are likely to be 2 to 200 W or more, as the regulatory power limits are less stringent than those placed on transmitters operating in the "unlicensed" bands available to the general public. In a large UAV, the risk of loss of control due to uplink interference or jamming can be mitigated by including completely independent hot standby radio systems in the design. Small UAVs, due to their light weight construction, low cost, small cost and widespread use, are potentially sensitive to EMC problems that are less likely to affect large UAVs. Large UAVs will not be discussed any further in this chapter.

11.2 Small UAVs

Putting aside historical model aircraft, where an ethanol engine and servo motors were controlled by an HF or VHF transmitter, the following broad characteristics apply to small UAVs considered in this section:

- Uplink control transmitters commonly operate in the 2.4-GHz band
- Uplink control transmissions are some form of spread spectrum digital modulation

- Downlink telemetry (for position, battery voltage and other status information) is not always present, but may also work on a 2.4-GHz link
- The UAV may accept a small payload, such as a camera and or gimbal, that can be installed by an unqualified user
- Some payloads may draw power from the UAVs main (only) battery
- Some payloads may incorporate radio transmitters

The 2.4-GHz frequency band is favored for control of small UAVs for a number of reasons: radio modules are readily available, spectrum access is open to the public (the 2.4-GHz band is one of the internationally harmonized industrial, scientific, and medical bands) and international regulations on transmission power are such that an inexpensive system can provide 1 km range comfortably under line-of-sight conditions. Furthermore, the wavelength lends itself to conveniently sized, inexpensive yet effective omnidirectional antennas.

To understand the interference scenario with small UAVs, it is necessary to have an appreciation of the remote control technology, as the greatest risk to safe flight of the UAV is loss of the control signal from the transmitter. While small UAVs incorporate varying capabilities for autonomous flight, an unexpected loss of the control signal will generally trigger an unexpected response and attendant risk to injury, property damage, or loss of the vehicle.

There are many and varied data rates and modulation schemes that are used in the 2.4-GHz band—Bluetooth and 802.11b/g Wi-Fi are ubiquitous examples. Most 2.4-GHz systems used in consumer and industrial applications, including those outside UAV control, are "spread spectrum" radio systems, where the transmitted signal's occupied bandwidth is larger than the baseband data bandwidth. These various systems have characteristics that suit different applications—for example, 802.11b/g Wi-Fi offers high data rates (up to approximately 54 Mbps) but has a larger occupied bandwidth and is more susceptible to interference from other users of the spectrum than, for example, Bluetooth (up to approximately 1.5 Mbps), which employs frequency hopping and occupies approximately the same amount of spectrum.

For small UAV control, frequency-hopping spread spectrum (FHSS) is favored, because it is more resistant to interference from other users of the 2.4-GHz spectrum than fixed frequency spread spectrum techniques. Even at the hobbyist end of the market, multiple techniques are usually incorporated into a single transmitter, so commands are transmitted over an FHSS scheme, but at each hop, direct sequence spread spectrum (DSSS) is used. DSSS mixes the baseband data (digitized commands to the UAV) with a code sequence—so each baseband data bit is represented by a sequence of code bits (Code Division Multiple Access—CDMA). The ratio of the number of bits transmitted per baseband bit is known as “spreading factor” or “processing gain”—a higher spreading factor allows data to be recovered at the receiver under worse signal to noise conditions than a lower spreading factor, but at the expense of reduced data rates for the same occupied RF bandwidth. UAV commands do not require high data rates, so it is possible to use high spreading factors to produce the most robust links possible. Most Wi-Fi uses a spreading factor of 11, small UAV controllers use spreading factors around 64, and GPS satellite signals (which are CDMA but not frequency hopping) use a processing gain of 1024, so that data can be extracted from very low amplitude signals. To gain an intuitive appreciation of the power of code spreading in extracting very weak signals from background noise, consider that a laboratory-grade spectrum analyzer, connected to a suitable antenna pointed at the sky, cannot display the signals from navigation satellites. It is merely detecting the amplitude of the RF energy present in the spectrum and has no capability to apply coding techniques to extract information from the received signals; the display looks very much like noise. In contrast, a small consumer-grade GPS receiver, which includes the necessary coding and processing circuits, can easily decode the data from the satellite signals (and very likely take photos and make phone calls into the bargain!). Code spreading radio communications technology is a game changer for EMC and no EMC engineering problem space brings this to the fore quite like small UAVs.

In an FHSS/DSSS/CDMA small UAV uplink, the control commands are split into small parcels of data and transmitted in short, narrow band bursts at discrete frequencies within the 2.4-GHz band—typically 36 channels, although other schemes are possible. The transmission channels for each parcel of data are chosen in a pseudorandom sequence that is known to the transmitter and the receiver, so once the receiver has locked onto the transmitter’s signal, it can track the sequence of short, comparatively narrow transmissions. In this way, a number of similar systems can occupy the same band without apparent interference, as they are time, code, and frequency sharing the spectrum. Since the data rate required to control a small UAV is low (that is, digitized control commands can be expected to be low bandwidth, as the operator’s manipulation of the controls is not rapid), the most effective use of the available spectrum is to optimize reliability with a high spreading factor, balanced against minimizing on-air time to maximize transmitter battery life.

Radio systems of the kind used in small UAVs are inherently robust, in terms of their ability to work with other transmitters operating on nearby frequencies, compared with traditional narrow band superheterodyne systems.

The ability of a radio receiver to function at its specified level of performance in the presence of an unwanted interfering signal is known as its selectivity; the specified level of performance may be signal to noise and distortion for an analog voice receiver or bit error rate for a digital receiver. For a traditional,

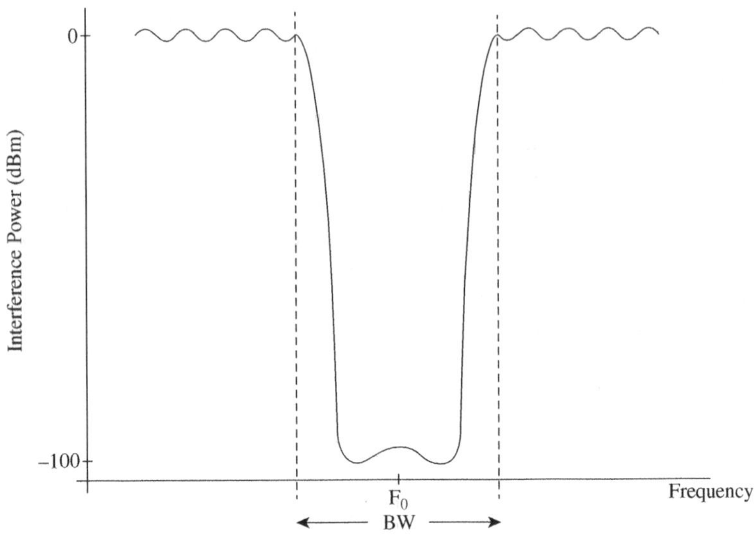

Figure 11.1 Typical selectivity curve for a traditional narrow band superheterodyne receiver.

fixed-frequency, narrow-band radio receiver of the kind used in aviation VHF band communications, for example, the selectivity curve looks much like the one shown in Figure 11.1. The selectivity curve gives a relative indication of how much power an unwanted signal must have at the input to the receiver in order to disrupt reception of the wanted signal.

Curves of this kind can be produced by testing the radio receiver with two signals simultaneously, usually introduced by coaxial connectors and adapters at the input to the receiver. The first signal is the wanted signal, set to an amplitude slightly more than the minimum necessary for the receiver to work properly and the second signal is the interfering signal, usually continuous wave (CW); the frequency of this signal is stepped or swept across a band wider than the receiver's operating channel. At each frequency, the interfering signal amplitude is adjusted so that the receiver produces its defined response and curve represents the level of RF power required at the input of the receiver under test that is necessary to degrade the receiver's performance below its rated specification. Actually, in many cases, the power level of the interfering signal outside the band is limited by the power output of the interference test generator rather than the receiver under test itself, particularly when the receiver under test uses good RF front end filters. Figure 11.1 shows that, outside the channel bandwidth for the receiver, a comparatively high amount of interference power is necessary to disrupt performance, but within the channel bandwidth, much less power is necessary. In EMC terms, this is a "front-door" interference scenario, where the unwanted interfering signal enters the victim system (radio receiver) by a pathway that is intended to receive other radio signals. When coming through the front door, very little power is required.

With a spread spectrum receiver, the selectivity curve produced by this kind of test is much shallower. In the case of an FHSS/DSSS combination signal of the kind used in small UAV uplinks, a CW interference signal may not bring about a reduction in receiver performance at all, so instead of the notch shown in Figure 11.1, the selectivity curve could be a horizontal line, or perhaps a shallow bathtub curve. Recall the earlier discussion about code spreading techniques: while the CW interfering signal is present in the receiver's bandwidth with much greater power than the wanted signal, it does not have the desired coding properties, so the receiver is less able to "see" it. Furthermore, since the CW interferer is, by definition, very narrow band (actually theoretically zero bandwidth according to Fourier), it only disrupts a small proportion of the entire 2.4 GHz control band—for example, it may disrupt one of the 36 frequency channels in a UAV control scenario. However, the control commands are transmitted many times, so the entire message is not embedded in one package at one frequency; as a result, the functional performance of the radio link is not affected by the CW interferer. To disrupt such a link, the interfering signal must either be so powerful that it drives the receiver into nonlinear compression, rendering it useless across the entire band, or be of sufficiently wide bandwidth to disrupt a high proportion, of the spread spectrum bandwidth.

In case of an analog receiver, the degradation of performance in the face of an interfering signal is gradual—the baseband signal (often audio) becomes noisier and more distorted, but remains intelligible until the interference is sufficiently strong to drown the wanted signal out completely. After removal or a reduction in level of the interfering signal, recovery of normal performance in the receiver is usually fairly rapid.

In the case of a digital radio communications link of the kind used in small UAV controls, failure modes are more complex, and are dependent on the nature of the interfering signal, the receiver's physical properties and the coding and error processing that the overall receiving system uses to extract the data. However, as a general characteristic, digital radio communications failure is nonlinear: the user of a digital radio link may observe small artifacts in the received data (e.g., pixilation in a video signal or, apparently, sluggish response to controls in a UAV), followed by an abrupt total loss of link. Moreover, recovery of the link may require a substantial reduction in interference power, because digital link establishment—the "locking on" of the receiver to the frequency and code of the transmitted signal—is more difficult than maintaining an established link. This process may take several seconds; for a UAV control signal, the practical implication may be that the link cannot be reestablished until the controller and UAV are brought together under much better propagation conditions and without the influence of an interfering signal.

11.3 Payloads

For small UAVs, one of the most popular payloads is video, and video acquisition is enhanced when the operator can see the video images in real time, so a live video downlink is a highly desirable feature in a UAV.

Figure 11.2 shows a possible small UAV/camera integration. The UAV and camera are made by different manufacturers; their spectra are respectively shown in Figures 11.3 and 11.4. The compact nature of the assembly brings the Wi-Fi camera transmitter into close proximity with a number of victim systems on the UAV itself, most importantly, the receiver for control

Figure 11.2 Small UAV with a Wi-Fi camera installed. There are unshielded wires on each of the four UAV landing legs. Two are wire antennas for the 2.4-GHz control receiver, one ribbon is for the compass module near the bottom of the leg and the other ribbon (not visible) is for an optional gimbal (not used in this installation). The camera's Wi-Fi transmitter antenna is integral with the camera module and is approximately 50 mm from each of the four legs and the body of the UAV.

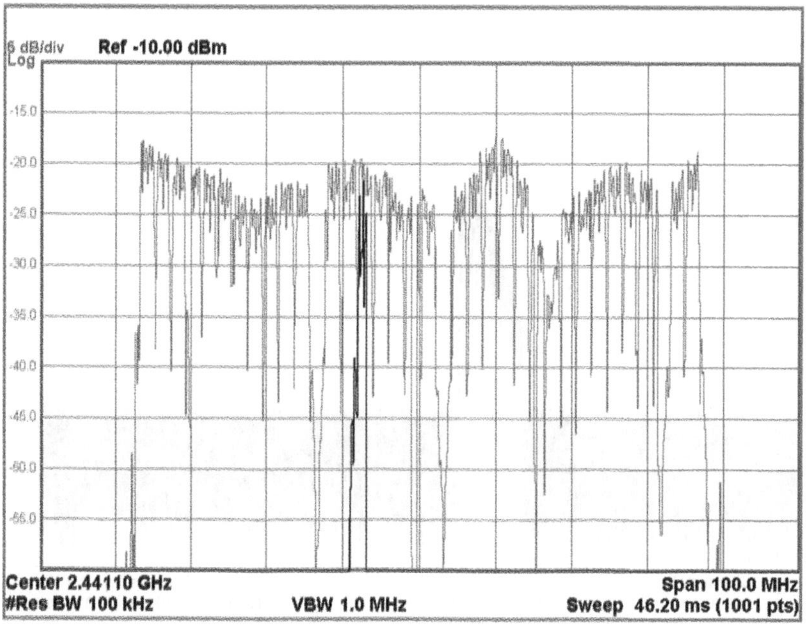

Figure 11.3 Example of FHSS/DSSS combination small UAV control signal. The light trace is built up over several minutes and shows most of the spectrum occupied by the controller and the dark trace is a single burst of data in one hop.

commands. This integration is one that may cause EMC problems for the UAV. The video downlink transmissions will almost certainly render one third of the FHSS control channels unusable, and depending on the in-band unwanted signal rejection characteristics of the receiver, possibly more than that. The difficulty with implementation of this kind is that the initial pre-flight checks of the system (especially verification of control functionality) may well indicate that the system will work, as the transmitter is very close to the UAV during such checks, but a dramatically reduced range may be observed once the UAV moves away from the controller.

In contrast to UAV control links, the video downlink is a high data rate and high bandwidth application. In this case, DSSS techniques with lower spreading factor and little or no frequency hopping are employed. This gives rise to a situation where the transmitting signal is "on the air" more often than a low duty-cycle FHSS transmission. Many readily available wireless video systems operate in the 2.4 GHz band, and this brings about a significant EMC challenge for the small UAV integrator/operator. Figure 11.3 shows the spectrum of a typical small UAV controller, and Figure 11.4 shows the spectrum of a typical Wi-Fi video "action" camera of the kind that could easily be installed on a small UAV.

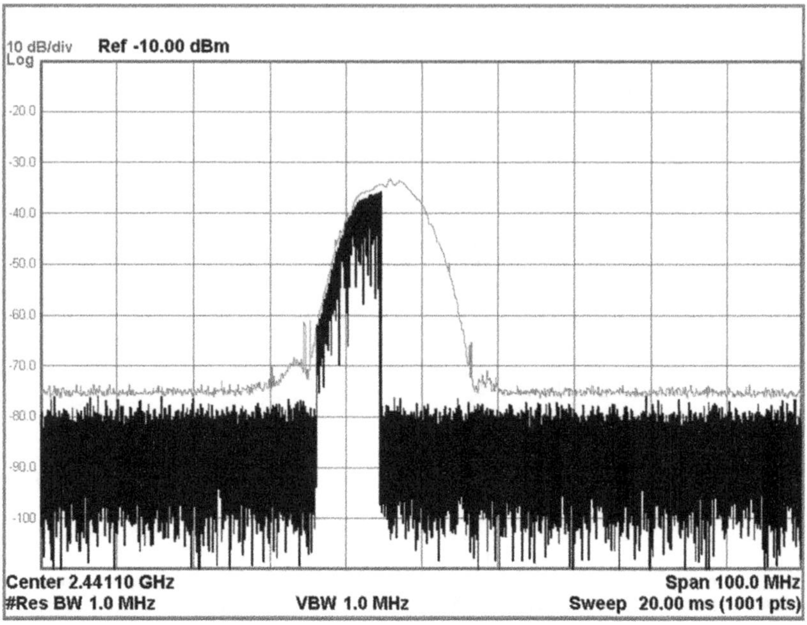

Figure 11.4 Example of a typical 2.4-GHz Wi-Fi camera signal. The light trace was built up over several seconds to indicate the total spectrum used by the camera; the dark trace is a single transmission burst.

Figure 11.5 shows the camera Wi-Fi signal superimposed over the UAV control signals; an amplitude offset has been applied for clarity of illustration. This figure shows that the camera occupies the same spectrum as approximately 12 of the 36 channels used by the UAV controller.

Incompatibilities between transmitting payloads and the operation of the UAV itself can be mitigated in a couple of ways:

- Payload transmissions in a completely different band from the UAV controls (e.g., 5.8-GHz band controls with 2.4-GHz band video)
- Payload transmissions integrated into the system under a single controller—this allows coordination of hopping spectrum use to ensure that payload transmissions occur at different channels and/or times to uplink control transmissions

The impact of a transmitting payload, or any payload for that matter, on the control uplink of a UAV can be evaluated by performing a desensitization test on the transmitter. A desensitization test provides an indication of the loss of sensitivity of a receiver due to the presence of a potentially interfering source. By using the actual control transmitter and comparing performance with and

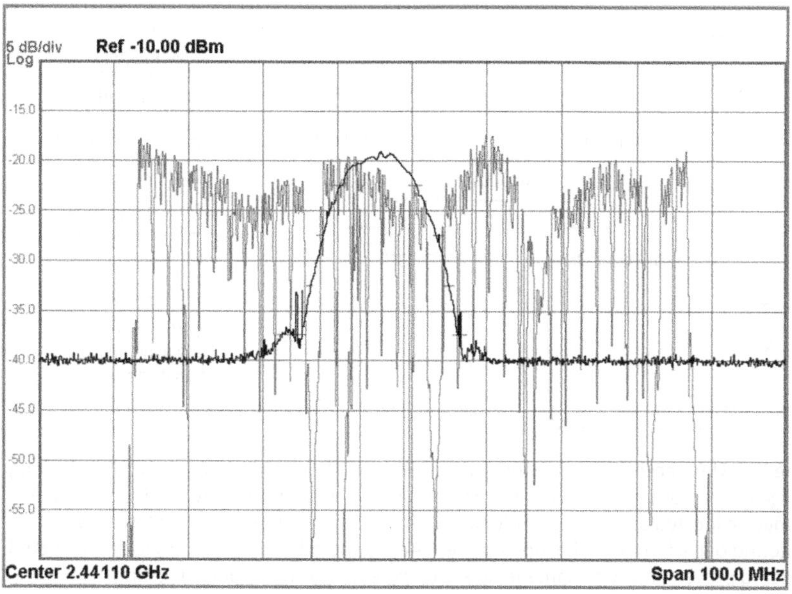

Figure 11.5 Composite of UAV controller spectrum (light traces) with Wi-Fi camera spectrum (dark trace) showing overlap in spectrum occupancy.

without a potential interferer, the system-level response of the UAV can be evaluated, which allows the UAV to employ any adaptive techniques it may have to manage interference—for example, it may adjust its use of hopping channels to avoid channels where it has detected interference.

The key steps are:

(i) Establish a test setup where the transmitter signal can be coupled to the receiver under test via variable coaxial attenuators. This usually entails placing the transmitter or receiver in a suitable shielded enclosure, in order to prevent leakage from the transmitter reaching the receiver with sufficient intensity to provide normal performance. The signal may be coupled from the output of the attenuators to the receiver under test via a suitable coupling antenna or other structure. An example of a suitable setup is shown in Figure 11.6.

(ii) Set the variable attenuators to zero and establish a baseline "normal performance" response from the receiver (in the case of a small UAV, a status lamp on the vehicle may provide a suitable indication; ensure that the UAV is suitably tethered as loss of signal during testing may elicit a "fly home" response). The highly nonlinear failure mode of spread spectrum radio links and the high level of redundancy in the transmissions means

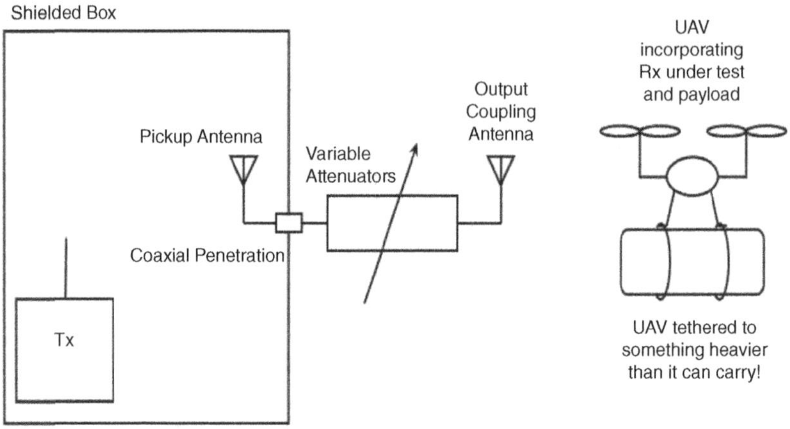

Figure 11.6 UAV sensitivity test setup example for a unidirectional radio control link. So long as the shielded box provides sufficient isolation between the transmitter and receiver, the UAV may be located at a conveniently close distance to the attenuators to allow easy monitoring of performance. Ensure that the UAV is tethered and that access to it is restricted in order to prevent injury or damage due to unexpected behavior during testing.

that a statistical approach may be beneficial to determining the baseline—for example, criteria for normal operation may be that the link must remain open for a defined period of time.

(iii) Increase the variable attenuation (typically in 10 and 1 dB steps) to determine the amount of attenuation required to break the control link as indicated by a change in behavior of the UAV from the baseline established at step (ii). Record this attenuation setting as the reference level A_{norm}. If the maximum available attenuation is reached without disrupting performance, it is likely that there is insufficient RF isolation between the transmitter and receiver in the setup at step (i) and further isolation (shielding and/or separation) is necessary.

(iv) Reset the variable attenuators to zero and verify normal performance has returned. Enable the potential interferer (in this example, the Wi-Fi camera transmitter) and monitor the UAV for normal performance. When the UAV cannot attain normal performance even with the variable attenuators set to zero, increase the efficiency of the transmitter—receiver coupling in the setup established at step (i), for example, by improving the gain of the coupling antennas or decreasing separation between the output coupling antenna and the UAV.

(v) Repeat the sensitivity test described at step (iii) and record the amount of attenuation required to break the link, A_{int}.

(vi) $D = A_{norm} - A_{int}$ is the desensitization of the control link due to the interferer. The number of dB difference between sensitivity with and without the interferer is related to the loss of range by the Friis free-space path loss equation

$$L = 20\log_{10}(4\pi d/\lambda)$$

where L is the path loss (dB), d is distance (m), and λ is wavelength (m), so that a 6-dB decrease in sensitivity is the same as a halving of the operational range of the UAV.

Refinements on this approach could be implemented with more exotic equipment, such as directional couplers and power meters to monitor signal strength, and continuously variable attenuators to provide a smoother transition in coupled signal level than the more commonly available step attenuators. Furthermore, if the UAV control link is part of an integrated duplex system (for example, if the UAV incorporates a telemetry downlink with the receiver integrated in the controller), then it will be necessary to employ circulators or directional couplers to separate the uplink and downlink sections so that the path losses can be manipulated separately. Notwithstanding these complexities, the method described above can be implemented with comparatively little specialist equipment and provides good insight into the degradation that a payload may cause to a control link for unidirectional systems or for duplex systems where the downlink component is not central to operation of the control system.

Such a test was performed on the example shown in Figure 11.2, using a setup shown in Figure 11.6. Despite the considerable overlap in spectrum occupancy shown in Figure 11.5 between the payload and the control radio links, there was no apparent loss of control sensitivity when the camera Wi-Fi was enabled (that is, the A_{norm} and A_{int} values were the same). Monitoring the spectrum while both the control link and the camera link were active confirmed that the UAV was not adaptively selecting channels (as expected for this uplink-only control system); it is concluded that there is sufficient redundancy in the control transmission of this particular UAV system to tolerate loss of approximately one third of its channels, some of the time (noting that neither the camera nor the control transmitter transmit continuously).

However, the tests also showed that even without the camera Wi-Fi operating, reestablishing the uplink after breaking it required an increase in signal level of approximately 20 dB above the level at which the link was broken. This is in stark contrast to an analog radio sensitivity result, where normal operation of the receiver is recovered when the signal is increased 3 dB or less above the level at which the signal is lost. In terms of the practical implication for UAV operation, this means that if control is lost, through either interference or flying the UAV out of range, a much stronger signal is necessary to regain lock. A possible scenario then is that a small UAV, on losing its control signal, enters "fly home" mode, and begins traveling back to its inertial/GPS reference location (the launch site if all is well). For most of this time, the UAV could remain unresponsive, as it would have to come well within range for sufficient signal strength to be available to obtain a lock again. If the UAV's programming does not include a "fly home" feature, and instead it enters a "soft landing" or other safe recovery procedure, loss of the UAV is possible, depending on where it is when the control signal is lost.

11.4 Small UAV Navigation and Control Systems

Small UAVs commonly incorporate satellite navigation, inertial navigation (solid state), and magnetic compass systems; the systems are normally used in a fused navigation solution that allows the UAV to return to home in the event of transmitter signal loss and fly predetermined waypoints for mission execution. Solid-state inertial sensors are used to help maintain the attitude of the UAV during flight and the compass can be used to determine the correct azimuthal orientation.

The sources of interference on the UAV itself include high current DC motors and motor controllers, processing circuits, and other sources related to payloads. High current DC sources can upset magnetic compass sensors, and such currents are certainly present in small UAV applications. The following strategies are important in ensuring that the magnetic compass sensor is not compromised by the UAV's high current loads:

- Locate the magnetic sensor as far away from the UAV's other circuits as possible, for example, on a small pole or a landing leg that does not incorporate any other electrical services
- Keep wires to and from batteries and engine speed controllers as short as possible and not close to compass sensors or compass sensor wiring
- Keep loop areas small, especially in high current wires associated with motors and motor drive circuits

While it is tempting to think of battery wires as "DC" and engine speed controller (ESC) wires as "AC," the truth is unfortunately more complex: the ESC module produces a set of AC outputs to drive the motors, and, in doing so, it pulls current in pulses from the DC battery—so the battery side of the ESC can have substantial AC content. ESCs employ some local energy storage (capacitors) to minimize voltage drop issues associated with potentially "long" wiring runs from the battery itself, but some AC (that is, RF interference currents) will be present on the battery wires themselves. Layout of wiring, as with any EMC application, is an important determinant of interference. Large loop areas are to be avoided, and, where possible, wires should be twisted together, to maximize the opportunities for RF currents traveling in common mode along the wires to cancel each other out. High currents flowing in loops give rise to high magnetic fields, so it is always desirable to keep loop areas small.

While space is always at a premium in a small UAV, the magnetic component of the fields produced by ESCs and the motors themselves decays rapidly with distance. So, even a small increase in separation between sensitive magnetic components and high intensity magnetic fields is worthwhile.

A further mitigation against compass interference is hardware or software filtering. Since the compass output is not expected to change rapidly (certainly less than a few tens of Hz), it is a suitable candidate for slow response filtering or averaging, either in software or with analog components (low pass filter) or both. By applying sufficient rejection of frequencies near and above the motor drive frequency range to the compass system, unwanted responses can be reduced to a level that provides acceptable performance.

11.5 Electromagnetic Environment for Small UAVs

Leaving the UAV itself and its payloads aside, the operating environment for a small UAV has a number of threats. Just as full-scale aircraft are sometimes restricted in their flight paths by high power transmitters, small UAVs should be flown with due consideration for environmental threats.

Radiated field immunity performance of a small UAV is generally not known to the operator, so a set of operating guidelines can only be based on best estimates. Given that the electronics inside the UAV are lightweight and inexpensive, it is reasonable to expect that the general immunity to back-door interference would be on the order of 10 to 30 V/m. So, prudence would dictate that small UAVs should not be flown in places where the field strength could exceed about 10 V/m. This level is less than that imposed by most human exposure protection standards, so there is no particular restriction on transmitters in the community producing fields of 10 V/m in publicly accessible places. That is, an unwittingly operated UAV in a public space could easily be exposed to a field strength of 10 V/m or more unless operational precautions are taken.

A nonexhaustive list of powerful RF sources that could be encountered by a small UAV in an urban or rural area is given below and estimated far-field strength distances for 10 V/m are given in Table 11.1:

- Cell phone base station transmitter
- Broadcast radio or television transmitter
- Wi-Fi access points
- Vehicle and fixed two-way radios
- Radar transmitters (airfield, harbor, ship, and pleasure craft systems)
- Amateur radio transmitter
- Terrestrial point-to-point microwave links
- Telemetry links

Some of these may exist at locations where transmitters are not expected, for example, telemetry radio links may be used to monitor apparently remote

Table 11.1 Some transmitter types and distances where 10 V/m may be expected (far-field assumptions)

Transmitter type	Typical power (W)	Typical main lobe gain (dB)	10 V/m distance in main beam (m)
Cell phone base station	30	18	23.8
Broadcast FM radio tower	150,000	0	212.1
Wi-Fi hotspot	1	10	1.7
Vehicle and fixed two-way radio	25	2.2	3.5
Small marine radar (pulse power)	2000	30	774.6
Telemetry link	1	5.2	1.0
Amateur radio shortwave transmitter	100	2.2	7.1

water tanks or pumping stations; even a small yacht can be equipped with a weather radar of 2 kW or more, married to a high gain antenna and cellular phone base station antennas can be concealed in clock towers and other structures to meet local amenity requirements. Moreover, many high-intensity RF threats are highly directional in their radiation, so a small UAV can fly quite happily apparently quite close to a structure, but could suffer abrupt loss of control on entering the main beam of the antenna.

While cellular base station antennas are notionally omnidirectional (or at least, a set of antennas form an omnidirectional system divided into sectors), the beams are usually quite shallow in the elevation direction, in order to spread the beam across the greatest possible service area (there is usually little radiation above the antenna as the intended service areas are on the ground around the base station). This means that a small UAV ascending in the vicinity of a cellular base station will not be exposed to significant field strengths until it is almost level with the antennas themselves. It is difficult to predict with any accuracy the region of maximum exposure in front of a base station antenna, as the beam tilt is electrically adjustable by the network operator; it is safer not to fly small UAVs in the vicinity of cellular base stations. In contrast, the fields around a powerful broadcast transmitter tower are more uniformly distributed in the region around the structure.

The distances given in Table 11.1 are not ranges at which failure of a small UAV is assured. Any particular UAV may tolerate greater than 10 V/m interference fields or may fail at lower field intensities. Radio frequency susceptibility in any case is a function of frequency, and the dimensions of small UAVs are such that they are not very efficient absorbers of VHF signals like those from broadcast FM radio towers (approximately 100 MHz, 3 m wavelength): a UAV that is 0.5 m across is approximately 1/6 of the wavelength of an FM tower signal so behaves more like a short dipole than a matched antenna. Moreover, the longest conducting path in a small UAV is usually less than the distance between extremities and is often confined to a small subset of the total vehicle, so the electrical length may be only 0.15 m. In contrast, a cell phone signal is near 1 or 2 GHz, with a wavelength 0.3 or 0.15 m, and it is possible that a small UAV could intercept the RF field with good coupling efficiency, giving rise to unwanted interference currents on power and control tracks and wires within the vehicle (i.e., back-door interference).

Radar sources have high-pulsed power levels and high gain antennas. In the case of a search radar (marine weather or airfield radar, for example), the antenna rotates, so the signal is radiated in all directions, and a UAV in the vicinity is likely to be illuminated regardless of its altitude. Radar pulses are typically very low duty cycle (for example, 1 μs pulse width and 1 ms pulse interval) so most electronic circuits have higher immunity to pulsed interference than continuous interference (the victim circuit may not have time to respond to the interference before the pulse transmission has ceased). So it is possible that a UAV may require perhaps 100 V/m of a pulsed interferer to bring about a failure, which occurs at 77 m from the source in the example marine radar given in Table 11.1.

A final word on directivity: any electronic equipment comprised of circuit boards, wires and other components, such as a UAV, will have directional properties in its ability to receive unwanted RF energy. The directionality of a system is related to its electrical size, that is, the ratio between the wavelength of impinging energy and the dimensions of the system. At wavelengths shorter than approximately twice the linear dimensions of the system, lobes and nulls in the directional response will make interference problems difficult to observe and repeat in operational environments. When the wavelength is much shorter than the dimensions of the system, the lobes can be narrow, making the amount of RF energy entering the system through unwanted pathways a strong function of azimuth and elevation. The field strength estimates given in this chapter are guidelines for safe operation, assuming the worst angle of arrival of the RF interferer; in practice much higher field strengths may be tolerated under some conditions. Generally, the directional properties of the UAV's tendency to receive RF radiation are not known to the operator and could only be determined by exhaustive laboratory testing or detailed computational electromagnetic simulation.

12

DC Magnetic Cleanliness Description for Spaceflight Programs

Pablo S. Narvaez

12.1 Magnetic Cleanliness Introduction

The goal of interplanetary missions carrying scientific magnetometers is to understand the origin and evolution of our solar system planets. Understanding our planets should provide knowledge needed to understand the origin of our own solar system and other planetary systems. The magnetometer instruments investigate a planet's magnetic fields and characterize the planet's internal magnetic field, its secular variation, and the external magnetic fields associated with currents in the magnetosphere. These measurements are typically performed by dual fluxgate magnetometers and scalar/vector helium magnetometers, as shown in Figure 12.1. The objectives of these measurements require high-accuracy magnetic field measurements and accurate attitude knowledge at the magnetometer sensors. The more challenging task is to meet a magnetic cleanliness requirement levied on typical spacecraft ranging from 0.2 to 10 nT static and 0.5 nT variable at the magnetometer sensor location.

To help achieve magnetometer science objectives, typical interplanetary missions should require that magnetic control be exercised over magnetic fields produced by other spacecraft subsystems and external sources such as mechanical equipment tools and ground support equipment used during integration and testing phase of a spacecraft assembly process. Most missions are low cost; hence, magnetic control plans are followed as best as is feasible under project cost constraints.

The Research was carried out at the Jet Propulsion Laboratory, California Institute of Technology under contract with the National Aerospace and Space Administration. U.S. government sponsorship.

Handbook of Aerospace Electromagnetic Compatibility, First Edition. Edited by Reinaldo J. Perez.

(a) (b)

Figure 12.1 Sketch of spacecraft and photos of scalar helium magnetometer and (b) fluxgate magnetometer sensor.

12.2 Magnetic Cleanliness and Control Philosophy

The degree of magnetic control required by today's typical spacecraft is essentially the same as that achieved on past missions that have flown scientific magnetometers except for the added emphasis placed on achieving a low spacecraft magnetic remnant field that is within manageable low cost constraints and available resources. The level of magnetic control is the result of concern about magnetic materials and devices and an effort to reduce or eliminate sources that may yield significant magnetic fields at the outboard and inboard magnetometer sensor locations, thus minimizing contamination of scientific data with spacecraft magnetic sources.

The required magnetic control is governed by the sensitivities of the magnetometer sensors whereby the magnetic field must be reduced to a level where science is not affected. The basic magnetic requirement for typical programs is for the spacecraft to limit its overall magnetic fields at the outboard magnetometer sensor location to levels from 0.2 nT up to about 10 nT. It is also required that the dynamic (or varying) field be less than 0.5 nT at the outboard magnetometer sensor location. This dynamic field requirement applies primarily to changes in the magnetic field over a period of tens of Hertz to several weeks and includes magnetic fields due to changing flow of currents in the spacecraft or to repositioned or moving magnetic devices such as mechanisms rotating the solar array panels.

In order to ensure that spacecraft magnetic fields are within the magnetometer science requirements, subsystem-level magnetic requirements are contained in this chapter for all typical major spacecraft subsystems. The magnetic cleanliness should be assessed at the subsystem level as well as at the system level. On any interplanetary program, emphasis on early testing of suspect hardware should be placed. Instruments with permanent magnets or solenoid devices should be tested or analyzed in order to identify problem areas that can be addressed early in the design stages. Recognizing that some instruments may have difficulty in achieving magnetic requirements due to the nature of their functions, their respective impact to magnetometer science should be assessed and changes implemented wherever necessary. Hardware with high power or high currents, likewise, should be given special attention.

A system-level magnetic compatibility test should be planned for the spacecraft in addition to the subsystem-level measurements. The system level test should determine the overall spacecraft magnetic field at the magnetometer sensor locations. Subsystem-level measurements should be used to determine the approximate magnetic field at the FGMs and scalar helium magnetometer by analysis prior to the system-level test. To improve the accuracy of this analysis, accurate magnetic characterization measurements of each subsystem should be made.

12.3 Magnetics Cleanliness Program Description

An interplanetary spacecraft magnetic control program is an effort to ensure that spacecraft has reasonably low magnetic field levels which are consistent with the overall magnetic cleanliness requirements at the magnetometer sensor locations and to maintain the flight hardware at its lowest magnetic field condition within reasonable cost constraints prior to launch. In order to achieve these goals, it is necessary to impose design requirements on hardware and to test and evaluate hardware to verify compliance with design requirements. A project's mission magnetic design requirements are based on these premises.

Each project should have a magnetics control engineer from the magnetics control review board or MCRB (consisting of members from all vendors delivering hardware) that should be made available in order to provide assistance to spacecraft-cognizant engineers regarding magnetic cleanliness matters on their respective instruments and to identify potential threats posed to the magnetometer instrument and to recommend solutions to these concerns. Assistance from the MCRB should be available to hardware-cognizant engineers for determining the most desirable hardware design approach for achieving spacecraft magnetic cleanliness goals. A set of guidelines are outlined in this chapter to assist each spacecraft-cognizant engineer in developing the best design which produces the least amount of magnetic fields. It is required that all hardware-cognizant engineers verify the adequacy of their magnetic control design as early as possible by having these parts assessed and magnetic measurements performed by the MCRB on representative magnetic devices planned for use in their subsystem. It is very important that this be done as soon as possible to allow for flexibility in the available solutions. The magnetics control engineer should work closely with cognizant engineers for those instruments that are found to be the worst-case sources of magnetic field generation. Advice as to what are the most desirable nonmagnetic materials should also be provided. Furthermore, the magnetics control engineer and the magnetics test facility should be available for evaluation and tests of parts, particularly those within close proximity to the magnetometer sensors.

The magnetic testing of spacecraft subsystems should be performed as early as possible. In testing, some of the subsystems may be energized in worst-case power-consuming modes to determine the magnetic fields due to their current flow. Magnetic tests consist of first measuring the subsystem, then demagnetizing the subsystem in a zero-field ambient and then measuring its residual magnetic field in a zero field (i.e., earth's field reduced from 50,000 nT to nearly zero in a working volume). This test should be performed just prior to delivery to the final spacecraft assembly area or integration and test (I&T) area. Subsequent to the demagnetization, any hardware removed from the spacecraft for reworking or modification must be remeasured prior to delivery and may be demagnetized if the residual magnetic field has changed. This post-demagnetization control should be continued from the moment the spacecraft leaves the I&T area en route to the launch site and launch pad. In order to ensure that the spacecraft was never subjected to high magnetized or perm conditions during transport to the launch site, small portable calibrated magnetic sensors should be placed, before spacecraft shipment, on the surfaces of the spacecraft and/or shipping containers. These calibrated magnetic sensors should easily magnetize in the presence of a strong magnetic field and, hence, determine the extent of the external magnetization exposure that the spacecraft instruments and other flight hardware has received. These sensors should then be measured upon arrival to the launch site to determine the extent of magnetization and whether a final demagnetization of any hardware is found to be necessary prior to launch.

Measurements of the effect of current loop fields and other dynamic effects at the magnetometer sensor locations caused by the operation of the spacecraft in various power-consuming modes are recommended during spacecraft level functional testing. Solar arrays should be analyzed and tested to ensure the absence of large current loops in its solar cell string (and circuits) design.

A comprehensive checklist of the magnetic cleanliness program is given in Table 12.1

Table 12.1 Magnetic cleanliness checklist

Magnetic cleanliness requirements for typical interplanetary spacecraft
1. The spacecraft total magnetic field contribution should not be greater than 10 nT DC and 0.5 nT AC (0–30 Hz) at the magnetometer sensor location for typical missions.
2. Magnetometer sensors should be mounted in a suitable location far from the spacecraft bus, preferably at the end of a long nonmagnetic boom usually 5–10 m on length.
3. A magnetic model should be produced using all measured subsystem magnetic fields. The model should indicate compliance with the overall maximum allowable spacecraft DC and AC magnetic field requirement at the magnetometer sensor location.
4. System-level magnetic field tests should be performed on the spacecraft to verify compliance with magnetic field specifications.
5. Magnetic control plan and design guidelines should be provided to all spacecraft project-cognizant engineers.
6. Magnetics working group should be formed to address all magnetic cleanliness issues.
7. A lead magnetics engineer should be identified at each contractor site and be responsible for the magnetics cleanliness program at each of these contractor sites.
8. The magnetometer experimenter team should provide instrumentation and training to the contractor in order to allow suitable magnetic measurement capability.
9. Lead magnetic engineer should be responsible for a magnetic bookkeeping data base that keeps track of all component magnetic moment measurements.
10. All significant DC and AC magnetic sources should be identified early in the project by the magnetics working group.
11. Components should undergo early magnetics characterization tests with results to be submitted to the magnetics working group.
12. Components should meet their respective magnetic field allocation requirement when measured at a distance of three times its largest dimension.
13. All magnetic specification exceedances should be submitted to the magnetic working group for waiver consideration and resolution.
14. For components exceeding their respective magnetic allocation, one of the following magnetic reduction techniques should be implemented when approved by the magnetics working group: (a) demagnetization, (b) self-compensation wherever identical dual parts are used, (c) magnet compensation using permanent magnets, (d) magnetic shielding of source within component, and (e) reduction of current loops.
15. Solar array wiring layout should employ backwiring of all cell strings with the intent to minimize all current loops.
16. Solar array wiring layout should be modeled using magnetics software tools in order to ensure current-generated magnetic fields meet specifications prior to selection of final wiring layout
17. Solar arrays should be measured and verified that the magnetic fields are within specifications.
18. Solar panel nearest magnetometer sensor should be nonmagnetic (i.e., no magnetic parts/components allowed).
19. Solar array/power distribution/battery/shunt system harnessing should be laid out with the goal of fully reducing all loop areas between these subsystems.

(*continued*)

Table 12.1 (*Continued*)

Magnetic cleanliness requirements for typical interplanetary spacecraft
20. All motors or other continuously moving magnetic parts should be considered for magnetic shielding with high permeability material after consulting with and with the approval of the magnetics working group.
21. The use of Invar or any soft magnetic material should be approved by the magnetics working group.
22. Battery should use cross-strapping of all cells to reduce loop area.
23. All power lines should have dedicated active and return lines and be wired side by side and twisted.
24. All heaters should be noninductive.
25. Chassis should not be allowed to be used as a return for power.
26. Connector pins should have active and return pins next to each other to minimize loop area in connectors.
27. Spacecraft should be monitored for magnetic fields when transported to the launch complex.

RECOMMENDED PROGRAMMATIC APPROACH

12.4 Early Magnetic Cleanliness Involvement

A consistent finding in spacecraft magnetic cleanliness programs is that early involvement of the magnetometer experimenter and the magnetic control engineer is imperative. History is replete with the excessive costs and adverse schedule impacts incurred by the need to implement material or location changes late in a project.

Early in the design process, the hardware-cognizant engineers should work with the magnetic control engineer to plan for testing hardware to resolve issues which may be of concern for magnetic compliance well in advance of subsystem/hardware design reviews.

Developmental testing should be performed for informational purposes only and pass/fail criteria are not applied. Typically, the testing consists of a "magnetic mapping" of the hardware in sufficient detail to identify those components which may preclude the magnetic compliance of the assembled unit. The detailed results of developmental testing should be provided in a brief report or interoffice memorandum (IOM) to the cognizant hardware engineer.

A typical approach is shown in Figure 12.2 whereby the MCRB identifies potential issues and evaluates impact prior to charting next path forward.

12.4.1 Magnetic Requirements Workshop

Communication of magnetic requirements is of major importance in implementing a magnetic cleanliness program on a large project. A useful tool for

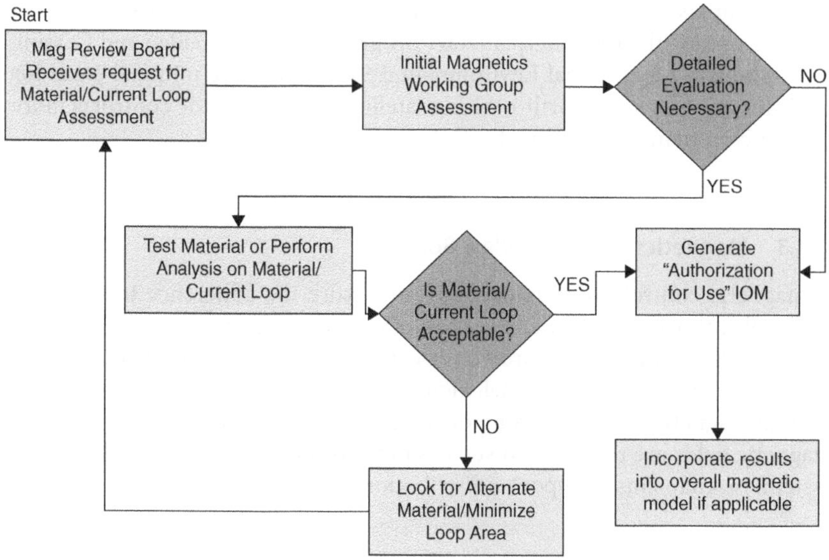

Figure 12.2 Typical developmental magnetic testing program approach.

the communication of the magnetic design rationale and requirements is a workshop conducted early in the design phase. Project management support is required to stress the potential savings in cost and schedule attainable by addressing issues upfront. Adverse schedule and cost impact of implementation of fixes after the design and fabrication phase are costly and usually not as effective.

The workshop should be attended by spacecraft hardware designers as well as science instrument and experimenters and cover such topics as the rationale for the requirements, material selection, compliance methods, available resources, test methods, and any lessons learned from previous magnetometer missions.

Questionnaires should be provided to both spacecraft cognizant hardware engineers and science instrument designers and experimenters early enough to have responses available in time for the workshop. The questionnaires should be designed by the magnetic control engineer to provide enough detail to flag "high tent poles" related to compliance.

12.4.2 Magnetics Control Plan

The magnetic control plan provides the overall philosophy and guidance necessary to ensure the integrity of the spacecraft magnetic characteristics. The full implementation of magnetic control can be a major cost and schedule driver and must, therefore, be tailored to mission-specific needs and resources.

The complexity of magnetic control plans has varied from the elaborate level of those used with large interplanetary programs, such as Galileo and Cassini, Geostationary Operational Environmental Satellites to the minimal plans for programs such as small earth-orbiting satellites where attitude control is more of a concern than scientific goals.

12.4.3 Magnetics Control Review Board

A magnetic control board is necessary to assure the adherence to the magnetic control plan, the implementation of necessary corrective measures, and the overall guidance to the project. It is important to have MCRB participation on the material review panel. Figure 12.3 is an overall suggested structure for this magnetics working group or MCRB. Major contributors to spacecraft magnetic fields are represented such as propulsion, radio frequency (RF) subsystem, mechanisms, and power distribution including solar arrays.

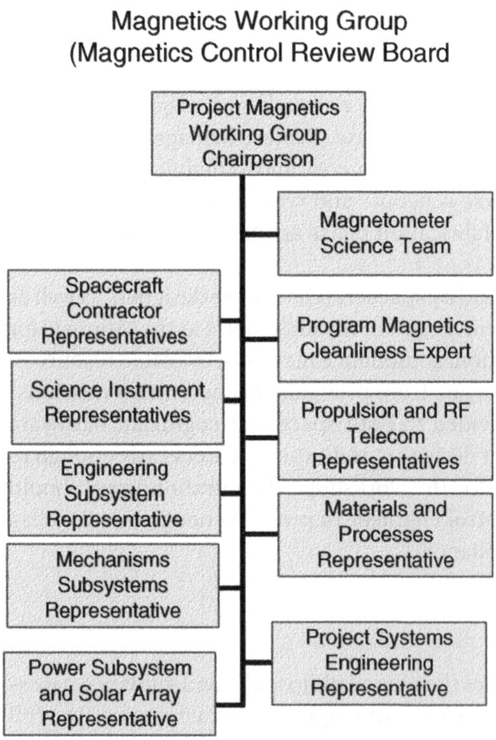

Figure 12.3 Recommended MCRB structure.

12.4.4 Materials Screening Panel

It is very important for a materials screening panel to be in place and active early in the design phase. "Nonmagnetic" is a very relative term and the control board must establish specific levels of acceptance. The acceptability of materials must be addressed in the final product (e.g., the adverse impact of work hardening on fabricated parts such as stainless steels). Subtleties such as the plating of surfaces must be monitored because industry practice frequently makes use of plating parts with ferromagnetic materials. This can be a significant problem for components close to a magnetometer sensor. Consequently, it is important to know the location of the material on the spacecraft as well as its composition.

Other considerations may be driven by the expected operational environment. For example, exposure to strong fields (spacecraft mission) can introduce unanticipated issues related to induced moments.

Compromises will be necessary, and the material screening group should work with the MCRB to assess and, if necessary, provide waivers. Waivers are implemented as a control and tracking mechanism but should not to be interpreted as a "sign of failure" by project management.

12.5 Design Requirements and Practices

In order to minimize magnetic fields produced by the spacecraft at the outboard magnetometer location to levels of less than 10 nT static (typical) and a varying field to less than 0.5 nT (typical), the primary effort should be to minimize as best as possible within the constraints of spacecraft subsystem designs the use of soft magnetic materials such as invar and Alloy 46 as well as hard magnetic materials such as ferromagnetic components (15-5 PH, 17-7 PH, etc.), particularly permanent magnets or solenoid devices. In addition, large current loops in a subsystem's wiring layout, particularly the solar array, should be avoided. All electronic boxes should strive for minimum loop areas in their respective printed circuit board (PCB) layout.

Because of the random orientation of magnetic sources in typical spacecraft subsystems, magnetic field requirements for all spacecraft subsystems should be based on probabilistic considerations. Unfavorable subsystem orientations could result in exceeding the desired field at the magnetometer sensor locations. To account for this, design margins should be added to the requirements. Since the magnetic field of a subsystem is directly related to its volume, mass, power consumption, and the number of magnetic sources therein, the design requirements stated in a magnetics control plan should reflect these differences in hardware.

Design requirements should be expressed in terms of maximum allowable radial magnetic field at 1 m from the approximate geometric center of the hardware and its related magnetic dipole moment expressed in Gauss centimeter^3 or milliamperes-meter^2. To assist hardware engineers in developing a subsystem with the lowest possible magnetic field, the following sections provide guidelines that describe the available means by which a subsystem magnetic field could be minimized.

12.5.1 Spacecraft Magnetic Design Requirements

All spacecraft flight hardware should be designed so that the maximum static magnetic field when extrapolated as the inverse cube of the distance to 1 m distance from the center of each subsystem does not exceed typical values given in Table 12.2 after demagnetization with a 50 Gauss field. The static magnetic field while the subsystem is energized should not exceed 120% of the level given in Table 12.2. Sinusoidal magnetic fields from DC to 1 kHz should be limited to 10 nT (peak-to-peak) at a distance of 1 m. This limit level is equivalent to a magnetic moment of 50 mA-m^2.

In order to meet typical science requirements at the outboard magnetometer sensor location, during test and integration and transportation to the launch site, the subsystems/spacecraft should not be exposed to a uniform external magnetic field exceeding 3 Gauss in order to minimize permanent magnetization. A typical 1 Gauss magnetization is considered statistically to be the worst magnetization level encountered during typical integration, shipment, and launch handling. Note that the Earth's magnetic field is 0.5 Gauss.

Vector changes in the magnetic field from whatever cause should be limited. This requirement applies to field changes caused by mode changes, moving or stepping devices, or changing current loop fields. Magnetic field changes caused by solenoid-type devices such as relays, or magnetic actuators which occur during special maneuvers, such as momentum offloading are acceptable if the actuation of the device can be correlated with telemetry data and operate in a very short time period.

12.5.2 Magnetometer Booms

Distance and the inverse cube relationship associated with dipole fields makes the use of boom-mounted magnetic sensors particularly attractive in that the spacecraft magnetic cleanliness requirements can be reduced significantly.

Although the use of booms is not by itself a complete solution to magnetic cleanliness concerns, their length is generally a practical trade-off between the cost of a strict magnetic cleanliness program and a tolerable level of background at the location of the magnetometer sensor. Figure 12.4 shows typical missions with long magnetometer booms:

Instead of a separate dedicated magnetometer boom, it may be possible to make use of the solar array or other structure. Scientists took advantage of the large size of the solar arrays of the Mars Global Surveyor mission and one arm as a boom to place the magnetometer sensors approximately 5 m from the main

Table 12.2 Typical maximum allowable radial magnetic field requirements for typical spacecraft hardware and space subsystems

Typical spacecraft subsystems Component/circuit	DC magnetic test required	Magnetic field limit at 1 m	Magnetic moment limit*
Command and data handling			
Central electronics unit	Yes	10 nT	50 mA-m^2
Attitude control electronics	Yes	10 nT	50 mA-m^2
Solid-state recorder	Yes	10 nT	50 mA-m^2
Telecommunications (Telecom)			
Traveling-wave tube amplifier	Yes	40 nT	200 mA-m^2
X-band small deep space transponder	Yes	10 nT	50 mA-m^2
Ka-band SDST	Yes	10 nT	50 mA-m^2
Ka-band solid-state power assembly	Yes	10 nT	50 mA-m^2
High-gain antenna, 2.5 m diameter	Yes	10 nT	50 mA-m^2
X-band medium-gain antenna	Yes	10 nT	50 mA-m^2
Low-gain antenna	Yes	10 nT	50 mA-m^2
Radio frequency passive components	Yes	10 nT	50 mA-m^2
Attitude control subsystem			
Star camera	Yes	10 nT	50 mA-m^2
Spinning sun sensor (SS)	Yes	10 nT	50 mA-m^2
Nutation damper	Yes	10 nT	50 mA-m^2
Power subsystem (PS)			
Li/ion battery	Yes	10 nT	50 mA-m^2
Li/ion battery electronics	Yes	10 nT	50 mA-m^2
Solar array switching unit	Yes	10 nT	50 mA-m^2
Power distribution and drive unit	Yes	10 nT	50 mA-m^2
Pyro initiation unit	Yes	10 nT	50 mA-m^2
Solar arrays	Yes	10 nT	50 mA-m^2
Propulsion subsystem			
Lateral rocket engine	Yes	10 nT	50 mA-m^2
Spin rocket engine	Yes	10 nT	50 mA-m^2
Latch valves	Yes	40 nT	200 mA-m^2
Thruster gimbals/actuators	Yes	10 nT	50 mA-m^2
Pressure transducers	Yes	10 nT	50 mA-m^2
Dual-mode engine module	Yes	10 nT	50 mA-m^2
Tanks	Yes	10 nT	50 mA-m^2
Payload instruments			
Typical science instruments	Yes	10 nT	50 mA-m^2

*1 mA-m^2 = 1 Gauss-cm^3.

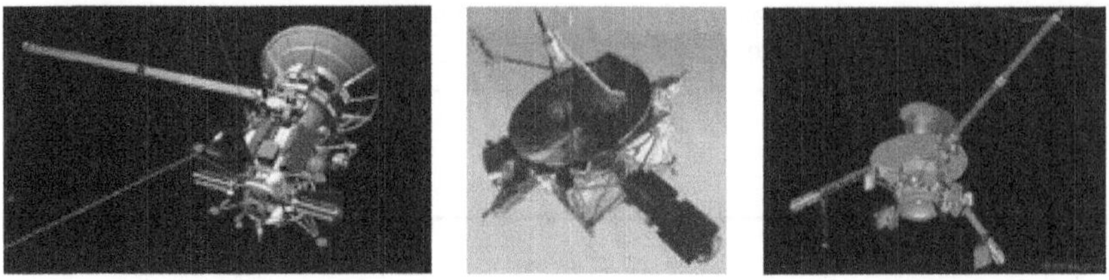

Cassini Ulysses Galileo

Figure 12.4 Spacecraft with dedicated magnetometer booms for magnetometer sensors.

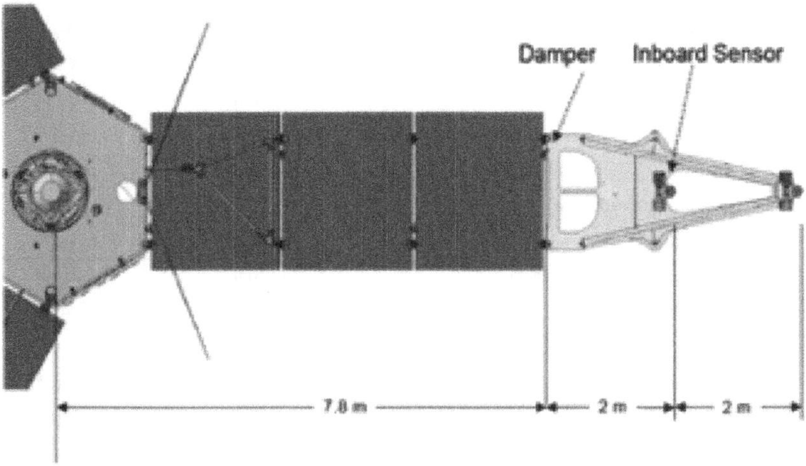

Figure 12.5 JUNO spacecraft using solar arrays as part of boom for magnetometer sensors.

body of the spacecraft. A similar approach was implemented on the Juno spacecraft as shown in Figure 12.5. With respect to magnetometer locations near the solar arrays, it is important to ensure that the array hardware, interlinks, and elements are nonmagnetic and the electrical distribution does not have uncompensated current loops.

12.6 Magnetic Assessment and Control

To provide assurance that subsystems will be able to satisfy the overall spacecraft magnetic requirements after their fabrication and assembly, the magnetics control engineer should address the adequacy of the magnetic design of subsystems based on descriptions and/or heritage of the hardware. As early as possible, each subsystem cognizant engineer should identify to the magnetics control engineer devices within his/her subsystem containing large current or permanent magnets, solenoids, and other known magnetic materials. The magnetics control engineer would then recommend which components should undergo informal developmental magnetic testing and proceed to obtain samples of these components, which may be measured by the magnetics control personnel. Any potential problems as determined by the assessment shall be reported by IOM and distribution to project personnel.

12.6.1 Magnetic Shielding

Although proven to be a good method for enclosing or shunting magnetic fields, whenever possible, the use of magnetic shielding should be avoided. It not only involves a mass penalty, it also introduces factors of uncertainty due to instability with time and temperature as well as with induced effects.

In order to separate the induced and "perm" moments, the hardware unit must be characterized both while in a near-zero field environment and again after exposure to the expected worse case environment.

12.6.1.1 Magnetic Shielding Limitations

Magnetic shields simply shunt the field lines by proving a low reluctance path, thus directing the field around the protected component. While this effect lends itself to containing the effects of a field source, particular care must be taken to avoid saturation of the shield. Saturation is evidenced by an obvious reduction in the shielding effectiveness. Figure 12.6 shows the region of optimum shielding performance below the B–H curve "knee" and where saturation occurs, above the B–H curve "knee":

Multilayer shields are attractive because of the advantage afforded by the multiplicative effect of nested layers (i.e., the shielding effectiveness increases as the product of the individual shielding factors). This can result in a significant weight saving over merely increasing the thickness of a shield to achieve a specific attenuation factor. It is recommended that a shield be designed, modeled, and then engineering model fabricated to ensure optimum performance is obtained prior to final design. Figure 12.7 is a sample unit.

12.6.1.2 Magnetic Shielding Materials

There are several commercial materials available for the magnetic shielding of spacecraft components. In addition to the traditional Ni/Fe-based materials such as u-metal and superpermalloy, there is now a class of amorphous metallic compounds, such as Metglass, which exhibit much higher permeability. The latter material is usually provided in the form of a very thin foil tape or sheet.

With the materials in the u-metal family, it is very important to ensure that a proper heat treatment follows any mechanical stress due to the fabrication of the shield. The fabricated part must be handled in such a manner as to avoid mechanical shock.

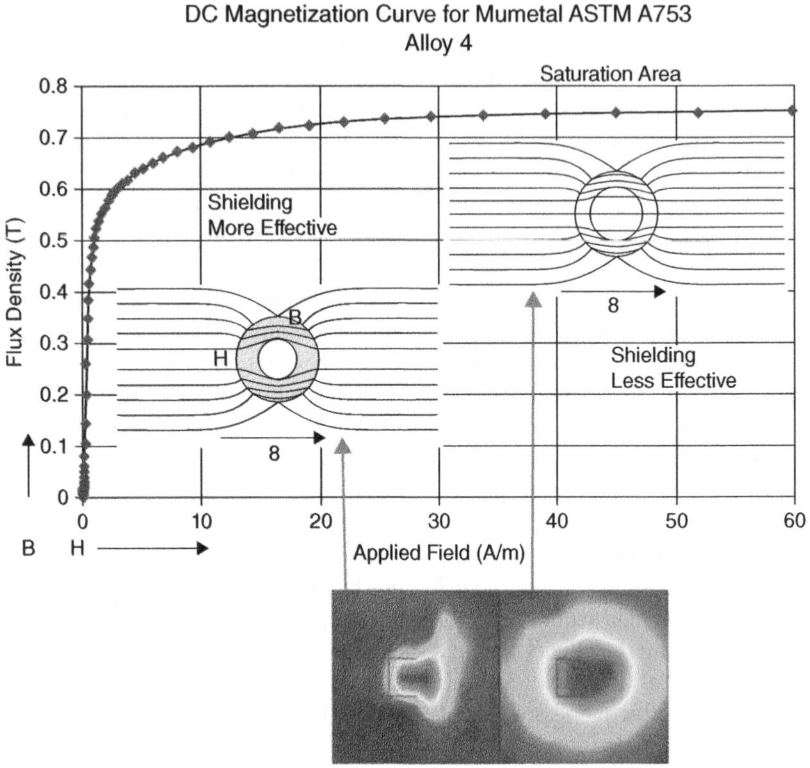

Figure 12.6 Shielding effectiveness is highest below knee and lowest at saturation.

12.6.2 Self-Compensation

Areas in which magnetic self-compensation can be most effective are telecommunications and propulsion. In cases where magnetic materials cannot be avoided, such as with traveling-wave tube amplifiers (TWTAs), RF isolators, solid state, and propulsion system latch valves, some degree of self-compensation can be accomplished by orienting the devices back to back as shown in Figure 12.8.

When self-compensation is not practical, it may be necessary to employ permanent magnets as compensating sources. However, this is more by art than by design and the implementation must be measured to verify its effectiveness. Typically, the compensation is most effective along the major axis and the orthogonal elements may remain only partially compensated. Care must be exercised in the location of the compensating magnets so as to not interfere with the performance of telecom devices such as the TWTAs and RF isolators.

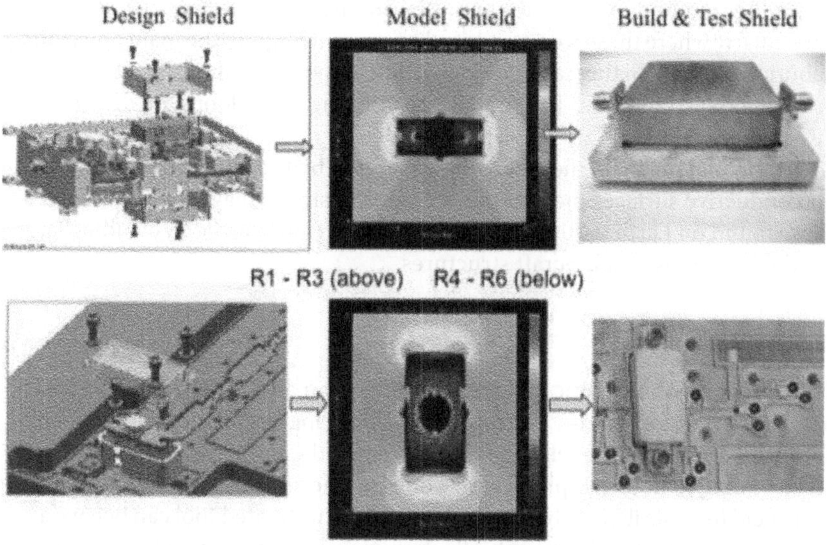

Figure 12.7 Mu-metal shielding evolution from initial design to implementation.

Depending upon the manufacturer and the design of the devices, distances in the order of 10 cm to 15 cm must be maintained. Since the compensation is to be effective at the magnetometer sensor location, there is flexibility in locating the magnets.

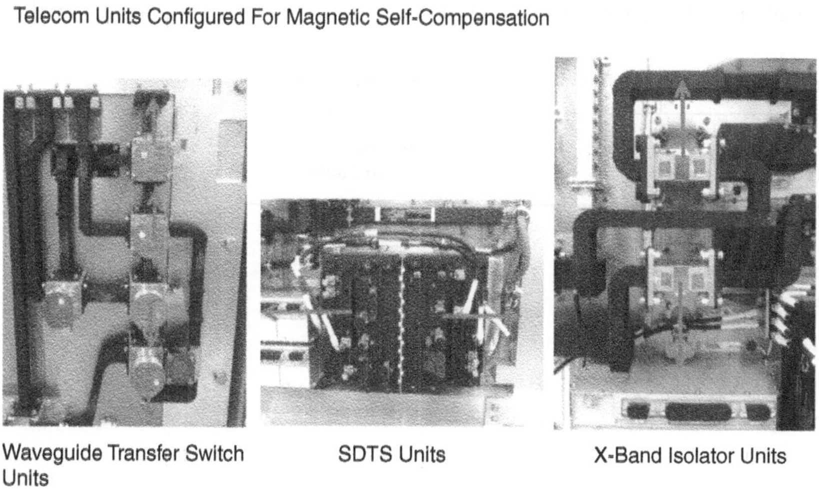

Figure 12.8 Examples of self-compensation attained by the orientation of components on a telecommunications panel.

Another method of compensation is by adding compensation current loops, particularly where these were produced by current loops in the main power distribution circuits. In this case, compensation loops can be introduced to reduce or cancel the stray fields. The size and positioning of the loops are determined by measurement.

Orthogonal magnetic fields can be generated by the eddy currents induced on conductive surfaces moving the presence of strong magnetic fields such as those found on Earth or Jupiter. Eddy currents are minimized by avoiding large conductive loops in spacecraft structures.

12.6.3 Tool Control

The proximity of magnetic hand tools can have an adverse effect on paramagnetic materials located on a spacecraft by inducing residual dipole moments. Table 12.3 illustrates the relative fields present at the surfaces of various tools. It is interesting to note that physical size does not necessarily correlate with magnetic field (e.g., scriber vs. hammer). In effect, an untreated tool can behave as a permanent magnet and adversely affect previously demagnetized components.

A procedure for the routine demagnetization and control of tools must be incorporated into the I&T procedures and a specific person should be designated with the responsibility to ensure compliance.

12.6.4 DC Current Loops

This is a not insignificant part of the allowable magnetic moment likely to be allocated to the spacecraft as a whole. In planning the cabling layout for the

Table 12.3 Magnetic tools and their approximate surface level

Tool type	Surface field level in Gauss
Holding screwdriver	30
Scissors	12
6-inch steel tape	6
Wire strippers	20
Long nose pliers	13
Scriber	100
Hammer	5
Hatchet	4
Steel scale	20
Phillips screwdriver	55

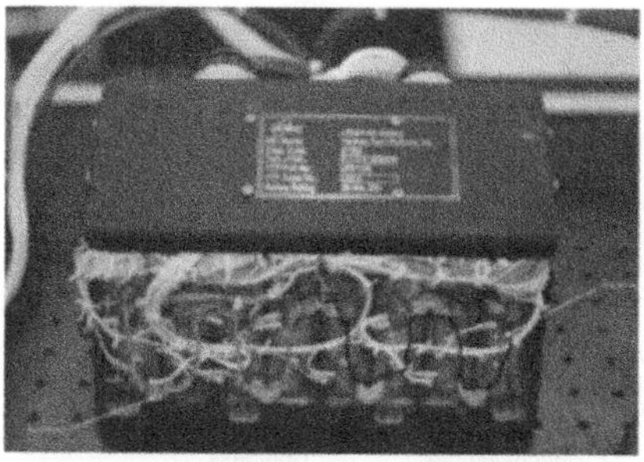

Figure 12.9 Typical battery wiring layout.

spacecraft, the designers should take care that no excessive and unnecessary current loops are formed.

Current loops associated with the spacecraft battery should receive particular attention. The mitigation of these loops is discussed in later chapters. Figure 12.9 is a typical battery layout. Cross-strapping reduces the net loop area and thus the net magnetic field.

12.6.5 Time-Variant Current Loops

Time-variant current loops are primarily associated with power, propulsion, and attitude control systems. Solar panels, power conditioners, and battery chargers all have associated with them large and variable currents. In the attitude control system, reaction wheels, magnetic torquers, and other actuators can generate large and time variable fields; in the propulsion system, latch valves, thrusters, and solenoids can also add large and variable components to the overall spacecraft generated field. Devices that involve mechanical motion must be carefully evaluated for shielding or mutual compensation (e.g., relays). Twisted pairs (not necessarily shielded) are extremely effective in eliminating the stray fields associated with electrical loads on the spacecraft.

12.6.6 Approach to Minimizing Fields from Thermoelectric and Eddy Currents

Thermal gradients across conductive surfaces can produce thermoelectric currents and associated magnetic fields. These fields can be significant if in close proximity to a magnetometer sensor. Therefore, it is important that magnetometer sensors not be mounted on highly conductive plates even though they are nonmagnetic. A temperature gradient as small as 1°C across a 0.5 cm thick aluminum plate can generate a field of several nanoTeslas at 3 cm above the surface.

Magnetic fields can be generated by the eddy currents induced on conductive surfaces moving the presence of strong magnetic fields such as those found on

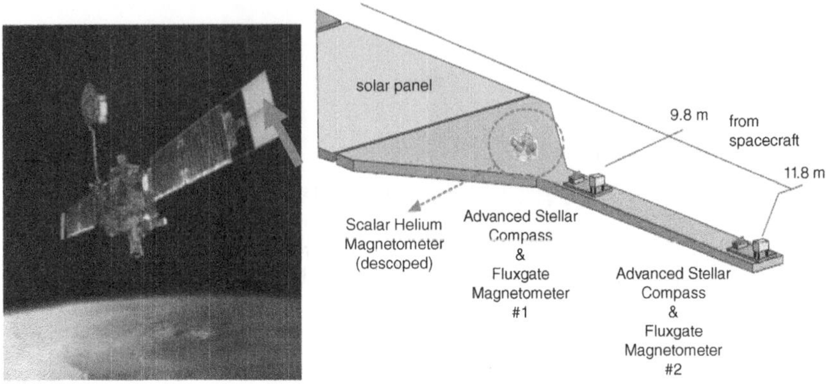

Figure 12.10 Mars global surveyor magnetometer.

Earth or Jupiter. Eddy currents are minimized by avoiding conductive loops in spacecraft structures and large conductive areas near magnetically sensitive sensors are to be avoided.

12.6.7 Magnetometer Booms

Depending upon the size of a spacecraft and its magnetic properties, nonmagnetic booms are used to position sensitive magnetometers beyond the influence of the spacecraft generated fields. Both Voyager spacecraft had 13 m booms and Galileo and Cassini used 11 m booms. In the case of Mars Reconnaissance Orbiter and Juno, the magnetometers were mounted on extensions added to the solar panels, see Figures 12.10 and 12.11. It must be noted that the latter approach imposes added requirements on the magnetic qualification of the

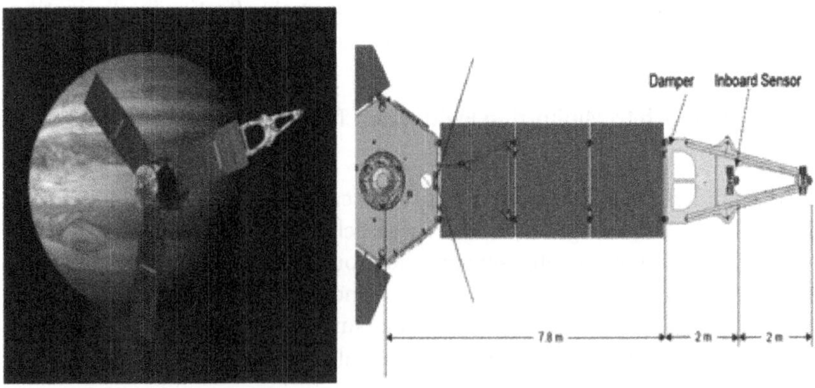

Figure 12.11 Juno spacecraft magnetometer.

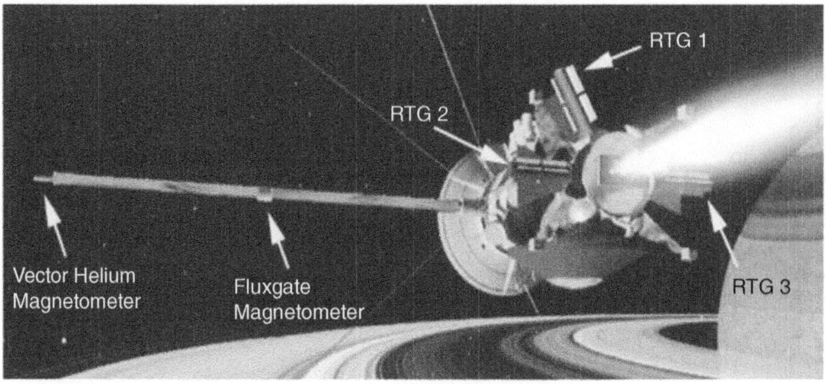

Figure 12.12 Cassini spacecraft and locations of magnetometers.

solar panel design. Cassini used a dedicated magnetometer boom as shown in Figure 12.12.

A linear plot of the Cassini spacecraft's field along the magnetometer boom provides further indication of the spacecraft's inverse cube fall off. This is shown in Figure 12.13. When the Cassini magnetometer boom was deployed, in-flight data showed the spacecraft's magnetic field gradually was reduced as approximately the inverse cube of the distance, as shown in the figure.

12.7 Magnetic Control Design Practices

In order to minimize magnetic fields produced by a spacecraft at the outboard magnetometer location to levels of less than the required specification limit for static and varying field, the primary effort should be to minimize as best as possible within the constraints of spacecraft subsystem designs, the use of soft magnetic materials such as kovar, invar, and Alloy 46 as well as hard magnetic materials such as ferromagnetic components, particularly permanent magnets, or solenoid devices. In addition, large current loops in a subsystem's wiring layout, particularly the solar array, should be avoided. All electronic boxes should strive for minimum loop areas in their respective PCB layout.

Because of the random orientation of magnetic sources in spacecraft subsystems, magnetic field requirements for all spacecraft subsystems are based on probabilistic considerations. Unfavorable subsystem orientations could result in exceeding the desired field at the magnetometer sensor locations. To account for this, design margins have been added to the requirements. Since the magnetic field of a subsystem is directly related to its volume, mass, power consumption, and the number of magnetic sources therein, the design requirements stated in this plan are intended to reflect these differences in hardware.

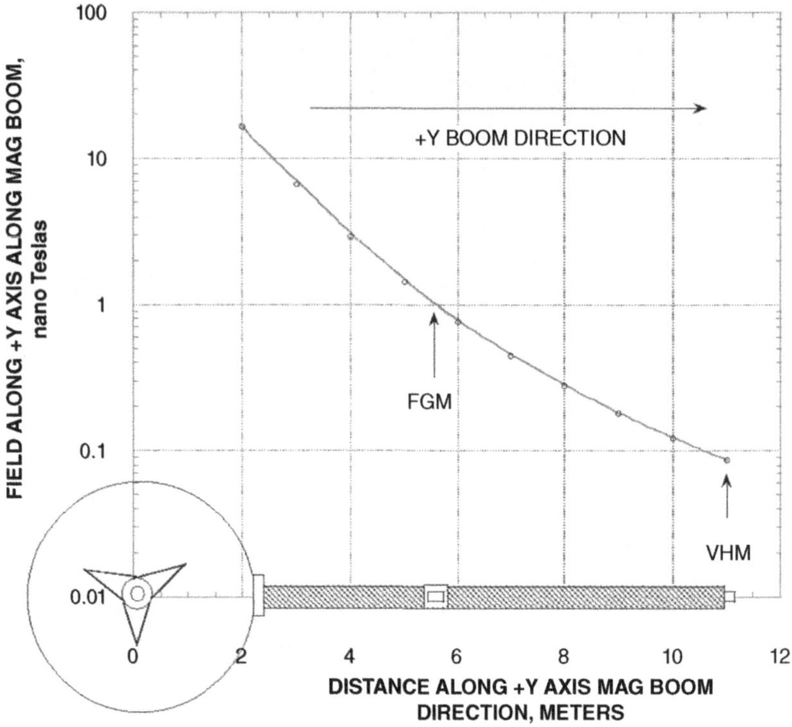

Figure 12.13 Inverse cube fall off along the magnetometer boom direction.

Design requirements are expressed in terms of maximum allowable radial magnetic field at 1 m from the approximate center of the hardware and its related magnetic dipole moment expressed in Gauss-centimeter^3 or milliamperes-meter^2. To assist hardware engineers in developing a subsystem with the lowest possible magnetic field, the following paragraphs provide guidelines that describe the available means by which a subsystem magnetic field could be minimized.

12.7.1 Design Practices and Guidelines

In order to minimize magnetic fields on spacecraft subsystems, the primary effort should be to minimize the use of ferromagnetic components in the design, particularly permanent magnets or solenoids devices and large current loops in the electrical wiring. The primary area of concern includes motors, latching relays, RF circulators and isolators, TWTAs, solenoid valves, actuators, and other devices functionally dependent on a magnetic field. Because "soft" magnetic materials pose a threat to the magnetometer science performance if highly magnetized, these materials should be used on a very limited basis or, if at all possible, eliminated altogether to prevent unintentional perming of the spacecraft due to external fields. All nonelectronic units should be able to withstand demagnetizing fields of up to 600 Gauss AC and units with electronics should be able to withstand up to 50 Gauss unless a waiver is issued for sensitive items such as the magnetometer sensors, Star Tracker, waves instrument, and associated electronics.

12.7.1.1 Selection of Nonmagnetic Materials

Spacecraft subsystems should utilize nonmagnetic materials in mechanical and electrical parts to the fullest extent possible. Nonmagnetic materials such as aluminum, brass, tungsten, fiberglass, magnesium, copper, gold and silver plating, and titanium are materials which are diamagnetic and paramagnetic having permeabilities less than 1. Paramagnetic materials usually do not show hysteresis and have permeabilities which are independent of temperature and field strength. When choosing materials, those that exhibit nonmagnetic properties should be chosen over ferromagnetic materials. For example, some stainless steel alloys are practically nonmagnetic (i.e., 303, 304). Carbon used in thermal blankets is nonmagnetic. There are different types of nonmagnetic wire and coaxial cables available (no wire with plated steel conductors nor shield wire with braided steel mesh shielding should be used). Table 12.4 is an example showing the results when an assembly changed its material from magnetic to non-magnetic material. The assembly changed its part from highly magnetic 440C to MP35M, reducing the magnetic field from 1290 to 5 nT. Table 12.5 provides a list of nonmagnetic materials.

12.7.1.2 Reduction of "Hard" Magnetic Materials

In a magnetics cleanliness program, the primary area of concern is with contributions from ferromagnetic materials, or "hard" magnetic materials, such as from permanent magnets or stainless steel alloys. "Hard" magnetic materials have properties similar to iron, such as nickel, cobalt, and all alloys of these elements. They exhibit hysteresis and have permeabilities that are dependent

Table 12.4 Ballscrew assembly magnetic test results in nanoTesla measured at 1 m material replacement—magnetic w/ nonmagnetic

Material	As received field @ 1 m (0 to p)	Immersed in 12 G field @ 1 m (0 to p)
440-C	574	1289
MP35M	< 1	5

Table 12.5 List of nonmagnetic materials

Aluminum	Gold	Molybdenum	Silver
Antimony	Germanium	Niobium	Tantalum
Beryllium	Indium	Osmium	Tin
Cadmium	Iridium	Palladium	Titanium
Chromium	Lead	Platinum	Tungsten
Columbium	Magnesium	Rhenium	Vanadium
Copper	Manganese	Ruthenium	Zinc
	Mercury	Silicon	Zirconium

on field strength and temperature. These "hard" magnetic materials have high coercivity, high retentivity, and low loss in magnetization due to mechanical treatment. They are difficult to demagnetize. Devices made up of these elements are expected to remain unaffected in the presence of expected ambient fields of 0.05 mT (0.5 Gauss) to 0.5 mT (5 Gauss); nevertheless, they should be used in limited numbers or, if possible, be avoided or demagnetized in strong fields prior to final assembly.

The most obvious sources of "hard" magnetic fields are primarily from ball bearings, motors, latching relays, RF circulators and isolators, TWTAs, solenoid valves, actuators, and other mechanical devices. Because some of these devices are functionally dependent on a magnetic field, the source cannot be eliminated or altered in any manner, which could directly affect that device's required function (other methods are used to reduce their fields without altering the source). Since more reliable mechanisms depend on magnetic actuation for switching and permanent magnets for latching, the use of relays should be limited only to the most critical functions that cannot be handled by solid state switching. The magnetic field of a permanent magnet in a latching relay can be minimized through the choice of the smallest, least magnetic relay adequate for the task.

Some "hard" magnetic materials listed in Table 12.6 typically do not make up an integral part of a subsystem's functional requirements and merely serve as a structural component. In this case, a suitable nonmagnetic material with similar characteristics should be used. In the case where "hard" magnetic materials must be used for a component, its magnetism should be eliminated or its field reduced to the point of becoming benign. Examples are the stainless steel radiation shields used in electronic assemblies or the stainless steel casing in an actuator. These materials usually incur their "hard" magnetic fields during fabrication, cold working, and annealing. If radiation shields must be used on circuit boards or if any other subsystem must utilize cold worked stainless steel (i.e., actuator casing), these parts should be checked for magnetic contamination prior to final assembly. If found magnetically contaminated, the units should be demagnetized in the presence of a zero-field ambient by exposing the

Table 12.6 "Hard" magnetic materials versus suitable nonmagnetic replacements

"Hard" magnetic stainless steel materials (difficult to demagnetize and highly magnetic)	Suitable nonmagnetic stainless steel materials
Ferritic stainless steels, CRES 400 series: 400, 400Cb, 409, 409Ni, 410S, 430, 434, 435Mod, 436, 439, 440A/B/C, 444	Austenitic stainless steels 201, Nitronic 30, 301, 304, 304L, 309S, 316, 316L, 321
Martensitic stainless steels, 400 series: 410, 410H, 420, 420HC	
Precipitation hardening stainless steels: 17-4PH, 15-5PH, 17-7PH, PH 15-7Mo	

part to fields of as high as 60 mT (600 Gauss) AC. Once these "hard" magnetic fields are reduced and found to be acceptable, the parts can then be assembled with the rest of the subsystem. Individually demagnetizing the "hard" magnetic material prior to final assembly is preferred because it enables the use of strong demagnetizing fields ordinarily not used in subsystem-level testing due to the threat of damaging sensitive electronic components. The required 5 mT (50 Gauss) demagnetization at the subsystem level is primarily aimed at reducing "soft" magnetic components.

Large mechanical structures that are unable to be measured inside a coil system will be magnetically surveyed with a Gaussmeter, and all items greater than 0.1 mT (1 Gauss) at point contact will be a candidate for demagnetization if deemed practical. High magnetic fields due to "hard" materials are generally from inserts, monoballs, and some non-titanium screws. To reduce fields from mechanical structures, large quantities of inserts, monoballs, and screws should be demagnetized in bulk before these items are used on mechanical structures (i.e., panels, hinges, struts, and main bus). Typical "hard" magnetic materials used on spacecraft are stainless steel 15-5 PH and 17-4 PH. These materials are a martensitic precipitation hardening stainless steel that provides high strength and corrosion resistance. They are used as radiation shields, damper mechanisms, and ball screw assemblies. Other "hard" magnetic materials are the 400 series ferritic stainless steels. This group of ferritic stainless steels contains 10.5% to 20% chromium for corrosion and oxidation resistant but are highly magnetic and difficult to demagnetize. On the other hand, nonmagnetic or low magnetic materials that provide a good replacement for the "hard" magnetic materials are the 200 and 300 series stainless steels. These austenitic stainless steels provide high strength and are oxidation resistant, but most importantly are nonmagnetic in the annealed condition. Depending on their composition, primarily nickel content, they become slightly magnetic when cold worked Nitronic 30 is also another nonmagnetic stainless steel material that is suitable for replacing highly magnetic "hard" magnetic materials.

12.7.1.3 Elimination of "Soft" Magnetic Materials

In past flight programs, it was determined that subsystems that used considerable amounts of "soft" magnetic materials (i.e., invar, kovar, nickel, and their alloys) had very unstable magnetic fields, particularly when these subsystems were removed from magnetically controlled areas. To maintain a magnetically stable spacecraft, the reduction or elimination of nonessential "soft" magnetic materials must be implemented by replacing these with proven nonmagnetic materials or efforts should be made to find a suitable nonmagnetic substitute with identical properties as those of the "soft" magnetic material. By nonessential, it is meant that the magnetic characteristics of the material are not a determinant on its selection. In the case where "soft" magnetic materials are essential for meeting an instrument's functional requirements such as the wide used of invar on camera structures, the invar material must first be demagnetized and then be monitored closely during shipment to the launch site in order to ensure that the material has not been magnetized. Next are given examples of the induced effects when soft magnetic materials are used. One example, in Figure 12.14, shows a nickel heat shield and the resulting induced magnetic moment when exposed to an external field of 12 Gauss. Figure 12.15 shows a magnetic model of the magnetic fields of the Main Engine Assembly (MEA) when exposed to an external 12 Gauss field. These induced effects are difficult to eliminate and predict from mission science data during postprocessing. Another example shows how mu-metal material can become an induced magnetic dipole moment in the presence of an external magnetic field is the Inertial Measurement Unit (IMU) as shown in Figure 12.16.

Figure 12.14 Main engine nickel heat shield.

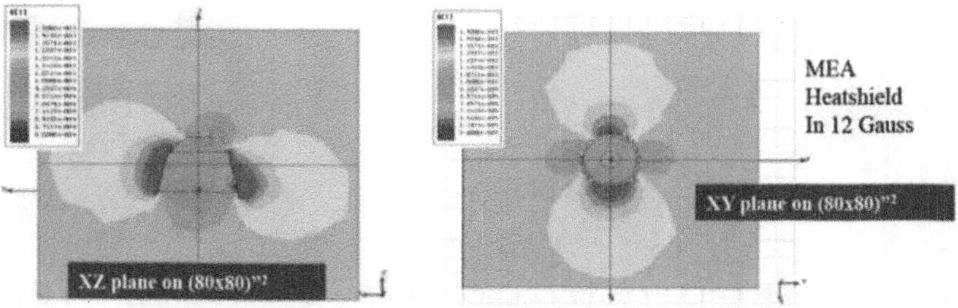

Figure 12.15 Nickel heat shield and induced effects from external magnetic fields.

12.7.1.4 Component Self-Compensation

Magnet and solenoid devices will be a serious problem where used in any quantity. It may be possible, where more than one is used, to pair them in such a way that their residual fields approximately cancel. This will not result in complete cancellation, but will result in a significant reduction in the field. For example, two TWTAs may be packaged sided by side and with opposite polarities in order to reduce their net magnetic field. In the case of banks of at least two magnetic latch relays, these should be arranged in back-to-back configuration to achieve maximum mutual compensation. Figure 12.17 is an example of isolator units mounted back to back to self-compensation their respective dipole moments (vector direction shown in "red").

12.7.1.5 Component with External Magnet(s) or Current Loops

Where only a single magnetic device is used, it is sometimes possible to reduce the field by means of an externally positioned and suitably oriented compensation magnet or current loop to partially or completely cancel the residual field, provided the field from the source is shown to remain stable in the presence of a 0.3 mT (3 Gauss) magnetization. This technique applies the principle of superposition. For example, the field due to a single latch valve can be reduced by the addition of a permanent compensating magnet, sized and positioned on the latch valve so that its magnetic field is equal and opposite to

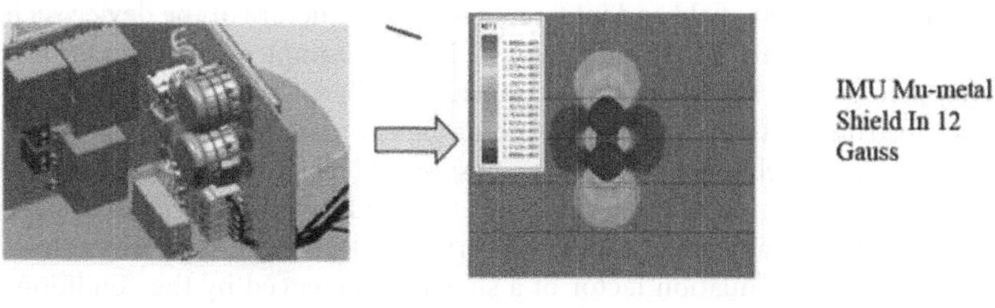

Figure 12.16 Inducted magnetic field effects due to the use of mu-metal material.

Figure 12.17 Magnetic self-compensation approach (back-to-back mounting).

the undesired field, thus partially cancelling the magnetic field of the magnet internal to the latch valve. Where current loop compensation is used, it must be observed that the magnetic moment of each offending current loop must be due to a steady current; the distance between the current loop and the compensating loop be as small as possible; and, the magnetic moment of the two loops must be equal and opposite. Any magnetic compensation practices must be evaluated by the magnetics control engineer. Because the magnetic field yielded by subsystems containing appreciable amounts of "soft" magnetic material are likely to be unstable and unpredictable, these subsystems should not be provided with magnetic compensation.

12.7.1.6 Magnetic Shielding of Source

Although it is not desirable to intentionally add magnetic shielding material to the spacecraft, in some cases it is unavoidable. Magnetic shielding material should be used carefully and in minimum amounts. It should be used to contain permanent magnets or slowly varying magnetic fields from sources such as stepper motors and stepping electromagnets. Transformers are also candidates for shielding. The shielding material should be used close to the source of the varying magnetic field and integrated into the structure of the device such as is shown in Figure 12.18 and Figure 12.19. For example, it requires far less material to directly shield a motor or transformer than it would be to shield the entire instrument that contains the motor. Shield material must be properly annealed, have a relatively complete enclosure, and all joints must overlap.

Once a shield is formed and fitted, the shielded instrument must be handled carefully, for any small shock will seriously degrade the effectiveness of the shield. The attenuation factor of a shield is influenced by the conditions such as external field intensities, mechanical and thermal shock, temperature, and stress. All applications of magnetic shielding must be cleared through the

Figure 12.18 Mu-metal shield to right used to shield isolator on the left.

MCRB. The caveat with using magnetic shielding is its soft magnetic characteristics. External magnetic fields will induce a dipole moment, and when this external field is removed, a residual (or remnant) magnetic moment remains. This is shown in the hysteresis curve in Figure 12.20.

12.7.1.7 Eliminate or Reduce Area of Current Loops

Current loops will produce magnetic fields proportional to the current, the area enclosed by the loop, and the number of wire turns in this loop. The solar array wiring layout should include minimum loop areas between all cell-to-cell wiring. The power distribution system wiring should be laid out such that current loops in the harness layout are minimized. The chassis should not be used

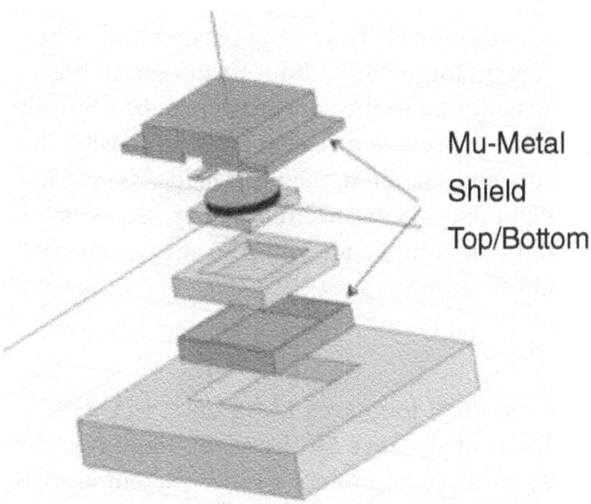

Figure 12.19 Mu-metal shield requires a full enclosure to ensure appropriate shielding.

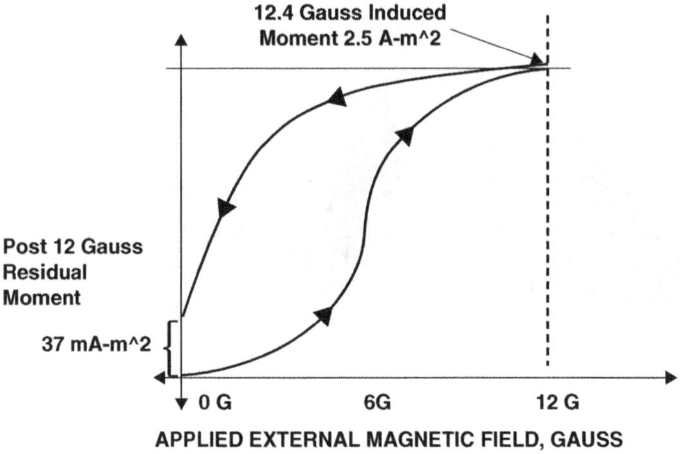

Figure 12.20 Hysteresis curve for soft magnetic material such as Mu-metal.

to intentionally conduct signal currents under normal conditions. Only fault and leakage currents are permissible through chassis grounds. All spacecraft bus subsystems should minimize current loops within their respective PCBs. A single PCB with a power line active and return grid layout having a diameter of 30 cm and 1 ampere of current yields a magnetic field of approximately 13 nT at a distance of 1 m. Considering that most subsystems have a dynamic magnetic field spec limit of 10 nT at 1 m, PCB with currents on the order of about 1 ampere, and loop area greater than 707 cm^2 could pose a potential magnetic cleanliness problem when energized. The effective area of current loops can be reduced by routing the active trace and its return together (side by side). Transformers and inductors with a minimum of air gaps are generally not a problem.

Circuit boards should use paired, side-by-side traces and twisted leads wherever possible and eliminate any current loops throughout its design. In power supplies, the current-carrying leads internal to the converter must be carefully routed and twisted wherever possible to provide cancellation of stray fields. To minimize lead lengths, and thus current-generated magnetic fields, the subsystem must observe an optimum layout of parts in the circuit board. To estimate if the PCB effective loop area will be in compliance with the subsystem-level spec limit, use the following equation:

$$|\mathbf{B}| = (2 \times 10^4\, n\, I\, A_{\text{eff}})/(r^2 + b^2)^{3/2}$$

where r^2 is the effective squared radius of loop area r^2 (in cm^2) = $(1/\pi)\, A_{\text{eff}}$; **B** is the magnetic field in nanoTeslas at distance "b" from the loop; b is the distance from PCB loop area along axis in centimeters; I is the current in PCB loop (amperes); and n is the number of turns in a PCB loop (use 1 turn).

12.7.1.8 Spacecraft Magnetometer Mounting Panel Parts Selection

A magnetically clean magnetometer mounting panel where the magnetometer sensors are located should be provided for spacecraft. To verify that all magnetometer mounting panel parts are magnetically clean, each component should be measured and verified to have the characteristics of a low residual field. Magnetometer mounting panel parts may not be considered nonmagnetic unless these have been magnetically inspected. For example, stainless steel 300 series parts may not be considered nonmagnetic unless these have been magnetically inspected. To ensure that no foreign magnetic materials are embedded (i.e., ferrous scraps obtained during fabrication) in the magnetometer panel parts, such as on plastics, titanium screws, epoxies, aluminum brackets, thermal blankets, connectors, and cables, all items should be tested. Requirements levied on magnetometer mounting panel parts are directly related to their distance to the magnetometer sensors (i.e., the closer the part is to the sensors, the more stringent the requirement becomes). For verification of these requirements, the most sensitive magnetometer equipment should be used. All thermal paints used on the magnetometer panel, particularly those within 10 cm of the sensors, will be tested to ensure that no ferrous impurities are present (in some cases, small steel beads are used for mixing paints). Steel wool or any other ferrous cleaning products should not be used on any magnetometer mounting panel part for cleaning purposes. Table 12.7 provides a list of requirements for the magnetometer mounting panel parts.

12.7.1.9 Spacecraft Solar Arrays and Panels

The solar arrays close to magnetometer sensors such as is shown in Figure 12.21, should be wired with the goal of minimizing the overall current loops areas. This is achieved by the process of backwiring each panel. Early in the program, an analysis of the solar arrays should be conducted in order to ensure the absence of large current loops. The inboard, center, and outboard panels of each solar array should be modeled by dividing each panel into circuits. These circuits should then further be divided into segments. Each individual

Table 12.7 Magnetometer mounting panel parts requirements

Magnetometer mounting panel part location	Post 100 Gauss magnetized condition	Post-demagnetized condition
Parts mounted on mounting panel	<10 nT at 2.54 cm	<2 nT at 2.54 cm
Parts mounted within 20 cm of mag sensors	<4 nT at 2.54 cm	<0.5 nT at 2.54 cm
Parts mounted within 10 cm of sensors	<1 nT at 2.54 cm	<0.2 nT at 2.54 cm

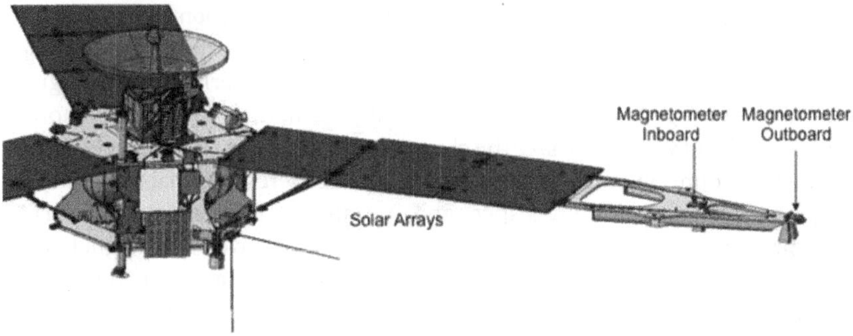

Figure 12.21 Typical spacecraft solar array and magnetometer locations.

segment is then modeled by assuming straight-line current elements flowing along the solar cell strings on the front of the panel. The backside wiring is also modeled assuming straight-line current segments representing each wire. These straight-line current elements eventually join to create a single large circuit element with a traced out loop area. The model will then determine the field produced by such traced out loop area. If the resulting loop area is too large, then the magnetics control engineer will work with the solar array vendor to devise ways of minimizing these loop areas, including rewiring the back panel.

To verify the model and ensure the absence of large loop areas, the solar panels can be measured by shining high-powered lights directly onto the front panel or by moving the panel outside and exposing this panel to direct sunlight. Measurements can be performed using a test magnetometer sensor positioned along the front surface of the panel and moved along a distance normal to the plane of the panel.

There are several methods to cancelling the effects of current loops within a solar array string. These are listed as follows (and shown in Figure 12.22):

1. Linear compensation
2. Paired linear string

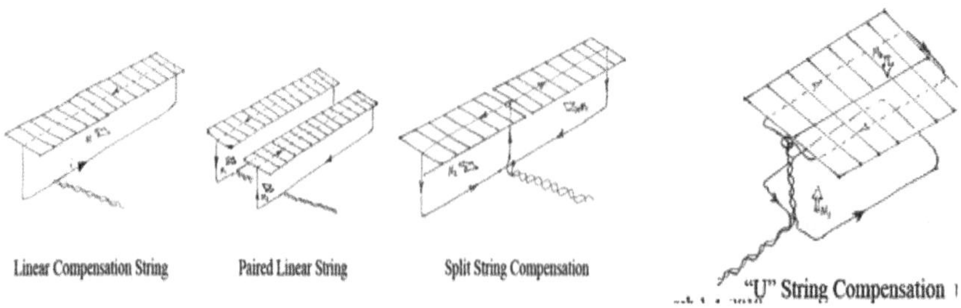

Figure 12.22 Various solar panel cell layout configurations for optimal field reduction.

Figure 12.23 Typical setup for magnetic field testing of solar panels.

3. Split string compensation
4. "U" string compensation (uses mirror image current loop nulling)

Measurements of the magnetic fields can be performed on a solar panel by injecting current into the array. Figure 12.23 is a sample picture of measurements conducted on a solar panel unit.

12.7.1.10 Spacecraft TWTAs

TWTA are typically highly magnetic. In the case where two TWTAs are employed, these units can be packaged back to back so as to cause a net cancellation of their total magnetic field. This will require that the TWTAs dipole moments be well understood so as to position one TWTA's polarity opposite to the other TWTA's polarity. In the case whereby cancellation by packaging is not feasible, then magnetic compensation can be employed by placing a permanent magnet on the surface of the TWTA and with its polarity opposite to the TWTA's polarity. Figure 12.24 is a sample TWTA compensation scheme implemented when two TWTAs were compensated using two magnets of equal strength but opposite polarity to the two TWTAs.

12.7.1.11 Spacecraft Battery Cell-to-Cell Wiring Layout

Careful layout of the battery's cell-to-cell wiring should be implemented with the goal of minimizing current loops. Figure 12.25 shows a picture of a battery and the improvement in field reduction attained with cross-strapping. During charging and discharging of the battery, load currents could potentially generate significant static as well as dynamic magnetic fields if the wiring layout contains significant loop areas such as is shown in Figure 12.26(a). Reduction of loop areas created by battery wiring harnesses can be accomplished by rerouting the active and returns as close as possible to each other. This approach is shown in Figure 12.26(b). The wiring scheme redirects one of the loop's current flow opposite to that of its counterpart, thus forcing a net field cancellation. Another simpler method is to cross-strap the cells so as to reduce the net loop area, as shown in Figure 12.26(c).

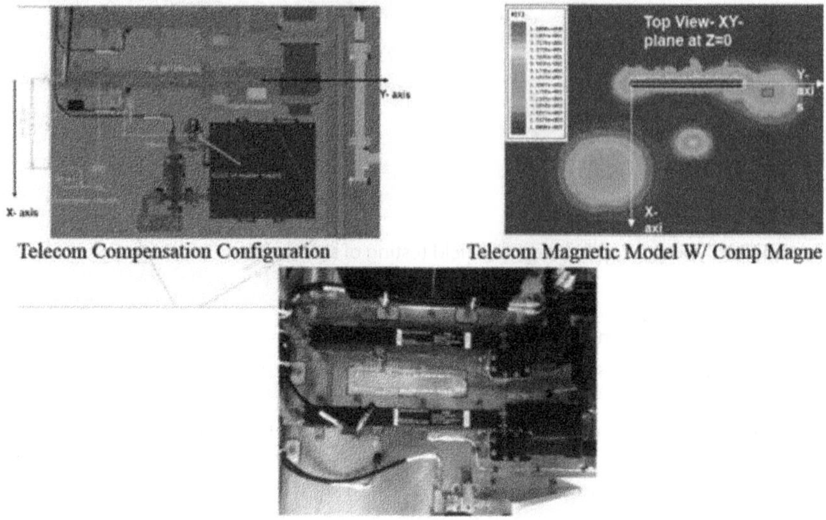

Figure 12.24 TWTA magnetic compensation using permanent magnets opposing TWTA dipole moment vector direction.

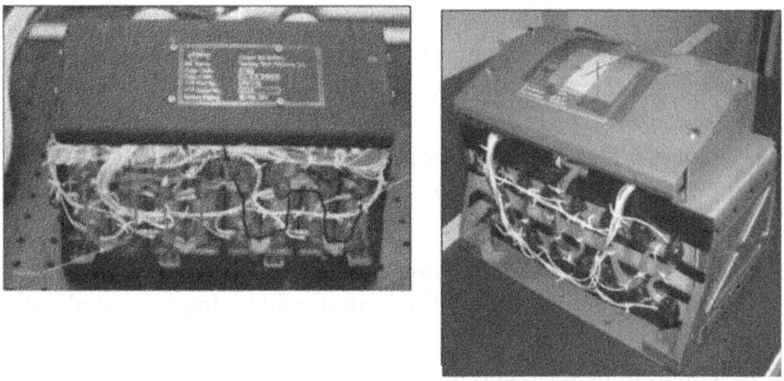

1 m Measurement Point

- Without cross strapping 0.44 nT
- With cross strapping 0.03 nT
  - Order of Magnitude of improvement

Figure 12.25 Battery wiring layout example/cross-strapping.

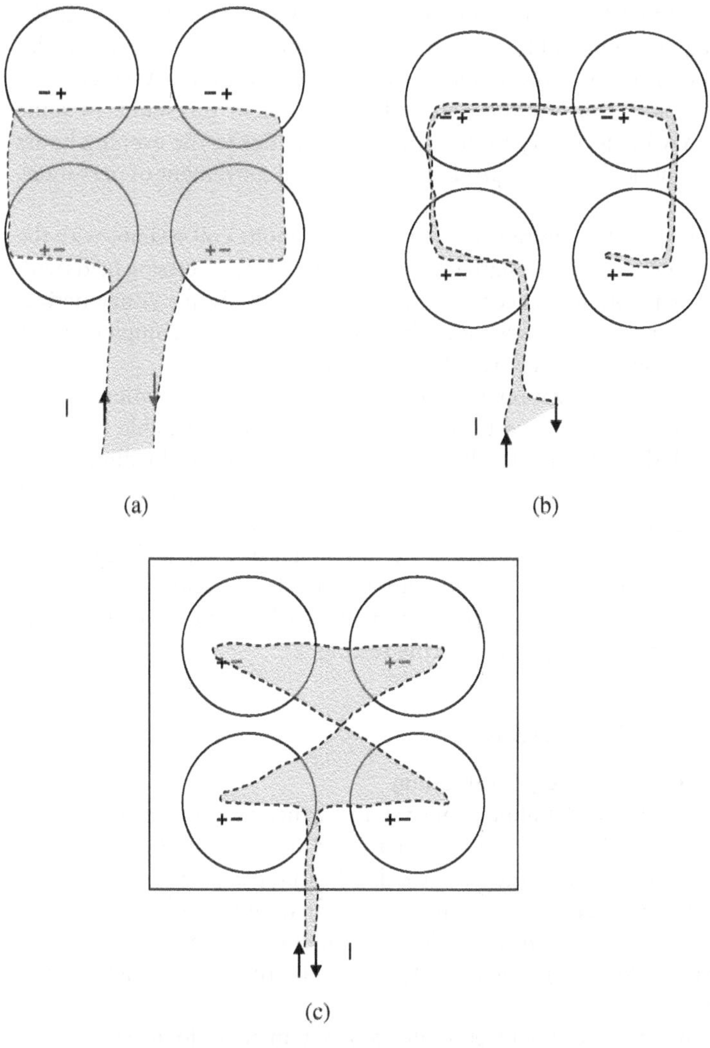

Figure 12.26 (a) Wrong way to string battery cells. Note large loop area around cell layout. (b) Right way to string battery cells. Note smaller loop area around cell layout. (c) Preferred way to string battery cells. Cross-strapping produces smaller loop area around cell layout.

12.8 Test Facilities Measurement and Methods

12.8.1 Test Environment

The facility for magnetic characterization measurements should include a capability to provide a uniform, near-zero (less than 500 nT) field environment within the device under test (DUT) volume. This is necessary to separate the true DUT magnetic moment from the magnetic moment induced by the Earth's magnetic field. In addition, the ambient magnetic conditions must be such that the noise level is well below the required measurement threshold. In some instances, it may be necessary to limit the test activities to the evening hours to avoid transient field changes produced by physical movement of personnel, trucks, cars, etc.

An additional facility requirement is to have Helmholtz coil systems with the capability to generate high-strength magnetizing and demagnetizing fields with sufficient uniform volume to accommodate the DUT hardware. A magnetization capability up to 25 Gauss is more than adequate and the complementing demagnetization source would be at least 50 Gauss.

Because the characterization process entails 360-degree rotational measurements about each DUT vertical axis, a nonmagnetic turntable is useful. The accommodation of flight hardware may involve the proof loading of this turntable and any test stands as well as providing Electrostatic Discharge or ESD grounding connections to ensure the DUT remains grounded throughout the test sequence. In some cases, the unit being tested can be bagged with nonmagnetic antistatic ESD bagging material, thereby lessening the need for a cleanroom environment. Optical instruments, such as spectrometers, may require a nitrogen purge capability during testing.

12.8.2 Helmholtz Coils Systems

12.8.2.1 Subsystem-/Assembly-Level Testing

Moderate-sized triaxial Helmholtz coil systems, 2 meters or more in diameter, can be used for assembly-level testing. Square coil systems have been designed to provide better access for larger test articles. Both types of systems are commercially available. Usually, the maximum size constraint is the entry opening to the test area. Commercially available constant current supplies are capable of providing the field-nulling currents. Typical coils are shown in Figure 12.27 through Figure 12.30.

The requirements for a small Helmholtz coil system are as follows:

1. Data acquisition system
2. Magnetometers able to measure from 1 nT up to about 100,000 nT
3. Power supplies to energize the Helmholtz coil system
4. Nonmagnetic and proof-loaded turntable
5. Three axes Helmholtz coil
6. Magnetizing (perm) and demagnetizing (deperm) Helmholtz coils

12.8.2.2 System-Level Test Facilities

There are several large coil systems capable of handling spacecraft testing in Europe, Asia, and the United States. Because of the need for very large test

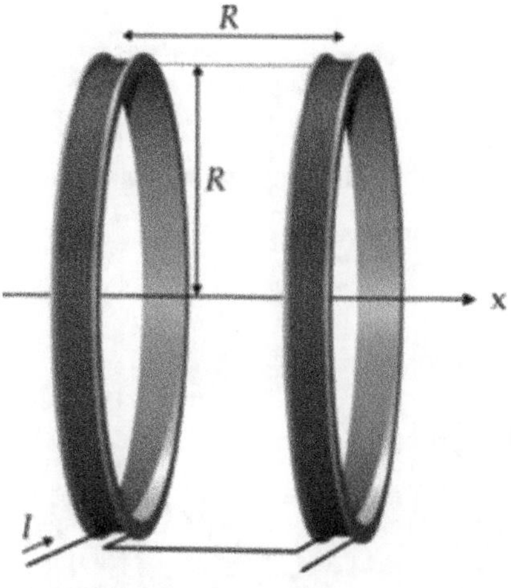

Figure 12.27 Classic Helmholtz coil pair (separation distance is equal to coil radius).

volumes, these coil systems are configured in multiple coil arrays of the Fanselau/Braunbeck or similar design.

The Goddard Space Flight Center or GSFC facility, shown in Figure 12.31, has a nominal diameter of 15 meters and incorporates a temporal correction system to null ambient field variations. It has been in operation since 1964.

12.8.3 DC Magnetic Tests Overview

All spacecraft subsystems identified by the magnetics control engineer as posing potential magnetic cleanliness problems will require magnetics tests (i.e., mag mapping). The subsystems tested must be of flight quality or magnetically equivalent to flight hardware. Tests are to be performed at locations having adequate magnetics test facilities. Tests should be closely monitored by a magnetics control engineer. A generic test procedure should be prepared and approved by the cognizant engineer and appropriate project representative. These magnetic tests are to be performed prior to delivery to Assembly Test and Launch Operations or ATLO for flight spacecraft buildup. The tests will verify that the spacecraft subsystems, when demagnetized, will satisfy the magnetic field requirements. Demagnetization of the hardware will ensure that the subsystem will have no adverse effect on the magnetometers. The tests should consist of the following sequence of tests in the order listed:

1. Calibration run with reference coil
2. "As received" mapping
3. 50-Gauss peak demagnetization in three axes
4. First post-demagnetization mapping
5. Magnetize with a uniform 25 (or 12) Gauss field in three axes
6. Perform magnetic mapping

Figure 12.28 Typical magnetic coil facilities.

7. Second 50-Gauss peak demagnetization in three axes
8. Second post-demagnetization mapping
9. Third 50-Gauss peak demagnetization in three axes (if necessary)
10. Third post-demagnetization mapping
11. Magnetic mapping while energized in worst-case mode (highest power-consuming mode)

Figure 12.29 DC magnetic data acquisition (left) and magnetometers/power rack (right).

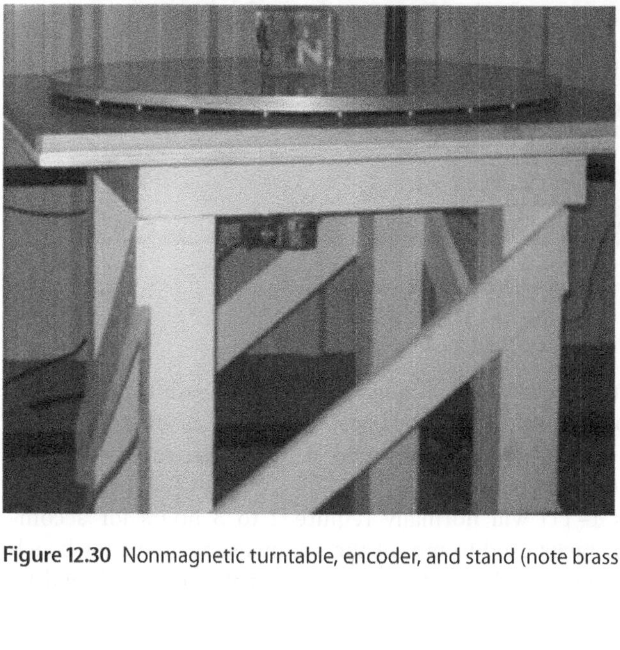

Figure 12.30 Nonmagnetic turntable, encoder, and stand (note brass ball bearings).

Figure 12.31 A 15-m diameter magnetic coils system.

Subsystems or chassis boards, which might be independently replaced during system operations, should be individually measured as in step 2. It is intended that these tests (1–11) be performed at the largest subsystem level (i.e., approaching complete instrument).

These tests (steps 1–11) will normally require 2 to 3 hours for accomplishment. Frames or racks used to mount the various subsystems for bench testing and handling and which will be used in magnetic testing should be nonmagnetic (i.e., aluminum, wood, brass, and copper are all acceptable) or placed at a distance from the spacecraft. For performing the tests, the subsystem or its nonmagnetic handling fixture should be capable of resting on three mutually orthogonal surfaces and on a fourth surface parallel to one of these three. For energized tests, cables should be sufficiently long so that they may be suspended about 1 m above the hardware during mapping and to allow the support equipment to be positioned at least 2.5 m from this support point. This will require cables at least 6 m in length. Hardware at all times should be handled and positioned by subsystem cognizant engineers or authorized spacecraft mechanical technicians. Mating and demating of connectors performed during energized tests should solely be done by certified technicians.

Transient or dynamic magnetic field measurements will not be performed on a subsystem unless the static magnetic field measurements or other evidence indicates that such tests will produce useful results. Changes in a magnetic field will be deduced from two static measurements made with the subsystem in different operating modes.

If, following this test, the subsystem is removed from the spacecraft assembly area or bonded stores (both locations are designated as magnetically controlled areas), that subsystem will require a retest, and possibly demagnetization.

12.8.3.1 DC Magnetic Test Methods and Procedures

Magnetic testing of spacecraft subsystems consists of measuring its incoming magnetic field, after various magnetic exposures, after demagnetizing the subsystem, and while it is energized in its highest power-consuming modes. The demagnetization process is then followed by a final measurement. The measurement of the residual magnetic field is known as "mapping." Magnetic mapping is concerned with static magnetic field determination.

12.8.3.1.1 Subsystem Static Magnetic Mapping Approach Magnetic field measurements on each instrument should be made at either 0.5, 1, 1.5, or 2 m from the rotation axis, selecting the largest distance such that the measured magnetic field is detectable. Only DC magnetic fields will be measured. Some subsystems will be measured powered off, and some will require both powered on and then off. Measurements will be made in a reasonably quiet ambient environment and should be repeated if external field changes are evident in the recorded data or the field measured at the start and end of a mapping rotation (or closure of field trace) differs by more than 2 nT. Measurements will be made in a low-field environment produced by bucking out the Earth's field with a 3-axis Helmholtz coil system (see Figure 12.32). A computer will regulate the current flowing in all three orthogonal Helmholtz coils such that an equal and opposite field component is produced in the direction of the Earth's east–west, vertical, and north–south field directions. Tests will not commence until the ambient magnetic field has been reduced to a homogeneous field that is less than 10 nT

Figure 12.32 Magnetic measurement the Helmholtz coil system.

achieved inside a 1 m^3 volume. Once an approximately zero field is achieved, a calibration coil with radius of 5 cm, current of 500 milliamperes, and 177 turns will be measured to verify the correct operation of the test magnetometer sensor. This will achieve 138 nT when the calibration coil is measured 1 m from the center of the coil and with its radial axis in parallel with the coil radial axis. The normal convention for magnetic field polarity will be a positive field into the outboard end of the sensor.

Magnetic mapping of spacecraft subsystems will be performed at a fixed distance from the approximate center of the test hardware as the hardware is rotated on a nonmagnetic turntable about a vertical axis through its approximate center of mass. By performing this operation about at least two orthogonal axes of the hardware, and with the assumption that the hardware magnetic field source is dipolar, the magnetic dipole moment can be determined. With a magnetic dipole moment, the maximum field is the radial component along the dipole axis. The maximum tangential component is one half the maximum radial component of the magnetic field with the maximum tangential component occurring normal to the dipole axis at its center. For greater sensitivity, the radial component of the field is measured and the maximum radial magnetic field component calculated from the maximum value measured in two or more orthogonal planes. It can be seen that the magnetic field determined by this method is a worst-case value that disregards the orientation of the dipole moment.

An overview of a typical test sequence is presented in Figure 12.33.

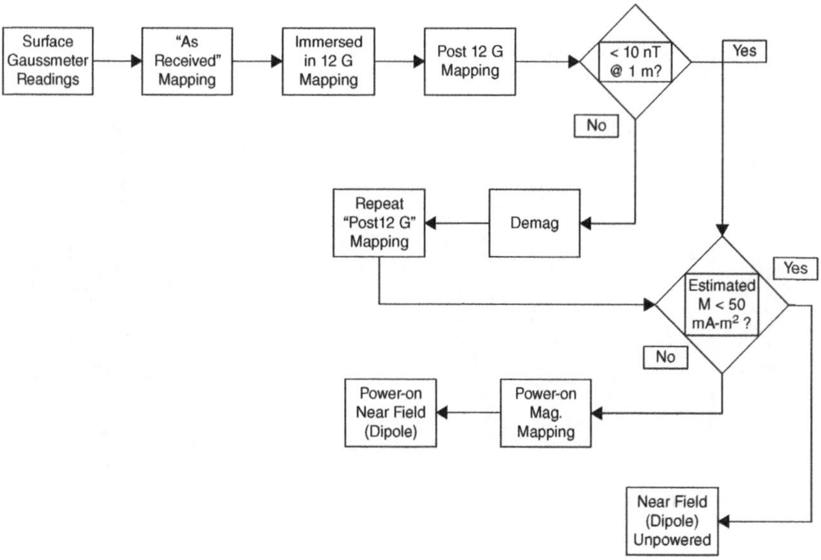

Figure 12.33 Typical DC magnetic mapping test sequence.

There are two basic types of DC magnetic measurements:

1. Magnetic mapping essentially characterizes the relative magnetism of the DUT in terms of the field seen at 1 m.
2. The near-field (dipole) measurement determines the dipole moment of the DUT and the results will be used in the modeling the net dipole moment of the spacecraft.

The magnetic qualification of Juno hardware may require testing in both an unpowered and a powered-on condition.

The unpowered magnetic mapping test involves a sequence of several simple steps:

(a) As received mag map of DUT
(b) Demagnetize DUT in an exponentially decaying AC mag field, 50-Gauss peak field
(c) Mag map DUT after demagnetization
(d) Immerse DUT in an external DC magnetic field, 25 (or 12) Gauss peak field
(e) Mag map DUT after magnetization
(f) Demagnetize DUT in an exponentially decaying AC field, 50-Gauss peak field
(g) If applicable, energize DUT and mag map in powered-on state

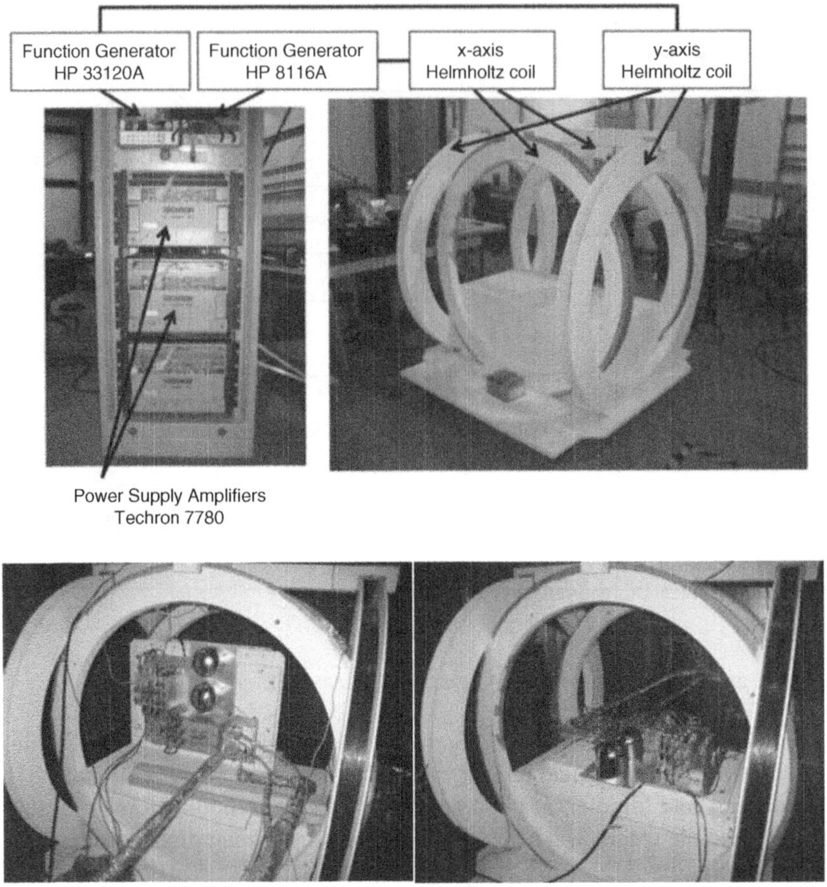

Figure 12.34 Magnetization and demagnetization process of a DUT.

The exposure to a 25 (or 12) Gauss environment is required to assess any induced field effects due to the presence of permeable materials in the DUT.

Demagnetization is required if the measured field value exceeds 10 nT at 1 m after exposure to the 25 Gauss environment. If the DUT returns to the "as received" level after being removed from the 25 Gauss environment, a demagnetizing step may not be required. Note that the "powered-on" testing does not include immersion in the 25-Gauss environment. The selected "powered-on" mode should be the highest current load condition. Figure 12.34 Figure 12.35 and Figure 12.36 are setups for magnetizing and demagnetizing of subsystems:

The following magnetic mapping steps can be applied for a typical subsystem and/or assembly:

"As Received"—Unpowered

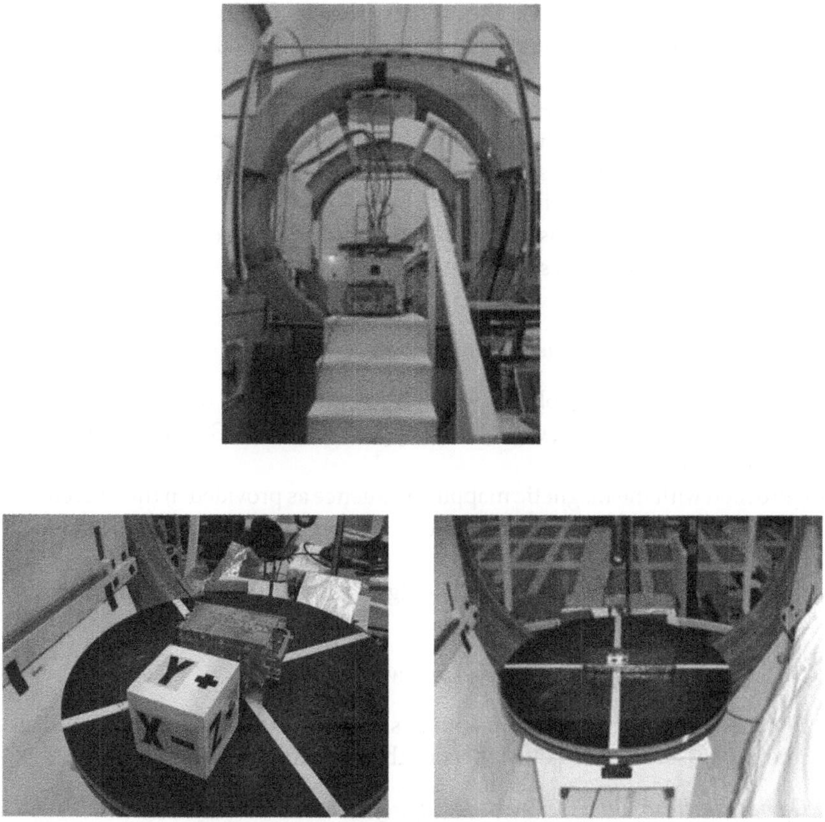

Figure 12.35 Example of a DUT on turntable for *Y*-axis rotation.

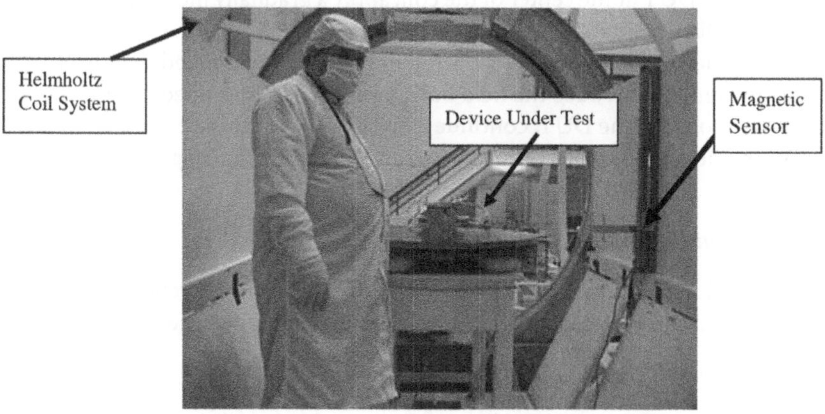

Figure 12.36 DUT inside a magnetic test Helmholtz coil system.

(a) Reduce the ambient field at the center of the Helmholtz coil system to zero. The zero-field environment settings should be maintained throughout the balance of the testing.
(b) Place DUT in the center of the Helmholtz coil system. Select and identify the reference axes for the measurement. Photograph the DUT in place.
(c) Using a handheld Gaussmeter, determine the polarity and field level of the selected axes at the surface.
(d) Orient DUT with X-axis up, the Y-axis toward the sensor, and proceed with the magnetic mapping sequence of the referenced DC Magnetic Measurement Manual.

Immersed in 12-Gauss Field—Unpowered

(a) Place a Gaussmeter probe in proximity to the DUT on the turntable.
(b) Using the Mag/Demag coil system, apply a 25- or 12-Gauss (DC) environment to the DUT.
(c) Proceed with the magnetic mapping sequence as provided in the referenced DC Magnetic Measurement Manual. Record the information on the data sheet.
(d) Reduce the 25- or 12-Gauss field (gradually) to zero and turn off the Mag/Demag power supply.

Post 25 (or 12) Gauss Exposure—Unpowered

(a) Proceed with the magnetic mapping sequence of the referenced DC Magnetic Measurement Manual. Record the information on the data sheet.

Demagnetization in a 50 Gauss Decaying Field

(a) Place a Gaussmeter probe in proximity to the DUT.
(b) With the DUT at the center of the coil, apply a gradually increasing DC field as the DUT is rotated.
(c) Continue rotation of the DUT while increasing the applied field up to 50 Gauss and then reduce the field back to zero with continued rotation.
(d) While rotating the DUT, continue steps b and c for three cycles.
(e) With the applied field reduced to zero, turn off the Mag/Demag power source.

Post 50 Gauss Demagnetization

(a) Proceed with the magnetic mapping sequence of the referenced DC Magnetic Measurement Manual. Record the information on the data sheet.

Powered-on Mapping

(a) Apply power to the DUT. Allow sufficient cable length (overhead drop) to accommodate the 360° rotation of the turntable.

(b) Proceed with the magnetic mapping sequence of the referenced DC Magnetic Measurement Manual. Record the information on the data sheet.

If the magnetic source is a true dipole at the center of rotation of the hardware, the plot of radial magnetic field versus rotation angle will appear as a pure sine wave on computer screen and X–Y plot output. If this is not the case, the dipole moment cannot be accurately determined. If the measured magnetic field appears to be that of a simple dipole, then the maximum radial field can be simply extrapolated to other distances from the center of the hardware as the inverse cube of the distance. Measurements at two different distances can also be compared to determine if the magnetic source is dipolar in nature. With the magnetic field due to higher order multipoles decreasing as the fourth and higher powers of the inverse distance, experience indicates that the measurement distance should be greater than 4 to 6 times the maximum hardware dimension. This assumes that the source is a dipole moment, so that calculations based on that fact can be made. With the larger items, measurements made at the larger distances frequently do not result in sufficient amplitude to be distinguished from the ambient noise disturbances. Near-field measurements can also be performed using the measurement approach by rotating unit about its vertical axis for each of six rotation angels each 30 degrees apart. See Figure 12.37. The goal of this measurement approach is to

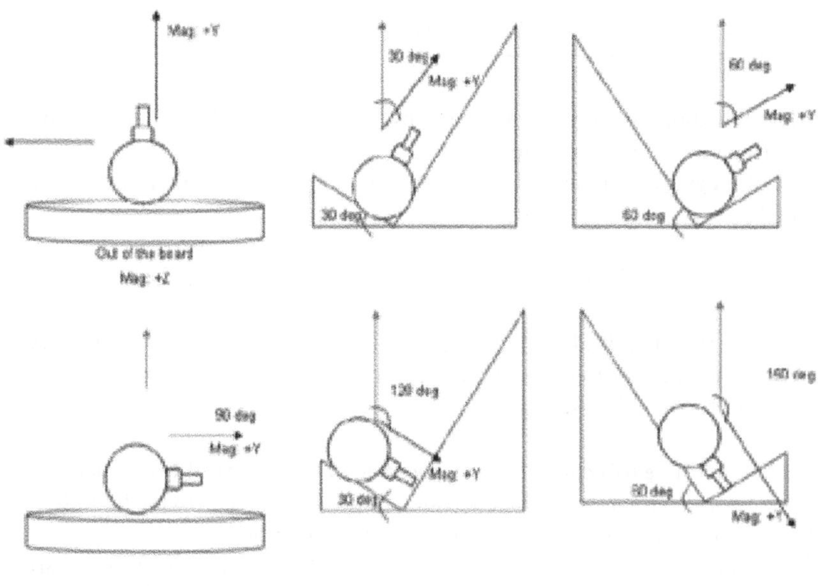

Figure 12.37 Near-field mapping angles.

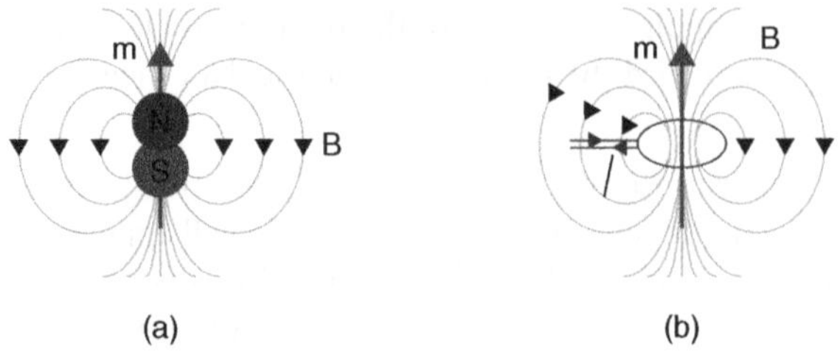

Figure 12.38 (a) Magnetic moment from material and (b) magnetic moment from current loop.

measure the pure dipole moment due to either a material or current loop as shown in Figure 12.38.

Once the moment's rectangular components (M_x, M_y, and M_z) are determined, the rectangular components of the field at the magnetometer sensor location can be found by the following equations:

$$\bar{B} = \frac{\mu_o M}{4\pi r^3}(2\cos\theta\hat{r} + \sin\theta\hat{\theta}) = \frac{\mu_o}{4\pi r^3}[3(\bar{M}\cdot\bar{r})\bar{r} - r^2\bar{M}]$$

$$B_x = \frac{3(M_x x + M_y y + M_z z)}{r^5}x - \frac{M_x}{r^3}$$

$$B_y = \frac{3(M_x x + M_y y + M_z z)}{r^5}y - \frac{M_y}{r^3}$$

$$B_z = \frac{3(M_x x + M_y y + M_z z)}{r^5}z - \frac{M_z}{r^3}$$

Magnetic dipoles can be represented by either material-based sources (permanent magnets) or current loops.

Radial magnetic field at a distance "r" from a magnetic moment can be calculated by

$$B_{radial} = \frac{2M\cos\theta}{r^3}$$

where M is the magnitude of the magnetic moment, theta is the angle between the moment and the direction of the field, and "r" is the distance from the location where the field is calculated.

12.8.3.1.2 Subsystem Demagnetization Processes Demagnetization of spacecraft subsystems, also referred to as deperming, is performed to provide flight hardware to be in its lowest and most stable magnetic condition.

Demagnetization is provided to erase the effect of prior exposures of hardware to low-level magnetic fields. Demagnetization is performed by exposing

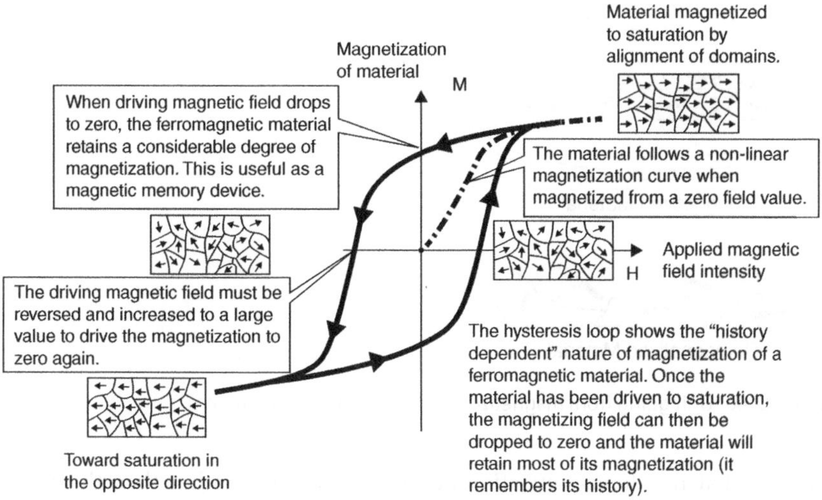

Figure 12.39 Demagnetization profile.

the hardware to a peak alternating magnetic field and decreasing this field exponentially. On small handheld nonelectronic components, it is possible to vary the orientation of the hardware in an approximate higher peak field (approximately 60 mT or 600 Gauss AC) by tumbling the object as the field decreases. As this is not possible with the large spacecraft instrument, it is necessary to demagnetize the hardware in each of three orthogonal axes or rotate the hardware about two orthogonal axes normal to the decreasing 5 mT (50 Gauss) peak demagnetizing field. This method has the advantage that the rate of change of magnetic intensity or induced voltages in the subsystem are kept at a minimum. Figure 12.39 shows the hysteresis process as the unit is being demagnetized.

The unit undergoing demagnetization drops its residual field to near zero as shown in Figure 12.40 with the dipole domains now showing a net zero effect:

Demagnetizing coils are shown in Figures 12.41 and 12.42.

For small spacecraft parts or nonelectronic subsystems, induced voltages, which might be caused by a 60-Hz demagnetizing field, are inconsequential. In this case, it is usually more convenient to demagnetize in a small coil or solenoid using 60 Hz power. For demagnetizing medium-sized structural parts, a small 60-Hz handheld tape demagnetizer could be used. For larger parts, such as the fully assembled spacecraft structure, the practice of using the lower frequency alternating field approach will be employed. Demagnetization will be performed by slowly rotating the hardware about an axis normal to the applied field as the unidirectional DC field is slowly decreased to zero. Successive, oppositely directed demagnetizing fields continue until the demagnetizing field is only a fraction of the Earth's ambient field value. This method

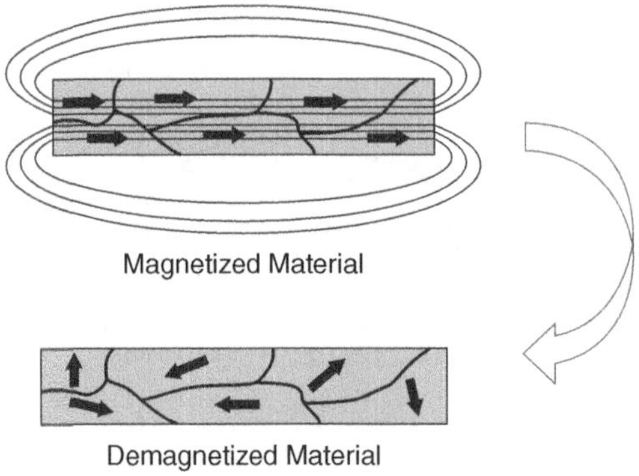

Figure 12.40 Unit going from magnetized to demagnetized condition.

requires that the demagnetization be performed twice rather than thrice to achieve the three-axis demagnetization. Figure 12.43 shows a graph of the demagnetizing field.

It has occasionally been found that some items such as invar will not demagnetize properly when demagnetized in an Earth's field environment. In fact, in some cases the hardware containing invar has been more magnetic following such demagnetization. In most cases, demagnetization in a low-field environment will permit a more thorough demagnetization of the hardware, which

Figure 12.41 Helmholtz coils demagnetizer.

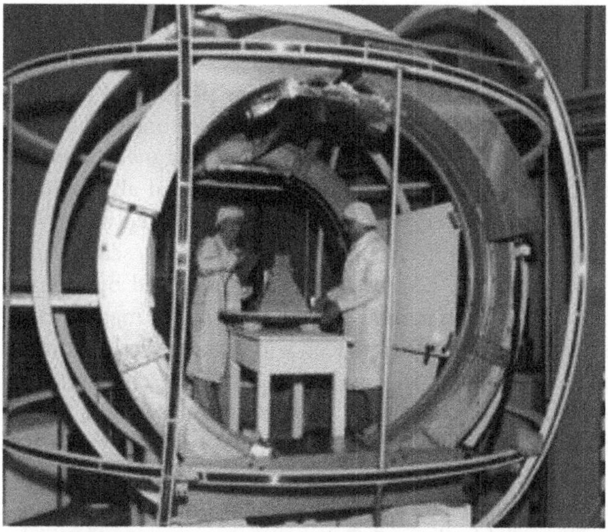

Figure 12.42 Large Helmholtz coil demagnetizer.

will remain even after return to the Earth's environment. Difficulty occurs if the residual field was produced by a field higher than the initial demagnetizing field (i.e., greater than 5 mT or 50 Gauss). The demagnetizing field only reduces the residual field in soft magnetic materials and is not expected to affect any permanent magnet devices in the subsystem such as magnets in motors.

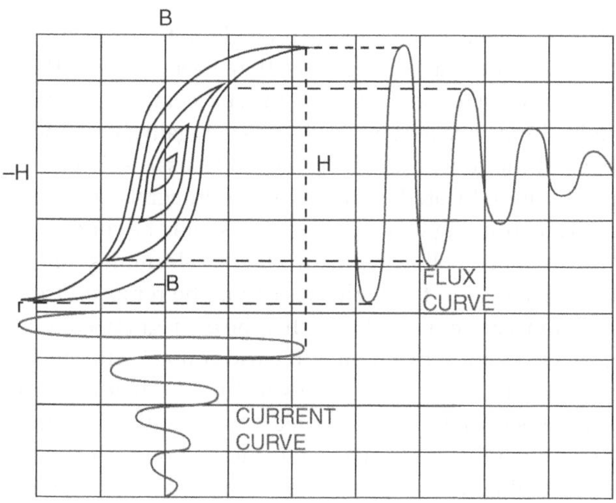

Figure 12.43 Coils driven with sinusoidal inputs producing slowly decaying fields.

12.8.4 Launch Site Survey and Control

Thorough magnetic surveys of all facilities used for spacecraft integration and storage spacecraft integration, and launch site should be performed prior to moving the spacecraft to such facilities. The surveys should be done well before spacecraft delivery to allow time for correction in case a magnetic cleanliness problem has been found. In addition, magnetic surveys of all mechanical vehicles and support structures used during spacecraft integration will be performed. These items include cherry pickers, scissors jacks, scaffolding, transport containers, and canisters. Results should be submitted by a technical report, and any noted highly magnetic sources capable of magnetizing spacecraft subsystems are to be removed, replaced with nonmagnetic sources, or demagnetized prior to introducing the spacecraft into the facility or near the mechanical devices. Surveys are performed by scanning all surfaces with a Gaussmeter probe such as the F. W. Bell Model 9500 Gaussmeter. Storage cabinets, workstations, nearby air-handlers, etc., can be quickly checked by scanning an object's surface and checking for any significant magnetic fields. If surfaces are determined to have fields above 5 Gauss and cannot cost effectively be demagnetized to less than 1 Gauss, these areas are to be noted and assessed to determine if they pose a significant threat to the spacecraft instruments.

Of particular importance are the magnetic surveys of the dynamic shaker coils used during environmental random vibration testing. A survey at this interface should be performed with the shaker coils powered off and then powered on. The environmental test lab dynamic shakers should be surveyed prior to spacecraft dynamics testing. If levels are higher than 1 Gauss at the shaker-to-spacecraft interface, then ramping down of the shaker coils should be employed.

In order to ensure that the spacecraft assembly tools do not cause subsystems of the spacecraft to become magnetized after they have received their final demagnetization, the tools planned for use during spacecraft integration should be first demagnetized and checked and demagnetized again if their maximum field on the surface of the tools exceeds 0.3 mT (3 Gauss). This will be done at a magnetic survey station located in the spacecraft integration area. A magnetics control tutorial and training session at each site should be conducted by the magnetics control engineer. Following the magnetics cleanliness check and demagnetization, any magnets in the toolboxes should be removed from the assembly area. In addition, care must be taken that the demagnetized tools do not acquire further magnetization by coming in close contact with highly magnetic tools. Repeated checks on assembly tools will not be made except on a selective basis. The transferring of unmeasured and potentially magnetized tools or support equipment into the spacecraft integration area should be avoided. Toolboxes containing nonmagnetic tools or previously demagnetized tools conforming to the less than 0.3 mT (3 Gauss) level should not be

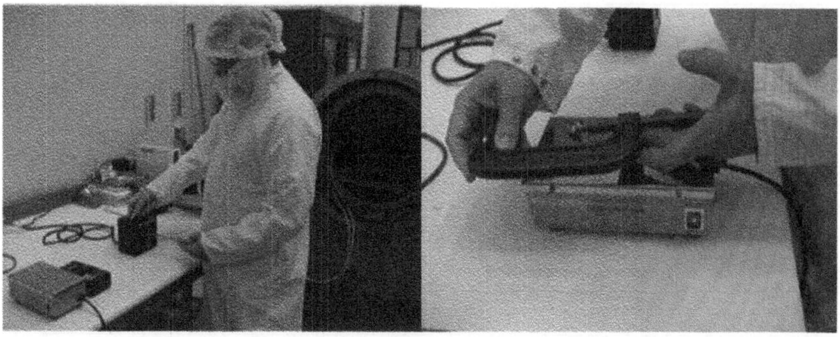

(a) (b)

Figure 12.44 Process of demagnetizing tools.

contaminated with magnetized particles. Good working practice includes the use of nonmagnetic (or degaussed) tools everywhere, especially anywhere near the magnetometer sensors. The danger of transferring magnetized particles to nonmagnetic materials again stresses the importance of testing all tools before use. Figure 12.44 presents pictures showing demagnetizing methods.

Good working practice includes the use of nonmagnetic (or degaussed) tools everywhere, especially anywhere along the magnetometer boom. The danger of transferring magnetized particles to nonmagnetized materials again stresses the importance of testing all tools before use.

12.9 Analytical Determination of Magnetic Fields

The total residual magnetic field will be estimated and updated at the magnetometer sensor locations by analysis using data from subsystem measurements and information from spacecraft-level measurements in order to verify that the spacecraft's magnetic condition satisfies the magnetic requirements. This analytical determination should be used to support a system-level measurement of the overall spacecraft. These calculations will be made for the flight spacecraft at the completion of the initial subsystem-level testing, at the pre-ship review, at the time of launch readiness review, and at launch. The calculated results should be updated with new measurement data as they become available.

The spacecraft's magnetic fields at the inboard and outboard sensor locations along the magnetometer mounting panel length should be calculated by summing all measured subsystems' extrapolated magnitudes of the dipolar moment. The extrapolation will be based on the relative orientation of the hardware and the location of the center of mass of the hardware in terms of spacecraft coordinates. From these measurements, extrapolation will be based

Error Checking
ok

Mode
AC

Total B field at magnetometer (nT) in SC coordinates			
Bx	By	Bz	Brms
1.7443123	0.0430663	-0.019045	1.7449478

	Dipole Moment mA-m^2			Results B field (nT) at magnetometer from DUT			
Name	x	y	z	Bx	By	Bz	Brms from DUT
Battery 1	**-9.9**	**1.2**	**-2.6**	0.0006403	9.812E-05	0.0001263	0.00066
Battery 2	**-9.9**	**1.2**	**-2.6**	0.0011744	-6.22E-05	0.0002833	0.0012096
CME	-0.2	-3.4	-3.7	0.0010649	1.377E-05	-1.57E-05	0.0010651
JADE Electron Sensor 060	3.2	0.6	-1.3	0.000433	-9.37E-05	-6.45E-06	0.000443
JADE Electron Sensor 180	3.2	0.6	-1.3	0.0002362	5.231E-08	-2.86E-06	0.0002363
JADE Electron Sensor 300	3.2	0.6	-1.3	0.0009719	0.0001995	-1.47E-05	0.0009923
JADE Ion Sensor	2.8	1.9	-0.8	0.0006666	-5.67E-05	-7.05E-06	0.0006691
JADE Electronics	-1.7	0.7	7.6	0.0100256	-0.000298	-0.000807	0.0100624
JEDI Sensor 180	-27.2	8.6	0.7	0.0057926	0.0004575	-6.41E-05	0.005811
JEDI Sensor 270	-27.2	8.6	0.7	0.0121903	0.0029652	-0.000225	0.0125478
JEDI Sensor 90	-27.2	8.6	0.7	0.0119751	-0.002913	-0.000221	0.0123262
JIRAM Optics	1.1	4.0	6.3	0	0	0	0
JIRAM Electronics	0.6	2.2	3.8	0	0	0	0
Junocam Optics	**-2.2**	**0.8**	**-0.1**	0.0010913	1.122E-05	1.422E-05	0.0010915
Junocam Electronics	**6.8**	**0.2**	**-0.2**	0.0004921	3.881E-05	-3.36E-05	0.0004948

Magnetometer location (m) in SC coord.		
x	y	z
10.026	0	1.7

Total B field at magnetometer (nT) in SC coordinates			
Bx	By	Bz	Brms
1.7443123	0.0430663	-0.019045	1.7449478

Figure 12.45 Magnetic spreadsheet summing all magnetic sources at the mag location.

on calculations or moments up to those causing fields to vary as the inverse cube of the distance. A subsystem's dipole moment may also be derived from measurements made with three triaxial magnetometers each at 1.5 m from the geometric center of the subsystem under test but in a zero-field region, such as a magnetic shielded area or an energized Helmholtz coil system.

Figure 12.45 shows a sample dipole model spreadsheet that takes all the subsystem-level dipole moment components and models the fields at the magnetometer sensor location. This method allows to the "book-keeping" of the overall magnetic cleanliness status of the spacecraft as new measurements are performed. Figure 12.45 is a sample of a spreadsheet that sums all fields at the magnetometer location and keeps track of the progress being made toward meeting a requirement at the magnetometer sensor location.

13

Spacecraft Charging

Robert C. Scully

Source: ESA (url: http://www.esa.int/spaceinimages/Images/2012/11/Space_Situational_Awareness_Space_Weather)

13.1 Introduction

Ionizing radiation, nonionizing radiation, plasma, manmade orbital debris, naturally occurring meteoroids of varying shapes and sizes, the geomagnetic field of the earth, and the gravitational field of the earth, are all components of the space environment that surrounds the earth. The relationships between these

Handbook of Aerospace Electromagnetic Compatibility, First Edition. Edited by Reinaldo J. Perez.

various components of the environment are complex and dynamic, and the results for spacecraft placed into orbit are harsh and unforgiving. As more and more everyday functions rely on satellite operations, knowledge of this environment and its effects on technological systems, including reliable modeling and forecasting methods, has become increasingly important. Understanding this environment and designing to survive it successfully is a steep challenge, requiring coordination among multiple engineering and scientific disciplines. Taken singly, any one component of the space environment presents unique hazards, and a full discussion of all of them, together with full consideration of their interactions, would be a monumental task.

The thrust of this handbook is to highlight the electromagnetic environmental effects of the space environment, and how these effects play a role in design, construction, deployment, and operation of spacecraft. In an effort to stay within the scope of the handbook, this chapter will focus solely on those aspects of the space environment most closely related to nonionizing electromagnetic effects, and the results of those effects, most of which manifest themselves in some form of electrostatic discharge (ESD) and the ensuing damage therefrom. To be specific, this chapter will focus on space plasma and its constituent elements, their interaction with the earth's geomagnetic field and the solar wind, and their aggregate effects and impacts on spacecraft, generally referred to as spacecraft charging.

Spacecraft charging and satellite anomalies are closely linked, one to the other. While spacecraft charging is a general physical phenomenon that cannot be totally avoided, it does not necessarily lead to anomalous behavior of the spacecraft. On the other hand, damaging anomalies can be caused by excessive charging. For both spacecraft engineering and operations, a key question is when, where, and why satellite charging and other environment-related anomalies occur, and how hazardous these phenomena may be. Although much work has been devoted to the understanding of the charging and other anomalous phenomena, and to protection against them, they still take place, sometimes with serious consequences. The development of smaller and more sensitive electronics and, most likely, the increasing use of spacecraft in the most hazardous orbits crossing high latitudes and radiation belts point out the necessity for intensified effort to understand both the origin and effects of these events.

Spacecraft charging may be generally defined as the accumulation of electrical charge by a spacecraft. This charge may reside on exterior surfaces, known as surface charging; it may be scattered throughout the interior of the spacecraft on conductive surfaces or embedded into dielectric materials, known as bulk charging, internal charging, or deep dielectric charging; or it may a combination of both surface and bulk charging. Exposed surfaces of the thermal protection system or external metallic structures, transparencies, antennas, solar array panels, internal support structures, electrical and electronic components, and wiring and cabling harness sheathing are all subject to the processes of spacecraft charging. The effects of spacecraft charging include physical spacecraft surface damage and erosion resulting from ESD events, degradation of thermal, optical, and electrical properties of spacecraft surface materials and solar arrays caused by surface contamination, and electrical anomalies in power generation, navigation, communication, and instrumentation subsystems caused by the increase of electrical fields related to the deposition of electrical charges and the coupling of ESD transients into spacecraft electronics.

All of these effects vary in degree and magnitude as a function of latitude, longitude, and altitude and are largely governed by the strength of the earth's own internal dipolar geomagnetic field and that of the external magnetic fields associated with the stream and flow of charged particles and ultraviolet, X-ray, and shorter wavelength radiation that make up the solar wind. Spacecraft that operate in low earth orbits and at moderate latitudes are likely to encounter dense plasma comprising particles exhibiting low energy levels. Spacecraft that operate in polar or highly elliptical orbits, or in geosynchronous altitudes, are likely to encounter much sparser plasma comprising particles exhibiting high energy levels. Spacecraft that pass near or through the Van Allen belts are exposed to concentrations of magnetically trapped, very high energy particles including both electrons and protons. These electromagnetic environments thus exhibit widely varying characteristics and dynamics, and correspondingly, the techniques for mitigation of spacecraft charging vary for different mission profiles. One of the lowest common denominators between these various environments, and a major focus of this chapter, is the threat of damage to the spacecraft from the effects of ESD.

The chapter is divided into sections, opening with this brief introduction. The second section provides historical background information regarding spacecraft charging. The third section provides a general description of the space environment and the various physical aspects of the electromagnetic environment known as the magnetosphere surrounding the earth out to several thousands of kilometers. This section also includes discussion of the space plasma environment itself, and how and why it varies as a function of latitude, longitude, and altitude. The fourth section provides discussion of the types of spacecraft charging that may occur and the roles of spacecraft geometry and physics that drive such effects. The fifth section explores the various damage mechanisms associated with spacecraft charging and its effects that might be expected to occur in poorly or unprotected spacecraft designs, and then follow up with a discussion of cost and mass effective means that may be employed to mitigate such damage effects and processes. The final section provides a summary of the chapter and contains a bibliography of related materials and published work that can facilitate collaboration and further study.

Engineering, manufacturing and assembly, and program management functions engaged to varying degrees in planning, designing, constructing, and operating spacecraft missions, are all impacted by spacecraft charging and its effects. The success of the efforts of these various organizations is, at least partially, predicated on understanding the spacecraft charging problem and gaining familiarity with ways and means available for its mitigation. It is hoped this chapter will in some small fashion facilitate that understanding.

13.2 Historical Background

In November of 1981, H. B. Garrett published [1] a 39-page document in which he provided a concise narrative not only of the science of spacecraft charging as it was known at the time, but also a thorough historical perspective on the topic, and an extensive bibliography. In his paper, Garrett separated the events over time in his historical perspective into four distinct phases. The first phase extended from 1924, marked by early electrostatic probe work [2] up to the launch of Sputnik in 1957. In this first phase, a great deal of the work was of course theoretical and focused on the potentials of interstellar dust grains, and estimates for potentials between satellites and the space environment. Spacecraft charging became recognized as a discipline with the advent of space-borne sensors and transducers in the early 1950s. One of the earliest technical papers [3] published that attempted to capture the effects of spacecraft charging discussed shifts in instrumentation readings that could most easily be traced to negative spacecraft potential.

The second phase, according to Garrett, stretched from 1957 to 1961, marked by the publication of the first review of spacecraft charging [4]. During the second phase, most of the elements of spacecraft charging as it was known in 1961 were identified, including not only confirmation through observation of rocket and satellite missions, but also confirmation of predicted levels in the ionosphere. Effects associated with photoemission and electron flux were recognized as dominant in the process of spacecraft charging, and the importance of relative motion in terms of velocity through a magnetic field were at least under serious consideration. Some areas that had yet to be addressed included secondary emission and backscatter effects, and non-Maxwellian distributions of energetic particles.

The third phase encompassed the foregoing shortcomings and ended in 1965 as marked by the near simultaneous publication of the first book concerned with spacecraft charging [5] and the seminal thesis by E. C. Whipple [6] titled "The equilibrium electric potential of a body in the upper atmosphere and in interplanetary space." As evidenced by these works, the spacecraft charging community had now fully realized the impact of spacecraft charging on plasma environment measurements and the importance of self-consistent solutions of particle trajectories and fields. Perhaps most important still was that quantitative measurements had finally become available for study.

The fourth phase, ranging from 1965 to 1980, saw the emergence of increasingly sophisticated models of surface charging, a healthy growth of measurements on-board various space-borne platforms, and a steadily improving understanding and definition of the plasma environment. Failure of an Air Force mission caused by ESD led to cooperation between the Air Force and NASA to develop the NASA Charging Analyzer Program (NASCAP). Launch of the Spacecraft Charging AT High Altitudes (SCATHA) satellite mission in 1979 into a near-synchronous elliptical orbit is considered by Garrett to mark closure of this phase. SCATHA was specifically designed to study the effects of spacecraft charging, and data from that mission included observation of arcing under multiple operating conditions, the response of a sampling of materials to both natural and artificial charging events, simultaneous data for plasma, magnetic and electric fields, surface potentials, and surface contamination.

The community is currently in a fifth phase. Design guidelines were developed by NASA in 1984 [7] and 1988 [8] and have since been replaced or updated and embellished in several documents [9–13]. The community has largely been successful in containing surface charging, and, perhaps because of this, is experiencing a growing awareness of internal charging caused by the penetration of energetic charged particles. Continued operation of the International Space Station, establishment of the Constellation Program and its successor the Multi-Purpose Crew Vehicle Program, the advent of the Commercial Crew Development Program, multiple research satellite missions to explore and measure aspects of the solar wind and its impact on the earth's magnetosphere, and a burgeoning interest in the use of high-voltage spacecraft electrical power systems employing solar arrays, have all been instrumental in growing and maintaining interest in spacecraft operations in plasma in the Low Earth Orbit (LEO) region.

13.3 General Description of the Near-Earth Electromagnetic Environment

The earth's atmosphere near the surface is composed of ~78% nitrogen (N_2), < 21% oxygen (O_2), and < 1% argon (Ar). The remaining ~0.04% is mostly carbon dioxide (CO_2), with trace amounts of neon (Ne), helium (He), methane (CH_4), krypton (Kr), and hydrogen (H_2). These constituents of the earth's atmosphere remain virtually unchanged up to an altitude of about 60 km, but above this level, the relative amounts and types of gases are no longer constant with altitude. Between 60 and 400 km, solar radiation comprising mostly energetic photons at ultraviolet and X-ray wavelengths begins to interact strongly with the atmospheric constituents. Energetic photons split molecules into component parts via photo-dissociation. Neutral species of oxygen, nitrogen, helium,

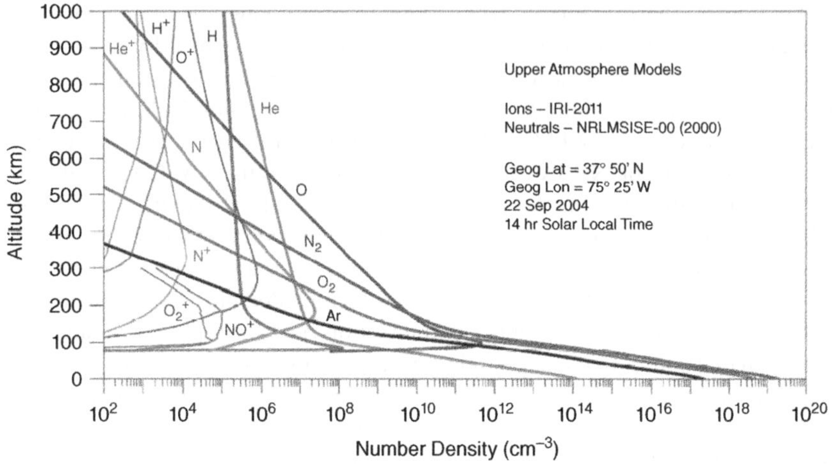

Figure 13.1 Model distributions of neutral and ionic constituents as a function of altitude for a mid-latitude location (Wallops Island, Va., USA) at 14 hour solar local time. Ion distributions below 80 km are not provided by the model. (*Source*: Pfaff [14])

and hydrogen are subject to photo-ionization, yielding approximately equal numbers of positively charged ions and negatively charged free electrons. These ions and electrons make up what is known as the plasma environment, and their densities are referred to as plasma densities. Plasma densities typically are far less than those corresponding to neutral species, as can be seen in Figure 13.1, but nonetheless, the plasma environment is noticeably electrically conductive.

The region of space within which these processes occur is known as the ionosphere, and it stretches well beyond 1000 km of altitude. The ionosphere has three separate regions, the D, E, and F layers, characterized by the types of photodissociation and photoionization processes that dominate. In the D layer, generally located between 60 and 90 km of altitude, neutral atmospheric species are relatively dense, and extreme ultraviolet radiation (EUV; 10–100 nm) has been greatly decreased by absorption at higher altitudes. Ultraviolet radiation (UV; 100–300 nm) is able to penetrate, however. Radiation at wavelengths greater than 112 nm ionizes O_2 and N_2 species, and Lyman-α radiation at 121.6 nm ionizes NO species. Even so, during the daytime in this region, plasma densities are far less than those of neutrals. The ionization process effectively ceases during the nighttime, almost all of the free electrons are recaptured by the ionized species, and the D layer almost disappears. This allows amplitude-modulated (AM; 540–1700 kHz) broadcast energy to reach up to the higher ionospheric layers and reflect back to the earth, facilitating long-range propagation. It is for this reason that AM broadcast stations are required to reduce their transmit power during the nighttime.

Between approximately 90 and 150 km of altitude is found the E layer of the ionosphere. Ionizing radiation composed of EUV, soft X-rays (1–10 nm), and Lyman-β radiation at 102.6 nm is able to penetrate into the E layer and react with O_2 and N_2 species. Photodissociation splits O_2, producing atomic O that is readily ionized by EUV energetic photons. Additional ionization may occur as a result of collisions between neutral species and photoelectrons released from primary interaction with incoming radiation. Eventually an equilibrium is achieved between ionization and recombination of ions with free electrons, leaving NO^+ and $O_2{}^+$ ions prevalent in the region, but still having density numbers far less than the surrounding neutral species. Like the D layer, the E layer tends to disappear at night with cessation of solar activity. At times, a region of unusually ionized atmospheric gas will form within the same altitude range from 90 to 150 km. Known as a sporadic E layer, such a region may last up to several hours and is able to reflect radio waves between 25 and 225 MHz. Amateur radio operators refer to this reflection sometimes as "skip," and very long distance communications may be possible when a sporadic E layer has formed.

The F layer of the ionosphere is generally located between 150 and 400 km. Atomic O, N_2, and O_2 are the dominant neutral species in this region of the ionosphere, but all are readily ionized by the strong EUV that is able to penetrate. Multiple reactions occur in this region of the ionosphere that affect the relationship between neutral and plasma species. Oxygen atoms are readily ionized by photon impact as previously described, but other species not so much. However, oxygen ions will readily exchange charge with some of these neutrals, effectively establishing a series of interactions yielding a mixed soup of molecular ions that tend to recombine with free electrons during the daytime and to a greater extent during the nighttime. In addition, ionized species will chemically react with free electrons, contributing to the ongoing recombination processes. Some of these dissociation and recombination processes occur very rapidly, while others occur comparatively slowly. At the relatively higher densities encountered at lower altitudes, recombination tends to dominate because neutral and charged particles are in closer proximity. The balance between the processes of ionization, dissociation, and recombination determines the total quantity of ionization present.

Directly related to solar weather activity and sunlight exposure, the F layer separates into the F1 and F2 layers. These two layers differ in that the F2 layer tends to form as a result of additional processes, including the upward diffusion of O+ ions, wind-driven ion flows, and upward plasma transport caused by the influence of electric fields associated with the ionization processes already described. Plasma number densities in the F2 layer can range between a maximum on the order of 10^6 cm^{-3} to a minimum on the order of 10^5 cm^{-3} near 300 km. The region of maximum densities is known as the F2 peak. Neutral and plasma number densities are observed to decrease slowly as altitude increases

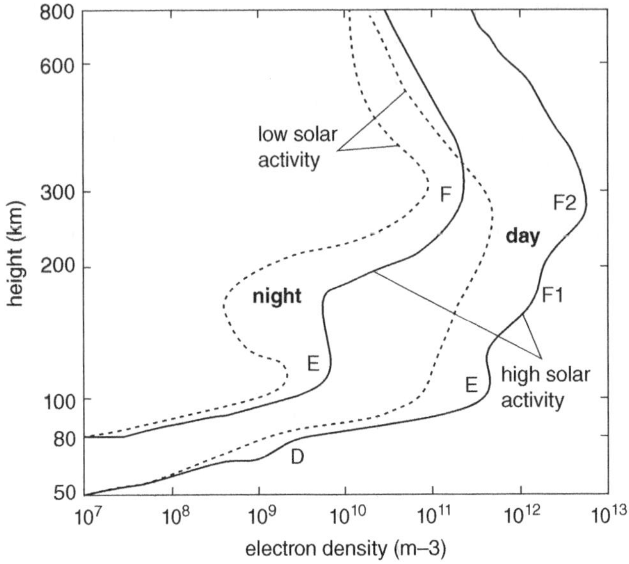

Figure 13.2 Representative electron density profiles for day and night, at solar minimum and maximum activity levels (similar to Figure 13.6.1 [15]).

above the F2 peak region. Figure 13.2 highlights the profiles of electron densities as a function of altitude and solar activity level.

Above the F2 peak region, sometimes referred to as the "topside" ionosphere, the charge exchange process is very active, resulting in a growing number of hydrogen nuclei and neutral oxygen atoms. Above about 1000 km, hydrogen nuclei begin to dominate. Helium nuclei are also present but except during geomagnetic storm activity, remain minor in number density in comparison with the hydrogen nuclei. This region is sometime referred to as the protonosphere, but more generally is known as the plasmasphere. The plasmasphere extends outward from the earth from roughly $\pm$ 60 degrees of magnetic latitude to about 4 earth radii (1 $R_E \sim 6380$ km) and forms a region about the earth that resembles a torus, bounded by magnetic field lines that emanate from and rotate with the earth. The outer extremity of the plasmasphere terminates with a rather sharp decrease in density at the plasmapause.

Two additional regions of high-energy charged particles known as the Van Allen belts, first discovered in 1958 [16], encircle the earth, as illustrated in the upper portion of Figure 13.3. The high-energy charged particles concentrated in the Van Allen radiation belts spiral around and move along the field lines of the surrounding geomagnetic field, similar to charged particle motions observed in cyclotrons. As the particles approach the polar regions, the magnetic field strength increases and causes the spiral paths followed by the charged particles to tighten. Eventually the field strength is sufficient to force the particle to

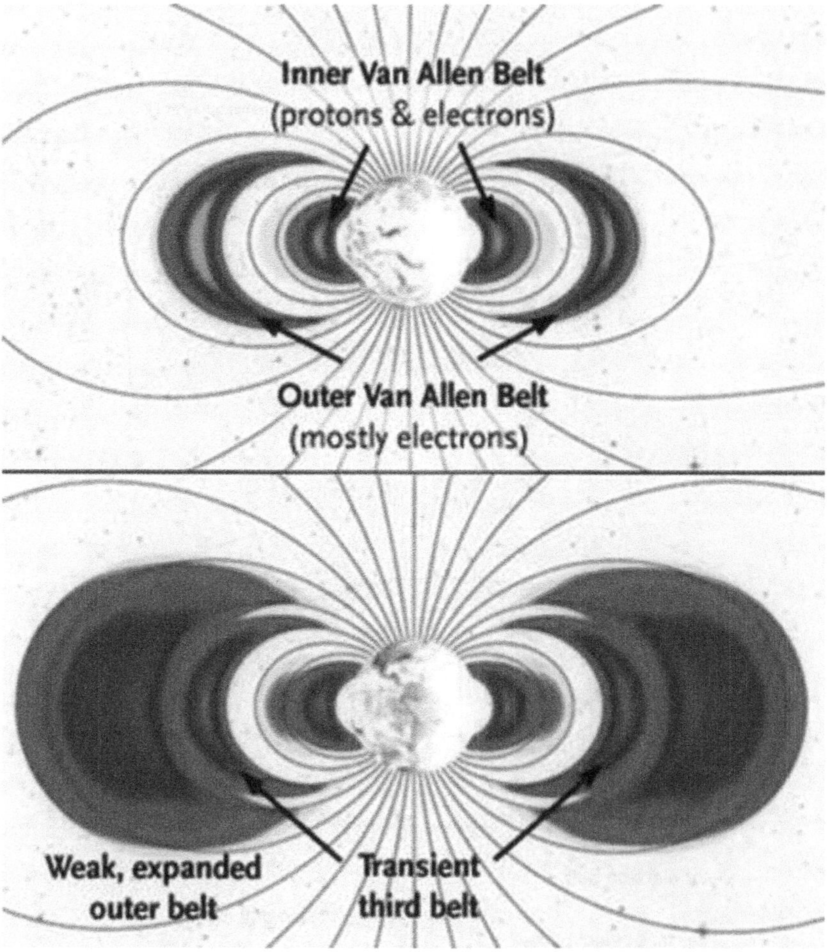

Figure 13.3 Illustration of the Van Allen belts showing a transient third belt (*Source*: NASA).

reverse direction. Thus, the particle is reflected between so-called mirror points and conjugate mirror points. Also, there is a slower longitudinal drift of the path around the earth that is westward for protons and eastward for electrons. Once a complete azimuthal rotation is made around the earth, the resulting toroidal surface that has been traced out is called a drift shell. A graphic illustrating the cyclotron motion and the mirror and conjugate mirror points, together with a schematic of such a drift shell, is shown in Figure 13.4.

As time has passed and more data have become available, scientists have observed that the belts expand and contract in concert with solar activity, merging at times, and even separating into three belts occasionally, as illustrated in the lower portion of Figure 13.3. The third belt appears following

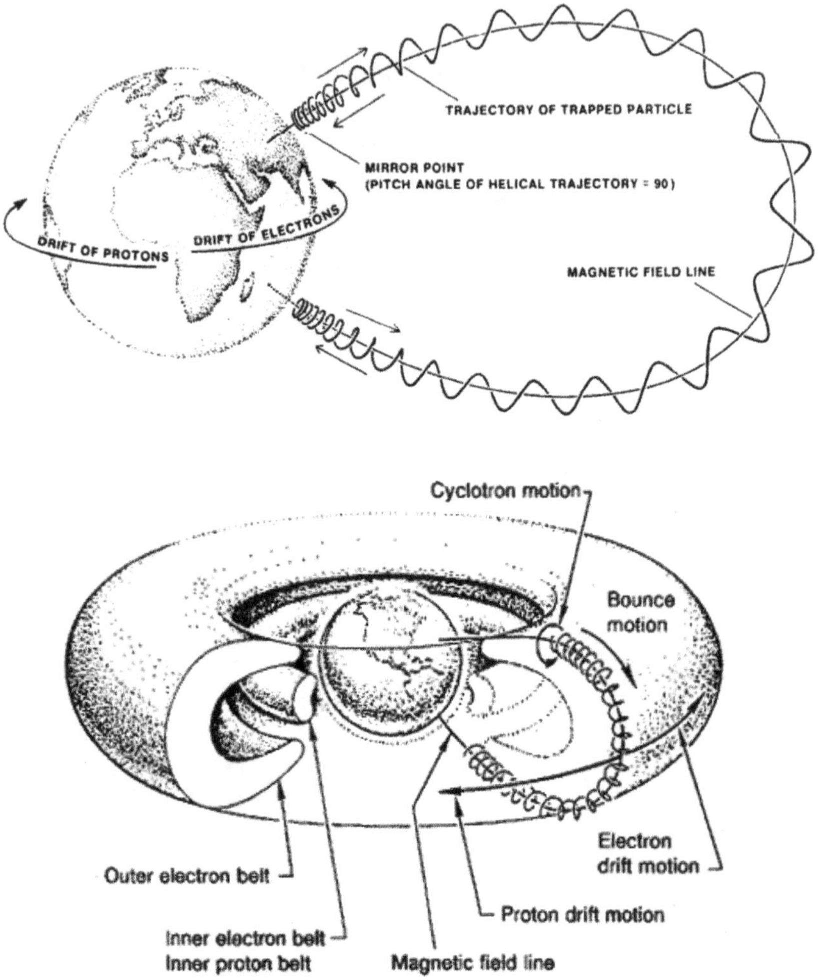

Figure 13.4 Illustration of cyclotron motion of charged particles along geomagnetic field lines [17] and illustration of the spiral path of charged particles along the geomagnetic lines of force, and the associated shells created by the longitudinal drift of these paths [18].

severe geomagnetic storms and can last for several months. The belts extend from ~1,500 km altitude out to ~6 R_E. These represent the most consistently present threat of satellite anomalies, especially to satellites in Middle Earth Orbit (MEO) and, to a lesser extent, Geosynchronous Earth Orbit (GEO). Satellites in the LEO, with altitudes up to ~2,000 km, experience high-energy particles from the radiation belts to a lesser degree than satellites at MEO and GEO, though high-inclination LEO satellites are exposed to the dynamic polar regions where geomagnetic activity can cause intense particle streams to concentrate. The quintessential example of geomagnetic activity is the geomagnetic storm, during which injected high-energy plasma acts to intensify the magnetic field strength of the earth's magnetosphere.

Nominally the inner belt, composed primarily of protons with energies up to 10 MeV and higher, ranges between about 650 and about 10,000 km in altitude, with the highest particle density found between 1.5 and 2.0 R_E. In like fashion, the outer belt, composed primarily of electrons with energies ranging from 10s of MeV to 100 MeV, ranges between about 13,500 and 58,000 km in altitude, with the highest particle density found between 4.0 and 5.0 R_E. A second region of high electron particle density is located near the same altitude as the highest proton density. Electron energies in this second region are on the order of 1 MeV. The belts do not really have distinct inner and outer boundaries, with near overlap between them.

Recently, NASA's Van Allen Probes mission (previously known as the Radiation Belt Storm Probes mission) has discovered a key interaction that occurs at the intersection of the charged particles at the plasmapause, the terminating edge of the plasmasphere, with the inner boundary of the outer Van Allen belt and the highly energetic electrons contained therein. Electrons in the outer radiation belt are weakly scattered by charged species in the plasmasphere, effectively preventing their transport inward toward the earth. Radiation belt electrons move very fast, spiraling along the earth's magnetic field lines of force. The Van Allen Probes mission data show that in the direction toward earth, even the most energetic electrons exhibit only a gentle, slow drift with a time scale on the order of months. Because this movement is so slow and weak it is easily countered by the plasmasphere particle scattering. This countering effect loses its effectivity during geomagnetic storms and periods of elevated solar activity, allowing the penetration of some highly energetic electrons in the slot region; a region of comparatively lower particle density between the inner and outer belts has been designated as the slot. The slot is found at about 2.5 R_E.

The inner Van Allen radiation belt approaches most closely to the earth's surface in the South Atlantic Anomaly (SAA), illustrated in Figure 13.5, dipping down to an altitude of about 200 km, and is farthest from the earth's surface over the northern Pacific. An increased flux of energetic particles exists in the SAA and can expose satellites to levels of radiation greater than in other parts of their orbits. The effect is directly caused by the nonconcentricity of the geocenter of the earth and the center of the geomagnetic dipolar field. The shape of the SAA has expanded steadily since its discovery in 1958, growing in the northern, northeastern, northwestern, and eastern directions, while remaining roughly the same to the south. At an altitude of approximately 500 km (300 mi), the SAA spans from – 50° to 0° geographic latitude and from – 90° to + 40° longitude. The shape and the particle density of the SAA also vary on a diurnal

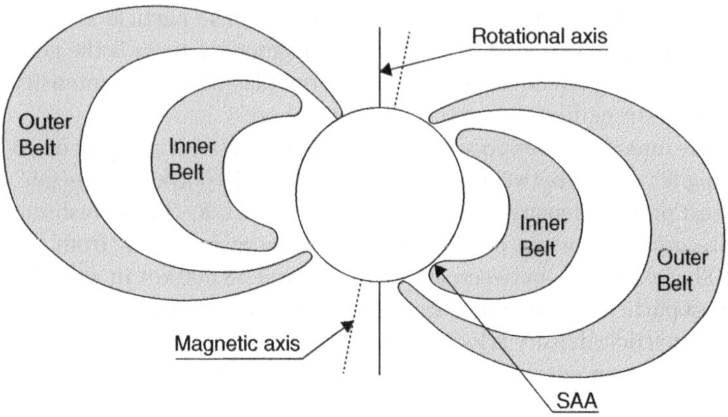

Figure 13.5 Illustration of how the close approach of the inner Van Allen belt is the source of the SAA. (*Source*: ESA)

basis, with the highest particle density observed around local noon, as illustrated in Figure 13.6. The highest intensity portion of the SAA is drifting to the west at a speed of about 0.3 degrees per year, very close to the estimated rotation differential between the earth's core and its surface.

The SAA is of great significance to astronomical satellites and other spacecraft that orbit the earth at several hundred kilometers altitude; these orbits

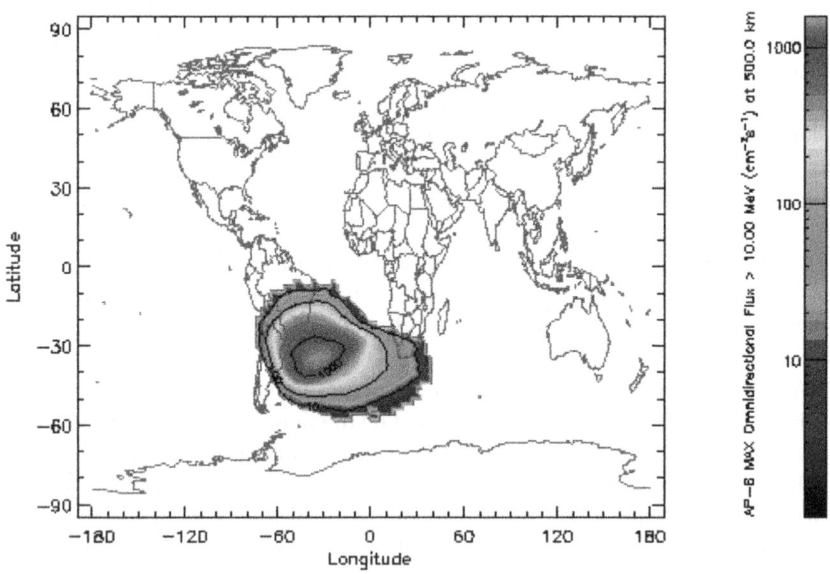

Figure 13.6 Illustration of a typical electron density profile in the SAA. (*Source*: NASA)

take satellites through the anomaly periodically, exposing them to several minutes of strong radiation, caused by the trapped protons in the inner Van Allen belt. The International Space Station, orbiting with an inclination of 51.6°, requires extra shielding to deal with this problem. The Hubble Space Telescope does not take observations while passing through the SAA. Astronauts are also affected by this region which is said to be the cause of peculiar "shooting stars" (phosphenes) seen in the visual field of astronauts. Passing through the SAA is thought to be the reason for the early failures of the Globalstar network's satellites. The PAMELA experiment, while passing through the SAA, detected antiproton levels that were orders of magnitude higher than expected. This suggests the Van Allen belt confines antiparticles produced by the interaction of the earth's upper atmosphere with cosmic rays. NASA has reported that modern laptops crashed when space shuttle flights passed through the anomaly. According to an article titled "Dragon enjoying ISS stay, despite minor issues – Falcon 9 investigation begins," published on NASA Spaceflight.com, Oct 19, 2012, "Overall, the SpX-1 Dragon is healthy. (However,) Dragon experienced a Single Event Effect (SEE) on the Trunk Remote Input/Output (RIO) Unit String B as ISS passed over the South Atlantic Anomaly. The team power cycled and re-synched the RIO string and recovered the unit."

The Van Allen radiation belts are primarily shaped and controlled by the earth's geomagnetic field. The earth's geomagnetic field is approximately dipolar for altitudes of up to about 4 or 5 R_E. The dipolar coordinate system is not aligned with the earth's geographic coordinate system, with the axis of the magnetic dipole field tilted about 11.5° with respect to the geographic north–south axis and its origin displaced by a distance of more than 500 km from the earth's geocenter. The lines of magnetic force associated with the earth's magnetic field leave the earth at nearly a 90° angle near the poles, with that angle steadily decreasing as one moves toward the equator. This change of angle is known as magnetic inclination. The lines also vary in their angle with respect to longitude, and this variation is known as declination. Declination is positive when magnetic north is east of true north. Finally, the lines of force vary in intensity, ranging between about 0.25 and 0.65 Gauss. In comparison, a common refrigerator magnet has a strength of about 100 Gauss. Figure 13.7 provides a graphic illustration of the dipolar internal magnetic field of the earth.

The geomagnetic field is composed of three major sources. The first of these is the internal field, or main field, thought to be caused by a combination of the convective motion of molten metals in the liquid outer core of the earth, a sort of magnetohydrodynamic dynamo, together with remnant fields of ferromagnetic materials trapped in the earth's lithosphere and upper mantle. The second source arises from currents associated with atmospheric tidal winds caused by differential solar heating or gravitational lunar forcing. These wind effects move the ionospheric plasma against the geomagnetic field lines, thus generating electric fields and currents just like a dynamo coil moving against

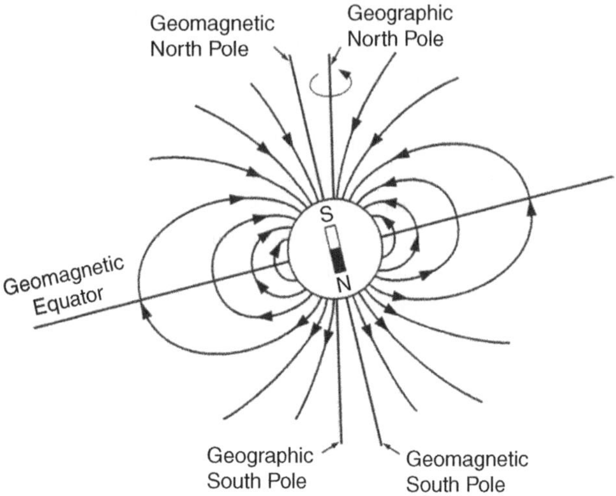

Figure 13.7 Dipolar magnetic field of the earth.

magnetic field lines. This wind-driven motion of the plasma tends to form vortices both north and south of the equator and a so-called electrojet along the equator itself, analogous to jet streams of air that flow from west to east about 10 km above the earth's surfaces. The third major source is composed of a number of current systems that exist in the earth's magnetosphere.

A magnetosphere is that area of space around a planet that is controlled and shaped by the planet's geomagnetic field and its interaction with the solar wind and the accompanying interplanetary magnetic field (IMF). The solar wind was named after Eugene Parker of the University of Chicago in 1958 [19] as a result of his efforts to derive the equilibrium structure of the sun's atmosphere, known as the corona. The sun's corona is held captive by the sun's gravity. In any atmosphere, the average velocity of atoms, molecules and ions depends on their temperature. Individual velocities are spread out around this average, and a few particles are always fast enough to "evaporate" and escape the restraining gravity. The sun's gravity is very strong, but the extremely high temperature of the corona essentially ejects the topmost layers of the corona. The flow of this ejecta was named "solar wind" and its existence was later confirmed by instruments aboard spacecraft.

The solar wind shapes the earth's magnetosphere and supplies energy to its many processes. Its density at the intersection with the earth's orbit is around 6 ions per cubic centimeter. The solar wind comprises neutral and ionized species together with free electrons and is essentially a neutral plasma, similar to, but different from, that surrounding the earth. The solar wind flows away from the sun with a mean speed of about 400 km/sec. Observations made by the Voyager 2 space probe indicate this same flow extends past the outermost planets. During moderate solar wind conditions, the magnetosphere terminates at roughly 10 R_E from the center of the planet on the sunward side. During turbulent magnetic storm conditions, it can be compressed to about 6 R_E. The solar wind generally flows around the geomagnetic field and consequently the magnetosphere stretches out to a distance of possibly 1000 R_E in the direction away from the sun.

Solar wind particles enter the magnetosphere through direct streaming along magnetic field lines at the cusps of the magnetosphere near the north and south magnetic poles, impacting the atmosphere and generating the aurora borealis and aurora australis, respectively. The solar wind also enters the magnetosphere both at the dayside magnetopause and at the flanks of the magnetotail at "reconnection sites"—regions where the IMF field lines intermittently connect to the earth's magnetic field, depending on the character of the solar wind density, speed, and IMF direction [20, 21].

The earth's magnetosphere prevents most of the particles from the sun carried in the solar wind from hitting the earth. The sun and other planets have magnetospheres, but the earth has the strongest one of all the rocky planets. The earth's magnetosphere is a highly dynamic structure that responds dramatically to solar wind variations. On the dayside of the earth, the magnetosphere is significantly compressed by the solar wind to a distance of approximately 65,000 km (40,000 mi). The earth's bow shock is about 17 km (11 mi) thick and is located about 90,000 km (56,000 mi) from the earth. The magnetopause is the boundary between the planet's magnetosphere and the solar wind. The location of the magnetopause is determined by the balance between the pressure of the dynamic planetary geomagnetic field and the dynamic pressure of the solar wind. As the solar wind pressure increases and decreases, the magnetopause moves inward and outward in response. The magnetopause exists at a typical distance of several hundred kilometers off the surface of the earth. The earth's magnetopause has been compared to a sieve, as it allows particles from the solar wind to enter. Kelvin–Helmholtz instabilities occur when large swirls of plasma travel along the edge of the magnetosphere in the magnetosheath, the plasma region between the bow shock and the magnetopause, at a different velocity from the magnetosphere. This results in magnetic reconnection, and as the magnetic field lines break and reconnect, solar wind particles are able to enter the magnetosphere. The solar wind drags out the night-side magnetosphere to possibly 1000 R_E; its exact length is not known. This extension of the magnetosphere is known as the magnetotail. Figure 13.8 illustrates the foregoing details.

The constantly shifting balance of pressures in the magnetosphere requires that the electron and ion species particles move in certain ways. The motions of these particles tend to follow various paths around and through the magnetosphere. Primary among these paths is the largest component of the external

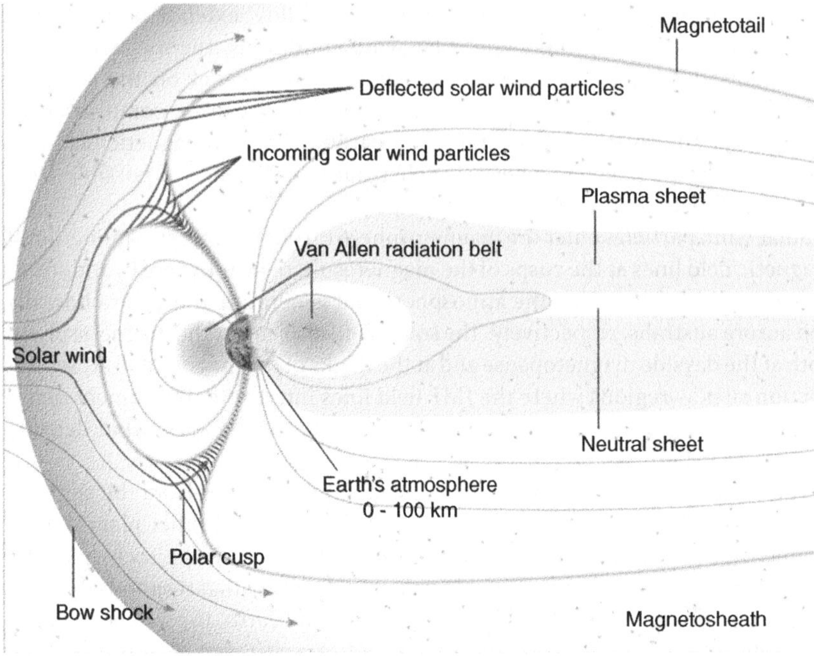

Figure 13.8 Schematic diagram of the earth's magnetosphere. (*Source*: NASA)

variations in earth's geomagnetic field, a ring of current circulating westward at a distance of 2–9 R_E. A charged particle moving with velocity v in a magnetic field of strength B will be acted on by the Lorentz force, causing the particle to spiral about the lines of force of the magnetic field. Consider a particle thus spiraling around a magnetic field line near the equator. The main magnetic field, originating inside the core, has a radial gradient and is larger closer to earth. The radius of curvature of the path *a* particle takes will thus be smaller closer to earth. Looking down from the north pole, a positive particle will circle clockwise and drift westward; an electron will do the opposite, and the combined motion of these charged particles constitutes what is known as the ring current. The strength of the ring current is characterized by the "DST" (disturbance storm time) index, an index of magnetic activity derived from a network of low- to mid-latitude geomagnetic observatories that measures the intensity of the globally symmetrical part of the equatorial ring current. Other currents that flow in the magnetosphere include the magnetopause current, the neutral sheet current, and various field-aligned currents. Figure 13.9 is a composite sketch illustrating the earth's magnetospheric currents.

The interaction of the geomagnetic field and the solar wind creates a uniquely hazardous electromagnetic environment for artificial earth orbiting satellites

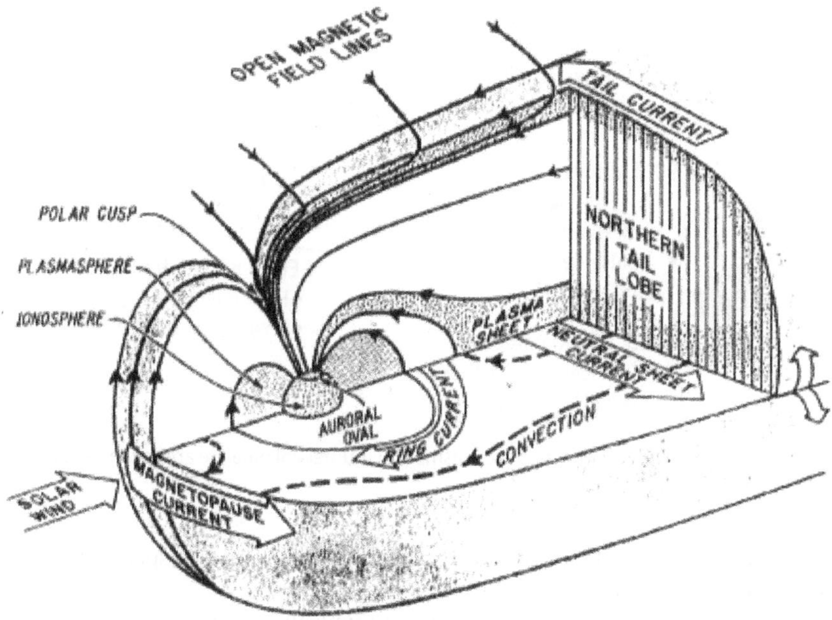

Figure 13.9 Composite sketch illustrating magnetospheric currents. (*Source*: NASA)

and other manned and unmanned space missions. Deep space ions, solar eclipses that occur each time a spacecraft enters the earth's shadow during orbit, electromagnetic interference from multiple sources, geomagnetic storms, plasma, radiation belts, and solar flares all present challenges to the spacecraft designer and operator. Effects on spacecraft operating in single or multiple orbital altitudes and regions about the earth are varied and can be dramatic, including deterioration of spacecraft surfaces from exposure to O^+ and the formation of surface oxides, ionizing radiation damage including single event upset and latch-up, total ionizing dose effects, displacement damage, and spacecraft charging. The last of these is the subject of the remaining material in this chapter.

13.4 Introduction to Spacecraft Charging

Spacecraft that either orbit the earth or travel away from the earth into deeper space and return, by definition operate in the plasma environment that surrounds the earth. Because of this spacecraft charging, naturally resulting from exposure of the spacecraft to the surrounding space plasma environment, cannot be avoided. Spacecraft charging is the common term used to describe both

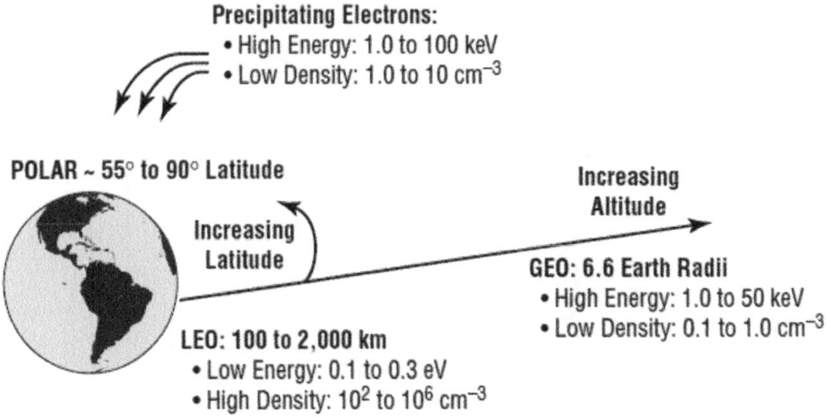

Figure 13.10 Properties of natural space plasma [22] (numerical data are approximate for illustration).

the accumulation and deposition of electrical charge in and on a spacecraft and its constituent materials.

Properties of the natural space plasma are typically described by specifying particle densities and energy levels. These properties are governed principally by latitude, longitude, altitude, the strength of the external magnetic fields associated with the stream and flow of charged particles and ultraviolet, X-ray, and shorter wavelength radiation that make up the solar wind, and the strength of the earth's own internal dipolar geomagnetic field. As depicted in Figure 13.10, the natural plasma encountered in orbital inclinations less than about 55° at LEO altitudes between roughly 90 km and 2000 km is relatively dense and exhibits an aggregate low energy. At orbital inclinations greater than about 55° at LEO altitudes, orbits sometimes referred to as Polar Earth Orbits, spacecraft will encounter the low density, high-energy environment of charged particles accelerated by the strong magnetic fields near the north and south magnetic poles. Many spacecraft missions are carried out in MEO altitudes roughly between 2000 km and 35,000 km. In this region are found high-energy charged particles concentrated in the Van Allen radiation belts that pervade the MEO environment. Because of the large number of satellites operating in this region of space and the decreasing density and increasing energy levels of charged particles encountered, it is considered a major environment of concern for internal charging effects.

In GEO altitudes, between 35,000 and 37,000 km, spacecraft frequently encounter high-energy, low-density plasma, much of which is associated with geomagnetic storms. These storms are a geophysical phenomenon by which electrons, protons, and other ions normally found at altitudes greater than GEO are accelerated toward earth, causing an increased flux of high-energy electrons. Such storms may inject high-energy charged particles in MEO and

LEO as well, complicating the dynamics of the environment and exacerbating its effects on orbiting spacecraft. The main phase of a geomagnetic storm, during which satellite anomalies may be expected to occur frequently, typically lasts between two and eight hours, with a gradual recovery to prestorm conditions that may last multiple days.

Engineers and scientists that constitute the spacecraft charging community by convention consider the ambient plasma environment to be neutral,

$$\varphi_p = 0 \tag{13.1}$$

If a spacecraft attains a potential greater than zero relative to the ambient plasma, it is said to be charged:

$$\varphi_s \neq 0 \tag{13.2}$$

Spacecraft charging is generally expressed as a differential relationship equating the time rate of change of the accumulated charge over a given surface area to the net current flowing both to and from the surface. The total charge transport is controlled by several factors, among them the type, conductivity, and thickness of the surface materials on the vehicle, the magnitude of the energy of the particles that make up the plasma environment in and through which the vehicle is passing, the latitude, longitude, and altitude of the spacecraft within the plasma environment, and the orbital location and velocity of the spacecraft with respect to the surrounding magnetic field structure dominated and created by the earth's geomagnetic field at lower altitudes and by the stream of photoelectrons and charged particles that make up the solar wind at higher altitudes.

Spacecraft charging may be simply expressed as the total flow of electrical current to and from the spacecraft to the surrounding plasma environment, or stated otherwise as the change in accumulated total electrical charge on the spacecraft as a function of time. Considered as a body with capacitance, the charging of a spacecraft can also be expressed as the capacitance of the body times the total change in its potential with respect to the plasma as a function of time.

$$\frac{dQ}{dt} = C\frac{d\varphi_s}{dt} = \sum_k I_k = I_E - \left[I_I + I_{SE} + I_{SI} + I_{BSE} + I_{PH}\right] \tag{13.3}$$

where

I_E = incident environmental electron current
I_I = incident environmental positive ion current
I_{SE} = secondary emitted electron current caused by I_E
I_{SI} = secondary emitted electron current caused by I_I
I_{BSE} = back-scattered electron current caused by I_E
I_{PH} = photoelectron current caused by exposure to sunlight

When total current flow stops, or the total amount of electrical charge stops changing as a function of time, the spacecraft has achieved equilibrium within its plasma environment, given simply as

$$\sum_k I_k = 0 \tag{13.4}$$

Despite the complexity of the interaction of spacecraft with the plasma environment, analysis can be facilitated greatly and with reasonable accuracy if the distribution of charged particles is assumed to be Maxwellian in nature. In fact, the most probable statistical distribution of charged particles in the unperturbed plasma environment is given by what is known as a Maxwell–Boltzmann distribution function, defined as

$$f(v_i) = n_i \left(\frac{m_i}{2\pi k T_i} \right)^{3/2} e^{(-m_i v_i^2 / 2kT_i)} \tag{13.5}$$

where

v_i = velocity of species i;
n_i = number density of species i;
m_i = mass of species i;
T_i = temperature of species i;
k = Boltzmann constant.

This is a so-called single Maxwellian distribution, typically used as a baseline for calculations of a spacecraft charging threat level. A double Maxwellian distribution, consisting of a superposition of two distributions characterized by different temperatures and densities, may be applied for worst-case environments such as those encountered in GEO and during solar storm activity. The associated charging current levels of these charged particles are often expressed as single or double Maxwellian distributions for analytical purposes.

The density and energy of electrons and positively charged ions composing the natural space plasma basically are the same. The mass of the positively charged ions, however, is orders of magnitude greater than the mass of electrons in all spacecraft orbits. Since their energies are equal, the electron current is greater than the ion current to spacecraft surfaces because lighter particles move faster, usually causing a net negative charge to spacecraft surfaces. Spacecraft charging occurs in all spacecraft orbits to varying degrees.

However, not all components of the charge balance equation can easily be expressed by a simple Maxwellian distribution. Such components include photoelectron current that, in some cases, is greater than the natural space plasma current. Photoelectron current consists of electrons liberated from spacecraft surface materials when sufficiently energized by solar radiation. Photoelectron current acts as a positive current since the electrons are leaving the spacecraft surfaces. Because photoelectron current is relatively large, variations in solar radiation caused by eclipse periods, temporal and seasonal changes in the sun's angle, and orbital changes in the spacecraft orientation to the sun must be considered. As an example, in a three-axis stabilized spacecraft, the solar arrays maintain a constant orientation relative to the sun while other instruments remain fixed on an object in space, such as the earth. Thus, different sides of the spacecraft body are illuminated by the sun at different points in the orbit. Because each variation in the sun's angle causes changes in the photoelectron current, determination of the equilibrium or floating potential must consider spacecraft orientation to the sun. Because of photoelectron currents, some surfaces may actually be positive by single-digit potentials.

Incident electrons and ions contribute to the total charge on the surface, and high-energy electrons can penetrate into dielectrics, depositing charge internally. Charged particles may leave the spacecraft surface as a result of reflection, backscatter, secondary emission electrons caused by absorption effects, or secondary emission being trapped and returned to a different location on the vehicle. All of these charge currents are captured diagrammatically as shown in Figure 13.11.

Note in Figure 13.11 the dashed line labeled "Sheath." This represents the edge of the Debye sheath that surrounds the vehicle in the plasma environment. The dimensions of the Debye sheath region are governed by the natural potential of the body with respect to the surrounding plasma. This potential will affect the space surrounding the body in such a manner that the potential decreases according to

$$V = V_0 e^{-|r|/\lambda_D} \tag{13.6}$$

where λ_D = the Debye length.

The Debye length is the scale length for the sheath and is given by

$$\lambda_D = \sqrt{\frac{\varepsilon_0 K T_e}{e^2 n}} \tag{13.7}$$

where

ε_0 = the permittivity of free space
K = Boltzmann's constant
T_e = the electron temperature in Kelvin
e = electron charge
n = the ambient plasma density.

The charged particles in the plasma environment will be attracted or repelled from the surface of the vehicle, depending on the sign of the vehicle's charge. Within the Debye sheath volume, charged particles of opposite sign to that of the vehicle will collect to the extent that outside of the Debye sheath, the volume

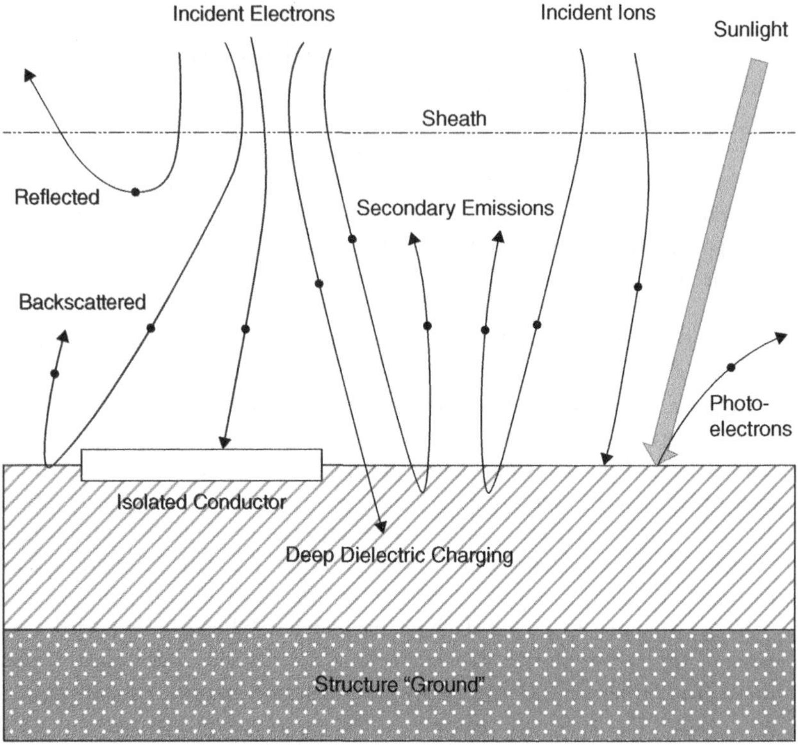

Figure 13.11 Balance of charge currents (similar to Wrenn and Sims [23, pp. 491–511]).

occupied by the vehicle and its sheath appears electrically neutral. When the vehicle is in LEO, the charged particles in the surrounding plasma have low energy and high density, such that an equilibrium state can be achieved over a very short distance or radius. As the energy level of the particles increases and their density decreases, the corresponding distance or radius becomes large. In LEO, the Debye sheath is quite thin, on the order of millimeters. In GEO, the reverse is true; the Debye sheath is quite large, on the order of 100s of meters. When the sheath is thin, the spacecraft potential is effectively charge limited, and the floating potential of the vehicle will be close to that of the plasma itself. When the sheath is thick, the potential is not charge limited, and can reach magnitudes of several tens of kV.

In addition to the physical issues already mentioned, motion of the spacecraft and the plasma constituents themselves through the plasma environment and the surrounding magnetic field potentials can result in changes in the vehicle's electric potential and anisotropic effects associated with dynamically changing particle densities and fluxes. Motion of a body across a magnetic field induces an electric field expressed as

$$\overrightarrow{E_m} = \overrightarrow{V_0} \times \vec{B} \tag{13.8}$$

where

V_0 = velocity of the body
B = the magnetic field.

If the body has low conductivity, this can create a potential difference between two positions on the spacecraft. This effect implies that large objects in strong magnetic field never can have the same potential at all points on the spacecraft without active potential control. Charged particles and fluxes may be accelerated, separated, or clumped together, resulting in environmental variations encountered by the spacecraft on a moment by moment basis.

13.5 Types of Spacecraft Charging

Spacecraft charging can generally be separated into two categories, known as surface charging and internal charging. Surface charging can be further broken down into absolute charging, sometimes referred to as structure or frame charging, and differential charging. Internal charging may also be referred to as deep dielectric charging, bulk charging, buried charging, or thick charging.

Surface charging is the buildup of electrical charge on the outer surfaces of the spacecraft resulting from interactions between spacecraft surfaces and space plasma, geomagnetic fields, and solar radiation. Unequal negative and positive currents to spacecraft surfaces produce an accumulation of charge on the exposed surfaces of a spacecraft. In addition, photons from solar illumination liberate electrons through the photoelectric effect, contributing to the total charging process. As a positive or negative charge accumulates, an electric field develops that acts to decelerate like-charged particles and accelerate oppositely charged particles. The charging process continues until the accelerated particles are collected rapidly enough to balance the currents. At this point, the spacecraft reaches its equilibrium charging level or "floating potential" and no more charge accumulates. The majority of charged particles affecting the charge level are electrons and ions, with energies ranging from as little as 1 eV to approximately 50 keV. Surface charging is a problem for high-inclination LEO satellites, primarily because of passage through downstreaming plasma in the auroral zones during periods of high geomagnetic activity. These same satellites will be affected in turn, and, to a much lesser degree, by passage through cold plasma from the F-region of the ionosphere at lower latitudes which peaks in free electron density between 300 and 400 km but extends to just above 1000 km. In GEO, surface charging occurs intermittently during geomagnetic substorms, which inject high-energy electrons from the plasma sheet inward to GEO, especially between 2000 and 0800 magnetic local time [24, pp. 23, 24].

Absolute charging describes the resultant electrical state of a spacecraft that has accumulated sufficient charge to establish a net potential different from that of the surrounding plasma environment. If the spacecraft is all conductive material, for example, a metal such as aluminum, the entire spacecraft will charge to the same potential. Since a spacecraft chassis is not normally "grounded" to the surrounding space plasma, it is free to collect charge and reach a floating potential relative to the space plasma. Although the absolute charge on a spacecraft should theoretically be very small, equaling only a few volts under "normal" conditions, geomagnetic-activity-induced and spacecraft-power-system-induced charging can charge spacecraft to tens, hundreds, or even thousands of volts. Although geomagnetic-activity-induced charging might be rare or mild for some orbits, spacecraft-power-system-induced charging to nearly the voltage of the power system, near 160 volts for the International Space Station for example, may take place every time a satellite's solar arrays are illuminated. Absolute charging typically takes place very rapidly, with characteristic periods on the order of microseconds during geomagnetic substorm activity.

Differential charging, with periods on the order of seconds to minutes, occurs more gradually than absolute charging. Dielectric materials used on many modern spacecraft (i.e., Kapton TM and Teflon TM) are poor distributors of accumulated charge and thereby maintain a portion of the charge deposited on them. Variation in the charged particle flux causes these surfaces to reach different floating potentials. The largest levels of differential charging will typically develop between sunlit and shaded surfaces because photoelectron current, often the largest source of positive current to a surface, tends to maintain the floating potential of sunlit surfaces near the plasma potential, while shaded dielectric surfaces can charge to large negative potentials. The rate of charge buildup is not constant, however, and there are times large differential potentials are more likely to build up, typically when the spacecraft moves from one plasma or illumination environment to another.

Internal charging is the buildup of charge on and within dielectric materials, such as circuit boards and wiring insulation, or on insulated floating conductors inside the spacecraft. Energetic electrons, with energies from approximately 10s of keV to several MeV, can penetrate the surface of the spacecraft and deposit charges inside. High-energy electrons associated with internal charging are concentrated in the inner and outer electron Van Allen belts, making high LEO orbits and GEO orbits similarly susceptible [25]. Photoemission and secondary emission currents are insignificant for internal charging, whereas internally conducted current is a major contributor to the overall current balance.

13.6 Potential Damage

The primary mechanism by which spacecraft charging disturbs mission activities is ESD that occurs when the accumulation of charges at a location is more rapid than its diffusion. This rapid accumulation will generate electric fields that exceed breakdown thresholds, resulting in ESD events. ESD resulting from spacecraft charging can take place between locations on or inside the spacecraft or between spacecraft surfaces and the surrounding plasma environment. At the breakdown threshold where the discharges occur, charges will move between the locations giving rise to a current, followed by an electromagnetic disturbance. Small discharges generally cannot be detected when the spacecraft is in space, but during the Meteosat-1 ground tests, small discharges on the spacecraft surface were observed as frequently as one discharge per second [26].

ESD radiates a transient broadband electromagnetic pulse that couples into spacecraft electronics and causes operational anomalies such as unintended logic changes, command errors, spurious signals, phantom commands, degraded sensor performance, and component failure. ESD may induce undesirable current spikes on electronics power busses leading to loss of solar array strings. Solar array strings are often affected by ESD events in LEO, with an accompanying loss of spacecraft electrical power. ESD may also induce transient currents in electrical and electronic power systems capable of shifting the vehicle electrical system reference and draining electrical power source current from these same systems. ESD can contribute to vehicle surface degradation and contamination that can alter material properties and can produce emissions that can interfere with sensitive optical instruments. Light flashes, such as arcs on solar arrays between interconnections and between the array and the surrounding plasma environment, are commonly observed discharge phenomena. Deep dielectric ESD is potentially more serious than surface ESD because the discharge event occurs within the spacecraft in close proximity to sensitive circuits where the probability is higher that the electromagnetic pulse will have a detrimental effect.

Besides generating electromagnetic pulses that can couple with spacecraft electronics, ESD events can lead to physical damage of surfaces. ESD events can produce localized heating and cause ion sputtering. In sputtering, large negative floating potentials of spacecraft surfaces accelerate positively charged ions to high energies, leading to the physical removal of surface atoms (i.e., sputtering) by the impacting ions. If the ESD events are repetitive, they can lead to ejection of sizable amounts of surface material from the discharge site. Ion sputtering and larger scale ejection of surface materials can create a source of contamination for other spacecraft surfaces. In addition to sputtering and ejection of materials, organic molecules outgassed from spacecraft surfaces can be ionized while still near the spacecraft by solar radiation and attracted to negatively charged surfaces. The more negative the floating potential of a surface is, the greater the probability of its contamination. Ultimately, loss of material can degrade spacecraft structural integrity, alter the electrical and thermal properties of spacecraft surface materials, and can even lead to spacecraft trajectory control and related navigational errors. ESD events may even stop after some period of time because of changes in the material caused by the discharges.

ESD can result in instantaneous and unexpected anomalies, but it can also cause detrimental effects through the accumulation of charge over long periods of time. The time scale for a significant level of internal charging can be days or longer and may even stretch from months to years. Once the rate at which the energetic electrons deposit on the surface or embed inside a bulk dielectric is greater than the rate at which the charge leaks out, the electric field begins to increase in magnitude. Once the generated electric field reaches the breakdown threshold of the dielectric, ESD will occur. ESD events inside spacecraft circuitry can lead to anomalies in spacecraft operation and can cause loss of mission or spacecraft. Recently, failures of solid-state power amplifiers (SSPAs) between 1996 and 2012 were analyzed [26] using data from the British telecommunications company Inmarsat amounting to 26 anomalies over 16 years. SSPAs amplify uplink signals received by a satellite and retransmit downlink signals, making them critical components for satellite operation. The authors found that most of these anomalies occurred during relatively quiet geomagnetic periods, but that a significant number of SSPAs experienced a high-energy electron flux in the two weeks prior to the anomaly. Also, the anomalies occurred in all local times, but most occurred either in the midnight-to-dawn sector or the local-noon-to-dusk sector. It is difficult to build reasonable statistics on such a small sample, but in their report the authors inferred that most of the SSPA anomalies were likely not caused by surface charging alone, but rather by internal charging of components by high-energy electrons or by the combined effects of surface and internal charging.

Internal charging depends on four factors: the plasma environment, the shielding thickness of the spacecraft, and the characteristics and geometry of the charged materials. The rate of charge accumulation may not be constant and generally depends on the plasma environment in which the spacecraft finds itself at any given time, as well as whether the spacecraft is in sunlight or eclipse. Also, note that the threshold electrical potential required to achieve breakdown varies depending on the material and the spacing between conductors within spacecraft components. The amount of internal or surface charge that may accumulate before ESD occurs may not even be comparable between different spacecraft, especially those traversing different plasma populations within the magnetosphere.

The Anik E1 and Anik E2 satellite anomalies are primary examples of problems caused by ESD. Both satellites were nearly identical in design and construction, and both were in GEO, exposed to very nearly the same plasma environment. The elevated fluence of high-energy electrons that led to ESD in both satellites actually started on January 13, 1994, and lasted for 10 days. Satellite systems failures were not reported until near the end of those 10 days, so that the eventual ESD events occurred after significant buildup of electrical charge on internal spacecraft components. Note the two satellites' momentum wheel control systems failed within nine hours of one another [27].

On June 2, 1973, perhaps the most significant event in spacecraft charging history was the catastrophic failure of the US Air Force Defense Space System Communication Satellite (DSCS) 9431 [28,29]. The satellite failed when power to its communication system was suddenly interrupted by a high-energy discharge caused by surface charge build up during a geomagnetic storm. Both 9431 and its sister satellite 9432 experienced a series of nuisance anomalies prior to the catastrophic event, but nothing that would or could have predicted the failure. The incident motivated a joint NASA and US Air Force investigation to explore and evaluate the spacecraft charging phenomenon, culminating in the SCATHA satellite mission, discussed previously in the historical section. SCATHA provided a tremendous amount of data, much of which is still used today for protective design efforts. Many other examples of spacecraft failures attributable to spacecraft charging exist, several of which are documented in NASA Reference Publication 1375 [30].

13.7 Ways and Means of Protection/Mitigation

Spacecraft charging and its detrimental effects can be controlled through careful design. LEO missions at low latitudes and low altitudes in general are operating in a relatively benign charging environment. However, spacecraft in LEO at low latitudes and low altitudes that also operate solar arrays will require specific design features to protect the arrays and prevent loss of electrical power. Spacecraft in polar orbits, in LEO at high altitudes that encounter some part of the Van Allen belts, and GEO operate in a much harsher charging environment. These missions will require some if not all of the protective features for general LEO operations, but also additional features that specifically the surface and internal charging threats they will encounter.

General ESD control guidelines summarized below are effective at all altitudes and are applicable both at the subsystem and system levels [18]:

1. All conducting elements are tied to a common electrical reference point either directly or through charge bleed-off resistors.
2. The exterior surfaces of the spacecraft are at least partially conductive.
3. The primary spacecraft structure, electronic components housings, and electric cable shields provide a physically and electrically continuous shielded surface around all electronics and wiring.

4. Electrical filters are employed to protect circuits from discharge induced upsets.
5. The electrical continuity of the spacecraft grounding system is intact with less than 2.5 mΩ resistance per joint.
6. The solar array panels and substrates are electrically grounded to structure.
7. Antistatic measures should be adapted in the design and fabrication of the solar array.
8. The aluminized surfaces of MLI blankets are electrically referenced to the structure in multiple locations such that spacing between the neighboring connections is not more than 1 m.
9. The conductive material on antenna reflector rear surfaces is referenced to structure.
10. Carbon fiber reinforced plastic composites are referenced to structure.
11. The return paths for all electrical signals including electrical power are through wires rather than structure.
12. Wire shields are not used as signal return lines except in the case of coaxial cables.
13. Pyrotechnic wiring is routed in separate bundles from all other wiring.
14. Twisted and shielded pair wires are used for all circuits, except for solar arrays.
15. Major charge collectors such as solar array substrates have hard wire return lines to structure through bleeder resistors.
16. All electronic elements should be shielded with Al or equivalent thickness so that the internal charging rate is benign.
17. Co-cured Tantalum foils may be used along with the composite face sheets to provide enhanced shielding against space radiation. For ESD control, however, the foil should be referenced to structure.
18. Dielectric materials having very high surface resistivity ($> 10^{12}$ Ω/sq.) should be avoided.
19. Dielectric materials should be static dissipative or leaky. This applies to circuit boards, wire insulation materials, and conformal coating materials.
20. Low pass filters and/or transient suppressors on interface circuits provide adequate protection against internal ESD. The filter or the protection circuit should be located at the device terminals.
21. Cable harness should be routed away from apertures. Wrapping of all harness, especially external harness, with copper tape and physically and electrically terminating the tape at connector backshells provides protection against energetic particles.
22. Thermal blankets provide very little protection against electron penetration. Additional Al shielding should be provided where external surfaces are covered only with MLI blankets.

Specific measures for spacecraft operating in LEO at low latitudes and low altitudes that also operate solar arrays may be found in NASA-STD-4005

[11] and NASA-HDBK-4006 [12]. In like fashion, measures and techniques for polar, high-altitude LEO, and GEO operations may be found in NASA-HDBK-4002A. General information applicable to the phenomenon of spacecraft charging may be located in NASA Reference Publications 1354 [31], 1375 [30], and 1390 [22].

13.8 Concluding Material

Spacecraft charging and satellite anomalies are closely linked, one to the other. While spacecraft charging is a general physical phenomenon that cannot be totally avoided, it does not necessarily lead to anomalous behavior of the spacecraft. On the other hand, damaging anomalies can be caused by excessive charging. For both spacecraft engineering and operations, a key question is when, where, and why satellite charging and other environment-related anomalies occur, and how hazardous these phenomena may be. The design and operation of a spacecraft involves many talented people with diverse backgrounds, ranging from manufacturing and assembly to engineering and program management. Many designers and builders of spacecraft are unfamiliar with the definition of spacecraft charging and the damage that spacecraft charging can cause. Understanding the problem is an important first step in getting everyone involved in addressing the appropriate spacecraft charging issues during spacecraft design and build phases.

References

1 H. B. Garrett, "The charging of spacecraft surfaces," *Rev. Geophys. Space Phys.*, vol. 19, no. 4, pp. 577–616, Nov. 1981.

2 I. Langmuir and K. B. Blodgett, "Currents limited by space charge between concentric spheres," *Phys. Rev.*, vol. 24, no. 1, 1924.

3 C. Y. Johnson and E. B. Meadows, "First investigation of ambient positive-ion composition to 219 km by Rocket-Borne spectrometer," *J. Geophys. Res.*, vol. 60, no. 2, pp. 193–203, 1955.

4 K. P. Chopra, "Interactions of rapidly moving bodies in terrestrial atmosphere," *Rev. Mod. Phy.*, vol. 33, no. 2, pp. 153–189, 1961.

5 S. F. Singer, ed. *Interactions of a Space Vehicles with an Ionized Atmosphere.* New York: Pergamon, 1965.

6 E. C. Whipple, "The equilibrium electric potential of a body in the upper atmosphere and in interplanetary space," NASA X-615-65-296, 1965.

7 C. K. Purvis, et al., "Design guidelines for assessing and controlling spacecraft charging effects," NASA-TP-2361, 1984.

8 W. G. Dunbar, "Design guide: Designing and building high voltage power supplies," AFWAL-TR-88-4143, vol. 2, Aug. 1988.

9 A. Whittlesey and H. B. Garrett, "Avoiding problems caused by spacecraft on-orbit internal charging effects," NASA-HDBK-4002, 1999.

*10 D. C. Ferguson and G. G. Hillard, "Low earth orbit spacecraft charging design guidelines," NASA TP-2003-212287, 2003.

11 D. C. Ferguson, "Low earth orbit spacecraft charging design standard," NASA-STD-4005, 2007.

*12 "Low earth orbit spacecraft charging design Handbook," NASA-HDBK-4006, 2007.

*13 "Mitigating in-space charging effects – A guideline," NASA-HDBK-4002A, 2011.

14 R. F. Pfaff Jr., "The near-earth plasma environment," *Space Sci. Rev.*, vol. 168, pp. 23–112, 2012.

15 J. K. Hargreaves, *The Solar-Terrestrial Environment*. Cambridge: Cambridge University Press, p. 209, 1992.

16 J. A. Van Allen, "The geomagnetically trapped corpuscular radiation," *J. Geophys. Res.*, vol. 64, no. 11, pp. 1683–1689, 1959.

17 W. N. Spjeldvik and P. L. Rothwell, "The earth's radiation belts," Report, AFGL-TR-83-0240, Air Force Geophysics Laboratory, Hanscom AFB, MA, Sep. 29, 1983.

18 M. F. McKay, D. S. McKay, and M. B. Duke, eds. "Space resources – Energy, power, and transport," Technologies Section, NASA SP-509, vol. 2, Lyndon B. Johnson Space Center, Houston, TX, 1992.

19 E. N. Parker, "Dynamics of the interplanetary gas and magnetic fields," *J. Astrophys.*, vol. 128, pp. 664–676, 1958.

20 J. R. Wertz and W. J. Larson, *Space Mission Analysis and Design*, 3rd ed. Space Technology Library, vol. 8. Dordrecht, Netherlands: Kluwer Academic Publishers, 1999.

21 M. G. Kivelson and C. T. Russell, *Introduction to Space Physics*. Cambridge, UK: Cambridge University Press, 1995.

22 K. L. Bedingfield, R. D. Leach and M. B. Alexander, "Spacecraft system failures and anomalies attributed to the natural space environment," NASA Reference Publication 1390, Aug. 1996.

23 R. N. DeWitt, D. Duston, and A. K. Hyder, eds. *The Behavior of Space Systems in the Space Environment. Series E: Applied Sciences*, vol. 245, NATO Advanced Science Institutes Series. Dordrecht, Netherlands: Kluwer Academic Publishers, 1993.

24 M. Romero and L. Levy, "Internal charging and secondary effects," in *The Behaviour of Systems in the Space Environment. Proc. NATO Advanced Study Institute 245*, Kluwer Academic Publishers, R. N. DeWitt et al., eds., pp. 565–580, 1993.

25 Tsoline Mikaelian, *Spacecraft Charging and Hazards to Electronics in Space*. Canada: York University, May. 2001.

*26 D. G. Hoge, "Meteosat spacecraft charging investigation," ESA N82-14265, 1980.

*27 H.-L. Lam, et al., "Anik-E1 and E2 satellite failures of January 1994 revisited," *Space Weather*, vol. 10, S10003, doi:10.1029/2012SW000811, 2012.
28 G. T. Inouye, "Spacecraft charging anomalies on the DSCS II launch 2 Satellites," in *Proceedings of the Spacecraft Charging Technology Conference*, NASA Lewis Research Center, pp. 829–852, Feb. 1977.
29 N. J. Stevens, A. Rosen and G. T. Inouye, "Communication satellite experience in the seventies," in *AIAA 25th Aerospace Sciences Conference*, Reno, NV, Jan. 1987.
*30 R. D. Leach and M. B. Alexander, "Failures and anomalies attributed to spacecraft charging," NASA Reference Publication 1375, Aug. 1995.
*31 J. L. Herr and M. B. McCollum, "Spacecraft environments interactions: protecting against the effects of spacecraft charging," NASA Reference Publication 1354, Nov. 1994.

Bibliography

W. Lohmeyer and K. Cahoy, "Data management of geostationary communications satellite telemetry and correlation to space weather," thesis, Massachusetts Institute of Technology, Feb. 2013.

H. B. Garrett and A. C. Whittlesey, "Spacecraft charging, an update," in *Proceedings, WESCON 96*, Oct. 22–24, Anaheim, CA, pp. 260–267, 1996.

E. C. Whipple, "Potentials of surfaces in space," *Rep. Prog. Phys*, vol. 44, pp. 1197–1250, Nov. 1981.

J. L. Barth, "Space and atmospheric environments: From low earth orbits to deep space," in *Protection of Materials and Structures from Space Environment, Space Technology Proceedings*, J. Kleiman and Z. Iskanderova, eds., vol. 5, pp. 7–29, 2003.

D. E. Hastings, "A review of plasma interactions with spacecraft in low earth orbit," *J. Geophys. Res.*, vol. 100, no. A8, pp. 14457–14483, Aug. 1995.

R. Grard, et al., "Spacecraft charging effects," *Space Sci. Rev.*, vol. 34, pp. 289–304, Aug. 1995.

R. Perez, "Electrostatic discharge analysis for spacecraft in geosynchronous orbit," in *Proceedings, Electrical Overstress – Electrostatic Discharge Symposium*: EOS-15, Sep. 28–30, Lake Buena Vista, FL, pp. 93-29–93-34, 1993.

D. A. Galvan, et al., *Satellite Anomalies: Benefits of a Centralized Anomaly Database and Methods for Securely Sharing Information among Satellite Operators*. Santa Monica, CA: RAND National Defense Research Institute, 2014.

M. Xapsos, "Modeling the space radiation environment," Section 1, IEEE Nuclear and Space Radiation Effects Conference Short Course, Ponte Vedra Beach, FL, Jul. 17, 2006.

T. W. Schlatter, "Atmospheric composition and vertical structure," in *Encyclopedia of Aerospace Engineering*. John Wiley & Sons, Ltd., Dec. 15, 2010.

*"Space engineering: Spacecraft charging," ECSS-E-ST-20-06C, *European Cooperation for Space Standardization*, ECSS Secretariat, ESA-ESTEC, Jul. 31, 2008.

*"Space engineering: Space environment," ECSS-E-ST-10-04C, *European Cooperation for Space Standardization*, ECSS Secretariat, ESA-ESTEC, Nov. 15, 2008.

*H. B. Garrett and A. C. Whittlesey, Hoboken, NJ: John Wiley & Sons, Inc., 2012.

R. C. Olsen, "Introduction to the space environment: PH 2514," Course text, Naval Postgraduate School, Jan. 2005.

S. T. Lai, ed. *Spacecraft Charging, Progress in Astronautics and Aeronautics*, vol. 237. Reston, VA: American Aeronautics and Astronautics, Inc., 2011.

*S. T. Lai, *Fundamentals of Spacecraft Charging: Spacecraft Interactions with Space Plasmas*. Princeton, NJ: Princeton University Press, 2012.

*A. Rosen, ed., *Spacecraft Charging by Magnetospheric Plasmas*. Progress in Astronautics and Aeronautics, vol. 47, New York, NY: American Aeronautics and Astronautics, 1975.

*H. B. Garrett and C. P. Pike, eds. *Space Systems and Their Interactions with Earth's Space Environment*. Progress in Astronautics and Aeronautics, vol. 71. New York, NY: American Aeronautics and Astronautics, 1980.

*D. Hastings and H. B. Garrett, *Spacecraft Environment Interactions. Cambridge Atmospheric and Space Science Series*. Cambridge, UK: Cambridge University Press, 1996.

C. W. Gordon, V. Canuto, and W. I. Axford, eds. *The Earth: 1 the Upper Atmosphere, Ionosphere, and Magnetosphere. Handbook of Astronomy, Astrophysics, and Geophysics*, vol. 1. New York, NY: Gordon and Breach Science Publishers, 1978.

K. G. Balmain, *Plasma Effects in EMC. Handbook of Electromagnetic Compatibility*, R. Perez eds. San Diego, CA: Academic Press, 1995.

S. E. Deforest, "Electrostatic potentials developed by ATS-5," in *Proceedings of the Sixth ESLAB Symposium*, Noordwijk, Netherlands, Sep. 26–29, pp. 263–276, 1972.

14

Analysis and Simulations of Space Radiation-Induced Single-Event Effects and Transients

Reinaldo J. Perez

14.1 Introduction

Spacecraft electronics are affected by the space radiation environment. Among the different types of radiation effects that can affect spacecraft electronics is the single-event transient (SET). The space environment is responsible for many of the SETs, which can upset the performance of the spacecraft avionics hardware. In this chapter, we first explore the origins of SETs and their effects as outlined by many scientists and researchers, many of which are mentioned in the references. This chapter also explores the modeling of a SET in digital and analog circuit, also as discussed by scientists who we have referenced. The chapter also addresses the concept of crosstalk that could develop among digital circuits in the present of a SET event. The chapter also provides a discussion of SET hardening. We then provide a discussion concerning propagation of a SET event at the local, subsystem, and system levels in a spacecraft using two different models, one of the models developed by this author, known as the State Transition Model. The final goal of the chapter is to provide a qualitatively methodology for assessing SETs and its effects so that spacecraft avionics engineers can develop either hardware or software countermeasures in their designs. SET is not a form of electromagnetic interference (EMI) in its origin, but semantically SET is very similar to EMI because they are both caused by a current source not previously accounted for. SET has the same effects as EMI, and it can cause interference problems in electronic circuits via multiple coupling mechanisms similar to EMI, and therefore makes such

The Research was carried out at the Jet Propulsion Laboratory, California Institute of Technology under contract with the National Aerospace and Space Administration. U.S. government sponsorship.

Handbook of Aerospace Electromagnetic Compatibility, First Edition. Edited by Reinaldo J. Perez.

Figure 14.1 The nature of the space radiation environment.

circuits incompatible. The chapter is the result of the contribution of many to this field over the years.

14.2 The Space Radiation Environment

The dominant energy source of the space radiation environment in the solar system is the Sun. The main effect of the Sun on the space environment is through its electromagnetic flux and the large number of charged particles it emits. The solar particle flux is composed basically of two main components: high-energy plasma ($E > 1$ MeV) bursts (e.g., from solar flares) and the lower energy ($E \approx 10$–100 eV) plasma, referred to as the solar wind. The high-energy plasma is primarily responsible for single-event effects (SEE). Heavy ions or protons striking sensitive junctions in semiconductors and depositing energy can cause SEE. These effects range from simple upsets on normal circuit operations to permanent circuit failures. Figure 14.1 shows the main sources of energetic particles that cause SEE events: solar energetic particles, cosmic ray protons and heavy ions, and protons and heavy ions from solar flares. Spacecraft electronics are affected by the space radiation environment [1].

14.3 Single-Event Effects

SEE is the general term used when space energetic particles are capable of creating an electrical response in an electronics device. All SEE events involve an electrical response observed following the generation of mobile electron-hole pairs by an energetic particle in a semiconductor device. The ionization caused by these energetic charged particles can cause either direct ionization or indirect ionization. Direct ionization, which is the most common, occurs when the incident charged particle creates electron-hole pairs. Indirect ionization occurs when the incident charged particle creates an energetic recoil charge particle, part of the nucleus of a target atom recoils, and then creates electron-hole pairs. Electronic devices that are susceptible to SEE have a level of immunity which is called "threshold LET." The threshold LET is defined as the minimum linear energy transfer (LET) that can cause a SEE.

The device SEE susceptibility to a specified type of charge particle is usually described in terms of a cross-section, which is defined as the number of errors observed divided by the fluence required to produce those errors. The single-event upset (SEU), a type of SEE, is defined by NASA as "radiation-induced errors in microelectronic circuits caused when charged particles (usually from the radiation belts or from cosmic rays) lose energy by ionizing, the medium through which they pass, leaving behind a wake of electron-hole pairs." SEUs are soft errors of a transient nature in ICs, but are generally nondestructive to the hardware or the IC. Often, a power reset or functional reset of the device is all that it takes for resuming normal operations of the device. SEUs can occur in analog, digital, optical, and mix analog/digital components, and they can also affect the interface circuitry (a common cause of failure propagation among circuits). SEUs can appear as transient signals in analog or digital circuitry or as erroneous bits in registers and digital gates. Other possibilities include multiple-bit SEUs, which are caused when a single ion induces SEUs in multiple bits simultaneously. When there are multiple-bit SEUs, there is a serious problem in the capability of single-bit error detection and correction (EDAC) of the system because it is very difficult to assign bits to fill in a word in different ICs; this is especially true in memory ICs. When a SEU is severe, it can cause a single-event functional interrupt (SEFI), in which an SEU in the IC, which provides control circuitry, places the circuit into a test mode, halt, or undefined state. A SEFI event interrupts normal operations of the circuitry and requires a power reset for the device to recover its functionality. A SEU in shift registers can also cause a SEFI if these registers control the address of the control logic, resulting in a misread of information in a device. The present trend in electronics development is in the decrease of electronic size scalability and increase functionalities, which will increase the susceptibility to SEU and SEFI.

Another type of SEE is the single-event latchup (SEL), which causes loss of device functionality due to a single-event-induced current state. SELs are hard failures in ICs and can be highly destructive because these can cause permanent damage to the device. The SEL can cause a high current effect in a given device, which may exceed the device specifications and will damage the device due to local joule heating. A latchup condition can destroy the device, stress the bus voltage, or damage the power supply. A latchup can be caused by heavy ion or by protons impacting very sensitive devices. A SEL can be cleared by a power reset or power strobing of the device. In a latchup condition, power must be removed quickly, otherwise a catastrophic failure can occur due to excessive heating or metallization or bond wire failure. SEL is strongly dependent on temperature and the threshold for latchup decreases at high temperature.

A single-event burnout (SEB) can cause the destruction of a transistor due to a high current surge in a power transistor when a heavy ion passing through the power transistor deposits enough charge to surge that current. SEB causes the device to fail permanently. SEBs can include burnout of power MOSFETs and frozen bits. Only SEB of n-channel power MOSFETs have been reported. A SEB can be triggered in a power MOSFET biased in the OFF state (i.e., blocking a high drain-source voltage) SEB susceptibility decreases with increasing temperature. SEB can also occur in bipolar junction transistors (BJTs) but are less common.

14.4 Single-Event Transient

In this chapter, we concentrate a lot in a type of SEE that can cause transients currents propagation, known as the SET. SET can not only cause propagation of transient currents within a system, but it can also couple to other electronics and systems via parasitic effects.

As electronic components have become smaller in device geometry, lower in operating voltage, and higher in complexity, their immunity to the space radiation environment has been diminished. SET on electronic hardware has become a major concern for spacecraft avionics designers. The spacecraft avionics must be designed such that no SET can cause an unrecoverable failure to the spacecraft. When a space charged particle strikes at a sensitive node such as the drain node in an IC, electron-hole pairs are created along an ionization track. A transient current pulse is generated following the drift and diffusion mechanisms. The current pulse results in an SET voltage generation at the particle hit node. Under favorable conditions, the pulse will propagate and cause soft errors in additional ICs and their derived digital circuits. The measured magnitude, pulse width, and shape of the transient pulse depend on the technology where the SET pulse is generated. Figure 14.2 illustrates a typical SET pulse and its parameters. Notice that transients can be very large and their pulse width varies considerably.

When a charge particle travels in space and strikes an electronic IC, it interact with the electrons and holes of inside the IC via electrostatic forces as shown in Figure 14.3 and discussed in [2–4]. Figure 14.4 shows the same illustration using a real IC gate. The three main types of interaction are excitation, ionization, and bremsstrahlung. In excitations, the charged particle transfers some of its energy to the electrons in the IC with insufficient energy to ionize them.

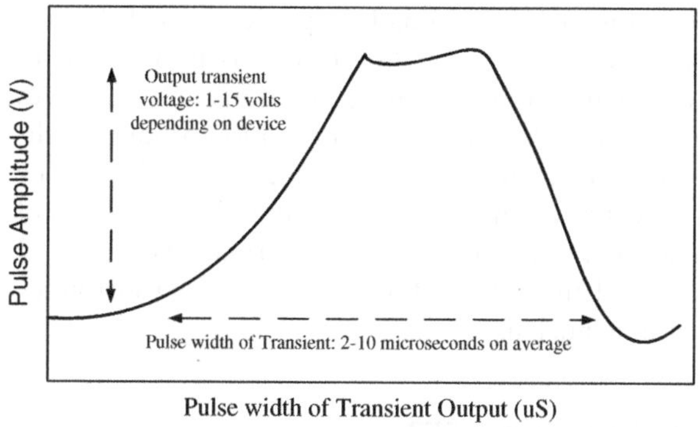

Figure 14.2 Typical measured parameters in a SET pulse.

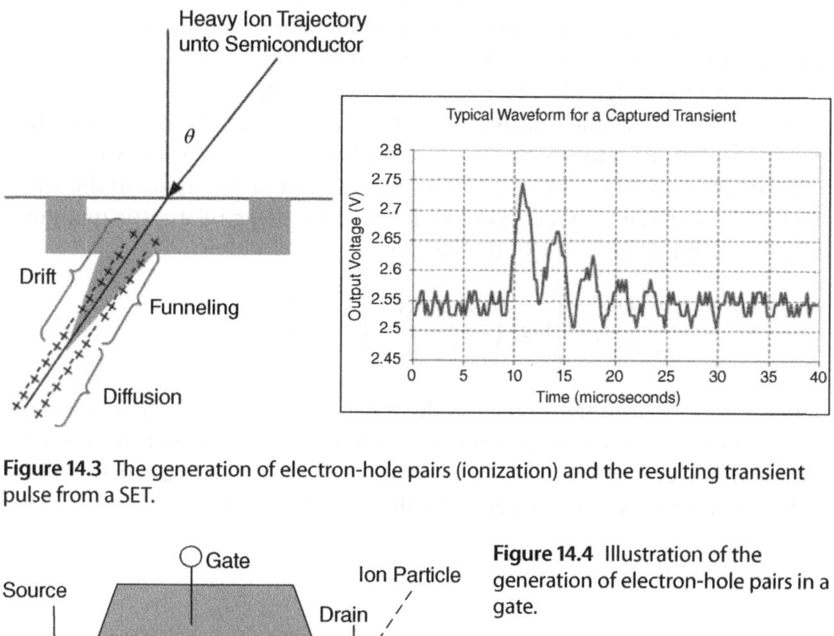

Figure 14.3 The generation of electron-hole pairs (ionization) and the resulting transient pulse from a SET.

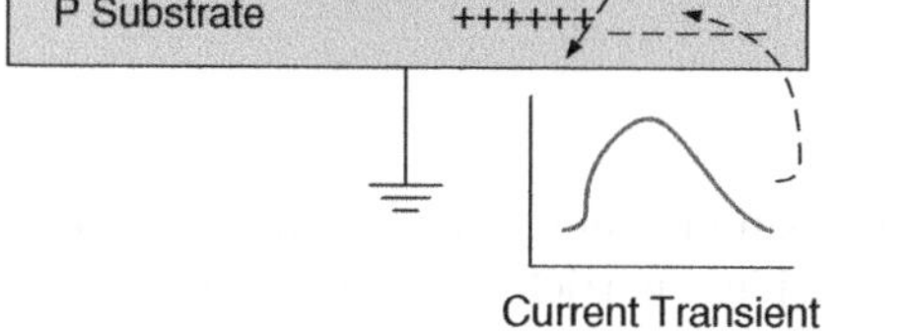

Figure 14.4 Illustration of the generation of electron-hole pairs in a gate.

When the electrons decay, they emit photons in a process called fluorescence. In ionization, the energy transferred to the electrons is sufficient to cause ionization within the IC material causing the generation of electro-holes pairs. Finally, if the electrons created by the original ionization produce additional ionization, these are called delta rays. Bremsstrahlung, the mechanism by which X-rays are produced, occurs when photons are emitted, in the form of electromagnetic radiation, when a fast-moving charged particle loses some of its energy upon being accelerated and deflected by the electric field surrounding the positively charged atomic nucleus.

14.5 Generation and Modeling a SET

Space charged particles such as photons, electrons, protons, and heavy ions are capable of ionization when these charged particles penetrate through semiconductor matter. The energy that is lost by an ionizing particle when it propagates through matter is called LET. LET is defined as the total energy that is lost per unit distance of travel and it is normalized by dividing the energy per unit distance by the density of the matter in order to obtain units that are KeV cm^2/mg. The normalization makes LET for a charged particle and energy about the same for different materials. Equation (14.1) defines LET using relativistic quantum mechanics and known as the Bethe formula,

$$LET = \frac{4\pi k_e^2 Z^2 e^4 n}{m_e c^2 \beta^2}\left[\log_e\left(\frac{2m_e c^2\beta^2}{I(1-\beta^2)}\right) - \beta^2\right] \tag{14.1}$$

where c is the speed of light m/s; e is the magnitude of the charge of the electron, in coulomb; I is the mean excitation energy of the absorber in Joules; k_e is the Coulomb constant Nm^2/C^2; m_e is electron rest mass in kg; n denotes the number of absorber electrons per unit value, electrons^{-3} or

$$n = \frac{N_A \rho \eta}{M}$$

where N_A is Avogadro's number, η is the number of electrons per molecule of absorber; M is the molecular mass of the absorber in kg/kmol; ρ is the density of the absorber, kg/m^3; Z is the atomic number of the incident particle;

$$\beta = vc^{-1} = \left(\frac{2E_k}{E_0}\right)^{1/2}$$

where v is the velocity (m/s), E_k is the kinetic energy of the incident particle (eV); $E_0 = mc^2$ = rest energy of the incident particle (eV); and m is the rest mass = eV/c^2.

The charge deposition for a striking particle with a given LET and with a penetration D is given by equation (14.2).

$$E_{\text{deposited}}(\text{eV}) = \text{LET}_{\rho} D(\text{cm})\rho \tag{14.2}$$

where ρ denotes density (g/cm^3). The charge deposited is given by equation (14.3).

$$Q = \frac{E_{\text{deposited}} \times e}{I_{\text{first}}} \tag{14.3}$$

where I_{first} is the first ionization energy of the atoms in the absorber, eV. It is the energy required to remove one or more electrons from a neutral atom to form a positive ion, for example, for silicon (Si) is 3.62 eV, for germanium (Ge) is 2.98 eV, and for gallium arsenide (GeAs) is 4.8 eV; *e* is the magnitude of electron charge, C; and *Q* is charge deposited, C (C is the charge in Coulombs).

Of great interest, however, is the calculation of the *upset rate* for a given IC as a function of LET. The upset rate is defined as the number of times an impacted IC causes the circuit to have an erroneous output. Therefore, in order to know if an IC is experiencing a SEE, its corresponding circuit output must be detected either at the board level or assembly level. An accurate method for calculating the upset rate (upsets/day) is the usage of the Weibull distribution [5] function *F*, which would have the form

$$\begin{aligned} F(\text{LET}) &= \sigma \left\{ 1 - e^{-\left[\frac{(\text{LET}-\text{LET}_{\text{threshold}})}{w}\right]^{s}} \right\} \quad \text{for LET} > \text{LET}_{\text{threshold}} \\ F(\text{LET}) &= 0 \quad \text{for LET} < \text{LET}_{\text{threshold}}, \end{aligned} \tag{14.4}$$

where in the above equation *w* is the width of the distribution and *s* is a shape parameter (*s* = 1 corresponds to an exponential distribution, *s* = 2 corresponds to a Rayleigh distribution, *s* = 4 approximates a normal distribution, while large "*s*" approaches the log-normal distribution). σ is the value of the limiting cross-section, with typical units of cross-section in cm^2 per bit.

In Table 14.1, we can observe a numerical application of the formulas previously derived for a series of common ICs in a given circuit application. The application is that of an input/output (IO) bus, for this particular example, and the environment is that of low earth geosynchronous orbit (1470 km at 53° inclination orbit) in a predominant proton environment. The "usage factor" for the ICs in Table 14.1 is hypothetical for a typical IO bus.

Table 14.1 SEE rate calculations for an earth 1470 km × 53° inclination orbit, in a proton-induced environment

Part type	σ (cm²/bit)	LET_{th}	Weibull "s"	Weibull "w"	#Bits/device	Bit SEE (rate/day)	Device SEE (rate /day)	Usage factor of IC in PCB
FPGA	2.3E−06	24.99	0.741	22.55	140	1.2E−08	1.6E−06	26%
256k SRAM	6.0E−06	29.44	3.923	15.91	262,144	1.7E−14	4.5E−09	80%
Op-Amp	1.0E−04	2.89	1.189	35.64	1	3.0E−03	3.0E−03	100%
512k SRAM Asyn	8.8E−09	20	5	100	262,144	2.0E−11	5.3E−06	40%
Octal transparent latch tristate outputs	2.5E−06	40	6.87	28.7	8	1.1E−08	8.4E−08	33%

14.6 Use of Upset Rates for Analyzing Vulnerabilities of Designs to SEE

For electronic designers, it is very important to potentially identify critical components in their design that are susceptible to SEE because such components now become part of a critical path to the failure of the whole design. The concept of critical paths is borrowed from management sciences [6], which states that the success of a project hinges mainly on a few critical paths and these paths are always the most constrained in terms of schedule, funding, and resources. Likewise, from an electronic hardware performance point of view, a critical path is made up of a series of electronic components whose failure are deemed critical because the hardware will fail to reach its design objectives. Failures in critical paths yield critical failures. In most cases, critical failures will result in failure to reach mission objectives, and in some cases, actual mission failures.

For each critical hardware assembly, critical circuits are first identified. Once the critical circuits are identified, then critical components are identified. These critical components constitute the critical path(s). This is then followed by assessing which of these critical components are most susceptible to SEE. This assessment is done via analysis. When the assessments are completed, several options can be pursued, ranging from doing nothing about it to recommending design changes.

We now illustrate a simple, but very illustrative example, of analyzing the SEE susceptibility (SET in this case) of several critical paths within an FPGA (only a very small partial design of the FPGA is shown). This approach can also be extended up to the board level. In this example, the FPGA itself has been deemed a critical component for this particular design. However, we intend to extend the SET susceptibility to several critical circuits within the FPGA. For the sake of the example, we will analyze only two of these critical circuits within the FPGA. There may be many critical circuits in the application, and in some cases all the circuits could be critical, though some more than others.

The FPGA under discussion is on a board used to fire several pyrotechnic devices for deployment functions (e.g., a solar array deployment). The FPGA interfaces with the main command and control bus and provides the timing signals and pulse signals for executing the enabling and firing commands on the enabling and firing circuits which are also part of the board.

Two of the circuits within the FPGA that were deemed to be critical are the *pulse generator* circuit and the *execute* circuit. These circuits constitute only 5% of the whole FPGA design. The *pulse generator* circuit (Figure 14.5) provides the pulse timing, pulse shaping, and pulse duration for the enabling and firing circuits. The *execute* circuit (Figure 14.6) guides and controls the execution of the pulse generator circuit.

One of the serious failures identified that could be caused by an SET event is *to drive an output shorter or longer than intended while firing.* The probabilities

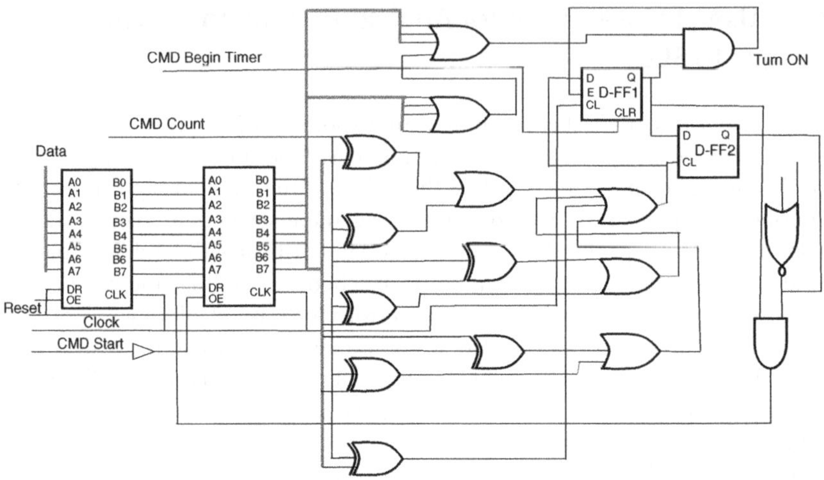

Figure 14.5 Pulse generator circuit.

for such an event to happen can be measured as the mean time between failures (MTBF).

There are four SET effects that could cause the above failure:

(1) SET will cause changes in the duration time in either of the two, eight bit registers in the pulse generator circuit in Figure 14.5. This would have to occur sometime between the issuance of an enable and a fire command. From Figure 14.5, each device (register) "normal" upset rate, per gate, at geosynchronous orbit = 2.1×10^{-5} upsets/day (from test data). For two whole

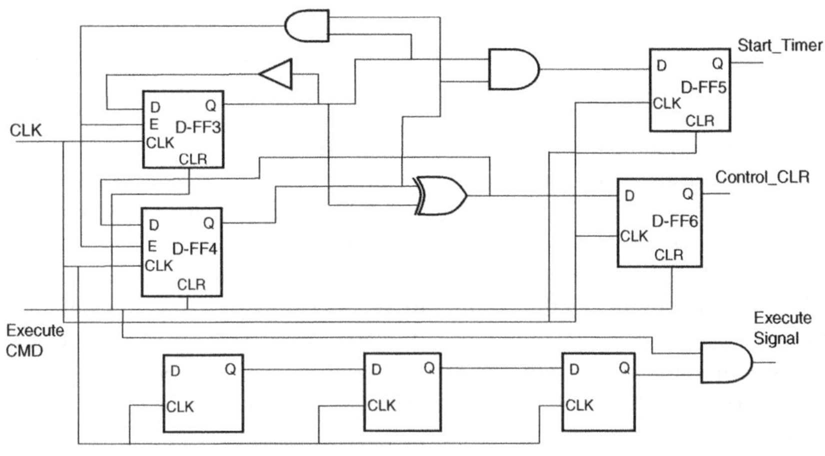

Figure 14.6 The execute circuit.

registers, the upset rate = 8 (outputs) × 2 (registers) × 2.1×10^{-5} upsets/day; MTBF = 1/ {$8 \times 2 \times 2.1 \times 10^{-5}$} = 2976 days/upset.

In a "high" solar flares environment the upset rate per register, per gate, at geosynchronous orbit = 4.9×10^{-2} upsets/day; and MTBF = 1/{$8 \times 2 \times 4.9 \times 10^{-2}$} = 1.27 days/upset.

(2) The master counter in Figure 14.5 gets a SET and skips over the duration. This would cause the fire command to last several tenths of milliseconds extra. SET would have to occur in one of the top six least significant bits of the master control.

Devices ("X-OR" and "OR" gates) "normal" upset rate, per gate, at geosynchronous orbit = 7.8×10^{-5} upsets/day. For the total of 13 outputs of the gates, the upset rate = 13 (outputs) × 7.8×10^{-5} upsets/day; MTBF = 1/ {$13 \times 7.8 \times 10^{-5}$} = 986 days/upset.

In a "high" solar flares environment upset rate, per gate, at geosynchronous orbit = 10.2×10^{-2} upsets/day; MTBF = 1/{$13 \times 10.2 \times 10^{-2}$} = 0.75 days/upset.

(3) CNTR_CLR in Figure 14.6 (Execute block) goes off during fire pulse and resets the master control before the counter reaches duration. D-FF6 would have to flip from low to high in the execute block.

Devices, such as D flip flops (D-FF) have "normal" upset rate, per gate, at geosynchronous orbit = 9.6×10^{-8} upsets/day (per test data). For two flip-flops (D-FF5, DFF6), the upset rate = 2 (outputs) × 9.6×10^{-8} upsets/day; MTBF = 1/ {$2 \times 9.6 \times 10^{-8}$} = 5,208,333 days/upset.

In a "high" solar flares environment upset rate, per gate, at geosynchronous orbit = 1.2×10^{-4} upsets/day; MTBF = 1/{$2 \times 1.2 \times 10^{-4}$} = 4166 days/upset.

(4) D-FF4 in Figure 14.6 gets an SET (low to high) during a fire pulse.

Devices (D-FF) "normal" upset rate, per gate, at geosynchronous orbit = 9.6×10^{-8} upsets/day. For the seven D-FF in Fig. 14.6, the upset rate = 1 × 7 (outputs) × 9.6×10^{-8} upsets/day

MTBF = 1/ (1 × 7{9.6×10^{-8}}) = 1,488,095 days/upset

In a "high" solar flares environment upset rate, per gate, at geosynchronous orbit = 1.2×10^{-4} upsets/day

MTBF = 1/ (1 × 7 {1.2×10^{-4}}) = 1190 days/upsets

After this analysis, we can proceed with assessing which components have the lowest MTBF as being the most susceptible. In the example described, it is the counter, especially in a high flare environment with the higher probability of getting an upset. We compare the MTBF of the most susceptible components with the mission function and assess the probabilities of having SET upsets during the particular mission function for those particular components. It is obvious from the example that no pyrotechnic functions should be attempted in the spacecraft during a solar flare event. As a following step, if a given mission function cannot be avoided, it is then important to assess the circuit behavior consequences of those upsets, and this assessment will result in further action, ranging from "we can live with it" to hardware redesigns.

Other options for improving space radiation immunity is boosting the common mode operating range of a part, choosing parts with higher ratings in critical designs, current limiting the power supply bias and even providing localized shielding.

14.7 Circuit Modeling of SETs

For modeling purposes (e.g., SPICE simulator), the current pulse generated is usually modeled using a double exponential waveform [7] as shown in equation (14.5). Q is the charge (positive or negative) deposited by the particle strike and given by equation (14.3), τ_α is the collection time constant of the p–n junction, and τ_β is the ion-track establishment time constant. The time constants τ_α and τ_β are dependent on process technology and can be taken conservatively as 0.1 and 0.05 nS, respectively.

$$I(t) = \frac{Q}{\tau_\alpha - \tau_\beta}\left(e^{-\frac{t}{\tau_\alpha}-\frac{t}{\tau_\beta}}\right) \tag{14.5}$$

A charge injection circuit for SPICE simulation is represented in Figure 14.7 as an independent current source, where $I(t)$ is given by equation (14.5).

An example on the use of such a current injection model is the circuit shown in Figure 14.8. In the figure, the induced transient current source is simulated at the output of a digital gate and propagating to another gate, where it will cause the wrong output on such gate.

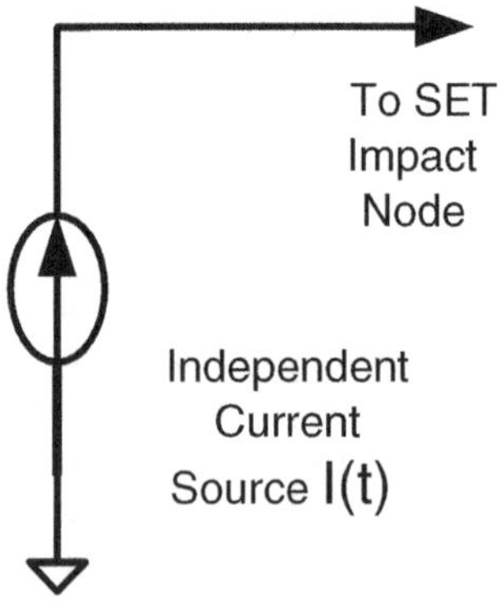

Figure 14.7 Charge injection circuit model.

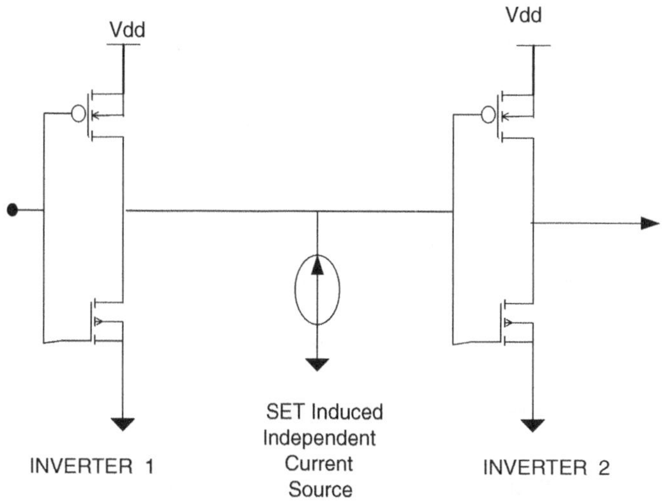

Figure 14.8 Modeling a SET-induced current transient inside a digital gate.

The main problem in modeling an independent current source, as shown in Figure 14.8, relates to the accuracy of the transient current used as the input stimulus. The model of the transient current source can affect considerably the circuit simulation accuracy. A typical example is the resulting SET current resulting from the device-level simulation of an unloaded device. In these cases, the circuit simulation inherits the inaccuracy of the improperly loaded device simulation.

The previous outlined limitations of circuit-level simulation can be overcome by using physically based device simulation to predict the response to ionizing radiation of the affected device. This approach is referred to as "mixed-mode" or "mixed-level" simulation, since the struck device is described by simulation in the device domain and the other devices by circuit models. The two simulation domains are tied together by the boundary conditions at the circuits, and the solution to both sets of equations is rolled into a single matrix solution. However, the main inconvenience of the mixed-level simulation approach is the increased CPU time compared with a full circuit-level (SPICE) approach. In addition, mixed-mode simulation becomes not tractable for complex circuits. Therefore, in this chapter, we advocate for the circuit-level simulation model as described in Figure 14.8 as sufficient for analyzing SET events.

Because we are often only interested in the behavior of an overall circuit to a SET event, a different approach should be followed. Rather than modeling the internals of the linear IC in order to assess the SET effects on the outputs of the IC, an easier approach is to model just the SET transient on the

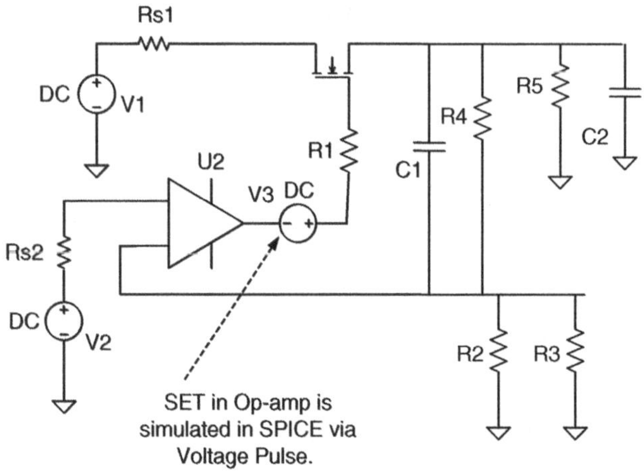

Figure 14.9 Modeling a pulse (V3) at the output of a SET susceptible device (U2).

IC output only. We can then propagate the behavior of the SET to the overall circuit. This approach is illustrated in Figure 14.9, where a transient pulse is modeled via SPICE on the output of U2. The transient is modeled as V3 using a pulse step function in SPICE. The magnitude of the voltage step function can be modeled from equation (14.5) and the use of a dummy resistive load in the V3 voltage source. We can then capture the pulse response effect at the circuit load (represented by C2 || R5). It is important to realize using the example of Figure 14.9 that a SET is capable of causing serious detrimental effects in a circuit's performance. In the example of Figure 14.9, it is shown that a SET transient can inadvertently turn-on the MOSFET and momentarily activate the circuit even if it was not supposed to. The inadvertent activation of the circuit can result in detrimental effects downstream from where the circuit is located.

14.8 SETs in Digital Devices

Advanced digital microelectronic technologies are often more susceptible to SEE more than other previous technologies developed over the past 30 years. Because of the factor related to technology scaling, there is now a greater sensitivity to SEU in memory-integrated circuits than ever before. Digital SETs in high-speed electronic ICs have also become a growing concern and specially in space systems. Digital SET (DSET) are transient voltages or currents that occur in a digital circuit that has been struck by a high-energy particle, even though the strike may not cause an upset in the circuit struck by an energetic particle [8]. The transient effect can propagate through subsequent circuitry and will eventually cause an SEU when the transient event reaches a latch or other memory element. There are mainly four criteria that must be met for a DSET to result in a circuit error: (1) the energetic charged particle strike must be able to generate a transient pulse capable of propagating through the circuit, (2) there must be an credible logic path through which the transient pulse can propagate to reach at a latch circuit or other memory element circuit, (3) the transient pulse must be of sufficient amplitude and duration to change the state of the latch circuit or memory element circuit, and (4) in synchronous logic electronics, the transient pulse must arrive at the latch circuit during a clock pulse in order to enable the latch. The probability is large that momentary glitches will be captured as valid data in combinational logic, and this probability increases greatly with frequency as the frequency of clock cycles increases. As logic circuits' speeds increase, it is also reassured that the ability of a given transient to propagate increases. Due to both, the greater capability of pulses to propagate in high-speed circuits and their higher probability of being captured by subsequent storage elements circuits such as in latches, digital transient pulses have become very common in deep submicron digital electronic circuits. Transients in digital circuits were first observed in the arithmetic logic unit of microprocessor [9]. Digital transient pulses due to strikes in clock logic circuits were reported in the early 1990s [10]. Research data have confirmed the existence of digital SEE transients in several technologies, such as high-speed GaAs FET [11], Si bipolar [12], and Si CMOS logic circuitry [13, 14]. Several experiments' results obtained during the measurements of the properties of digital SETs in specialized test fixtures have been done in an effort to understand the mechanisms of transient effects which can help in the development of mitigation techniques [15–19]. Circuit analyses and modeling techniques have been applied for the purpose of estimating the logic error rates caused by digital SETs [20–23], but few device-level simulations have been performed to research the physical mechanisms involved in DSETs.

Typical sequential elements in the core logic of electronic circuits are the latch, a domino cell, or a register file cell. State changes can occur in the core logic of digital circuits similarly to memory elements. In sequential logic (like in SRAM), the soft error rate has been found to be independent of the clock frequency of the circuit [24]. For example, the latch state of a gate can be flipped by the charge deposited by a particle strike on a gate node regardless of the state of the clock signal, as shown in Figure 14.10.

Flip-flop circuits and sequential logic circuits are main types of logic circuits. Technology has been scaling down, considerably going to smaller features, and therefore flip-flops have become more susceptible to many types of soft errors. The decrease in supply voltage for flip-flops and in their node capacitances has made flip-flops more susceptible to SEE. Flip-flops circuits are similar in design to memory cells, since they both have feedback loops of cross-coupled inverter

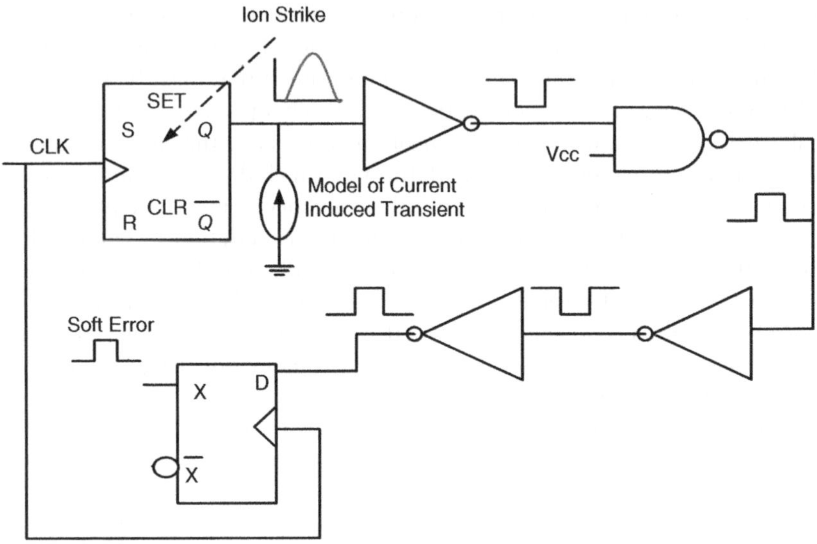

Figure 14.10 Illustration of the SET effect in a synchronous circuit.

pairs. The soft error sensitivity of flip-flops and memory circuits is determined by the critical charge that can accumulate in those devices and the collection efficiency. For example, in an SRAM cell, the charge accumulation is mainly the same for the two storage nodes, because the memory cell is symmetrical. In the case of flip-flops, the inverters are normally sized differently and they also have different fan-outs, which makes the flip-flop circuit asymmetric compared with the SRAM. The individual storage nodes in flip-flops circuits also have a different critical charge per flip-flop when compared with SRAM cells and their SET sensitivity can vary by several orders of magnitude.

Any node which is part of a combinational circuit can be susceptible by an SET event and the node is capable of generating a voltage transient which can then propagate through several nodes of the combinational circuits as shown in Figures 14.11 and 14.12.

A SET event can cause errors in the gates if the SET event gets latched by a sequential element, such as in a memory cell. In combinational logic components, some transient events will be latched and others will not be latched depending on the circuit topology. However, even for the cases where such transient events are latched, some of the data will not be transmitted as errors during software operation. The rationale for this lack of software errors is that a transient event in a logic circuit may not be captured because it could be hidden by one of the following three phenomena: (i) logical hidden response, which occurs when a charge particle strikes part of the combinational circuits but it is

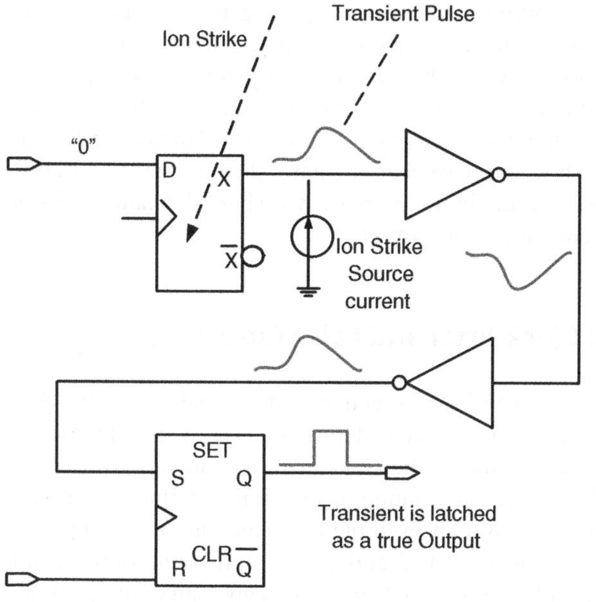

Figure 14.11 Transient propagation in a flip-flop where transient causes a nonintended output.

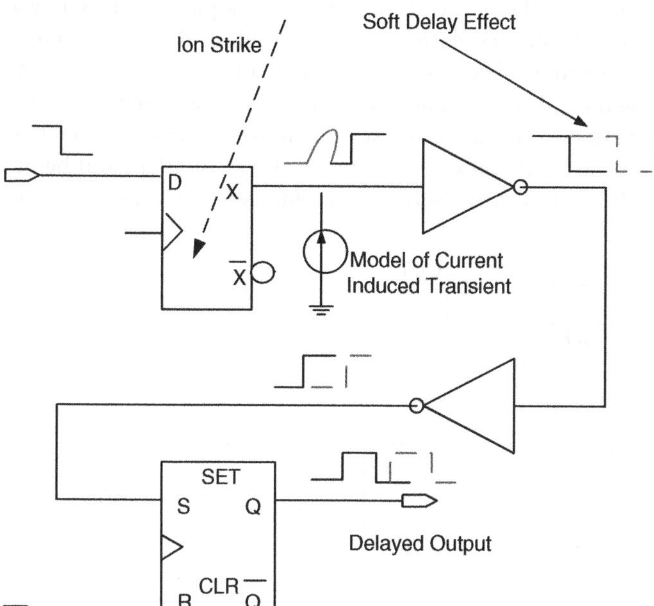

Figure 14.12 Transient propagation in a flip-flop where transient causes a nonintended delay in the output.

not capable of affecting the output because subsequent gates outputs are completely determined by other input values from those gates; (ii) latching-window hidden response, which occurs when the pulse resulting from a charge particle strike reaches a latch but not at the clock transition time at which the latch can captures its input value. Therefore, the error bit cannot be latched, and there will be no soft error which would have been the natural consequence; and (iii) electrical hidden response occurs in pulse transients where the bandwidths is larger than the cutoff frequency of the circuit.

14.9 SET-Induced Clock Jitter and False Clock Pulse

One of the consequences of a SET is the generation of clock jitters [8]. The clock jitter occurs when the charge particle deposits its charge into the nodes of a clock circuit at the same time as the clock edge is available. As a result of this scenario, the clock edge can change in position and the wrong data can be stored. Figure 14.13 shows an example of a clock jitter in a flip-flop configuration where the signals IN, OUT, and CLK denote input, output, and clock signals, respectively, of the flip-flop. Figure 14.13 gives an example of the clock of the flip-flop when the clock is not being affected (i.e., in the absence of a single-event-induced transient affecting the clock). This figure shows the clock pulse, when there is either a "jitter" or a "false clock pulse" induced by a transient event. As can be observed, the jitter will result in the output signal OUT to be delayed, as can be observed in the figure. For the case of the clock jitter, if the output signal "OUT" is connected to another storage element in the circuit, the storage of the wrong data will result when the output signal that has been delayed arrives during the next setup time of the sequential input being received. It is also shown in Figure 14.13 that an energetic particle strike can

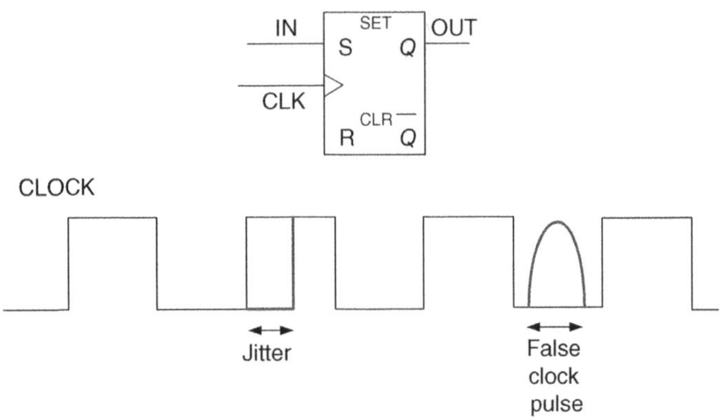

Figure 14.13 SET-induced jitter and a false clock pulse present in a clock signal.

also create a "false clock" pulse on the nodes of clock circuit even in the absence of any clock pulse [8]. If the false pulse that was generated is of enough large magnitude and enough width, the pulse can be taken for a real clock signal and can cause a wrong latched behavior in combinatorial circuits.

14.10 Designing Digital Circuits for *SET* Survivability

Two of the most common type of digital circuits are field-programmable gate arrays (FPGA) and applications-specific interface circuits (ASIC). These ICs are used in large-scale digital designs, and they are often configured to perform complex communications, control, signal processing, and command functions in spacecraft. In FPGA and ASIC, a straightforward approach to get around SET is to use triple-majority redundancy for all critical FPGA registers. Triple-majority voting refers to a FPGA register implementation technique in which each FPGA register is implemented by three flip-flops or latches whose votes (two out of three) determine the state of the register. Any failed flip-flop is out-voted by the other two. Triple-majority voting can also be applied to complete designs or parts of circuits, not just registers.

The inclusion of triple-majority voting can be done directly with very-high-speed-integrated hardware description language (VHDL) source code, which does not require too great of an effort. There are also synthesis tools that can replace any flip-flop with a triple-majority voting set of three flip-flops, without the need for rewriting the VHDL source code.

Some FPGA technologies, such as the SX-S family from Actel (now Microsemi), include triple-majority voting flip-flops on silicon [25]. This technique should be applied to all crucial designs (including data paths, status, and configuration registers), specifically in those applications for which no power cycling or reset is possible. An example of triple-majority implementation using an FPGA shift register in an Actel FPGA is shown in Figure 14.14. Notice that if one flip-flop is accidentally flipped to the wrong state, the other two outvote it and the correct pulse is propagated to the rest of the circuit. The three "D" flip-flops must receive the same inputs and the outputs are compared with select the "two of three" and the result is then passed to the output holding register in the diagram.

Figure 14.15 shows the basic R-cell functionality of an Actel RT54SX-S FPGA. Figure 14.16 illustrates a simplified representation of how the "D" flip-flop in the R-cell is implemented in the SX architecture. The flip-flop consists of a master and a slave latch gated by opposite edges of the clock. Each latch is constructed by feeding back the output of the input stage. The potential problem in a space environment is that either of the latches can change state when hit by an ion with enough LET energy. To achieve SET hardness, the D flip-flop is enhanced (Figure 14.16). Both the master and slave latches are actually implemented with

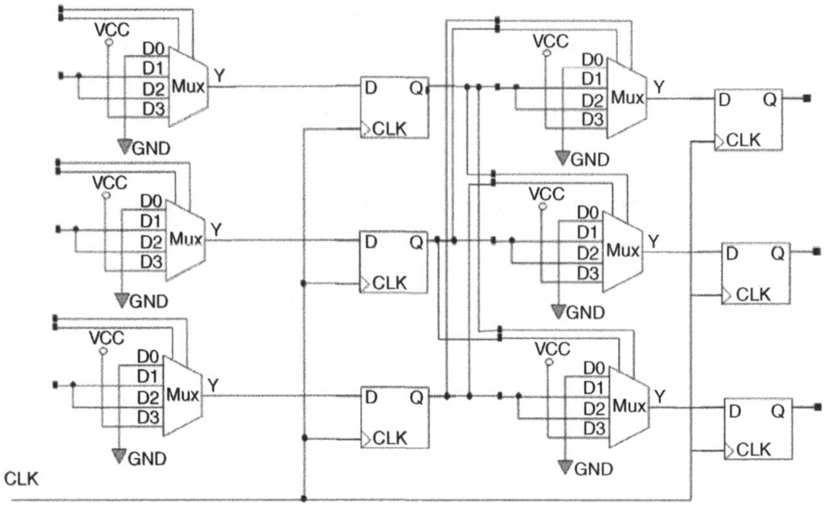

Figure 14.14 Triple-majority voting using separate voters. MUX and flip-flop combine into a single S module. Actel FPGA implementation (with permission from Actel).

three latches. The feedback path of each of the three latches is voted with the outputs of the other two latches, hence the name triple-majority voting. If one of the three latches is struck by an ion charge and starts to change state, the voting with the other two latches prevents the change from feeding back and permanently latching. The layout must be such that a single ion can only strike one latch. This approach can be extended to higher level designs within an FPGA, for example, Figure 14.17 shows a conceptual design on the implementation of

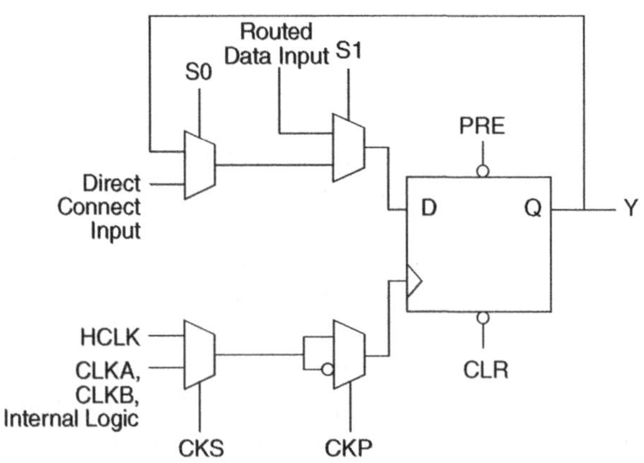

Figure 14.15 R-cell functional diagram in an RT54SX-S FPGA (with permission from Actel).

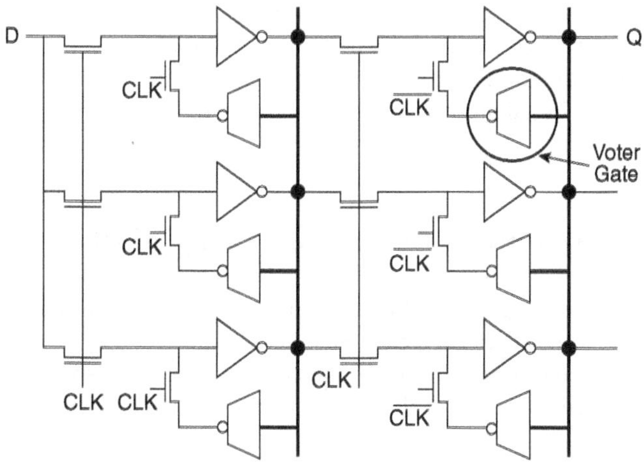

Figure 14.16 RT54SX-S R-cell implementation of D flip-flop using voter gate logic (with permission from Actel).

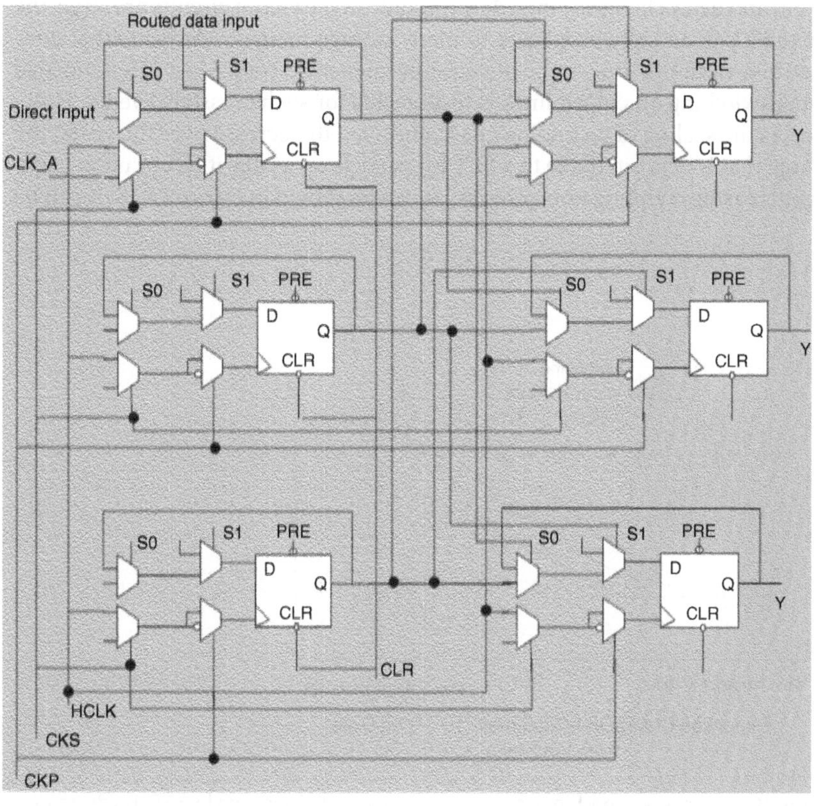

Figure 14.17 RT54SX triple-majority implementation of an FPGA shift register.

a triple-majority voting in the design of an FPGA shift register. The output "Y" is the result of the majority voting of three samples of the "Direct Input" of the circuit. While majority voting is by far the best approach for eliminating SET in FPGA logic, care must be taken to minimize effects on the majority voter output block.

14.11 Crosstalk Noise from SET Events and Delay Effects

Design techniques for hardening against SEUs can produce electronic designs that can reduce, and in some cases eliminate, SETs in CMOS circuits. Due to the scaling issues in the electronics, coupling effects can increase if spacing between interconnect is reduced and if there is increased thickness-to-width ratio of interconnects.

The electrical noise interaction caused by parasitic coupling between interconnects of circuits, known as crosstalk, can produce detrimental effects in electronic circuits, such as positive and negative glitches, overshoot, undershoot, and delay changes. Since the crosstalk on the victim circuits are large, the crosstalk can actually propagate to many storage memory elements that connect to a victim IC line and can cause permanent errors. Most often, normal signal switching on culprit lines is responsible for such crosstalk events. However, as the scaling of technology has continued to decrease over the years, the charge that is deposited due to a SET hit particle on a culprit IC circuit is creating increasing coupling noise effects, as can be shown in Figure 14.18. It can be

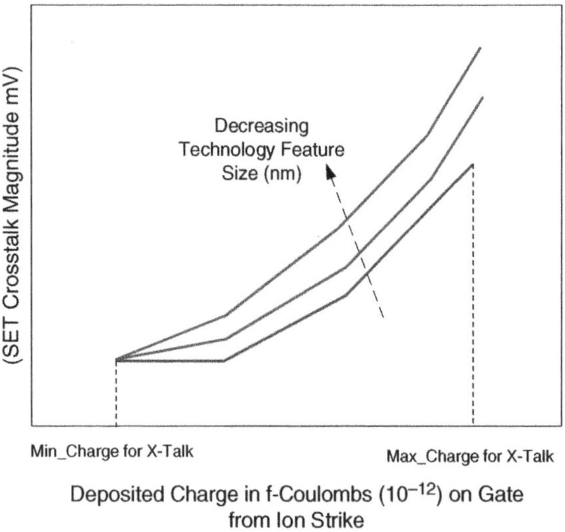

Figure 14.18 Direct relationship between increased crosstalk induced by SET and decrease in IC technology feature size.

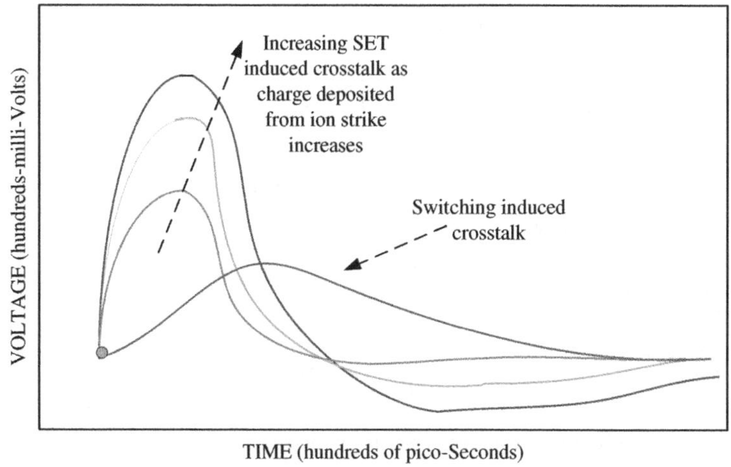

Figure 14.19 Comparison of crosstalk induced from SET and from normal switching.

shown that SETs by an aggressor IC circuit are capable of creating larger noise effects than those transients induced by the normal signal switching crosstalk as shown in Figure 14.19.

During the IC layout optimization process used by EDA tools, even though a given net may pass the normal crosstalk noise check, the net may still pose a susceptible scenario if SETs are not properly considered during the design process. Present EDA tools lack this capability. It is much easier to address SETs effects during the design process than afterwards when issues are found and modifying the design would cause serious schedule and cost problems.

For the circuits which are susceptible to experience coupling effects among its interconnects, as found in the design process, a SET pulse which has been generated on a circuit node can affect multiple logic paths in the circuit due to the strong coupling among wires, vias, and even board traces. Figure 14.20 shows the aggressor and the susceptible circuit pair along with its drivers and receivers. In this scenario, when the inputs of both drivers are held at logic 1, the outputs are normally held at logic 0. A SET ion strike at the drain of OFF transistor in the inverter driver can cause the output to go to logic 1 for some pulse duration. The transient voltage created in the culprit circuit can then affect the susceptible circuit through parasitic coupling capacitance, and this will induce a SET crosstalk noise on the susceptible circuit. The coupling effects produced by SET strikes in a circuit can violate the noise margins of gates connected to other affected parts in the circuit, and this will result in logic errors as shown in Figure 14.20. Large detrimental effects may occur if the susceptible circuits are of such importance such as in clock circuits.

The SETs generated at the culprit circuits may also cause increased signal delays on a neighboring susceptible circuit, such as in switching circuits. This

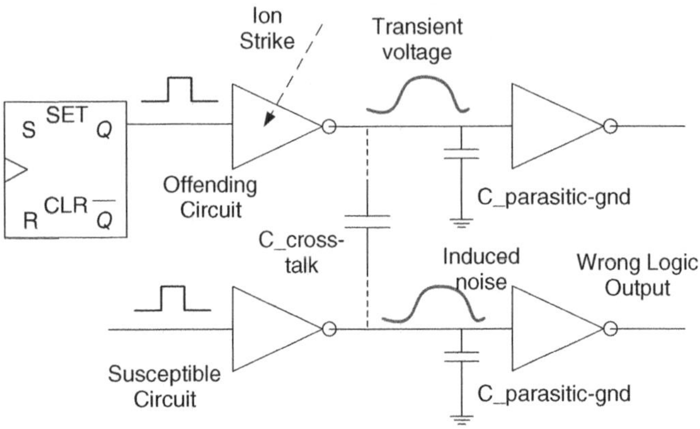

Figure 14.20 SET-induced crosstalk capable of causing wrong logic outputs.

effect can be named SET-induced crosstalk delay. In the example shown in Figure 14.21, a SEU particle strikes the output node of the culprit driver and causes a voltage transient to be generated in the culprit circuit. The transient then couples into the susceptible circuit during switching, through capacitance coupling, and causing a signal slowdown on the susceptible circuit. The increase in interconnect delay due to the SET coupling can affect circuit performance. The delay changes caused by this noise coupling can violate the setup or hold time requirements of logic storage circuits which are connected to these receivers.

Due the SET-induced coupling effects, a SET pulse generated on a circuit node can propagate beyond to the propagation path that exists between the

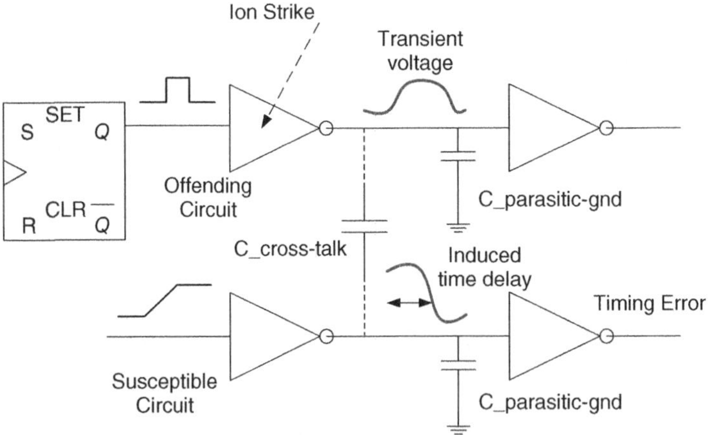

Figure 14.21 SET-induced crosstalk capable of causing wrong timing delays.

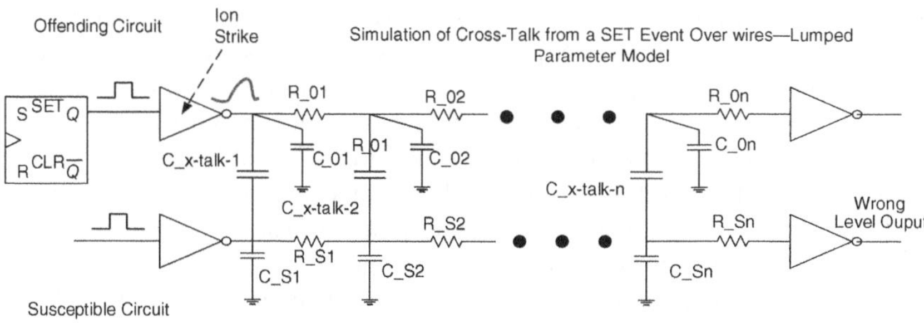

Figure 14.22 Simulation of crosstalk and SET propagation through adjacent circuits.

affected node and latch that may reside in the circuit. Therefore, the coupling effects among interconnects in multiple circuits can cause SEU transients to propagate within electronically unrelated circuit paths, and this can increase the SET susceptibility of circuits to SET, as shown in Figure 14.22.

SETs can induce crosstalk effects on a neighboring electronic component copper path and can induce logic noise level as small as 100 μV on technologies 90 nm and lower. As the technology scaling goes down, there can be an increase in crosstalk effects due to decreased spacing and increased thickness to width ratio of the copper interconnects among circuits. With enough coupling, a single-event pulse can easily spread from one part of the circuit to unrelated parts of combinatorial logic circuit causing single-event crosstalk noise effects.

14.12 SET in Voltage Regulators

The effects of transients in linear devices depend greatly on the circuit application conditions, and this situation makes it difficult to draw conclusion from tests on a specific circuit configuration to other applications with different configurations [26–35]. Linear devices in this configuration are, for example, the regulators. Regulators are used over a very wide range of input voltages, with input/output voltage differences up to 40 V. The wide range of supply voltages and variation in output loading conditions that we find in regulators are critical parameters that need to be addressed in determining SETs effects in regulators. This flexibility, however, increases the complexity of radiation testing, the analysis of SET, and the evaluation of data for regulators concerning SET. In studying SETs, the first step is to define the parameters for the amplitude and duration of output voltage response. For voltage regulators, SETs with small amplitude or short duration are treated with no much concern, but in reality, these transients can induce unacceptable noise levels in critical circuits. SET with large amplitude can produce voltage conditions that can cause damage to other circuits or produce noise effects that disrupt normal circuit functions. It must be remembered that the SETs from these regulators are actually occurring even when there is a presence of large capacitor at the device output. This is because the low output impedance and high current capability of these regulators can easily overcome the expected filtering effect of bypass capacitors inside the regulators, and therefore can allow transients to propagate in a network that at first thought may have been immune from such SET effects when a first-order analysis was done by circuit designers.

Because voltage regulators are capable of working under a wide range of load conditions, it is often difficult to deal with SET in voltage regulators, which are intended for use in a wide range of applications. The SET effect on regulators depends on several factors, such as input/output voltage, load capacitance, load current, and operating temperature. All these variables are interdependent of each other and this creates a complex problem for the solution of SET events in voltage regulators.

Just as it is with normal operation in regulators, the transient output voltage from a SET event inside the regulator also depends on the difference between input and output voltage. Therefore, charge generation and collection inside the regulator from a SET strike also depend on voltage conditions, increasing with voltage. These effects are often difficult to model with circuit analysis programs, such as those in EDA tools, because the modeling accuracy is limited, and it does not incorporate the voltage dependence of charge collection in the internal transistor structures.

SET-induced transients from voltage regulators can affect the other circuits fed by the regulator in a variety of ways, increasing the difficulty for engineers of characterizing and defining transient events. Although usually the concern is with transients of high amplitude, voltage regulators can also be used in the regulation of voltage for other circuit cards where small transients can interfere with the normal operation in those cards and can induce noise that is greater than the expected interfering with the operations of those cards. Since the threshold LET for low-amplitude transients is very low, it is very likely that protons can induce low-amplitude transients in addition to heavy ions.

As discussed in [36] and [37], the circuit designs used in these regulators are complex, using sophisticated design techniques that rely on close matching of various internal transistors. However, there are common ways to reduce the susceptibility of regulators to SETs which do not require a deep understanding of the detailed circuit design. These measures are post design corrective measures and address the circuits outside the regulators. For example, it is quite possible to limit the maximum transient amplitude by adding a circuit that will clamp the voltage between the output of the amplifier and the input to the Darlington output stage, reducing the output voltage transient to mV. The duration of the transients can also be reduced.

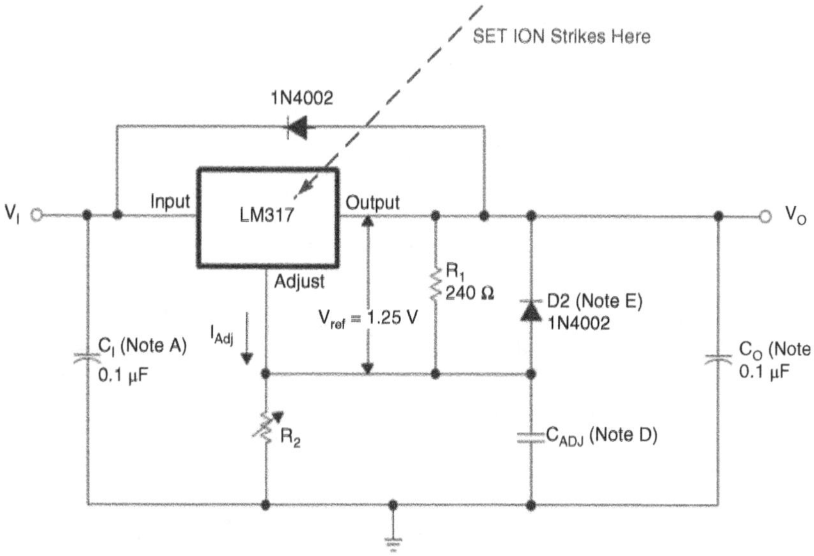

Figure 14.23 An adjustable voltage regulator susceptible to SET.

Figure 14.23 shows an example application of an adjustable voltage regulator (LM317) capable of supplying up to 1.5 A with an adjustable output voltage V_o that can range from 1.25 to 37 V. The output voltage is given by

$$V_0 = V_{ref}\,(1 + R_2/R_1) + (I_{ADJ} \times R_2) \tag{14.6}$$

where I_{adj} is usually around 50 μA in most applications, and it is often ignored. V_{ref} is defined in Figure 14.23. The LM 317 regulator (with 26 BJT transistors inside) is susceptible to SET, and a SET will cause V_{ref} to go to zero volts, hence, from equation (14.1), $I_{ADJ} \times R_2 = 0$. A SET can last from a few microseconds up to tens of microseconds; hence, $V_0 = 0$ should last also for the duration of the transient. However, in the circuit of Figure 14.23, capacitor C_0 will improve the transient response and C_{ADJ} will greatly improve ripple rejection as it prevents amplification of the ripple as the output voltage is adjusted higher. In this example, SET propagation across the circuit has been shown to have little effect, but that is not always the case.

14.13 SET Propagation through Multiple Circuits

Transient "glitches" can now be rigorously modeled at the IC level and address its effects within the IC. EDA tools are not capable of doing this modeling but

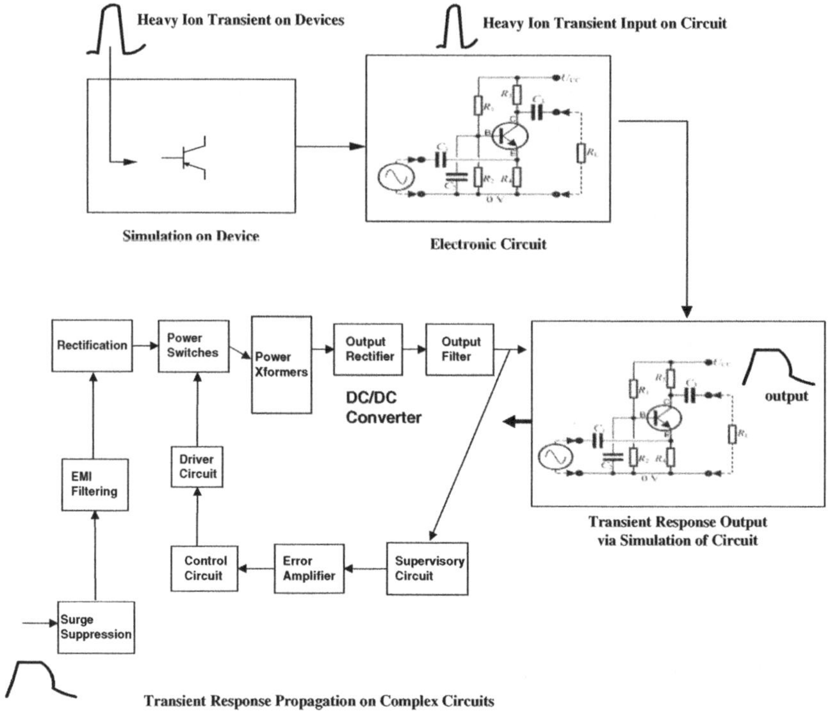

Figure 14.24 The effect of SET propagation from device level to whole assembly.

charge transport model tools are capable of providing preliminary data for use in EDA tools. This approach is illustrated in Figure 14.24, where an ion strike causes a glitch in a transistor gate. The glitch causes the logic to go from logic 0 to logic 1 at the output of the gate. Figure 14.24 illustrates how such as glitch will ripple through downstream logic causing even more errors on multiple circuits of ever-increasing complexity. Even, in clock circuits, as previously shown, a SET can cause momentary glitches which are often recoverable in the next clock cycle but such errors can temporarily interrupt critical function in a space vehicle which can trigger internal alarms and cause emergency "safing" procedures to be executed. For example, work by this author has shown (nonpublishable) that a few such events in critical circuits in resonant power supplies can cause the fault management software in a spacecraft to execute power-on resets in addition to other measures which can put the spacecraft in "safe mode." To address this issue without any design changes in the resonant power supplies, the culprit circuits need to be identified and correction of the fault management software needs to be implemented to account for the glitch effects. The goal is to "screen out" the effect of these glitches via software.

14.14 SET Hardening of Interconnects

In PCB layout, some of the standard techniques used to decrease crosstalk among circuits are wire sizing, spacing among circuits, and wiring. Driver sizing should also be an effective technique for decreasing SET crosstalk noise and delay. For a fixed wire width, if a wire's spacing to its neighbors is increased, its coupling capacitance decreases while ground capacitance increases. Therefore, the reduction of spacing between wires and traces is an effective way in the reduction of SET crosstalk noise and delay, even though we pay some penalty due to use of routing resources. The wire-sizing approach can also be used in mitigation of SET crosstalk. As a wire width is changed, its resistance and capacitances also change. The larger wire size (i.e., width) causes a reduction in wire resistance and an increased in ground capacitance, all of which contribute to susceptible circuits' decreased susceptibility to noise. The wire-sizing approach, however, may not help the SET-induced crosstalk delay. This is because the increased ground capacitance of susceptible wire will also cause an increase in delay if such capacitance is not also decreased.

Not all crosstalk mitigation approaches may be applicable to SET crosstalk noise and reduction. In crosstalk analysis, driver sizing can also be used to mitigate crosstalk effects. In case of an aggressor driver-sizing method, if the driver is sized down, its effective trans conductance decreases. As a result, it cannot transition as fast due to its large resistance. Finally, the amount of noise it induces on the victim line decreases.

14.15 Modeling Subsystem- and System-Level Effects from SET

The rigorous implementation of SET propagation effects as shown in Figure 14.24 is difficult to implement because it requires costly and time-consuming SPICE-like simulations on multiple circuits with ever-increasing complexities. Rather than providing a quantitative estimate of SET propagation across electronics of interest, a qualitative approach is pursued. In the qualitative approach the interest is mainly on the behavioral assessment of the circuits and the behavioral performance of such circuits. The SET effects are described in terms of behavioral responses which are first postulated at the circuit level and then extrapolated to higher level of complexities, such as the circuit card and then subsystem levels.

There are two methods to address subsystem- and system-level effects from SET events in a behavioral approach. The first approach is the *Tabular model*. The Tabular model is quite good, and it has advantages and very few disadvantages. The main advantage of the Tabular model is that it emphasizes the outcome, including potential failures, without the need for transitional stages.

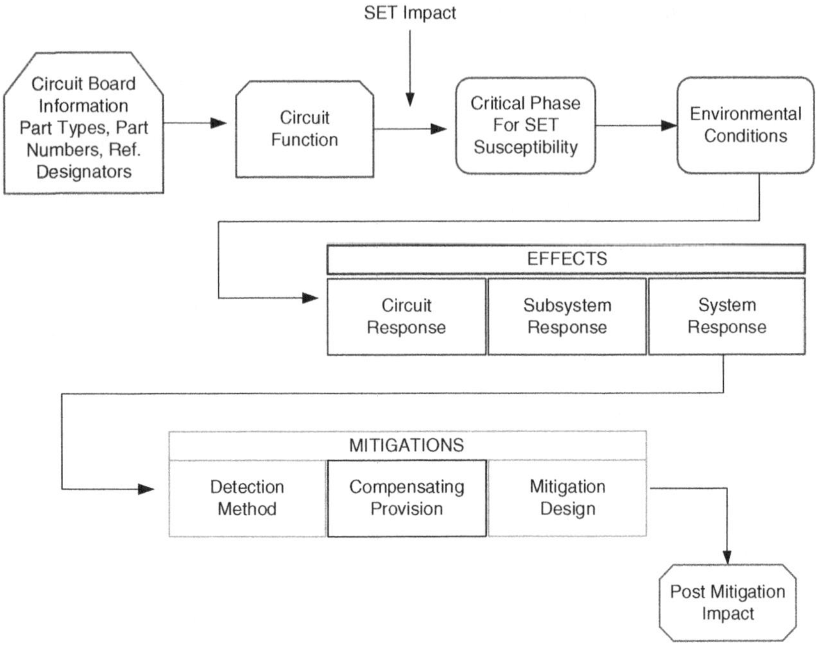

Figure 14.25 Flowchart on the use of the Tabular model for assessing SET effects.

The Tabular model is the most common approach used in the industry when performing SET analyses, and a block diagram of the *Tabular model* is illustrated in Figure 14.25.

An example of a Tabular model addressing only the effects of a SET in an LM158A linear amplifier, which is part of a larger electronics assembly, is shown in Table 14.2.

The second approach is developed by this author; and it is identified as the SET *state transition model*. This author believes that the *state transition model* is much more revealing from a software engineering point of view and more rigorous when addressing SET propagation. The state transition model shows the states and the transitional probabilities to go from a fault-free state of the system to the actual failure state in multiple transitions. Figure 14.8 shows an example of the *state transition model*. The *state transition model* has the unique advantage that it can be mathematically, logically, and discretely represented via software models, and can be made into an intricate part of a fault management system.

The *tabular model* of Figure 14.25, which is the model used in the space industry, has the advantage that it allows for the analyst to propose mitigation effects to a SET event. The mitigation effect can be made in a qualitative manner. First, the analyst makes an assumption as to where in the hardware the

Table 14.2 Example of modeling SET propagation via a tabular model

Ref. no	SH	Sec	Qty	Part number	Part function	SET perturbation assessed	SET circuit assessment	SET subsystem assessment	SET system assessment
U2B	1	D4	2	LM158A	Main loop error amp	$\leq \pm 1$ V transient for 15 μsec on output	Transients of this magnitude and duration due to the inherent delays of the system will have little effect on the output voltage. No effect at NHA.	No effect since transient will be filtered out	No effect
						Transients of greater magnitudes than + 1 V transient for 15 μsec on output	Direct testing and analysis shows that for pulses on the input pin of the UC1864J shorter than 17.5 μsec, the 5 V main output will not trip, but may transition above the required point of regulation.	Regulation holds but at a higher voltage	No effect that can be observed if small
						Transients of magnitudes greater than − 1 V including transients to ground output	Due to the high gain of the feedback system at this point, a small negative transient at this point may cause a dip in the output voltage, possibly beyond acceptable regulation limits. Always voltages will remain within about 0.5 V below nominal.	Small period for out of regulation	Drop output voltage by no more than 0.5 V may affect low voltage circuits temporarily

SET event can be detected. Often that detection is not really observable (i.e., by flight software and/or fault protection software) until the SET event affects a subsystem or system functionality. The analyst then postulates a compensating provision that can be activated in software, hardware, or both in order to diminish, and if possible eliminate, the detrimental effect of the SET event. This compensating provision is most often achieved by a system-level response (e.g., use of redundancy in the system). The analyst then explores the possibility, and can even suggest, if a design change is needed to address the effects of the SET, if the compensating provision is not adequate or the proposed design change is actually simple to implement. Once the SET effects have been mitigated either through a design change or an adequate compensating provision a post-mitigation impact is postulated at the mission level. As can be observed, the *tabular model* is hardware oriented and it is actually performed during the design stages of avionic hardware. The main goal in the *tabular model* is to identify potential hardware design changes that will diminish the effects of an SET.

The *state transition model* is a new model developed by this author and it is proposed for use in fault protection software development. The *state transition model* is software oriented. The main goal of this *model* is to greatly improve the efficiency of fault protection software which is part of the flight software in a spacecraft. In the diagram of Figure 14.26, a SET which occurs inside the pulse

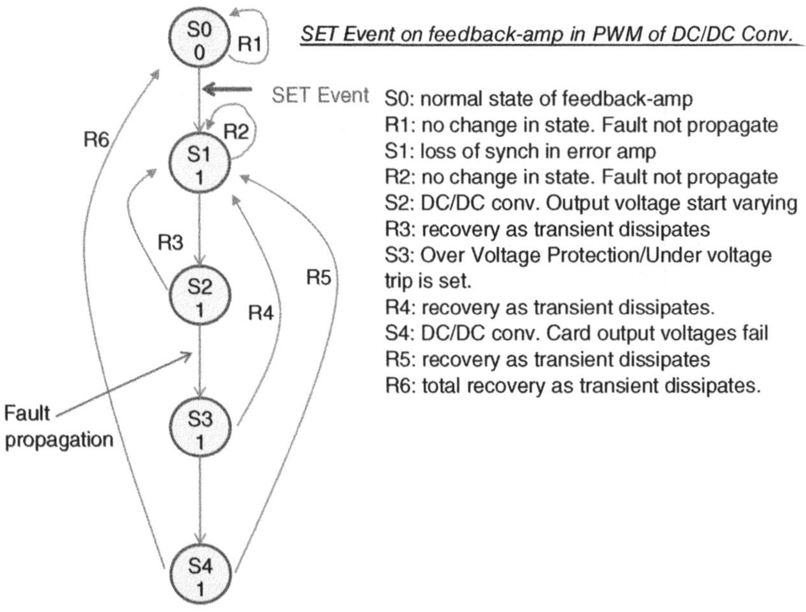

Figure 14.26 Example on the use of state transition diagram for modeling SET effects.

width modulators (PWM) of a DC/DC converter is tracked through a series of states nodes. The S0 node is not only the starting node but it is the node which identifies the PWM as working correctly. After a SET event, the *state transition model* shows the states S1 through S4, NOT as failure modes (as in the *tabular model*), but rather as behavioral states of the PWM. These behavioral states can be monitored by a fault management software which can then diagnose, very accurately, the SET propagation location, which is a capability that the *tabular model* could not assess. The *state transition* model also has another unique capability it allows for feedback among the transition states. These feedbacks are known as recovery modes (shown as recovery "R1–R6" modes in Figure 14.8). The recovery modes allow for SET annealing in case that at some point the hardware cleared the SET effect. Therefore, it allows for the fault management software to estimate the end location of the SET effect if annealing occurs.

14.16 Analyses and Protection for SET for Electronic Devices

14.16.1 Introduction

There are many methods to diminish the effects of SETs in the many different types of linear devices. The easiest and sometimes the most effective way to protect linear devices against SET is by filtering the output of the linear devices in order to eliminate the propagation of SET to other linear devices. For several applications, filtering may not be the correct approach, and other techniques will need to be used. In some electronic devices, the susceptibility to SET and the transient behavior resulting from the SET are very much functions of the input and bias conditions of the electronic devices. A very simple way to decrease the transients in these electronic devices is to use input and biasing processes that are less susceptible to SETs. As with many other transient events, a powerful means to eliminate possible transient effects in digital electronics is to use a synchronous design. A few other mitigation methods in digital electronics that may be used for mitigating SET effects are triple-majority voting, oversampling, and/or use software in an intelligent fashion. We now outline the effects of SET for different classes of electronic devices.

14.16.2 Voltage Comparators

It has been found in experimental research that the effect of a SET in a voltage comparator is a transient pulse at the output of the comparator that can have a worst-case signature of a rail-to-rail change of state at the comparator output and the duration of the transient output is in the order of a few microseconds. In general, it has been noticed in experimental data that the lower the comparator differential input voltage, the greater is the device sensitivity to a SET event.

14.16.3 Operational Amplifiers

When an operational amplifier gets impacted by an SET, its output experiences an output glitch. Experimental research over many years and extensive flight data from spacecraft missions have demonstrated that a great variety of transient outputs have been recorded (such as positive-going transients, negative-going transients, or bipolar transients, with either short or long duration). The worst-case transient that has been observed has an amplitude to the power supply rail of the operational amplifiers and with a duration of tens of microseconds normally. The SETs can be very difficult to mitigate in an analog circuit, specifically if the bandwidth of those circuits is wide compared with the SET transient pulse width. Therefore, the analysis of potential subsystem- and system-level effects and anomalies induced by a SET should be performed. If an operational amplifier is used in critical circuits, a triple-majority voting technique and extensive filtering should be used where appropriate.

14.16.4 Voltage References

The effect of a SET in a voltage reference circuit is an output transient pulse. The best approach to mitigate such transient effects is by the addition of low pass filtering at the device output of the voltage reference circuit.

14.16.5 Voltage Regulators

The effect of a SET is an output transient pulse. SETs in voltage regulators, however, are generally filtered out by the large output capacitors used in most typical applications of voltage regulators. Therefore, there are no specific remediation used in most voltage regulators to eliminate or decrease the effects of SET.

14.16.6 MOSFET Drivers

There are not much experimental data available concerning SET for MOSFET drivers. MOSFET drivers are generally considered not to be very susceptible to SET. However, the use of MOSFET driver which could theoretically allow a destructive failure mode, like short circuit, due to a SET on the MOSFET driver should be avoided.

14.16.7 Analog-to-Digital/Digital-to-Analog Converters (ADC/DAC)

There are two possible mechanisms for SETs to propagate in ADCs. One of the most well-known mechanisms for SEE in ADC/DAC is the SEU. The outcome at the output due to a SEU is most often observed as just a spread in the distribution of the digital output for a given analog input. In these cases, a comparator in the converter is the most likely to have been hit and this causes the output to be shifted by a bit.

The other common mechanism that needs to be addressed is the case when the analog input is a rapidly varying input (in the time scale of a transient), and in this scenario a SET on the analog input of the ADC can be propagated through the entire chain of the ADC signaling and component. The SET manifests itself as the digital output of the ADC.

For DACs, the response to a SET is much simpler. Since on the output of the device we have an analog signal, the SET is manifested as the output transient of the analog output. It should be noted that these changes in the analog output of DACs are in addition to any SEU events that may be occurring. An upset can occur in the digital input latches of the DAC and that changes the state of the affected latch, causing a change in the analog output.

14.16.8 Line Drivers/Receivers/Transceivers

Line drivers and receivers are used for the transmission of data between two locations. Either at the source or at the end of the data transmission line, transients can be generated in the form of glitches in the data lines. At the end of transmission line, SETs can place transients on the data line, that the receiver would have to see as valid data, for the error to propagate. A receiver can have a SET on its input side that glitch can then be interpreted as valid data. The primary mitigation for this class of parts is via software with data EDAC.

14.16.9 Sample and Hold Devices

Sample and hold devices are designed to sample analog inputs and hold this information for use in ADC. The typical response of this type of device to SET would be having a transient form on the analog input of the device that the sample and hold circuitry that will use it cannot distinguish from the correct data. This means that any transient generated in the input would be locked into the output data. However, by their very nature, SETs are transient in nature, so oversampling, redundant sampling, and majority voting can be used to counter these SET effects.

14.16.10 Timers

Timer devices are designed to produce pulsed output at specified time intervals. SETs can affect the output of timers by either placing glitches on the output pulse train or, even worst, by adding or removing pulses from the pulse train. Depending on the speed of the timer, glitches may or may not be a concern.

However, the addition or the lack of pulses can affect system performance for those systems not designed to deal with these events.

14.16.11 Pulse Width Modulators

There are three different types of SETs that have been identified in PWMs: (1) both outputs go to a low output state condition for a period of time correlated with the soft start capability or the shutdown capability of the device. The time it takes for the duty cycle to increase from 0% to DC max after the SET event is equal to the time it takes to discharge and recharge the soft start capacitor in the soft start input circuit; (2) the second type of SET affecting PWM has an anomaly much shorter in duration. These short anomalies come in two forms. In the first type of anomaly, both complementary outputs return to the low reference. This scenario lasts less than one clock period, after which the PWM would return to normal output amplitude and frequency. The second form of upset manifests itself as a toggling of the outputs but not related to the clock. The correct function is restored before the next clock cycle begins; and (3) the third type of SET causes a phase shift of the clock circuit. The outputs follow the change in the clock phase. This scenario also affects the device frequency output. Therefore, depending on how the device is used in a circuit, this sort of SET upset can affect more than one function of the device.

For DC/DC converters, which use PWM the two last types of SETs, do not affect the operations of the DC/DC converters. This is due to the short duration of the event. However, the first type of SET could have an impact on the application depending on the soft start circuitry characteristics and design. The longer the duration of the soft start, the higher the impact on the application. It could be very critical on PWM devices where the user could not use the soft start circuitry. After shutdown, the device never starts again. The PWMs that do not implement the soft start and/or shutdown functions are not sensitive to these types of events. Figure 14.27 shows an example of a PWM in a DC/DC converter where SET could cause a detrimental behavior in the converter itself.

14.16.12 Hybrid Devices

This classification of devices is generally assumed to be many types of devices. For example, there are linear devices that are of the hybrid design. Therefore, hybrid devices cover a large range of other types of devices, from simple devices such as optocoupler to more complex devices such as complex DC/DC converter, regulators, microcontrollers, etc. For each of these examples SETs are widely different.

Optocouplers can have their outputs susceptible to transients, which is typical of many types of linear devices, that transient varies widely with the application biasing. An oscillator can have either SETs as output glitches or extra or

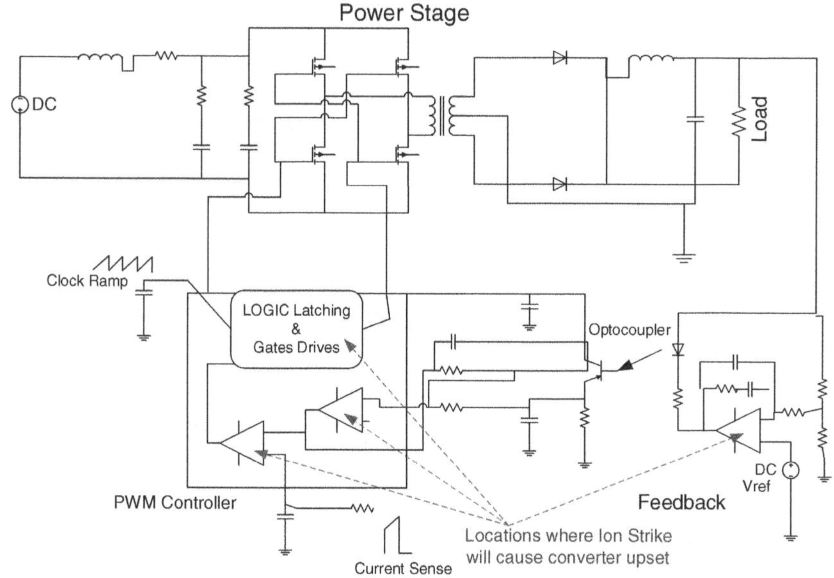

Figure 14.27 The effect of set in PWM of a DC/DC converter.

missing pulses, depending on which device within the oscillator has suffered the initial SET. DC/DC converters as previously stated can have simple transients on their outputs if the SET is generated in one of the devices near the output. Under some more severe case of SET susceptibility, DC/DC converters can have output voltage dropouts, where the output voltage typically drops to zero. Though these dropouts can be short in durations (a few microseconds), it requires a reset of the converter to recover the normal output voltage.

Hybrid devices need to be selected very carefully for SET effects. A process needs to be established for the selection of linear devices that are either not susceptible or least susceptible to SET. If a hybrid is selected that has unknown SET characteristics and utilized in an important system, radiation characterization for SETs will be required via radiation testing.

14.17 SEE Testing of Spacecraft Hardware Electronics

The IC's upset rates can cause actual damage to the component itself or to the system in a very strong radiation environment (e.g., solar flare). Therefore, prior testing of an IC for SEE is often done in electronic space systems design. This is a standard industry practice. IC testing is necessary because most commercially fabricated electronic components are not subjected to any kind of

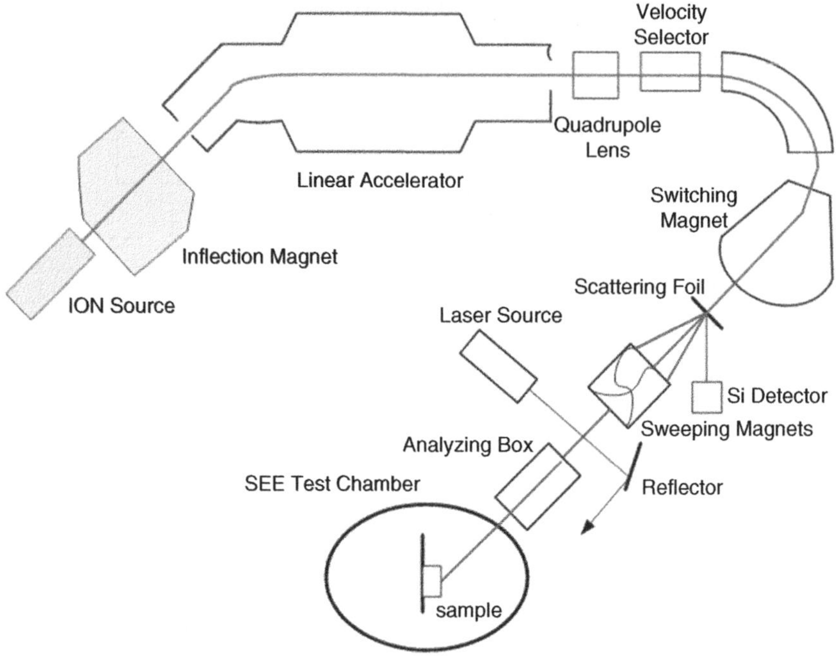

Figure 14.28 Layout of a high-energy particle accelerator.

radiation survivability. SEE testing involves exposing the IC device to an ion beam from a particle accelerator as shown in Figure 14.28. Due to the short range of the charge ions, the IC must be de-lidded in order for the particles to reach the sensitive regions within the device. Tests are performed on de-lidded parts in a vacuum environment (simulating space). Several ions are used in testing and measuring the number of errors produced by the IC, and the total particle fluence is used to determine the cross-section at various LET values. A sample of the types of ions used in SEE testing is shown in Table 14.3. Single-event testing is relatively straightforward for memory circuits. It is easy to define the internal conditions and to test the entire storage array of a memory circuit.

Very large scale integrated (VLSI) devices, such as microprocessors and random logic, are much more difficult to test. Bias conditions play a major role in SEUs, particularly for complex circuits. In order to interpret results, one must know which regions of a device involve internal storage cells and how many of them are being exercised during the test. For example, test results for some types of microprocessors have shown an order of magnitude increase in cross-section for a given LET when the device is exercised by operations that use

Table 14.3 Typical list of ions used in SEE testing by particle accelerators

Typical ions used for SEE testing				
Ion	Atomic number	Energy (MeV)	LET ($MeV\text{-}cm^2/mg$)	Range (μm)
Li	7	44	0.45	>100
C	12	105	1.39	202
F	19	150	3.02	133
Si	28	195	7.7	81
Cl	35	210	11.5	63
Ni	58	255	27	40.3
Br	79	285	37.3	36.4
Ag	107	300	53.1	30.9
I	127	230	59.9	30.7
Au	197	345	82.3	27.9

cache memory compared with noncache results. The LET can vary with incident angles and it is different for different devices.

The types of SEE that have been observed during testing in different types of ICs are shown in Table 14.4 [37].

14.18 Conclusions

This chapter addresses the physics of SET and its effects in digital electronics with a few examples. The chapter shows that SET events can be modeled in electronic circuits by first calculating the induced charge from a SET event and then using SPICE-like modeling tools to analyze the effects of such induced charge in the circuits themselves. The chapter discusses the modeling of an induced charge transient inside a digital gate and the modeling of an induced charge transient in a circuit. The chapter shows that transient propagation can affect multiple circuits in a chain, which is where the main problem lies in terms of detrimental effects. The chapter also addresses the concept of crosstalk that could develop among digital circuits in the present of a SET event. The chapter also provides a brief discussion of SET hardening, a subject that needs much more coverage in the future. Finally, the paper focuses on SET propagation effects models and discusses two models, the tabular and the state transition models. The state transition model is a new methodology which uses state transition diagrams for tracking the effects of SET events through an electronics

Table 14.4 Sample device types, sensitive areas, and SEE

Device type	Sensitive area of device	SEE effects
Memories	Memory cells	Bit flips
	Control logic	Bit flips if sequential, transients if combinatorial
Combinatorial logic	Combinatorial logic	Transients
Sequential logic	Sequential logic	Bit flips
FPGAs	Combinatorial logic	Transients
	Sequential logic	Bit flips
Microprocessors	Registers, cache, sequential control logic	Bit flips
ADC, DAC	Analog portion	Transients
	Digital portion	Bit flips or transients depending on design
Linear ICs	Analog area	Transients
Photodiodes	Photodiode	Transients
Optocouplers	Photodiode	Transients
Power MOSFETs	MOS	SEB, SEGR
FET drivers	MOS	Transients
Analog switches	Analog portion	Momentarily switched state
Op-Amps, comparators, and Voltage regulators and references	Analog portion	Transients

assembly and was developed by this author in order to be easily incorporated with fault management software. Overall, the chapter shows that SETs are a potential risk for space missions now more than ever because of the multiple ways its detrimental effects can be realized.

References

1 R. Das Sawato, "Killer electrons from outer space," *IEEE Spectrum Magazine*, www.spectrum.ieee.org/sep07/5498.

2 B. E. Pritchard, G. M. Swift, and A. H. Johnston, "Radiation effects predicted, observed, and compared for spacecraft systems," in *IEEE NSREC 2002 Data Workshop Proceedings*, pp. 7–17, 2002.

3 Eugene Normand, "Single event effects in avionics," *Boeing Radiation Effects Lab, Boeing Phantom Works*, Dec. 16, 1998.

4 R. Perez, "Methods for spacecraft avionics protection against space radiation in the form of single event transients," in *IEEE Trans on Electromagnetic Compatibility*, Aug. 2008.

5 W. Mendehall and T. Sincich, *Statistics for Engineers and Sciences*, 4th ed. Englewood Cliff, NJ: Prentice Hall, 1995.

6 Robert Wysocki, *Effective Project Management*, 7th ed. Indianapolis, IN: John Wiley and Sons, Inc., 2013.

7 G. C. Messenger and M. S. Ash, *Single Event Phenomena*. New York: Chapman and Hall, 1997.

8 Salahattin Sayil and Juyu Wang, "Single event soft errors in CMOS logic," in *IEEE Proceedings*, Mar-Apr. 2012.

9 R. Koga and W. A. Kolasinski, "Effects of heavy ions on microcircuits in space: Recently investigated upset mechanisms," *IEEE Trans. Nucl. Sci.*, vol. 34, pp. 46–51, Feb. 1987.

10 J. F. Leavy, L. F. Hoffman, R.W. Chovan, and M. T. Johnson, "Upset due to a single particle caused propagated transient in a bulk CMOS microprocessor," *IEEE Trans. Nucl. Sci.*, vol. 38, pp. 1493–1499, Dec. 1991.

11 P. W. Marshall, C. J. Dale, T. R. Weatherford, M. L. Macchia, and K. A. LaBel, "Particle-induced mitigation of SEU sensitivity in high data rate GaAs HIGFET technologies," *IEEE Trans. Nucl. Sci.*, vol. 42, pp. 1844–1849, Dec. 1995.

12 R. A. Reed, M. A. Carts, P. W. Marshall, C. J. Marshall, S. Buchner, M. La Macchia, B. Mathes, and D. Mc. Morrow, "Single event upset cross sections at various data rates," *IEEE Trans. Nucl. Sci.*, vol. 43, pp. 2862–2867, Dec. 1996.

13 N. Seifert, X. Zhu, D. Moyer, R. Mueller, R. Hokinson, N. Leland, M. Shade, and L. Massengill, "Frequency dependence of soft error rates for sub-micron CMOS technologies," in *IEDM Tech. Dig.*, pp. 323–326, Dec. 2001.

14 J. Benedetto, P. Eaton, K. Avery, D. Mavis, M. Gadlage, and T. Turflinger, "Heavy ion induced digital single-event transients in deep submicron processes," *IEEE Trans. Nucl. Sci.*, vol. 51, pp. 3480–3485, Dec. 2004.

15 S. Buchner, M. Baze, D. Brown, D. McMorrow, and J. Melinger, "Comparison of error rates in combinational and sequential logic," *IEEE Trans. Nucl. Sci.*, vol. 44, pp. 2209–2216, Dec. 1997.

16 M. P. Baze and S. P. Buchner, "Attenuation of single event induced pulses in CMOS combinational logic," *IEEE Trans. Nucl. Sci.*, vol. 44, pp. 2217–2223, Dec. 1997.

17 M. Nicolaidis and R. Perez, "Measuring the width of transient pulses induced by ionising radiation," in *Proc. Int. Reliability Physics Symp.*, pp. 56–59, Apr. 2003.

18 P. Eaton, J. Benedetto, D. Mavis, K. Avery, M. Sibley, M. Gadlage, and T. Turflinger, "Single event transient pulse width measurements using a variable

temporal latch technique," *IEEE Trans. Nucl. Sci.*, vol. 51, pp. 3365–3368, Dec. 2004.
19 M. J. Gadlage, R. D. Schrimpf, P. H. Eaton, and T. L. Turflinger, "Single-event transient pulse widths in digital microcircuits," *IEEE Trans. Nucl. Sci.*, vol. 51, pp. 3285–3290, Dec. 2004.
20 H. T. Nguyen and Y. Yagil, "A systematic approach to SER estimation and solutions," in *Proc. Int. Reliability Physics Symp.*, pp. 60–70, Apr. 2003.
21 S. E. Diehl-Nagle, J. E. Vinson, and E. L. Petersen, "Single event upset rate predictions for complex logic systems," in *IEEE Trans. Nucl. Sci.*, vol. 31, pp. 1132–1138, Dec. 1984.
22 N. Kaul, B. L. Bhuva, and S. E. Kerns, "Simulation of SEU transients in CMOS ICs," *IEEE Trans. Nucl. Sci.*, vol. 38, pp. 1514–1520, Dec. 1991.
23 M. P. Baze, S. Buchner, W. G. Bartholet, and T. A. Dao, "An SEU analysis approach for error propagation in digital VLSI CMOS ASICs," *IEEE Trans. Nucl. Sci.*, vol. 42, pp. 1863–1869, Dec. 1995.
24 Véronique Ferlet-Cavrois, Lloyd W. Massengill, and Pascale Gouker, "Single event transients in digital CMOS—A review," *IEEE Trans. Nucl. Sci.*, Jun. 2013.
25 Actel RT54SX-S Rad. Tolerant FPGA for Space Applications, ACTEL Corp, Nov. 2002.
26 R. Koga, S. D. Pinkerton, S. D. Moss, D. C. Mayer, S. D. LaLumendiere, S. F. Hansel, K. B. Crawford, and S. D. Crain, "Observation of single event upsets in analog microcircuits," *IEEE Trans. Nucl. Sci.*, vol. 40, no. 6, pp. 1838–1844, Dec. 1993.
27 R. Koga, S. H. Penzin, K. B. Crawford, S. D. Crain, S. D. Moss, S.D. Pinkerton, S. LaLumendiere, and M. C. Maher, "Single event upset (SEU) sensitivity dependence of linear integrated circuits (ICs) on bias conditions," *IEEE Trans. Nucl. Sci.*, vol. 44, no. 6, pp. 2325–2332, Dec. 1997.
28 R. Ecoffet, S. Duzellier, P. Tastet, C. Aicardi, and M. Labrunee, "Observation of heavy ion induced transients in linear circuits," in *Proc. IEEE NSREC '94 Data Workshop*, pp. 72–77, 1994.
29 R. L. Pease, A. L. Sternberg, L. W. Massengill, R. D. Schrimpf, S. Buchner, M. Savage, J. Titus, and T. Turflinger, "Critical charge for single-event transients (SETs) in bipolar linear circuits," *IEEE Trans. Nucl. Sci.*, vol. 48, no. 6, pp. 1966–1972, Dec. 2001.
30 P. Adell, R. D. Schrimpf, H. J. Barnaby, R. Marenc, C. Chatry, P. Calvel, C. Barillot, and O. Mion, "Analysis of single event transients in analog circuits," *IEEE Trans. Nucl. Sci.*, vol. 47, no. 6, pp. 2616–2623, Dec. 2000.
31 S. Buchner and D. Mc. Morrow, "Single event transients in linear integrated circuits," in *Proc. Nuclear and Space Radiation Effects Conf. Short Course*, pt. 4, 2005.
32 A. H. Johnston, G. M. Swift, T. F. Miyahira, and L. D. Edmonds, "A model for single event transients in comparators," *IEEE Trans. Nucl. Sci.*, vol. 47, no. 6, pp. 2624–2633, Dec. 2000.

33 S. Buchner, D. McMorrow, A. L. Sternberg, L. W. Massengill, R. L. Pease, and M. Maher, "Single event transient (SET) characterization of a LM119 voltage comparator: An approach to SET model validation using a pulsed laser," in *Proc. RADECS '01*, pp. 431–437, 2001.

34 R. Pease, A. L. Sternberg, Y. Boulghassoul, L. W. Massengill, S. Buchner, D. McMorrow, D. S. Walsh, G. L. Hash, S. D. LaLumendiere, and S. C. Moss, "Comparison of SETs in bipolar linear circuits generated with an ion microbeam, laser light and circuit simulation," *IEEE Trans. Nucl. Sci.*, vol. 49, no. 6, pp. 3163–3170, Dec. 2002.

35 R. J. Widlar, "New developments in IC voltage regulators," *IEEE J. Solid-State Circuits*, vol. 6, no. 1, pp. 2–7, 1971.

36 S. Cuk and R. D. Middlebrook, "Basics of switched-mode power conversion: Topologies, magnetics, and control," in *Advances in Switching Mode Power Conversion*. Pasadena, CA: TESLAco, vol. 2, 1981.

37 R. W. Erickson, *DC-DC Power Converters*. in Wiley Encyclopedia of Electrical and Electronics Engineering. New York: Wiley, 1988.

Index

Handbook of Aerospace Electromagnetic Compatibility, First Edition. Edited by Reinaldo Perez.

Printed and bound by CPI Group (UK) Ltd, Croydon, CR0 4YY

17/09/2026

14964286-0003